an introduction to

STATISTICAL COMMUNICATION THEORY

an introduction to

STATISTICAL COMMUNICATION THEORY

JOHN B. THOMAS
Department of Electrical Engineering
Princeton University

John Wiley & Sons, Inc. New York London Sydney Toronto

 *Library of
Congress Catalog Card
Number: 68–8108
SBN 471 85893 5
Printed in the
United States
of America*

Preface

This book is intended to be a first treatment of statistical communication theory and communication systems at a senior-graduate level. The only formal prerequisite is a knowledge of elementary calculus; however, some familiarity with linear systems and transform theory will be helpful.

Chapters 1 and 2 and the first parts of 3 and 4 are basic in the sense that this material is used freely as background throughout the rest of the book. Sections 3.12–3.17 and 4.9–4.12 and all of Chapters 5, 6, 7, and 8 are optional and can be assigned as time permits and as the instructor desires. All eight chapters can be covered easily in two semesters. Furthermore, many teachers may wish to cover only the first part of Chapter 6 since this chapter treats nonlinear filters with random inputs in somewhat more detail than is conventional.

Chapter 1 is introductory and contains no substantial technical material.

Chapter 2 is an elementary introduction to probability theory at a nonrigorous and nonabstract level. It is essential to the remainder of the book but may be skipped (or reviewed hastily) by any student who has taken a one-semester undergraduate course in probability. The material on density function representation at the end of this chapter (Sections 2.24 and 2.25) is used only in Chapter 6 and can be skipped if only Part 1 of Chapter 6 is to be covered.

Chapter 3 is a brief treatment of random processes and spectral analysis. It includes a discussion of Markov processes (Sections 3.12 and 3.13) which is used only in Chapter 8 and an introduction to shot noise (Sections 3.14–3.17) which is not subsequently used explicitly.

Chapter 4 considers linear systems with random inputs. It includes a reasonable amount of material on narrow-band systems and on the representation of random processes. In addition it includes material on the noise figure of linear systems (Sections 4.9–4.12) which is not used subsequently.

Chapter 5 treats the least mean-squared-error linear filter at an elementary level but in some detail.

Chapter 6 is a rather comprehensive treatment of nonlinear systems with random inputs. This material is not generally available in organized form at so elementary a level and in such detail. Part 3 of this chapter is, I believe, of particular interest in that it treats general (nonlinear) least mean-squared-error filtering.

Chapter 7 treats modulation theory in a comprehensive but elementary fashion. Those who wish to spend less time on the subject can skip most of Part 1 and concentrate on Parts 2 and 3.

Chapter 8 gives an introduction to information theory in the Shannon sense. Both discrete and continuous signals are treated. All of the principal results of elementary information theory are included.

A number of appendices cover material that is either outside the mainstream of a chapter's development or is too detailed to treat in the text without loss of continuity. In most cases, the appendices can be ignored if desired.

John B. Thomas

Princeton, New Jersey

November 1968

Contents

an introduction to
STATISTICAL COMMUNICATION THEORY

1

1

Introduction

1.1 Background

The field of electrical communications has expanded tremendously since World War II. With this growth has come a realization of the need for significant changes in the emphasis and approach to communications problems. Techniques have been developed which are based on the pioneer work of Wiener and Shannon and which rely essentially on probabilistic concepts and statistical techniques both for the development of a better understanding of communication system performance (analysis) and for the design of communication systems (synthesis).

In a general way the need for statistical communication theory arose with the invention of the triode in the early 1900's and with the consequent apparent possibility of producing unlimited amplification of signals. It soon became obvious, of course, that random fluctuations, or noise, provided an inherent limitation to system performance. In 1918 Schottky showed that at least part of the noise in triode amplifiers, the *shot noise*, was a fundamental attribute of the triode; that is, it resulted from the discrete nature of the electronic charges and their random arrival at the anode. In addition another fundamental source of noise was soon discovered. The effect that we call *thermal conductor noise* was first measured by Johnson in 1925 and frequently is called *Johnson noise*. This is the noise that arises in any conductor because of the random excitation of the electron gas from thermal agitation of the molecules of the structure.

For the next twenty years a good deal of work was directed toward a more complete understanding, both theoretically and practically, of the quantitative relationships involved in noises and their underlying mechanisms. By 1940 the point had been reached where good quantitative studies had been made of the important types of noises that arise in electrical devices and where these could be related to the performance of the systems involving these devices. As far as signal analysis was concerned, systems were characterized in terms of their response to sinusoids, square waves, step functions, and other standard signals.

At this time, under the impetus of World War II and the increased need for automatic fire control, radar, and sonar, another major breakthrough took place in the field of communications—and this was in many ways more revolutionary than the invention of the triode. This new step forward has associated with it the names of Wiener and Shannon among many others. The basic idea developed was that signals, if they are to convey information, must have associated with them certain characteristics which must be indeterminate by the receiver *before reception.* As a consequence the signals to be considered in the analysis and synthesis of communication systems should be statistical in nature rather than the deterministic standard signals used heretofore. Thus signals, as well as noises, are to be treated as random processes. The behavior of these random processes in a system can be described in terms of their average properties which frequently are known or can be estimated accurately. The determination of such average properties comprises the fields of probability and statistics, which are essential mathematical background to the study of communication systems.

1.2 Communication Systems

A communication system can be defined in a loose way as a system designed to transmit information in the form of messages or data. This definition is vague since, among other shortcomings, it does not define precisely the meaning of the term *information.* The usual problem in a communication system is to eliminate or minimize the effect of unwanted disturbances (noises) and to maximize the amount of information (?) transmitted for a given system complexity, cost, amount of power used, and so on.

A typical communication system could be represented by the block diagram of Figure 1.1. An information source produces messages which are processed, and possibly corrupted, before transmission. The transmission system itself

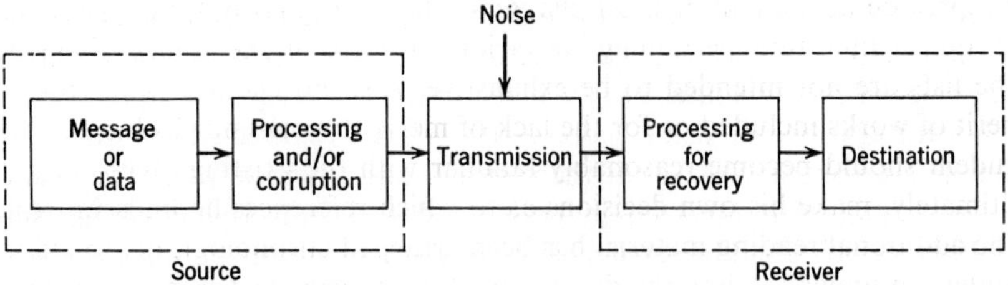

Figure 1.1 *A typical communication system*

introduces additional disturbances. The received message then is processed in an attempt to recover it in its original form before passing it on to its destination. Thus the problem is to study the effect of the communication system on the message. To do this the characteristics of the system must be known. In turn, these characteristics are determined, or at least constrained, by (1) the physical environment; (2) the nature of the communication process; (3) practical factors of cost, size, and reliability.

1.3 Statistical Communication Theory

A working definition of *statistical communication theory* would be that it is the application of probability and statistics to the analysis and synthesis of communication systems. For our purposes we might subdivide this field as follows:

1. *A mathematical background* in linear analysis, elementary probability and statistics, and random processes (Chapters 2, 3, and part of 4 of this book).
2. *System analysis*—analysis of linear and nonlinear systems with random inputs. (Chapters 4, 6, and 7 of this book.)
3. *System synthesis*—constructive procedures to obtain systems satisfying some statistical criterion of performance. Two major areas here are signal detection and signal extraction (waveform estimation). (Chapter 5 and parts of 6 of this book.)
4. *Information theory*—development of the concept of information measure and its application to communication systems. (Chapter 8 of this book.)
5. *Noise physics*—a study of the characteristics of physically realizable random processes based on a knowledge of the mechanisms generating them. (Sections 3.14–3.17 of this book.)

In the next chapter we proceed to develop some of the elementary concepts of probability and statistics in order to provide part of the required mathematical background for this study.

1.4 Additional Reading

We provide at the end of most chapters a list of books and, in some cases, periodical literature pertaining to various aspects of the topics discussed. The lists are not intended to be exhaustive. Also no claim is made for the merit of works included or for the lack of merit of works not included. Each student should become reasonably familiar with the existing literature and, ultimately, make his own decisions as to which references he finds valuable. The additional reading material has been arranged chronologically to aid the student in understanding the development of the various topics, particularly in Chapters 5, 6, and 7.

2

Probability and Random Variables

2.1 Introductory Remarks

It was pointed out in Chapter 1 that a fundamental aspect of statistical communication theory is the concept that signals (and noises) can be treated as random phenomena and described in statistical terms. The starting point for such a description is the theory of probability. Although a rigorous treatment of probability requires considerable mathematical sophistication and some knowledge of measure and integration theory, it is possible to develop an elementary theory using very simple concepts. Such an approach will be given in this chapter.

It will be convenient to use some of the language and elementary ideas from set theory; therefore a brief introduction to this subject will be our first topic.

2.2 Elements of Set Theory

We start with the undefined notion of a *set*. This notion is intuitively familiar and will be taken simply to mean *a collection, an aggregate, a class, or a family of any objects whatsoever*. Each of the objects that make up the set is called an *element*, or *member*, or *point* of the set.

It will be convenient in most cases to denote sets by upper-case letters and elements of sets by lower-case letters. Thus we write A to denote a given set and

$$a \in A$$

to denote that a is an element of the given set. If the element a does not belong to the set A, we write

$$a \notin A.$$

It is conventional to specify a set in one of two ways:

1. By listing the elements of the set between braces; or
2. By listing some property common to all the elements of the set.

For example, let I be the set of all integers and consider the set B with elements 1, 2, and 3. We write:

$$B = \{1, 2, 3\}$$

or

$$B = \{b \in I / 0 < b < 4\}.$$

The last expression is read "B is the set of all integers b which are greater than zero and less than four."

It should be noted that the order of the elements is immaterial and that no element can appear more than once. Thus the set $\{1, 2, 3\}$ and the set $\{3, 1, 2\}$ are the same, and the collection $\{1, 1, 2, 3, 1\}$ is a set only if the repeated elements are deleted.

We shall use equality in the sense of identity. Two sets A and B are equal if, and only if, they contain exactly the same elements. In this case we write

$$A = B.$$

If the elements in the two sets A and B are not the same, we write

$$A \neq B$$

and say that A and B are *distinct*.

If every element of B is also an element of A, we say that *B is contained in A* or *B is a subset of A* and write

$$B \subseteq A.$$

We might also write

$$A \supseteq B$$

and say that A contains B. If $B \subseteq A$ and at least one element of A is not in B, we call B a *proper subset* of A and say *A properly contains B*

$$A \supset B$$

or *B is properly contained in A*

$$B \subset A.$$

Since it may be necessary to consider a particular set which will turn out to have no elements, it is convenient to define the *void* or *empty* or *null* set $\varnothing$ as the set that contains no elements.

In most discussions of sets we will have in mind a particular class of objects that constitute an all-embracing set. The sets with which we concern ourselves then will be sets of elements from this fixed set. This fixed set is known as the *universal set* U or the *sample space* S. We shall use the two terms interchangeably. Then, for every set A, we have

$$\varnothing \subseteq A \subseteq S$$

and, for every element a, we have

$$a \in S \qquad \text{and} \qquad a \notin \varnothing.$$

It is sometimes convenient to attach a geometric significance to the subsets of a given sample space. By this convention we associate areas with sets and points with elements. We indicate the sample space S by a rectangle and subsets of the space by areas within the rectangle. Figure 2.1 illustrates the

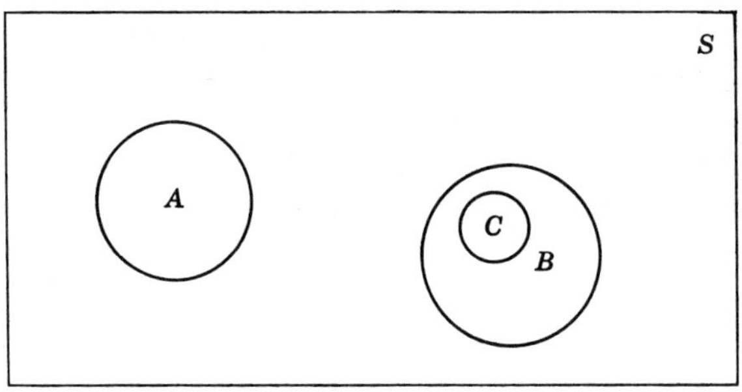

Figure 2.1 *A Venn diagram.*

technique. Here A, B, and C are subsets of the sample space S. In addition, C is a proper subset of B and we could write

$$C \subset B.$$

These diagrams are called *Venn diagrams.*

2.3 Algebra of Sets

We consider in this section some of the ways in which sets can be combined with one another and some of the properties of such combinations.

The *intersection* of two sets A and B is denoted by

$$A \cap B$$

and is the set of all elements common to both A and B; that is,

$$A \cap B = \{x/x \in A \text{ and } x \in B\}.$$

The *union* of two sets A and B is denoted by $A \cup B$ and is the set of all elements in A or B or both; that is,

$$A \cup B = \{x/x \in A \text{ or } x \in B\}.*$$

The Venn diagram of Figure 2.2 shows the union and intersection of A and B for a typical situation.

The *complement* of the set A is denoted by A' (or A^c or $\bar{A}$) and is the set of all elements not in A; that is,

$$A' = \{x/x \notin A\}.$$

Some useful relationships involving the complement are

$$(A')' = A$$

$$\varnothing' = S \qquad S' = \varnothing$$

$$A \cup A' = S \qquad A \cap A' = \varnothing$$

If two sets A and B have no elements in common they are said to be *disjoint* or *exclusive* and

$$A \cap B = \varnothing.$$

It has already been emphasized that the elements of a set can be anything whatsoever; in particular, they can be sets themselves. To avoid confusion, it is conventional to call

a *set of sets* a *class* (of sets)

and

a *set of classes* a *family* (of classes).

A set will be called a *finite set* if it is empty or contains n elements, for n a positive integer; otherwise, the set will be called an *infinite set.*

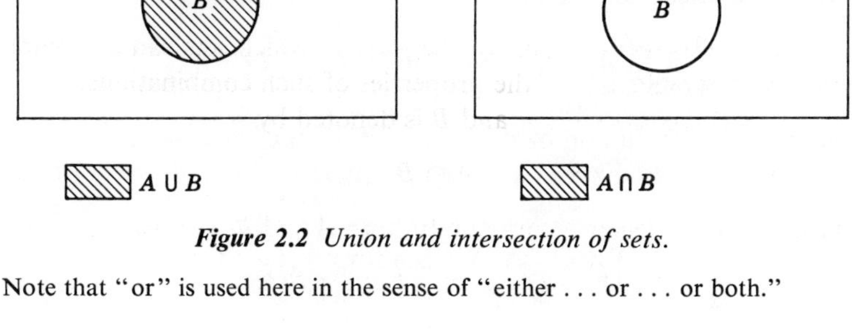

Figure 2.2 *Union and intersection of sets.*

* Note that "or" is used here in the sense of "either . . . or . . . or both."

Based on the previous definitions, all subsets of the universal set S form an algebraic system, for which the following theorems hold. For completeness, some relationships that have been given previously are repeated.

$$
\begin{aligned}
A \cup \varnothing &= A & \qquad A \cup S &= S \\
A \cap \varnothing &= \varnothing & \qquad A \cap S &= A \\
A \cup A &= A & \qquad S \cup \varnothing &= S \\
A \cap A &= A & \qquad S \cap \varnothing &= \varnothing.
\end{aligned}
$$

If $A \subset B$, then $A \cup B = B$ and $A \cap B = A$.
For all A, B; $A \subseteq A \cup B$ and $A \cap B \subseteq A$.
If $B \subset C$, then $A \cap B \subset A \cap C$.
If $A \subset C$ and $B \subset C$, then $A \cup B \subset C$.
If $A \subset C$ and $B \subset D$, then $A \cup B \subset C \cup D$.

The commutative law:

$$A \cap B = B \cap A$$
$$A \cup B = B \cup A.$$

The distributive law:

$$A \cap (B \cup C) = (A \cap B) \cup (A \cap C)$$
$$A \cup (B \cap C) = (A \cup B) \cap (A \cup C).$$

The associative law:

$$(A \cup B) \cup C = A \cup (B \cup C) = A \cup B \cup C$$
$$(A \cap B) \cap C = A \cap (B \cap C) = A \cap B \cap C.$$

The ideas of union and intersection may be extended in a rather obvious fashion to n sets. If $A_1, A_2, \ldots, A_n$ are subsets of S, the union of $A_1, A_2, \ldots, A_n$ is the set of all elements that belong to at least one of the sets $A_1, A_2, \ldots, A_n$ and may be denoted by

$$A_1 \cup A_2 \cup \cdots \cup A_n \qquad \text{or} \qquad \bigcup_{i=1}^{n} A_i.$$

Similarly, let $\mathscr{E}$ be a class of subsets of S. The set of all $x \in S$ such that $x \in A$ *for at least one* $A \in \mathscr{E}$ can be written as

$$\bigcup_{A \in \mathscr{E}} A = \{x \in S / x \in A \text{ for some } A \in \mathscr{E}\}$$

and can be called the union of the sets of $\mathscr{E}$. In the same way, the set of all $x \in S$ such that $x \in A$ *for every* $A \in \mathscr{E}$ is the intersection of the sets of $\mathscr{E}$ and is written

$$\bigcap_{A \in \mathscr{E}} A = \{x \in S / x \in A \text{ for all } A \in \mathscr{E}\}.$$

An infinite set is said to be *countable* or *denumerable* if its elements can be indexed by the natural numbers (positive integers)—that is, placed in one-to-one correspondence with them. Otherwise an infinite set is said to be *noncountable* or *nondenumerable*. As an example, the set of natural numbers is a countable set; the set of all real numbers is not.

We proceed now to develop a concept of probability based on *relative frequency*. Later we will see that the relations developed can be more rigorously obtained from an axiomatic approach.

2.4 Probability and Relative Frequency

We begin with the idea of a *random experiment*. For each performance of this experiment there are N possible outcomes or events $A, B, \ldots, N$, only one of which actually occurs. The use of the word *random* implies that it is not possible to predict with certainty which of the N outcomes will occur in a given experiment.

We consider a sequence of n of these random experiments. In this sequence, the outcome A will occur n_A times, the outcome B will occur n_B times, and so on. Since one of the $A, B, \ldots, N$ outcomes must occur in each experiment, it follows that

$$n_A + n_B + \cdots + n_N = n. \tag{2.4-1}$$

It is convenient to divide both sides of Equation 2.4-1 by n to obtain

$$\frac{n_A}{n} + \frac{n_B}{n} + \cdots + \frac{n_N}{n} = 1. \tag{2.4-2}$$

Here the term n_A/n is called the *relative frequency* or *frequency ratio* of the event A.

Now let us assume that we repeat this sequence of experiments many times for different values of n, say $n_1, n_2, \ldots$. In this way, we will obtain the frequency ratios

$$\frac{n_{A1}}{n_1}, \frac{n_{A2}}{n_2}, \ldots$$

for the event A. Our intuitive concept of probability is based on the observation that, for many kinds of experiments, the frequency ratios for the event A tend to cluster more or less about some fixed point as the number of experiments in each sequence becomes large. We call the value of this fixed point the *probability of the event A* and denote it by $P(A)$.

Perhaps the simplest example of a random experiment is coin tossing. Here the outcome of each experiment is one of the two events "heads" (H) or "tails" (T). Thus a typical sequence of forty experiments, obtained by actually tossing a coin forty times, is

TTHHTTHHHHTHTTHTHHTTTHTHTHTHHTTHTHTHTTHH.

Suppose the relative frequency of the event "tails" is calculated after each experiment in the sequence and plotted versus the number of experiments which have occurred by that time. The resulting plot is shown in Figure 2.3(*a*). The striking feature of this graph is the way in which, after large initial fluctuations, the relative frequency n_T/n approaches and tends to remain in the neighborhood of one-half as n becomes larger. Let us now toss the coin again forty times to obtain the new sequence

HTTHHHTHTTTTTHHHTHTTTHTHTTHTHTTTHHHHTHHH.

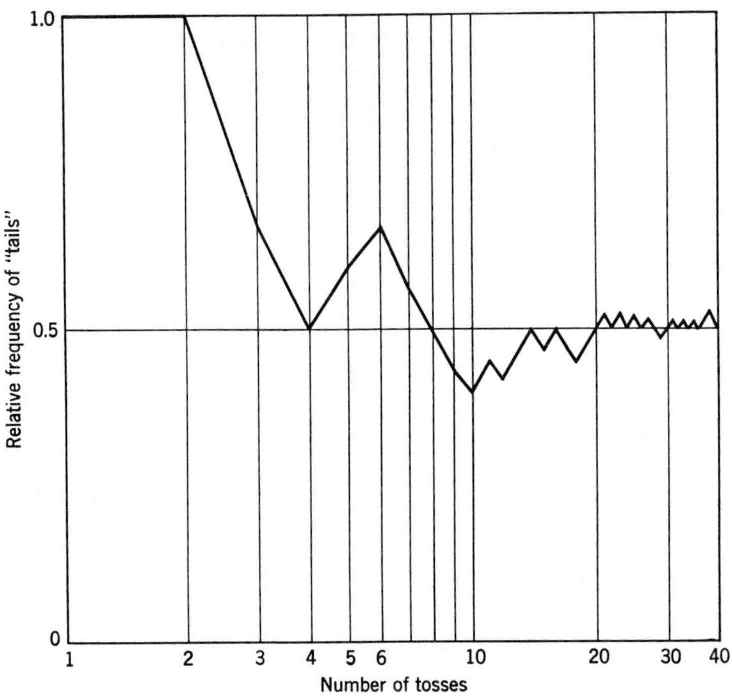

Figure 2.3(a) *Relative frequency of the occurrence of "tails" in coin tossing—a "typical" sequence of forty tosses.* (*Courtesy of Gwendolyn E. Thomas.*)

The relative frequency of tails for this sequence is shown in Figure 2.3(*b*) and, again, tends toward the neighborhood of one-half. In this case, however, it may not be so clear that the tendency is toward exactly one-half. In fact it appears that the relative frequency may be tending toward some number such as 0.54 instead of 0.50.

Let us now consider what the relative frequency looks like if the coin is tossed forty times and tails is obtained on *each* toss *except* the seventh. Such a plot is shown in Figure 2.3(*c*). In this case it is clear that the curve is tending toward unity. We have assumed that the coin is unbiased and has no memory; consequently a head or a tail is equally likely on each toss, regardless of the results of previous tosses. Thus any sequence of forty heads and tails is as likely to occur as any other. In particular, each of the three sequences corresponding to Figures 2.3(*a*), (*b*), and (*c*) is equally likely. Why then do we feel intuitively that the first two sequences are "typical" and the last one is not?

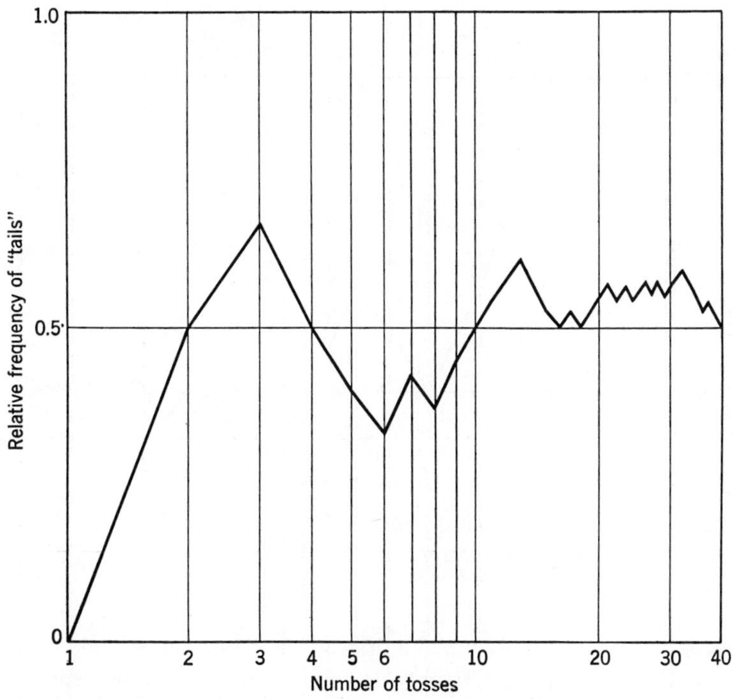

Figure 2.3(b) *Relative frequency of the occurrence of "tails" in coin tossing—a "typical" sequence of forty tosses. (Courtesy of Gwendolyn E. Thomas.)*

The answer to this question lies in the fact that there are many possible distinct sequences similar to the first two and relatively few similar to the third. Any sequence containing approximately the same number of heads and tails will produce a relative frequency plot somewhat similar to the first two while either heads or tails must predominate in any sequence producing a relative frequency plot similar to Figure 2.3(*c*). Let us look briefly then at the number of *distinct* sequences of forty heads and tails containing exactly r tails where $r \leq 40$. If $r = 40$, it is clear that only one distinct sequence exists, namely the sequence of forty tails. If $r = 39$, then there are forty

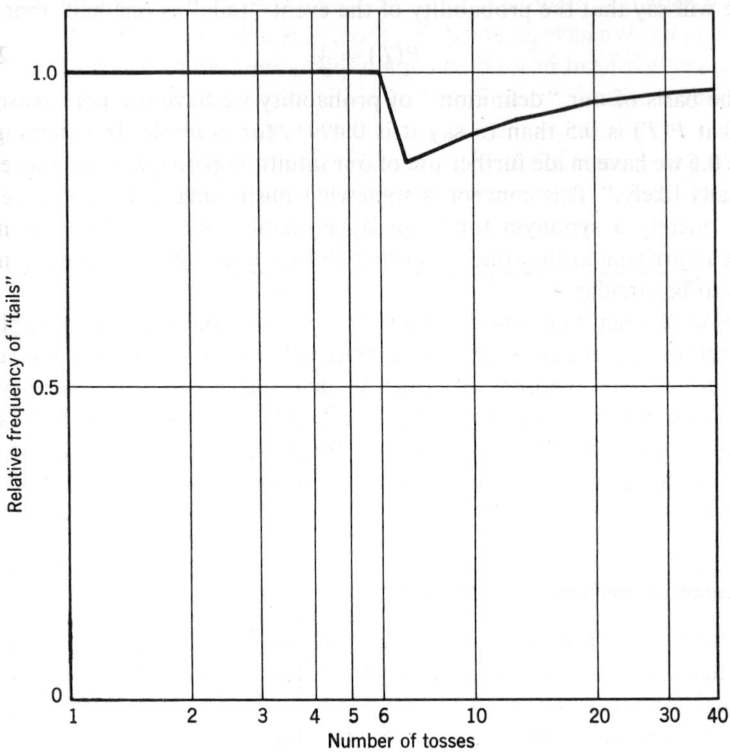

Figure 2.3(c) Relative frequency of the occurrence of "tails" in coin tossing—an "untypical" sequence of forty tosses.

distinct sequences: all tails but the first, all tails but the second, and so on. We will show in Section 2.7 that the number of distinct sequences with r tails is

$$\frac{40!}{(40-r)!r!}, \qquad r = 0, 1, 2, \ldots, 40.$$

This number is relatively small for r near zero or for r near forty. On the other hand, it is a very large number for any value of r near twenty. For $r = 20$, we have, for example,

$$\frac{40!}{20!20!} \approx 13.78 \times 10^{10}.$$

Thus there are an enormous number of sequences similar to Figures 2.3(a) and (b) and relatively few similar to Figure 2.3(c). Thus we feel that the first two sequences are "typical" and the last one is not.

We will say that the probability of the event "tails" is one-half, that is,

$$P(T) = \tfrac{1}{2}. \tag{2.4-3}$$

On the basis of our "definition" of probability we have no more reason to say that $P(T)$ is 0.5 than to say it is 0.499^+, for example. In assigning the value 0.5 we have made further use of our intuition, perhaps in the concept of "equally likely." This concept is somewhat misleading as it is, in a certain sense, merely a synonym for "equally probable" which we are trying to define in arriving at the conclusion that $P(T)$ is one-half. Thus the argument tends to be circular.

We have seen that our "definition" of probability is not completely satisfactory; nevertheless, it can be extremely useful, particularly in those cases where the random experiments involved relate to some physical situation in which probabilities can be assigned or assumed on the basis of information in addition to the relative frequencies. Such, certainly was the case for the coin-tossing experiment where the idea of "equally likely" entered.

2.5 Some Elementary Properties of Probability

We shall proceed now to develop some of the properties of probability on the basis that the relative frequency of an event, for a large number of events, is a reasonable approximation to the probability of the event. We shall assume that the same relationships exist for probabilities of events as for relative frequencies of the same events.

We consider the case described previously where n random experiments have been performed, and where the outcome of each experiment is one of the events $A, B, \ldots, N$. It is apparent that, for any event A,

$$0 \leq \frac{n_A}{n} \leq 1. \tag{2.5-1}$$

On the basis of our definition of probability, we conclude that

$$0 \leq P(A) \leq 1, \tag{2.5-2}$$

or, in words, the *probability of an event is a nonnegative number less than or equal to one*.

Two special events are easily distinguished. The first of these is the *certain event* S, the event that occurs on every experiment. For this case, $n_S = n$ and

$$P(S) = 1. \tag{2.5-3}$$

Thus *the probability of a certain event is one*. It should be emphasized that the converse is not necessarily true. An event with probability one is not certain to occur but practically certain; in other words, the frequency ratio is very nearly one.

The second special event is the *impossible event* O, the event which cannot occur. Here $n_O = 0$ and

$$P(O) = 0. \tag{2.5-4}$$

A similar statement holds for the converse of Equation 2.5-4 as held for the converse of Equation 2.5-3.

Let us generalize our random experiment to include cases where more than one outcome is possible on each experiment. Specifically we consider cases where only A or B or both or neither can occur. The event that *one or both* of the events A and B have occurred is $A \cup B$, the union of A and B (also written as $A + B$). Furthermore, the event that *both* A and B have occurred is $A \cap B$, the *intersection* of A and B (also written AB). In the same way we introduce n_{AB} and $n_{A\cup B}$ where n_{AB} is, for example, the number of times that both A and B occur out of the sequence of n experiments.

In any given experiment the outcome must be one of the following four events:

1. The event A only, occurring n_1 times in n trials.
2. The event B only, occurring n_2 times in n trials.
3. The event AB, occurring $n_3 = n_{AB}$ times in n trials.
4. The event that neither A nor B has occurred, occurring n_4 times in n trials.

It must be true that

$$n_1 + n_2 + n_3 + n_4 = n. \tag{2.5-5}$$

We can now write the frequency ratios of the various events:

$$\frac{n_A}{n} = \frac{n_1 + n_3}{n}, \tag{2.5-6}$$

$$\frac{n_B}{n} = \frac{n_2 + n_3}{n}, \tag{2.5-7}$$

$$\frac{n_{AB}}{n} = \frac{n_3}{n}, \tag{2.5-8}$$

$$\frac{n_{A\cup B}}{n} = \frac{n_1 + n_2 + n_3}{n}. \tag{2.5-9}$$

From Equations 2.5-6 through 2.5-9, we obtain the relationship:

$$\frac{n_{A\cup B}}{n} = \frac{n_A}{n} + \frac{n_B}{n} - \frac{n_{AB}}{n}, \tag{2.5-10}$$

from which we deduce that

$$P(A \cup B) = P(A) + P(B) - P(AB). \tag{2.5-11}$$

If A and B are mutually exclusive, then $P(AB) = 0$ and

$$P(A \cup B) = P(A) + P(B). \tag{2.5-12}$$

It is convenient at this point to introduce the idea of *conditional occurrence*. We say that the relative frequency of the event A on the condition that B has occurred is $n_3/(n_2 + n_3)$ and that the probability associated with this relative frequency is the conditional probability $P(A/B)$, sometimes called the probability of A given B. Similarly the relative frequency of B given A is $n_3/(n_1 + n_3)$ and the corresponding probability is $P(A/B)$, sometimes called the probability of B given A. It is apparent that

$$P(AB) = P(A)P(B/A) = P(B)P(A/B). \tag{2.5-13}$$

The foregoing relationships can be extended to the case where more than two events can occur. Consider the events A, B, and C. If B is replaced in Equation 2.5-11 by $B \cup C$, we have

$$P(A \cup B \cup C) = P(A) + P(B) + P(C) - P(AB) - P(AC) - P(BC) + P(ABC), \tag{2.5-14}$$

where use has been made of the identity

$$A(B \cup C) = AB + AC - ABC. \tag{2.5-15}$$

Here $A \cup B \cup C$ means that at least one of the events A, B, or C has occurred and ABC means that all of the events A, B, and C have occurred. In a similar way, Equation 2.5-14 may be extended to the union of any number of events.

In a manner analogous to that of the previous paragraph, let B in Equation 2.5-13 be replaced by BC. Then

$$P(ABC) = P(A)P(B/A)P(C/AB), \tag{2.5-16}$$

and so on.

2.6 Independence

Let us consider the events A and B, both assumed to have probabilities different from zero. If

$$P(A/B) = P(A), \tag{2.6-1}$$

we say that A and B are *statistically independent*. Note that Equations 2.6-1 and 2.5-13 imply that

$$P(B/A) = P(B), \tag{2.6-2}$$

which could have been taken equally well as the definition of independence. This concept can be extended as before to cases where more than two variables are involved. Note that Equation 2.6-1 or 2.6-2 implies that Equation 2.5-13 can be rewritten for the independent case as

$$P(AB) = P(A)P(B) \tag{2.6-3}$$

which also could have been taken as the definition of independence.

Three events A, B, and C are mutually independent if

$$\begin{aligned} P(AB) &= P(A)P(B) \\ P(AC) &= P(A)P(C) \\ P(BC) &= P(B)P(C) \end{aligned} \tag{2.6-4}$$

and if

$$P(ABC) = P(A)P(B)P(C). \tag{2.6-5}$$

The last condition is necessary since the first three do not ensure that such events as AB and C are independent as they would be if

$$P\{(AB)(C)\} = P(AB)P(C) = P(A)P(B)P(C). \tag{2.6-6}$$

Example 2.1 Two dice are thrown. The events A, B, and C are taken to be

Event A—"odd face on first die."
Event B—"odd face on second die."
Event C—"sum of the faces odd."

The probabilities for fair dice are

$$\begin{array}{lll} P(A) = \frac{1}{2} & P(A/B) = \frac{1}{2} & P(A/C) = \frac{1}{2} \\ P(B) = \frac{1}{2} & P(B/A) = \frac{1}{2} & P(B/C) = \frac{1}{2} \\ P(C) = \frac{1}{2} & P(C/A) = \frac{1}{2} & P(C/B) = \frac{1}{2}. \end{array}$$

However, A, B, and C cannot occur simultaneously and it is apparent that the events AB and C are not independent.

• •

For N events A_i to be statistically independent, it is necessary that, for all combinations $1 \le i \le j \le k \le \cdots \le N$,

$$\begin{aligned} P(A_iA_j) &= P(A_i)P(A_j) \\ P(A_iA_jA_k) &= P(A_i)P(A_j)P(A_k) \\ &\vdots \\ P(A_1A_2\cdots A_N) &= P(A_1)P(A_2)\cdots P(A_N). \end{aligned} \tag{2.6-7}$$

2.7 Permutations and Combinations

We proceed now with some ideas concerning possible arrangements and subsets that can be formed from given sets. We begin with the notion of a *sample*. A sample of size r will be taken to be an *ordered* set of r elements obtained in some way from a universal set of n elements. Two possible policies to follow are to *sample with replacement* of the elements drawn and to *sample without replacement*.

Consider first the case of sampling with replacement. There are n possible elements any one of which can be chosen on the first drawing to form the first element in the sample. Since the chosen element is replaced prior to the second drawing, there are n possible elements available to be chosen as the second element in the sample. Thus, after r drawings, the sample (of length r) could consist of any one of

$$n \cdot n \cdot n \cdots n = n^r \tag{2.7-1}$$

distinct possible samples.

In sampling without replacement it is apparent that $r \leq n$. After the first drawing, there are only $n - 1$ elements available from which to choose and, after r drawings, only $n - r$. Thus the number of possible samples of length r in this case is only

$$n(n-1)(n-2)\cdots(n-r+1) = \frac{n!}{(n-r)!}. \tag{2.7-2}$$

An (ordered) sample taken without replacement (so that the same element does not occur twice) is often called a *permutation*. The symbol P_r^n is sometimes used to indicate the number of permutations of n things taken r at a time:

$$P_r^n = \frac{n!}{(n-r)!}. \tag{2.7-3}$$

Suppose we sample without replacement to obtain a sample of size n from a set of n elements. Then the number of such samples is just $n!$ since

$$\frac{n!}{(n-n)!} = n!.$$

Furthermore this is the number of ways of *ordering* the n elements since each sample is a unique ordering and the $n!$ possible samples exhaust the possibilities.

A *combination* is a set of elements without repetition and without regard to order. Thus the set $\{1, 2\}$ and the set $\{2, 1\}$ are different permutations but only one combination. Out of n elements the possible permutations of size r are

$$\frac{n!}{(n-r)!}.$$

We have seen that r elements can be ordered in $r!$ ways so that there are $r!$ permutations containing any particular set of r elements. Therefore the number of possible combinations of n elements taken r at a time must be

$$\frac{n!}{(n-r)!} \div r! = \frac{n!}{(n-r)!r!}. \tag{2.7-4}$$

This quantity is frequently denoted by C_r^n or by $\binom{n}{r}$ and is often called the binomial coefficient since it occurs in the binomial expansion

$$(a + b)^n = \sum_{r=0}^{n} \binom{n}{r} a^{n-r} b^r. \tag{2.7-5}$$

It follows directly from Equation 2.7-4 that

$$\binom{n}{r} = \binom{n}{n-r} \tag{2.7-6}$$

and that

$$\binom{n}{n} = 1, \binom{n}{0} = 1. \tag{2.7-7}$$

Example 2.2 What is the probability that a bridge hand will contain all thirteen card values: A, K, Q, . . ., 3, 2?

The number of possible bridge hands is given by

$$\binom{52}{13}.$$

The number of possible bridge hands with all thirteen card values is

$$4^{13}.$$

Thus the required probability is

$$4^{13} \div \binom{52}{13} \approx 0.0001057.$$

• •

2.8 An Axiomatic Approach to Probability

We have based our notions of probability on the concept of relative frequency and on the intuitive feeling that this relative frequency has some kind of natural limit. From a mathematical point of view, it would have been more satisfactory to have postulated a set of axioms from which the theory of probability could be derived. This resulting theory would then be useful if it could be related in some meaningful way to reality. We proceed with the axiomatic approach as follows.

We begin with the certain event S of Section 2.5 and identify it with the *sample space* S of Section 2.2. The events $A, B, C, \ldots$ are subsets of the space. We define a function $P(\cdot)$ which associates with each event A a real number called the probability of A. This function $P(\cdot)$ is required to satisfy the following axioms:

I. $P(S) = 1$.
II. $P(A) \geq 0$.
III. $P(A \cup B) = P(A) + P(B) \quad$ if $\quad AB = 0$.

It will be noted that these three axioms are just Equations 2.5-3, 2.5-2, and 2.5-12 respectively. The system of sets $A, B, C, \ldots$, together with the assignment of the numbers $P(A), P(B), \ldots$, is often called a field of probability.*

It is possible to build an entire treatment of probability on these three axioms and to derive many useful theorems from them [2]. It should be emphasized, however, that they comprise a mathematical structure and are not inherently useful in the practical problems of relating the subsets of the sample space to real events and of assigning probabilities to these events.

2.9 Random Variables, Discrete and Continuous

A *random variable* x is a real-valued function defined on the events of our probability system (the elements of the sample space S). In other words, for each event there is a real number that is the corresponding value of the random variable. Strictly speaking, the random variable x should be written as $x(\cdot)$ where the argument is the event being considered.

The random variable has been introduced in order to map each element of the sample space onto the real line. Thus subsets of the sample space become sets of points on the real line. For example, on each throw of a die, a number (the number of dots on the top face) results. Each of these numbers $1, 2, \ldots, 6$ may be taken to be the values of the random variable defined in this case. It should be emphasized, however, that any other number assignment could have been made. Thus the event "two dots appear on top face" could have been assigned the number unity and all other events the number zero. In this case the sum of the values of the random variable in repeated trials becomes the number of "twos" obtained. The situation may be more obvious for the coin-tossing example in which the outcomes are H and T rather than numbers. A set of values frequently used to define a random variable in this case are unity for H and zero for T.

The definition of a random variable has left the probabilities unchanged. In the coin-tossing case just considered, if the coin is fair, then the probability of obtaining the value of unity or zero for the random variable on each experiment is still one-half in each case. The practical advantage that has accrued from the use of the random-variable concept is that mathematical manipulation has been transferred to the real line with a consequent gain in the analytical tools available.

In the examples just considered, the random variable has taken on only a discrete set of values. In such cases we say that we are dealing with a *discrete random variable* sometimes denoted as x_i where the subscript i indexes the possible values of x_i.

* Actually a somewhat more rigorous approach is required to define a field of probability, and some ideas from measure theory must be used. This topic is discussed in more detail in references 1 and 2.

Example 2.3 Let us consider the random experiment of tossing a coin three times. We will presume that the coin is weighted so that $P(H) = \frac{2}{3}$ and $P(T) = \frac{1}{3}$. The random variable x_i, $i = 0, 1, 2, 3$, will be taken to be the number of heads produced in the three tosses. Thus

$$x_0 = 0 \qquad x_2 = 2$$
$$x_1 = 1 \qquad x_3 = 3.$$

The sample space and probabilities may be tabulated as:

$S =$	$\{(HHH)$,	(HHT),	(HTH),	(THH),	(HTT),	(THT),	(TTH),	$(TTT)\}$
$P =$	$\frac{8}{27}$	$\frac{4}{27}$	$\frac{4}{27}$	$\frac{4}{27}$	$\frac{2}{27}$	$\frac{2}{27}$	$\frac{2}{27}$	$\frac{1}{27}$
$x_i =$	3	2	2	2	1	1	1	0

• •

For the *continuous random variable* x, we assume that a continuum of values is permissible over some interval (a, b) or set of intervals. We include the case where the interval is infinite. For this continuous random variable it is not possible to speak of the probability of a single selected value since this probability is zero. The situation is analogous to considering the area under a curve. The area under any point is zero but the area under any segment of nonzero length may be nonzero. Thus we are led to consider not the probability that a random variable x takes on some value a—that is, $P(a)$—but rather the probability that it lies in some interval $a \leq x \leq b$—that is, $P(a \leq x \leq b)$.

A problem has now arisen that may not be immediately apparent. We cannot set up a one-to-one correspondence between the values of the continuous random variable and the events $A, B, C, \ldots$ even if we consider an infinite number of events since the points in a continuum are not denumerable. The problem can be resolved, however, in an elementary fashion.

Suppose we divide the real line into two parts by the number a; that is, we consider values of the random variable $x \leq a$ and $x > a$. With some event A we can associate the values of the random variable $x \leq a$. In other words A is the specific set of sample points s for which $x(s) \leq a$. Thus the probability $P(A)$ of the event A becomes the probability $P(x \leq a)$. It is conventional to denote this probability by $F(a)$ and write

$$F(a) = P(x \leq a). \tag{2.9-1}$$

If the function $F(a)$ is given for all values of a $(-\infty < a < \infty)$, then the values of x as a random variable have been defined. This function $F(x)$ is called the *distribution function* of x or the *cumulative distribution function* (c.d.f.). It can be defined in an analogous manner for discrete random variables and will be considered in more detail in Section 2.10.

Example 2.4 It is known that a random variable x can take on all values in the interval (0, 1). Furthermore, the probability that x lies in any interval Δx is proportional to the length of Δx.

Since $P(0 \leq x \leq 1)$ must be unity, it follows that $P(x \leq a) = a$ for $0 \leq a \leq 1$. Therefore

$$F(a) = \begin{cases} 0, & a < 0, \\ a, & 0 \leq a \leq 1, \\ 1, & a > 1. \end{cases}$$

This cumulative distribution function can be plotted as shown in Figure 2.4.

• •

At this point a word on notation is in order. We have made no real distinction between the random variable itself and the values that it assumes. It is conventional in many treatments to use the notation X for the former and x for the latter. Thus the probability that the random variable lies in some interval dx is more appropriately written as

$$P(x \leq X \leq x + dx).$$

2.10 Distribution Functions and Densities

We have defined the cumulative distribution function $F(a)$ by Equation 2.9-1. A number of basic properties of $F(a)$—which will often be written as $F(x)$—follow directly from its definition:

1. *$F(a)$ is nondecreasing* (monotonicity); that is, $F(a + h) \geq F(a)$ if $h > 0$.
   The function $F(a)$ is a probability by definition. Let us associate the event A with the probability $P(x \leq a)$ and the event B with the probability $P(x \leq a + h)$ where $h > 0$. Then $B \supseteq A$ and $P(B) \geq P(A)$, or $F(a + h) \geq F(a)$.

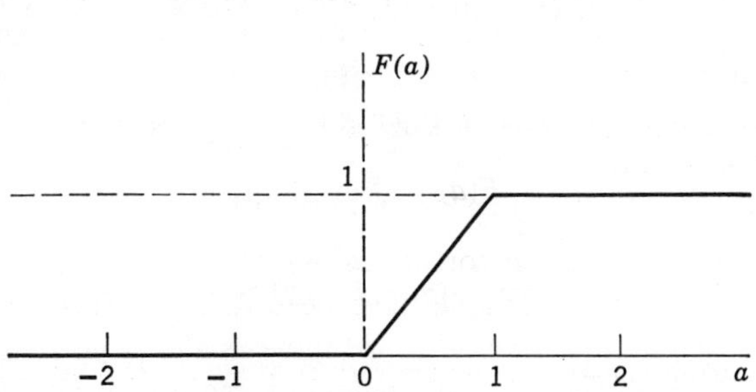

Figure 2.4 *A cumulative distribution function for a continuous random variable.*

2. *F(a) is right-continuous;* that is,

$$\lim_{\substack{h \to 0 \\ h > 0}} F(a + h) = F(a).$$

From the definitions of events A and B in property 1, we have

$$P(a < x \le a + h) = F(a + h) - F(a)$$

or

$$\lim_{\substack{h \to 0 \\ h > 0}} P(a < x \le a + h) = P(\varnothing) = 0$$

or

$$0 = \lim_{\substack{h \to 0 \\ h > 0}} F(a + h) - F(a).$$

3. $\lim_{a \to -\infty} F(a) = 0$; $\lim_{a \to \infty} F(a) = 1$.

If we assume that the random variable can take on only finite values,* then

$$\lim_{a \to -\infty} F(a) = P(\varnothing) = 0$$

and

$$\lim_{a \to \infty} F(a) = P(S) = 1.$$

We shall assume that the random variable takes on only finite values and shall use the notation:

$$\lim_{a \to -\infty} F(a) = F(-\infty)$$

$$\lim_{a \to \infty} F(a) = F(\infty)$$

$$\lim_{\substack{h \to 0 \\ h > 0}} F(a + h) = F(a^+) = F(a + 0)$$

$$\lim_{\substack{h \to 0 \\ h > 0}} F(a - h) = F(a^-) = F(a - 0).$$

* Suppose that the random variable can assume the values ∞ and $-\infty$ with nonzero probability. Then we have

$$\lim_{a \to \infty} F(a) < F(\infty) = 1$$

and

$$\lim_{a \to -\infty} F(a) = F(-\infty) > 0.$$

This situation could occur when a random variable has a nonzero probability of assuming the value zero; its logarithm would have a nonzero probability of assuming the value $-\infty$.

There are a number of corollaries to these properties which either are obvious or follow immediately from the previous discussion:

1. $0 \leq F(a) \leq 1.$ (2.10-1)
2. $F(b) - F(a) = P(a < x \leq b).$ (2.10-2)
3. $F(a) - F(a^-) = P(x = a).$ (2.10-3)
4. $F(a^-) = P(x < a).$ (2.10-4)

It is apparent from corollary 3 that the c.d.f. is continuous for the continuous random variable x. For the discrete random variable x_i, the c.d.f. is constant except for jumps at the points x_i, these jumps being equal to the probability $P(x_i)$. These characteristics are illustrated in the second example of Section 2.9 and in the following example.

Example 2.5 In the Example 2.3 of Section 2.9, a random variable x_i $(i = 0, 1, 2, 3)$ was defined on the outcomes of a random experiment. The c.d.f. for this random variable is shown in Figure 2.5. The crosses on the graph are used to indicate the right-continuity of the function.

• •

It can be shown [3] that if a function $F(a)$ of a real variable a satisfies the basic properties 1, 2, and 3, then there is a random variable x (or x_i) for which $F(a)$ is the c.d.f. In other words we could have started with these three properties as the definition of a cumulative distribution function [4].

For the discrete random variable x_i, we have seen that

$$F(a) - F(a^-) = P(x = a).$$

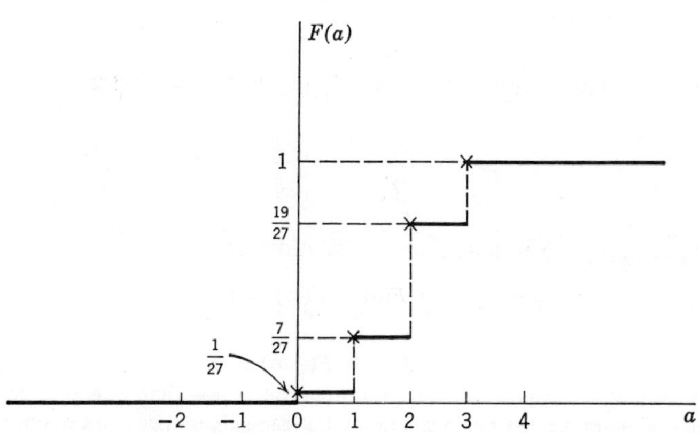

Figure 2.5 *A cumulative distribution function for a discrete variable.*

Let us call this probability $f(a)$ so that*

$$P(x_i) = f(x_i).$$

It follows that $f(x_i)$ has the following two properties:

$$f(x_i) \geq 0 \tag{2.10-5}$$

$$\sum_i f(x_i) = 1. \tag{2.10-6}$$

We say that a function $f(x_i)$ defined for the set of x_i is a *discrete density function* (d.f.) if it satisfies Equations 2.10-5 and 2.10-6. A function $f(x)$ is a *continuous density function* (d.f.) if it is defined on x and satisfies the two conditions:

$$f(x) \geq 0 \tag{2.10-7}$$

and

$$\int_{-\infty}^{\infty} f(x)\, dx = 1. \tag{2.10-8}$$

It will often be convenient to use the symbol x to denote either the continuous or the discrete random variable. It will be apparent from the context as to which situation exists and whether the sum of Equation 2.10-6 or the integral of Equation 2.10-8 is to be used. If we are willing to use a somewhat more general concept of the integral (the Riemann-Stieltjes integral), then both equations can be expressed in a single form. In addition the case is easily handled where the random variable is mixed; that is, where it assumes a continuum of values but also has a nonzero probability of assuming some distinct values. From the applied point of view, this last situation is conveniently treated using the Dirac delta function. The Riemann-Stieltjes integral and the Dirac delta function are discussed briefly in Appendices A and B respectively.

Example 2.6 The following function $f(x)$ is to be tested as a density function:

$$f(x) = \begin{cases} e^x, & x \leq 0, \\ 0, & x > 0. \end{cases}$$

It is apparent that $f(x)$ is nonnegative. In addition

$$\int_{-\infty}^{\infty} f(x)\, dx = \int_{-\infty}^{0} e^x\, dx = 1.$$

* Again it should be emphasized that the notation is less than adequate (although conventional). If X_i is the random variable and x_i the values that it takes on, then we mean that

$$P(x_i) = P(X_i = x_i) = f(x_i).$$

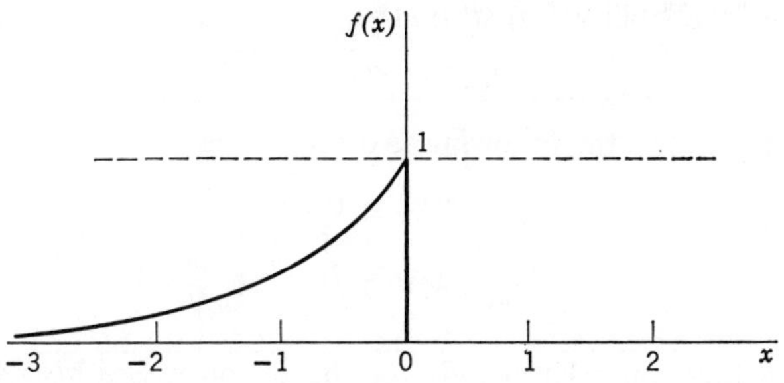

Figure 2.6 *A continuous density function.*

Thus this $f(x)$ is an example of a continuous density function and may be plotted as shown in Figure 2.6.

• •

Example 2.7 Let us define a function $b(k; n, p)$ as

$$b(k; n, p) = \binom{n}{k}(p)^k(1 - p)^{n-k}$$

for $k = 0, 1, 2, \ldots, n$. In this case let $n = 2$ and $p = \frac{1}{2}$ so that

$$b(k; 2, \tfrac{1}{2}) = \frac{2!}{(2 - k)!k!}(\tfrac{1}{2})^2.$$

Again it is apparent that $b(k; 2, \frac{1}{2})$ is nonnegative and that

$$\sum_{k=0}^{2} b(k; 2, \tfrac{1}{2}) = \tfrac{1}{4} + \tfrac{2}{4} + \tfrac{1}{4} = 1.$$

This function $b(k; 2, \frac{1}{2})$—or, in general, $b(k; n, p)$—is a discrete density. The general form is called the *binomial distribution* and will be discussed in more detail in Section 2.15. The plot of $b(k; 2, \frac{1}{2})$ is shown in Figure 2.7.

• •

It is clear from the previous discussion that the c.d.f. $F(a)$ and the d.f. $f(x)$ are related in the discrete case by

$$F(a) = P(x \leq a) = \sum_{x_i \leq a} f(x_i) \tag{2.10-9}$$

and in the continuous case by

$$F(a) = \int_{-\infty}^{a} f(x)\, dx. \tag{2.10-10}$$

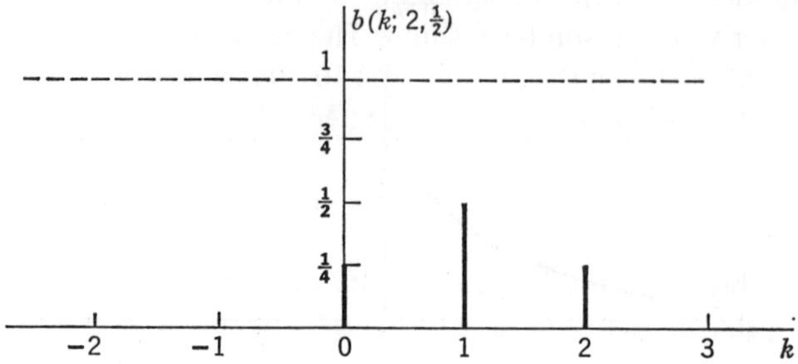

Figure 2.7 *A discrete density function.*

We prove that the function $F(a)$ defined by either Equation 2.10-9 or 2.10-10 is a c.d.f. by showing that it has the basic properties 1, 2, and 3 developed at the beginning of this section. The proof will be given only for the continuous case; extension to the discrete case is obvious.

First Property

1. *$F(a)$ is nondecreasing:*

$$F(a + h) = F(a) + \int_a^{a+h} f(x)\, dx, \qquad h > 0.$$

Since $f(x)$ is nonnegative, the last term of this expression must be nonnegative and

$$F(a + h) \geq F(a), \qquad h > 0.$$

Second Property

2. *$F(a)$ is right-continuous:*

$$\begin{aligned} F(a^+) - F(a) &= \lim_{\substack{h \to 0 \\ h > 0}} \int_a^{a+h} f(x)\, dx \\ &= \lim_{\substack{h \to 0 \\ h > 0}} hf(a) \qquad \text{(by the Mean-Value Theorem)} \\ &= 0. \end{aligned}$$

Third Property

3. *$F(-\infty) = 0$; $F(\infty) = 1$:*

$$F(-\infty) = \lim_{a \to -\infty} \int_{-\infty}^{a} f(x)\, dx = 0$$

$$F(\infty) = \lim_{a \to \infty} \int_{-\infty}^{a} f(x)\, dx = 1 \qquad \text{(from Equation 2.10-8)}.$$

The properties of $F(a)$ corresponding to Equations 2.10-1 through 2.10-4 are easily derived in a similar fashion. The derivations are left to the reader.

Since we have shown that Equation 2.10-10 is a legitimate definition of the c.d.f. $F(a)$, it is apparent from direct differentiation* that

$$\frac{dF(x)}{dx} = f(x) \tag{2.10-11}$$

except possibly at points of discontinuity of $f(x)$. Usually we will be dealing with well-behaved distribution and density functions, and Equation 2.10-11 can be applied without difficulty.

We have mentioned previously that the probability of occurrence of a single point value of a continuous random variable is zero. We have

$$\begin{aligned} P(x = a) &= F(a) - F(a^-) = \lim_{\substack{h\to 0\\ h>0}} \int_{a-h}^{a} f(x)\, dx \\ P(x = a) &= \lim_{\substack{h\to 0\\ h>0}} hf(a - h) = 0. \end{aligned} \tag{2.10-12}$$

It is apparent that, in this case, the probability of a set of values of x occurring is equal to the corresponding area under the density function curve. For this probability to be nonzero, it is necessary that an interval in x of finite length be considered. Thus, instead of considering the probability that a continuous random variable assumes a single value, we should consider the probability that the random variable lies in some small interval dx and write

$$P(x \le x \le x + dx) = f(x)\, dx \tag{2.10-13}$$

from the Mean-Value Theorem.

2.11 The Transformation of Random Variables

We consider now the problem that arises when a random variable x is transformed to a new random variable y through a functional relationship $y = h(x)$. The usual problem is, given the distribution of x, to find the distribution of y.

We begin with the idea of a *monotonic function*. A function $h(x)$ is *monotonically increasing* if $h(x_2) > h(x_1)$ when $x_2 > x_1$. Similarly a function $h(x)$ is

* Let

$$G(u) = \int_{a(u)}^{b(u)} H(x, u)\, dx.$$

Then the reader will recall that the derivative of this definite integral is given by

$$\frac{dG}{du} = \int_a^b \frac{\partial H}{\partial u}\, dx + H(b, u)\frac{db}{du} - H(a, u)\frac{da}{du}$$

if $H(x, u)$ is a continuous function of x and u.

monotonically decreasing if $h(x_2) < h(x_1)$ when $x_2 > x_1$. It is apparent that for monotonic functions there is a one-to-one correspondence between $h(x)$ and x; that is, for each value of x, there is one, and only one, value of $h(x)$.

Suppose $y = h(x)$ is monotonically increasing and differentiable. Let the random variable x have a density $f(x)$ and a c.d.f. $F(x)$ while the random variable y has a density $g(y)$ and a c.d.f. $G(y)$. If $h(a) = b$, then it is clear that

$$P(y \leq b) = P(x \leq a) \tag{2.11-1}$$

or

$$G(b) = F(a). \tag{2.11-2}$$

We may differentiate Equation 2.11-2 with respect to b to yield

$$\frac{dG(b)}{db} = \frac{dF(a)}{da}\frac{da}{db}. \tag{2.11-3}$$

Since x and y are related one to one, a unique inverse exists for h and

$$a = h^{-1}(b) \tag{2.11-4}$$

where the notation h^{-1} is used to denote the relationship

$$h(a) = h[h^{-1}(b)] = b. \tag{2.11-5}$$

Thus Equation 2.11-3 becomes

$$g(b) = f[h^{-1}(b)]\frac{d[h^{-1}(b)]}{db}$$

or, replacing b by y,

$$g(y) = f[h^{-1}(y)]\frac{d[h^{-1}(y)]}{dy}. \tag{2.11-6}$$

As a matter of convenience, we shall frequently write Equation 2.11-6 in the form

$$g(y) = f(x)\frac{dx}{dy} \tag{2.11-7}$$

but it will be understood that Equation 2.11-6 is what we actually have in mind.

Example 2.8 Let $y = e^x$ and let x be uniformly distributed in (0, 1). Then, from Equation 2.11-7, we have

$$\begin{aligned} g(y) &= f(x)\, e^{-x} = e^{-x}, && 0 \leq x \leq 1, \\ g(y) &= 1/y, && 0 \leq x \leq 1. \end{aligned}$$

For $0 \leq x \leq 1$, it follows that $1 \leq g \leq 2.718\ldots$; hence

$$g(y) = \begin{cases} 1/y, & 1 \leq y \leq 2.718\ldots, \\ 0, & \text{elsewhere.} \end{cases}$$

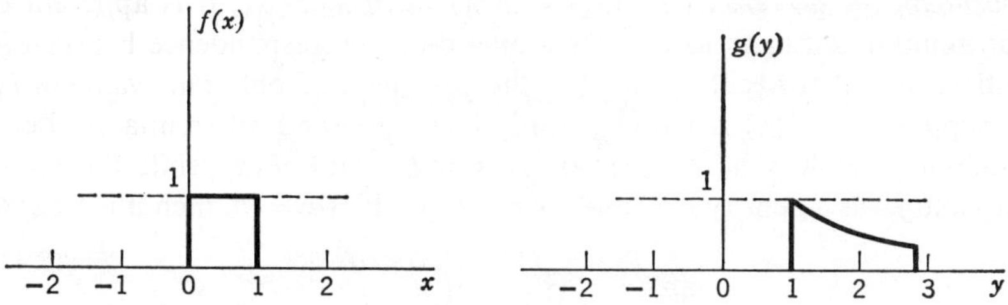

Figure 2.8 *A plot of the densities $f(x)$ and $g(y)$ of Example 2.8.*

It is clear that $g(y) \geq 0$ and that $\int_1^{2.718\ldots} 1/y\, dy = 1$; therefore $g(y)$ is a density. The two densities are plotted in Figure 2.8.

• •

Example 2.9 Let $y = x^3$ and let x be uniformly distributed in (0, 1). We find the density $g(y)$ directly from the c.d.f.

$$P(y \leq b) = P(x^3 \leq b) = P(x \leq b^{1/3})$$

or

$$P(x \leq b^{1/3}) = \begin{cases} 0, & b < 0, \\ b^{1/3}, & 0 \leq b \leq 1, \\ 1, & b > 1. \end{cases}$$

$$g(b) = \frac{dP(x \leq b^{1/3})}{db} = \begin{cases} 0, & b < 0, \\ \frac{1}{3}b^{-2/3}, & 0 \leq b \leq 1, \\ 0, & b > 1. \end{cases}$$

We plot $G(y)$ and $g(y)$ in Figure 2.9. It is left to the interested reader to show that the same result is obtained when Equation 2.11-7 is used.

• •

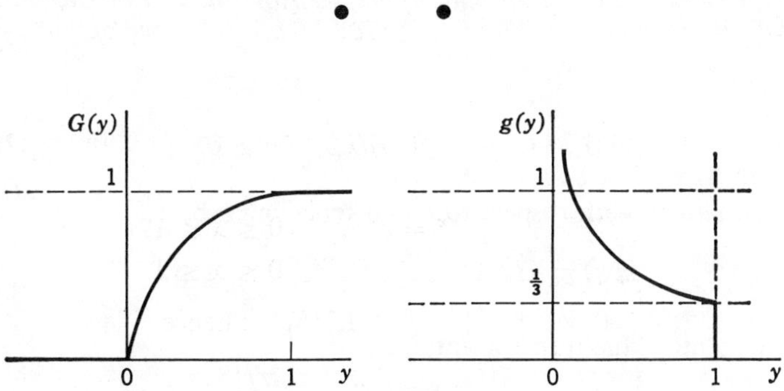

Figure 2.9 *A cumulative distribution function and the corresponding density function.*

We take up next the case where $y = h(x)$ is *monotonically decreasing*. Now we have, for $b = h(a)$, that

$$P(y \le b) = P(x \ge a)$$

or

$$G(b) = 1 - F(a)$$

or, on differentiating,

$$g(b) = -f[h^{-1}(b)]\frac{d[h^{-1}(b)]}{db}.$$

However, the derivative $d[h^{-1}(b)]/db$ is always negative, and we can write

$$g(b) = f[h^{-1}(b)]\left|\frac{d[h^{-1}(b)]}{db}\right|. \tag{2.11-8}$$

This expression is identical to Equation 2.11-6 and hence we can write in general

$$g(y) = f(x)\left|\frac{dx}{dy}\right| \tag{2.11-9}$$

when $y = h(x)$ is a one-to-one continuously differentiable function of x.

For the case where $y = h(x)$ is not monotone, we must consider the problem in more general terms. Now several values of $y = h(x)$ may correspond to a single value of x, and the inverse h^{-1} is multivalued. We now write, for $b = h(a)$,

$$P(y \le b) = P(\text{set of } x \text{ corresponding to } y \le b)$$

and we must take into account the fact that, for each value of x, a set of values of y may exist. Thus

$$G(y) = \int_{\{x/y \le b\}} f(x)\,dx$$

where the notation $\{x/y \le b\}$ means simply that we integrate over all x for which $y \le b$. We now illustrate this procedure with several examples.

Two common transformations that will be encountered later are the *square-law transformation* and the *absolute-value transformation.* For the former

$$y = x^2 \tag{2.11-10}$$

and

$$G(b) = \int_{-\sqrt{b}}^{\sqrt{b}} f(x)\,dx, \qquad b \ge 0, \tag{2.11-11}$$

or, differentiating* with respect to b and replacing b by y,

$$g(y) = \frac{f(\sqrt{y}) + f(-\sqrt{y})}{2\sqrt{y}}, \qquad y \ge 0. \tag{2.11-12}$$

For the absolute-value transformation,

$$y = |x| \tag{2.11-13}$$

* See footnote on page 27.

and

$$G(b) = \int_{-b}^{b} f(x)\,dx, \qquad b \geq 0, \tag{2.11-14}$$

or

$$g(y) = f(y) + f(-y), \qquad y \geq 0. \tag{2.11-15}$$

Example 2.10 Let $y = x^2$ and let $f(x)$ be given by

$$f(x) = \frac{1}{\sqrt{2\pi}} e^{-x^2/2}, \qquad -\infty < x < \infty.$$

This density is called the *normal* or *gaussian* density and will be encountered extensively in noise problems. We have, from Equation 2.11-12,

$$g(y) = \frac{1}{\sqrt{2\pi}\, 2\sqrt{y}} (e^{-y/2} + e^{-y/2}), \qquad 0 < y < \infty$$

or

$$g(y) = \frac{1}{\sqrt{2\pi}} y^{-1/2} e^{-y/2}, \qquad 0 < y < \infty.$$

This density is called the *chi-squared distribution* with one degree of freedom [5]. Both it and the normal distribution are plotted in Figure 2.10.

• •

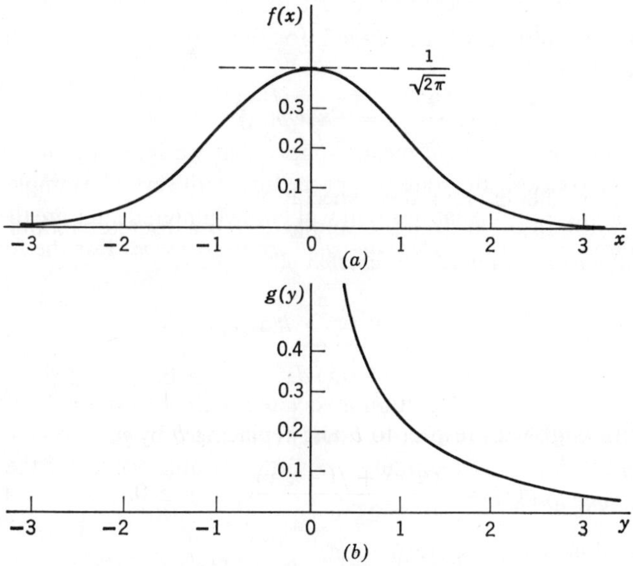

Figure 2.10 *The normal and chi-squared distribution: (a) the normal distribution; (b) the chi-squared distribution with one degree of freedom.*

2.12 Expectation

In the random experiments that motivate our study of probability, a given outcome cannot be predicted but its probability of occurrence is presumed to be available. We seek now some way of determining the "average" behavior of these outcomes.

Suppose that we are offered the opportunity of playing a game with the following payoff: the probability of winning \$1 is $p_1 = 0.3$, of winning \$3 is $p_2 = 0.1$, and of losing \$1 is $p_3 = 0.6$. A reasonable question to ask is whether or not it is profitable to play the game. In other words, what are the average winnings or *expectation* per game? Assume that the game is played n times and we

win \$1,	m_1 times,
win \$3,	m_2 times,
lose \$1,	m_3 times,

where $n = m_1 + m_2 + m_3$. In n games, the total winnings will be $m_1 + 3m_2 - m_3$ and the average winnings per game will be

$$E = \frac{m_1}{n} + 3\frac{m_2}{n} - \frac{m_3}{n} \tag{2.12-1}$$

or, approximately,

$$E = p_1 + 3p_2 - p_3. \tag{2.12-2}$$

If the random variable x_i was defined, for this game, as the dollars won on each game, then

$$E = \sum_i x_i p(x_i) \tag{2.12-3}$$

would be a formula defining this expectation.

We generalize from this simple example. If x_i is a discrete random variable, we define the *expectation of the function* $h(x_i)$ by

$$E\{h(x_i)\} = \sum_i h(x_i)f(x_i) \tag{2.12-4}$$

where $f(x_i)$ is the density function associated with the random variable and the x_i are the values of the discrete random variable for which $f(x_i)$ is defined. In the same way, if x is a continuous random variable, we define the *expectation of the function* $h(x)$ by

$$E\{h(x)\} = \int_{-\infty}^{\infty} h(x)f(x)\,dx \tag{2.12-5}$$

where $f(x)$ is the density function associated with the random variable x.

As previously mentioned and as discussed in Appendix A for the case where the random variable can take on both discrete and continuous characteristics, we make use of the Stieltjes integral to write the expectation of $h(x)$ as

$$E\{h(x)\} = \int_{-\infty}^{\infty} h(x)\, dF(x) \tag{2.12-6}$$

where $F(x)$ is the c.d.f. corresponding to the density $f(x)$. Of course this last equation includes Equations 2.12-4 and 2.12-5 as special cases and could be used as the general definition of the expectation of $h(x)$. However, it will often be convenient to use the other forms when there is no confusion as to the nature of the random variable.

The expectation has a number of rather obvious properties:

1. $E\{h_1(x) + h_2(x)\} = E\{h_1(x)\} + E\{h_2(x)\}$. (2.12-7)
2. $E\{Ch(x)\} = CE\{h(x)\}$. (2.12-8)

These two are *linearity* properties and are a consequence of the fact that the expectation is a *linear operator*. In addition:

3. $E\{C\} = C; \qquad E\{1\} = 1$. (2.12-9)
4. $E\{h(x)\} \geq 0, \quad \text{if} \quad h(x) \geq 0 \quad \text{for all } x$. (2.12-10)

These properties are obtained in an elementary way from the definition of expectation.

2.13 Moments

One of the principal problems that will arise later is that of characterizing an arbitrary distribution or density function. The *moments* of a distribution will turn out to be one of the most convenient ways for such characterization.

For the discrete case, we define the rth *moment about the origin* for a given density function $f(x_i)$ as

$$u'_r = E\{x_i^r\} = \sum_i x_i^r f(x_i) \tag{2.13-1}$$

and, for the continuous case,

$$u'_r = E\{x^r\} = \int_{-\infty}^{\infty} x^r f(x)\, dx. \tag{2.13-2}$$

In particular, the first moment about the origin is called the *mean*. It is usually written as u and, for the continuous case, we have

$$u = u'_1 = E\{x\} = \int_{-\infty}^{\infty} xf(x)\, dx. \tag{2.13-3}$$

This is just the centroid of the area under the curve $f(x)$.

The rth *moment about the mean* is defined for the discrete case and for the density function $f(x_i)$ as

$$u_r = E\{(x_i - u)^r\} = \sum_i (x_i - u)^r f(x_i). \tag{2.13-4}$$

For the continuous case, we have

$$u_r = E\{(x - u)^r\} = \int_{-\infty}^{\infty} (x - u)^r f(x)\, dx. \tag{2.13-5}$$

In particular, the second moment about the mean is called the *variance*, is usually written as σ^2, and is given for the continuous case by

$$\sigma^2 = u_2 = E\{(x - u)^2\} = \int_{-\infty}^{\infty} (x - u)^2 f(x)\, dx. \tag{2.13-6}$$

Moments have a number of interesting properties. Some of the most obvious and most useful are:

1. *The first moment about the mean is zero.*

$$u_1 = E\{x - u\} = E\{x\} - u = u - u = 0,$$

$$\boxed{u_1 = 0.} \tag{2.13-7}$$

2. *The second moment about the origin is equal to the variance plus the square of the mean.*

$$\sigma^2 = E\{(x - u)^2\} = E\{x^2\} - 2uE\{x\} + u^2,$$

$$\sigma^2 = u_2' - u^2,$$

$$\boxed{u_2' = \sigma^2 + u^2.} \tag{2.13-8}$$

3. *The least second moment is the variance.* We find the value of b for which $E\{(x - b)^2\}$ is a minimum. Expanding,

$$E\{(x - b)^2\} = E\{x^2\} - 2bu + b^2.$$

We take the partial derivative with respect to b and equate the result to zero:

$$\frac{\partial}{\partial b}[E\{x^2\} - 2bu + b^2] = -2u + 2b = 0$$

$$\boxed{b = u.}$$

Check for minimum: the second derivative is

$$\frac{\partial}{\partial b}(-2u + 2b) = 2 \qquad \text{positive!}$$

4. *There is a general relationship between moments about the mean (central moments) and moments about the origin.* From the binomial expansion:

$$(x - u)^r = \sum_{s=0}^{r} (-1)^s \binom{r}{s} x^{r-s} u^s.$$

We write

$$E\{(x - u)^r\} = \sum_{s=0}^{r} (-1)^s \binom{r}{s} u^s E\{x^{r-s}\}$$

or

$$\boxed{u_r = \sum_{s=0}^{r} (-1)^s \binom{r}{s} u^s u'_{r-s}.} \tag{2.13-9}$$

For the case $r = 2$, we have

$$u_2 = \sigma^2 = u'_2 - 2u^2 + u^2$$

or

$$u'_2 = \sigma^2 + u^2$$

as was obtained in property 2.

It is apparent that the mean u is a measure of the central tendency or center of gravity of a density. On the other hand, the variance σ^2 is a measure of the dispersion of the density about the mean. A density with a small variance tends to be concentrated about the mean. The positive square root σ of the variance is often used instead of the variance and is called the *standard deviation.*

One of the most commonly encountered transformations of random variables is the *linear transformation*

$$y = a + bx \tag{2.13-10}$$

where a and b are constants and x and y are random variables. In this case the means and variances of x and y are simply related. Let u_y be the mean of y, and u_x the mean of x. Then

$$u_y = E\{y\} = a + bu_x$$

$$\boxed{u_y = a + bu_x.} \tag{2.13-11}$$

In the same way, the variance of y is

$$\begin{aligned}\sigma_y^2 &= E\{(y - u_y)^2\} \\ &= E\{(a + bx - a - bu_x)^2\} \\ &= b^2E\{(x - u_x)^2\} = b^2\sigma_x^2\end{aligned}$$

$$\boxed{\sigma_y^2 = b^2\sigma_x^2.} \tag{2.13-12}$$

2.14 The Tchebycheff Inequality

As mentioned in the previous section, the variance σ^2, or the standard deviation σ, is closely related to the idea of the dispersion of the distribution about the mean. This relationship is shown clearly by the following very useful and very general inequality:

The Tchebycheff Inequality: For any density with finite mean u and variance σ^2,

$$P(|x - u| \geq \lambda\sigma) \leq \frac{1}{\lambda^2} \tag{2.14-1}$$

for any real $\lambda > 0$.

Proof: Define a random variable y by

$$y = \begin{cases} 0 & \text{for} \quad |x - u| < \lambda\sigma, \\ \lambda^2\sigma^2 & \text{for} \quad |x - u| \geq \lambda\sigma. \end{cases} \tag{2.14-2}$$

The expected value of y is given by

$$E\{y\} = 0[P(y = 0)] + \lambda^2\sigma^2[P(y = \lambda^2\sigma^2)]$$

or

$$E\{y\} = \lambda^2\sigma^2P(|x - u| \geq \lambda\sigma). \tag{2.14-3}$$

It is apparent from Equation 2.14-2 that

$$y \leq (x - u)^2$$

or that

$$E\{y\} \leq E\{(x - u)^2\} = \sigma^2.$$

Substituting from Equation 2.14-3 for $E\{y\}$, we obtain

$$\lambda^2\sigma^2P(|x - u| \geq \lambda\sigma) \leq \sigma^2$$

which is the inequality desired.

2.15 Generating Functions

We consider now two functions which are closely related to the moments of a distribution. These will be useful later in the representation and manipulation of moment relationships. They are the *moment-generating function* $M(t)$ and the *characteristic function* $F(jv)$.

The moment-generating function (m.g.f.) $M(t)$ is defined for a random variable x with density function $f(x)$ as

$$M(t) = E\{e^{tx}\}. \tag{2.15-1}$$

Since e^{tx} has the power series expansion

$$e^{tx} = 1 + tx + \frac{t^2x^2}{2!} + \cdots + \frac{t^nx^n}{n!} + \cdots, \tag{2.15-2}$$

we may write Equation 2.15-1 as

$$M(t) = 1 + tE\{x\} + \frac{t^2}{2!}E\{x^2\} + \cdots + \frac{t^n}{n!}E\{x^n\} + \cdots \tag{2.15-3}$$

or as

$$M(t) = 1 + tu_1' + \frac{t^2}{2}u_2' + \cdots + \frac{t^n}{n!}u_n' + \cdots. \tag{2.15-4}$$

It is apparent that the moments u_i' are easily found as follows:

$$M(0) = 1 \tag{2.15-5}$$

$$\left.\frac{dM(t)}{dt}\right|_{t=0} = u_1' \tag{2.15-6}$$

$$\vdots$$

$$\left.\frac{d^nM(t)}{dt^n}\right|_{t=0} = u_n'. \tag{2.15-7}$$

It can be shown that the m.g.f. uniquely specifies the distribution $f(x)$; that is, no two distinct distributions correspond to the same m.g.f.

The characteristic function $F(jv)$ is defined for a random variable x with density function $f(x)$ as

$$F(jv) = E\{e^{jvx}\} \tag{2.15-8}$$

where $j = \sqrt{-1}$. For the continuous random variable x, this becomes

$$F(jv) = \int_{-\infty}^{\infty} f(x)\, e^{jvx}\, dx \tag{2.15-9}$$

and for the discrete random variable x_i,

$$F(jv) = \sum_i f(x_i)\, e^{jvx_i}. \tag{2.15-10}$$

Equation 2.15-9 is the Fourier transform of $f(x)$ (actually, the inverse Fourier transform). Consequently, from the theory of Fourier integrals [6], this equation may be inverted to give

$$f(x) = \frac{1}{2\pi}\int_{-\infty}^{\infty} F(jv)\, e^{-jvx}\, dv. \tag{2.15-11}$$

The function $F(jv)$ is uniquely determined by $f(x)$ through Equation 2.15-9 and $f(x)$ is uniquely determined by $F(jv)$ through Equation 2.15-11.

Equation 2.15-8 may be expanded as in the case of the m.g.f. to yield

$$F(jv) = 1 + jvu_1' + \frac{(jv)^2}{2!} u_2' + \cdots + \frac{(jv)^n}{n!} u_n' + \cdots \tag{2.15-12}$$

or

$$\left.\frac{d^n F(jv)}{dv^n}\right|_{v=0} = (j)^n u_n'. \tag{2.15-13}$$

A number of specific examples of both the characteristic function and the moment-generating function will be given in the next sections where certain special distributions that are commonly encountered are considered in some detail.

2.16 The Binomial Distribution

This distribution $b(k; n, p)$ has been given previously as Example 2.7 of Section 2.10 and a special case is plotted in Figure 2.7. The importance of the binomial distribution in our studies lies in the fact that both the Poisson distribution and the normal distribution can be considered to be limiting forms of the binomial distribution. We proceed now to derive the form of this distribution.

In many of the simplest situations where probability concepts are to be applied, it happens that there is a sequence of repeated independent trials with only two possible outcomes on each trial and with the probabilities of these two outcomes fixed throughout the trials. Such sequences are called *Bernoulli trials.* Perhaps the commonest example is a sequence of coin tossings, where a typical sequence might be

$$HTTHTHH\cdots.$$

It is conventional to call one of the two outcomes S (success) with probability of occurrence given by p and the other outcome F (failure) with probability $q = 1 - p$. Since the individual trials are independent by definition, the probabilities multiply and the probability of any given sequence is the product obtained when S is replaced by p and F by $q = 1 - p$. Thus, for the previous example, the probability P of the sequence is

$$P = p(1 - p)(1 - p)p(1 - p)pp\cdots, \tag{2.16-1}$$

or, for a sequence of length n,

$$P = p^k(1 - p)^{n-k} \tag{2.16-2}$$

where k is the number of successes and $(n - k)$ is the number of failures.

Example 2.11 Consider the successive tosses of a true die. Let throwing a two be considered success and all other outcomes failure. Then the probability P of the Bernoulli sequence $F2FFFF2$, where F stands for any number 1, 3, 4, 5, 6, is

$$P = (\tfrac{1}{6})^2(\tfrac{5}{6})^5.$$

• •

Suppose that we are interested in the probability of a given number of successes (say k successes) in n Bernoulli trials. One way to obtain k successes would be to have the sequence

$$\underbrace{SSS\cdots S}_{k}\,\underbrace{FFF\cdots F}_{n-k}$$

where the first k trials are successes and the last $(n - k)$ are failures. The probability P of this sequence has already been given by Equation 2.16-2 and is

$$P = p^k(1 - p)^{n-k}. \tag{2.16-2}$$

It is apparent that any other sequence containing exactly k successes will be acceptable. In fact there are available n cells or slots into which k successes can be put (the remaining cells each have a failure placed in them in each case). The number of different ways in which this can be done is just the number of combinations of n things taken k at a time or

$$\binom{n}{k} = \frac{n!}{(n - k)!k!}. \tag{2.16-3}$$

Each of these $\binom{n}{k}$ sequences has a probability given by Equation 2.16-2. Consequently the probability of exactly k successes in n Bernoulli trials is

$$b(k; n, p) = \binom{n}{k}p^k(1 - p)^{n-k} \tag{2.16-4}$$

where p is the probability of success in a single trial.

The quantity $b(k; n, p)$ is called the *binomial distribution.* It is the kth term in the binomial expansion of $(p + q)^n$

$$(p + q)^n = \sum_{k=0}^{n} \binom{n}{k} p^k q^{n-k} = \sum_{k=0}^{n} b(k; n, p). \tag{2.16-5}$$

It is apparent that $q = 1 - p$ and, consequently, that

$$[p + (1 - p)]^n = \sum_{k=0}^{n} b(k; n, p) = 1. \tag{2.16-6}$$

Since, in addition, $b(k; n, p)$ is nonnegative, it is a density function.

The mean u of the binomial distribution may be found as follows:

$$u = E\{k\} = \sum_{k=0}^{n} kb(k; n, p) \tag{2.16-7}$$

or

$$u = \sum_{k=0}^{n} k \frac{n(n-1)\cdots(n-k+1)}{k(k-1)\cdots(3)(2)(1)} p^k(1 - p)^{n-k}. \tag{2.16-8}$$

Since the term for which $k = 0$ is zero, we may write

$$u = np \sum_{k=1}^{n} \binom{n-1}{k-1} p^{k-1}(1 - p)^{(n-1)-(k-1)}. \tag{2.16-9}$$

If we change the index of summation by placing $i = k - 1$, then Equation 2.16-9 becomes

$$u = np \sum_{i=0}^{n-1} \binom{n-1}{i} p^i(1 - p)^{(n-1)-i} \tag{2.16-10}$$

or

$$u = np(p + 1 - p)^{n-1} = np \tag{2.16-11}$$

so that we conclude that the mean u is given by np. This result is intuitively reasonable. In a large number of trials n with probability p of success on each trial, we would expect approximately np successes.

If other moments of the binomial distribution are desired, it will be profitable to calculate the m.g.f. and obtain the moments by differentiation. The m.g.f. $M(t)$ is given by

$$M(t) = E\{e^{tk}\} = \sum_{k=0}^{n} e^{tk} \binom{n}{k} p^k(1 - p)^{n-k} \tag{2.16-12}$$

or

$$M(t) = \sum_{k=0}^{n} \binom{n}{k} (p\, e^t)^k(1 - p)^{n-k} \tag{2.16-13}$$

$$M(t) = (p\, e^t + 1 - p)^n. \tag{2.16-14}$$

We note that $M(0) = 1$ as required by Equation 2.15-4. Also the mean u can be found from

$$u = \left.\frac{dM(t)}{dt}\right|_{t=0} = n(p + 1 - p)^{n-1}p = np$$

as was found previously by direct calculation. The second moment about the origin is

$$u_2' = \left.\frac{d^2M(t)}{dt^2}\right|_{t=0}.$$

We have

$$\frac{d^2M(t)}{dt^2} = n(n-1)(p\,e^t + 1 - p)^{n-2}p^2\,e^{2t} + n(p\,e^t + 1 - p)^{n-1}p\,e^t$$

or

$$u_2' = n(n-1)p^2 + np = \sigma^2 + u^2. \tag{2.16-15}$$

Consequently the variance σ^2 is given by

$$\sigma^2 = np(1-p). \tag{2.16-16}$$

For large n, it will often be convenient to work with the normalized random variable

$$y = \frac{k}{n} \tag{2.16-17}$$

which cannot exceed unity. For example, if k is the number of heads observed in n tosses of a coin, then y is the proportion of heads. The mean of y is

$$E\{y\} = E\left\{\frac{k}{n}\right\} = \frac{np}{n} = p \tag{2.16-18}$$

and the variance is

$$\sigma_y^2 = \frac{np(1-p)}{n^2} = \frac{p(1-p)}{n}. \tag{2.16-19}$$

This last expression indicates that, as n becomes large, the distribution of y clusters more and more about the mean p. A question of some interest is: How close to the mean p is y for large n? Since y is a random variable, the question can be answered only in a probabilistic sense. We have, from the Tchebycheff inequality,

$$P(|y - p| \geq \epsilon) \leq \frac{\sigma_y^2}{\epsilon^2} \tag{2.16-20}$$

for any arbitrary $\epsilon > 0$. Since σ_y^2 is given by Equation 2.16-19, we write

$$\lim_{n\to\infty} P(|y - p| \geq \epsilon) = \lim_{n\to\infty} \frac{p(1-p)}{n\epsilon^2} = 0 \tag{2.16-21}$$

for $\epsilon > 0$. Thus, in a probabilistic sense, the random variable y can be made to approach the mean p arbitrarily closely by making the number of trials n large enough. This relationship is called *Bernoulli's Theorem* and is a special case of the *Law of Large Numbers* [6]. The type of convergence given by Equation 2.16-21 is called *convergence in probability*; that is, we say that y converges to p in probability. This subject will be discussed further in Section 3.5.

2.17 The Poisson Distribution

As previously mentioned, this distribution can be obtained as a limiting approximation to the binomial distribution. Suppose the situation exists where n is very large and p is very small but the product $np = \lambda$ is of reasonable size. More precisely, we want n to approach infinity and p to approach zero in such a way that the product λ exists and is nonzero. For this case we seek an approximation to the binomial distribution in a form more amenable to calculation and manipulation.

Consider the term $b(0; n, p)$:

$$b(0; n, p) = \binom{n}{0} p^0 (1 - p)^n = (1 - p)^n. \tag{2.17-1}$$

Replace p by λ/n so that

$$\lim_{n \to \infty} b(0; n, p) = \lim_{n \to \infty} \left(1 - \frac{\lambda}{n}\right)^n = e^{-\lambda} \tag{2.17-2}$$

since

$$\lim_{x \to \infty} \left(1 + \frac{1}{x}\right)^x = e. \tag{2.17-3}$$

Thus, for large n, the term $b(0; n, p)$ is given approximately by $e^{-\lambda}$ where $\lambda = np$. In the same way

$$b(1; n, p) = \binom{n}{1} p (1 - p)^{n-1} = \frac{np}{1 - p} b(0; n, p) \tag{2.17-4}$$

and

$$\lim_{n \to \infty} b(1; n, p) = \lim_{n \to \infty} \frac{\lambda}{1 - \lambda/n} e^{-\lambda} = \lambda e^{-\lambda}. \tag{2.17-5}$$

We continue and find that

$$\lim_{n \to \infty} b(2; n, p) = \frac{\lambda^2}{2} e^{-\lambda}. \tag{2.17-6}$$

Finally

$$\lim_{n \to \infty} b(k; n, p) = p(k; \lambda) = \frac{\lambda^k}{k!} e^{-\lambda}, \qquad k = 0, 1, 2, \ldots \tag{2.17-7}$$

if p approaches zero in such a way that $np = \lambda$. For large n, Equation 2.10-7 is called the *Poisson approximation to the binomial distribution* and, in the limit, is the *Poisson distribution*.

We note that

$$\sum_{n=0}^{\infty} p(k; \lambda) = e^{-\lambda} \sum_{n=0}^{\infty} \frac{\lambda^k}{k!} \tag{2.17-8}$$

or, from the power series expansion of e^{λ},

$$\sum_{n=0}^{\infty} p(k; \lambda) = e^{-\lambda} e^{\lambda} = 1. \tag{2.17-9}$$

Thus, since $p(k; \lambda)$ is also nonnegative, it is a density function if defined for $k = 0, 1, 2, \ldots$.

We would suppose that the mean of the Poisson distribution is given by $\lambda = np$. By direct calculation,

$$\begin{aligned} E\{k\} &= \sum_{k=0}^{\infty} k \frac{\lambda^k}{k!} e^{-\lambda} \\ &= \lambda e^{-\lambda} \sum_{k=1}^{\infty} \frac{\lambda^{k-1}}{(k-1)!}. \end{aligned} \tag{2.17-10}$$

Again, we let $i = k - 1$ and find that

$$\begin{aligned} E\{k\} &= \lambda e^{-\lambda} \sum_{i=0}^{\infty} \frac{\lambda^i}{i!} = \lambda e^{-\lambda} e^{\lambda} \\ &= \lambda \end{aligned} \tag{2.17-11}$$

as expected. The second moment may be found in a similar fashion:

$$\begin{aligned} E\{k(k-1)\} &= \sum_{k=0}^{\infty} k(k-1) \frac{\lambda^k}{k!} e^{-\lambda} \\ &= \lambda^2 e^{-\lambda} \sum_{k=2}^{\infty} \frac{\lambda^{k-2}}{(k-2)!} \end{aligned} \tag{2.17-12}$$

or

$$E\{k^2\} - E\{k\} = \lambda^2 e^{-\lambda} e^{\lambda} = \lambda^2. \tag{2.17-13}$$

Equation 2.17-13 may be rewritten as

$$u_2' - u = \lambda^2 = u_2' - \lambda$$

or, since $u_2' = \sigma^2 + u^2$,

$$\sigma^2 = \lambda^2 - \lambda^2 + \lambda = \lambda. \tag{2.17-14}$$

The variance and the mean are both equal to λ.

These moments could have been found from the m.g.f. which is given by

$$M(t) = E\{e^{tk}\} = \sum_{k=0}^{\infty} e^{tk}\, e^{-\lambda} \frac{\lambda^k}{k!} \tag{2.17-15}$$

and

$$M(t) = e^{-\lambda} \sum_{k=0}^{\infty} \frac{(\lambda\, e^t)^k}{k!} = e^{-\lambda}\, e^{\lambda e t}$$

or, finally,

$$M(t) = e^{\lambda(e t - 1)} \tag{2.17-16}$$

is the moment-generating function for the Poisson distribution. We have

$$M(0) = 1, \tag{2.17-17}$$

$$\left.\frac{dM(t)}{dt}\right|_{t=0} = e^{\lambda(e^t - 1)}\lambda\, e^t\big|_{t=0} = \lambda, \tag{2.17-18}$$

$$\frac{d^2M(t)}{dt^2} = e^{\lambda(e^t - 1)}\lambda^2\, e^{2t} + e^{\lambda(e^t - 1)}\lambda\, e^t,$$

$$u_2' = \lambda^2 + \lambda \tag{2.17-19}$$

as before.

2.18 The Normal or Gaussian Distribution

We can show [6] that if x is binomially distributed; that is, if

$$f(x) = \binom{n}{x} p^x(1 - p)^{n-x}, \qquad x = 0, 1, 2, \ldots, \tag{2.18-1}$$

then, in the limit as n becomes large, the density of x approaches

$$f(x) \sim \frac{1}{\sqrt{np(1 - p)}\sqrt{2\pi}} \exp\left[-\frac{(x - np)^2}{2np(1 - p)}\right]. \tag{2.18-2}$$

In the limit as $n \to \infty$, the random variable x is continuous $(-\infty, \infty)$ and

$$\lim_{n\to\infty} f(x) = \frac{1}{\sqrt{2\pi\sigma^2}} \exp\left[-\frac{1}{2}\left(\frac{x - u}{\sigma}\right)^2\right], \qquad -\infty < x < \infty \tag{2.18-3}$$

where

$$u = np,$$
$$\sigma^2 = np(1 - p).$$

As discussed by Feller [1], the approximation is amazingly good for small n ($n \geq 5$), provided p is not too far from $\frac{1}{2}$.

Equation 2.18-3 is called the *normal density or distribution* (sometimes called the *gaussian distribution*) with mean u and variance σ^2. For the moment, we will take $u = 0$ and $\sigma^2 = 1$ and consider the *unit normal distribution*

$$f(x) = \frac{1}{\sqrt{2\pi}} e^{-x^2/2}, \qquad -\infty < x < \infty. \tag{2.18-4}$$

This density has already been plotted in Figure 2.10(a). It is nonnegative and we can show that the area under the curve is equal to unity; that is, that

$$\frac{1}{\sqrt{2\pi}} \int_{-\infty}^{\infty} e^{-x^2/2}\, dx = 1. \tag{2.18-5}$$

Let us denote the integral of Equation 2.18-5 by J and write

$$\begin{aligned} J^2 &= \frac{1}{2\pi} \int_{-\infty}^{\infty} e^{-x^2/2}\, dx \int_{-\infty}^{\infty} e^{-y^2/2}\, dy \\ &= \frac{1}{2\pi} \int_{-\infty}^{\infty} e^{-(x^2+y^2)/2}\, dx\, dy. \end{aligned}$$

We may change to polar coordinates by the substitution

$$\begin{aligned} r^2 &= x^2 + y^2 \\ r\, dr\, d\theta &= dx\, dy \end{aligned}$$

to obtain

$$J^2 = \frac{1}{2\pi} \int_0^{2\pi} \int_0^{\infty} r\, e^{-r^2/2}\, dr\, d\theta. \tag{2.18-6}$$

This expression may be integrated directly to yield

$$\begin{aligned} J^2 &= \frac{1}{2\pi} \int_0^{2\pi} \{-e^{-r^2/2}\}_0^{\infty}\, d\theta \\ &= \frac{1}{2\pi} \int_0^{2\pi} d\theta = 1. \end{aligned} \tag{2.18-7}$$

Since J must be positive, we conclude that $J = 1$ and that Equation 2.18-5 is valid. It is apparent that the same result is obtained for the general form of Equation 2.18-3, since, after the substitution

$$y = \frac{x - u}{\sigma}, \tag{2.18-8}$$

Equation 2.18-5 is obtained as the integral of the general normal distribution.

The cumulative distribution function for the unit normal distribution is given by

$$F(a) = \frac{1}{\sqrt{2\pi}} \int_{-\infty}^{a} e^{-x^2/2}\, dx. \tag{2.18-9}$$

Since the integrand is even,

$$F(-a) = 1 - F(a). \tag{2.18-10}$$

For the case $a = -a = 0$, we have

$$F(0) = \tfrac{1}{2}. \tag{2.18-11}$$

Since the normal c.d.f. cannot be evaluated directly, it is necessary to refer to tables which have been obtained by some numerical technique. One often finds tables, not of the c.d.f., but of the *error integral* (or *error function*) $\theta(t)$ instead, where

$$\theta(t) = \operatorname{erf} t = \frac{2}{\sqrt{\pi}} \int_0^t e^{-y^2}\, dy. \tag{2.18-12}$$

Let us make the substitution

$$y = \frac{x}{\sqrt{2}}. \tag{2.18-13}$$

Then Equation 2.18-12 becomes

$$\theta(t) = \frac{2}{\sqrt{2\pi}} \int_0^{\sqrt{2}t} e^{-x^2/2}\, dx. \tag{2.18-14}$$

It is apparent from Equation 2.18-11 that

$$\theta(t) = 2F(\sqrt{2}t) - 1 \tag{2.18-15}$$

relates the c.d.f. for the unit normal distribution to the error function. As a matter of convenience, Table 2.1 gives the values of the density $f(x)$ and the c.d.f. $F(x)$ for a range of values of x. (A more complete listing is given in Table 1 in Appendix I.) For the normal distribution with nonzero mean u and variance σ^2, the linear change of variable of Equation 2.18-8 may be employed and the data of Table 2.1 used.

Example 2.12 Find the c.d.f. $F_1(1.2)$ for the normal distribution with mean of 0.6 and variance of 9.0.

We want to find

$$F_1(1.2) = \frac{1}{\sqrt{2\pi 9}} \int_{-\infty}^{1.2} \exp\left[-\frac{1}{2}\left(\frac{x - 0.6}{3}\right)^2\right] dx.$$

We let

$$y = \frac{x - 0.6}{3}, \qquad dx = 3dy.$$

Table 2.1 The Unit Normal Distribution[a,b]

x	$f(x)$	$F(x)$
0.0	0.3989	0.5000
0.2	0.3910	0.5793
0.4	0.3683	0.6554
0.6	0.3332	0.7257
0.8	0.2897	0.7881
1.0	0.2420	0.8413
1.2	0.1942	0.8849
1.4	0.1497	0.9192
1.6	0.1109	0.9452
1.8	0.0790	0.9641
2.0	0.0540	0.9773
2.2	0.0355	0.9861
2.4	0.0224	0.9918
2.6	0.0136	0.9953
2.8	0.0080	0.9974
3.0	0.0044	0.9987
3.5	0.0009	0.9998
4.0	0.0001	0.9999^+
4.5	0.00002	0.9999^+

[a] $f(x)$ = unit normal density $(1/\sqrt{2\pi})\, e^{-x^2/2}$.
[b] $F(x)$ = unit normal c.d.f. $(1/\sqrt{2\pi}) \int_{-\infty}^{x} e^{-y^2/2}\, dy$.

Then

$$F_1(1.2) = \frac{1}{\sqrt{2\pi}} \int_{-\infty}^{(1.2-0.6)/3} e^{-y^2/2}\, dy$$

$$= F\left(\frac{1.2 - 0.6}{3}\right) = F(0.2) = 0.5793$$

from Table 2.1. Thus

$$F_1(1.2) = 0.5793.$$

• •

The moments for the unit normal distribution are found most conveniently from the m.g.f. $M(t)$ where

$$M(t) = \frac{1}{\sqrt{2\pi}} \int_{-\infty}^{\infty} e^{tx}\, e^{-x^2/2}\, dx. \tag{2.18-16}$$

This integral is typical of many integrals connected with the normal distribution. The general procedure to follow is to complete the square in the variable of integration and then to make a linear change in variable. We have

$$M(t) = e^{t^2/2} \frac{1}{\sqrt{2\pi}} \int_{-\infty}^{\infty} e^{-(x-t)^2/2}\, dx. \tag{2.18-17}$$

We make the substitution

$$y = x - t$$

and obtain

$$M(t) = e^{t^2/2} \frac{1}{\sqrt{2\pi}} \int_{-\infty}^{\infty} e^{-y^2/2}\, dy \tag{2.18-18}$$

or

$$M(t) = e^{t^2/2} \tag{2.18-19}$$

for the result. In the same way, we find the characteristic function $F(jv)$ to be

$$F(jv) = \frac{1}{\sqrt{2\pi}} \int_{-\infty}^{\infty} e^{jvx}\, e^{-x^2/2}\, dx \tag{2.18-20}$$

or

$$F(jv) = e^{-v^2/2}. \tag{2.18-21}$$

From Equation 2.18-19 we write

$$M(t) = 1 + \frac{t^2}{2} + \frac{1}{2!}\left(\frac{t^2}{2}\right)^2 + \cdots + \frac{1}{n!}\left(\frac{t^2}{2}\right)^n + \cdots. \tag{2.18-22}$$

The nth moment about the origin u_n' is the coefficient of $t^n/n!$ in this power series expansion of $M(t)$. There are no odd powers of t in Equation 2.18-22 and for n even, the $n/2$th term is

$$\frac{1}{(n/2)!}\left(\frac{t^2}{2}\right)^{n/2} = \frac{n(n-1)\cdots[(n/2)+1]}{2^{n/2}} \frac{t^n}{n!}. \tag{2.18-23}$$

Consequently, the moments are

for n odd: $$u_n' = 0,$$

for n even: $$u_n' = u_n = \frac{n(n-1)\cdots[(n/2)+1]}{2^{n/2}}. \tag{2.18-24}$$

In particular,

$$u_1' = u = 0, \tag{2.18-25}$$

$$u_2 = \sigma^2 = 1. \tag{2.18-26}$$

Let us consider again the linear transformation of Equation 2.18-8 where y is a random variable with unit normal distribution. Then x is given by

$$x = u + \sigma y \tag{2.18-27}$$

and the density of x is easily obtained as

$$f(x) = \frac{1}{\sqrt{2\pi\sigma^2}} \exp\left[-\frac{1}{2}\left(\frac{x-u}{\sigma}\right)^2\right], \qquad -\infty < x < \infty. \tag{2.18-28}$$

We have immediately from Equation 2.18-27 that

$$E\{x\} = u_x = u + \sigma E\{y\} = u \tag{2.18-29}$$

and

$$E\{(x - u_x)^2\} = \sigma^2\sigma_y^2 = \sigma^2 \tag{2.18-30}$$

as expected for the normal distribution of Equation 2.18-28 with mean u and variance σ^2. In the same way, the characteristic function $F(jv)$ is

$$F(jv) = \exp\left[\tfrac{1}{2}(-\sigma^2 v^2 + 2juv)\right]. \tag{2.18-31}$$

2.19 Limit Theorems

In this section we will consider very briefly three limit theorems of fundamental importance in the study of sequences of random variables and of sums of random variables. These three theorems are:

1. Bernoulli's Theorem.
2. The Law of Large Numbers.
3. The Central Limit Theorem.

Bernoulli's Theorem

This theorem has already been discussed at the end of Section 2.16. The theorem states that:

> The relative frequency y with which an event occurs in n Bernoulli trials converges in probability to the probability p of occurrence in a single trial; that is,
>
> $$\lim_{n\to\infty} P(|y - p| \geq \epsilon) = 0 \tag{2.19-1}$$
>
> for any $\epsilon > 0$; $y = k/n$ where k is the number of successes in n trials.

This theorem was proved in Section 2.16 by an application of Tchebycheff's Inequality. It is a special case of the Law of Large Numbers.

The Law of Large Numbers

This theorem states that:

> The average y of n identically distributed independent random variables x_i with the same mean u and variance σ^2 converges in probability to the mean u.

Let $x_1, x_2, \ldots, x_n$ be independent random variables with the same distribution and with mean u and variance σ^2 (they could be n independent observations of the same random variable). Define the average y as

$$y = \frac{x_1 + x_2 + \cdots + x_n}{n}. \tag{2.19-2}$$

If Tchebycheff's Inequality is applied to the random variable y, then

$$P(|y - u_y| \geq \lambda\sigma_y) \leq \frac{1}{\lambda^2}, \qquad \lambda > 0 \tag{2.19-3}$$

where u_y is the mean and σ_y^2 is the variance of y. If we let $\epsilon = \lambda\sigma_y$, then Equation 2.19-3 becomes

$$P(|y - u_y| \geq \epsilon) \leq \frac{\sigma_y^2}{\epsilon}. \tag{2.19-4}$$

We now find the mean u_y and the variance σ_y^2 for substitution in Equation 2.19-4.

The mean of y is

$$u_y = \frac{1}{n}\sum_i E\{x_i\} = \frac{nu}{n} = u \tag{2.19-5}$$

and the variance is

$$\sigma_y^2 = E\{(y - u_y)^2\} = \frac{1}{n^2}E\left\{\left[\sum_i (x_i - u)\right]^2\right\}. \tag{2.19-6}$$

This last expression is more conveniently written as

$$\sigma_y^2 = \frac{1}{n^2}E\left\{\sum_i (x_i - u)\sum_j (x_j - u)\right\}. \tag{2.19-7}$$

The double sum contains n terms where the indices are equal—that is, for $i = j$, and $(n^2 - n)$ terms where $i \neq j$. For the cases where $i = j$, each term is

$$E\{(x_i - u)^2\} = \sigma^2 \tag{2.19-8}$$

and there are n of these. For the cases where $i \neq j$, each term is

$$E\{(x_i - u)(x_j - u)\} = E\{x_i x_j\} - u^2. \tag{2.19-9}$$

Since x_i and x_j are independent, we can write

$$E\{x_i x_j\} = E\{x_i\}E\{x_j\} = u^2. \tag{2.19-10}$$

(This subject will be discussed in more detail in Section 2.20.) Consequently, all of the cross terms of Equation 2.19-7 are zero and σ_y^2 becomes

$$\sigma_y^2 = \frac{1}{n^2}n\sigma^2 = \frac{\sigma^2}{n}. \tag{2.19-11}$$

Now, if Equations 2.19-5 and 2.19-11 are substituted in Equation 2.19-4, the result, in the limit of large n, is

$$\lim_{n \to \infty} P(|y - u| \geq \epsilon) \leq \frac{\sigma^2}{n\epsilon} = 0 \tag{2.19-12}$$

since $\epsilon > 0$. Thus we say that y converges in probability to u as was to be proved.

The Central Limit Theorem

This is a very general limit theorem, which states that sums of independent random variables tend to be normally distributed even though the summands are not. More specifically:

> Let $x_1, x_2, \ldots, x_n$ be a series of independent random variables having arbitrary distributions, means $u_1, u_2, \ldots, u_n$, and variances $\sigma_1^2, \sigma_2^2, \ldots, \sigma_n^2$, respectively. We form the new (normalized) random variables y_n where
>
> $$y_n = \frac{\sum_{i=1}^{n} x_i - \sum_{i=1}^{n} u_i}{\left(\sum_{i=1}^{n} \sigma_i^2\right)^{1/2}}, \qquad n = 1, 2, \ldots, n \tag{2.19-13}$$
>
> so that the mean of y_n is zero and its variance is unity. Then, under very general conditions, the distribution of y_n approaches in the limit the unit normal distribution.

The necessary conditions under which this theorem holds are difficult to state [6], but a *sufficient* set of conditions are that two positive numbers m and M exist such that

$$\sigma_i^2 > m > 0 \tag{2.19-14}$$

and

$$E\{|x_i - u_i|^3\} < M \tag{2.19-15}$$

for all $i = 1, 2, \ldots.$ These conditions ensure that no one term in the sum dominates.

It is relatively easy to prove the Central Limit Theorem for the special case where the x_i are independent random variables with the same distribution and with mean u and variance σ^2. The proof consists of showing that the moment-generating function for y_n approaches in the limit the moment-generating function $e^{t^2/2}$ for the unit normal distribution. We proceed as follows:

We form y_n from Equation 2.19-13. For our special case it can be written as

$$y_n = \sum_{i=1}^{n} \frac{x_i - u}{\sqrt{n}\,\sigma}. \tag{2.19-16}$$

Let $M(t)$ be the moment-generating function for the random variable $(x_i - u)/\sigma$; that is,

$$M(t) = E\{e^{t(x_i - u)/\sigma}\}. \tag{2.19-17}$$

Then, if the symbol $M_n(t)$ is used to denote the m.g.f. for y_n, it follows from Equation 2.19-16 that

$$M_n(t) = E\{e^{ty_n}\} = E\left\{\exp\left(\frac{t}{\sqrt{n}}\sum_i \frac{x_i - u}{\sigma}\right)\right\}. \tag{2.19-18}$$

Since the x_i are independent, the expectation of the product is equal to the product of the expectations (see Section 2.20) and Equation 2.19-18 becomes

$$M_n(t) = \prod_{i=1}^{n} M\left(\frac{t}{\sqrt{n}}\right) = \left[M\left(\frac{t}{\sqrt{n}}\right)\right]^n. \tag{2.19-19}$$

It has been pointed out previously that the m.g.f. $M(t)$ may be expanded as

$$M(t) = 1 + tu_1' + \frac{t^2}{2!}u_2' + \cdots + \frac{t^n}{n!}u_n' + \cdots \tag{2.19-20}$$

or

$$M\left(\frac{t}{\sqrt{n}}\right) = 1 + \frac{t}{\sqrt{n}}u_1' + \frac{t^2}{2n}u_2' + \cdots + \frac{t^n}{n!(n)^{n/2}}u_n' + \cdots. \tag{2.19-21}$$

Since the terms $(x_i - u)/\sigma$ have zero means and variances of unity, Equation 2.19-21 becomes

$$M\left(\frac{t}{\sqrt{n}}\right) = 1 + 0 + \frac{t^2}{2n} + \cdots \text{ higher-order terms.} \tag{2.19-22}$$

If n is taken large enough, then the function $M(t/\sqrt{n})$ is approximated by

$$M\left(\frac{t}{\sqrt{n}}\right) \approx 1 + \frac{t^2}{2n} \approx e^{t^2/2n} \tag{2.19-23}$$

and it follows from Equation 2.19-19 that the m.g.f. for y_n is, in the limit of large n,

$$M_n(t) \approx (e^{t^2/2n})^n = e^{t^2/2} \tag{2.19-24}$$

which was to be proved.

The Central Limit Theorem holds under much weaker conditions and even, in certain cases, for dependent random variables; however, the interested reader is referred elsewhere for further discussion [6].

2.20 Bivariate Distributions

Thus far we have been concerned with single random variables and their distributions. We call these *univariate* cases. In the last section we encountered for the first time the expression $E\{x_i x_j\}$; here we were interested in statistical relations between two random variables or in the *bivariate* case. In general, when we are concerned with more than one random variable, we talk about *multivariate* statistics, distributions, and so on. For our purposes, it will be sufficient to consider bivariate problems. The extension to more than two random variables will usually be obvious.

The joint or bivariate cumulative distribution function $F(a_1, a_2)$ for two real numbers a_1 and a_2 is defined on the random variables x_1 and x_2 as

$$F(a_1, a_2) = P(x_1 \leq a_1 \text{ and } x_2 \leq a_2). \tag{2.20-1}$$

Here we use the subscripts 1 and 2 to distinguish the two random variables, which may be discrete or continuous. It will usually be apparent from the context as to which is the case. This c.d.f. $F(a_1, a_2)$ has the following properties:

1. $F(a_1, a_2)$ is monotone nondecreasing in both variables.
2. If *either* a_1 or a_2 approaches $-\infty$, then $F(a_1, a_2) \to 0$.
3. If *both* a_1 and a_2 approach ∞, then $F(a_1, a_2) \to 1$.

For the discrete case, we consider the *bivariate* or *joint density* $f(x_{i1}, x_{j2})$ as a function defined on the discrete random variables x_{i1} and x_{j2} and satisfying

$$1.\ f(x_{i1}, x_{j2}) \geq 0. \tag{2.20-2}$$

$$2.\ \sum_{i,j} f(x_{i1}, x_{j2}) = 1. \tag{2.20-3}$$

It is apparent that each bivariate density is related to a bivariate c.d.f. by

$$F(a_1, a_2) = \sum_{\substack{x_{i1} \leq a_1 \\ x_{j2} \leq a_2}} f(x_{i1}, x_{j2}). \tag{2.20-4}$$

For the continuous case, we consider the *bivariate* or *joint density* $f(x_1, x_2)$ as a function defined on the continuous random variables x_1 and x_2 and satisfying

$$1.\ f(x_1, x_2) \geq 0. \tag{2.20-5}$$

$$2.\ \int_{-\infty}^{\infty} \int_{-\infty}^{\infty} f(x_1, x_2)\, dx_1\, dx_2 = 1. \tag{2.20-6}$$

It is apparent again that each bivariate density is related to a bivariate c.d.f. by

$$F(a_1, a_2) = \int_{-\infty}^{a_2} \int_{-\infty}^{a_1} f(x_1, x_2)\, dx_1\, dx_2 \tag{2.20-7a}$$

and that, for continuous and differentiable $F(a_1, a_2)$,

$$f(x_1, x_2) = \frac{\partial^2}{dx_1\, \partial x_2} [F(x_1, x_2)]. \tag{2.20-7b}$$

As in the univariate case, an expectation operator can be defined. For discrete random variables, the expectation of the function $h(x_1, x_2)$ is given by

$$E\{h(x_{i1}, x_{j2})\} = \sum_{x_{i1}, x_{j2}} h(x_{i1}, x_{j2}) f(x_{i1}, x_{j2}) \tag{2.20-8}$$

and, in the continuous case, by

$$E\{h(x_1, x_2)\} = \int_{-\infty}^{\infty} \int_{-\infty}^{\infty} h(x_1, x_2) f(x_1, x_2)\, dx_1\, dx_2. \tag{2.20-9}$$

As a matter of convenience, we shall use the notation for the continuous random variable in the future even when we deal with the discrete case.

The expectation operator has the same properties in the bivariate case as were developed in Section 2.12 for univariate random variables:

1. $E\{h_1(x_1, x_2) + h_2(x_1, x_2)\} = E\{h_1(x_1, x_2)\} + E\{h_2(x_1, x_2)\}$. (2.20-10)
2. $E\{Ch(x_1, x_2)\} = CE\{h(x_1, x_2)\}$. (2.20-11)
3. $E\{C\} = C$. (2.20-12)
4. $E\{h(x_1, x_2)\} \geq 0 \quad \text{if} \quad h(x_1, x_2) \geq 0$. (2.20-13)

We come now to the concept of *independence*, which was originally defined in Section 2.6. In terms of our present notation, the definition of Equation 2.6-3 is as follows: Two random variable x_1 and x_2 are independent if

$$P(x_1 \leq a_1 \text{ and } x_2 \leq a_2) = P(x_1 \leq a_1) P(x_2 \leq a_2) \tag{2.20-14}$$

for all a_1 and a_2. In other words, Equation 2.20-14 states that

$$F(a_1, a_2) = F_1(a_1) F_2(a_2) \tag{2.20-15}$$

where $F(\cdot, \cdot)$ is the joint c.d.f. of x_1 and x_2, $F_1(\cdot)$ is the c.d.f. of x_1, and $F_2(\cdot)$ is the c.d.f. of x_2. Equation 2.20-15 implies that the bivariate density can be factored in the independent case, or that

$$f(x_1, x_2) = f_1(x_1) f_2(x_2) \tag{2.20-16}$$

where $f(\cdot, \cdot)$ is the bivariate density of x_1 and x_2, $f_1(\cdot)$ is the (univariate) density of x_1, and $f_2(\cdot)$ is the (univariate) density of x_2.

A fifth property of the expectation can be written when x_1 and x_2 are independent:

5. $E\{h_1(x_1) h_2(x_2)\} = E\{h_1(x_1)\} E\{h_2(x_2)\}$. (2.20-17)

The proof is straightforward, since Equation 2.20-17 can be rewritten from Equation 2.20-16 as

$$\int_{-\infty}^{\infty} \int_{-\infty}^{\infty} h_1(x_1) h_2(x_2) f_2(x_1, x_2)\, dx_1\, dx_2$$

$$= \int_{-\infty}^{\infty} h_1(x_1) f_1(x_1)\, dx_1 \int_{-\infty}^{\infty} h_2(x_2) f_2(x_2)\, dx_2.$$

In particular, if the random variables x_1 and x_2 are independent, then

$$E\{x_1, x_2\} = E\{x_1\}E\{x_2\}. \tag{2.20-18}$$

This was the relationship used several times in Section 2.19.

We proceed now to define moments for bivariate distributions. We define the *rs*th moment about the origin u'_{rs} for the distribution $f(x_1, x_2)$ as

$$\begin{aligned} u'_{rs} &= E\{x_1^r x_2^s\} \\ &= \int_{-\infty}^{\infty}\int_{-\infty}^{\infty} x_1^r x_2^s f(x_1, x_2)\, dx_1\, dx_2. \end{aligned} \tag{2.20-19}$$

In particular, the mean of x_1 is

$$u'_{10} = u_{x_1} = E\{x_1\} \tag{2.20-20}$$

and the mean of x_2 is

$$u'_{01} = u_{x_2} = E\{x_2\}. \tag{2.20-21}$$

As in the univariate case, we find it convenient to define also the *rs*th moment about the mean as

$$u_{rs} = E\{(x_1 - u_{x_1})^r(x_2 - u_{x_2})^s\}. \tag{2.20-22}$$

In particular, the variance of x_1 is

$$u_{20} = \sigma_{x_1}^2 = E\{(x_1 - u_{x_1})^2\} \tag{2.20-23}$$

and the variance of x_2 is

$$u_{02} = \sigma_{x_2}^2 = E\{(x_2 - u_{x_2})^2\}. \tag{2.20-24}$$

In addition, there is a third possibility, the second mixed moment u_{11}, called the *covariance* and sometimes written $\sigma_{x_1x_2}$:

$$u_{11} = \sigma_{x_1x_2} = E\{(x_1 - u_{x_1})(x_2 - u_{x_2})\}. \tag{2.20-25}$$

We note that the covariance is zero if x_1 and x_2 are independent, since, in that case,

$$u_{11} = E\{x_1 - u_{x_1}\}E\{x_2 - u_{x_2}\} = 0. \tag{2.20-26}$$

The reader will recall that, for the univariate case, the variance σ^2 and the second moment about the origin u'_2 were related by

$$u'_2 = \sigma^2 + u^2$$

where u was the mean of the distribution. In the bivariate case a similar relationship can be derived by expanding Equation 2.20-25:

$$\sigma_{x_1x_2} = E\{x_1x_2\} - u_{x_1}u_{x_2} - u_{x_2}u_{x_1} + u_{x_1}u_{x_2}$$

or

$$u'_{11} = \sigma_{x_1x_2} + u_{x_1}u_{x_2}. \tag{2.20-27}$$

We can normalize the covariance and define the *correlation coefficient* $\rho_{x_1x_2}$ by

$$\rho_{x_1x_2} = \frac{\sigma_{x_1x_2}}{\sigma_{x_1}\sigma_{x_2}}. \tag{2.20-28}$$

It can be shown that this quantity is bounded in magnitude by unity; that is,

$$-1 \leq \rho \leq 1. \tag{2.20-29}$$

Let us consider the nonnegative function

$$[(x_1 - u_{x_1})a + (x_2 - u_{x_2})b]^2 \geq 0 \tag{2.20-30}$$

where a and b are real variables. Taking the expectation of this function, we have

$$\sigma_{x_1}^2 a^2 + 2u_{11}ab + \sigma_{x_2}^2 b^2 \geq 0. \tag{2.20-31}$$

The left side of this equation is a homogeneous quadratic form in a and b and could be written as

$$Aa^2 + B2ab + Cb^2$$

where A and C are nonnegative. The condition that this expression be nonnegative is that it have no real roots, or that $AC - B^2$ be nonnegative; that is,

$$\sigma_{x_1}^2\sigma_{x_2}^2 - u_{11}^2 \geq 0 \tag{2.20-32}$$

whence

$$\rho^2 = \frac{u_{11}^2}{\sigma_{x_1}^2\sigma_{x_2}^2} \leq 1 \tag{2.20-33}$$

as was to be proved.

If x_1 and x_2 are linearly related, then we have equality in Equation 2.20-29. Let

$$x_1 = bx_2$$

where b is an arbitrary real nonzero number. It is clear that

$$\rho_{x_1x_2} = \frac{E\{(bx_2 - bu_{x_2})(x_2 - u_{x_2})\}}{|b|\sigma_{x_2}\sigma_{x_2}}$$

or

$$\rho_{x_1x_2} = \begin{cases} 1, & \text{if } b > 0, \\ -1, & \text{if } b < 0. \end{cases} \tag{2.20-34}$$

We consider now the mean and variance of a linear combination of random variables. Let y be defined by

$$y = \sum_i a_i x_i \tag{2.20-35}$$

where the x_i are random variables with means u_{x_i} and variances $\sigma_{x_i}^2$. The mean of y is

$$u_y = \sum_i a_i u_{x_i}$$

as has been established. The variance, on the other hand, is more complicated because of the presence of cross terms which may not be zero. We have

$$\sigma_y^2 = E\{(y - u_y)^2\} = E\left\{\left[\sum_i a_i(x_i - u_{x_i})\right]^2\right\} \tag{2.20-36}$$

or

$$\sigma_y^2 = \sum_i a_i^2\sigma_{x_i}^2 + \sum_i \sum_{\substack{j \\ i \neq j}} a_i a_j \sigma_{x_i x_j}. \tag{2.20-37}$$

If all the x_i are uncorrelated; that is, if all

$$\sigma_{x_i x_j} = 0, \qquad i \neq j,$$

then Equation 2.20-37 reduces to

$$\sigma_y^2 = \sum_i a_i^2 \sigma_{x_i}^2. \tag{2.20-38}$$

A sufficient condition for this to be true is that the x_i be independent as already discussed in conjunction with Equation 2.20-17. However, random variables may be uncorrelated but not independent; thus, although independence implies Equation 2.20-38, the converse is not necessarily true.

Suppose we are given the bivariate density $f(x_1, x_2)$. The probability of x_1 assuming some value less than or equal to a_1, independent of the value of x_2 is sometimes called the *marginal* c.d.f. of x_1 and is given by

$$F(a_1, \infty) = F_1(a_1) = \int_{-\infty}^{a_1} \int_{-\infty}^{\infty} f(x_1, x_2)\, dx_2\, dx_1. \tag{2.20-39}$$

In the same way, we define the marginal c.d.f. of x_2 as

$$F(\infty, a_2) = F_2(a_2) = \int_{-\infty}^{a_2} \int_{-\infty}^{\infty} f(x_1, x_2)\, dx_1\, dx_2. \tag{2.20-40}$$

The joint characteristic function $F(v_1, v_2)$ and the joint moment-generating function $M(t_1, t_2)$ are defined in a manner analogous to that of the univariate case:

$$\begin{aligned} F(v_1, v_2) &= E\{e^{j(v_1x_1 + v_2x_2)}\} \\ &= \int_{-\infty}^{\infty} \int_{-\infty}^{\infty} e^{j(v_1x_1 + v_2x_2)} f(x_1, x_2)\, dx_1\, dx_2 \end{aligned} \tag{2.20-41}$$

and

$$M(t_1, t_2) = E\{e^{t_1x_1 + t_2x_2}\}. \tag{2.20-42}$$

It would be more accurate to write $F_{x_1x_2}(v_1, v_2)$ instead of $F(v_1, v_2)$, but this kind of notation will be avoided unless absolutely necessary. It will usually be clear from the context as to what random variables are involved. As in one dimension we can invert Equation 2.20-41 to give

$$f(x_1, x_2) = \frac{1}{(2\pi)^2}\int_{-\infty}^{\infty}\int_{-\infty}^{\infty} e^{-j(v_1x_1+v_2x_2)}F(v_1, v_2)\,dv_1\,dv_2. \quad (2.20\text{-}43)$$

The problem of functional transformations is considerably more complicated for bivariate distributions although the principles are the same as in Section 2.11. Suppose that

$$y_1 = h_1(x_1, x_2) \quad (2.20\text{-}44)$$

and

$$y_2 = h_2(x_1, x_2) \quad (2.20\text{-}45)$$

are both one-to-one continuously differentiable transformations of (x_1, x_2) into (y_1, y_2). It is clear that

$$P\begin{pmatrix} a_1 \le x_1 \le a_1 + da_1 \\ \text{and} \\ a_2 \le x_2 \le a_2 + da_2 \end{pmatrix} = P\begin{pmatrix} b_1 \le y_1 \le b_1 + db_1 \\ \text{and} \\ b_2 \le y_2 \le b_2 + db_2 \end{pmatrix} \quad (2.20\text{-}46)$$

where $h_1(a_1, a_2) = b_1$ and $h_2(a_1, a_2) = b_2$. In other words, we have

$$f(x_1, x_2)\,dx_1\,dx_2 = g(y_1, y_2)\,dy_1\,dy_2 \quad (2.20\text{-}47)$$

where $f(x_1, x_2)$ is the joint density of the x's and $g(y_1, y_2)$ is the joint density of the y's. This expression may be written as

$$f(x_1, x_2) = g(y_1, y_2)|J(y/x)| \quad (2.20\text{-}48)$$

where $J(y/x)$ is the *Jacobian* of the transformation of Equations 2.20-44 and 2.20-45 and is given by the determinant

$$J(y/x) = \begin{vmatrix} \dfrac{\partial y_1}{\partial x_1} & \dfrac{\partial y_1}{\partial x_2} \\ \dfrac{\partial y_2}{\partial x_1} & \dfrac{\partial y_2}{\partial x_2} \end{vmatrix}. \quad (2.20\text{-}49)$$

This is just a problem in coordinate transformation and is discussed in more detail in Appendix C, where the general n-dimensional case is considered. Note that in the univariate case, we have

$$J(y/x) = \frac{dy}{dx} \quad (2.20\text{-}50)$$

which agrees with the results of Section 2.11. As pointed out in Appendix C, Equation 2.20-48 may be written as

$$g(y_1, y_2) = f(x_1, x_2)|J(x/y)|. \quad (2.20\text{-}51)$$

Example 2.13 Suppose $y_1 = x_1 + x_2$ where x_1 and x_2 are independent random variables with densities $f(x_1, x_2) = f_1(x_1)f_2(x_2)$. Find the density $g_1(y_1)$.

Let $y_2 = x_2$ for convenience and form

$$J(y/x) = \begin{vmatrix} 1 & 1 \\ 0 & 1 \end{vmatrix} = 1.$$

Then $g(y_1, y_2) = f_1(x_1)f_2(x_2)$ or

$$g_1(y_1) = \int_{-\infty}^{\infty} g(y_1, y_2)\, dy_2$$

$$= \int_{-\infty}^{\infty} f_1(y_1 - y_2)f_2(y_2)\, dy_2.$$

This last expression is called the *convolution* of f_1 and f_2 and is sometimes written

$$g_1(y_1) = f_1(y_1)*f_2(y_1).$$

• •

In the general case where the transformations are not one to one, we start with Equation 2.20-47 and have

$$F_y(b_1, b_2) = \iint_{\{(x_1,x_2)/(y_1 \le b_1, y_2 \le b_2)\}} f(x_1, x_2)\, dx_1\, dx_2 \qquad (2.20\text{-}52)$$

where $F_y(b_1, b_2)$ is the joint c.d.f. of the y's and integration is over that (x_1, x_2) region where $y_1 \le b_1$ and $y_2 \le b_2$. Choosing the proper regions of integration may be difficult. An excellent and comprehensive discussion with many examples is given in reference [7, Chap. 5].

2.21 The Bivariate Normal Distribution

The normal distribution is of such importance in problems to be considered later that we devote this section and the next to a discussion of some of the elementary properties of the bivariate and multivariate normal distributions.

Suppose we are given the random variables x_1 and x_2 with densities $f_1(x_1)$ and $f_2(x_2)$ respectively. It is not in general a simple problem to construct an appropriate joint density $f(x_1, x_2)$. For the case where x_1 and x_2 are independent, we have immediately

$$f(x_1, x_2) = f_1(x_1)f_2(x_2). \qquad (2.21\text{-}1)$$

For normally distributed random variables with means u_1 and u_2 and variances σ_1^2 and σ_2^2, this becomes

$$f(x_1, x_2) = \frac{1}{2\pi\sigma_1\sigma_2} \exp\left\{-\frac{1}{2}\left[\frac{(x_1 - u_1)^2}{\sigma_1^2} + \frac{(x_2 - u_2)^2}{\sigma_2^2}\right]\right\}. \qquad (2.21\text{-}2)$$

The characteristic function $F(v_1, v_2)$ is given by Equation 2.20-41. This expression may be integrated by completing the square, first in one variable of integration and then in the other as in previous examples. Let us rewrite Equation 2.20-41 with the substitution of Equation 2.21-2:

$$F(v_1, v_2) = E\{e^{jv_1x_1}\}E\{e^{jv_2x_2}\} \tag{2.21-3}$$

or

$$= \exp\left[j(u_1v_1 + u_2v_2) - \tfrac{1}{2}(\sigma_1^2v_1^2 + \sigma_2^2v_2^2)\right] \tag{2.21-4}$$

from Equation 2.18-3. An examination of Equation 2.21-4 shows that the second part of the exponent is a quadratic form with the cross terms missing. Let us form a new characteristic function given by

$$F(v_1, v_2) = \exp\left(j\sum_{i=1}^{2} u_iv_i - \tfrac{1}{2}\sum_{i=1}^{2}\sum_{j=1}^{2}\sigma_{ij}v_iv_j\right) \tag{2.21-5}$$

where the notation means that $\sigma_{ii} = \sigma_i^2$ and $\sigma_{ij}(i \neq j)$ is the covariance of x_i and x_j. Here we have added the cross terms. We will now find the joint density $f(x_1, x_2)$ from Equation 2.20-43.

The algebra is considerably simplified if we take the means to be zero so that

$$u_1 = u_2 = 0.$$

A simple linear change of variable later will allow for nonzero means. We now consider Equation 2.20-43 with the substitution of $F(v_1, v_2)$ from Equation 2.21-5. We complete the square in the exponent, first in v_1 and then in v_2. The exponent becomes

$$-\tfrac{1}{2}\sigma_1^2\left[\left(v_1 + \frac{\sigma_{12}v_2}{\sigma_1^2} + j\frac{x_1}{\sigma_1^2}\right)^2 + \frac{\sigma_1^2\sigma_2^2 - \sigma_{12}^2}{\sigma_1^4}\left(v_2 - j\frac{\sigma_{12}x_1 - \sigma_1^2x_2}{\sigma_1^2\sigma_2^2 - \sigma_{12}^2}\right)^2\right]$$
$$- \frac{1}{2\sigma_1^2}\left[x_1^2 + \frac{(\sigma_{12}x_1 - \sigma_1^2x_2)^2}{\sigma_1^2\sigma_2^2 - \sigma_{12}^2}\right].$$

Let us make the change of variables

$$w = \sigma_1\left(v_1 + \frac{\sigma_{12}v_2}{\sigma_1^2} + j\frac{x_1}{\sigma_1^2}\right)$$

$$z = \left(\frac{\sigma_1^2\sigma_2^2 - \sigma_{12}^2}{\sigma_1^2}\right)^{1/2}\left(v_2 - j\frac{\sigma_{12}x_1 - \sigma_1^2x_2}{\sigma_1^2\sigma_2^2 - \sigma_{12}^2}\right).$$

After these changes, the exponent becomes

$$-\tfrac{1}{2}w^2 - \tfrac{1}{2}z^2 - \tfrac{1}{2}\frac{\sigma_2^2x_1^2 - 2\sigma_{12}x_1x_2 + \sigma_1^2x_2^2}{\sigma_1^2\sigma_2^2 - \sigma_{12}^2}$$

and Equation 2.20-43 can be written as

$$f(x_1, x_2) = \frac{1}{2\pi\sqrt{\sigma_1^2\sigma_2^2 - \sigma_{12}^2}}e^{-d/2}\left[\frac{1}{\sqrt{2\pi}}\int_{-\infty}^{\infty} e^{-w^2/2}\,dw\right]\left[\frac{1}{\sqrt{2\pi}}\int_{-\infty}^{\infty} e^{-z^2/2}\,dz\right] \tag{2.21-6}$$

where, for convenience, we have used d to denote

$$d = \frac{\sigma_2^2 x_1^2 - 2\sigma_{12} x_1 x_2 + \sigma_1^2 x_2^2}{\sigma_1^2 \sigma_2^2 - \sigma_{12}^2}.$$

Each of the two terms in brackets in Equation 2.21-6 is unity since each is the integral of the unit normal density. Let us now introduce the correlation coefficient defined by Equation 2.20-20 or, in our notation,

$$\rho = \frac{\sigma_{12}}{\sigma_1 \sigma_2}. \tag{2.21-7}$$

Equation 2.21-6 can be rewritten as

$$f(x_1, x_2) = \frac{1}{2\pi\sigma_1\sigma_2(1-\rho^2)^{1/2}} \exp\left[-\frac{1}{2(1-\rho^2)}\left(\frac{x_1^2}{\sigma_1^2} - 2\rho\frac{x_1 x_2}{\sigma_1\sigma_2} + \frac{x_2^2}{\sigma_2^2}\right)\right] \tag{2.21-8}$$

which is the bivariate normal distribution with means zero, variances σ_1^2 and σ_2^2, and correlation coefficient ρ. Note that when $\rho = 0$, the joint density factors into the product of two univariate normal densities. Thus *two uncorrelated normal random variables are independent*, although zero correlation does not imply independence in the general nonnormal case.

If we replace x_1 by $x_1 - u_1$ and x_2 by $x_2 - u_2$, then Equation 2.21-8 becomes the bivariate normal distribution with means u_1 and u_2.

2.22 The Multivariate Normal Distribution

Our purpose in this section is to generalize the normal distribution to the multivariate case. At this point it will be convenient to introduce some matrix concepts in order to simplify the notation. As we proceed we will use the bivariate case as an illustration.

Let us denote the random variables $x_1, x_2, \ldots, x_n$ by the column matrix (also called a column vector) $\mathbf{x}$, where

$$\mathbf{x} = \begin{bmatrix} x_1 \\ x_2 \\ \vdots \\ x_n \end{bmatrix}. \tag{2.22-1}$$

The *transpose* $\mathbf{x}'$ of the matrix is obtained by interchanging rows and columns so that $\mathbf{x}'$ is the row matrix (row vector) given by

$$\mathbf{x} = [x_1, x_2, \ldots, x_n]. \tag{2.22-2}$$

We will define a *covariance matrix* Λ by

$$\Lambda = [\sigma_{ij}] = \begin{bmatrix} \sigma_{11} & \sigma_{12} & \cdots & \sigma_{1n} \\ \sigma_{21} & \sigma_{22} & \cdots & \sigma_{2n} \\ \vdots & & & \\ \sigma_{n1} & \sigma_{n2} & \cdots & \sigma_{nn} \end{bmatrix} \tag{2.22-3}$$

with determinant $|\Lambda|$ or $|\sigma_{ij}|$. Here the element σ_{ij} is given by

$$\sigma_{ij} = E\{(x_i - u_i)(x_j - u_j)\} \tag{2.22-4}$$

and $\sigma_{ij} = \sigma_{ji}$. Note that it might be more correct (but less conventional) to refer to Λ as the variance-covariance matrix. In the same way we define the column matrix $\mathbf{u}$ of the means of $x_1, x_2, \ldots, x_n$ by

$$\mathbf{u} = \begin{bmatrix} u_1 \\ u_2 \\ \vdots \\ u_n \end{bmatrix}. \tag{2.22-5}$$

In the bivariate case, the covariance matrix is

$$\Lambda = \begin{bmatrix} \sigma_{11} & \sigma_{12} \\ \sigma_{21} & \sigma_{22} \end{bmatrix} = \begin{bmatrix} \sigma_1^2 & \sigma_1\sigma_2\rho \\ \sigma_1\sigma_2\rho & \sigma_2^2 \end{bmatrix} \tag{2.22-6}$$

and the characteristic function $F(v_1, v_2)$ of Equation 2.21-5 can be written as

$$F(\mathbf{v}) = e^{j\mathbf{u}'\mathbf{v} - (\mathbf{v}'\Lambda\mathbf{v})/2} \tag{2.22-7}$$

where $\mathbf{v}$ is the column matrix

$$\mathbf{v} = \begin{bmatrix} v_1 \\ v_2 \end{bmatrix}$$

or, in general,

$$\mathbf{v} = \begin{bmatrix} v_1 \\ v_2 \\ \vdots \\ v_n \end{bmatrix}. \tag{2.22-8}$$

Not only is Equation 2.22-7 more compact than Equation 2.21-5, but it represents the n-variate characteristic function (the case where the index of summation in Equation 2.21-5 takes on the values $1, 2, \ldots, n$).

Let us denote the *inverse* of the matrix Λ, or $[\sigma_{ij}]$, by Λ^{-1}, or $[\sigma^{ij}]$. For the bivariate case, the determinant of Λ is

$$|\Lambda| = |\sigma_{ij}| = \sigma_1^2\sigma_2^2(1 - \rho^2) \tag{2.22-9}$$

and the inverse Λ^{-1} is

$$\Lambda^{-1} = [\sigma^{ij}] = \begin{bmatrix} \dfrac{\sigma_2^2}{|\sigma_{ij}|} & -\dfrac{\sigma_1\sigma_2\rho}{|\sigma_{ij}|} \\ -\dfrac{\sigma_1\sigma_2\rho}{|\sigma_{ij}|} & \dfrac{\sigma_1^2}{|\sigma_{ij}|} \end{bmatrix}$$

$$= \begin{bmatrix} \dfrac{1}{\sigma_1^2(1-\rho^2)} & \dfrac{-\rho}{\sigma_1\sigma_2(1-\rho^2)} \\ \dfrac{-\rho}{\sigma_1\sigma_2(1-\rho^2)} & \dfrac{1}{\sigma_2^2(1-\rho^2)} \end{bmatrix}. \tag{2.22-10}$$

The determinant of this inverse is

$$|\sigma^{ij}| = \frac{1}{|\sigma_{ij}|} = \frac{1}{\sigma_1^2\sigma_2^2(1-\rho^2)}. \tag{2.22-11}$$

The bivariate normal density—Equation 2.21-8 with the substitution $(x_i - u_i)$ for x_i—can now be written in matrix notation as

$$f(\mathbf{x}) = \frac{|\sigma^{ij}|^{1/2}}{2\pi} e^{-[(\mathbf{x}-\mathbf{u})'\Lambda^{-1}(\mathbf{x}-\mathbf{u})]/2} \tag{2.22-12}$$

It can be shown that, for the n-variate case, the joint density $f(\mathbf{x})$ corresponding to the characteristic function $F(\mathbf{v})$ of Equation 2.22-7 is given by

$$f(\mathbf{x}) = \frac{|\sigma^{ij}|^{1/2}}{(2\pi)^{n/2}} e^{-[(\mathbf{x}-\mathbf{u})'\Lambda^{-1}(\mathbf{x}-\mathbf{u})]/2}. \tag{22.2-13}$$

This will be taken to be the *n-variate normal density*. A more complete discussion in terms of the moment-generating function $M(\mathbf{t})$ rather than the characteristic function is given in reference [8, Chap. 9].

As a matter of interest, we note that the n-variate normal moment-generating function can be written as

$$M(\mathbf{t}) = e^{\mathbf{u}'\mathbf{t} + (\mathbf{t}'\Lambda\mathbf{t})/2} \tag{2.22-14}$$

where $\mathbf{t}$ is a column matrix

$$\mathbf{t} = \begin{bmatrix} t_1 \\ t_2 \\ \vdots \\ t_n \end{bmatrix}. \tag{2.22-15}$$

Since this m.g.f. is defined by

$$M(\mathbf{t}) = E\{e^{\mathbf{t}'\mathbf{x}}\} = E\left\{\exp\left(\sum_{i=1}^{n} t_i x_i\right)\right\} \tag{2.22-16}$$

it follows by direct differentiation that

$$E\{x_i^a x_j^b x_k^c \cdots\} = \left. \frac{\partial^{a+b+c+\cdots} M(\mathbf{t})}{\partial t_i^a \, \partial t_j^b \, \partial t_k^c \cdots} \right|_{\substack{t_m = 0 \\ \text{all } m}} . \tag{2.22-17}$$

Thus, for the normal distribution, in the univariate case,

$$\left. \frac{\partial m(t_1)}{\partial t_1} \right|_{t_1 = 0} = E\{x_1\}; \tag{2.22-18}$$

in the bivariate case,

$$\left. \frac{\partial^2 M(t_1, t_2)}{\partial t_1 \, \partial t_2} \right|_{t_1 = t_2 = 0} = E\{x_1 x_2\}; \tag{2.22-19}$$

in the trivariate case,

$$\begin{aligned} \left. \frac{\partial^3 M(t_1, t_2, t_3)}{\partial t_1 \, \partial t_2 \, \partial t_3} \right|_{\substack{t_1 = t_2 \\ = t_3 = 0}} &= E\{x_1 x_2 x_3\} \\ &= u_1 E\{x_2 x_3\} + u_2 E\{x_3 x_1\} \\ &\quad + u_3 E\{x_1 x_2\} - 2u_1 u_2 u_3; \end{aligned} \tag{2.22-20}$$

in the quadrivariate case,

$$\begin{aligned} \left. \frac{\partial^4 M(t_1, t_2, t_3, t_4)}{\partial t_1 \, \partial t_2 \, \partial t_3 \, \partial t_4} \right|_{\substack{t_1 = t_2 = \\ t_3 = t_4 = 0}} &= E\{x_1 x_2 x_3 x_4\} \\ &= E\{x_1 x_2\} E\{x_3 x_4\} + E\{x_1 x_3\} E\{x_2 x_4\} \\ &\quad + E\{x_1 x_4\} E\{x_2 x_3\} - 2u_1 u_2 u_3 u_4. \end{aligned} \tag{2.22-21}$$

These relationships will be of considerable use later in treating the problem of nonlinear operations on normal random variables.

2.23 Linear Transformations on Normal Random Variables

One of the interesting and useful properties of normal random variables is that they are invariant to linear transformations. In other words, a linear transformation on a normal random variable, or a set of normal random variables, results in a normal random variable, or a set of normal random variables.

Let us represent an arbitrary linear transformation on the set of normal random variables $x_1, x_2, \ldots, x_n$ by

$$\mathbf{y} = A\mathbf{x} \tag{2.23-1}$$

where $\mathbf{y}$ is a column matrix

$$\mathbf{y} = \begin{bmatrix} y_1 \\ y_2 \\ \vdots \\ y_n \end{bmatrix} \tag{2.23-2}$$

and A is a matrix

$$A = \begin{bmatrix} a_{11} & a_{12} & \cdots & a_{1n} \\ a_{21} & a_{22} & \cdots & a_{2n} \\ \vdots & & & \\ a_{n1} & a_{n2} & \cdots & a_{nn} \end{bmatrix}. \tag{2.23-3}$$

Thus, in the bivariate case, we have

$$y_1 = a_{11}x_1 + a_{12}x_2,$$

$$y_2 = a_{21}x_1 + a_{22}x_2.$$

The column matrix of the means of $\mathbf{y}$ is given by

$$\mathbf{u}_y = E\{\mathbf{y}\} = A\mathbf{u} \tag{2.23-4}$$

and the covariance matrix by

$$\begin{aligned} \Lambda_y &= E\{(\mathbf{y} - \mathbf{u}_y)(\mathbf{y} - \mathbf{u}_y)'\} \\ &= E\{A(\mathbf{x} - \mathbf{u})(\mathbf{x} - \mathbf{u})'A'\} \end{aligned} \tag{2.23-5}$$

Since

$$\Lambda = E\{(\mathbf{x} - \mathbf{u})(\mathbf{x} - \mathbf{u})'\}, \tag{2.23-6}$$

then Equation 2.23-5 becomes

$$\Lambda_y = A\Lambda A'. \tag{2.23-7}$$

We have already written (Equation 2.22-7) the characteristic function of $\mathbf{x}$ as

$$F(\mathbf{v}) = e^{j\mathbf{u}'\mathbf{v} - (\mathbf{v}'\Lambda\mathbf{v})/2}.$$

Suppose we make the substitution

$$\mathbf{v} = A'\mathbf{w} \tag{2.23-8}$$

where A is the linear transformation matrix and $\mathbf{w}$ is a column matrix defined by Equation 2.22-8; that is,

$$\mathbf{w} = \begin{bmatrix} w_1 \\ w_2 \\ \vdots \\ w_n \end{bmatrix}. \tag{2.23-9}$$

With this substitution Equation 2.22-7 becomes

$$F(\mathbf{v}) = e^{j(A\mathbf{u})'\mathbf{w} - [\mathbf{w}'(A\Lambda A')\mathbf{w}]/2} \tag{2.23-10}$$

or, with the use of Equations 2.23-4 and 2.23-7,

$$F(\mathbf{v}) = e^{j\mathbf{u}_y'\mathbf{w} - (\mathbf{w}'\Lambda_y\mathbf{w})/2}. \tag{2.23-11}$$

However, this is the characteristic function of n normally distributed random variables with mean matrix $\mathbf{u}_y$ and covariance matrix Λ_y. We have that the characteristic function of $\mathbf{y}$ is defined by

$$F_y(\mathbf{w}) = E\{e^{j\mathbf{w}'\mathbf{y}}\}. \tag{2.23-12}$$

But, since $\mathbf{y} = A\mathbf{x}$ from Equation 2.23-1, we can rewrite this as

$$F_y(\mathbf{w}) = E\{e^{j\mathbf{w}'A\mathbf{x}}\} = E\{e^{j\mathbf{v}'\mathbf{x}}\} \tag{2.23-13}$$

or

$$F_y(\mathbf{w}) = F(\mathbf{v}). \tag{2.23-14}$$

We compare Equations 2.23-11 and 2.23-14 and conclude that $\mathbf{y}$ has an n-variate normal characteristic function and, hence, is normally distributed with means $\mathbf{u}_y$ and covariance matrix Λ_y. Thus we have proved that linear transformations on normal random variables yield normal random variables.

2.24 Series Expansions of Density Functions, Hermite Polynomials

Various attempts have been made to produce a general representation of density functions. One method which will have considerable utility later, particularly in the study of nonlinear systems in Chapter 6, involves the expression of the density function as an infinite series of orthogonal terms. We begin with the unit normal distribution of Equation 2.18-4 and denote this density by $\alpha(x)$ where

$$\alpha(x) = \frac{1}{\sqrt{2\pi}} e^{-x^2/2}, \qquad -\infty < x < \infty. \tag{2.24-1}$$

Suppose now that we differentiate this function n times to obtain

$$\frac{d}{dx}[\alpha(x)] = -x\alpha(x) \tag{2.24-2}$$

$$\frac{d^2}{dx^2}[\alpha(x)] = (x^2 - 1)\alpha(x) \tag{2.24-3}$$

$$\frac{d^3}{dx^3}[\alpha(x)] = -(x^3 - 3x)\alpha(x) \tag{2.24-4}$$

$$\vdots$$

$$\frac{d^n}{dx^n}[\alpha(x)] = (-1)^n H_n(x)\alpha(x) \tag{2.24-5}$$

where $H_n(x)$ is obviously a polynomial of degree n and is defined by Equation 2.24-5 or, after rearranging, by the *generalized Rodrigues' formula*

$$H_n(x) = (-1)^n e^{x^2/2} \frac{d^n}{dx^n}(e^{-x^2/2}). \tag{2.24-6}$$

The polynomial $H_n(x)$ is called a *Hermite polynomial* and has a number of interesting properties [5]. A *generating function* for $H_n(x)$ can be obtained as follows. From Equation 2.24-1, we write

$$\alpha(x - t) = \frac{1}{\sqrt{2\pi}} e^{-(x-t)^2/2} = \alpha(x)\, e^{tx - t^2/2}. \tag{2.24-7}$$

Now, if we expand $\alpha(x - t)$ in a power series, we have

$$\alpha(x - t) = \sum_{n=0}^{\infty} (-1)^n \frac{t^n}{n!} \frac{d^n}{dx^n} [\alpha(x)] \tag{2.24-8}$$

or, after using Equations 2.24-6 and 2.24-7, we can write a generating function

$$e^{tx - t^2/2} = \sum_{n=0}^{\infty} \frac{t^n}{n!} H_n(x). \tag{2.24-9}$$

Thus $H_n(x)$ is the coefficient of $t^n/n!$ in a power series expansion of $\exp(tx - \frac{1}{2}t^2)$. It is easy to show that

$$H_n(x) = x^n - \frac{n(n-1)}{2 \cdot 1!} x^{n-2} + \frac{n(n-1)(n-2)(n-3)}{2^2 \cdot 2!} x^{n-4} + \frac{n(n-1)(n-2)(n-3)(n-4)(n-5)}{2^3 \cdot 3!} x^{n-6} + \cdots. \tag{2.24-10}$$

Thus we have

$$H_0(x) = 1 \tag{2.24-11}$$

$$H_1(x) = x \tag{2.24-12}$$

$$H_2(x) = x^2 - 1 \tag{2.24-13}$$

$$H_3(x) = x^3 - 3x \tag{2.24-14}$$

$$H_4(x) = x^4 - 6x^2 + 3 \tag{2.24-15}$$

$$H_5(x) = x^5 - 10x^3 + 15x \tag{2.24-16}$$

$$\vdots$$

It is clear that $\alpha(x)$ as given by Equation 2.24-1 satisfies the differential equation

$$\frac{d}{dx}[\alpha(x)] + x\alpha(x) = 0. \tag{2.24-17}$$

On differentiating once, we obtain

$$\frac{d^2}{dx^2}[\alpha(x)] + x\frac{d}{dx}[\alpha(x)] + \alpha(x) = 0 \tag{2.24-18}$$

or, after n times,

$$\frac{d^{n+1}}{dx^{n+1}}[\alpha(x)] + x\frac{d^n}{dx^n}[\alpha(x)] + n\frac{d^{n-1}}{dx^{n-1}}[\alpha(x)] = 0. \qquad (2.24\text{-}19)$$

We now substitute Equation 2.24-5 into this last expression and find that

$$H_{n+1}(x) - xH_n(x) + nH_{n-1}(x) = 0 \qquad (2.24\text{-}20)$$

is a recurrence relationship connecting three successive Hermite polynomials. From this expression, all higher-degree Hermite polynomials can be found from $H_0(x)$ and $H_1(x)$. It can also be shown that $H_n(x)$ satisfies the differential equation

$$\frac{d^2w(x)}{dx^2} - x\frac{dw(x)}{dx} + nw(x) = 0. \qquad (2.24\text{-}21)$$

The $H_n(x)$ are orthogonal over the interval $(-\infty, \infty)$ with respect to the weighting function $\alpha(x)$; that is,

$$\int_{-\infty}^{\infty} \alpha(x)H_m(x)H_n(x)\,dx = n!\,\delta_{mn} \qquad (2.24\text{-}22)$$

where δ_{mn} is the Kronecker delta given by

$$\delta_{mn} = \begin{cases} 0, & m \neq n, \\ 1, & m = n. \end{cases} \qquad (2.24\text{-}23)$$

This relation is easily proved by using Equation 2.24-6 to rewrite the left side of Equation 2.24-22 as

$$I = \int_{-\infty}^{\infty} \alpha(x)H_m(x)H_n(x)\,dx = (-1)^n \int_{-\infty}^{\infty} H_m(x)\frac{d^n}{dx^n}[\alpha(x)]\,dx. \qquad (2.24\text{-}24)$$

Now we let

$$\begin{aligned} u &= H_m(x) & dv &= \frac{d^n}{dx^n}[\alpha(x)]\,dx \\ du &= \frac{d}{dx}[H_m(x)] & v &= \frac{d^{n-1}}{dx^{n-1}}[\alpha(x)] \end{aligned} \qquad (2.24\text{-}25)$$

and integrate by parts to obtain

$$I = (-1)^{n+1}\int_{-\infty}^{\infty} \frac{d}{dx}[H_m(x)]\frac{d^{n-1}}{dx^{n-1}}[\alpha(x)]\,dx. \qquad (2.24\text{-}26)$$

After m integrations by parts and for $m < n$, we have

$$I = (-1)^{n+m}\int_{-\infty}^{\infty} \frac{d^m}{dx^m}[H_m(x)]\frac{d^{n-m}}{dx^{n-m}}[\alpha(x)]\,dx. \qquad (2.24\text{-}27)$$

We now integrate once more and have, for $m \neq n$, $I = 0$ since

$$\frac{d^{m+1}}{dx^{m+1}} [H_m(x)] = 0 \tag{2.24-28}$$

as H_m is a polynomial of degree m. For the case where $m = n$,

$$I = \int_{-\infty}^{\infty} \alpha(x) H_n^2(x)\, dx = \int_{-\infty}^{\infty} \alpha(x) \frac{d^n}{dx^n} [H_n(x)]\, dx. \tag{2.24-29}$$

However, since $H_n(x)$ is a polynomial of degree n,

$$\frac{d^n}{dx^n} [H_n(x)] = n! \tag{2.24-30}$$

and

$$I = n! \int_{-\infty}^{\infty} \alpha(x)\, dx = n! \tag{2.24-31}$$

and Equation 2.24-22 is proved.

2.25 The Gram-Charlier and Edgeworth Series

Suppose that we are given an arbitrary density function $f(x)$ and wish to represent it by the series

$$f(x) = \sum_{n=0}^{\infty} a_n \alpha(x) H_n(x), \qquad -\infty < x < \infty. \tag{2.25-1}$$

After multiplying both sides of this equation by $H_m(x)$ and integrating from $-\infty$ to ∞, we obtain

$$a_n = \frac{1}{n!} \int_{-\infty}^{\infty} f(x) H_n(x)\, dx \tag{2.25-2}$$

since the orthogonality relationship of Equation 2.24-22 applies. This formal expansion of $f(x)$ as given by Equation 2.25-1 is called the *Gram-Charlier series* and may be a useful way to express the density $f(x)$. For the normalized gaussian case, only the first term is required, of course.

If we substitute the series expression for $H_n(x)$ as given by Equation 2.24-10 into Equation 2.25-2, we have

$$a_n = \frac{1}{n!} \left[u_n' - \frac{n(n-1)}{2 \cdot 1!} u_{n-2}' + \frac{n(n-1)(n-2)(n-3)}{2^2 \cdot 2!} u_{n-4}' + \cdots \right] \tag{2.25-3}$$

where u_n' is the nth moment of $f(x)$ about the origin and has been defined previously by Equation 2.13-2. If we normalize $f(x)$ to zero mean and unit variance before using Equations 2.25-1 and 2.25-3, then

$$a_0 = 1 \tag{2.25-4}$$

$$a_1 = 0 \tag{2.25-5}$$

$$a_2 = \tfrac{1}{2}(u_2 - 1) = 0 \tag{2.25-6}$$

$$a_3 = \tfrac{1}{6}u_3 \tag{2.25-7}$$

$$a_4 = \tfrac{1}{24}(u_4 - 6u_2 + 3) = \tfrac{1}{24}(u_4 - 3) \tag{2.25-8}$$

$$\vdots$$

where u_n is the nth moment of $f(x)$ about the mean as defined by Equation 2.13-5. (Of course in this case $u_n' = u_n$ since the mean of $f(x)$ has been taken to be zero.) Thus, for an arbitrary density $f(x)$ with zero mean and unit variance, we have the representation

$$f(x) = \alpha(x)\{1 + \tfrac{1}{6}u_3H_3(x) + \tfrac{1}{24}(u_4 - 3)H_4(x) + \cdots\}. \tag{2.25-9}$$

As a matter of interest we might note at this point that the cumulative distribution function $F(x)$ corresponding to the density $f(x)$ of Equation 2.25-1 is easy to obtain. We have

$$\begin{aligned} F(x) &= \int_{-\infty}^{x} f(y)\,dy = \int_{-\infty}^{x} \sum_{n=0}^{\infty} a_n\alpha(y)H_n(y)\,dy \\ &= \sum_{n=0}^{\infty} a_n \int_{-\infty}^{x} \alpha(y)H_n(y)\,dy = -\sum_{n=0}^{\infty} a_n\alpha(x)H_{n-1}(x) \end{aligned} \tag{2.25-10}$$

where we have used Equation 2.24-5 to write

$$\alpha(x)H_n(x) = -\frac{d}{dx}\left\{(-1)^{n-1}\frac{d^{n-1}}{dx^{n-1}}[\alpha(x)]\right\} = -\frac{d}{dx}[\alpha(x)H_{n-1}(x)]. \tag{2.25-11}$$

Thus far we have ignored the vital question as to whether or not the series of Equation 2.25-1 has any useful properties. One justification for its form can be made as follows. Let us suppose that the random variable x is the result of a superposition of independent effects so that

$$x = x_1 + x_2 + \cdots + x_n \tag{2.25-12}$$

where the x_i are independent random variables. Then, from the Central Limit Theorem of Section 2.19, we would expect x to be normally distributed in the limit as $n \to \infty$ and to be approximately normal for large n. Therefore the cumulative distribution function and the density function of the normalized x, for large n, might be approximated by

$$F(x) = \beta(x) + R(x) \tag{2.25-13}$$

and

$$f(x) = \alpha(x) + r(x) \tag{2.25-14}$$

where $\beta(x)$ and $\alpha(x)$ are the cumulative distribution function and density function of the unit normal distribution and $R(x)$ and $r(x)$ are correction terms. It is clear that both Equations 2.25-1 and 2.25-10 are of this form where the remainder is approximated by a power series in the derivatives of the unit normal density.

There are at least two different kinds of uses to which the representation of Equation 2.25-1 might be put. In the first place, the series might be used in a formal fashion to simplify the analysis of certain problems. Such will be the case in Chapter 6 in the treatment of instantaneous nonlinearities. For such problems it would be necessary to know that at least the series of Equations 2.25-1 and 2.25-10 converge and do represent $f(x)$ and $F(x)$. In the second place, the series might be used to determine the approximate form of a density (or cumulative distribution) by using the first few terms in Equation 2.25-9 together with estimates of the first few moments $u_3, u_4, \ldots, u_n$. In this case, we would be interested not in the convergence itself but in its rate of convergence. Actually we would like to know if a very few terms would give a good approximation to $f(x)$.

Unfortunately these series do not necessarily converge even for well-behaved distributions, although sufficient conditions for convergence are known [6]. Furthermore it is not true that each term in Equation 2.25-1 is smaller than the preceding term. Thus, if a finite number n terms is to be used, it is not necessarily true that the first n of these should be selected, and in principle these should be an optimum regrouping of terms for each application. A commonly used regrouping is called the *Edgeworth series* [6], but further discussion is somewhat beyond the scope of this treatment.

PROBLEMS

1. Indicate by cross-hatching the following sets on the Venn diagram.
   (a) $B \cap C$
   (b) $A \cap B'$
   (c) $(A \cup B') \cap C'$
   (d) $(A' \cap B)' \cup C$

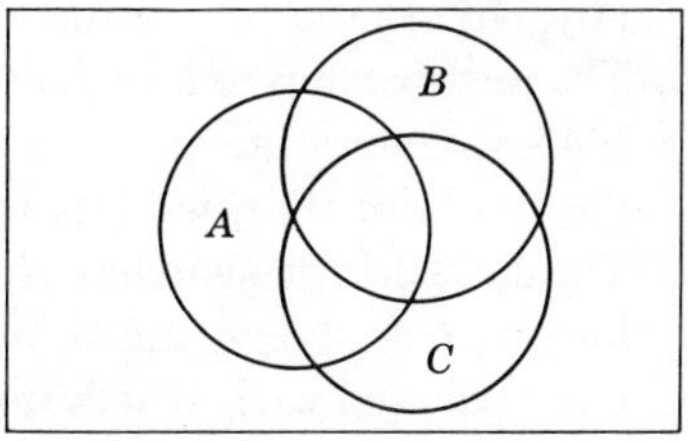

2. Let A be the set of positive even integers, let B be the set of positive integers divisible by 3, and let C be the set of positive odd integers. Describe the following sets.
   (a) $A \cup C$
   (b) $A \cap B$
   (c) $A \cap C$
   (d) $(A \cup B) \cap C$
   (e) $A \cup (B \cap C)$
3. Using the basic definitions of set equality, intersection, and union, prove the distribution law

$$A \cap (B \cup C) = (A \cap B) \cup (A \cap C).$$

4. If $\binom{n}{13} = \binom{n}{6}$, what is n?

   If $\binom{13}{r} = \binom{13}{r-3}$, what is r?

5. Three identical coins, marked A, B, and C for identification, are tossed simultaneously. If the order in which the coins fall is noted, how many ways can the coins fall assuming each coin has a "head" and "tail" side.
6. How many distinguishable arrangements are there of ten persons (a) in a row, (b) in a ring?
7. Three dice are rolled. Let A be the event that at least one ace appears and let B be the event that no two dice show the same value. Find $P(A)$, $P(B)$, and $P(AB)$. Are A and B statistically independent? Prove your answer.
8. Consider a game that consists of two successive trials. The first trial has outcomes A or B and the second outcomes C or D. The probabilities for the four possible outcomes of the game are as follows:

Outcome	AC	AD	BC	BD
Probability	$\frac{1}{3}$	$\frac{1}{6}$	$\frac{1}{6}$	$\frac{1}{3}$

   Are A and C statistically independent? Prove your answer.

9. Consider a family with two children of different ages. Assume that each child is as likely to be a boy as it is to be a girl. What is the conditional probability that both children are boys, given that (a) the older child is a boy, (b) at least one of the children is a boy?
10. The membership of a certain club is composed of 30 men and 20 women. If 40 percent of the men and 60 percent of the women play bridge, what is the conditional probability that a member who plays bridge is a man?
11. The network of switches A, B, C, D are connected across the power lines x, y as shown. Each switch has probability p of not closing when operated, and each switch functions independently of the other switches.

What is the probability that the circuit from x to y will fail to close when all four switches are operated? Will the addition of bus m change this probability? If so, what is the new probability?

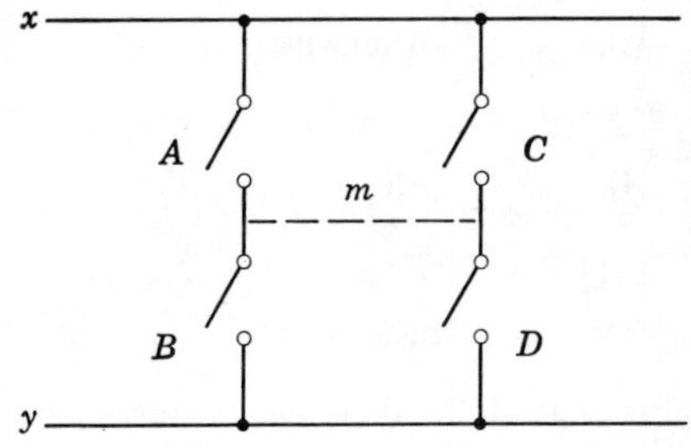

12. The three letters in the word "cat" are permuted in all possible distinct ways and then one permutation is picked at random. What is the probability that it spells a recognizable word in the English language?
13. What is the probability of throwing a "4" before a "7" with two dice?
14. Find the value of A which makes f a density function.

$$f(x) = \begin{cases} Ax, & \text{for } 0 < x < 4, \\ A(8 - x), & \text{for } 4 \leq x \leq 8, \\ 0, & \text{otherwise.} \end{cases}$$

Plot the c.d.f. and density function. What is the probability that $x \leq 6$?

15. Which of the following are density functions?

(a) $$f(x) = \begin{cases} \dfrac{1}{b - a}, & \text{for } a < x < b, \\ 0, & \text{otherwise.} \end{cases}$$

(b) $$f(x) = \begin{cases} \dfrac{1}{\pi} e^{-|x|}, & \text{for } -3 < x < 1, \\ 0, & \text{otherwise.} \end{cases}$$

(c) $$f(x) = \begin{cases} |x|, & \text{for } -1 < x < 1, \\ 0, & \text{otherwise.} \end{cases}$$

16. Plot the c.d.f. and density function for the random variable (a) the sum of the outcomes of two dice, (b) the product of the outcomes of two dice.

17. Which of the following are discrete density functions:

(a) $$p(x) = \begin{cases} \frac{1}{12}, & \text{for } x = 0, \\ \frac{7}{12}, & \text{for } x = 1, \\ \frac{1}{4}, & \text{for } x = 3, \\ 0, & \text{otherwise.} \end{cases}$$

(b) $$p(x) = \begin{cases} 2^{-|x|}, & \text{for } x = \ldots, -1, 0, 1, 2, \ldots, \\ 0, & \text{otherwise.} \end{cases}$$

(c) $$p(x) = \begin{cases} \frac{2}{3}(\frac{1}{3})^x, & \text{for } x = 0, 1, 2, \ldots, \\ 0, & \text{otherwise.} \end{cases}$$

18. A random variable n is defined as the number of times a fair coin is tossed until a tail appears.
    (a) Find and sketch the d.f. and c.d.f. associated with this random variable.
    (b) Find the mean of n and its second moment about the origin and about the mean.
19. If the random variable x has a density function

$$f(x) = \begin{cases} \frac{1}{2}, & -1 < x \leq 0, \\ \frac{1}{4}(2 - x), & 0 < x < 2, \\ 0, & \text{otherwise,} \end{cases}$$

and the random variable y is given by $y = |x|$, find $g(y)$, the density function of y.

20. If the random variable x has a density function

$$f(x) = \begin{cases} \dfrac{1}{\pi}, & \text{for } -\dfrac{\pi}{2} < x < \dfrac{\pi}{2}, \\ 0, & \text{otherwise.} \end{cases}$$

and the random variable y is given by $y = \cos x$, find $g(y)$, the density function of y.

21. A random variable is said to be uniformly distributed in (a, b) if

$$f(x) = \begin{cases} \dfrac{1}{b - a}, & \text{for } a < x < b, \\ 0, & \text{otherwise.} \end{cases}$$

Find the mean and variance of such a uniformly distributed random variable.

22. If the random variable x is the product of the two outcomes of a pair of dice, what is the mean of x? What is the most probable value of x?

23. Bill, Fred, and Al take turns tossing a coin, in that order, until a head appears. The man who first tosses a head keeps the coin. Find a set of entrance fees to make the game fair; that is, give equal expected gain for all players.
24. Find the moment-generating function and characteristic function for a random variable uniformly distributed in the interval (a, b).
25. Using Table 2.1 plot $p(|x| \geq \lambda)$ for $\lambda \leq 3$ for the unit normal distribution (Equation 2.18-4). On the same graph plot the upper bound given by Tchebycheff's Inequality (Equation 2.14-1) and compare the two plots.
26. A book with 100 pages has 100 misprints distributed at random. Assuming that the misprints follow a Poisson distribution with λ equal to the average number of misprints per page, estimate the probability that a page contains at least two misprints.
27. An ordinary die is thrown seven times. What is the most probable number of sixes?
28. A submarine launches n torpedos at a ship. Each torpedo has probability $\frac{1}{2}$ of hitting the ship independently of the other torpedos. What is the minimum number of torpedos that must be launched so that the probability of at least one hit is at least 0.8?
29. An insurance company writes a policy for A dollars which must be paid if some event E occurs. If E has probability p of occurring, what premium C must the company charge so that the expected profit is (a) 10 percent of A, (b) 10 percent of C?
30. The random variable x is uniformly distributed in $(-\pi/2, \pi/2)$. If $y = A \sin x$, find the density, the mean, and the variance of y.
31. A die is rolled 1000 times. What is the probability that the sum of the outcomes lies between 3534 and 3466? You may use the Central Limit Theorem.
32. If x is a random variable with an exponential density function

$$f(x) = A\, e^{-2A|x|}, \qquad -\infty < x < \infty,$$

find the nth moment of x.
33. A certain random variable is normally distributed with an average value of 10 and a standard deviation of unity. What is the probability of the random variable assuming a value (a) less than 11, (b) between 9 and 11?
34. Let x be a random variable with a binomial distribution with $n = 100$, $p = \frac{1}{5}$. *Estimate* the probability that $x \leq 25$.
35. The random variable x is normally distributed with zero mean and variance σ^2. Find the mean of the random variable $y = |x|$.
36. Let x be the random variable which takes on the values $-2, -1, 1, 2$ each with probability $\frac{1}{4}$. Let the random variable y be given by $y = x^2$. Find the correlation coefficient of x and y. Are x and y statistically independent?

37. If the random variables b and c are uniformly distributed on (0, 1) and independent, what is the probability that the roots of

$$x^2 + 2bx + c = 0$$

are real?

38. The two random variables x and y have the joint density function

$$f(x, y) = \begin{cases} e^{-(x+y)}, & \text{for } x \geq 0,\ y \geq 0, \\ 0, & \text{otherwise.} \end{cases}$$

Verify that $f(x, y)$ is a density function. Find the probability that $x < 3$, $y < 1$. Are x and y independent?

39. A red die and a green die are thrown. Let x be the random variable which is the outcome of the red die and let y be the random variable which is the larger of the two outcomes. Find (a) the joint probability density function of x and y, (b) the mean of x, and (c) the mean of y.

40. A certain marksman finds that his shots at a target are distributed about the center of the target in a bivariate normal distribution with zero means, unit variances, and no correlation; that is, the x and y coordinates of any shot are independent random variables with zero means and unit variances. What is the radius of the circle centered on the target such that the probability is 0.95 that a shot falls inside the circle?

41. The exponent in a bivariate normal distribution is

$$-\tfrac{1}{6}[4(x + 1)^2 - 2(x + 1)(y - 2) + (y - 2)^2].$$

What are the means, variances, and correlation coefficient of the random variables x and y?

42. The random variables x and y have a bivariate normal distribution. Show that the curves for which the distribution is a constant are ellipses on an x, y coordinate plane.

43. The covariance matrix Λ of three random variables x_1, x_2, x_3 is given by

$$\Lambda = \begin{bmatrix} \sigma_{11} & 0 & \sigma_{13} \\ 0 & \sigma_{22} & 0 \\ \sigma_{31} & 0 & \sigma_{33} \end{bmatrix}.$$

The linear transformation $\mathbf{y} = A\mathbf{x}$ is made where

$$A = \begin{bmatrix} 1 & 0 & 0 \\ 0 & 2 & 0 \\ 1 & 0 & 1 \end{bmatrix}.$$

Find the covariance matrix for the random variables y_1, y_2, y_3.

44. Assume that x is a nonnegative random variable with density $f(x)$ and mean u. Derive Markov's inequality

$$P(x \geq c) \leq \frac{u}{c}, \qquad c > 0.$$

45. Let us define for the random variable x the rth *absolute moment* ν_r about the origin by

$$\nu_r = E\{|x - u|^r\}, \qquad u = E\{x\}.$$

Prove the general inequality

$$P(|x - u| \geq \lambda \nu_r^{1/r}) \leq \frac{1}{\lambda^r}$$

for real $\lambda > 0$. For the special case where $r = 2$, show that this reduces to the *Tchebycheff Inequality.*

REFERENCES

1. W. Feller, *An Introduction to Probability Theory and Its Applications*, Wiley, New York, 1957.
2. A. A. Kolmogorov, *Foundation of the Theory of Probability*, Chelsea, New York, 1954.
3. M. Loeve, *Probability Theory*, 3rd ed., Van Nostrand, Princeton, N. J., 1963.
4. E. Lukacs, *Characteristic Functions*, Hafner, New York, 1960.
5. M. G. Kendall and A. Stuart, *The Advanced Theory of Statistics*, Vol. I, 2nd ed., Hafner, New York, 1963.
6. H. Cramer, *Mathematical Methods of Statistics*, Princeton Univ. Press, Princeton, N. J., 1946.
7. G. P. Wadsworth and J. G. Bryan, *Introduction to Probability and Random Variables*, McGraw-Hill, New York, 1960.
8. A. M. Mood, *Introduction to the Theory of Statistics*, McGraw-Hill, New York, 1950.

ADDITIONAL READING

(1950) Arley, N., and K. R. Buch, *Introduction to the Theory of Probability and Statistics*, Wiley (Science Editions), New York.

(1955) Cramer, H., *The Elements of Probability Theory and Some of Its Applications*, Wiley (Science Editions), New York.

(1960) Parzen, E., *Modern Probability Theory and Its Applications*, Wiley, New York.

(1963) Freeman, H., *Introduction to Statistical Inference*, Addison-Wesley, Reading, Mass.

3

Random Processes and Spectral Analysis

3.1 Definition

From an intuitive point of view, we view a *random* or *stochastic process* as a process developing in time and controlled by probabilistic laws. For further clarification, we consider the following example. It has been well known for many years that a noise voltage develops across the ends of a resistor. The mean-squared value of this *thermal noise* voltage v is given by

$$v^2 = 4kTR\,\Delta f\,(\text{volts})^2 \tag{3.1-1}$$

where

k = Boltzmann's constant = 1.37×10^{-23} watt-sec/degK,
T = absolute temperature, degK,
R = resistance, ohms,
Δf = measurement bandwidth, cps.

Suppose that we consider the ensemble (set) of all possible resistors having the same given resistance R ohms, maintain all of these resistors at the same temperature T, and display the thermal noise voltage of each on an ideal (noiseless) oscilloscope with bandwidth Δf. The result will be something similar to Figure 3.1 where each resistor is shown together with the thermal noise voltage produced across its terminals.

We will say that the (infinite) set of waveforms $v^{(1)}(t)$, $v^{(2)}(t), \ldots$, comprise a random process $v(t)$, or $\{v(t), -\infty < t < \infty\}$. An inherent part of this random process is the physical foundation from which it arises and from which it derives its statistical properties. With any random process, there must be

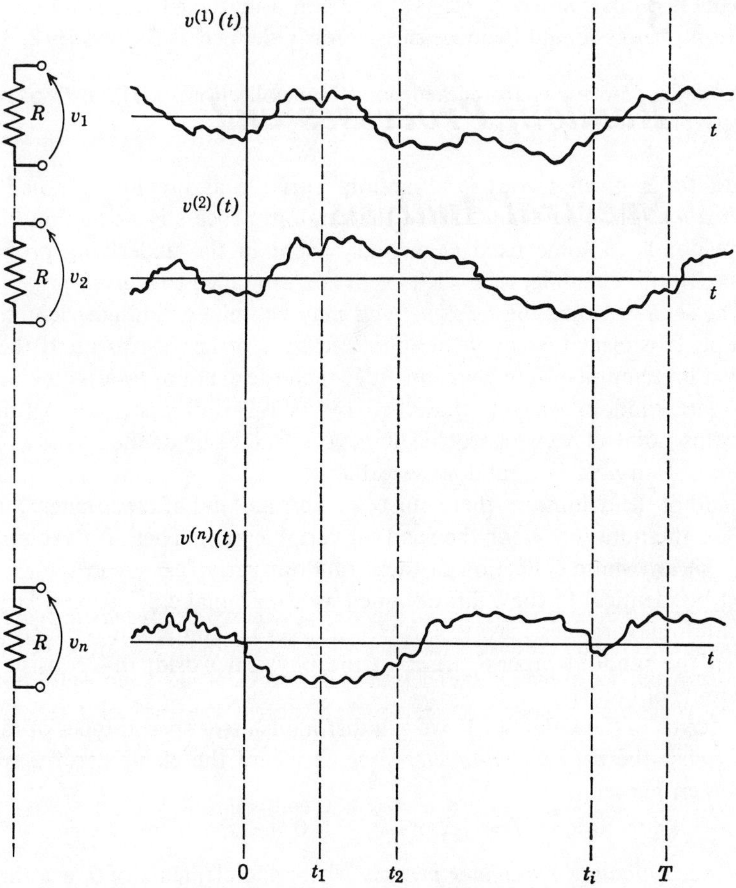

***Figure 3.1** A physical model for a random process $v(t)$*

connected, explicitly or implicitly, the laws that govern its statistical behavior. On the basis of this simple example, we define:

> A *random process* (or *stochastic process*) is an ensemble, or set, of functions of some parameter (usually taken to be time) together with a probability measure by which we can determine that any member, or group of members, has certain statistical properties.

We will call any member of the ensemble a *realization* (or *sample function*) of the process, for example, $v^{(1)}(t)$, or $v^{(2)}(t)$, in our example. At any time, say t_1, the realizations of $v(t)$ are just the realizations of a random variable which we will denote by $v(t_1)$.

This last observation suggests that, from a mathematical point of view, a random process could be more satisfactorily defined as follows [1, 2, 3, 4]:

> A *random process* is an indexed family (or collection) $\{x(t), t \in T\}$ of random variables.

Thus, for a given t, $x(t)$ is a random variable as previously noted. Each *realization* (or *sample function*) of the random process is a function of t and corresponds to some fixed elementary event in the underlying probability space. In our example, each such event corresponded to a given resistor.

The set T is called the *index set* and may be finite or infinite, denumerable or not. It is clear that only the infinite case is of interest to us. If there are only a finite number n of elements in T, then the random process is simply an n-variate random variable described by its n-variate distribution function. For this point of view, a stochastic process might be defined as the limit as $n \to \infty$ of an n-variate random variable.

In all of the definitions there must be a certain kind of *consistency*. Suppose we use the notation x_i for the random variable $x(t_i)$. Then, for example, the c.d.f. of any finite collection of these random variables, say $(x_1, x_2, \ldots, x_n)$, must be identical to the c.d.f. obtained as a marginal c.d.f. from a larger set of random variables, say $(x_1, x_2, \ldots, x_n, x_{n+1})$, containing our given set. When the random process arises in the physical world, this consistency is inherent, of course.

In terms of the index set T, we can distinguish two special types of random processes: the *discrete parameter process* where the elements of T are the denumerable set

$$T = \{\ldots, -2, -1, 0, 1, 2, \ldots\}$$

and the *continuous parameter process* where the elements of T are the non-denumerable set

$$T = \{t\colon -\infty < t < \infty\}.$$

We have already discussed extensively in Chapter 2 both *discrete* and *continuous* random variables. Thus we may conveniently distinguish four classes of random processes:

1. A *discrete random process with discrete parameter*; also sometimes called a *discrete random series* or a *discrete random sequence.* Here the random variable takes on only discrete values at any time t and the parameter t assumes only discrete values.
2. A *discrete random process with continuous parameter*; also sometimes called a *discrete random process.* Here the random variable is discrete but the parameter t assumes a continuum of values.
3. A *continuous random process with discrete parameter*, or a *continuous random series*, or a *continuous random sequence.*

4. A *continuous random process with continuous parameter*; sometimes called just a *continuous random process*.

For added clarity we now illustrate each of these classes by an example.

Example 3.1 Let us toss a coin N times. We define a random variable $x(i)$ by associating the value 1 with the occurrence of a head and 0 with the occurrence of a tail at each toss, and by indexing the number of the toss with i Thus $x(i) = 1$ if a head occurred on the ith toss and $x(i) = 0$ if a tail occurred.

There are 2^N possible distinct sequences of zeroes and ones of length N. Each sequence is a *realization* or a *sample function* of the random process We might call the jth sample function $x^{(j)}(i)$, but it will often be convenient to avoid this added complexity and call it simply $x(i)$ when no confusion will result.

We can assume that the probability of a head or of a tail on each toss is one-half and that the outcomes are all independent. The set (or *ensemble*) of the 2^N possible sample functions and this probability law comprise the *discrete random process with discrete parameter* or *discrete random series*. This process is illustrated in Figure 3.2.

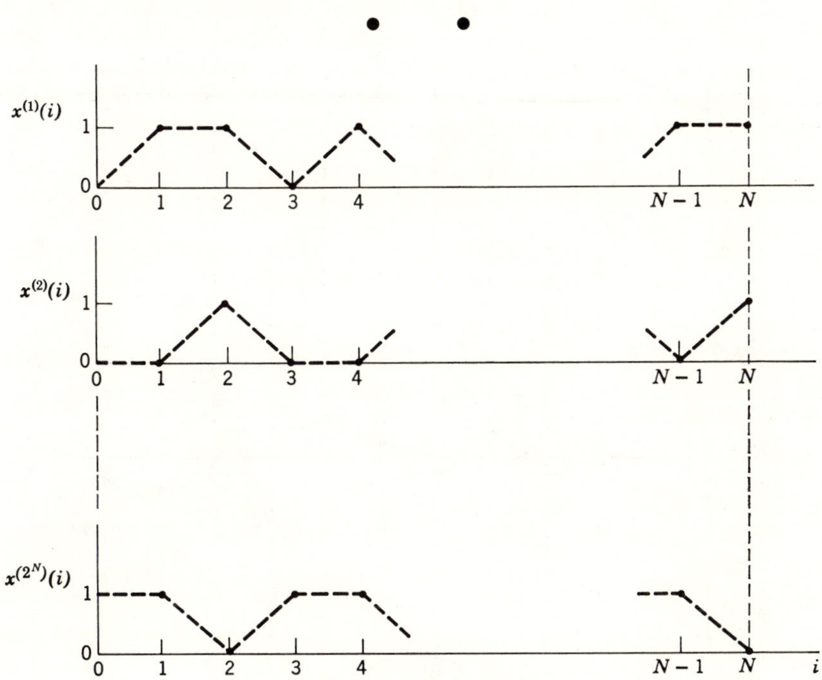

Figure 3.2 *A discrete random series $x(i)$*

Example 3.2 Figure 3.1 illustrates a *continuous random process with continuous parameter.* Here the sample space (range of the random variable) is $(-\infty, \infty)$ and the process could be defined for any interval T (or set of intervals) of the parameter t.

• •

Example 3.3 Suppose we sample the random process of Figure 3.1 at equally spaced intervals t_0 as shown in Figure 3.3. The result is a *continuous random process with discrete parameter.* The range of the random variable $v(nt_0)$ is the same as that of any $v(t)$ in the previous example, but the parameter time takes on only the discrete values nt_0.

• •

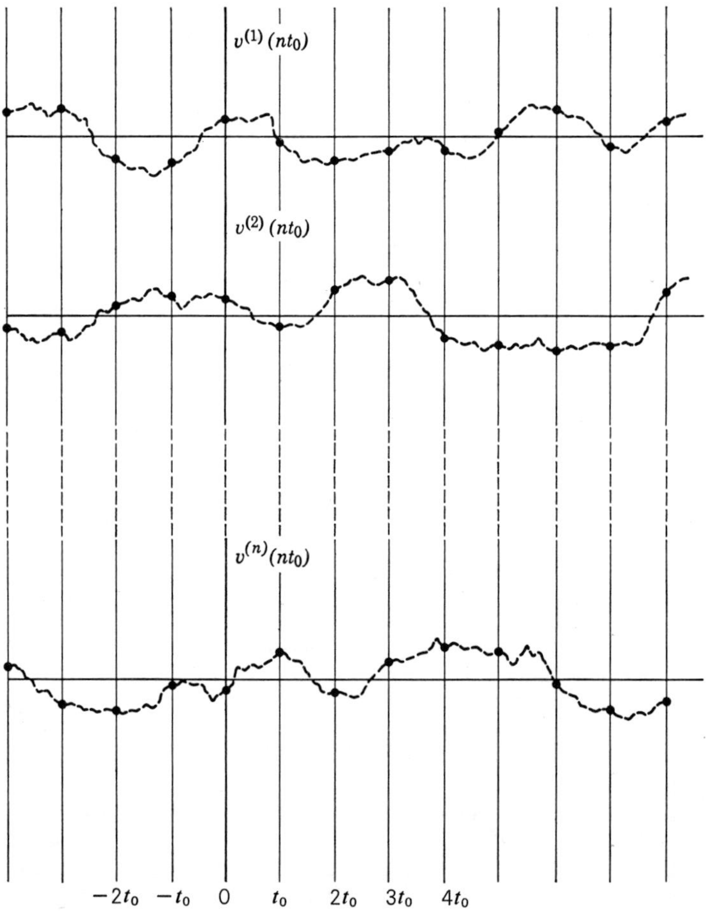

Figure 3.3 *A continuous random series* $v(nt_0)$

Example 3.4 Let us pass the random process $v(t)$ of Figure 3.1 through a *hard clipper* to produce the process $x(t)$ defined by

$$x(t) = \begin{cases} 1, & v(t) > 0, \\ -1, & v(t) < 0. \end{cases}$$

This process is illustrated in Figure 3.4 and is an example of a *discrete random process with continuous parameter*.

• •

Thus far our discussions have been in terms of a real-valued random variable resulting in *real-valued random processes.* It would have been possible to have defined the random variable of the process as complex and to have considered *complex-valued random processes.* Such processes will be useful in

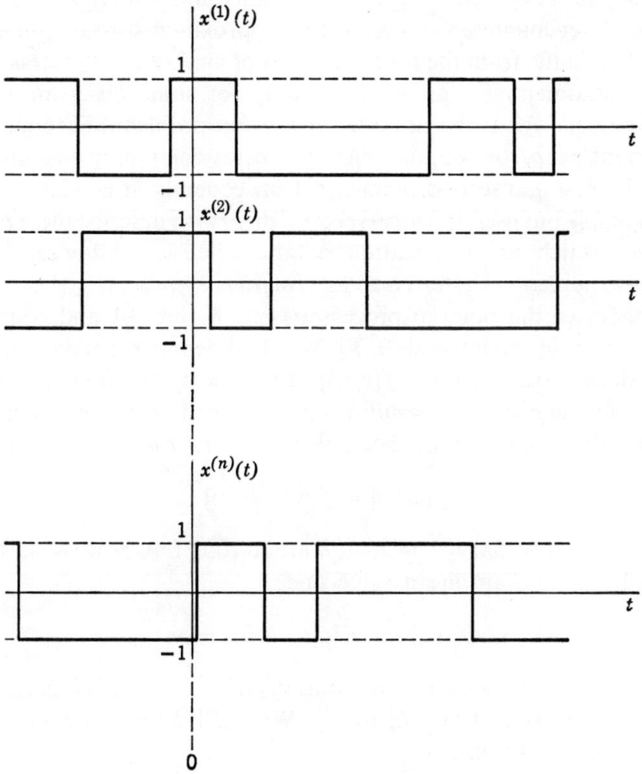

Figure 3.4 *A discrete random process $x(t)$*

some situations and, for our purposes, can be treated as linear combinations of real processes; that is, the complex-valued random process $x(t)$ can be written as

$$x(t) = x_1(t) + jx_2(t) \tag{3.1-2}$$

where $j = \sqrt{-1}$ and $x_1(t)$ and $x_2(t)$ are real random processes.

3.2 Stationarity

In a general way, we say that a random process is *stationary* if its statistical properties are invariant to time translation. This invariance implies that the underlying physical mechanism producing the process is not changing with time. The thermal noise random process described in the previous section and illustrated in Figure 3.1 is an example of a stationary process so long as the resistance values and the environmental temperature remain constant.

Stationary processes are of great importance for two reasons. First, they are frequently encountered in practice or approximated to a high degree of accuracy. (Actually, from the practical point of view, it is not necessary that a process be stationary for all time but only for some observation interval which is long enough to be suitable for a given problem.) Second, many of the important properties of stationary processes commonly encountered are described by first and second moments. Consequently, it is relatively easy to develop a simple but useful theory (spectral theory) to describe these processes.

Processes which are not stationary are called *nonstationary*, although they are also sometimes referred to as *evolutionary* processes [1].

Refer again to the random process $v(t)$ of Figure 3.1 and consider any arbitrary time t_1 in the interval $(0, T)$. We shall denote the univariate density of the random variable $v(t_1)$ by $f[v(t_1)]$. This density $f(\cdot)$ is frequently called the *first-order amplitude probability density* when it relates to a random process. We define a process to be *stationary to order one in* $(0, T)$ if

$$f[v(t_1)] = f[v(t_1 + h)] \tag{3.2-1}$$

for all t_1 and h such that t_1 and $t_1 + h$ are in $(0, T)$. A process stationary to order one has a constant mean value since

$$E\{v(t_1)\} = E\{v(t_1 + h)\} = u. \tag{3.2-2}$$

Now let us consider all joint distributions of the form $f_2[v(t_1), v(t_2)]$ where t_1 and t_2 are arbitrary times in $(0, T)$. We shall often write this bivariate density as $f_2(v_1, v_2)$ so that

$$f_2(v_1, v_2)\, dv_1\, dv_2 = \text{Prob}\begin{Bmatrix} v(t_1) < v(t_1) \le v(t_1) + dv(t_1) \\ v(t_2) < v(t_2) \le v(t_2) + dv(t_2) \end{Bmatrix} \tag{3.2-3}$$

where the right side of Equation 3.2-3 is the joint probability that v takes on values in the interval $(v_1, v_1 + dv_1)$ at time t_1 and values in the interval $(v_2, v_2 + dv_2)$ at time t_2. We define a process as *stationary to order two in* $(0, T)$ if

$$f_2[v(t_1), v(t_2)] = f_2[v(t_1 + h), v(t_2 + h)] \tag{3.2-4}$$

for all t_1, t_2, and h such that t_1, t_2, $t_1 + h$, and $t_2 + h$ are in $(0, T)$. For a process that is stationary to order two, we have

$$E\{v(t_1)v(t_2)\} = E\{v(t_1 + h)v(t_2 + h)\} = E\{v_1v_2\}. \tag{3.2-5}$$

Let us define the function $R(t_1, t_2)$ by

$$R(t_1, t_2) = E\{v(t_1)v(t_2)\}. \tag{3.2-6}$$

Then, it follows from Equation 3.2-5 that

$$R(t_1, t_2) = R(t_1 + h, t_2 + h) \tag{3.2-7}$$

for a process stationary to order two. Now place $h = -t_1$ so that Equation 3.2-7 becomes

$$R(t_1, t_2) = R(0, t_2 - t_1) = R(t_2 - t_1). \tag{3.2-8}$$

The function $R(t_1, t_2)$ is called the *correlation function* or the *autocorrelation function* of the random process. When the process $v(t)$ has zero mean, $R(t_1, t_2)$ is often called the *covariance function* or *autocovariance function.* For processes which are stationary to order two, the autocorrelation function has an argument which depends not on the times t_1 and t_2 but only on their difference $t_2 - t_1$. In fact, a similar argument shows that, in this case, the second-order density depends only on the time difference and could be written

$$f_2(v_1, v_2) = f_2(v_1, v_2; t_2 - t_1). \tag{3.2-9}$$

We proceed in the same way to define a process as *stationary to order n in* $(0, T)$ if

$$f_n[v(t_1), \ldots, v(t_n)] = f_n[v(t_1 + h), \ldots, v(t_n + h)] \tag{3.2-10}$$

for all $t_1, \ldots, t_n$, and h such that $t_1, \ldots, t_n, t_1 + h, \ldots, t_n + h$ are in $(0, T)$. It is clear that Equation 3.2-10 implies

$$E\{v(t_1)\cdots v(t_n)\} = E\{v(t_1 + h)\cdots v(t_n + h)\} \tag{3.2-11}$$

and that this function depends only on the time differences $t_2 - t_1, \ldots, t_n - t_1$.

It is clear that a process which is stationary to order n is stationary to all orders less than n. For example, consider the case where $n = 2$; then

$f_2[v(t_1), v(t_2)]$ and $f_2[v(t_1 + h), v(t_2 + h)]$ are equal and we can equate and integrate both of these densities to give

$$\int_{-\infty}^{\infty} f_2[v(t_1), v_2; \tau]\, dv_2 = \int_{-\infty}^{\infty} f_2[v(t_1 + h), v_2; \tau]\, dv_2 \qquad (3.2\text{-}12)$$

where $\tau = t_2 - t_1$ and where v_2 is the variable of integration, standing for $v(t_2)$ and $v(t_2 + h)$ respectively. But Equation 3.2-12 may be written as

$$f[v(t_1)] = f[v(t_1 + h)] \qquad (3.2\text{-}13)$$

where $f(\cdot)$ is the first-order amplitude probability distribution. The process is stationary to order one from Equation 3.2-1.

A process that is stationary to all orders will be called *strictly stationary* or simply *stationary* if no confusion will exist.

3.3 Correlation Functions

Again we consider the random process $v(t)$ and write v_1 for $v(t_1)$ and v_2 for $v(t_2)$ when convenient. We have already defined the *autocorrelation function* $R(t_1, t_2)$ of the process by Equation 3.2-6

$$R(t_1, t_2) = E\{v(t_1)v(t_2)\} \qquad (3.2\text{-}6)$$

if the random process is real. For a complex random process we write

$$R(t_1, t_2) = E\{v(t_1)v^*(t_2)\} \qquad (3.3\text{-}1)$$

where $v^*(t_2)$ is the complex conjugate of $v(t_2)$. As stated previously, we shall be concerned principally with real processes and shall always assume that we are dealing with such processes unless we specifically state otherwise.

For processes that are stationary at least to order two, it was shown in Section 3.2 that the autocorrelation function can be written as $R(t_2 - t_1)$. Since we are interested in developing a theory involving only first and second moments, it will be convenient at this point to define another kind of stationarity. A process $v(t)$ will be said to be *stationary in the wide sense* [4] or *covariance stationary* [1] if

$$1.\ E\{|v(t_i)|^2\} < \infty, \qquad \text{for all } t_i, \qquad (3.3\text{-}2)$$

$$2.\ E\{v(t_1)v(t_2)\} = R(t_2 - t_1), \qquad \text{for all } t_1 \text{ and } t_2. \qquad (3.3\text{-}3)$$

It is apparent that a process that is stationary to order two is also wide-sense stationary if its second moments exist. However, the converse may not be true.

Example 3.5 Let $\ldots, x_{-1}, x_0, x_1, \ldots = \{x_n\}$ be a sequence of mutually independent random variables each with zero mean and variance σ^2. Then the process $\{x_n\}$ is wide-sense stationary since $E\{|x_n|^2\} < \infty$ and the autocorrelation function $R(\cdot)$ is given by

$$R(n) = \begin{cases} \sigma^2, & n = 0, \\ 0, & n \neq 0. \end{cases}$$

However, the process is not strictly stationary (to any order) unless all of the x_i have a common distribution function.

• •

The autocorrelation function of a wide-sense stationary random process has a number of interesting properties that will be of considerable use later. For convenience, we define the time difference $t_2 - t_1 = \tau$ so that

$$R(\tau) = E\{v(t)v(t + \tau)\}. \tag{3.3-4}$$

1. *The mean-squared value of the random process is given by R(0).*

$$\boxed{R(0) = E\{[v(t)]^2\}.} \tag{3.3-5}$$

2. *The autocorrelation function of a wide-sense stationary random process is even.*

$$R(-\tau) = E\{v(t)v(t - \tau)\}.$$

Let $t - \tau = y$ so that

$$R(-\tau) = E\{v(y + \tau)v(y)\} = R(\tau)$$

$$\boxed{R(-\tau) = R(\tau).} \tag{3.3-6}$$

3. *The autocorrelation function of a wide-sense stationary random process is a maximum at the origin.*

Form the nonnegative quantity

$$E\{[v(t) \pm v(t + \tau)]^2\} \geq 0,$$

$$E\{[v(t)]^2\} + E\{[v(t + \tau)]^2\} \pm 2E\{v(t)v(t + \tau)\} \geq 0,$$

$$2R(0) \pm 2R(\tau) \geq 0,$$

$$\boxed{R(0) \geq |R(\tau)|.} \tag{3.3-7}$$

It should be emphasized, however, that the equality may hold.

Example 3.6 Let us consider the random process

$$e(t) = A \sin(\omega t + \theta)$$

where A, ω, and θ are possible random variables. We form the autocorrelation function

$$R(t, t + \tau) = E\{e(t)e(t + \tau)\}.$$

On substituting for $e(t)$, we have

$$R(t, t + \tau) = E\{A^2 \sin(\omega t + \theta) \sin[\omega(t + \tau) + \theta]\}.$$

We make use of the trigonometric identity

$$\sin x \sin y = \tfrac{1}{2} \cos(x - y) - \tfrac{1}{2} \cos(x + y)$$

to write

$$R(t, t + \tau) = \tfrac{1}{2} E\{A^2 \cos \omega\tau - A^2 \cos(2\omega t + 2\theta + \omega\tau)\}.$$

At this point, it is necessary to specify the nature of A, ω, and θ. We consider two cases.

Case I—A and ω are constants and θ is a random variable. Then we can write

$$R(t, t + \tau) = \frac{A^2}{2} \cos \omega\tau - \frac{A^2}{2} E\{\cos(2\omega t + 2\theta + \omega\tau)\}.$$

Now suppose that θ is uniformly distributed in $(0, 2\pi)$ so that the last term becomes

$$-\frac{A^2}{2} \frac{1}{2\pi} \int_0^{2\pi} \cos(2\omega t + 2\theta + \omega\tau)\, d\theta = 0.$$

In this case

$$R(t, t + \tau) = R(\tau) = \frac{A^2}{2} \cos \omega\tau$$

and the process $e(t)$ is wide-sense stationary.

Case II—ω and θ are constants and A is a random variable. Then we write

$$R(t, t + \tau) = \tfrac{1}{2}[\cos \omega\tau - \cos(2\omega t + 2\theta + \omega\tau)]E\{A^2\}$$

where $E\{A^2\}$ is determined from the distribution of A. In general, in this case, $R(t, t + \tau)$ is not a function of τ only and the process is not wide-sense stationary.

• •

In the definition of wide-sense stationarity, nothing was said about the dependence of $E\{v(t)\}$ on the parameter t. It will often be convenient to impose the added condition that

$$E\{v(t)\} = u = \text{const.} \tag{3.3-8}$$

In this case, the process can be centered; that is, a new process $\{v(t) - u\}$ can be defined with zero mean. Now the autocorrelation function

$$R(\tau) = E\{[v(t) - u][v(t + \tau) - u]\} \tag{3.3-9}$$

of the new process is a true covariance (second moment about the mean) and is properly called either a correlation function or a covariance function.

Suppose that we have two random processes $u(t)$ and $v(t)$ with autocorrelation functions $R_{uu}(t_1, t_2)$ and $R_{vv}(t_1, t_2)$ respectively. Then we may define the *cross-correlation functions*

$$R_{uv}(t_1, t_2) = E\{u(t_1)v(t_2)\} \tag{3.3-10}$$

and

$$R_{vu}(t_1, t_2) = E\{v(t_1)u(t_2)\}. \tag{3.3-11}$$

Now the correlation properties of the two random processes may be defined by the correlation matrix

$$\begin{bmatrix} R_{uu}(t_1, t_2) & R_{uv}(t_1, t_2) \\ R_{vu}(t_1, t_2) & R_{vv}(t_1, t_2) \end{bmatrix}. \tag{3.3-12}$$

If this matrix can be written as

$$\begin{bmatrix} R_{uu}(t_2 - t_1) & R_{uv}(t_2 - t_1) \\ R_{vu}(t_2 - t_1) & R_{vv}(t_2 - t_1) \end{bmatrix} \tag{3.3-13}$$

then we say that the random processes are each wide-sense stationary and, in addition, are jointly wide-sense stationary. The cross-correlation function in this case is not generally even, as was true for the autocorrelation function, nor does it necessarily have a maximum at the origin. On the other hand, we have

$$\begin{aligned} R_{uv}(\tau) &= E\{u(t)v(t + \tau)\} \\ R_{uv}(-\tau) &= E\{u(t)v(t - \tau)\} \\ &= E\{u(y + \tau)v(y)\} \end{aligned}$$

$$\boxed{R_{uv}(-\tau) = R_{vu}(\tau)} \tag{3.3-14}$$

which, for $u = v$, becomes Equation 3.3-6.

Also, suppose we form the nonnegative quantity

$$E\{[u(t) \pm v(t + \tau)]^2\} \geq 0$$

where $u(t)$ and $v(t)$ are individually and jointly wide-sense stationary. Then, on expanding, we have

$$R_{uu}(0) + R_{vv}(0) \pm 2R_{uv}(\tau) \geq 0$$

or

$$\boxed{|R_{uv}(\tau)| \leq \tfrac{1}{2}[R_{uu}(0) + R_{vv}(0)].} \tag{3.3-15}$$

For $u = v$, this reduces to Equation 3.3-7. Also, it follows immediately from Equation 2.20-29 that

$$\boxed{|R_{uv}(\tau)|^2 \leq R_{uu}(0)R_{vv}(0).} \tag{3.3-16}$$

3.4 Time Averages and Ergodicity

In practical problems involving random processes, what will generally be available to the observer is not the random process but one of its sample functions or realizations. Thus, in an experimental study of thermal noise, one will work with a given resistor and with the thermal voltage produced by it. In such cases, the quantities that are easily measured are various time averages, and an important question to answer is: Under what circumstances can these time averages be related to the statistical properties of the process?

We define the *time average* of the random process $v(t)$ by

$$A\{v(t)\} = \lim_{T \to \infty} \frac{1}{2T} \int_{-T}^{T} v(t)\, dt. \tag{3.4-1}$$

Other notations often are used instead of $A\{v(t)\}$ and it will sometimes be convenient to use the symbol $\overline{v(t)}$ for this time average. The *time autocorrelation function* $\mathscr{R}(\tau)$, or $\mathscr{R}_{vv}(\tau)$, of the random process $v(t)$ is defined by

$$\mathscr{R}(\tau) = A\{v(t)v(t + \tau)\} = \lim_{T \to \infty} \frac{1}{2T} \int_{-T}^{T} v(t)v(t + \tau)\, dt \tag{3.4-2}$$

and the *time cross-correlation function* $\mathscr{R}_{uv}(\tau)$ by

$$\mathscr{R}_{uv}(\tau) = A\{u(t)v(t + \tau)\}. \tag{3.4-3}$$

It is intuitively reasonable to suppose that, for stationary processes, time averages should be equal to expectations; for example,

$$E\{v(t)v(t + \tau)\} = A\{v(t)v(t + \tau)\}. \tag{3.4-4}$$

A heuristic argument to support this claim would go as follows. Let us divide the parameter t of the random process $v(t)$ into long intervals of T length. If these intervals are long enough (compared to the time scale of the underlying physical mechanism), then the statistical properties of the process in one interval T should be very similar to those in any other interval. Furthermore, a new random process could be formed in the interval $(0, T)$ by using as sample functions the segments of length T from a single sample function of the original process. This new process should be statistically indistinguishable from the original process and its ensemble averages would correspond to time averages of the sample function from the original process.

The foregoing is intended as a very crude justification of the condition of *ergodicity* [1, 4]. A random process is said to be *ergodic* if time averages of sample functions of the process can be used as approximations to the corresponding ensemble averages or expectations. A further discussion of ergodicity is beyond the scope of this treatment but this condition will be assumed to exist for all stationary processes unless explicitly stated otherwise. In this case, time averages and expectations can be interchanged at will and, in particular,

$$E\{v(t)\} = A\{v(t)\} = u \tag{3.4-5}$$

and

$$R(\tau) = \mathscr{R}(\tau). \tag{3.4-6}$$

3.5 Convergence of Random Variables

In the application of the theory of random processes we shall frequently encounter sequences of random variables of the form

$$x_1, x_2, \ldots, x_n$$

and we shall be interested in the convergence of such sequences to a random variable x. In Section 2.19, we have already discussed the concept of *convergence in probability*. Although there are a number of types of convergence of random variables which are commonly encountered [4], we shall consider only *convergence in probability* and *convergence in the mean.*

As mentioned earlier, we say that a sequence of random variables x_n—that is, $x_1, x_2, \ldots, x_n$—*converges in probability* to the random variable x if

$$\lim_{n\to\infty} P(|x_n - x| \geq \epsilon) = 0 \tag{3.5-1}$$

for arbitrary $\epsilon > 0$. We sometimes write Equation 3.5-1 as

$$p \lim x_n = x \tag{3.5-2}$$

or

$$x_n \to x \qquad \text{i.p.} \tag{3.5-3}$$

We say that the sequence of random variables x_n *converges in the mean* to the random variable x if

$$\lim_{n\to\infty} E\{|x_n - x|^2\} = 0. \tag{3.5-4}$$

This is sometimes written as

$$\underset{n\to\infty}{\text{l.i.m.}}\ x_n = x \tag{3.5-5}$$

where l.i.m. is read "limit in the mean."

If a sequence x_n converges in the mean to x, then it also converges in probability to x although the converse is not necessarily true. This statement is easily proved using Tchebycheff's Inequality as follows: we have from Section 2.14

$$P(|y - u_y| \geq \epsilon) \leq \frac{\sigma_y^2}{\epsilon^2} \tag{3.5-6}$$

for $\epsilon > 0$. Let us make the linear change of variable $z = y - u_y$ so that Equation 3.5-6 becomes

$$P(|z| \geq \epsilon) \leq \frac{E\{z^2\}}{\epsilon^2}. \tag{3.5-7}$$

Now let $|z| = |x_n - x|$ and we have

$$P(|x_n - x| \geq \epsilon) \leq \frac{E\{|x_n - x|^2\}}{\epsilon^2}. \tag{3.5-8}$$

If x_n converges in the mean to x, then, from Equation 3.5-4, we have

$$\lim_{n\to\infty} P(|x_n - x| \geq \epsilon) = 0 \tag{3.5-9}$$

and x_n also converges in probability to x, as was to be proved.

Example 3.7 We now give an illustration that convergence in probability does not imply convergence in the mean. Consider Cauchy's distribution where the density of x_n is

$$f(x_n) = \frac{n}{\pi}\frac{1}{1 + n^2x^2}, \qquad -\infty < x < \infty.$$

As $n \to \infty$, this density becomes the discrete density $\epsilon(x)$ where

$$\epsilon(x) = \begin{cases} 1, & x = 0, \\ 0, & x \neq 0. \end{cases}$$

We have

$$\begin{aligned} \lim_{n\to\infty} P(|x_n - 0| > \epsilon) &= \int_{-\infty}^{-\epsilon} \epsilon(x)\,dx + \int_{\epsilon}^{\infty} \epsilon(x)\,dx \\ &= 0 \end{aligned}$$

and x_n converges in probability to zero. On the other hand

$$\begin{aligned} E\{|x_n - x|^2\} &= E\{|x_n|^2\} \\ &= \frac{n}{\pi}\int_{-\infty}^{\infty} \frac{x^2}{1 + n^2x^2}\,dx \end{aligned}$$

which is infinite even for finite n, so that x_n does not converge in the mean to zero.

• •

Example 3.8 Let us define the sequence x_n by

$$x_n = \begin{cases} 1, & \text{with probability } a^n,\ a < 1, \\ 0, & \text{with probability } 1 - a^n. \end{cases}$$

Then x_n converges in the mean to zero since

$$E\{|x_n - 0|^2\} = a^n \cdot 1 + (1 - a^n) \cdot 0 = a^n$$

and

$$\lim_{n \to \infty} E\{|x_n - 0|^2\} = 0.$$

Consequently, x_n also converges in probability to zero.

• •

3.6 Fourier Transforms

In this section, we review briefly the concept of the Fourier transform [5, 6, 7]. It is presumed that the reader will already have been exposed to transform techniques; however, a review of the background material on Fourier series and an introduction to Fourier transforms is given in Appendix D.

Let $x(t)$ be a real-valued function of the real variable t and assume that

$$\int_0^T |x(t)|^2\, dt < \infty \qquad \text{for some } (0, T). \tag{3.6-1}$$

A function that satisfies Equation 3.6-1 is said to be of *integrable square* (or square integrable) for $0 \le t \le T$. If $x(t)$ is of integrable square for $-\infty < t < \infty$, then from the theory of Fourier transforms [5], there is a function $X(\omega)$ such that

$$\frac{1}{2\pi} \int_{-\infty}^{\infty} |X(\omega)|^2\, d\omega = \int_{-\infty}^{\infty} |x(t)|^2\, dt \tag{3.6-2}$$

and the functions $x(t)$ and $X(\omega)$ are related by

$$x(t) = \underset{A \to \infty}{\text{l.i.m.}}\ \frac{1}{2\pi} \int_{-A}^{A} X(\omega)\, e^{j\omega t}\, d\omega \tag{3.6-3}$$

and

$$X(\omega) = \underset{A \to \infty}{\text{l.i.m.}} \int_{-A}^{A} x(t)\, e^{-j\omega t}\, dt. \tag{3.6-4}$$

Here the notation l.i.m. of Equation 3.6-4 means "limit in the mean" and is defined by

$$\lim_{A \to \infty} \int_{-\infty}^{\infty} \left| X(\omega) - \int_{-A}^{A} x(t)\, e^{-j\omega t}\, dt \right|^2 d\omega = 0. \tag{3.6-5}$$

This is the analog for real variables of the "convergence in the mean" defined in Section 3.5 on an expectation basis for random variables. We call

$x(t)$ and $X(\omega)$ a *Fourier-transform pair.* The function $X(\omega)$ is usually called the (direct) Fourier transform of $x(t)$, and $x(t)$ is called the (inverse) Fourier transform of $X(\omega)$.

If the transform $X(\omega)$ *is absolutely integrable*, that is, if

$$\int_{-\infty}^{\infty} |X(\omega)|\, d\omega < \infty, \tag{3.6-6}$$

then Equation 3.6-3 becomes

$$x(t) = \frac{1}{2\pi}\int_{-\infty}^{\infty} X(\omega)\, e^{j\omega t}\, dt \tag{3.6-7}$$

and, if $x(t)$ is absolutely integrable so that

$$\int_{-\infty}^{\infty} |x(t)|\, dt < \infty, \tag{3.6-8}$$

then Equation 3.6-4 becomes

$$X(\omega) = \int_{-\infty}^{\infty} x(t)\, e^{-j\omega t}\, dt. \tag{3.6-9}$$

Example 3.9 Find the Fourier transform of

$$x(t) = A\, e^{-b|t|}, \qquad -\infty < t < \infty,$$

where A and b are positive constants. The function $x(t)$ is both square integrable and absolutely integrable since

$$\int_{-\infty}^{\infty} |A\, e^{-b|t|}|^2\, dt = \frac{A^2}{b} < \infty$$

and

$$\int_{-\infty}^{\infty} |A\, e^{-b|t|}|\, dt = \frac{2A}{b} < \infty.$$

Thus, from Equation 3.6-9, we have

$$X(\omega) = A\int_{-\infty}^{0} e^{bt}\, e^{-j\omega t}\, dt + A\int_{0}^{\infty} e^{-bt}\, e^{-j\omega t}\, dt$$

$$X(\omega) = \frac{2Ab}{b^2 + \omega^2}, \qquad -\infty < \omega < \infty.$$

The inverse transform $x(t)$ is obtained from Equation 3.6-7 as

$$x(t) = \frac{1}{2\pi}\int_{-\infty}^{\infty} \frac{2Ab}{b^2 + \omega^2}\, e^{j\omega t}\, d\omega$$

$$= \frac{Ab}{\pi}\int_{-\infty}^{\infty} \frac{\cos \omega t}{b^2 + \omega^2}\, d\omega.$$

This expression may be evaluated by contour integration; however, it is a standard form given in tables of integrals [8] and we may write

$$x(t) = \begin{cases} A\,e^{bt}, & t < 0 \\ A\,e^{-bt}, & t > 0 \end{cases}$$

as we had originally.

• •

As a matter of convenience, some of the most commonly encountered Fourier-transform pairs are given in Table 3.1.

Table 3.1. Common Fourier-Transform Pairs[a,b]

Number	$f(t)$	$F(\omega)$
1	$u(t-a) - u(t-b)$ $a < b$	$\dfrac{e^{-j\omega a} - e^{-j\omega b}}{j\omega}$
2	$u(t+a) - u(t-a)$ $a > 0$	$2a\dfrac{\sin a\omega}{a\omega}$
3	$e^{-bt}u(t)$ $b > 0$	$\dfrac{1}{b + j\omega}$
4	$e^{-b\lvert t\rvert}$ $b > 0$	$\dfrac{2b}{b^2 + \omega^2}$
5	$\dfrac{1}{b^2 + t^2}$ $b > 0$	$\dfrac{\pi}{b}\,e^{-b\lvert\omega\rvert}$
6	1	$2\pi\,\delta(\omega)$
7	$\delta(t)$	1
8	$\dfrac{1}{\sqrt{2\pi}}\,e^{-t^2/2}$	$e^{-\omega^2/2}$
9	$\dfrac{d^n}{dt^n}[f(t)]$ if $f(\infty) = f(-\infty) = 0$	$(j\omega)^n F(\omega)$

[a] $f(t) = (1/2\pi)\int_{-\infty}^{\infty} F(\omega)\,e^{j\omega t}\,d\omega$; $F(\omega) = \int_{-\infty}^{\infty} f(t)\,e^{-j\omega t}\,dt$.

[b] a, b are real.

$$u(t) = \text{unit step function} = \begin{cases} 0, & t < 0, \\ 1, & t > 0. \end{cases}$$

$$\delta(t) = \text{Dirac delta function} = \begin{cases} \infty, & t = 0, \\ 0, & t \neq 0, \end{cases} \qquad \int_{-\infty}^{\infty} \delta(t)\,dt = 1.$$

3.7 Integrals of Random Processes

There will be many places in the study of communication theory where integrals of random processes will arise. For example, in Section 3.4 we have already studied time averages. As another example, it may be desirable to obtain the Fourier transform $X(\omega)$ of the random process $x(t)$. Formally, we could write

$$X(\omega) = \int_{-\infty}^{\infty} x(t)\, e^{-j\omega t}\, dt. \tag{3.7-1}$$

It is apparent that $X(\omega)$ is a random variable if it exists; that is, for each sample function $x^{(i)}(t)$, a realization $X^{(i)}(\omega)$ of the random variable $X(\omega)$ results provided the sample function satisfies Equation 3.6-1. The probability law governing the random variable $X(\omega)$ is determined by the underlying probability laws of the random process and by the transformation of Equation 3.7-1. As a matter of fact, $X(\omega)$ determined by Equation 3.7-1 is actually a new random process with (real) parameter ω instead of t. It is apparent that $X(\omega)$ will be, in general, a complex random process.

Suppose we are given that $y(t)$ is a random process which is the result of a general linear operation $h(t, \tau)$ on the random process $x(t)$; that is,

$$y(t) = \int_a^b h(t, \tau)x(\tau)\, d\tau \tag{3.7-2}$$

where (a, b) is some interval, finite or infinite. It has been shown [4] that this integral exists if

$$\int_a^b E\{|h(t, \tau)x(\tau)|\}\, d\tau = \int_a^b |h(t, \tau)|\, E\{|x(\tau)|\}\, d\tau < \infty. \tag{3.7-3}$$

Furthermore, in this case, we can write

$$E\{y(t)\} = E\left\{\int_a^b h(t, \tau)x(\tau)\, d\tau\right\} = \int_a^b h(t, \tau)E\{x(\tau)\}\, d\tau \tag{3.7-4}$$

and the interchange of the order of integration and expectation operation is justified.

In this development, we have ignored some of the more subtle conditions which the random processes must satisfy for these results to hold. A more complete discussion requires some background in measure theory and is beyond the scope of this treatment. Nevertheless these results hold for reasonably well-behaved random processes such as those that arise in the physical world.

3.8 Power Spectra

As mentioned previously, a considerable theory (spectral theory) involving second moments only has been developed for random processes which are at least wide-sense stationary. In this section we proceed to develop some of the more elementary aspects of this theory.

We begin with the random process $x(t)$ assumed to be at least wide-sense stationary so that its autocorrelation function can be written as

$$R(\tau) = E\{x(t)x(t + \tau)\}. \tag{3.8-1}$$

In addition we will often assume that the process has a constant mean of zero so that the variance σ^2 of the process is a constant given by

$$\sigma^2 = R(0). \tag{3.8-2}$$

We note in passing that the restriction to zero mean is not necessary. Actually the process could have any constant mean u and we could work with the new process $y(t) = x(t) - u$.

We define the *power spectral density* $\varphi(\omega)$ of the random process $x(t)$ to be the Fourier transform of the autocorrelation function $R(\tau)$ of the process:

$$\varphi(\omega) = \int_{-\infty}^{\infty} R(\tau)\, e^{-j\omega\tau}\, d\tau. \tag{3.8-3}$$

Then the autocorrelation function is the inverse Fourier transform of $\varphi(\omega)$, or

$$R(\tau) = \frac{1}{2\pi}\int_{-\infty}^{\infty} \varphi(\omega)\, e^{j\omega\tau}\, d\omega. \tag{3.8-4}$$

It was pointed out in Section 3.6 that $R(\tau)$ must be absolutely integrable for Equation 3.8-3 to hold. An example was given in Section 3.3 where the autocorrelation function was of the form

$$R(\tau) = \cos \omega_0 \tau. \tag{3.8-5}$$

For this case, the power spectral density $\varphi(\omega)$ does not properly exist. However, it can be expressed in terms of the Dirac delta functions discussed in Appendix B since

$$\begin{aligned}\varphi(\omega) &= \int_{-\infty}^{\infty} \cos \omega_0 \tau\, e^{-j\omega\tau}\, d\tau \\ &= \frac{1}{2}\int_{-\infty}^{\infty} e^{-j(\omega-\omega_0)\tau}\, d\tau + \frac{1}{2}\int_{-\infty}^{\infty} e^{-j(\omega+\omega_0)\tau}\, d\tau \\ &= \pi\delta(\omega - \omega_0) + \pi\delta(\omega + \omega_0)\end{aligned} \tag{3.8-6}$$

from pair number 6 of Table 3.1. Here the power spectral density consists of two delta functions, one at $\omega = \omega_0$ and the other at $\omega = -\omega_0$. For most of the practical problems that we will encounter, either the power spectral density will be well defined or it can be written in terms of delta functions.

Let us assume at this point that we are dealing with an ergodic process $x(t)$ so that the time autocorrelation function is equal to the autocorrelation function, that is,

$$\mathscr{R}(\tau) = R(\tau). \tag{3.8-7}$$

We would like to relate the power spectral density $\varphi(\omega)$ to a time average through the transform relation of Equation 3.8-3. Using this equation and Equation 3.8-7, we can write

$$\varphi(\omega) = \int_{-\infty}^{\infty} \left[\lim_{T \to \infty} \frac{1}{2T} \int_{-T}^{T} x(t)x(t + \tau)\, dt \right] e^{-j\omega\tau}\, d\tau. \tag{3.8-8}$$

Suppose now that we interchange the order of integration to obtain

$$\varphi(\omega) = \lim_{T \to \infty} \frac{1}{2T} \int_{-T}^{T} x(t) \int_{-\infty}^{\infty} x(t + \tau)\, e^{-j\omega\tau}\, d\tau\, dt. \tag{3.8-9}$$

Let us define a function $X_T(\omega)$ by

$$X_T(\omega) = \int_{-T}^{T} x(t)\, e^{-j\omega t}\, dt. \tag{3.8-10}$$

Thus in the limit as $T \to \infty$, $X_T(\omega)$ becomes $X(\omega)$, the Fourier transform of $x(t)$. Now Equation 3.8-9 can be written, after the change in variable $t + \tau = y$, as

$$\varphi(\omega) = \lim_{T \to \infty} \frac{1}{2T} \int_{-T}^{T} x(t)\, e^{j\omega t} \int_{-\infty}^{\infty} x(y)\, e^{-j\omega y}\, dy\, dt \tag{3.8-11}$$

or

$$\varphi(\omega) = \lim_{T \to \infty} \frac{1}{2T} X_T(-\omega) X_T(\omega) = \lim_{T \to \infty} \frac{1}{2T} |X_T(\omega)|^2. \tag{3.8-12}$$

The results seem reasonable and they agree with the power spectrum concept developed for a square-integrable function in Section 7.8; however, they are *incorrect*. Unfortunately, the procedure that we have just followed is not justifiable. The reason is that the $X_T(\omega)$ are random variables which may not exist in the limit for most sample functions of the process; hence the interchange of limiting operations and order of integration was not permissible. It will now be shown [9, 10] that $\varphi(\omega)$ is actually given by

$$\varphi(\omega) = \lim_{T \to \infty} E\left\{ \frac{|X_T(\omega)|^2}{2T} \right\}. \tag{3.8-13}$$

What happens is that, although the mean of $|X_T(\omega)|^2/2T$ exists, its variance does not approach zero in the limit and hence the right side of Equation 3.8-12 does not converge in the mean to the power spectral density $\varphi(\omega)$.

We now proceed to show that the power spectral density $\varphi(\omega)$, as defined by Equation 3.8-3, is also given by Equation 3.8-13. We begin with $X_T(\omega)$ as defined by Equation 3.8-10, note that $X_T(-\omega) = X_T^*(\omega)$, and form

$$E\left\{\frac{1}{2T} X_T(\omega)X_T(-\omega)\right\} = E\left\{\frac{1}{2T}\int_{-T}^{T} x(t)\, e^{-j\omega t}\, dt \int_{-T}^{T} x(s)\, e^{j\omega s}\, ds\right\}$$

$$= \frac{1}{2T}\int_{-T}^{T}\int_{-T}^{T} E\{x(t)x(s)\}\, e^{-j\omega t}\, dt\, e^{j\omega s}\, ds. \qquad (3.8\text{-}14)$$

If $x(t)$ is a random process which is at least wide-sense stationary, then

$$E\{x(t)x(s)\} = R(t - s). \qquad (3.8\text{-}15)$$

Let us now make the following change of variables in the double integral of Equation 3.8-14:

$$t - s = u, \qquad (3.8\text{-}16a)$$

$$s = v. \qquad (3.8\text{-}16b)$$

The corresponding regions of integration are shown in Figure 3.5. If we integrate first with respect to v, we note that we must consider two regions of values of u, namely $u < 0$ and $u > 0$. Equation 3.8-14 becomes

$$E\left\{\frac{1}{2T} X_T(\omega)X_T(-\omega)\right\} = \frac{1}{2T}\int_{-2T}^{0} R(u)\, e^{-j\omega u}\int_{-T-u}^{T} dv\, du$$

$$+ \frac{1}{2T}\int_{0}^{2T} R(u)\, e^{-j\omega u}\int_{-T}^{T-u} dv\, du$$

$$= \int_{-2T}^{0}\left(1 + \frac{u}{2T}\right)R(u)\, e^{-j\omega u}\, du$$

$$+ \int_{0}^{2T}\left(1 - \frac{u}{2T}\right)R(u)\, e^{-j\omega u}\, du \qquad (3.8\text{-}17)$$

or, finally,

$$E\left\{\frac{1}{2T} X_T(\omega)X_T(-\omega)\right\} = \int_{-2T}^{2T}\left(1 - \frac{|u|}{2T}\right)R(u)\, e^{-j\omega u}\, du$$

$$= \varphi_T(\omega) \qquad (3.8\text{-}18)$$

where the definite integral is denoted by $\varphi_T(\omega)$. We see immediately that

$$\boxed{\lim_{T\to\infty} E\left\{\frac{1}{2T} X_T(\omega)X_T(-\omega)\right\} = \lim_{T\to\infty} \varphi_T(\omega) = \varphi(\omega)} \qquad (3.8\text{-}19)$$

where $\varphi(\omega)$ is defined by Equation 3.8-3, as was to be proved.

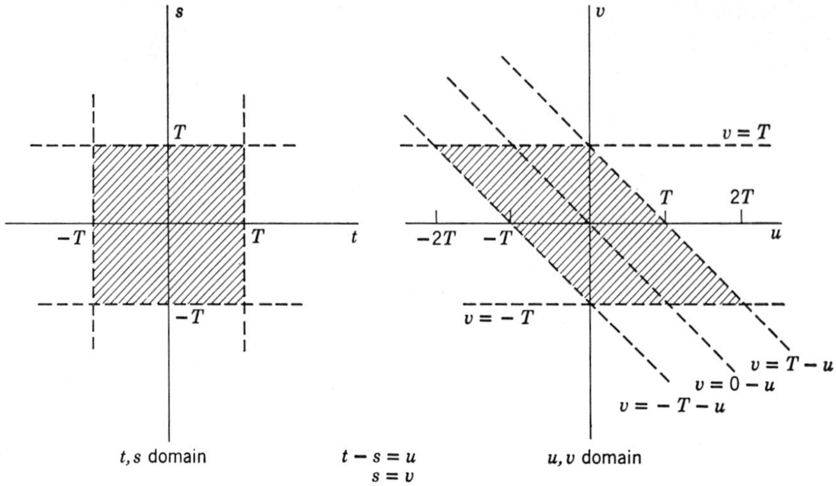

***Figure 3.5** Regions of integration for Equation 3.8-17*

We now show that the inverse transform relationship of Equation 3.8-4 is valid. Let us define a function $R_T(u)$ by

$$R_T(u) = \begin{cases} \left(1 - \dfrac{|u|}{2T}\right) R(u), & |u| \leq 2T, \\ 0, & |u| > 2T, \end{cases} \tag{3.8-20}$$

so that, from Equation 3.8-18,

$$\varphi_T(\omega) = E\left\{\frac{1}{2T} X_T(\omega) X_T(-\omega)\right\} = \int_{-\infty}^{\infty} R_T(u)\, e^{-j\omega u}\, du \geq 0. \tag{3.8-21}$$

We now multiply both right and left sides of this equation by $[1 - (|\omega|/\Omega)]$ $e^{j\omega t}$ and integrate in the variable ω from $-\Omega$ to Ω. We have

$$\int_{-\Omega}^{\Omega} \left(1 - \frac{|\omega|}{\Omega}\right) e^{j\omega t}\, \varphi_T(\omega)\, d\omega$$

$$= \int_{-\Omega}^{\Omega} \int_{-\infty}^{\infty} R_T(u)\, e^{-j\omega u}\, du \left(1 - \frac{|\omega|}{\Omega}\right) e^{j\omega t}\, d\omega$$

$$= \int_{-\infty}^{\infty} R_T(u) \int_{-\Omega}^{\Omega} e^{-j\omega(u-t)} \left(1 - \frac{|\omega|}{\Omega}\right) d\omega\, du. \tag{3.8-22}$$

The inner integral on the right side is just the Fourier transform of a triangular pulse and is easily calculated so that

$$\int_{-\Omega}^{\Omega} \left(1 - \frac{|\omega|}{\Omega}\right) e^{j\omega t} \varphi_T(\omega)\, d\omega = \int_{-\infty}^{\infty} R_T(u) \left\{\frac{\sin\left[\frac{1}{2}\Omega(u - t)\right]}{\frac{1}{2}\Omega(u - t)}\right\}^2 \Omega\, du. \tag{3.8-23}$$

As in Section 3 of Appendix B, it can be shown that this transform is also one of the approximating functions for the delta function. Specifically, we have

$$\lim_{\Omega\to\infty}\left\{\frac{\sin\left[\tfrac{1}{2}\Omega(u-t)\right]}{\tfrac{1}{2}\Omega(u-t)}\right\}^2\Omega = 2\pi\,\delta(u-t) \tag{3.8-24}$$

where $\delta(u - t)$ is the Dirac delta function. Thus, in the limit as $\Omega \to \infty$, Equation 3.8-23 becomes

$$\begin{aligned}\lim_{\Omega\to\infty}\int_{-\infty}^{\infty}\left(1-\frac{|\omega|}{\Omega}\right)e^{j\omega t}\varphi_T(\omega)\,d\omega &= \int_{-\infty}^{\infty}\varphi_T(\omega)\,e^{j\omega t}\,d\omega\\ &= 2\pi R_T(t)\end{aligned} \tag{3.8-25}$$

and, on allowing $T \to \infty$, we have

$$\begin{aligned}\lim_{T\to\infty} R_T(t) = R(t) &= \lim_{T\to\infty}\frac{1}{2\pi}\int_{-\infty}^{\infty}\varphi_T(\omega)\,e^{j\omega t}\,d\omega\\ &= \frac{1}{2\pi}\int_{-\infty}^{\infty}\varphi(\omega)\,e^{j\omega t}\,d\omega\end{aligned} \tag{3.8-26}$$

as was to be proved.

Thus, for an ergodic (and therefore stationary) random process $x(t)$, the autocorrelation function and the power spectral density are related by

$$\mathscr{R}(\tau) = R(\tau) = \frac{1}{2\pi}\int_{-\infty}^{\infty}\varphi(\omega)\,e^{j\omega\tau}\,d\omega \tag{3.8-27}$$

where

$$\mathscr{R}(\tau) = \lim_{T\to\infty}\frac{1}{2T}\int_{-T}^{T}x(t)x(t+\tau)\,dt \tag{3.4-2}$$

$$R(\tau) = E\{x(t)x(t+\tau)\} \tag{3.8-1}$$

and

$$\varphi(\omega) = \lim_{T\to\infty} E\left\{\frac{|X_T(\omega)|^2}{2T}\right\} = \int_{-\infty}^{\infty}R(\tau)\,e^{-j\omega\tau}\,d\tau.$$

(3.8-3) and (3.8-19)

The power spectral density $\varphi(\omega)$ of the random process $x(t)$ has a number of interesting properties that follow directly from its definition and from the properties of the autocorrelation function $R(\tau)$. For convenience, we list the following:

1. *The integral of the power spectral density is the average power in the random process.* This property follows directly from Equation 3.8-4 since

$$R(0) = E\{x^2(t)\} = \frac{1}{2\pi}\int_{-\infty}^{\infty}\varphi(\omega)\,d\omega. \tag{3.8-28}$$

2. *The power spectral density is real.* Since $R(\tau)$ is an even function, we can write

$$\begin{aligned}\varphi(\omega) &= \int_{-\infty}^{\infty} R(\tau)(\cos \omega\tau - j \sin \omega\tau)\, d\tau \\ &= \int_{-\infty}^{\infty} R(\tau) \cos \omega\tau\, d\tau \\ &= 2\int_{0}^{\infty} R(\tau) \cos \omega\tau\, d\tau. \end{aligned} \tag{3.8-29}$$

   Since $R(\tau)$ is real for a real random process, then $\varphi(\omega)$ is real also.
3. *The power spectral density is nonnegative.* This property is apparent from Equation 3.8-13. Also it is interesting to note that, when properly normalized, the autocorrelation function and the power spectral density can be related as a density function and its associated characteristic function. Let us write the *normalized autocorrelation function* $\rho(\tau)$ (compare Equation 2.20-28) as

$$\rho(\tau) = \frac{R(\tau)}{R(0)}. \tag{3.8-30}$$

   Then we may write

$$\rho(\tau) = \int_{-\infty}^{\infty} \frac{\varphi(\omega)}{2\pi R(0)} e^{j\omega\tau}\, d\omega \tag{3.8-31}$$

   or

$$\rho(v) = E\{e^{j\omega v}\} \tag{3.8-32}$$

   where $\rho(v)$ is the characteristic function corresponding to the density $\Psi(\omega)$

$$\Psi(\omega) = \frac{\varphi(\omega)}{2\pi R(0)} = \frac{\varphi(\omega)}{\int_{-\infty}^{\infty} \varphi(\omega)\, d\omega}. \tag{3.8-33}$$

   Note that $\Psi(\omega)$ is a true density since

$$\Psi(\omega) \geq 0, \tag{3.8-34}$$

$$\int_{-\infty}^{\infty} \Psi(\omega)\, d\omega = 1. \tag{3.8-35}$$

   Using the Riemann-Stieltjes integral as discussed in Appendix A, we can write Equation 3.8-31 as

$$\rho(v) = \int_{-\infty}^{\infty} e^{j\omega v}\, dF(\omega) \tag{3.8-36}$$

   where

$$F(\omega) = \int_{-\infty}^{\omega} \Psi(x)\, dx \tag{3.8-37}$$

is a c.d.f. with $F(-\infty) = 0$ and $F(\infty) = 1$. Also we see $\Psi(\omega)$ is given by

$$\Psi(\omega) = \frac{1}{2\pi}\int_{-\infty}^{\infty} \rho(\tau)\, e^{-j\omega\tau}\, d\tau \tag{3.8-38}$$

4. *The power spectral density is an even function of* ω; that is, $\varphi(\omega) = \varphi(-\omega)$. From the definition

$$\varphi(-\omega) = \int_{-\infty}^{\infty} R(\tau)\, e^{j\omega\tau}\, d\tau. \tag{3.8-39}$$

After the change of variable $x = -\tau$ and after noting that $R(-x) = R(x)$, we have

$$\varphi(-\omega) = \int_{-\infty}^{\infty} R(x)\, e^{-j\omega x}\, dx = \varphi(\omega). \tag{3.8-40}$$

In Section 3.3, we introduced the concept of cross-correlation function. If two processes $u(t)$ and $v(t)$ are wide-sense stationary and jointly wide-sense stationary, then the cross-correlation function $R_{uv}(\tau)$ is defined as

$$R_{uv}(\tau) = E\{u(t)v(t + \tau)\}. \tag{3.8-41}$$

In analogy with Equation 3.8-3, we define the *cross-spectral density* $\varphi_{uv}(\omega)$ as the Fourier transform of the cross-correlation function $R_{uv}(\tau)$:

$$\varphi_{uv}(\omega) = \int_{-\infty}^{\infty} R_{uv}(\tau)\, e^{-j\omega\tau}\, d\tau. \tag{3.8-42}$$

It was pointed out earlier that the power spectral density $\varphi(\omega)$ was real since $R(\tau)$ was an even function. The cross-spectral density is not necessarily real since the cross-correlation function $R_{uv}(\tau)$ obeys $R_{uv}(\tau) = R_{vu}(-\tau)$. However, direct substitution of this last relation in Equation 3.8-42 shows that

$$\varphi_{uv}(\omega) = \varphi_{vu}(-\omega) = \varphi_{vu}^{*}(\omega) \tag{3.8-43}$$

where the asterisk denotes the complex conjugate.

Example 3.10 We consider the random process $x(t)$ defined by

$$x(t) = u(t) + v(t)$$

where $u(t)$ and $v(t)$ are individually and jointly wide-sense stationary. The autocorrelation function of $x(t)$ is given by

$$\begin{aligned} R_{xx}(\tau) &= E\{[u(t) + v(t)][u(t + \tau) + v(t + \tau)]\} \\ &= R_{uu}(\tau) + R_{vv}(\tau) + R_{uv}(\tau) + R_{vu}(\tau). \end{aligned}$$

We find the power spectral density $\varphi_{xx}(\omega)$ by taking the Fourier transform term by term.

$$\varphi_{xx}(\omega) = \varphi_{uu}(\omega) + \varphi_{vv}(\omega) + \varphi_{uv}(\omega) + \varphi_{vu}(\omega).$$

If $u(t)$ and $v(t)$ are uncorrelated with zero means, then $R_{uv}(\tau)$, $\varphi_{uv}(\omega)$, $R_{vu}(\tau)$, and $\varphi_{vu}(\omega)$ are zero and the power spectral density of the sum is the sum of the power spectral densities. Also, since $\varphi_{xx}(\omega)$, $\varphi_{uu}(\omega)$, and $\varphi_{vv}(\omega)$ are all real, it follows from Equation 3.8-43 that

$$\begin{aligned}\text{Re}\,[\varphi_{uv}(\omega)] &= \text{Re}\,[\varphi_{vu}(\omega)] = \tfrac{1}{2}[\varphi_{uv}(\omega) + \varphi_{vu}^*(\omega)] \\ &= \tfrac{1}{2}[\varphi_{xx}(\omega) - \varphi_{uu}(\omega) - \varphi_{vv}(\omega)].\end{aligned}$$

• •

3.9 An Example of the Calculation of Autocorrelation Functions and Power Spectral Densities

In this section we illustrate some of the concepts of previous sections by computing directly both the power spectral density and the autocorrelation function of a random process and showing that they form a Fourier transform pair.

Consider the random process $x(t)$, a typical sample function of which is shown in Figure 3.6(*a*). This process might arise in certain computers or switching systems. The process has only two possible values $+1$ or -1. It can be considered as a set of connected rectangular pulses of width b. The sign of each pulse is independent of the signs of all other pulses, and positive and negative pulses are equally likely to occur. Suppose we consider a single centered pulse $f(t)$ of positive sign as shown in Figure 3.6(*b*). The Fourier transform of this pulse is (pair 2 of Table 3.1)

$$F(\omega) = \int_{-b/2}^{b/2} e^{-j\omega t}\,dt = \frac{2}{\omega}\sin\frac{\omega b}{2} \tag{3.9-1}$$

or

$$F(\omega) = b\,\frac{\sin \omega b/2}{\omega b/2}. \tag{3.9-2}$$

A given sample function of $x(t)$ is simply a given sum of these pulses properly shifted and with the proper sign. The process $x(t)$ could be represented as

$$x(t) = \sum_{n=-\infty}^{\infty} a_n f(t - nb) \tag{3.9-3}$$

where a_n is a random variable equally likely to be $+1$ or -1. Furthermore a_i and a_j are independent if $i \neq j$, and

$$E\{a_i\} = 0, \tag{3.9-4}$$

$$E\{a_i a_j\} = \begin{cases} 1, & i = j, \\ 0, & i \neq j. \end{cases} \tag{3.9-5}$$

With this notation, a typical sample function is shown in Figure 3.6(*c*).

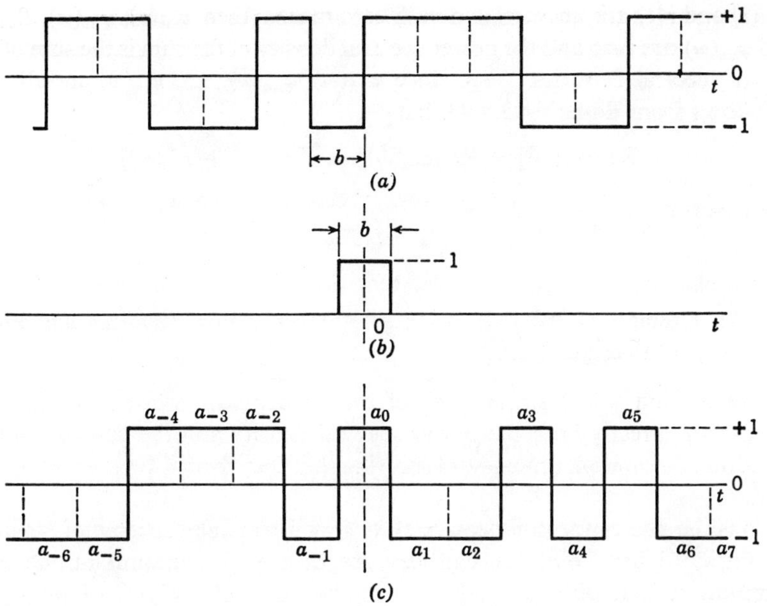

Figure 3.6 *A random process: (a) a typical sample function; (b) basic pulse; (c) a centered sample function*

Consider the Fourier transform of a section of the random process extending from $-T$ to T or from $n = -N$ to $n = N$ where $2T = (2N + 1)b$:

$$X_T(\omega) = \sum_{n=-N}^{N} a_n F(\omega)\, e^{-jn\omega b} = F(\omega) \sum_{n=-N}^{N} a_n\, e^{-jn\omega b} \tag{3.9-6}$$

since translation by nb in the time domain is equivalent to multiplication by $e^{-jn\omega b}$ in the frequency domain. We can easily show this as follows:

$$F(\omega) = \int_{-\infty}^{\infty} f(t)\, e^{-j\omega t}\, dt,$$

$$F(\omega)\, e^{-jn\omega b} = \int_{-\infty}^{\infty} f(t)\, e^{-j\omega(t+nb)}\, dt.$$

Let $t + nb = x$ so that $t = x - nb$:

$$F(\omega)\, e^{-jn\omega b} = \int_{-\infty}^{\infty} f(x - nb)\, e^{-j\omega x}\, dx = \mathscr{F}[f(t - nb)]$$

From Equation 3.8-13, we can write the power spectral density $\varphi(\omega)$ as

$$\varphi(\omega) = \lim_{T\to\infty} E\left\{\frac{|X_T(\omega)|^2}{2T}\right\}. \tag{3.8-13}$$

With the substitution of Equation 3.9-6 into this form, we have

$$\varphi(\omega) = \lim_{N\to\infty} E\left\{\frac{1}{(2N+1)b}|F(\omega)|^2\left|\sum_{n=-N}^{N} a_n\, e^{jn\omega b}\right|^2\right\}$$

or

$$\varphi(\omega) = \lim_{N\to\infty} \frac{1}{(2N+1)b}|F(\omega)|^2 \sum_{m=-N}^{N}\sum_{n=-N}^{N} E\{a_m a_n\}\, e^{jm\omega b}\, e^{-jn\omega b}. \qquad (3.9\text{-}7)$$

But we have that $E\{a_m a_n\} = \delta_{ij}$, so that Equation 3.9-7 can be written as

$$\varphi(\omega) = \frac{|F(\omega)|^2}{b} \lim_{N\to\infty} \frac{1}{2N+1} \sum_{n=-N}^{N} e^{jn\omega b}\, e^{-jn\omega b}$$

or

$$\varphi(\omega) = \frac{|F(\omega)|^2}{b} = b\,\frac{\sin^2 \omega b/2}{(\omega b/2)^2}. \qquad (3.9\text{-}8)$$

This is the power spectral density of $x(t)$ by direct calculation and is plotted in Figure 3.7(*a*). We now consider the direct computation of the autocorrelation function.

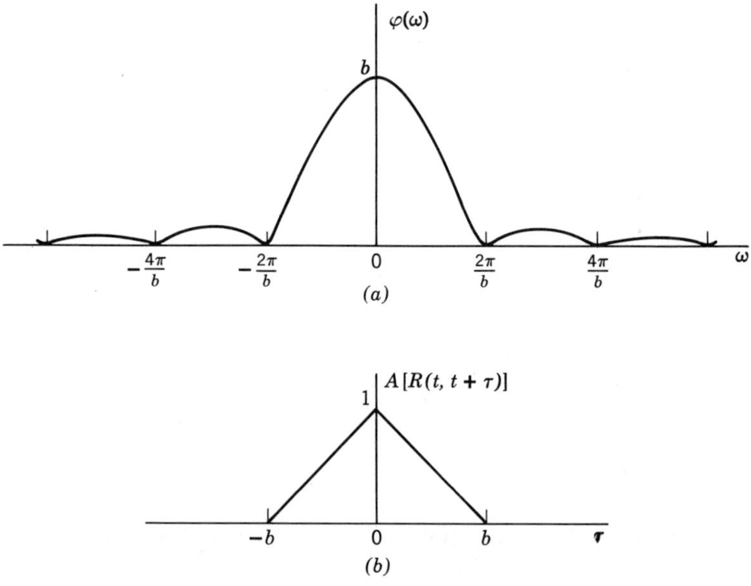

Figure 3.7 *Power spectral density and time average of the autocorrelation function of the random process of Figure 3.6:* (*a*) *the power spectral density;* (*b*) *the time average of the autocorrelation function*

The autocorrelation function of $x(t)$ can be written as

$$R(t, t + \tau) = E\{x(t)x(t + \tau)\}.$$

It follows directly from the definition of the random process that this function is zero unless t and $t + \tau$ lie in the same interval between switching instants. We have

$$R(t, t + \tau) = \begin{cases} 1, & nb \leq t, t + \tau \leq (n + 1)b, \\ 0, & \text{elsewhere.} \end{cases} \tag{3.9-9}$$

Thus the autocorrelation function depends on t as well as τ and the process $x(t)$ is not wide-sense stationary even though the power spectral density $\varphi(\omega)$ exists and is not a function of time.

Suppose we consider the time average $A[R(t, t + \tau)]$ of $R(t, t + \tau)$ where

$$A[R(t, t + \tau)] = \lim_{T \to \infty} \frac{1}{2T} \int_{-T}^{T} R(t, t + \tau)\, dt.$$

It is clear that this time average will be zero if $|\tau| > b$ since, in that case, t and $(t + \tau)$ will not lie in the same basic interval and $R(t, t + \tau)$ will be zero. For $0 \leq \tau < b$ there will be $2N + 1$ basic intervals of length less than b where $R(t, t + \tau)$ is nonzero. In fact, the nonzero region in each basic interval will be

$$nb < t < (n + 1)b - \tau. \tag{3.9-10}$$

Now the time average becomes, for $0 \leq \tau < b$,

$$A[R(t, t + \tau)] = \lim_{N \to \infty} \frac{1}{(2N + 1)b} (2N + 1) \int_{nb}^{(n+1)b - \tau} dt = 1 - \frac{\tau}{b}.$$

Thus the time average for all τ is

$$A[R(t, t + \tau)] = \begin{cases} 1 - \dfrac{|\tau|}{b}, & |\tau| \leq b, \\ 0, & |\tau| > b, \end{cases} \tag{3.9-11}$$

as shown in Figure 3.7(b). This last expression and Equation 3.9-8 form a Fourier-transform pair, that is,

$$\varphi(\omega) = \int_{-\infty}^{\infty} A[R(t, t + \tau)]\, e^{-j\omega\tau}\, d\tau. \tag{3.9-12}$$

This expression is easily evaluated to yield

$$\varphi(\omega) = \int_{-b}^{b} e^{-j\omega\tau}\, d\tau - \frac{2}{b} \int_{0}^{b} \tau \cos \omega\tau\, d\tau$$

$$= \frac{2}{\omega^2 b} (1 - \cos \omega b) = b \frac{\sin^2 \omega b/2}{(\omega b/2)^2}$$

in agreement with Equation 3.9-8.

Let us now consider a new random process $y(t)$ which is similar to the one we have just treated except that it is not centered at $t = 0$. Instead, the process has an initial phase angle θ which is a random variable uniformly distributed in the interval $(-b/2, b/2)$. In other words, θ has a density function $p(\theta)$ given by

$$p(\theta) = \begin{cases} \dfrac{1}{b}, & -\dfrac{b}{2} \leq \theta \leq \dfrac{b}{2}, \\ 0, & \text{elsewhere.} \end{cases} \tag{3.9-13}$$

A typical sample function of $y(t)$ is shown in Figure 3.8(*a*). This new process with random phase is wide-sense stationary as will now be demonstrated.

The power spectral density calculation proceeds exactly as in the previous case except that the shifted functions $f(t - nb)$ are replaced by $f(t - nb - \theta)$ where θ is the random variable just defined. It will be left to the reader to show that the same results are obtained as before and that $\varphi_y(\omega)$ is given by Equation 3.9-8. We now consider the direct calculation of the autocorrelation function.

Refer to Figure 3.8(*b*) which shows a typical sample function. We use the random variable notation

$$y_{1i} = y(t), \qquad t \text{ arbitrary}; \; i = 1, 2,$$
$$y_{2j} = y(t + \tau), \qquad j = 1, 2.$$

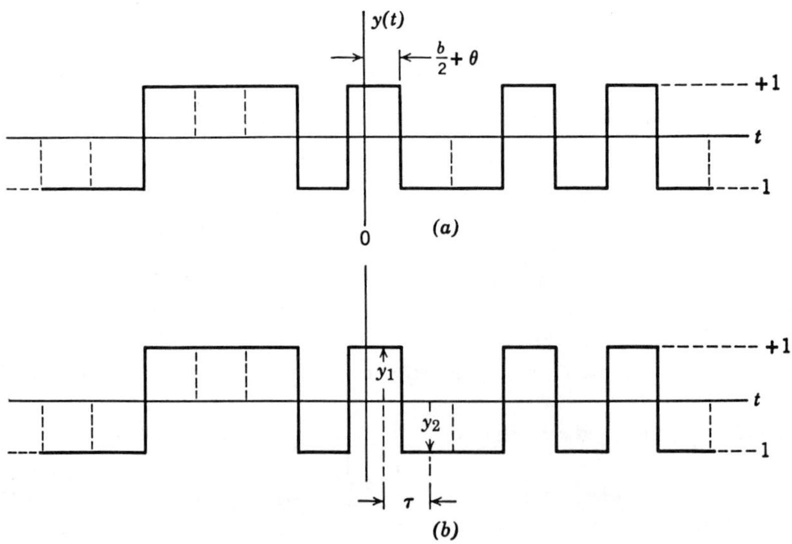

***Figure 3.8** A random process that is wide-sense stationary: (a) a typical sample function, θ is a random variable; (b) notation for the calculation of the autocorrelation function*

Now the autocorrelation function is given by

$$R(\tau) = E\{y_{1i}y_{2j}\} = \sum_{i=1}^{2}\sum_{j=1}^{2} y_{1i}y_{2j}f_2(y_{1i}, y_{2j}) \tag{3.9-14}$$

where f_2 is the joint density of y_{1i} and y_{2j}. This expression may be rewritten as

$$R(\tau) = P(1, 1; \tau) + P(-1, -1; \tau) - P(1, -1; \tau) - P(-1, 1; \tau) \tag{3.9-15}$$

where $P(1, 1; \tau)$ is the probability that y_{1i} has the value $+1$ and y_{2j} has the value $+1$, and so on. From the conditional probability relations of Chapter 2, we write

$$\begin{aligned} R(\tau) = {} & P(1)P(1/1; \tau) + P(-1)P(-1/-1; \tau) \\ & - P(1)P(-1/1; \tau) - P(-1)P(1/-1; \tau). \end{aligned} \tag{3.9-16}$$

From symmetry we have that

$$P(1/1; \tau) = P(-1/-1; \tau), \tag{3.9-17}$$

$$P(-1/1; \tau) = P(1/-1; \tau). \tag{3.9-18}$$

In other words, the random process has been defined in such a manner that its properties are symmetric with respect to sign. With this substitution, the autocorrelation function is given by

$$R(\tau) = P(1/1; \tau) - P(-1/1; \tau) \tag{3.9-19}$$

since

$$P(1) = P(-1) = \tfrac{1}{2}. \tag{3.9-20}$$

We now compute $R(\tau)$ from Equation 3.9-19 using Figure 3.8(b) as a model. There are at least two ranges to be considered: $0 \leq \tau \leq b$ and $b \leq \tau \leq 2b$. *For* $0 \leq \tau \leq b$ *we have*

$$\begin{aligned} P(-1/1; \tau) = {} & (\text{probability that a shift time will occur}) \times \\ & (\text{probability of a change in sign at this shift}) \end{aligned} \tag{3.9-21}$$

or

$$P(-1/1; \tau) = \frac{\tau}{b} \times \frac{1}{2} = \frac{\tau}{2b} \tag{3.9-22}$$

and, consequently,

$$P(1/1; \tau) = 1 - \frac{\tau}{2b}. \tag{3.9-23}$$

For $b \le \tau \le 2b$ we have either one or two shifts in this interval since less than one or more than two is impossible. The probabilities are

$$\text{probability of two shift times is } \frac{\tau - b}{b},$$

and consequently,

$$\text{probability of one shift time is } 1 - \frac{\tau - b}{b} = \frac{2b - \tau}{b}.$$

Also the probability of a shift in sign at a shift time is $\frac{1}{2}$.

In the range $b \le \tau \le 2b$ the samples x_{1i} and x_{2j} can have different signs in three ways:

1. One shift time with a shift in sign occurring.
2. Two shift times with a shift in sign occurring on the first.
3. Two shift times with a shift in sign occurring on the second.

Then

$$P(-1/1;\tau) = \frac{1}{2}\left(\frac{2b - \tau}{b}\right) + \frac{1}{4}\left(\frac{\tau - b}{b}\right) + \frac{1}{4}\left(\frac{\tau - b}{b}\right) = \frac{1}{2}. \quad (3.9\text{-}24)$$

In the range $b \le \tau \le 2b$ the samples x_1 and x_2 can have the same sign in three different ways:

1. One shift time with no shift in sign occurring.
2. Two shift times with no shift in sign occurring.
3. Two shift times with two shifts in sign occurring.

Then

$$P(1/1;\tau) = \frac{1}{2}\left(\frac{2b - \tau}{b}\right) + \frac{1}{4}\left(\frac{\tau - b}{b}\right) + \frac{1}{4}\left(\frac{\tau - b}{b}\right) = \frac{1}{2}. \quad (3.9\text{-}25)$$

Hence $R(\tau)$ is zero in the range $b < \tau < 2b$. An extension of this argument will show $R(\tau)$ to be zero for all $\tau > b$ so that, since $R(\tau)$ is even:

$$R(\tau) = \begin{cases} 1 - \dfrac{|\tau|}{b}, & -b \le \tau \le b, \\ 0, & \text{elsewhere.} \end{cases} \quad (3.9\text{-}26)$$

We might have concluded originally that $R(\tau) = 0$ for $\tau > b$ since the changes in sign at the shift times are independent and since the shift times are b seconds apart. The autocorrelation function of Equation 3.9-26 is shown in Figure 3.7(b).

3.10 The Gaussian Process

A random process $x(t)$ is called a *gaussian random process* if, for every finite set of parameter values $\{t_i\}$, the set of random variables $\{x(t_i)\}$ have a multivariate normal distribution as discussed in Section 2.22; that is, if the joint density $f(\mathbf{x})$ is given by

$$f(\mathbf{x}) = \frac{|\sigma^{ij}|^{1/2}}{(2\pi)^{n/2}} e^{-[(\mathbf{x}-\mathbf{u})'\Lambda^{-1}(\mathbf{x}-\mathbf{u})]/2} \tag{3.10-1}$$

where

$$\mathbf{x} = \begin{bmatrix} x(t_1) \\ x(t_2) \\ \vdots \\ x(t_n) \end{bmatrix} = \begin{bmatrix} x_1 \\ x_2 \\ \vdots \\ x_n \end{bmatrix}; \tag{3.10-2}$$

$\mathbf{u}$ is the mean matrix

$$\mathbf{u} = \begin{bmatrix} E\{x(t_1)\} \\ E\{x(t_2)\} \\ \vdots \\ E\{x(t_n)\} \end{bmatrix} = \begin{bmatrix} u_1 \\ u_2 \\ \vdots \\ u_n \end{bmatrix}; \tag{3.10-3}$$

Λ is the covariance matrix

$$\Lambda = [\sigma_{ij}] = [E\{(x_i - u_i)(x_j - u_j)\}]; \tag{3.10-4}$$

and $|\sigma^{ij}|$ is the determinant of the inverse of Λ:

$$|\sigma^{ij}| = |\Lambda^{-1}|. \tag{3.10-5}$$

For the case where the process is wide-sense stationary and has a constant mean u, the terms of the covariance matrix become

$$\begin{aligned} \sigma_{ij} &= E\{(x_i - u)(x_j - u)\} \\ &= R_{xx}(t_j - t_i) - u^2. \end{aligned} \tag{3.10-6}$$

Thus not only the covariance matrix but also the joint densities to all orders are a function of the time differences only. Consequently, a gaussian random process which is wide-sense stationary and has a constant mean is also strictly stationary since the joint densities are invariant to a time translation.

In particular, a gaussian random process has a first-order density $f(x_1)$ given by

$$f(x_1) = \frac{1}{\sqrt{2\pi\sigma_1^2}} \exp\left[-\frac{1}{2}\left(\frac{x_1 - u_1}{\sigma_1}\right)^2\right] \tag{3.10-7}$$

where x_1 is $x(t)$ at any time t_1; $u_1 = E\{x_1\}$; and $\sigma_1^2 = E\{(x_1 - u_1)^2\}$. Also the bivariate density $f_2(x_1, x_2)$ is

$$f_2(x_1, x_2) = \frac{\exp\left[-\dfrac{\sigma_2^2(x_1 - u_1)^2 - 2\sigma_1\sigma_2\rho(x_1 - u_1)(x_2 - u_2) + \sigma_1^2(x_2 - u_2)^2}{2\sigma_1^2\sigma_2^2(1 - \rho^2)}\right]}{2\pi\sigma_1\sigma_2(1 - \rho^2)^{1/2}} \tag{3.10-8}$$

where ρ has been previously defined as

$$\rho = \frac{\sigma_{12}}{\sigma_1\sigma_2} = \frac{E\{(x_1 - u_1)(x_2 - u_2)\}}{\sigma_1\sigma_2}. \tag{3.10-9}$$

For the case where the process is wide-sense stationary with zero mean (and, hence, is strictly stationary), this density becomes

$$f_2(x_1, x_2) = \frac{1}{2\pi\sigma^2[1 - \rho^2(\tau)]^{1/2}} \exp\left\{-\frac{x_1^2 - 2\rho(\tau)x_1x_2 + x_2^2}{2\sigma^2[1 - \rho^2(\tau)]}\right\} \tag{3.10-10}$$

where $\sigma^2 = \sigma_1^2 = \sigma_2^2$, $\tau = t_2 - t_1$, and

$$\rho(\tau) = \frac{R(\tau)}{R(0)} = \frac{E\{x(t_1)x(t_2)\}}{E\{x^2(t)\}}. \tag{3.10-11}$$

The form of Equation 3.10-10 will be of considerable utility later.

3.11 White Noise

It will often be convenient to consider a random process $x(t)$ with a constant spectral density N_0:

$$\varphi(\omega) = N_0, \qquad -\infty < \omega < \infty. \tag{3.11-1}$$

Such a process is often called *white noise*. It is not realizable since its mean-squared value (power) is not finite:

$$R(0) = E\{x^2(t)\} = \frac{N_0}{2\pi}\int_{-\infty}^{\infty} d\omega = \infty. \tag{3.11-2}$$

On the other hand it is very useful in many types of analysis and often can be postulated where the actual process has an approximately constant spectral density over a frequency range much greater than the system bandwidth.

For the white noise of Equation 3.11-1 we have the Fourier-transform pair

$$R(\tau) = \frac{1}{2\pi}\int_{-\infty}^{\infty} N_0\, e^{j\omega\tau}\, d\omega = N_0\,\delta(\tau) \tag{3.11-3}$$

and

$$\varphi(\omega) = \int_{-\infty}^{\infty} N_0\,\delta(\tau)\, e^{-j\omega\tau}\, d\tau = N_0 \tag{3.11-4}$$

as illustrated in Figure 3.9 and as given by pair number 7 of Table 3.1. We often consider the case where the power spectral density is constant but bandlimited. Here the autocorrelation function and power spectral density are related by

$$R(\tau) = \frac{1}{2\pi}\int_{-\omega_b}^{\omega_b} N_0\, e^{j\omega\tau}\, d\omega$$

$$= \frac{\omega_b N_0}{\pi}\frac{\sin \omega_b\tau}{\omega_b\tau} \tag{3.11-5}$$

and

$$\varphi(\omega) = \frac{\omega_b N_0}{\pi}\int_{-\infty}^{\infty} \frac{\sin \omega_b\tau}{\omega_b\tau}\, e^{-j\omega\tau}\, d\tau$$

$$= \begin{cases} N_0, & |\omega| \le \omega_b, \\ 0, & |\omega| > \omega_b, \end{cases} \tag{3.11-6}$$

as illustrated in Figure 3.9 and as given by pair number 2 of Table 3.1. It is clear that this last process becomes white noise as $\omega_b \to \infty$; for example, the autocorrelation function becomes

$$\lim_{\omega_b \to \infty} \frac{\omega_b N_0}{\pi}\frac{\sin \omega_b\tau}{\omega_b\tau} = N_0\,\delta(\tau).$$

This last expression is one of many approximations to the delta function as discussed in more detail in Appendix B.

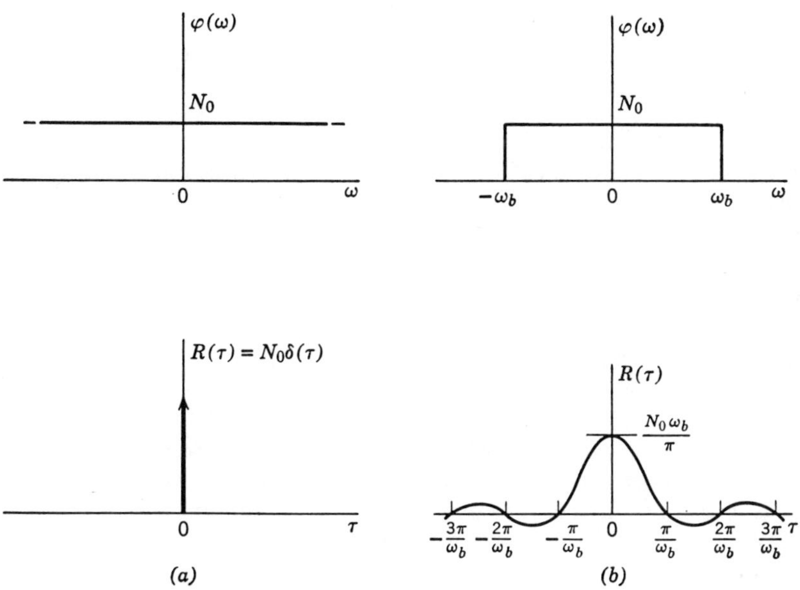

Figure 3.9 Power spectral densities and autocorrelation functions of white noise: (a) white noise; (b) bandlimited white noise

Suppose now that the bandlimited process $x(t)$ defined by Equation 3.11-6 is sampled at the fixed time interval $\Delta = \pi/\omega_b$. The result is the discrete-time (discrete-parameter) process

$$\{x(n\Delta)\} = \{\ldots x(-2\Delta), x(-\Delta), x(0), x(\Delta), x(2\Delta), \ldots\}.$$

It follows from Equation 3.11-5 that the random variables $x(n\Delta)$ are uncorrelated; that is,

$$E\{x(m\Delta)x(n\Delta)\} = N_0\omega_b\, \delta_{mn}/\pi.$$

If, in addition, the original process $x(t)$ is gaussian, then all random variables $x(n\Delta)$ are normally distributed and *independent*. Thus, in this case, the random process $\{x(n\Delta)\}$ is just a sequence of independent, normal, identically distributed random variables. In subsequent discussions it will be convenient to change notation slightly and consider the process $\{x_t\}$ where

$$\{x_t\} = \{\ldots x_{-2}, x_{-1}, x_0, x_1, x_2, \ldots\}.$$

Thus the sampling interval Δ has been suppressed.

3.12 Markov Processes

From one point of view the simplest discrete-time random process $\{x_t\}$ is a sequence of independent, identically distributed random variables such as we have just discussed. It is not necessary that the random variables be normally distributed of course. Let the density function of the (identically distributed) random variables be denoted by $f(x_m)$. Then for arbitrary times $t_l < t_m < t_n$ it is clear that

$$f_c(x_n/x_m, x_l, \ldots) = f(x_n) \tag{3.12-1}$$

where $f_c(\cdot)$ is a conditional density function. What this last equation indicates is that the random variable x_n is independent of all previously occurring random variables comprising the process $\{x_t\}$.

A logical generalization of Equation 3.12-1 is to assume that a given random variable x_n is statistically related to some finite number of previously occurring random variables. This is the essential idea in the definition of Markov processes. More specifically, if

$$f_c(x_n/x_{n-1}, x_{n-2}, \ldots) = f_{c1}(x_n/x_{n-1}), \tag{3.12-2}$$

then we say that $\{x_n\}$ is a *discrete-time first-order Markov process.* Here $f_c(\cdot)$ and $f_{c1}(\cdot)$ are conditional density functions and the occurrence of x_n is conditioned only on the immediately preceding random variable x_{n-1}. In general, $\{x_n\}$ is a *discrete-time kth-order* Markov process if

$$f_c(x_n/x_{n-1}, \ldots, x_{n-k}, \ldots) = f_{ck}(x_n/x_{n-1}, \ldots, x_{n-k}). \tag{3.12-3}$$

Actually we have been defining *homogeneous* Markov processes. In the general case, the conditional density $f_{ck}(\cdot)$ on the right side of Equation

3.12-3 would depend on the time t_n (and hence on the index n) as well as the index k. In this and subsequent discussions we will consider only the homogeneous case. In addition we will restrict our attention to situations where the random variable x_n can take on only a finite set of possible values; that is, it is a finite discrete random variable. These discrete-time, finite, discrete Markov processes sometimes are called *finite Markov chains.*

Let us restrict our attention to the first-order Markov chain where the outcome (realization of the random variable) at any time t_{n+1} depends only on the outcome at the immediately preceding time t_n. Thus for each pair (x_n, x_{n+1}) we can define the transition probability as the conditional probability $p_{ij}^{n,n+1}$ where

$$p_{ij}^{n,n+1} = P(x_{n+1} = \Psi_j/x_n = \Psi_i). \tag{3.12-4}$$

Here the Ψ_i are the possible realizations of the random variables x_{n+1} and x_n. This notation indicates that the transition probabilities depend not only on the initial and final outcome but also on the time at which the transition takes place. As mentioned previously, we shall consider only the *homogeneous* case where the transition probabilities are *stationary* so that this last equation is just the one-step transition probability and can be rewritten as

$$p_{ij} = P(\Psi_j/\Psi_i). \tag{3.12-5}$$

For convenience we shall arrange these transition probabilities into a $q \times q$ square *transition matrix* $\boldsymbol{T}$

$$\boldsymbol{T} = [p_{ij}] = \begin{bmatrix} p_{11} & p_{12} & \cdots & p_{1q} \\ p_{21} & p_{22} & \cdots & p_{2q} \\ \vdots & & & \\ p_{q1} & & \cdots & p_{qq} \end{bmatrix} \tag{3.12-6}$$

where q is the number of possible realizations of x_i. It is clear that each element of the matrix satisfies the conditions

$$p_{ij} \geq 0, \qquad i, j = 1, 2, \ldots, q, \tag{3.12-7}$$

$$\sum_{j=1}^{q} p_{ij} = 1, \qquad i = 1, 2, \ldots, q. \tag{3.12-8}$$

Square matrices whose elements satisfy Equations 3.12-7 and 3.12-8 sometimes are called *Markov matrices* or *stochastic matrices.*

If Ψ_i is the particular realization of x_n, we shall say that the stochastic process is in *state* σ_i at time t_n. (In Section 8.8 we shall extend this concept of state to account for the nth-order as well as the first-order Markov process.) It is clear that the process has q states, the number of realizations of any x_i. Let the unconditional probability of state σ_i occurring at time t_n be denoted by $p_i^{(n)}$; that is,

$$P(x_n = \Psi_i) = p_i^{(n)}. \tag{3.12-9}$$

Also denote all state probabilities at time t_n by the *state distribution vector* $\mathbf{\Pi}^{(n)}$ where

$$\mathbf{\Pi}^{(n)} = [p_1^{(n)}, p_2^{(n)}, \cdots, p_q^{(n)}]. \tag{3.12-10}$$

The *initial* state distribution vector $\mathbf{\Pi}^{(0)}$ will be written as $\mathbf{\Pi}$ and is given by

$$\mathbf{\Pi}^{(0)} = \mathbf{\Pi} = [p_1^{(0)}, p_2^{(0)}, \cdots, p_q^{(0)}]. \tag{3.12-11}$$

This is equivalent to assuming that the random process x_t starts at $t = t_0$.

Let us now consider r-step transition probabilities and define

$$p_{ij}^{(r)} = P(x_{r+m} = \Psi_j/x_m = \Psi_i), \qquad \text{all } m. \tag{3.12-12}$$

Then the one-step transition probability has already been defined since

$$p_{ij}^{(1)} = p_{ij}. \tag{3.12-13}$$

The absolute state probability at time $t = t_1$ is

$$p_j^{(1)} = \sum_{i=1}^{q} p_i^{(0)} p_{ij} \tag{3.12-14}$$

or, in matrix notation,

$$\mathbf{\Pi}^{(1)} = \mathbf{\Pi}\boldsymbol{T}. \tag{3.12-15}$$

In the same way, we continue the process to obtain

$$p_k^{(2)} = \sum_{j=1}^{q} p_j^{(1)} p_{jk} = \sum_{j=1}^{q} \sum_{i=1}^{q} p_i^{(0)} p_{ij} p_{jk}. \tag{3.12-16}$$

This last expression may be rearranged to give

$$p_k^{(2)} = \sum_{i=1}^{q} p_i^{(0)} \sum_{j=1}^{q} p_{ij} p_{jk} \tag{3.12-17}$$

or

$$\mathbf{\Pi}^{(2)} = \mathbf{\Pi}^{(1)}\boldsymbol{T} = \mathbf{\Pi}\boldsymbol{T}^2. \tag{3.12-18}$$

It is clear that, in the general case, we obtain

$$\boxed{\mathbf{\Pi}^{(r)} = \mathbf{\Pi}\boldsymbol{T}^r.} \tag{3.12-19}$$

Note that Equation 3.12-19 may be rewritten as

$$\mathbf{\Pi}^{(r+s)} = \mathbf{\Pi}\boldsymbol{T}^{r+s} = (\mathbf{\Pi}\boldsymbol{T}^s)\boldsymbol{T}^r \tag{3.12-20}$$

or

$$\boxed{\mathbf{\Pi}^{(r+s)} = \mathbf{\Pi}^{(s)}\boldsymbol{T}^r.} \tag{3.12-21}$$

Thus $\boldsymbol{T}^r$ is the required r-step transition matrix with elements $p_{ij}^{(r)}$ as given by Equation 3.12-12. For a homogeneous Markov chain, this matrix is obviously independent of m, the time index in Equation 3.12-12.

That the matrix $\boldsymbol{T}^r$ is also a stochastic matrix is easily shown. Let **1** be a *column* vector of 1's. Then

$$\boldsymbol{T}\mathbf{1} = \mathbf{1} \tag{3.12-22}$$

and

$$\boldsymbol{T}^2\mathbf{1} = \boldsymbol{T}\mathbf{1} = \mathbf{1} \tag{3.12-23}$$

or

$$\boldsymbol{T}^r\mathbf{1} = \mathbf{1} \tag{3.12-24}$$

and $\boldsymbol{T}^r$ is stochastic.

3.13 Regularity and Ergodicity of Finite Markov Chains

Suppose that the state probability vector $\boldsymbol{\Pi}^{(r)}$ approaches a limiting distribution independent of the initial state distribution; that is,

$$\boxed{\lim_{r\to\infty} \boldsymbol{\Pi}^{(r)} = \mathbf{P}} \tag{3.13-1}$$

where **P** is a fixed probability vector with elements p_j so that

$$\mathbf{P} = [p_1, p_2, \cdots, p_q]. \tag{3.13-2}$$

Then it is clear that the unconditional state probability vector has reached *equilibrium*. We will call $p_i = p(\Psi_i)$ the *stationary* probability distribution of state σ_i. A Markov chain with the property of Equation 3.13-1 is called *regular*.

It may not be easy to find the limit of the left side of Equation 3.13-1. On the other hand, suppose that we consider the equation

$$\mathbf{Q}\boldsymbol{T} = \mathbf{Q} \tag{3.13-3}$$

subject to the condition that

$$\sum_{i=1}^{q} q_i = 1, \qquad q_i \geq 0 \tag{3.13-4}$$

where the q_i are the elements of the row vector **Q**. A set of q independent linear relationships for the q_i results and may be solved simultaneously for the q_i. It is clear that **Q** is a stationary distribution in the sense that, if it is chosen as the initial distribution $\boldsymbol{\Pi}$, all subsequent $\boldsymbol{\Pi}^{(r)}$ will be equal to **Q**; that is,

$$\begin{aligned} \boldsymbol{\Pi}^{(1)} &= \boldsymbol{\Pi}\boldsymbol{T} = \mathbf{Q}\boldsymbol{T} = \mathbf{Q} \\ \boldsymbol{\Pi}^{(2)} &= \boldsymbol{\Pi}^{(1)}\boldsymbol{T} = \mathbf{Q}\boldsymbol{T} = \mathbf{Q} \\ &\vdots \end{aligned} \tag{3.13-5}$$

It can be shown [11, 12] that, if a stationary limiting distribution **P** exists, then this is the only limiting distribution. Furthermore, it is clear that **P** = **Q** and that **P** is found by solving

$$\mathbf{P}T = \mathbf{P} \tag{3.13-6}$$

subject to the condition that

$$\sum_{i=1}^{q} p_i = 1, \qquad p_i \geq 0. \tag{3.13-7}$$

The concept of regularity has been defined in terms of an eventual statistical equilibrium of states independent of the initial state distribution. It seems intuitively reasonable to suppose that, in a regular Markov chain, it must be possible to go from any one state to any other state in a finite number of steps. If this were not true, then it should be possible to pick initial state distributions that affect all subsequent state distributions. The condition that any state can be reached from any other state in a finite number of steps is just the condition that T^r have no zero elements for some positive finite r. Let us now use Equations 3.12-19 and 3.13-1 to write

$$\lim_{r\to\infty} \mathbf{\Pi}^{(r)} = \mathbf{\Pi} \lim_{r\to\infty} T^r = \mathbf{P} \tag{3.13-8}$$

for a regular Markov chain. Here $\mathbf{\Pi}$ is *any* initial distribution vector. This equation is just q linear equations, a typical one being

$$\sum_{i=1}^{q} \Pi_i \left[\lim_{r\to\infty} p_{ij}^{(r)} \right] = p_j, \qquad j = 1, 2, \ldots, q. \tag{3.13-9}$$

Since this expression must be true for any set of Π_i such that

$$\sum_{i=1}^{q} \Pi_i = 1, \tag{3.13-10}$$

it is clear that

$$\lim_{r\to\infty} p_{ij}^{(r)} = p_j, \qquad j = 1, 2, \ldots, q \tag{3.13-11}$$

or, in matrix notation,

$$\boxed{\lim_{r\to\infty} T^r = A} \tag{3.13-12}$$

where A is a stochastic matrix with identical rows, each row being the stationary distribution vector **P**. A regular Markov chain may be defined [11] by either Equation 3.13-1 or 3.13-12.

The foregoing has been intended as only the most elementary discussion of Markov processes. A number of excellent and more detailed treatments exist. Those of Feller [13] and Parzen [1] are particularly clear although elementary. A detailed treatment of finite Markov chain is given by Kemeny, Snell, and Knapp [12]. An advanced and detailed treatise on Markov processes is that of Dynkin [14]. The elementary Markov chains of this section will be used in Chapter 8 as models of information sources.

3.14 Shot Noise

In this section we begin a brief study of the *shot-noise process.* One of the simplest examples of such a process occurs in the temperature-limited vacuum-tube diode. Because of its underlying importance as well as its analytical simplicity, this case will now be treated in some detail.

Consider a diode connected to a resistive load R_L and with proper plate and filament voltages to be operated temperature limited as shown in Figure 3.10(*a*). Since the electrons possess a discrete charge and since they are emitted randomly from the cathode, the plate current $i(t)$ and, hence, the voltage across the load will fluctuate. This is the *shot effect* and the fluctuating component is called the *shot noise.* The effect was first predicted and treated theoretically by Schottky [15] in 1918.

In the temperature-limited case the electron emissions are independent and the current may be considered as the superposition of all of the current pulses due to the individual electrons. A given electron (assumed to be emitted at time $t = 0$) will produce a current pulse of some form $f(t)$ as shown in Figure 3.10(*b*) where t_T is the transit time and the shape of $f(t)$ depends on tube geometry, potential distribution, and electron velocity. Note that time t_T is very short (in this respect the figure is very misleading)

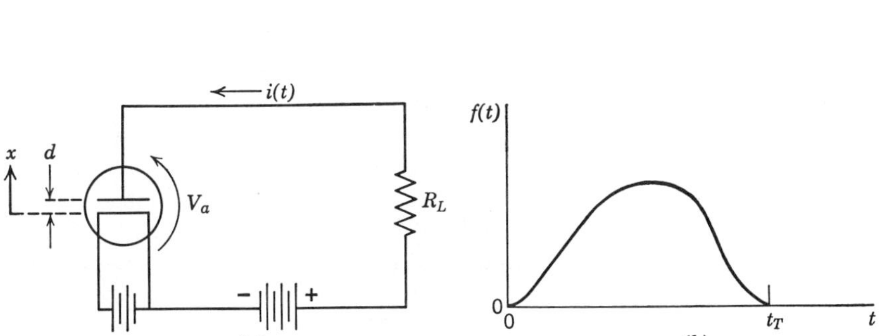

Figure 3.10 *The temperature-limited diode:* (*a*) *the diode circuit;* (*b*) *the current pulse* $f(t)$ *due to a single electron*

so that $f(t)$ is almost like a δ-function. From the properties of the electron, we have

$$\int_{-\infty}^{\infty} f(t)\,dt = \int_{0}^{t_T} f(t)\,dt = e \text{ (electronic charge).} \tag{3.14-1}$$

Consider the diode current in a long interval $-T \leq t \leq T$ where T is very large compared to t_T. If N electrons are emitted in the time $2T$, then the diode current $i_T(t)$ is just the linear superposition of the N currents due to each electron so that

$$i_T(t) = \sum_{i=1}^{N} f(t - t_i) \tag{3.14-2}$$

where t_i is the emission time of the ith electron. Note that N is a function of the length of the interval $(-T, T)$ and will sometimes be written $N(T)$.

The time average value of this current is just

$$A[i_T(t)] = \langle I \rangle = \lim_{T\to\infty} \frac{1}{2T} \int_{-T}^{T} i_T(t)\,dt. \tag{3.14-3}$$

On substituting Equation 3.14-2 into this last expression, we have

$$\langle I \rangle = \lim_{T\to\infty} \frac{1}{2T} \int_{-T}^{T} \sum_{i=1}^{N} f(t - t_i)\,dt. \tag{3.14-4}$$

Interchanging the order of integration and summation, we obtain

$$\langle I \rangle = \lim_{T\to\infty} \frac{1}{2T} \sum_{i=1}^{N} \int_{-\infty}^{\infty} f(t - t_i)\,dt \tag{3.14-5}$$

$$= \lim_{T\to\infty} \frac{1}{2T} \sum_{i=1}^{N} e = \lim_{T\to\infty} \frac{1}{2T} Ne \tag{3.14-6}$$

or

$$\langle I \rangle = \langle n \rangle e \tag{3.14-7}$$

where $\langle n \rangle$ is the average number of electrons per second and is given by

$$\langle n \rangle = \lim_{T\to\infty} \frac{N}{2T}. \tag{3.14-8}$$

Suppose now that the diode circuit of Figure 10(a) is only one of a very large number with identical characteristics. Then the current given by Equation 3.14-2 can be considered as one of the sample functions or realizations of a random process $i_T(t)$. In this case both N, the number of electrons emitted in the interval $(-T, T)$, and t_i, the emission time of the ith electron, will be random variables defined on the ensemble of diode currents which are the sample functions. It is apparent that each realization of N is just the number of electrons involved in that particular sample function. Ultimately we are interested in the process $i(t)$ obtained when T becomes infinitely large

$$i(t) = \lim_{T\to\infty} i_T(t). \tag{3.14-9}$$

We may write the mean current or ensemble average as

$$E\{i_T(t)\} = I_T = \int_{-\infty}^{\infty}\int_{-\infty}^{\infty}\cdots\int_{-\infty}^{\infty} i_T(t)W(t_1, t_2, \ldots, t_N, N)\, dt_1\, dt_2 \cdots dt_N\, dN \tag{3.14-10}$$

where W is the joint distribution (density function) of the N emission times t_i and N itself. In order to proceed further, it is now necessary to determine this joint density W. From the conditional probability relationship

$$P(AB) = P(A)P(B/A) \tag{3.14-11}$$

we can rewrite W as

$$W(t_1, \ldots, t_N, N) = W_N(N)W_C[(t_1, \ldots, t_N)/N] \tag{3.14-12}$$

where W_C is a conditional probability density. We now proceed to find the two density functions on the right side of Equation 3.14-12, beginning with $W_N(N)$.

3.15 Random Events in Time

Let the random variable k be the number of events in an interval of length t. We assume that:

1. The probability of a single event in a very small time interval Δt is proportional to Δt so that

$$P[\text{one electron emitted in } (t, t + \Delta t)] = \lambda\, \Delta t. \tag{3.15-1}$$

2. The probability of more than one event in Δt is negligible so that

$$P[\text{no electrons emitted in } (t, t + \Delta t)] = 1 - \lambda\, \Delta t. \tag{3.15-2}$$

3. The events are independent.

From these assumptions it is possible to show that k is Poisson distributed with mean λt. The Poisson distribution is discussed in Section 2.17. We proceed as follows.

Let the probability of k events in time t be denoted* by $p_k(t)$. In the small time interval Δt, there will be either no events or one event since we have assumed that the probability of more than one event is negligible. Therefore

* For consistency it would be better to denote this probability by $p_k[k(t)]$ but the notation given above is conventional.

the probability of k events in the time interval $(t + \Delta t)$ is equal to the product of the probabilities of k events in time t and no events in time Δt plus the product of the probabilities of $(k - 1)$ events in time t and one event in time Δt. In other words,

$$p_k(t + \Delta t) = p_k(t)(1 - \lambda\,\Delta t) + p_{k-1}(t)\lambda\,\Delta t, \qquad k \geq 1. \tag{3.15-3}$$

From the definition of derivative, we have

$$\frac{dp_k(t)}{dt} = \lim_{\Delta t \to 0} \frac{p_k(t + \Delta t) - p_k(t)}{\Delta t}. \tag{3.15-4}$$

Thus Equation 3.15-3 may be rewritten as

$$\frac{dp_k(t)}{dt} = \lambda[p_{k-1}(t) - p_k(t)]. \tag{3.15-5}$$

This equation may be solved by induction as follows: let $k = 0$, then

$$\frac{dp_0(t)}{dt} = -\lambda p_0(t). \tag{3.15-6}$$

This is an ordinary homogeneous differential equation in the dependent variable $p_0(t)$ with solution

$$p_0(t) = C_1\, e^{-\lambda t} \tag{3.15-7}$$

where C_1 is an arbitrary constant determined from the boundary condition

$$p_0(0) = 1 = C_1. \tag{3.15-8}$$

Thus, for $k = 0$, we have

$$p_0(t) = e^{-\lambda t}. \tag{3.15-9}$$

Now let $k = 1$ and make use of this last result to write Equation 3.15-5 as

$$\frac{dp_1(t)}{dt} + \lambda p_1(t) = \lambda\, e^{-\lambda t}. \tag{3.15-10}$$

This is an ordinary nonhomogeneous differential equation with solution

$$p_1(t) = C_2\, e^{-\lambda t} + \lambda t\, e^{-\lambda t}. \tag{3.15-11}$$

The appropriate boundary condition now is

$$p_1(0) = 0 = C_2 \tag{3.15-12}$$

so that

$$p_1(t) = \lambda t\, e^{-\lambda t}. \tag{3.15-13}$$

We continue in the same way and find that

$$\boxed{p_k(t) = e^{-\lambda t}\frac{(\lambda t)^k}{k!}, \qquad k = 0, 1, \ldots.} \tag{3.15-14}$$

This result may be checked by direct substitution into Equation 3.15-5. It is the Poisson distribution of Equation 2.17-7 with the parameter λ replaced by λt. Thus the random variable k with density function given by Equation 3.15-14 has a mean

$$E\{k\} = \lambda t \tag{3.15-15}$$

and a variance

$$E\{(k - \lambda t)^2\} = \lambda t. \tag{3.15-16}$$

Both of these results follow from Equations 2.17-11 and 2.17-14.

Note that $p_k(t)$ is the probability that exactly k events occur in a time interval t. The term $W_N(N)$ of Equation 3.14-12 is the probability that exactly N electrons are emitted in the interval $2T$. On setting up the correspondences

$$t \sim 2T, \tag{3.15-17}$$

$$\lambda \sim \frac{E\{N\}}{2T}, \tag{3.15-18}$$

and hence

$$\lambda t \sim E\{N\} = \bar{N}, \tag{3.15-19}$$

we have

$$W_N(N) = \frac{e^{-\bar{N}}(\bar{N})^N}{N!}, \qquad N = 0, 1, \ldots \tag{3.15-20}$$

where $\bar{N}$ has been written for $E\{N\}$ in order to simplify notation. It should be kept in mind that $W_N(N)$ is a discrete density defined on the nonnegative integers.

The next problem is to determine the conditional density $W_C[(t_1, \ldots, t_N)/N]$ of Equation 3.14-12. Let us use Equation 3.14-11 and construct the probability

$$P\begin{bmatrix}\text{one electron is emitted in each of the}\\ \text{intervals } (t_1, t_1 + dt_1)\cdots(t_N, t_N + dt_N)\\ \text{given that } N \text{ electrons are emitted in}\\ (-T, T) \text{ where all } t_i \in (-T, T)\end{bmatrix}$$

$$= \frac{P\begin{bmatrix}\text{no electrons emitted in the intervals } (-T, t_1),\\ (t_1 + dt_1, t_2)\cdots(t_N + dt_N, T) \text{ and one electron}\\ \text{emitted in each of the intervals } (t_1, t_1 + dt_1),\\ (t_2, t_2 + dt_2)\cdots(t_N, t_N + dt_N)\end{bmatrix}}{P[\text{exactly } N \text{ electrons emitted in } (-T, T)] = W_N(N)} \tag{3.15-21}$$

or

$$P(\cdots) = \frac{[e^{-\lambda(t_1+T)}\, e^{-\lambda(t_2 - t_1 - dt_1)} \cdots e^{-\lambda(T - t_N - dt_N)}][\lambda\, dt_1\, e^{-\lambda dt_1} \cdots \lambda\, dt_N\, e^{-\lambda dt_N}]}{\dfrac{(\lambda 2T)^N}{N!}\, e^{-\lambda 2T}}$$

where λ has been written for $\bar{N}/2T$. This expression is readily reduced to

$$P(\cdots) = \frac{N!}{(2T)^N}\, dt_1 \cdots dt_N. \tag{3.15-22}$$

But the probability $P(\cdots)$ can also be written as

$$P(\cdots) = \sum_{\substack{\text{all orderings}\\ \text{of the } x_i}} p(x_1, x_2, \ldots, x_N)\, dx_1\, dx_2 \cdots dx_N \tag{3.15-23}$$

where

$$p(x_1, x_2, \ldots, x_N)\, dx_1\, dx_2 \cdots dx_N = P\begin{bmatrix}\text{the emission time } t_i \text{ falls in the interval} \\ (x_i, x_i + dx_i) \text{ for each } i = 1, 2, \ldots, N\end{bmatrix}. \tag{3.15-24}$$

Since there are exactly $N!$ orderings of N objects (see Section 2.7), we easily calculate Equation 3.15-23 to be

$$P(\cdots) = N!p(t_1, t_2, \ldots, t_N)\, dt_1\, dt_2 \cdots dt_N. \tag{3.15-25}$$

By direct comparison of Equations 3.15-22 and 3.15-25 we conclude that

$$p(t_1, t_2, \ldots, t_N) = W_1(t_1)W_1(t_2)\cdots W_1(t_N) \tag{3.15-26}$$

where each univariate density $W_1(t_i)$ is given by

$$W_1(t_i) = \begin{cases} \dfrac{1}{2T}, & -T \leq t_i \leq T, \\ 0, & \text{elsewhere,} \end{cases} \qquad i = 1, 2, \ldots, N. \tag{3.15-27}$$

The results of the last two equations are intuitively reasonable. The emission times of the electrons are independent and each is as likely to occur in any one small interval as in any other.

3.16 The Mean and Autocorrelation Function of Shot Noise

We now return to the evaluation of Equation 3.14-10. It follows from Equations 3.15-1 and 3.15-26 that

$$W(t_1, \ldots, t_N, N) = W_N(N)W_1(t_1)\cdots W_1(t_N) \tag{3.16-1}$$

where the $W_1(t_i)$ are the uniform distributions of Equation 3.15-27. Therefore Equation 3.14-10 may be written as

$$I_T = \int_{-\infty}^{\infty}\int_{-T}^{T}\cdots\int_{-T}^{T}\sum_{i=1}^{N} f(t - t_i)W_N(N)W_1(t_1)\cdots W_1(t_N)\, dt_1\cdots dt_N\, dN. \tag{3.16-2}$$

Let us now bring each integral with respect to each t_j inside the summation. It is clear that $(N - 1)$ of these (where $i \neq j$) will be of the form

$$\int_{-T}^{T} W_1(t_j)\, dt_j = \frac{1}{2T}\int_{-T}^{T} dt_j = 1. \tag{3.16-3}$$

Thus Equation 3.16-3 may be rewritten as

$$I_T = \int_{-\infty}^{\infty} W_N(N)\left[\sum_{i=1}^{N} \frac{1}{2T}\int_{-T}^{T} f(t - t_i)\, dt_i\right] dN \tag{3.16-4}$$

or

$$I_T = \frac{e}{2T}\int_{-\infty}^{\infty} NW_N(N)\, dN = \frac{e\overline{N}}{2T} = \lambda e \tag{3.16-5}$$

where $\overline{N}$ is the mean value of N and is given by

$$\overline{N} = E\{N\} = \int_{-\infty}^{\infty} NW_N(N)\, dN \tag{3.16-6}$$

and λ is the mean number of electrons per unit of time and is

$$\lambda = \frac{\overline{N}}{2T} \tag{3.16-7}$$

as previously defined by Equation 3.15-18. Notice that the result of Equation 3.16-5 is independent of T so that we have

$$\boxed{E\{i_T(t)\} = I_T = E\{i(t)\} = I = \lambda e} \tag{3.16-8}$$

as the final result for the mean value of the shot current process $i(t)$.

We now proceed to calculate the autocorrelation function of $i(t)$. We begin by forming

$$E\{i_T(t)i_T(t + \tau)\} = \int_{-\infty}^{\infty}\cdots\int_{-\infty}^{\infty}\left[\sum_{i=1}^{N} f(t - t_i)\right]\left[\sum_{j=1}^{N} f(t + \tau - t_j)\right] \times W(t_1, \ldots, t_N, N)\, dt_1\cdots dt_N\, dN. \tag{3.16-9}$$

We now make the substitution of Equation 3.16-1 for W and write $E\{i_T(t)i_T(t + \tau)\}$

$$= \int_{-\infty}^{\infty} W_N(N)\left[\sum_{i=1}^{N}\sum_{j=1}^{N}\left(\frac{1}{2T}\right)^N \int_{-T}^{T}\cdots \int_{-T}^{T} f(t - t_i)f(t + \tau - t_j)\, dt_1\cdots dt_N\right] dN. \tag{3.16-10}$$

For each of the N terms where $i = j$ in the double summation, we have

$$\frac{1}{2T}\int_{-T}^{T} f(t - t_i)f(t + \tau - t_i)\, dt_i$$

and for each of the $N^2 - N$ terms where $i \neq j$, we find

$$\frac{1}{2T}\int_{-T}^{T} f(t - t_i)\, dt_i \frac{1}{2T}\int_{-T}^{T} f(t + \tau - t_j)\, dt_j.$$

Thus Equation 3.14-48 becomes (approximately)

$$\begin{aligned} E\{i_T(t)i_T(t + \tau)\} \\ = \int_{-\infty}^{\infty} W_N(N)\Bigg[\frac{N}{2T}\int_{-T}^{T} f(t - t_i)f(t + \tau - t_i)\, dt_i + \frac{N^2 - N}{(2T)^2} \\ \times \int_{-T}^{T} f(t - t_i)\, dt_i \int_{-T}^{T} f(t + \tau - t_j)\, dt_j\Bigg]\, dN. \quad (3.16\text{-}11) \end{aligned}$$

In the limit as $T \to \infty$, we obtain the autocorrelation function $R(\tau)$ of $i(t)$. Note that

$$\lim_{T\to\infty} \int_{-T}^{T} f(t - t_i)\, dt_i = \lim_{T\to\infty} \int_{-T}^{T} f(t + \tau - t_j) = e$$

where e is the electronic charge. Also we have

$$\lim_{T\to\infty} \int_{-T}^{T} f(t - t_i)f(t + \tau - t_i)\, dt_i = \int_{-\infty}^{\infty} f(t)f(t + \tau)\, dt.$$

We may now write

$$\begin{aligned} R(\tau) &= \lim_{T\to\infty} E\{i_T(t)i_T(t + \tau)\} \\ &= \lim_{T\to\infty} \int_{-\infty}^{\infty} f(t)f(t + \tau)\, dt \frac{1}{2T}\int_{-\infty}^{\infty} NW_N(N)\, dN \\ &\quad + \lim_{T\to\infty} \left(\frac{e}{2T}\right)^2 \int_{-\infty}^{\infty} (N^2 - N)W_N(N)\, dN. \quad (3.16\text{-}12) \end{aligned}$$

Note that $W_N(N)$ is actually a discrete density and

$$\int_{-\infty}^{\infty} NW_N(N)\, dN = \sum_{N=0}^{\infty} N\frac{(\bar{N})^N}{N!} e^{-\bar{N}} = \bar{N}, \quad (3.16\text{-}13)$$

$$\int_{-\infty}^{\infty} (N^2 - N)W_N(N)\, dN = \sum_{N=0}^{\infty} N(N - 1)\frac{(\bar{N})^N}{N!} e^{-\bar{N}} = (\bar{N})^2, \quad (3.16\text{-}14)$$

from the properties of the Poisson distribution as given by Equations 2.7-11 and 2.17-13 respectively. Thus we have finally

$$R(\tau) = E\{i(t)i(t+\tau)\} = \lambda \int_{-\infty}^{\infty} f(t)f(t+\tau)\,dt + \lambda^2 e^2. \quad (3.16\text{-}15)$$

It is convenient to define a new shot current $i_a(t)$ with zero mean

$$i_a(t) = i(t) - \lambda e \quad (3.16\text{-}16)$$

with autocorrelation function

$$R_a(\tau) = E\{i_a(t)i_a(t+\tau)\} = R(\tau) - \lambda^2 e^2 \quad (3.16\text{-}17)$$

or

$$\boxed{R_a(\tau) = \lambda \int_{-\infty}^{\infty} f(t)f(t+\tau)\,dt.} \quad (3.16\text{-}18)$$

When τ is set equal to zero, this last expression is known as *Campbell's Theorem*. Thus the shape of the autocorrelation function is determined only by the shape of the basic current pulses due to the individual electrons.

Let us consider briefly the power spectral density $\varphi_a(\omega)$ where

$$\varphi_a(\omega) = \int_{-\infty}^{\infty} R_a(\tau)\, e^{-j\omega\tau}\, d\tau. \quad (3.16\text{-}19)$$

This expression is easily evaluated as follows:

$$\varphi_a(\omega) = \lambda \int_{-\infty}^{\infty} \int_{-\infty}^{\infty} f(t)f(t+\tau)\,dt\, e^{-j\omega\tau}\, d\tau. \quad (3.16\text{-}20)$$

We make the linear change in variable $\tau = \sigma - t$ to obtain

$$\varphi_a(\omega) = \lambda \int_{-\infty}^{\infty} f(\sigma)\, e^{-j\omega\sigma}\, d\sigma \int_{-\infty}^{\infty} f(t)\, e^{j\omega t}\, dt \quad (3.16\text{-}21)$$

or

$$\boxed{\varphi_a(\omega) = \lambda F(\omega)F(-\omega) = \lambda |F(\omega)|^2} \quad (3.16\text{-}22)$$

where $F(\omega)$ is the Fourier transform of the basic pulse shape.

The low-frequency behavior of the power spectra density is given by

$$\lim_{\omega \to 0} \varphi_a(\omega) = \lambda \left| \int_{-\infty}^{\infty} f(t)\, dt \right|^2 = \lambda e^2 \quad (3.16\text{-}23)$$

from Equation 3.16-21. Note that this last expression may be written as

$$\boxed{\lim_{\omega \to 0} \varphi_a(\omega) = e(\lambda e) = eI} \quad (3.16\text{-}24)$$

which is often called the *Schottky Formula*.

Example 3.11 We consider a parallel-plane diode as shown in Figure 3.10(a) operated temperature limited and with cathode-anode spacing d and plate voltage V_a. Since the diode is operated temperature limited, there is no space charge and Laplace's Equation is satisfied in the cathode-anode space

$$\nabla^2 V = 0 = \frac{\partial^2 V}{\partial^2 x}.$$

The solution is

$$V = ax + b.$$

At $x = 0$, $V = 0$ so that $b = 0$; also at $x = d$, $V = V_a$ so that $a = V_a/d$. Therefore we have

$$V = \frac{V_a}{d} x, \qquad 0 \leq x \leq d$$

for the potential distribution in the cathode-anode space.

The energy $\mathscr{E}$ gained by an electron in moving through a potential V is

$$\mathscr{E} = eV$$

and the corresponding power flow is

$$\frac{\partial \mathscr{E}}{\partial t} = e\frac{dV}{dt} = \frac{eV_a}{d}\frac{dx}{dt} = V_a i.$$

Thus the current flow due to a single electron is

$$i = f(t) = \frac{e}{d}\frac{dx}{dt} = \frac{e}{d} v_x$$

where v_x is the electron velocity. The electron acceleration a_x is a constant since

$$a_x = \frac{e}{m}\frac{V_a}{d},$$

hence

$$v_x = \frac{e}{m}\frac{V_a}{d} t$$

where the additive constant of integration is zero from our assumption that v_x is zero at $t = 0$; that is, we assume that the electron is emitted with zero initial velocity* from the cathode. Thus the current is

$$i = f(t) = \frac{e^2 V_a}{md^2} t.$$

* This assumption of zero initial velocity is not correct since the actual velocity is a random variable with a Maxwellian distribution. For details see [16].

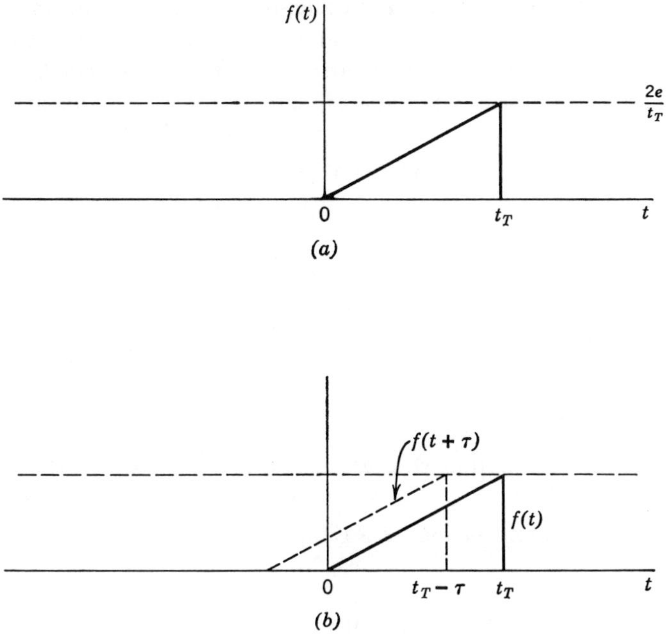

Figure 3.11 *The current in a parallel-plane diode: (a) the current pulse due to a single electron; (b) calculation of the autocorrelation function*

If we call the transit time t_T, then the average value of the velocity v_x is given by

$$\frac{d}{t_T} = \frac{1}{2}\frac{e}{m}\frac{V_a}{d}\, t_T.$$

We may now write the current due to a single electron as

$$i = f(t) = \begin{cases} \dfrac{2e}{t_T^2}\, t, & 0 \le t \le t_T, \\ 0, & \text{elsewhere}, \end{cases}$$

as shown in Figure 3.11(a).

The autocorrelation function is now easily found from Equation 3.16-18 and Figure 3.11(b). For the region $0 \le \tau \le t_T$ we see that

$$R_a(\tau) = \lambda \int_0^{t_T-\tau} \frac{2e}{t_T^2}\, t\, \frac{2e}{t_T^2}(t+\tau)\, dt = \frac{4e^2\lambda}{t_T^4}\left(\frac{t^3}{3} + \frac{t^2\tau}{2}\right)_{t=0}^{t=t_T-\tau}$$

$$= \frac{4}{3}\frac{\lambda e^2}{t_T}\left(1 - \frac{\tau}{t_T}\right)^2\left(1 + \frac{1}{2}\frac{\tau}{t_T}\right).$$

Since $I = \lambda e$ and since $R_a(\tau)$ is even, we have

$$R_a(\tau) = \begin{cases} \dfrac{4}{3}\dfrac{eI}{t_T}\left(1 - \dfrac{|\tau|}{t_T}\right)^2\left(1 + \dfrac{1}{2}\dfrac{|\tau|}{t_T}\right), & 0 \le |\tau| \le t_T, \\ 0, & \text{elsewhere.} \end{cases}$$

This function is plotted in Figure 3.12(*a*). The power spectral density $\varphi_a(\omega)$ is obtained by direct transformation or from Equation 3.16-22. We have

$$\begin{aligned} \varphi_a(\omega) &= 2\int_0^\infty R(\tau)\cos\omega\tau\,d\tau \\ &= \frac{8eI}{t_T^2}\int_0^{t_T}\left(\frac{t_T}{3} - \frac{\tau}{2} + \frac{1}{6}\frac{\tau^3}{t_T^2}\right)\cos\omega\tau\,d\tau \end{aligned}$$

or, finally,

$$\varphi_a(\omega) = \frac{4eI}{(\omega t_T)^4}[(\omega t_T)^2 + 2(1 - \cos\omega t_T - \omega t_T\sin\omega t_T)].$$

This spectrum is plotted in Figure 3.12(*b*).

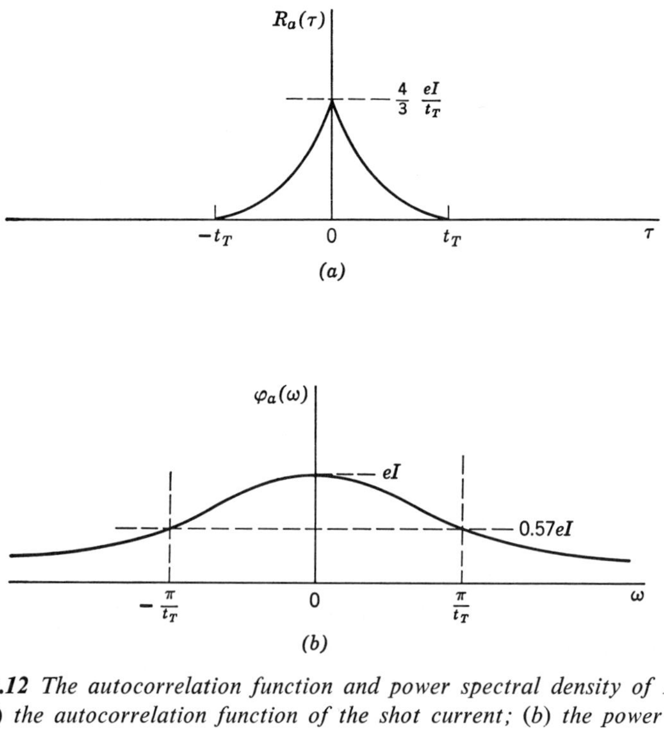

Figure 3.12 *The autocorrelation function and power spectral density of Example 3.11: (a) the autocorrelation function of the shot current; (b) the power spectral density of the shot current*

At low frequencies ($\omega \ll \pi/t_T$) the spectrum is nearly constant at a value of eI. Thus, at *low frequencies*, the shot-noise power in a bandwidth $\Delta f = f_2 - f_1$ is approximately

$$P_N \approx \frac{1}{2\pi}\int_{-\omega_2}^{-\omega_1} \varphi(0)\, d\omega + \frac{1}{2\pi}\int_{\omega_1}^{\omega_2} \varphi(0)\, d\omega$$

or

$$P_N \approx \frac{1}{\pi}\int_{\omega_1}^{\omega_2} eI\, d\omega = 2eI(f_2 - f_1)$$

$$\boxed{P_N \approx 2eI\,\Delta f.}$$

Some practical idea of the magnitude of t_T and hence of the frequency range over which $\varphi(\omega)$ is essentially constant can be obtained by considering the Sylvania type 5722 noise-generating diode. This is a seven-pin miniature tube with an anode-cathode spacing of 0.0375 in. When operated as recommended, it has a transit angle of 58 deg at 500 Mc/sec, or a transit time t_T of approximately 300 $\mu\mu$sec. Hence the frequency at which $\varphi(\omega)$ has dropped to 0.57 of its value at zero frequency is approximately

$$f = \frac{1}{2t_T} \approx \frac{10^{12}}{600} \approx 1.6 \times 10^9 \text{ cycles/sec}$$

or

$$f \approx 1600 \text{ Mc/sec.}$$

The 5722 is used as a noise generator at frequencies to 400 or 500 Mc/sec and has a reasonably flat power spectral density up to this range of frequencies.

• •

3.17 The Amplitude Distribution of Shot Noise

As defined in Section 2.15, the characteristic function $F(v)$ of a random variable x is given by

$$F(v) = E\{e^{jvx}\} = \int_{-\infty}^{\infty} W_1(x)\, e^{jvx}\, dx \tag{3.17-1}$$

where $W_1(x)$ is the density function of x. As in Section 2.19 we consider the case where x is made up of a sum of *independent* random variables; that is,

$$x = x_1 + x_2 + \cdots + x_n = \sum_{i=1}^{n} x_i. \tag{3.17-2}$$

Then the characteristic function becomes

$$F(v) = E\{e^{jvx}\} = E\left\{\exp\left(jv \sum_{i=1}^{n} x_i\right)\right\} = \prod_{i=1}^{n} F_i(v) \tag{3.17-3}$$

where $F_i(v)$ is the characteristic function of x_i; that is,

$$F_i(v) = E\{e^{jvx_i}\}. \tag{3.17-4}$$

If the x_i are identically distributed so that each has the same characteristic function, say $F_1(v)$, then Equation 3.17-3 becomes

$$F(v) = [F_1(v)]^N. \tag{3.17-5}$$

Let us now consider the shot current $i(t)$ of Equation 3.14-9 and find

$$F(v) = E\{e^{jvi(t)}\} = \lim_{T\to\infty} E\{e^{jvi_T(t)}\} \tag{3.17-6}$$

where $i_T(t)$ is given by Equation 3.14-2. Note that $i_T(t)$ is a sum of independent random variables $f(t - t_i)$ and that the number N of the terms in the sum is also a random variable with density function $W_N(N)$ given by Equation 3.15-20. We can write

$$F(v) = \lim_{T\to\infty} \sum_{N=0}^{\infty} W_N(N)F(v/N) \tag{3.17-7}$$

where $F(v/N)$ is the characteristic function for given N. It follows from Equation 3.17-3 that

$$F(v) = \lim_{T\to\infty} \sum_{N=0}^{\infty} W_N(N)[F_1(v)]^N \tag{3.17-8}$$

where $F_1(v)$ is the characteristic function of one of the $f(t - t_k)$ and can be written as

$$F_1(v) = E\{e^{jvf(t-\sigma)}\} = \frac{1}{2T}\int_{-T}^{T} e^{jvf(t-\sigma)}\, d\sigma \tag{3.17-9}$$

from Equation 3.15-27.

We may use Equation 3.15-20 to rewrite Equation 3.17-8 as

$$F(v) = \lim_{T\to\infty} \sum_{N=0}^{\infty} \frac{e^{-\bar{N}}(\bar{N})^N}{N!}[F_1(v)]^N = \lim_{T\to\infty} e^{-\bar{N}} \sum_{N=0}^{\infty} \frac{[\bar{N}F_1(v)]^N}{N!} \tag{3.17-10}$$

or

$$F(v) = \lim_{T\to\infty} e^{-\bar{N}} e^{\bar{N}F_1(v)} = \lim_{T\to\infty} e^{\bar{N}[F_1(v)-1]}. \tag{3.17-11}$$

We now use Equation 3.17-9 to obtain finally

$$F(v) = \lim_{T\to\infty} \exp\left\{\frac{\bar{N}}{2T}\int_{-T}^{T} [e^{jvf(t-\sigma)} - 1]\, d\sigma\right\} \tag{3.17-12}$$

or

$$\boxed{F(v) = \exp\left\{\lambda\int_{-\infty}^{\infty} [e^{jvf(\sigma)} - 1]\, d\sigma\right\}.} \tag{3.17-13}$$

In general this last expression is difficult to evaluate. For the high density case ($\lambda \to \infty$), however, we can show that $F(v)$ is the characteristic function of a gaussian random variable. For convenience, we introduce the normalized current

$$x(t) = \frac{i(t) - \lambda e}{\sqrt{\lambda}} \tag{3.17-14}$$

with zero mean; that is,

$$E\{x\} = \frac{E\{i\} - \lambda e}{\sqrt{\lambda}} = \frac{\lambda e - \lambda e}{\sqrt{\lambda}} = 0 \tag{3.17-15}$$

and autocorrelation function $R_x(\tau)$

$$R_x(\tau) = E\{x(t)x(t + \tau)\} = \frac{R(\tau) - \lambda^2 e^2}{\lambda} \tag{3.17-16}$$

or

$$R_x(\tau) = \int_{-\infty}^{\infty} f(t)f(t + \tau)\, d\tau. \tag{3.17-17}$$

We now note that the characteristic function of x is

$$\begin{aligned} F_x(v) &= E\{e^{jvx}\} = E\{e^{jv(i-\lambda e)/\sqrt{\lambda}}\} \\ &= e^{-j\sqrt{\lambda}ev}E\{e^{ji(v/\sqrt{\lambda})}\} = e^{-j\sqrt{\lambda}ev}F\left(\frac{v}{\sqrt{\lambda}}\right) \end{aligned} \tag{3.17-18}$$

where $F(v)$ is given by Equation 3.17-13. Thus

$$F_x(v) = \exp\left\{\lambda\int_{-\infty}^{\infty}\left[e^{j(v/\sqrt{\lambda})f(\sigma)} - 1 - j\frac{v}{\sqrt{\lambda}}f(\sigma)\right] d\sigma\right\} \tag{3.17-19}$$

since

$$e^{j\sqrt{\lambda}ev} = \exp\left[\lambda\int_{-\infty}^{\infty} j\frac{v}{\sqrt{\lambda}}f(\sigma)\, d\sigma\right]. \tag{3.17-20}$$

On expanding the exponential inside the integral of Equation 3.17-19 in a power series, we obtain

$$F_x(v) = \exp\left\{\lambda\int_{-\infty}^{\infty}\sum_{k=2}^{\infty}\frac{j^k}{k!}\left(\frac{1}{\sqrt{\lambda}}\right)^k [vf(\sigma)]^k\, d\sigma\right\}. \tag{3.17-21}$$

On allowing λ to become large, we have

$$\lim_{\lambda \to \infty} F_x(v) = \exp\left[\int_{-\infty}^{\infty} \frac{(j)^2}{2} v^2 f^2(\sigma)\, d\sigma\right]$$

$$= \exp\left[-\frac{v^2}{2}\int_{-\infty}^{\infty} f^2(\sigma)\, d\sigma\right]. \tag{3.17-22}$$

We see from Equation 3.17-17 that

$$\lim_{\lambda \to \infty} F_x(v) = e^{-(v^2/2)R_x(0)}. \tag{3.17-23}$$

As shown in Section 2.18, this is the characteristic function of a gaussian random variable with zero mean and variance $R_x(0)$. Thus the shot current has a limiting first-order probability amplitude distribution that is gaussian or normal.

The second-order distribution may be found in the same way. We are interested in the joint density of $i(t)$ and $i(t + \tau)$ with corresponding characteristic function $F(v_1, v_2)$ given by

$$F(v_1, v_2) = \lim_{T \to \infty} E\{e^{j[v_1 i_T(t) + v_2 i_T(t+\tau)]}\}. \tag{3.17-24}$$

As in Equation 3.17-7, this may be written as

$$F(v_1, v_2) = \lim_{T \to \infty} \sum_{N=0}^{\infty} \frac{e^{-\bar{N}}(\bar{N})}{N!} \prod_{i=1}^{n} \frac{1}{2T}\int_{-T}^{T} e^{j[v_1 f(t-t_i) + v_2 f(t+\tau-t_i)]}\, dt_i. \tag{3.17-25}$$

We now follow the previous development of Equations 3.17-10 through 3.17-13 to obtain

$$\boxed{F(v_1, v_2) = \exp\left\{\lambda\int_{-\infty}^{\infty} [e^{j[v_1 f(\sigma) + v_2 f(\sigma+\tau)]} - 1]\, d\sigma\right\}.} \tag{3.17-26}$$

Again this expression is difficult to evaluate except in the limit of large λ. As in Equation 3.17-14 we define a normalized current $x(t)$ with joint characteristic function $F_x(v_1, v_2)$ where

$$F_x(v_1, v_2) = E\{e^{j[v_1 x(t) + v_2 x(t+\tau)]}\}$$

$$= E\left\{\exp\left[j\left(v_1 \frac{i_1 - \lambda e}{\sqrt{\lambda}} + v_2 \frac{i_2 - \lambda e}{\sqrt{\lambda}}\right)\right]\right\}$$

$$= e^{-j\sqrt{\lambda} e(v_1 + v_2)} F\left(\frac{v_1}{\sqrt{\lambda}}, \frac{v_2}{\sqrt{\lambda}}\right) \tag{3.17-27}$$

where i_1 has been written for $i(t)$ and i_2 for $i(t + \tau)$ and where $F(v_1, v_2)$ is given by Equation 3.17-26. Thus we have

$$F_x(v_1, v_2) = \exp\left\{\lambda \int_{-\infty}^{\infty} \sum_{k=2}^{\infty} \frac{j^k}{k!}\left(\frac{1}{\sqrt{\lambda}}\right)^k [v_1 f(\sigma) + v_2 f(\sigma + \tau)]^k \, d\sigma\right\}. \quad (3.17\text{-}28)$$

On passing to the limit we have

$$\lim_{\lambda \to \infty} F_x(v_1, v_2) = \exp\left\{-\frac{1}{2}\int_{-\infty}^{\infty} [v_1 f(\sigma) + v_2 f(\sigma + \tau)]^2 \, d\sigma\right\}. \quad (3.17\text{-}29)$$

On using Equation 3.17-17 we can rewrite this last expression as

$$\lim_{\lambda \to \infty} F_x(v_1, v_2) = \exp\{-\tfrac{1}{2}[v_1^2 R_x(0) + 2v_1 v_2 R_x(\tau) + v_2^2 R_x(0)]\}. \quad (3.17\text{-}30)$$

As shown by Equation 2.22-7, this is the characteristic function for the bivariate normal distribution with mean of zero and covariance function $R_x(\tau)$. Thus the second-order distribution of the shot current is jointly normal in the limit of high density.

PROBLEMS

1. Consider the random process $e(t) = A \sin(\omega t + \theta)$ where A and ω are constants and θ is a random variable. What are necessary and sufficient conditions on the random variable θ such that the random process $e(t)$ is wide-sense stationary? Give a specific example of such a random variable.
2. Consider the random process $y(t) = x \cos t$ where x is a normally distributed random variable with zero mean and unit variance. What is the probability density function of $y(0)$? What is the joint density function of $y(0)$ and $y(\pi)$? Is the random process stationary?
3. Give at least two reasons why the following matrix cannot be a correlation matrix of two jointly wide-sense stationary processes:

$$\begin{bmatrix} A^2 \cos \tau & 2A^2 \cos \dfrac{3\tau}{2} \\ 2A^2 \cos \dfrac{3\tau}{2} & A^2 \sin 2\tau \end{bmatrix}.$$

4. Consider two stationary random processes having sample functions $x(t)$ and $y(t)$. Find the autocorrelation function of the sum of these two processes assuming: (a) that the processes are correlated, (b) that the processes are uncorrelated, (c) that the processes are uncorrelated and have zero means.

5. Consider the random process $e_1(t) = A \sin(\omega t + \theta)$ where A and ω are constants and θ is a uniformly distributed random variable on $(-\pi, +\pi)$. Let $e_2(t) = |e_1(t)|$. Find the autocorrelation function of $e_2(t)$ and the cross-correlation function of $e_1(t)$ and $e_2(t)$.
6. Consider the constant random process $x(t) = a$ where a is a random variable uniformly distributed on $(-1, +1)$. Find the autocorrelation function and time autocorrelation function of this random process. Is the process ergodic?
7. If $x(t)$ is a bounded periodic function with fundamental period P, show that

$$\lim_{T\to\infty} \frac{1}{2T}\int_{-T}^{+T} x(t)x(t+\tau)\,dt = \frac{1}{P}\int_0^P x(t)x(t+\tau)\,dt.$$

8. Find the time autocorrelation function of the periodic process $x(t)$.

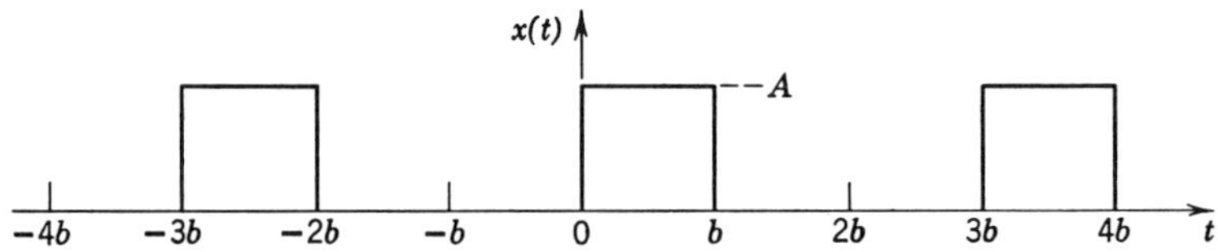

9. Find the time cross-correlation function of the two periodic processes $x(t)$ and $y(t)$.

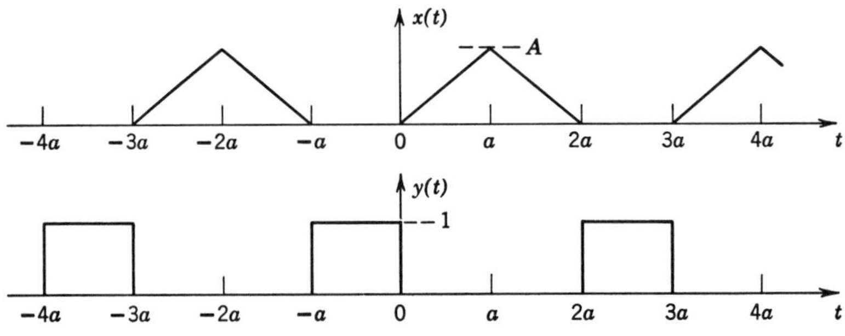

10. Consider two different periodic functions $x(t)$ and $y(t)$ having the same fundamental period. Show that their time cross-correlation function depends only on those harmonics that are present in both $x(t)$ and $y(t)$.
11. Is the following function the power spectral density of a physical process? If so, what is the autocorrelation function of the process? If not, what condition does it violate?

$$\frac{A^2\omega^2}{a^2 + \omega^2}.$$

12. Find the Fourier transform of $1/(b^2 + t^2)$ using contour integration.
13. Consider a wide-sense stationary random process having the power spectral density $\varphi(\omega)$ as shown. Find the autocorrelation function of the random process.

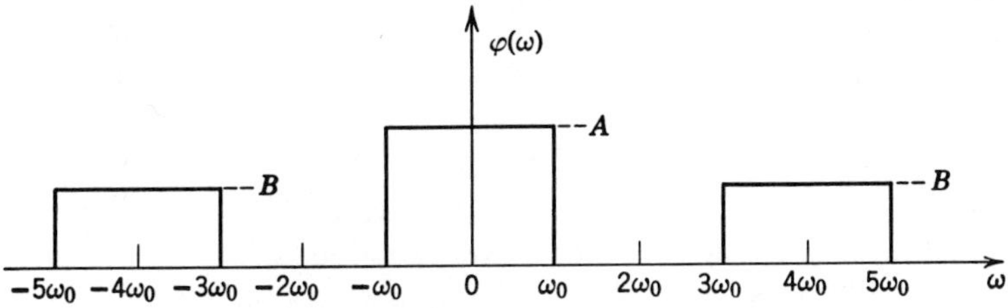

14. Consider the random process $x(t)$ which has the following discrete density function:

$$f(x) = \begin{cases} 1 - \dfrac{1}{t^2}, & x = 0, \quad t > 1, \\ \dfrac{1}{2t^2}, & x = t, \quad t > 1, \\ \dfrac{1}{2t^2}, & x = -t, \ t > 1, \\ \frac{1}{2}, & x = 1, \quad t \leq 1, \\ \frac{1}{2}, & x = -1, \ t \leq 1. \end{cases}$$

Find the mean and variance of $x(t)$. Is the process stationary?

15. A random process $x(t)$ has an autocorrelation function as shown in the figure. Find the power spectral density of this process. As b becomes large, what does the process tend toward?

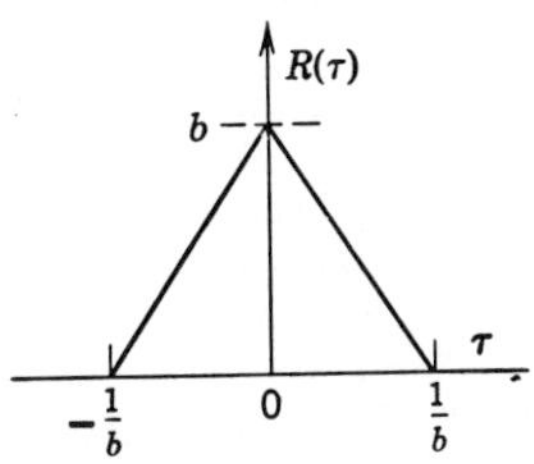

16. Consider the following sequences of random variables:

$$\text{(a)} \quad f(x_k) = \begin{cases} 2^{-(k+1)}, & \text{for } x_k = -2^k, \\ 1 - 2^{-k}, & \text{for } x_k = 0, \\ 2^{-(k+1)}, & \text{for } x_k = 2^k, \end{cases} \qquad \text{for } k = 1, 2, \ldots.$$

$$\text{(b)} \quad f(x_k) = \begin{cases} 2^{-(2k+1)}, & \text{for } x_k = -2^k, \\ 1 - 2^{-2k}, & \text{for } x_k = 0, \\ 2^{-(2k+1)}, & \text{for } x_k = 2^k, \end{cases} \qquad \text{for } k = 1, 2, \ldots.$$

(c) $$f(x_k) = \begin{cases} \frac{1}{2}k^{-1/2}, & \text{for } x_k = -k, \\ 1 - k^{-1/2}, & \text{for } x_k = 0, \\ \frac{1}{2}k^{-1/2}, & \text{for } x_k = k, \end{cases} \qquad \text{for } k = 2, 3, \ldots.$$

Which of these sequences converge to zero, (a) in probability, (b) in the mean?

17. Consider the random process:

$$y(t) = A \sin t + B \cos t$$

where A and B are independent random variables with zero means and equal variances σ^2. Show that the process is wide-sense stationary but not strictly stationary if the density function of A differs from the density function of B.

REFERENCES

1. E. Parzen, *Stochastic Processes*, Holden-Day, San Francisco, 1962.
2. M. Rosenblatt, *Random Processes*, Oxford Univ. Press, New York, 1962.
3. M. S. Bartlett, *An Introduction to Stochastic Processes with Special Reference to Methods and Applications*, Cambridge Univ. Press, Cambridge, England, 1955.
4. J. L. Doob, *Stochastic Processes*, Wiley, New York, 1953.
5. E. C. Titchmarsh, *Introduction to the Theory of Fourier Integrals*, Oxford Univ. Press, New York, 1948.
6. W. Kaplan, *Operational Methods for Linear Systems*, Addison-Wesley, Reading, Mass., 1962.
7. M. E. Van Valkenburg, *Network Analysis*, Prentice-Hall, Englewood Cliffs, N. J., 1955.
8. R. S. Burington, *Handbook of Mathematical Tables and Formulas*, Handbook Publ., Sandusky, Ohio, 1940.
9. D. Middleton, *An Introduction to Statistical Communication Theory*, McGraw-Hill, New York, 1960.
10. W. B. Davenport, Jr., and W. L. Root, *An Introduction to the Theory of Random Signals and Noise*, McGraw-Hill, New York, 1958.
11. S. Karlin, *A First Course in Stochastic Processes*, Academic Press, New York, 1966.
12. J. G. Kemeny, J. L. Snell, A. W. Knapp, *Denumerable Markov Chains*, Van Nostrand, Princeton, N. J., 1966.
13. W. Feller, *An Introduction to Probability Theory and Its Applications*, Wiley, New York, 1957.
14. E. B. Dynkin, *Markov Processes*, Vols. I and II, Academic, New York, 1965.
15. W. Schottky, "Theory of Shot Effect," *Ann. Phys.*, **57**, 1918, 541–568.
16. W. R. Bennett, *Electrical Noise*, McGraw-Hill, New York, 1960, Sect. 4.2.

4

Linear Systems with Random Inputs

4.1 Introduction

The problems that will be most frequently encountered in statistical communication theory involve the study of the action of some system on a random process. In this chapter we treat the analysis of linear systems subjected to such random inputs. Linear systems are treated first for a variety of reasons:

1. They are the simplest systems to analyze.
2. They are commonly encountered.
3. A procedure has been developed to synthesize *optimum* linear systems; a knowledge of linear system analysis is necessary in order to apply this theory.

We start with some definitions concerning systems in general and linear systems in particular [1, 2, 3].

4.2 Linear Systems

We shall define a (*physical*) *system* as a (physical) unit that converts an input function $x(t)$ into an output function $y(t)$ as indicated schematically in Figure 4.1(a). It is not necessary that there be a single input and output; we can consider systems with an arbitrary number of inputs and outputs as shown in Figure 4.1(b). In this case it is convenient to consider the vector-valued input $\mathbf{x}(t)$ and output $\mathbf{y}(t)$. Note that the number of components of $\mathbf{x}(t)$ need not equal the number of components of $\mathbf{y}(t)$. We call $x(t)$ the *input* or *forcing function* and $y(t)$ the *output* or *response* of the system. Although in most cases the parameter t (usually time) will be continuous, it need not

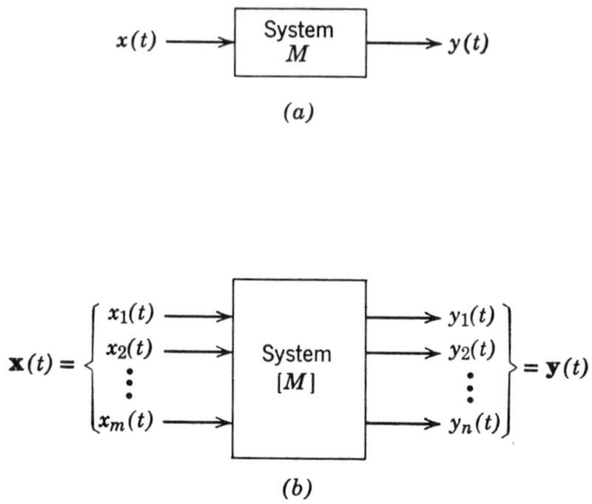

Figure 4.1 *Schematic representation of a system: (a) A one-input, one-output system; (b) an m-input, n-output system*

necessarily be so; more generally we take t to belong to some set T and include the case where $T = \{t_i; i = \cdots -1, 0, 1, \ldots\}$.

We require that the output $y(t)$ of the system be determined completely by the input $x(t)$, the system characteristics, and a set of initial conditions, and write

$$y(t) = M[x(t)] \tag{4.2-1}$$

where M is the *system operator*. By an operator we mean an assignment of a member of the set of possible outputs $\{y(t)\}$ to each member of the set of possible inputs $\{x(t)\}$. Thus the system operator completely characterizes the system for a given class of inputs $\{x(t)\}$.

An operator L is called a *linear operator* for some class of inputs $\{x(t)\}$, and hence the system described by L is a *linear system* for that class of inputs, if

$$L[a_i x_i(t) + a_j x_j(t)] = a_i y_i(t) + a_j y_j(t) \tag{4.2-2}$$

for all $x_i(t)$ and $x_j(t) \in \{x(t)\}$ where the a_i, a_j are arbitrary constants and

$$y_i(t) = L[x_i(t)]$$

is the output for arbitrary input $x_i(t) \in \{x(t)\}$.

An operator M is called *time invariant* (or *fixed parameter*) for some class of inputs $\{x(t)\}$, and hence the system described by M is a *time-invariant system* (or *fixed-parameter system*) for that class of inputs, if

$$y_i(t) = M[x_i(t)] \underset{\text{(implies)}}{\Rightarrow} y_i(t + t_0) = M[x_i(t + t_0)] \tag{4.2-3}$$

for all $x_i(t) \in \{x(t)\}$ and for all t_0.

An operator M is said to be *causal*, and hence the system described by M is a *causal system*, if there is no output prior to there being an input; that is if,

$$x(t) = 0, t < t_0 \Rightarrow y(t) = M[x(t)] = 0, t < t_0. \tag{4.2-4}$$

A *linear* system L is said to be *stable* if its response to any bounded input is bounded; that is, if

$$|x(t)| < M \Rightarrow |y(t)| < MI \tag{4.2-5}$$

where I is a constant.

Example 4.1 Let the operator M be defined by

$$y(t) = \int_{-\infty}^{t} x(s)\, ds$$

for all $x(t)$ for which this integral exists. It is apparent from a direct application of Equation 4.2-2 that the operator thus defined is linear since

$$\int_{-\infty}^{t} [a_1x_1(s) + a_2x_2(s)]\, ds = a_1y_1(t) + a_2y_2(t).$$

The operator is also time invariant and causal.

• •

Example 4.2 Consider the linear differential operator of order n with constant coefficients

$$M = N(p) = a_np^n + a_{n-1}p^{n-1} + \cdots + a_1p + a_0$$

where $p = d/dt$ and the a_i are constants. We have

$$y(t) = N(p)x(t)$$

and this operator is linear, time invariant, and causal.

• •

4.3 Description of Linear Systems

Let us consider an arbitrary time-invariant linear system L with input $e^{j\omega t}$ and corresponding steady-state output $y(t)$:

$$y(t) = L(e^{j\omega t}). \tag{4.3-1}$$

Since the system is time invariant and linear, we have

$$L[e^{j\omega(t+t_0)}] = L(e^{j\omega t}\, e^{j\omega t_0}) = e^{j\omega t_0}L(e^{j\omega t}) = e^{j\omega t_0}y(t). \tag{4.3-2}$$

On setting $t = 0$, we obtain

$$L(e^{j\omega t_0}) = y(0)\, e^{j\omega t_0}. \tag{4.3-3}$$

In other words, if the input to the linear, time-invariant system is an exponential $e^{j\omega t}$, then the steady-state output is also of the same form except for a constant of proportionality or scale factor.

The quantity $y(0)$ in general is complex and is a function of the frequency ω. Let us denote this fact by using the symbol $H(\omega) = y(0)$ so that

$$H(\omega) = A(\omega) + jB(\omega) \tag{4.3-4}$$

and

$$L(e^{j\omega t}) = H(\omega)\, e^{j\omega t} \tag{4.3-5}$$

or

$$H(\omega) = \frac{L(e^{j\omega t})}{e^{j\omega t}}. \tag{4.3-6}$$

This quantity $H(\omega)$ is called the *complex transfer function* of the system or, sometimes, just the *system function*. It completely describes a linear, time-invariant system.

Let us denote by $h(t, \tau)$ the response of a linear system to a delta-function input $\delta(t - \tau)$ occurring at $t = \tau$:

$$L[\delta(t - \tau)] = h(t, \tau). \tag{4.3-7}$$

From the properties of the delta function (see Appendix B) we can write

$$x(t) = \int_{-\infty}^{\infty} x(\tau)\, \delta(t - \tau)\, d\tau \tag{4.3-8}$$

for an arbitrary function $x(t)$. From the linearity property of L, it follows that

$$L[x(t)] = \int_{-\infty}^{\infty} x(\tau) L[\delta(t - \tau)]\, d\tau = y(t) \tag{4.3-9}$$

or

$$y(t) = \int_{-\infty}^{\infty} x(\tau) h(t, \tau)\, d\tau. \tag{4.3-10}$$

This last expression relates the output of a linear system to its input through the *impulse response* $h(t, \tau)$ of the system. Note that $h(t, \tau)$ is not a single function of time but a family of functions indexed by the parameter τ.

Suppose now that the system L is not only linear but time invariant so that Equation 4.3-7 becomes

$$L[\delta(t - \tau)] = h(t - \tau) \tag{4.3-11}$$

and so that Equation 4.3-10 becomes

$$y(t) = \int_{-\infty}^{\infty} x(\tau) h(t - \tau)\, d\tau. \tag{4.3-12}$$

This relation is called the *convolution* of $x(t)$ and $h(t)$ and is sometimes written

$$y(t) = x(t) * h(t). \tag{4.3-13}$$

Let us denote the Fourier transforms of $y(t)$, $x(t)$, and $h(t)$ by $Y(\omega)$, $X(\omega)$, and $H(\omega)$, respectively. Then, by direct transformation, Equation 4.3-12 becomes

$$\begin{aligned} Y(\omega) &= \int_{-\infty}^{\infty} x(\tau)\left[\int_{-\infty}^{\infty} h(t-\tau)\, e^{-j\omega t}\, dt\right] d\tau \\ &= H(\omega)\int_{-\infty}^{\infty} x(\tau)\, e^{-j\omega\tau}\, d\tau \\ &= H(\omega)X(\omega). \end{aligned} \tag{4.3-14}$$

The convolution of two functions is the product of their Fourier transforms. Suppose the input is a delta function $\delta(t)$. Then the output of the linear time-invariant system is $h(t)$. It follows from Equations 4.3-12 and 4.3-13, by setting $y(t) = h(t)$ and $x(t) = \delta(t)$, that

$$\mathscr{F}[h(t)] = H(\omega).$$

Thus a linear time-invariant system is uniquely specified by a single function, its *impulse response* $h(t)$, or its (*complex*) *transfer function* $H(\omega)$, where the two are a Fourier-transform pair

$$H(\omega) = \int_{-\infty}^{\infty} h(t)\, e^{-j\omega t}\, dt \tag{4.3-15}$$

and

$$h(t) = \frac{1}{2\pi}\int_{-\infty}^{\infty} H(\omega)\, e^{j\omega t}\, d\omega. \tag{4.3-16}$$

We have already defined a *causal* system by Equation 4.2-4. Let us assume that our linear, time-invariant system $h(t)$ is causal. Since $h(t)$ is the response to a delta function at $t = 0$, we must have

$$h(t) = 0, \qquad t < 0 \tag{4.3-17}$$

and

$$y(t) = \int_{-\infty}^{t} x(\tau)h(t-\tau)\, d\tau = \int_{0}^{\infty} x(t-\tau)h(\tau)\, d\tau \tag{4.3-18}$$

for a causal, linear, time-invariant system. A system with the property given by Equations 4.3-17 is frequently called *physically realizable.*

We have defined the *stability* of a linear system by Equation 4.2-5. We can write

$$|y(t)| = \left|\int_{-\infty}^{\infty} x(t-\tau)h(\tau)\, d\tau\right|.$$

If the input $x(t)$ is bounded by M, then

$$|y(t)| \leq M\int_{-\infty}^{\infty} |h(\tau)|\, d\tau = MI.$$

Thus, if the impulse response is absolutely integrable, the system is stable.

4.4 Random Inputs [4, 5]

We consider the linear, time-invariant system $h(t)$ with input $x(t)$ assumed to be an ergodic random process so that time and ensemble averages can be interchanged if desired. The system $h(t)$ operating on the random process $x(t)$ defines a new process $y(t)$ as discussed in Section 3.7 where

$$y(t) = \int_{-\infty}^{\infty} x(t - \tau)h(\tau)\, d\tau. \tag{4.4-1}$$

The mean value u_y of the output process is given by

$$u_y = E\{y(t)\} = \int_{-\infty}^{\infty} E\{x(t - \tau)\}h(\tau)\, d\tau$$

or

$$u_y = u_x \int_{-\infty}^{\infty} h(\tau)\, d\tau \tag{4.4-2}$$

since $x(t)$ is stationary with mean u_x. The autocorrelation function of the output is

$$\begin{aligned} R_y(\tau) &= E\{y(t)y(t + \tau)\} \\ &= E\left\{\int_{-\infty}^{\infty} x(t - \alpha)h(\alpha)\, d\alpha \int_{-\infty}^{\infty} x(t + \tau - \beta)h(\beta)\, d\beta\right\} \\ &= \int_{-\infty}^{\infty}\int_{-\infty}^{\infty} E\{x(t - \alpha)x(t + \tau - \beta)\}h(\alpha)h(\beta)\, d\alpha\, d\beta \\ &= \int_{-\infty}^{\infty}\int_{-\infty}^{\infty} R_x(\tau - \beta + \alpha)h(\alpha)h(\beta)\, d\alpha\, d\beta \end{aligned} \tag{4.4-3}$$

where the autocorrelation function of the input is

$$R_x(\tau) = E\{x(t)x(t + \tau)\}. \tag{4.4-4}$$

We might note in passing that the mean-squared value of the output process is given by

$$R_y(0) = \int_{-\infty}^{\infty}\int_{-\infty}^{\infty} R_x(\alpha - \beta)h(\alpha)h(\beta)\, d\alpha\, d\beta. \tag{4.4-5}$$

Let us take the Fourier transform of Equation 4.4-3 in order to obtain the output power spectral density $\varphi_y(\omega)$ from the relation

$$\varphi_y(\omega) = \int_{-\infty}^{\infty} R_y(\tau)\, e^{-j\omega\tau}\, d\tau. \tag{4.4-6}$$

If we make the change of variable $\gamma = \tau - \beta + \alpha$, then

$$\varphi_y(\omega) = \int_{-\infty}^{\infty} h(\alpha)\, e^{j\omega\alpha}\, d\alpha \int_{-\infty}^{\infty} h(\beta)\, e^{-j\omega\beta}\, d\beta \int_{-\infty}^{\infty} R_x(\gamma)\, e^{-j\omega\gamma}\, d\gamma$$

or

$$\varphi_y(\omega) = H^*(\omega)H(\omega)\varphi_x(\omega) = |H(\omega)|^2\varphi_x(\omega) \qquad (4.4\text{-}7)$$

is the relation connecting the input and output power spectral densities. Because of its simplicity, this last relation is of great utility in analyzing linear systems subjected to random inputs.

Example 4.3 We consider a low-pass *RC* filter as shown in Figure 4.2 subjected to a white-noise input of spectral density $N_0/2$. The filter has a complex transfer function

$$H(\omega) = \frac{1}{1 + j\omega RC} = \frac{1}{1 + j\omega/\beta}$$

where $\beta = 1/RC$. Its impulse response is

$$h(t) = \frac{1}{2\pi}\int_{-\infty}^{\infty} \frac{e^{j\omega t}}{1 + j\omega/\beta}\, d\omega$$

$$= \begin{cases} 0, & t < 0, \\ \beta e^{-\beta t}, & t > 0. \end{cases}$$

The input spectral density is

$$\varphi_x(\omega) = \frac{N_0}{2}$$

and the input autocorrelation function is

$$R_x(\tau) = \frac{N_0}{2}\,\delta(\tau)$$

from pair number 7 of Table 3.1. The output spectral density is

$$\varphi_y(\omega) = |H(\omega)|^2\varphi_x(\omega) = \frac{N_0/2}{1 + \omega^2/\beta^2}$$

and the output autocorrelation function is

$$R_y(\tau) = \frac{1}{2\pi}\int_{-\infty}^{\infty} \frac{N_0/2}{1 + \omega^2/\beta^2}\, e^{j\omega\tau}\, d\omega$$

$$= \frac{N_0\beta}{4}\, e^{-\beta|\tau|}.$$

The various quantities are illustrated in Figure 4.2.

• •

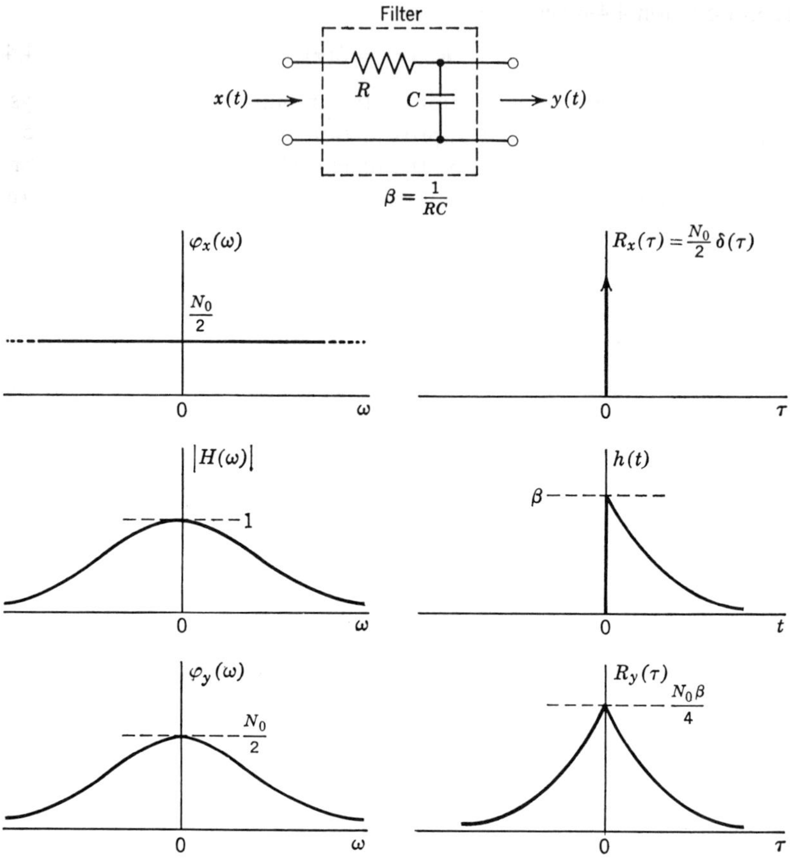

Figure 4.2 *White noise through a low-pass filter*

Let us write an expression for the cross-correlation between the input and output of a linear system. We have

$$\begin{aligned}R_{xy}(\tau) &= E\{x(t)y(t+\tau)\}\\ &= E\left\{x(t)\int_{-\infty}^{\infty} x(\sigma)h(t+\tau-\sigma)\,d\sigma\right\}\\ &= \int_{-\infty}^{\infty} E\{x(t)x(\sigma)\}h(t+\tau-\sigma)\,d\sigma\\ &= \int_{-\infty}^{\infty} R_x(\sigma-t)h(t+\tau-\sigma)\,d\sigma. \qquad (4.4\text{-}8)\end{aligned}$$

Suppose that the input $x(t)$ is white noise with unity spectral density so that

$$R_x(\tau) = \delta(\tau). \qquad (4.4\text{-}9)$$

Then Equation 4.4-8 becomes

$$R_{xy}(\tau) = h(\tau) \tag{4.4-10}$$

and the cross-correlation function is the impulse response of the system. This result suggests that an approximation to the impulse response of a linear system can be obtained by the scheme shown in Figure 4.3. A broadband input $x(t)$ and the shifted output $y(t + \tau)$ are multiplied and averaged. The result is

$$\frac{1}{2T}\int_{-T}^{T} x(t)y(t + \tau)\, dt \approx KR_{xy}(\tau) = Kh(\tau) \tag{4.4-11}$$

where K is a scale factor.

The cross-spectral density $\varphi_{xy}(\omega)$ can be found by direct transformation of Equation 4.4-8:

$$\varphi_{xy}(\omega) = \int_{-\infty}^{\infty}\int_{-\infty}^{\infty} R_x(\sigma - t)h(t + \tau - \sigma)\, d\sigma\, e^{-j\omega\tau}\, d\tau \tag{4.4-12}$$

or, after interchanging the order of integration,

$$\varphi_{xy}(\omega) = \int_{-\infty}^{\infty} R_x(\sigma - t)\int_{-\infty}^{\infty} h(t + \tau - \sigma)\, e^{-j\omega\tau}\, d\tau\, d\sigma$$

$$= \int_{-\infty}^{\infty} R_x(\sigma - t)\, e^{-j\omega(\sigma - t)}H(\omega)\, d\sigma \tag{4.4-13}$$

or, finally,

$$\varphi_{xy}(\omega) = H(+\omega)\varphi_x(\omega). \tag{4.4-14}$$

We could also write

$$\varphi_{xy}(-\omega) = \int_{-\infty}^{\infty} R_x(t - \sigma)\, e^{-j\omega(t-\sigma)}H(-\omega)\, d\sigma$$

$$= H(-\omega)\varphi_x(\omega). \tag{4.4-15}$$

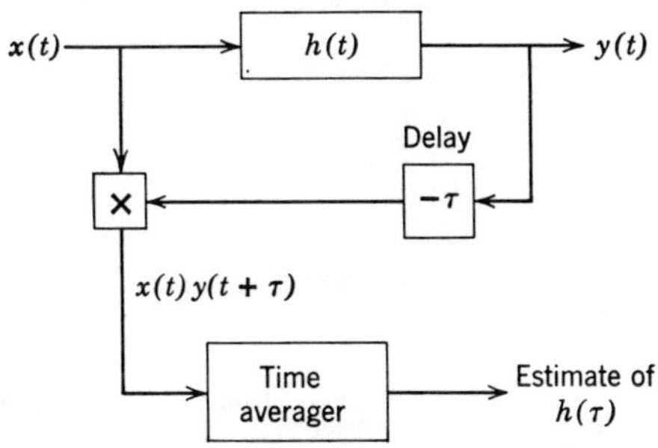

Figure 4.3 *A method for approximating the impulse response of a linear, time-invariant system*

4.5 Complete Statistical Description

The procedures of the previous section can be extended in an obvious way to find higher moments of the output of a linear system in terms of the input moments; thus the third output moment is

$$E\{y(t_1)y(t_2)y(t_3)\}$$
$$= \iiint_{-\infty}^{\infty} E\{x(\alpha)x(\beta)x(\gamma)\}h(t_1 - \alpha)h(t_2 - \beta)h(t_3 - \gamma)\, d\alpha\, d\beta\, d\gamma. \quad (4.5\text{-}1)$$

In the same fashion, all higher-order output moments can, in principle, be found if they exist. It is apparent, however, that the amount of work involved will become prohibitively large for even the first few moments. Thus there seems to be no practical way to characterize completely the output of a linear system subjected to an arbitrary random input.

It was shown in Section 2.23 that the result of a linear operation on a normal random variable was another normal random variable. *In other words, if a linear system has a gaussian input, the output will be gaussian also.* This is an extremely important result since, in this case, a knowledge of only the first two moments is required to completely characterize the output $y(t)$. This knowledge is given by Equations 4.4-2 and 4.4-3, which are now sufficient descriptions of the statistics of linear systems with gaussian inputs.

4.6 Representation of Random Processes

It will often be convenient to represent a random process $x(t)$ as an infinite series of functions, each having certain properties that facilitate a given analysis. Perhaps the most frequently encountered representation of this type is the Fourier series which has been discussed in some detail in Appendix D in conjunction with Fourier transforms. Although the Fourier series arises naturally in the expansion of periodic deterministic functions, it is of considerable utility in the representation of random processes.

Let us represent the random process $x(t)$ in the interval $(-T/2, T/2)$ by the Fourier series

$$x(t) = \frac{a_0}{2} + \sum_{n=1}^{\infty} \left(a_n \cos \frac{2\pi nt}{T} + b_n \sin \frac{2\pi nt}{T}\right) \quad (4.6\text{-}1)$$

or by the complex form

$$x(t) = \sum_{n=-\infty}^{\infty} c_n\, e^{2\pi njt/T} \quad (4.6\text{-}2)$$

where $j = \sqrt{-1}$, as usual, and the Fourier coefficients are related by

$$a_n = c_n + c_{-n}, \quad (4.6\text{-}3)$$

$$b_n = j(c_n - c_{-n}). \quad (4.6\text{-}4)$$

These coefficients a_n, b_n, and c_n are random variables, of course, and are given by

$$a_n = \frac{2}{T}\int_{-T/2}^{T/2} x(t) \cos \frac{2\pi nt}{T}\, dt, \tag{4.6-5}$$

$$b_n = \frac{2}{T}\int_{-T/2}^{T/2} x(t) \sin \frac{2\pi nt}{T}\, dt, \tag{4.6-6}$$

and

$$c_n = \frac{1}{T}\int_{-T/2}^{T/2} x(t)\, e^{-2\pi njt/T}\, dt \tag{4.6-7}$$

where the integrals are to be interpreted in the sense discussed in Section 3.7. In other words, for a given sample function, Equations 4.6-5 through 4.6-7 yield a given realization of the random variables a_n, b_n, and c_n. For convenience, we shall use the symbol

$$\omega_1 = \frac{2\pi}{T} \tag{4.6-8}$$

in the future.

An important question is the meaning of the equality in Equation 4.6-1 or 4.6-2; that is, in what sense do the right sides of these equations converge to the random process $x(t)$? We proceed to answer this question in the next paragraphs.

A stationary random process $x(t)$ will be called *periodic* with period T if its autocorrelation function $R(\tau)$ is periodic with this period. In this case, $R(\tau)$ may be expanded in a Fourier series

$$R(\tau) = R(\tau + T) = \sum_{n=-\infty}^{\infty} r_n\, e^{jn\omega_1\tau} \tag{4.6-9}$$

where the coefficient r_n is given by

$$r_n = \frac{1}{T}\int_{-T/2}^{T/2} R(\tau)\, e^{-jn\omega_1\tau}\, d\tau. \tag{4.6-10}$$

Let us form the partial sum

$$x_N(t) = \sum_{n=-N}^{N} c_n\, e^{jn\omega_1 t}. \tag{4.6-11}$$

Then we have

$$E\{|x(t) - x_N(t)|^2\} = R(0) - 2E\{x(t)x_N(t)\} + E\{|x_N(t)|^2\}. \tag{4.6-12}$$

The last two terms in this expression may be identified as follows:

$$E\{x(t)x_N(t)\} = \sum_{n=-N}^{N} e^{jn\omega_1 t} E\{c_n x(t)\}$$

which becomes, from Equation 4.6-7,

$$E\{x(t)x_N(t)\} = \sum_{n=-N}^{N} e^{jn\omega_1 t} \frac{1}{T}\int_{-T/2}^{T/2} R(\tau - t)\, e^{-jn\omega_1 \tau}\, d\tau$$

$$= \sum_{n=-N}^{N} \frac{1}{T}\int_{-T/2}^{T/2} R(\tau - t)\, e^{-jn\omega_1(\tau - t)}\, d\tau$$

$$= \sum_{n=-N}^{N} r_n. \tag{4.6-13}$$

In the same way, we form

$$E\{|x_N(t)|^2\} = \sum_{m=-N}^{N} \sum_{n=-N}^{N} E\{c_m c_n^*\}\, e^{j\omega_1(m-n)t}. \tag{4.6-14}$$

The expression $E\{c_m c_n^*\}$ may be written as follows from Equations 4.6-7 and 4.6-9:

$$E\{c_m c_n^*\} = \frac{1}{T^2}\int_{-T/2}^{T/2}\int_{-T/2}^{T/2} \sum_{k=-\infty}^{\infty} r_k\, e^{jk\omega_1(t-u)}\, e^{j\omega_1(nu - mt)}\, du\, dt.$$

This equation can be rearranged to give

$$E\{c_m c_n^*\} = \frac{1}{T^2}\sum_{k=-\infty}^{\infty} r_k \int_{-T/2}^{T/2} e^{j\omega_1(k-m)t}\, dt \int_{-T/2}^{T/2} e^{j\omega_1(n-k)u}\, du$$

$$= \sum_{k=-\infty}^{\infty} r_k\, \delta_{km}\, \delta_{nk}$$

$$= r_m\, \delta_{mn} \tag{4.6-15}$$

where δ_{mn} is the Kronecker delta given by

$$\delta_{mn} = \begin{cases} 1, & m = n, \\ 0, & m \neq n. \end{cases}$$

Thus the Fourier coefficients c_n are uncorrelated for a periodic random process. We substitute Equation 4.6-15 into 4.6-14 to obtain

$$E\{|x_N(t)|^2\} = \sum_{n=-N}^{N} r_n. \tag{4.6-16}$$

Thus, in the limit, Equation 4.6-12 becomes

$$\lim_{N\to\infty} E\{|x(t) - x_N(t)|^2\} = R(0) - 2R(0) + R(0) = 0. \quad (4.6\text{-}17)$$

The Fourier series expansion of a periodic random process converges mean square to the process. Mean-square convergence has been discussed in Section 3.5.

It follows from Equations 4.6-3 and 4.6-4 that the complex coefficients c_k are related to the coefficients a_k and b_k by

$$c_k = \tfrac{1}{2}(a_k - jb_k), \quad (4.6\text{-}18)$$

$$c_{-k} = \tfrac{1}{2}(a_k + jb_k). \quad (4.6\text{-}19)$$

Since $R(\tau)$ is a real function (for a real random process), it is clear from Equations 4.6-10 and 4.6-15 that the imaginary part of $E\{c_m c_n^*\}$ is zero:

$$\text{Im } E\{c_m c_n^*\} = 0 = \tfrac{1}{4}[E\{a_m b_n\} - E\{a_n b_m\}], \quad (4.6\text{-}20)$$

$$\text{Im } E\{c_m c_{-n}^*\} = 0 = \tfrac{1}{4}[E\{a_m b_n\} + E\{a_n b_m\}], \quad (4.6\text{-}21)$$

or

$$E\{a_m b_n\} = E\{a_n b_m\} = E\{a_k b_k\} = 0. \quad (4.6\text{-}22)$$

We see that the Fourier series expansion of a periodic random process has two very interesting properties:

1. It converges mean square to the process.
2. Its coefficients a_n, b_n, and c_n are uncorrelated.

The first of these properties allows us to use the expansion with some confidence since it ensures that such difficulties as we encountered in defining the power spectral density will not exist. The second property will greatly simplify problems where we are interested in the statistics of the coefficients. This will be particularly true when the process $x(t)$ is gaussian. Since Equations 4.6-5 through 4.6-7 are then linear operations on gaussian processes, the coefficients a_n, b_n, and c_n will be normal random variables. Normal random variables which are uncorrelated are also statistically independent and, thus, the joint densities of these coefficients will be products of univariate densities.

Of course most processes in which we are interested will not be periodic and, strictly speaking, Fourier-series representation cannot be applied. We can avoid this problem by assuming that the processes are periodic with period $T \to \infty$. Actually it can be shown that the coefficients are approximately uncorrelated for any process whose spectral density is approximately constant over intervals large compared to $1/T$. Thus, for the white-noise case, where the spectral density is some constant K for all frequencies ω, the coefficients are always uncorrelated. This is easily shown by using Equation 4.6-7 to form

$$E\{c_m c_n^*\} = \frac{1}{T^2}\int_{-T/2}^{T/2}\int_{-T/2}^{T/2} E\{x(s)x(t)\}\, e^{-j\omega_1(mt-ns)}\, dt\, ds. \tag{4.6-23}$$

But, we have postulated that

$$E\{x(s)x(t)\} = K\,\delta(t-s) \tag{4.6-24}$$

so that Equation 4.6-23 becomes

$$E\{c_m c_n^*\} = \frac{K}{T^2}\int_{-T/2}^{T/2} e^{-j\omega_1(m-n)t}\, dt$$

or

$$E\{c_m c_n^*\} = \frac{K}{T}\,\delta_{mn} \tag{4.6-25}$$

as was to be proved.

We digress briefly at this point to obtain some further statistical properties of the Fourier coefficients which will be useful later. It is apparent from Equations 4.6-10 and 4.6-15 that

$$\lim_{T\to\infty} Tr_n = \lim_{T\to\infty} TE\{c_n c_n^*\} = \varphi(n\omega_1). \tag{4.6-26}$$

We also have, from Equations 4.6-18 and 4.6-19 that

$$c_n c_n^* = c_n c_{-n} = |c_n|^2 = \tfrac{1}{4}(a_n^2 + b_n^2). \tag{4.6-27}$$

Thus,

$$\lim_{T\to\infty} TE\{a_n^2 + b_n^2\} = 4\varphi(n\omega_1). \tag{4.6-28}$$

Now let us use Equation 4.6-7 to form

$$\lim_{T\to\infty} T^2E\{c_n^2\} = \lim_{T\to\infty}\int_{-T/2}^{T/2}\int_{-\infty}^{\infty} R(t-s)\, e^{-jk\omega_1(t+s)}\, dt\, ds. \tag{4.6-29}$$

We make the linear change of variable $y = (t - s)$ to obtain

$$\lim_{T\to\infty} T^2E\{c_n^2\} = \varphi(k\omega_1)\lim_{T\to\infty}\int_{-T/2}^{T/2} e^{-2jk\omega_1 s}\, ds = 0 \tag{4.6-30}$$

since $\omega_1 = 2\pi/T$. Now, from Equations 4.6-18 and 4.6-19 we have

$$4E\{c_n^2\} = E\{a_n^2 - b_n^2\} - 2jE\{a_n b_n\} = 0 \tag{4.6-31}$$

or, in the limit,

$$4E\{c_n^2\} = E\{a_n^2\} - E\{b_n^2\} = 0 \tag{4.6-32}$$

and

$$\lim_{T\to\infty} TE\{a_n^2\} = \lim_{T\to\infty} TE\{b_n^2\} = 2\varphi(n\omega_1). \tag{4.6-33}$$

The mean values of the coefficients are given by

$$E\{a_n\} = \frac{2}{T}\int_{-T/2}^{T/2} E\{x(t)\} \cos n\omega_1 t \, dt, \tag{4.6-34}$$

$$E\{b_n\} = \frac{2}{T}\int_{-T/2}^{T/2} E\{x(t)\} \sin n\omega_1 t \, dt, \tag{4.6-35}$$

and

$$E\{c_n\} = \frac{1}{T}\int_{-T/2}^{T/2} E\{x(t)\} e^{-jn\omega_1 t} \, dt. \tag{4.6-36}$$

If the process $x(t)$ has a zero mean, then

$$E\{a_n\} = E\{b_n\} = E\{c_n\} = 0. \tag{4.6-37}$$

We have seen that the Fourier series, when used to represent a periodic random process or in the limit as $T \to \infty$, has two interesting (and useful) properties, namely uncorrelated coefficients and mean-square convergence.

For a general (nonperiodic) random process, these properties do not necessarily hold, although we frequently assume that they do. A more general approach is to start with an arbitrary orthogonal expansion in the interval (a, b) of the form

$$x(t) = \sum_{n=1}^{\infty} X_i \varphi_i(t), \qquad a \leq t \leq b \tag{4.6-38}$$

where

$$\int_a^b \varphi_i(t)\varphi_j(t) \, dt = \delta_{ij}. \tag{4.6-39}$$

We now attempt to choose the functions $\varphi_i(t)$ so that the random variables X_i are uncorrelated; that is, we require that

$$E\{X_i X_j\} = \frac{\delta_{ij}}{\lambda_j} \tag{4.6-40}$$

where $E\{X_i^2\} = 1/\lambda_i$. It is apparent from the orthogonality condition of Equation 4.6-39 that the coefficients X_i are given by

$$X_i = \int_a^b x(t)\varphi_i(t) \, dt \tag{4.6-41}$$

and that

$$E\{X_i X_j\} = \int_a^b \int_a^b R(t - s)\varphi_i(t)\varphi_j(s) \, dt \, ds. \tag{4.6-42}$$

Let us assume that the functions $\varphi_i(t)$ are the solutions of the integral equation

$$\varphi_i(t) = \lambda_i \int_a^b R(t - s)\varphi_i(s) \, ds. \tag{4.6-43}$$

Note that the function $R(t - s)$ (called the *kernel*) is given and the function $\varphi_i(t)$ is unknown. It is clear that $\varphi_i(t) = 0$ is a solution to this equation. It can be shown [6] that there is a set of functions $\varphi_i(t)$ (called the eigenfunctions) and a corresponding set of values λ_i (called the eigenvalues) which also are solutions. If the right side of this equation is now substituted into Equation 4.6-42, we have

$$E\{X_iX_j\} = \frac{1}{\lambda_j}\int_a^b \varphi_i(t)\varphi_j(t)\,dt \tag{4.6-44}$$

or, from Equation 4.6-39,

$$E\{X_iX_j\} = \frac{\delta_{ij}}{\lambda_j}.$$

Thus the coefficients in the expansion of Equation 4.6-38 are uncorrelated if the orthogonal functions are chosen as the solutions to the integral equation (4.6-43). Furthermore it can be shown that the right side of Equation 4.6-38 converges mean square to the random process $x(t)$ under very weak restrictions on the autocorrelation function $R(t - s)$. This expansion (called the Karhunen-Loeve expansion) has been used extensively in problems involving statistical design and optimization. One of its disadvantages is that the orthogonal functions depend on the autocorrelation function, and, furthermore, may be hard to find from Equation 4.6-43. Further details are beyond the scope of this treatment [4, 5].

We are now in a position to consider problems involving random process representation by orthogonal expansions. In the next section, we will treat narrow-band processes by means of the Fourier series.

4.7 Narrow-Band Systems

In many communication systems, the situation will arise where broad-band noise is passed through a narrow-band linear filter to produce a narrow-band random process. By a *narrow-band process* we mean one whose spectral density is concentrated in a region Δf_m which is narrow compared to a center frequency f_m as shown in Figure 4.4. In radio receivers, for example, the RF stage is a tunable filter-amplifier with a bandwidth which may be very small compared to its center frequency. In such situations, the input process frequently will be gaussian or nearly gaussian and, thus, the output will be gaussian also. Furthermore, the narrow-band filters frequently will be followed by such nonlinear operations as rectification, squaring, and clipping. It is pertinent, therefore, to examine the properties of narrow-band noise, particularly those properties that will be useful in determining the effect of certain nonlinear operations.

Suppose a broad-band noise is passed through a narrow-band filter with an adjustable bandwidth and the output is displayed on an oscilloscope. As the

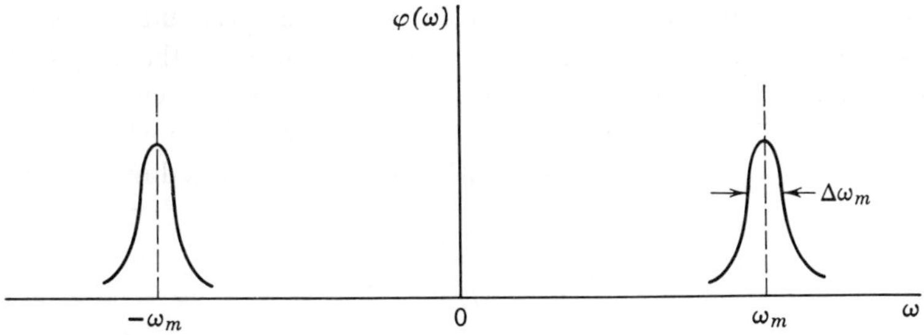

Figure 4.4 *The spectral density of a narrow-band process:* $\omega_m = 2\pi f_m$; $\Delta\omega_m = 2\pi\,\Delta f_m$

filter bandwidth is decreased, the observed waveform increasingly takes on the appearance of a sinusoid of random amplitude and phase. This suggests that the output might be represented in the form

$$f(t) = E(t)\cos\,[\omega_m t + \phi(t)] \tag{4.7-1}$$

where $\omega_m/2\pi = f_m$ is the center frequency of the filter and $E(t)$ and $\phi(t)$ are random processes which will be called the *envelope function* and the *phase function*, respectively.

We now make use of the Fourier series as discussed in Section 4.6 to represent the narrow-band process $f(t)$ in the interval $(-T/2, T/2)$:

$$f(t) = \sum_{k=1}^{\infty} [a_k \cos k\omega_1 t + b_k \sin k\omega_1 t] \tag{4.7-2}$$

where $\omega_1 = 2\pi/T$ and where the d-c term has been dropped for convenience. After some algebraic manipulation of Equation 4.7-2, we will relate the envelope function $E(t)$ and the phase function $\phi(t)$ to the Fourier coefficients a_k and b_k. From a knowledge of the statistical properties of these coefficients, we will determine the statistical properties of $E(t)$ and $\phi(t)$.

We can rewrite Equation 4.7-1 in the form

$$f(t) = E(t)\cos\phi(t)\cos\omega_m t - E(t)\sin\phi(t)\sin\omega_m t \tag{4.7-3}$$

since

$$\cos(x + y) = \cos x\cos y - \sin x \sin y.$$

We also write $\cos k\omega_1 t$ and $\sin k\omega_1 t$ as

$$\begin{aligned}\cos k\omega_1 t &= \cos\,[(k\omega_1 - \omega_m)t + \omega_m t]\\ &= \cos(k\omega_1 - \omega_m)t\cos\omega_m t - \sin(k\omega_1 - \omega_m)t\sin\omega_m t\end{aligned} \tag{4.7-4}$$

and

$$\begin{aligned}\sin k\omega_1 t &= \sin\,[(k\omega_1 - \omega_m)t + \omega_m t]\\ &= \sin(k\omega_1 - \omega_m)t\cos\omega_m t + \cos(k\omega_1 - \omega_m)t\sin\omega_m t\end{aligned} \tag{4.7-5}$$

since

$$\sin(x + y) = \sin x\cos y + \cos x\sin y.$$

We may now use Equations 4.7-4 and 4.7-5 to connect the alternate expressions for $f(t)$ as given by Equations 4.7-2 and 4.7-3. Thus $f(t)$ can be represented as

$$f(t) = e_c(t) \cos \omega_m t + e_s(t) \sin \omega_m t \tag{4.7-6}$$

where the two functions $e_c(t)$ and $e_s(t)$ are

$$e_c(t) = E(t) \cos \phi(t) \tag{4.7-7}$$

or

$$e_c(t) = \sum_{k=1}^{\infty} [a_k \cos \omega_k t + b_k \sin \omega_k t] \tag{4.7-8}$$

and

$$e_s(t) = -E(t) \sin \phi(t) \tag{4.7-9}$$

or

$$e_s(t) = \sum_{k=1}^{\infty} [b_k \cos \omega_k t - a_k \sin \omega_k t] \tag{4.7-10}$$

and where, for convenience, we have introduced the notation

$$\omega_k = k\omega_1 - \omega_m. \tag{4.7-11}$$

The foregoing procedure has related the envelope and phase functions to the Fourier coefficients since, from Equations 4.7-7 and 4.7-9,

$$E(t) = [e_c^2(t) + e_s^2(t)]^{1/2} \tag{4.7-12}$$

and

$$\phi(t) = \arc\tan \frac{-e_s(t)}{e_c(t)}. \tag{4.7-13}$$

The function $e_c(t)$ and $e_s(t)$ are linear combinations (Fourier series) of a_k's and b_k's given by Equations 4.7-8 and 4.7-10. These coefficients are found from the integrals

$$a_k = \frac{2}{T} \int_{-T/2}^{T/2} f(t) \cos k\omega_1 t \, dt, \tag{4.7-14}$$

$$b_k = \frac{2}{T} \int_{-T/2}^{T/2} f(t) \sin k\omega_1 t \, dt. \tag{4.7-15}$$

If $f(t)$ is a gaussian process, then a_k and b_k are normal random variables (result of a linear operation on a gaussian process). Then, from Equations 4.7-8 and 4.7-10, it follows that $e_c(t)$ and $e_s(t)$ are gaussian processes (result of a linear operation on normal random variables). It follows also from the discussion of Section 4.6 that the coefficients a_k, b_k are uncorrelated if the interval of expansion T is allowed to become infinite.

Let us check to see if $e_c(t)$ and $e_s(t)$ are correlated. We have

$$E\{e_c(t)e_s(t)\}$$

$$= \sum_{k=1}^{\infty} \sum_{m=1}^{\infty} [E\{a_k b_m\} \cos \omega_k t \cos \omega_m t + E\{b_k b_m\} \sin \omega_k t \cos \omega_m t - E\{a_k a_m\} \cos \omega_k t \sin \omega_m t - E\{b_k a_m\} \sin \omega_k t \sin \omega_m t]$$

which can be written as

$$E\{e_c(t)e_s(t)\} = 0 + \sum_{k=1}^{\infty} E\{b_k^2\} \sin \omega_k t \cos \omega_k t - \sum_{k=1}^{\infty} E\{a_k^2\} \sin \omega_k t \cos \omega_k t - 0.$$

We have already shown in Equation 4.6-33 that $E\{b_k^2\} = E\{a_k^2\} = (2/T)\varphi(k\omega_1)$ and hence that $e_c(t)$ and $e_s(t)$ are uncorrelated in the limit as $T \to \infty$. Since $e_c(t)$ and $e_s(t)$ are also gaussian random variables, they are independent and the joint density $f_2(e_c, e_s)$ can be written as the product of the univariate densities of e_c and e_s, respectively:

$$f_2(e_c, e_s) = f_1(e_c)f_1(e_s). \tag{4.7-16}$$

If the process $f(t)$ has zero mean, it follows from Equations 4.7-8, 4.7-10, and 4.6-37 that

$$E\{e_c(t)\} = E\{e_s(t)\} = 0. \tag{4.7-17}$$

The variance of e_c and e_s can be found directly from Equations 4.7-6, 4.7-8, and 4.7-10 to be

$$\begin{aligned} E\{e_c^2(t)\} = E\{e_s^2(t)\} &= \frac{1}{2} \sum_{k=1}^{\infty} E\{a_k^2 + b_k^2\} \\ &= \sum_{k=1}^{\infty} E\{a_k^2\} = \sum_{k=1}^{\infty} E\{b_k^2\} \\ &= E\{f^2(t)\} = \sigma^2. \end{aligned} \tag{4.7-18}$$

Thus, if $f(t)$ is a gaussian process, the joint density of Equation 4.7-16 is

$$f_2(e_c, e_s) = \frac{1}{2\pi\sigma^2} e^{-(e_c^2 + e_s^2)/2\sigma^2}. \tag{4.7-19}$$

We will now find the joint density $g_2(E, \phi)$ of E and ϕ, the envelope and phase functions, by using the functional relationships of Equations 4.7-7 and 4.7-9 and the transformation procedure of Section 2.20. We write

$$g_2(E, \phi) = f_2(e_c, e_s) \left| J\left(\frac{e_c, e_s}{E, \phi}\right) \right| \tag{4.7-20}$$

where the magnitude of the Jacobian of the transformation is given by

$$\left| J\left(\frac{e_c, e_s}{E, \phi}\right) \right| = \left\| \begin{matrix} \cos\phi & -E\sin\phi \\ -\sin\phi & -E\cos\phi \end{matrix} \right\| = E \tag{4.7-21}$$

since

$$e_c = E\cos\phi,$$
$$e_s = -E\sin\phi.$$

Now Equation 4.7-20 can be written as

$$g_2(E, \phi) = \begin{cases} \dfrac{E}{2\pi\sigma^2} e^{-E^2/2\sigma^2}, & 0 \le E < \infty,\ 0 \le \phi < 2\pi, \\ 0, & \text{elsewhere.} \end{cases} \tag{4.7-22}$$

This is the joint density of the envelope and phase functions. Note that E is the square root of the sum of the squares of two gaussian random variables.

The univariate densities $g_1(E)$ and $g_1(\phi)$ can be obtained by integrating the joint density $g_2(E, \phi)$ with respect to the proper variable. We have

$$g_1(E) = \int_0^{2\pi} g(E, \phi)\, d\phi \tag{4.7-23}$$

or

$$g_1(E) = \begin{cases} \dfrac{E}{\sigma^2} e^{-E^2/2\sigma^2}, & 0 \le E < \infty, \\ 0, & -\infty < E < 0. \end{cases} \tag{4.7-24}$$

This distribution is called the Rayleigh distribution and is plotted in Figure 4.5 in the normalized form:

$$f(x) = x\, e^{-x^2/2} u(x) \tag{4.7-25}$$

where E/σ is written as x, $\sigma g_1(E)$ is written as $f(x)$, and $u(x)$ is the unit step function. The value of x for which $f(x)$ is a maximum is found from

$$\frac{d}{dx} f(x) = 0 = (1 - x^2)\, e^{-x^2/2}$$

or

$$x = 1. \tag{4.7-26}$$

This maximum is

$$f_{\max} = f(1) = e^{-1/2} \approx 0.606. \tag{4.7-27}$$

The expected value of x is

$$E\{x\} = \int_0^{\infty} x f(x)\, dx = \int_0^{\infty} x^2 e^{-x^2/2}\, dx.$$

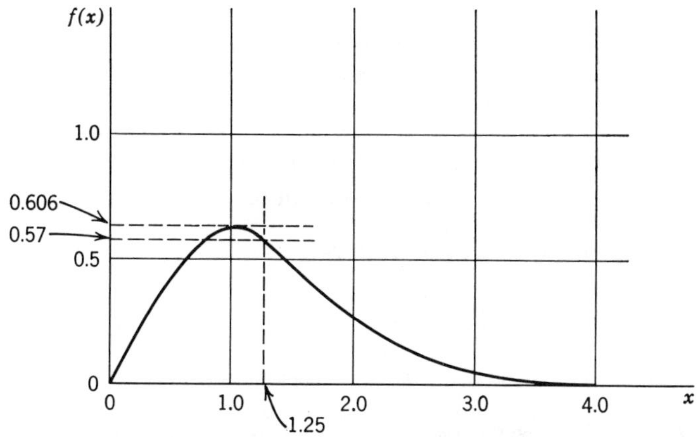

Figure 4.5 *The normalized Rayleigh distribution:* $f(x) = x\,e^{-x^2/2}u(x)$

This can be rewritten as

$$E\{x\} = \sqrt{\frac{\pi}{2}}\left(\frac{1}{\sqrt{2\pi}}\int_{-\infty}^{\infty} x^2\,e^{-x^2/2}\,dx\right).$$

The quantity in brackets has a value of unity since it is the variance of the unit normal distribution; therefore

$$E\{x\} = \sqrt{\frac{\pi}{2}} \approx 1.25. \tag{4.7-28}$$

These various quantities are indicated in Figure 4.5.

The distribution of ϕ is found from

$$\begin{aligned} g_1(\phi) &= \int_0^{\infty} g(E, \phi)\,dE \\ &= \frac{1}{2\pi\sigma^2}\int_0^{\infty} E\,e^{-E^2/2\sigma^2}\,dE \\ &= -\frac{1}{2\pi}\,e^{-E^2/2\sigma^2}\bigg|_0^{\infty} = \frac{1}{2\pi} \end{aligned} \tag{4.7-29}$$

or

$$g_1(\phi) = \begin{cases} \dfrac{1}{2\pi}, & 0 \le \phi \le 2\pi, \\ 0, & \text{elsewhere.} \end{cases} \tag{4.7-30}$$

The phase is uniformly distributed in the interval $(0, 2\pi)$ as might be expected since there is no physical basis to prefer one phase angle over another.

It is apparent from Equations 4.7-22, 4.7-24, and 4.7-30 that

$$g_2(E, \phi) = g_1(E)g_1(\phi) \tag{4.7-31}$$

and hence that the random variables E and ϕ at any time t are statistically independent. It does not follow, however, that the random processes $E(t)$ and $\phi(t)$ are statistically independent. Joint densities involving two times t_1 and t_2 may not factor. It can be shown [4, 5] that the joint densities

$$f_4(E_1, E_2, \phi_1, \phi_2) \neq f_2(E_1, E_2)f_2(\phi_1, \phi_2) \tag{4.7-32}$$

where $E_i = E(t_i)$ and $\phi_i = \phi(t_i)$.

4.8 Narrow-Band Systems with Added Sinusoids

It may be more realistic in many cases to consider a sinusoidal signal added to the input of the narrow-band filter discussed in the Section 4.7. Then the output would be

$$g(t) = f(t) + A \cos(\omega_m t + \Psi) \tag{4.8-1}$$

where $f(t)$ is the narrow-band process given by Equations 4.7-1 and 4.7-2, $f_m = \omega_m/2\pi$ is the "center frequency" of the filter, Ψ is an arbitrary angle in $(0, 2\pi)$, and A is a constant. The output $g(t)$ may be represented in the same form as Equation 4.7-1 and written as

$$g(t) = R(t) \cos[\omega_m t + \theta(t)] \tag{4.8-2}$$

where $R(t)$ and $\theta(t)$ are new envelope and phase functions.

In Equation 4.7-6 the noise process $f(t)$ was written in a form that defined the quadrature components $e_c(t)$ and $e_s(t)$:

$$f(t) = e_c(t) \cos \omega_m t + e_s(t) \sin \omega_m t. \tag{4.8-3}$$

It is clear that the new output $g(t)$ can be written as

$$g(t) = e_c'(t) \cos \omega_m t + e_s'(t) \sin \omega_m t \tag{4.8-4}$$

where

$$e_c'(t) = e_c(t) + A \cos \Psi, \tag{4.8-5a}$$

$$e_s'(t) = e_s(t) + A \sin \Psi, \tag{4.8-5b}$$

and $e_c(t)$ and $e_s(t)$ have been defined by the Fourier series of Equations 4.7-8 and 4.7-10, respectively. As before, the envelope and phase functions are given in terms of these quadrature components by

$$R(t) = \{[e_c'(t)]^2 + [e_s'(t)]^2\}^{1/2} \tag{4.8-6}$$

and

$$\theta(t) = \arc\tan \frac{-e_s'(t)}{e_c'(t)}. \tag{4.8-7}$$

It is clear that $e_c'(t)$ and $e_s'(t)$ are gaussian if the filter input is gaussian and that their joint density is

$$f_2(e_c', e_s') = \frac{1}{2\pi\sigma^2} \exp\left[-\frac{(e_c' - A\cos\Psi)^2 + (e_s' - A\sin\Psi)^2}{2\sigma^2}\right] \tag{4.8-8}$$

since both $e_c'(t)$ and $e_s'(t)$ have mean equal to $A\cos\Psi$ and $A\sin\Psi$ respectively. As before, the joint density of R and θ is

$$g_2(R, \theta) = f_2(e_c', e_s')\left| J\left(\frac{e_c', e_s'}{R, \theta}\right)\right|. \tag{4.8-9}$$

Since

$$e_c' = R\cos\theta, \tag{4.8-10}$$

$$e_s' = -R\sin\theta, \tag{4.8-11}$$

the Jacobian of the transformation is R. Also the exponent of Equation 4.8-8 can be rewritten in terms of R and θ as follows:

$$(e_c' - A\cos\Psi)^2 + (e_s' - A\sin\Psi)^2 = R^2 + A^2 - 2AR\cos(\theta - \Psi). \tag{4.8-12}$$

Now Equation 4.8-9 becomes

$$g_2(R, \theta) = \begin{cases} \dfrac{R}{2\pi\sigma^2} \exp\left[-\dfrac{R^2 + A^2 - 2AR\cos(\theta - \Psi)}{2\sigma^2}\right], & 0 \le R < \infty,\ 0 \le \theta \le 2\pi, \\ 0, & \text{elsewhere,} \end{cases} \tag{4.8-13}$$

where σ^2 has been defined previously by Equation 4.7-18 and is the variance of the gaussian random process $f(t)$.

This joint density $g_2(R, \theta)$ is not so simply integrated as was $g_2(E, \phi)$, treated in the previous section. The univariate density for R is

$$\begin{aligned} g_1(R) &= \int_0^{2\pi} g_2(R, \theta)\, d\theta \\ &= \frac{R}{2\pi\sigma^2} e^{-(R^2 + A^2)/2\sigma^2} \int_0^{2\pi} e^{[AR\cos(\theta - \Psi)]/\sigma^2}\, d\theta \\ &= \frac{R}{\sigma^2} e^{-(R^2 + A^2)/2\sigma^2} I_0\left(\frac{AR}{\sigma^2}\right) \end{aligned} \tag{4.8-14}$$

where $I_0(x)$ is the modified Bessel function of the first kind and order zero. This last relationship follows from

$$J_n(z) = \frac{(j)^{-n}}{2\pi} \int_{\Psi}^{\Psi + 2\pi} e^{jz\cos(\theta - \Psi)}\, e^{jn(\theta - \Psi)}\, d\theta \tag{4.8-15}$$

where $J_n(z)$ is the Bessel function of the first kind and order n. The modified Bessel function of the first kind and order n (also called a hyperbolic Bessel function) is denoted by $I_n(x)$ and is given by

$$I_n(x) = (-j)^n J_n(jx). \tag{4.8-16}$$

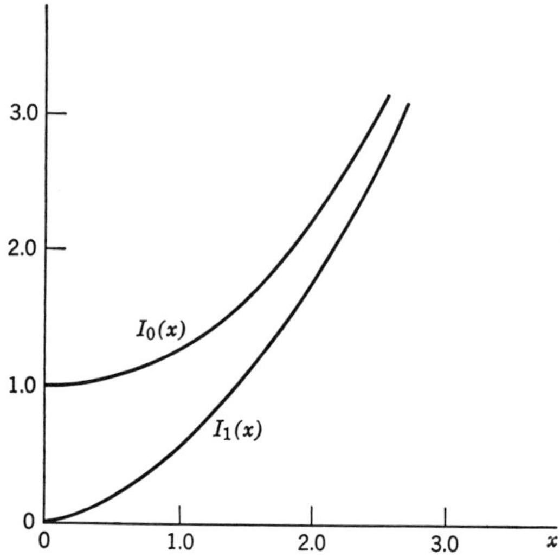

Figure 4.6 *Modified Bessel functions of the first kind*

Thus $I_0(x)$ is

$$I_0(x) = J_0(jx). \tag{4.8-17}$$

This function is plotted in Figure 4.6 for positive argument and is seen to be monotonically increasing. Further discussion of its properties and tabulations of its values will be found in the literature [6, 7, 8].

The density function $g_1(R)$ has been plotted in Figure 4.7. Again a normalized form has been used:

$$f(x) = x\, e^{-(x^2+k^2)/2} I_0(kx) \tag{4.8-18}$$

where R/σ is written as x, $\sigma g_1(R)$ as $f(x)$, and A/σ as the parameter k. The modified Bessel function $I_0(x)$ has a series expansion, valid for large x, of the form [9]:

$$I_0(x) = \frac{e^x}{\sqrt{2\pi x}}\left(1 + \frac{1}{8x} + \frac{9}{128x^2} + \cdots\right). \tag{4.8-19}$$

Therefore, for the case where $kx = AR/\sigma^2 \gg 1$, Equations 4.8-14 and 4.8-18 can be written as

$$g_1(R) \to \frac{1}{\sigma}\sqrt{\frac{R}{2\pi A}}\, e^{-[(R-A)/\sigma]^2/2} \tag{4.8-20}$$

and

$$f(x) \to \sqrt{\frac{x}{2\pi k}}\, e^{-(x-k)^2/2}. \tag{4.8-21}$$

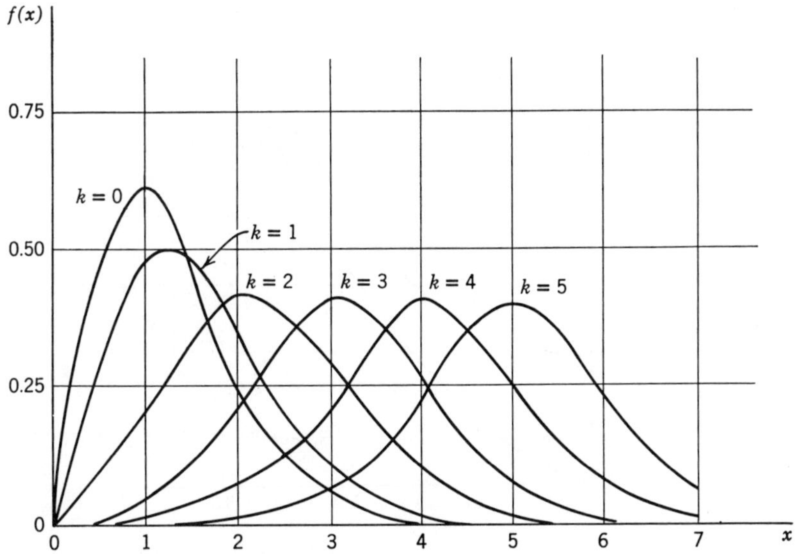

Figure 4.7 *The probability amplitude distribution of a sinusoid in narrow-band gaussian noise:* $f(x) = x\,e^{-(x^2+k^2)/2} I_0(kx)$

It is apparent from these equations and from the curves given in Figure 4.7 that the envelope distribution is approximately gaussian in the vicinity of $R = A$ when the sine-wave magnitude A is large compared to the square root of the noise power σ^2.

It is clear from Figure 4.7 that the mean value of the envelope is a function now of the parameter $k = A/\sigma$, which is, in turn, a measure of the signal-to-noise ratio. This mean value is given by

$$E\{x\} = E\left\{\frac{R}{\sigma}\right\} = \int_0^\infty xf(x)\,dx$$

$$= e^{-k^2/2} \int_0^\infty x^2 e^{-x^2/2} I_0(kx)\,dx \qquad (4.8\text{-}22)$$

or

$$E\{x\} = \sqrt{\frac{\pi}{2}}\, e^{-k^2/4} \left[\left(1 + \frac{k^2}{2}\right) I_0\left(\frac{k^2}{4}\right) + \frac{k^2}{2} I_1\left(\frac{k^2}{4}\right)\right] \qquad (4.8\text{-}23)$$

where $I_1(x)$ is the modified Bessel function of the first kind and order one and is shown in Figure 4.6 for positive arguments. The value of $E\{x\}$ as given by Equation 4.8-23 is plotted on Figure 4.8 as a function of $k^2 = A^2/\sigma^2$. Also plotted on the same figure and shown by the dashed curve is the case for the sinusoid alone; that is, where $R = A$.

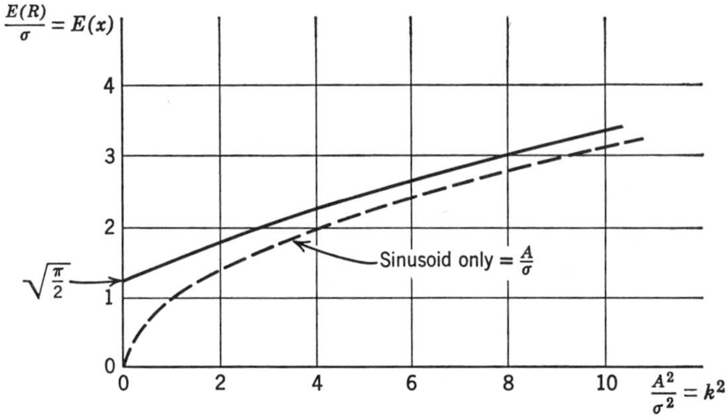

Figure 4.8 *The expected value of the envelope of a sinusoid in narrow-band gaussian noise*

The results of the preceding paragraph are a special case of a more general development [5, Chap. 9]. The rth moment of the envelope R can be written from Equation 4.8-14 as

$$E\{R^r\} = \int_0^\infty \frac{R^{r+1}}{\sigma^2} e^{-(R^2+A^2)/2\sigma^2} I_0\left(\frac{AR}{\sigma^2}\right) dR \qquad (4.8\text{-}24)$$

or

$$E\{R^r\} = (2\sigma^2)^{r/2}\Gamma\left(\frac{r}{2}+1\right) {}_1F_1\left(-\frac{r}{2}; 1; -\frac{A^2}{2\sigma^2}\right) \qquad (4.8\text{-}25)$$

where $\Gamma(x)$ is the gamma function [6, Chap. 12] defined by

$$\Gamma(x) = \int_0^\infty y^{x-1} e^{-y}\, dy \qquad (4.8\text{-}26)$$

and ${}_1F_1(\alpha; \beta; \pm y)$ is the confluent hypergeometric function [5, 7] given by

$$\begin{aligned} {}_1F_1(\alpha; \beta; \pm y) &= 1 + \frac{\alpha}{\beta}\frac{(\pm y)}{1!} + \frac{\alpha(\alpha+1)}{\beta(\beta+1)}\frac{(\pm y^2)}{2!} + \cdots \\ &\quad + \frac{\alpha(\alpha+1)(\alpha+2)\cdots(\alpha+n-1)}{\beta(\beta+1)(\beta+2)\cdots(\beta+n-1)}\frac{(\pm y^n)}{n!} + \cdots. \end{aligned} \qquad (4.8\text{-}27)$$

It is the solution to the differential equation

$$y\frac{d^2F}{dy^2} + (\beta - y)\frac{dF}{dy} - \alpha F = 0. \qquad (4.8\text{-}28)$$

For the case where $r = 1$, Equation 4.8-25 becomes

$$E\{R\} = (2\sigma^2)^{1/2}\Gamma\left(\frac{3}{2}\right){}_1F_1\left(-\frac{1}{2}; 1; -\frac{A^2}{2\sigma^2}\right). \tag{4.8-29}$$

Since [6, Chap. 12]

$$\Gamma\left(\frac{3}{2}\right) = \frac{\sqrt{\pi}}{2}$$

and [5, App. A]

$${}_1F_1\left(-\frac{1}{2}; 1; -y\right) = e^{-y/2}\left[(1 + y)I_0\left(\frac{y}{2}\right) + yI_1\left(\frac{y}{2}\right)\right], \tag{4.8-30}$$

then Equation 4.8-29 reduces to Equation 4.8-23 as expected. For the case where $r = 2$, we have

$$E\{R^2\} = 2\sigma^2\Gamma(2){}_1F_1\left(-1; 1; -\frac{A^2}{2\sigma^2}\right). \tag{4.8-31}$$

But [6, Chap. 12]

$$\Gamma(2) = 1 \tag{4.8-32}$$

and, from Equation 4.8-27,

$${}_1F_1(-1; 1; -y) = 1 + y \tag{4.8-33}$$

so that

$$E\{R^2\} = 2\sigma^2 + A^2 = \sigma^2(2 + k^2) \tag{4.8-34}$$

where, as before, $k = A/\sigma$. The variance of R is

$$\sigma_R^2 = E\{R^2\} - [E\{R\}]^2 \tag{4.8-35}$$

or

$$\frac{\sigma_R^2}{\sigma^2} = 2 + k^2 - \frac{\pi}{2}e^{-k^2/2}\left[\left(1 + \frac{k^2}{2}\right)I_0\left(\frac{k^2}{4}\right) + \frac{k^2}{2}I_1\left(\frac{k^2}{4}\right)\right]^2 \tag{4.8-36}$$

from Equations 4.8-23 and 4.8-34. This last relationship is plotted in Figure 4.9.

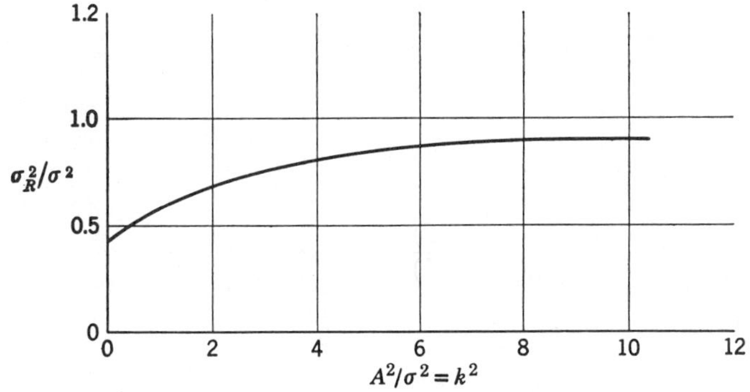

Figure 4.9 *The normalized power of a sinusoid in narrow-band gaussian noise*

The distribution of the phase function θ is found from

$$g_1(\theta) = \int_0^\infty g_2(R, \theta)\, dR$$

$$= \frac{1}{2\pi\sigma^2} \int_0^\infty R \exp\left[-\frac{R^2 + A^2 - 2AR\cos(\theta - \Psi)}{2\sigma^2}\right] dR. \quad (4.8\text{-}37)$$

Let us complete the square in R in the exponent to obtain

$$g_1(\theta) = \frac{1}{2\pi\sigma^2} \exp\left\{-\frac{A^2}{2\sigma^2}[1 - \cos^2(\theta - \Psi)]\right\}$$

$$\times \int_0^\infty R \exp\left\{-\frac{1}{2}\left[\frac{R - A\cos(\theta - \Psi)}{\sigma}\right]^2\right\} dR. \quad (4.8\text{-}38)$$

At this point it is convenient to introduce the notation

$$b = \frac{A\cos(\theta - \Psi)}{\sigma} \quad (4.8\text{-}39)$$

and make a linear change in variable

$$\frac{R}{\sigma} - b = y \quad (4.8\text{-}40)$$

in Equation 4.8-38 to obtain

$$g_1(\theta) = \frac{1}{2\pi} \exp\left[-\frac{A^2 \sin^2(\theta - \Psi)}{2\sigma^2}\right] \int_{-b}^\infty (y + b)\, e^{-y^2/2}\, dy. \quad (4.8\text{-}41)$$

The first term in this integral may be evaluated directly while the second may be rearranged to yield

$$g_1(\theta) = \exp\left[-\frac{A^2 \sin^2(\theta - \Psi)}{2\sigma^2}\right]\left(\frac{e^{-b^2/2}}{2\pi} + \frac{1}{\sqrt{2\pi}} \int_{-b}^\infty e^{-y^2/2}\, dy\right) \quad (4.8\text{-}42)$$

or

$$g_1(\theta) = \frac{e^{-A^2/2\sigma^2}}{2\pi} + \frac{A\cos(\theta - \Psi)}{2\pi\sigma} \exp\left[-\frac{A^2 \sin^2(\theta - \Psi)}{2\sigma^2}\right] \Phi(b) \quad (4.8\text{-}43)$$

where $\Phi(b)$ is the cumulative distribution function for the unit normal distribution and is given by

$$\Phi(b) = \frac{1}{\sqrt{2\pi}} \int_{-\infty}^b e^{-y^2/2}\, dy. \quad (4.8\text{-}44)$$

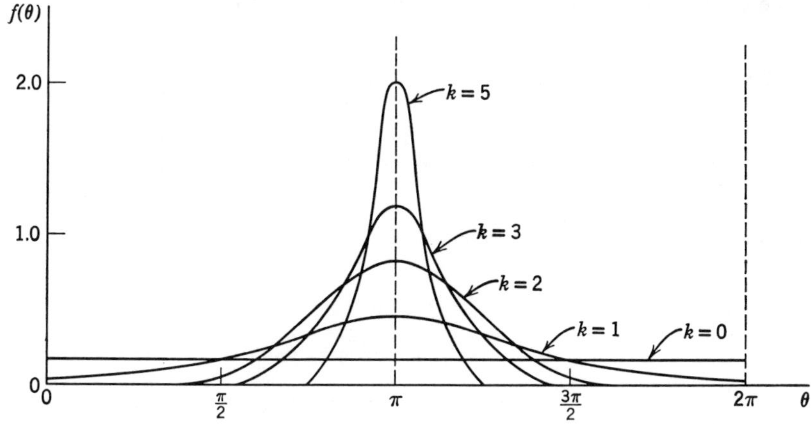

Figure 4.10 The density function of the phase of a sinusoid in narrow-band gaussian noise

If, as in Equation 4.8-18, we let $A/\sigma = k$, then Equation 4.8-43 becomes

$$f(\theta) = \frac{e^{-k^2/2}}{2\pi} + \frac{k\cos\theta}{2\pi}\exp\left[-\frac{k^2\sin^2(\theta - \Psi)}{2}\right]\Phi[k\cos(\theta - \Psi)].$$

This equation is plotted in Figure 4.10 with k as the parameter and with $\Psi = \pi$ for convenience. As $k \to 0$ the effect of the sinusoid becomes negligible and the distribution becomes uniform as expected. For strong signal conditions, $k \gg 1$, the distribution approaches a delta function at $\theta = \pi$.

4.9 Maximum-Power-Transfer Theorems

In the next several sections we will investigate the effects of noise in linear amplifiers. Before treating this subject, however, we review some of the elementary theorems connected with maximum-power transfer in linear systems.

We consider the system illustrated in Figure 4.11 where E_S is a signal source, Z_S is the source impedance, and Z_L is the load impedance. In many problems of communications, it is desirable to transfer maximum power to the load. At least three situations can exist with respect to the impedance; each of these will now be considered separately.

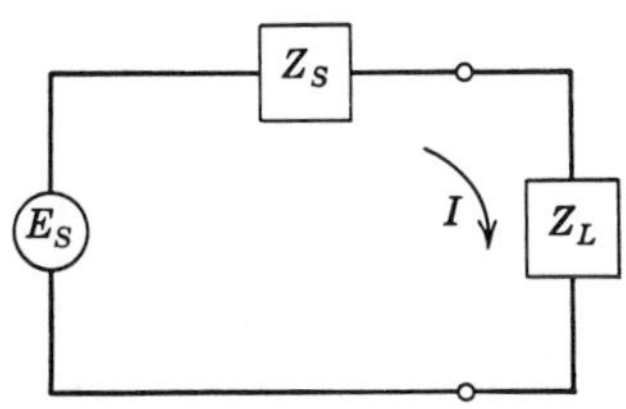

Figure 4.11 A linear system

Case I—Source Impedance Fixed and Load Impedance Variable

Let us denote the real and imaginary parts of the impedances by

$$Z_S = R_S + jX_S \tag{4.9-1}$$

and

$$Z_L = R_L + jX_L. \tag{4.9-2}$$

The power P_L dissipated in the load is

$$P_L = |I|^2 R_L$$

$$= \frac{|E|^2}{(R_S + R_L)^2 + (X_S + X_L)^2} R_L. \tag{4.9-3}$$

If this power P_L is to be a maximum, then the following relationships must be satisfied simultaneously:

$$\frac{\partial P_L}{\partial R_L} = \frac{\partial P_L}{\partial X_L} = 0. \tag{4.9-4}$$

Taking first the partial derivative with respect to R_L, we have

$$R_S^2 - R_L^2 + (X_S + X_L)^2 = 0 \tag{4.9-5}$$

and, from the partial derivative with respect to X_L,

$$X_S + X_L = 0 \tag{4.9-6}$$

or

$$X_L = -X_S, \tag{4.9-7}$$

$$R_L = R_S. \tag{4.9-8}$$

Thus, for fixed-source impedance and maximum-power transfer, the load impedance should be the complex conjugate of the source impedance:

$$\boxed{Z_L = Z_S^*.} \tag{4.9-9}$$

The maximum power dissipated in the load is obtained from Equation 4.9-3:

$$(P_L)_{\max} = \frac{|E|^2}{4R_L} = \frac{|E|^2}{4R_S}$$

which is equal to the power P_S dissipated in the source; thus the efficiency is

$$\eta = \frac{P_L}{P_S + P_L} \times 100 = 50\%.$$

Case II—Load Impedance Fixed and Source Impedance Variable

For this case, maximum-power-transfer conditions are apparent from an inspection of Equation 4.9-3. Since R_L is fixed,

$$X_L = -X_S$$

as before, and

$$R_S = 0$$

since $R_S \geq 0$. The efficiency then is

$$\eta = \frac{P_L}{P_L + P_S} \times 100 = 100\%. \tag{4.9-10}$$

This situation is not realizable, of course, since the source resistance can never be zero. However, the source resistance should be made as small as possible.

Case III—Source Impedance Fixed and the Phase Angle of Load Impedance Fixed

The power dissipated in the load is

$$P_L = |I|^2|Z_L| \cos \theta \tag{4.9-11}$$

where θ is the fixed phase angle associated with the load impedance Z_L. Equation 4.9-11 can be rewritten as

$$P_L = \frac{|E|^2|Z_L| \cos \theta}{(R_S + |Z_L| \cos \theta)^2 + (X_S + |Z_L| \sin \theta)^2}$$

$$= \frac{|Z_L|\,|E|^2 \cos \theta}{|Z_L|^2 + |Z_L|2[R_S \cos \theta + X_S \sin \theta] + (R_S^2 + X_S^2)}. \tag{4.9-12}$$

If this expression is now differentiated with respect to $|Z_L|$ and the result set equal to zero, we obtain

$$|Z_L| = [R_S^2 + X_S^2]^{1/2} = |Z_S|.$$

The magnitude of the load impedance should be made equal to the magnitude of the fixed-source impedance.

4.10 Noise in Linear Amplifiers

Linear amplifiers are an integral part of most communication systems where signal levels must be increased. The performance of these amplifiers is limited ultimately by the noises that arise within them and prevent arbitrarily large amplification. Some of these noise sources, such as thermal sources, are fundamental in nature and arise in all amplifiers; others, such as flicker noise, depend on the devices being used in the amplifier. The common sources of noise and their characteristics are discussed in a number of texts [10, 11, 12].

The problem to be considered in this section is the characterization of (linear) amplifier performance in the presence of small signals. In such situations one of the most frequently used measures of the ability to distinguish a signal from a noise background is the signal-to-noise power ratio S/N at some point in the system.

Let us define an *improvement factor* I for a system or an *operating noise figure* $1/I$ of the system as the ratio of the actual signal-to-noise power ratios at the input and output. Thus

$$I = \frac{(S/N)_o}{(S/N)_i} \tag{4.10-1}$$

where $(S/N)_o$ is the signal-to-noise power ratio at the output of the system and $(S/N)_i$ is the signal-to-noise power ratio at the input of the system. It is clear that

$$I \leq 1 \tag{4.10-2}$$

since, in an ideal (noiseless) amplifier, no noise would be added in the amplifier and the input signal and noise would be amplified by the same amount as they passed through the system, yielding an improvement factor of unity. This case illustrates a fundamental property of linear systems (see Equation 4.2-2)—that input noise and noise generated internally can be isolated and treated individually.

It was pointed out in Section 4.9 that the power obtained from a source depends on both the load and source impedances. The *available power* P_a from a source is the power delivered to a matched load; hence it is the maximum power that the particular source can deliver. As shown by Equation 4.9-3, it is

$$P_a = \frac{|E|^2}{4R_S} \tag{4.10-3}$$

where R_S is the source resistance and E is the source voltage. For a thermal noise source (see Equation 3.1-1) with mean-squared voltage $4kTR\,\Delta f$, this available power in a band Δf is

$$P_a = \frac{4kTR\,\Delta f}{4R} = kT\,\Delta f. \tag{4.10-4}$$

It is interesting to note that this available power depends on the resistor temperature but not on the resistor value.

An arbitrary linear amplifier is shown in Figure 4.12 together with a driving source E_S with impedance Z_S and a load Z_L. The impedance Z_i is the *input impedance* of the amplifier and Z_o is the *output impedance*. The amplifier

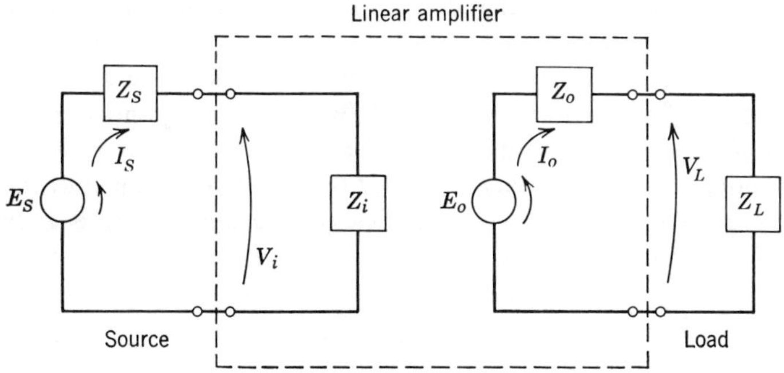

***Figure 4.12** An arbitrary linear amplifier*

source E_o is the open-circuit output voltage. It is apparent that the generator has an available power given by

$$P_{a1} = \frac{|E_S|^2}{4R_S} \tag{4.10-5}$$

where R_S is the real part of Z_S. The amplifier has an available power

$$P_{a2} = \frac{|E_o|^2}{4R_o} \tag{4.10-6}$$

where R_o is the real part of Z_o. The *available power gain* G_a of the amplifier is defined as

$$G_a = \frac{P_{a2}}{P_{a1}} = \frac{|E_o|^2}{|E_S|^2}\frac{R_S}{R_o}. \tag{4.10-7}$$

The available power gain does not depend on the amplifier load but it does depend on the source impedance; thus it is appropriate to talk of the available power gain for a given source. Also, although the input impedance of the amplifier does not appear explicitly, it does influence the open-circuit output voltage E_o for a given source.

The available power gain G_a is not the only gain that can be defined. The *power insertion gain* G_I (or *voltage insertion gain* V_I) is commonly used and is defined by

$$G_I\ (\text{or } V_I) = \frac{\text{load power (or voltage) when amplifier is connected}}{\text{load power (or voltage) when amplifier is removed}}. \tag{4.10-8}$$

In terms of the notation of Figure 4.12, the power insertion gain is

$$G_I = \frac{|E_o|^2/[(R_o + R_L)^2 + (X_o + X_L)^2]}{|E_S|^2/[(R_S + R_L)^2 + (X_S + X_L)^2]}. \tag{4.10-9}$$

For the case where $Z_L = \infty$ (open circuit) and $Z_S = 0$, this becomes

$$G_I = \frac{|E_o|^2}{|E_S|^2}. \tag{4.10-10}$$

Up to this point the discussion of this section has been in terms of the instantaneous power at some unspecified frequency f. In general, the source voltage, the noise voltages, the amplifier gain, and the various impedances will be functions of frequency. Also, at least the noise voltages will be typical realizations of random processes and, at any instant of time, will be random variables. For ergodic random processes, it will be convenient to consider expected values (or, equivalently, time averages) so that the available power from a random source is

$$P_a = \frac{E(E_S^2)}{4R_S} = \frac{A[E_S^2(t)]}{4R_S} = \frac{R_S(0)}{4R_S} \tag{4.10-11}$$

where $R_S(\tau)$ is the autocorrelation function of the process. In order to account for frequency dependence, it is conventional to define [4] the *incremental available power** P_a of a random source as the available power in a frequency band df which is centered on the frequency f. Thus

$$P_a = \frac{2\varphi(f)df}{4R_S} \tag{4.10-12}$$

where $\varphi(f)$ is the power spectral density related to $R_S(0)$ by

$$R_S(0) = \frac{1}{2\pi}\int_{-\infty}^{\infty} \varphi(f)d(2\pi f). \tag{4.10-13}$$

For the thermal noise source of Equation 4.10-4, this incremental available power is

$$P_a = kTdf = \frac{\varphi(f)df}{2R_S} \tag{4.10-14}$$

or the thermal noise power spectral density of a resistor R_S is

$$\varphi(f) = 2kTR_S. \tag{4.10-15}$$

In view of the simplicity of the expressions involving thermal noise, it is often desirable to assume that all of the source noise is thermal noise arising in the resistive component R_S of the source impedance. In this case the noisiness of the source can be expressed as an *effective noise temperature* T_e defined from Equation 4.10-14 as

$$T_e = \frac{P_a}{kdf} \tag{4.10-16}$$

* It would be desirable to introduce another symbol (say $\mathscr{P}_a$) to denote the incremental available power defined in terms of spectral densities. However, in the interests of simplicity of notation, the same symbols as already defined will be used for incremental quantities.

where P_a is the incremental available power from the source. Note that this temperature is not directly related to the actual source temperature. It is defined, however, in terms of a bandwidth *df*. The *relative noise temperature* t of a source is just the noise temperature referred to some *standard noise temperature* T_o, which is often taken to be 290° K (62.6° F):

$$t = \frac{T_e}{T_o}. \tag{4.10-17}$$

The relative power gain G_a of the amplifier of Figure 4.12 is given by Equation 4.10-7. It follows from Equation 4.10-12 that the incremental available power gain G_a is

$$G_a = \frac{\varphi_o(f)\, R_S}{\varphi_S(f)\, R_o} \tag{4.10-18}$$

where $\varphi_S(f)$ and $\varphi_o(f)$ are the power spectral densities from the source and from the open-circuited amplifier output, respectively. Suppose now that the amplifier has a complex transfer function $H(\omega)$ (see Equation 4.3-15). It follows from Equation 4.4-7 that $\varphi_o(f)$ is related to $\varphi_S(f)$ by

$$\varphi_o(f) = \varphi_S(f)\left|\frac{Z_i}{Z_i + Z_S}\, H(\omega)\right|^2 \tag{4.10-19}$$

and that Equation 4.10-18 can be written as

$$G_a = \left|\frac{Z_i}{Z_i + Z_S}\, H(\omega)\right|^2 \frac{R_S}{R_o}. \tag{4.10-20}$$

As mentioned previously the amplifier available power gain, whether on an incremental basis or not, depends on the source impedance. Actually, as shown by Equation 4.10-20, the important quantity in this regard is the ratio of impedances in the input circuit.

4.11 Noise Figure

The term *noise figure* was first introduced by Friis [13] as a measure of amplifier noisiness. The *noise figure* F of an amplifier is defined as the ratio of incremental available signal-to-noise power ratio of the amplifier source to incremental available signal-to-noise power ratio of the amplifier output:

$$F = \frac{S_{as}/N_{as}}{S_{ao}/N_{ao}}. \tag{4.11-1}$$

It is understood that the same source is used for calculating both signal-to-noise power ratios. This noise figure can be written in an equivalent form which is sometimes more useful. Let N_{aos} be used to denote the incremental available noise power out of the amplifier due only to the noise from the source. In other words N_{aos} is given by

$$N_{aos} = G_a N_{as}. \tag{4.11-2}$$

Also

$$S_{ao} = G_a S_{as} \tag{4.11-3}$$

since all of the signal output originates in the source. If these substitutions are made in Equation 4.11-1, then the noise figure can be written as

$$F = \frac{N_{ao}}{N_{aos}}. \tag{4.11-4}$$

Since the amplifier is linear the total incremental available output noise power can be separated into two components N_{aos}, due to the source, and N_{aoi}, arising within the amplifier:

$$N_{ao} = N_{aos} + N_{aoi}. \tag{4.11-5}$$

It follows from Equations 4.11-3 and 4.11-4 that

$$F = 1 + \frac{N_{aoi}}{N_{aos}} \tag{4.11-6}$$

so that $F \geq 1$. The size of the second term in Equation 4.11-6 depends on the relative amount of internal noise produced in the amplifier. A good (relatively noiseless) amplifier will have a noise figure near unity whereas a noisy amplifier will have a large noise figure. In terms of the effective noise temperature T_e of the source (Equation 4.10-16), the noise figure can be written as

$$F = 1 + \frac{N_{aoi}}{G_a k T_e \, df}. \tag{4.11-7}$$

If T_e is taken to be the standard noise temperature T_o, then the resulting F is called the *standard noise figure* F_o of the amplifier. It is apparent that N_{aoi} can be eliminated from Equation 4.11-7 to yield

$$F = 1 + \frac{T_o}{T_e}(F_o - 1). \tag{4.11-8}$$

The quantity F is sometimes called the *spot noise figure* since it is defined in terms of a given frequency. The *average noise figure* $\bar{F}$ is defined as a ratio of total available noises; that is,

$$\bar{F} = \frac{\int_0^\infty N_{ao} df}{\int_0^\infty N_{aos} df} \tag{4.11-9}$$

or

$$\bar{F} = \frac{\int_0^\infty F G_a N_{as} df}{\int_0^\infty G_a N_{as} df}. \tag{4.11-10}$$

If the incremental available noise power N_{as} from the source is not a function of frequency, then the average noise figure becomes

$$\bar{F} = \frac{\int_0^\infty FG_a df}{\int_0^\infty G_a df}. \tag{4.11-11}$$

It is sometimes convenient to introduce the idea of the *equivalent noise bandwidth* β of an amplifier. If the input of the amplifier consists of thermal noise from a resistor R, then the available output noise power N_a is

$$N_a = kT \int_0^\infty G_a df \tag{4.11-12}$$

where G_a is the available power gain of the amplifier and might be written as $G_a(f)$. The equivalent noise bandwidth β is defined as

$$\beta = \frac{N_a}{kTG_a(f_o)} \tag{4.11-13}$$

where f_o is some characteristic frequency of the passband of the amplifier. Note that β is the width of a rectangular passband of height $G_a(f_o)$ and total area $\int_0^\infty G_a(f)df$. The situation is illustrated in Figure 4.13 which shows the available power gain for a bandpass amplifier and its equivalent noise bandwidth. In this case an obvious choice for the characteristic frequency f_o is the center frequency. However, it may not always be clear as to the value of f to take as f_o, in which case the equivalent noise bandwidth will not be uniquely defined. A way out of this difficulty is to use

$$\beta = \frac{N_a}{KTG_{a\ \max}} \tag{4.11-14}$$

as a definition of β, where $G_{a\ \max}$ is the maximum value of the available power gain of the amplifier. For the example illustrated in Figure 4.13, the definitions of Equations 4.11-13 and 4.11-14 are equivalent. In many practical

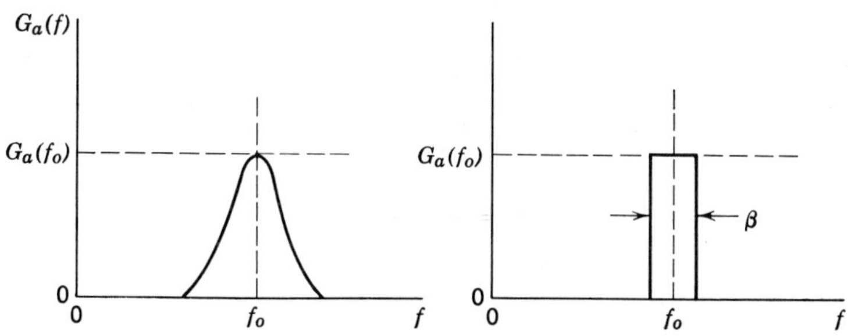

Figure 4.13 *Equivalent noise bandwith* β: $\int_0^\infty G_a(f)\, df = \beta G_a(f_o)$

cases it will turn out that the equivalent noise bandwidth is a good approximation to the usual 3-dB bandwidth.

Example 4.4 Consider the network shown in Figure 4.14(*a*) where $V(t)$ is an ideal voltage source. Thus all noises in the system are thermal noises in the resistors. The noise figure of the network is given by Equation 4.11-6 as

$$F = 1 + \frac{N_{aoi}}{N_{aos}}.$$

The incremental available noise power arising within the amplifier is

$$N_{aoi} = kT_R df$$

where T_R is the effective noise temperature of resistor R as given by Equation 4.10-16. In the same way, the incremental available noise power arising from the source is

$$N_{aos} = kT_{R_S} df$$

and the noise figure is

$$F = 1 + \frac{T_R}{T_{R_S}} = 1 + \frac{R_S}{R}.$$

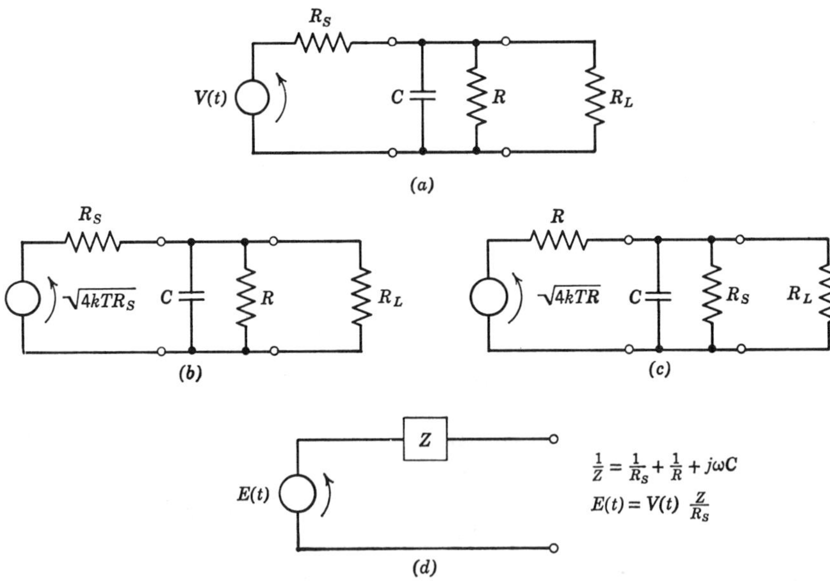

Figure 4.14 *An RC network: (a) network; (b) source noise circuit; (c) internal noise circuit; (d) Thevenin equivalent circuit for network*

The problem may be looked at in another way. The source noise may be considered to arise in a voltage source $\sqrt{4kTR_S}$ in series with a noiseless resistor R_S as shown in Figure 4.14(*b*). The internal noise may be considered to arise in a voltage source $\sqrt{4kTR}$ in series with a noiseless resistor R as shown in Figure 4.14(*c*). The Thevenin equivalent circuit for the network is given by Figure 4.14(*d*). It is apparent from this last figure that the incremental available power (to a matched load Z^*) is

$$P_{ao} = \frac{|E(t)|^2}{4ReZ}.$$

Thus

$$N_{aos} = 4kTR_S \frac{|Z|^2}{R_S^2} \frac{1}{ReZ}$$

and

$$N_{aoi} = 4kTR \frac{|Z|^2}{R^2} \frac{1}{ReZ}$$

or

$$F = 1 + \frac{N_{aoi}}{N_{aos}} = 1 + \frac{R_S}{R}$$

as before. It should be clear, however, that the first method of solving this problem must be used with discretion since the pertinent noisy elements are not always in parallel. Note that the noise figure does not depend on the load impedance, as an examination of Equation 4.11-6 reveals.

● ●

The principal types of noise that arise in electronic amplifiers are discussed in the references [10, 11, 12], and a number of examples of the calculation of noise figure are given there.

4.12 Noise in Cascaded Stages

In many amplifiers involving electronic devices, there will be a number of cascaded stages that are more or less isolated from each other. In each stage there will be internal noises as well as noises from the preceding stages. One of the useful features of our definition of noise figure is the way in which the noise figure of individual stages can be combined to form an overall noise figure.

A cascade of two stages is shown in Figure 4.15. The overall (incremental) available power gain G_a is

$$G_a = \frac{P_{ao2}}{P_{as}} \tag{4.12-1}$$

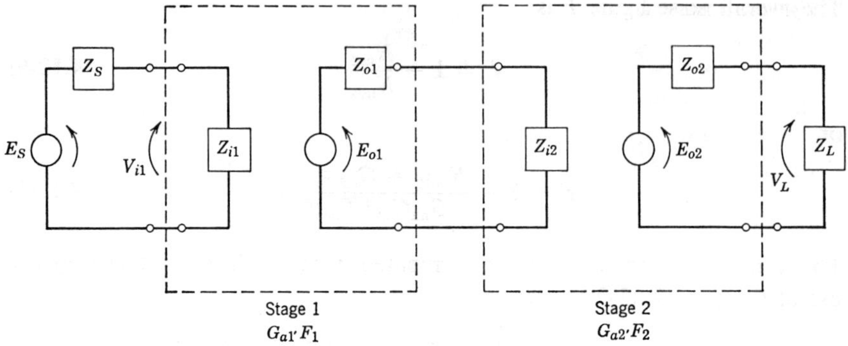

***Figure 4.15** A two-stage linear amplifier*

where P_{ao2} is the (incremental) available power from Stage 2 and P_{as} is the (incremental) available power from the source. However, Equation 4.12-1 can be written as

$$G_a = \frac{G_{a2}P_{ao1}}{P_{as}} = \frac{G_{a2}G_{a1}P_{as}}{P_{as}}$$
$$= G_{a1}G_{a2} \tag{4.12-2}$$

where G_{a1} and G_{a2} are the (incremental) available power gains of Stages 1 and 2, respectively, *with the given source impedances.* In other words G_{a2} is calculated with the given source and Stage 1 driving Stage 2, and G_{a1} is calculated with the given source. It is apparent that, for N stages, the overall (incremental) available power gain is

$$G_a = \prod_{n=1}^{N} G_{an} \tag{4.12-3}$$

where each G_{an} is calculated for the given combination of stages and sources preceding it.

It is clear that, for the two stages of Figure 4.15, the incremental available noise power due to the source is

$$N_{aos} = G_{a1}G_{a2}kT_{es}df \tag{4.12-4}$$

where T_{es} is the effective source temperature. The total incremental available noise power N_{aoi} due to internal sources has two components, one arising from second stage internal noises and one from first stage internal noises:

$$N_{aoi} = N_{aoi2} + G_{a2}N_{aoi1}. \tag{4.12-5}$$

The overall noise figure F is

$$F = 1 + \frac{N_{aoi}}{N_{aos}} \tag{4.12-6}$$

or

$$F = 1 + \frac{N_{aoi2} + G_{a2}N_{aoi1}}{G_{a1}G_{a2}kT_{es}df}. \tag{4.12-7}$$

The terms N_{aoi2} and N_{aoi1} can be eliminated from this last expression by the use of Equation 4.11-7 to yield

$$F = 1 + \frac{(F_2 - 1)G_{a2}kT_{e1} + G_{a2}(F_1 - 1)G_{a1}kT_{es}}{G_{a1}G_{a2}kT_{es}}$$

or

$$F = F_1 + \frac{F_2 - 1}{G_{a1}}\frac{T_{e1}}{T_{es}} \tag{4.12-8}$$

where T_{e1} is the effective noise temperature of Stage 1 with the given source. If the two noise temperatures are equal, then

$$F = F_1 + \frac{F_2 - 1}{G_{a1}}. \tag{4.12-9}$$

In the same way, for N cascaded stages, the overall noise figure (for all of the system at the same temperature) is

$$F = F_1 + \sum_{n=2}^{N} \frac{F_n - 1}{\prod_{m=1}^{n-1} G_{am}}. \tag{4.12-10}$$

Suppose that the mth stage of a cascade of N stages has a high (incremental) available power gain

$$G_{am} \gg 1. \tag{4.12-11}$$

Then, unless succeeding stages have abnormally low gains, the contributions to F from succeeding stages will be small and F will depend principally on the first m stages. This fact will lead, in many practical situations, to considerable simplification of noise figure calculations.

Let us return to the two-stage system of Figure 4.15. A question of some interest is which amplifier should be placed first in the string. If Stage 1 is first as shown in Figure 4.15, then the overall noise figure is

$$F_{12} = F_1 + \frac{F_2 - 1}{G_{a1}}. \tag{4.12-12}$$

On the other hand, if Stage 2 is placed first, then

$$F_{21} = F_2 + \frac{F_1 - 1}{G_{a2}}. \tag{4.12-13}$$

It is desirable to have as low a noise figure as possible. Assume that the first combination is best; that is,

$$F_{12} < F_{21} \tag{4.12-14}$$

or

$$\frac{F_1 - 1}{1 + (1/G_{a1})} < \frac{F_2 - 1}{1 - (1/G_{a2})}. \tag{4.12-15}$$

If this equation is satisfied, then Stage 1 should precede Stage 2; otherwise the stages should be reversed.

It is sometimes convenient [14] to define the *noise measure M* of a linear system as

$$M = \frac{F - 1}{1 - (1/G_a)}. \tag{4.12-16}$$

With this definition, Equation 4.12-15 becomes

$$M_1 < M_2. \tag{4.12-17}$$

The stage with the lowest noise measure should be placed first. If M_{12} is defined as

$$M_{12} = \frac{F_{12} - 1}{1 - (1/G_{a1}G_{a2})} \tag{4.12-18}$$

then

$$M_{12} = M_1 + (M_2 - M_1)\frac{G_{a2} - 1}{G_{a1}G_{a2} - 1}. \tag{4.12-19}$$

If both G_{a1} and G_{a2} are greater than unity so that

$$0 < \frac{G_{a2} - 1}{G_{a1}G_{a2} - 1} \leq 1, \tag{4.12-20}$$

then, for $M_1 < M_2$,

$$M_1 \leq M_{12} \leq M_2. \tag{4.12-21}$$

For the special case where $M_1 = M_2$, it follows that $M_{12} = M_1 = M_2$ and either stage could be placed first.

It is fitting to close this section with a word of warning about noise-figure calculation. The whole subject requires careful attention and is full of pitfalls. This is particularly true in dealing with amplifiers that are not unilateral. In the negative resistance amplifiers such as the maser and the parametric amplifier, there is a noise contribution due to load noise amplified by the negative output resistance of the device. There are some difficulties involved here in noise performance specification which do not seem to be completely resolved.

PROBLEMS

1. Consider the system described by the equation

$$y(t) = a_1 x(t + a_2) + a_3 x\left(\frac{t}{a_4}\right) + a_5|x(t)| + \frac{a_6}{t} x(t) + a_7$$

where $x(t)$ is the input, $y(t)$ is the output, and the a_i are constants. What are the necessary and sufficient conditions on the a_i for the system to be (a) linear, (b) time invariant, (c) causal, (d) stable?

2. Let $x(t)$ be an input with finite energy to a stable, linear, time-invariant system $h(t)$, that is,

$$\int_{-\infty}^{\infty} |h(\tau)|^2 \, d\tau < +\infty \quad \text{and} \quad \int_{-\infty}^{\infty} |x(t)|^2 \, dt < +\infty.$$

Show that the output $y(t)$ has finite energy.

3. Consider the following low-pass filter. If $x(t)$ is a sample function of a random process with a flat spectral density (white noise) $N_0/2$, then find the spectral density and mean-square value of y.

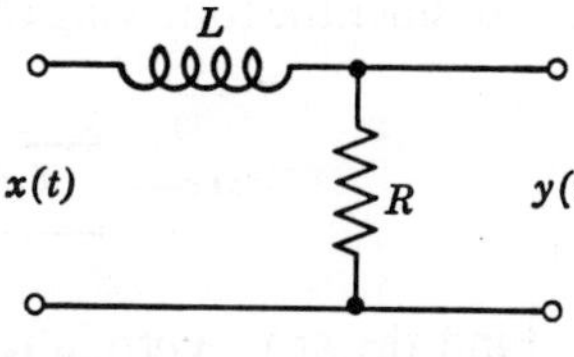

4. Consider the following bridge circuit with white noise with spectral density N_0 as the input.

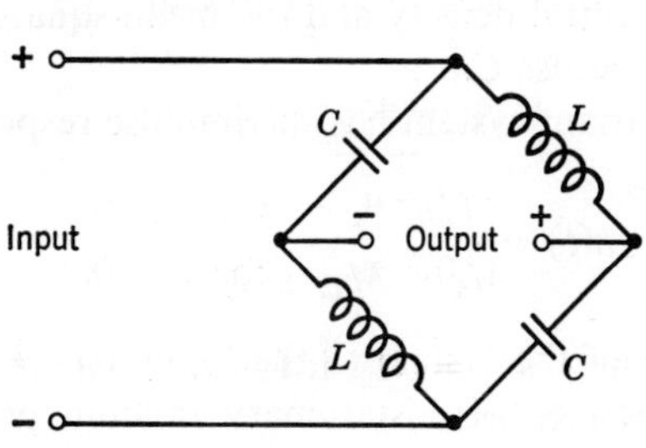

Find the spectral density of the output.

5. Show that if the impulse response of a linear system is not absolutely

integrable, that is, if

$$\int_{-\infty}^{\infty} |h(\tau)|\, d\tau$$

is not finite, then the system is unstable. *Hint:* Use the convolution integral with a properly chosen input.

6. Consider the following linear system composed of a delay line, subtractor, and ideal integrator.

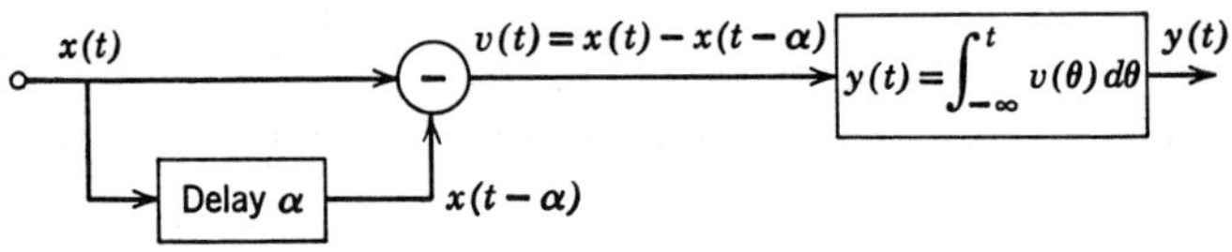

Find the impulse response and transfer function of the system. Prove that the system is stable.

7. Consider the following two linear, time-invariant systems: Express the cross-correlation function $R_{y_1y_2}(\tau)$ of $y_1(t)$ and $y_2(t)$ in terms of $h_1(t)$, $h_2(t)$, and the cross-correlation function $R_{x_1x_2}(\tau)$ of the inputs $x_1(t)$ and $x_2(t)$.

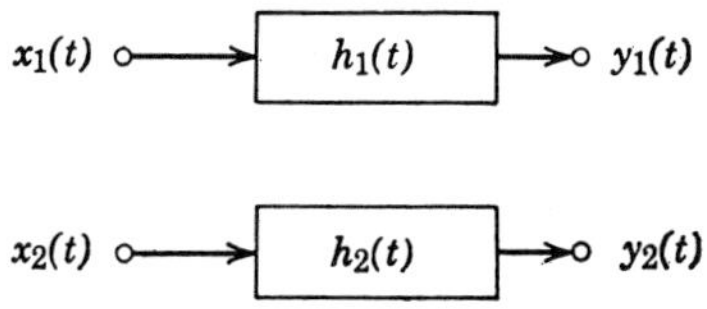

8. Consider the following cascade of two linear, time-invariant systems:

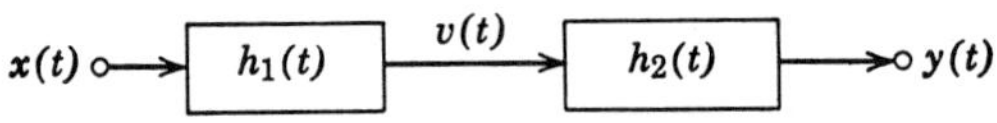

Find the cross-correlation function $R_{vy}(\tau)$ of $v(t)$ and $y(t)$ in terms of $h_1(t)$, $h_2(t)$, and the autocorrelation function $R_{xx}(\tau)$ of the input $x(t)$.

9. Consider a resistor R at temperature T in parallel with a capacitor C. Find the power spectral density and the mean-square value of the noise voltage developed across C.

10. A linear, time-invariant system has the impulse response

$$h(t) = \begin{cases} 2e^{-4t}, & \text{for } t > 0, \\ 0, & \text{for } t < 0. \end{cases}$$

Find the response of the system to the input $x(t) = 10 \cos 4t$.

11. If the sample functions of a stationary random process are periodic, that is, $y(t) = y(t + nT)$, $n = \ldots, -1, 0, 1, 2, \ldots$, show that the process is periodic. Conversely, if a random process $x(t)$ is periodic show that $x(t)$ converges in the mean to $x(t + nT)$.

12. Let $E(t)$ be the envelope function of a narrow-band gaussian process with variance σ^2. Find the mean and variance of $E(t)$ in terms of σ.

13. Consider a linear system with fixed-source impedance Z_S and fixed-load reactance X_L. Find R_L for maximum-power transfer.

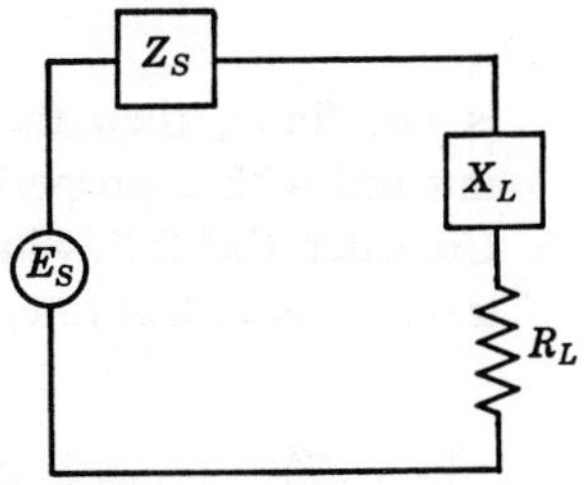

14. Find the available power gain G_a of the following attenuator using the approximation $a \ll 1$.

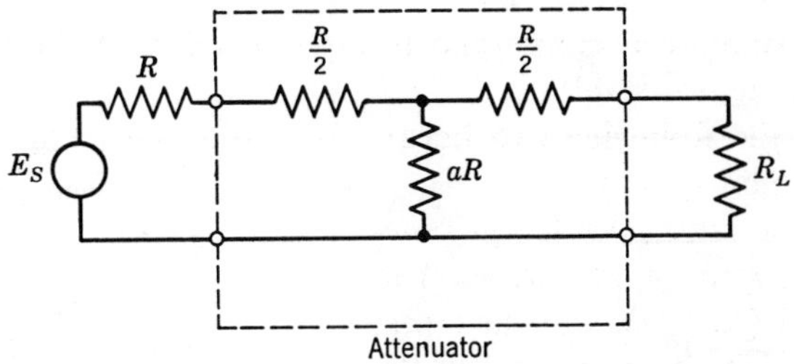

15. The following linear amplifier is a small signal equivalent circuit for a transistor amplifier.

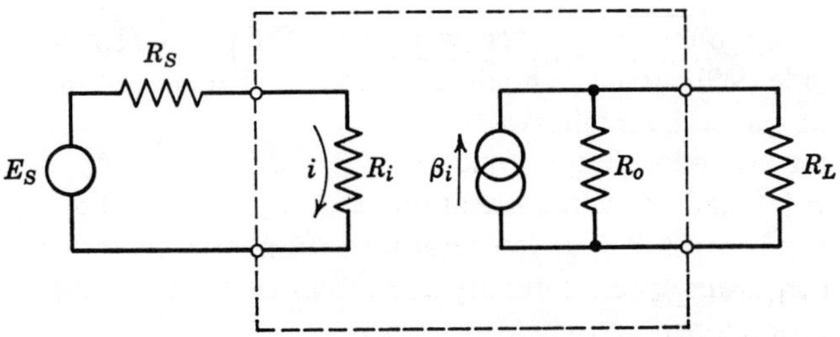

Express the available power gain G_a and the power insertion gain G_I in terms of R_S, R_i, R_o, R_L, and β. Evaluate G_a and G_I for the parameters $R_S = 5k$, $R_i = 100\Omega$, $R_o = 20k$, $R_L = 2k$, and $\beta = 30$.

16. An amplifier has an incremental available noise power N_{aoi} which is directly proportional to the temperature T_a of the amplifier, that is, $N_{aoi} = N_i T_a$. The source noise is thermal noise with an effective noise temperature T_e and the incremental available power gain G_a is a constant A. If the noise figure F is to be less than 2, what is the allowable temperature range for T_a?

17. The incremental available internal noise power N_{aoi} and available power gain G_a of an amplifier are the following functions of frequency:

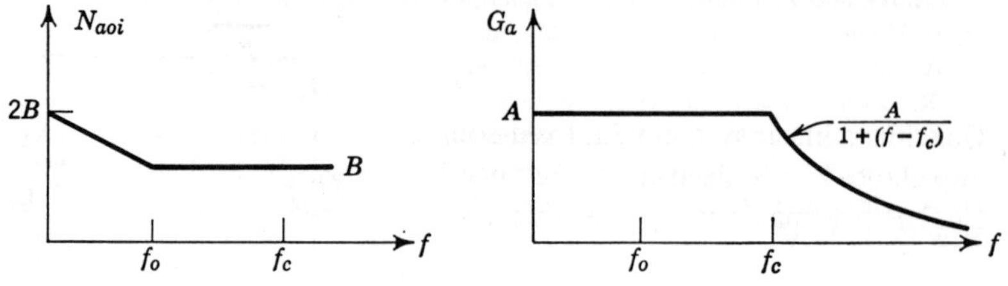

If the source noise is thermal noise with an effective noise temperature T_e, find and sketch the noise figure (spot noise figure) as a function of frequency.

18. Two amplifier stages have the following noise figures and available power gains.

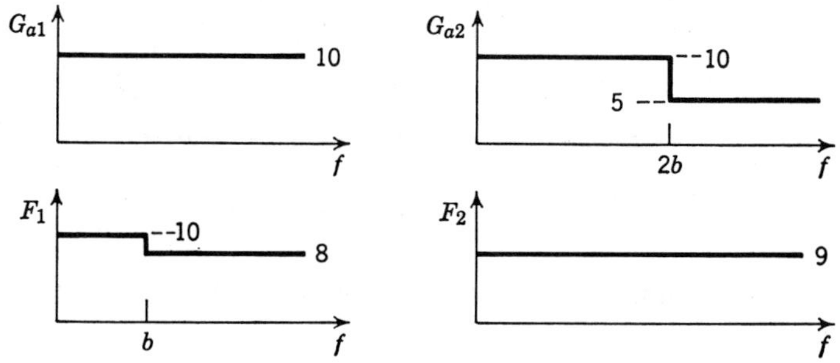

Find the noise measure M for each stage. The two stages are connected in cascade. Which stage should be placed first if the overall spot noise figure at f_o is to be minimized?

REFERENCES

1. M. E. Van Valkenburg, *Network Analysis*, Prentice-Hall, Englewood Cliffs, N. J., 1955.
2. W. Kaplan, *Operational Methods for Linear Systems*, Addison-Wesley, Reading, Mass., 1962.
3. D. K. Cheng, *Analysis of Linear Systems*, Addison-Wesley, Reading, Mass., 1959.
4. W. B. Davenport, Jr., and W. L. Root, *An Introduction to the Theory of Random Signals and Noise*, McGraw-Hill, New York, 1958.
5. D. Middleton, *An Introduction to Statistical Communication Theory*, McGraw-Hill, New York, 1960.
6. E. T. Whittaker and G. N. Watson, *A Course in Modern Analysis*, Cambridge Univ. Press, New York, 1927, Chap. 11.

7. P. M. Morse and H. Feshbach, *Methods of Theoretical Physics*, McGraw-Hill, New York, 1953.
8. E. Jahnke and F. Emde, *Tables of Functions*, Dover, New York, 1947.
9. H. B. Dwight, *Tables of Integrals*, Macmillan, New York, 1947.
10. D. A. Bell, *Electrical Noise*, Van Nostrand, Princeton, N. J., 1960.
11. W. R. Bennett, *Electrical Noise*, McGraw-Hill, New York, 1960.
12. D. K. C. MacDonald, *Noise and Fluctuations*, Wiley, New York, 1962.
13. H. T. Friis, "Noise Figures of Radio Receivers," *Proc. IRE*, **32**, 1944, 419–422.
14. H. A. Haus and R. B. Adler, *Circuit Theory of Linear Noisy Networks*, Wiley, New York, 1959.

5

Optimum Linear Systems

5.1 Optimum Systems

The discussion of the previous chapters has been concerned with the analysis of signals and systems. We study now a much more difficult problem: the *synthesis* or *design* of systems that are *optimum* in some sense. We also might consider the design of optimum signals, that is, signals that are best for some given transmission problem. Actually an important aspect of this latter subject will be discussed in Chapter 8 in the study of *information theory*, where the *encoding* of messages in some optimum form is considered. In this chapter, we restrict our attention to *system design*. Since we are interested principally in systems with random inputs and with criteria of optimization which are statistical in nature, it might be appropriate to call this area, including signal design, by the general term *statistical design* or *statistical optimization theory*.

To proceed with the design of an optimum system, at least three factors must be available:

1. *Specification of the input*—The input may be random or a combination of random and nonrandom signals and noises. At least some minimum knowledge of the characteristics of the input must be available.
2. *Restrictions on system design*—In general, it is necessary that some restricted class of systems be considered—for example, linear, time-varying systems. In addition, if the systems are to be built, they must be physically realizable. (An unrealizable system may be useful, however, as a criterion against which to measure other systems.) In practical situations, cost and complexity must be taken into account.
3. *A criterion of optimization*—A measure of the "goodness" of the system must be adopted. This measure must lead to nontrivial and unique systems and must be meaningful in the sense that it is a reasonable criterion to apply to the particular problem. Most important of all, the criterion must yield equations that are mathematically tractable.

It should be apparent that all three of these factors are interrelated and that useful results will be obtained only if a certain compatibility exists among them.

This chapter will be restricted to the study of optimum *linear* systems since a large class of optimization problems of this type have been solved and since such problems are relatively simple. We begin with the *matched filter* based on a signal-to-noise ratio criterion.

PART 1—THE MATCHED FILTER

5.2 The Matched Filter

It was pointed out in Chapter 4 that the concept of signal-to-noise ratio is useful in the evaluation of system performance, particularly in linear systems. In fact, the *noise figure* of a linear system was defined in terms of the input and output signal-to-noise power ratios.

We now consider a problem of statistical design which is illustrated in Figure 5.1. The following conditions prevail:

1. The input consists of a known signal $s(t)$ and an *additive random* noise $n(t)$. Additional information about these inputs will be supplied as needed. The corresponding output signal and noise are $s_o(t)$ and $n_o(t)$ respectively.
2. The system is linear and time invariant with impulse response $h(t - \tau)$ as discussed in Sections 4.2 and 4.3.
3. The criterion of optimization will be that the output signal-to-noise power ratio R_o be a maximum. Since $n_o(t)$ is random, its mean-squared value $E\{n_0^2(t)\}$ will be used as the output noise power. Thus the criterion is that

$$R_o = \frac{s_0^2(t_1)}{E\{n_0^2(t_1)\}} = \text{maximum} \tag{5.2-1}$$

   at some time t_1.

The problem is to find the system $h(t)$ that accomplishes the maximization of Equation 5.2-1. This optimum system will be called a *matched filter* for reasons that will become clear as the derivation proceeds.

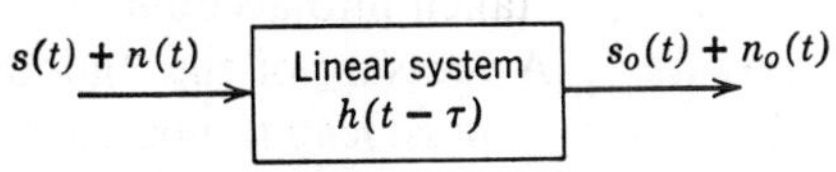

Figure 5.1 *The matched filter*

Since the filter is linear with impulse response $h(t)$, the output signal $s_o(t_1)$ is given by Equation 4.3-18 as

$$s_o(t_1) = \int_{-\infty}^{t_1} s(\tau)h(t_1 - \tau)\, d\tau \tag{5.2-2}$$

and the output noise $n_o(t_1)$ by

$$n_o(t_1) = \int_{-\infty}^{t_1} n(\tau)h(t_1 - \tau)\, d\tau. \tag{5.2-3}$$

The upper limit in the last two integrals is taken to be t_1 rather than ∞ as would be the case if $h(t)$ were physically realizable, that is, if

$$h(t) = 0, \qquad t < 0. \tag{5.2-4}$$

It should be made clear that we desire a physically realizable filter, but it may very well turn out (and will do so in general) that the optimum filter is unrealizable. We will deal with this problem later.

The signal-to-noise output power ratio of Equation 5.2-1 becomes

$$R_o = \frac{\left|\int_{-\infty}^{t_1} s(\tau)h(t_1 - \tau)\, d\tau\right|^2}{\int_{-\infty}^{t_1}\int_{-\infty}^{t_1} R_n(\tau, \sigma)h(t_1 - \tau)h(t_1 - \sigma)\, d\tau\, d\sigma} \tag{5.2-5}$$

where $R_n(\tau, \sigma)$ is the autocorrelation function of the input noise $n(t)$ and is given by

$$R_n(\tau, \sigma) = E\{n(\tau)n(\sigma)\}. \tag{5.2-6}$$

This function must be known in order to proceed. Now only the filter $h(t)$ is unknown in Equation 5.2-5.

The maximization of Equation 5.2-5 with respect to $h(t)$ is particularly simple for the case where the input noise is white; that is, where $n(t)$ is stationary with autocorrelation function $R_n(\tau) = (N_o/2)\,\delta(\tau)$ and power spectral density $\varphi_n(\omega) = N_o/2$. We now consider this special case in detail.

5.3 The Matched Filter for White Noise

For the white noise case, the denominator of Equation 5.2-5 becomes

$$\int_{-\infty}^{t_1}\int_{-\infty}^{t_1} R_n(\tau, \sigma)h(t_1 - \tau)h(t_1 - \sigma)\, d\tau\, d\sigma$$

$$= \frac{N_o}{2}\int_{-\infty}^{t_1}\int_{-\infty}^{t_1} \delta(\tau - \sigma)h(t_1 - \tau)h(t_1 - \sigma)\, d\tau\, d\sigma$$

$$= \frac{N_o}{2}\int_{-\infty}^{t_1} h^2(t_1 - \tau)\, d\tau. \tag{5.3-1}$$

At this point, let us suppose that the condition of physical realizability (Equation 4.3-17) has been met so that $h(t) = 0$ for $t < 0$. Now we can replace the upper limit t_1 in the integral by ∞. We will have to be careful later to ensure that we do not violate our new assumption on $h(t)$. In this case Equation 5.2-5 reduces to

$$R_o = \frac{\left|\int_{-\infty}^{\infty} s(\tau)h(t_1 - \tau)\, d\tau\right|^2}{(N_o/2)\int_{-\infty}^{\infty} h^2(t_1 - \tau)\, d\tau}. \tag{5.3-2}$$

It is useful to convert this last expression to an integral in the frequency domain by writing the complex transfer function $H(\omega)$ of the system where

$$H(\omega) = \int_{-\infty}^{\infty} h(t)\, e^{-j\omega t}\, dt \tag{5.3-3}$$

and

$$h(t) = \frac{1}{2\pi}\int_{-\infty}^{\infty} H(\omega)\, e^{j\omega t}\, d\omega. \tag{5.3-4}$$

If Equation 5.3-4 is substituted into Equation 5.3-2, the result is

$$R_o = \frac{(1/2\pi)\left|\int_{-\infty}^{\infty} H(\omega)S(\omega)\, e^{j\omega t_1}\, d\omega\right|^2}{(N_o/2)\int_{-\infty}^{\infty} |H(\omega)|^2\, d\omega} \tag{5.3-5}$$

where

$$|H(\omega)|^2 = H(\omega)H(-\omega) = H(\omega)H^*(\omega) \tag{5.3-6}$$

and

$$S(\omega) = \int_{-\infty}^{\infty} s(t)\, e^{-j\omega t}\, dt \tag{5.3-7}$$

is the Fourier transform of the known input signal $s(t)$.

The value of $H(\omega)$ which makes Equation 5.3-5 a maximum at time t_1 can be found by applying the *Schwarz Inequality* developed in Appendix E to the numerator of the equation:

$$\left|\int_{-\infty}^{\infty} H(\omega)S(\omega)\, e^{j\omega t_1}\, d\omega\right|^2 \le \int_{-\infty}^{\infty} |H(\omega)|^2\, d\omega \int_{-\infty}^{\infty} |S(\omega)|^2\, d\omega. \tag{5.3-8}$$

It follows that Equation 5.3-5 can be written as

$$R_o \le \frac{2}{N_o}\frac{1}{2\pi}\int_{-\infty}^{\infty} |S(\omega)|^2\, d\omega. \tag{5.3-9}$$

The right side of this inequality does not depend on the filter transfer function $H(\omega)$ but only on the signal and noise energies which must be presumed to be given. Consequently the ratio R_o will be a maximum when $H(\omega)$ is chosen so that the equality holds. For this case $H(\omega)$ assumes its optimum value $H^\dagger(\omega)$ or

$$H^\dagger(\omega)S(\omega)\, e^{j\omega t_1} = H^\dagger(\omega)H^\dagger(-\omega) = S(\omega)S(-\omega). \tag{5.3-10}$$

Hence the matched-filter transfer function is

$$\boxed{H^{\dagger}(\omega) = S(-\omega)\, e^{-j\omega t_1} = S^{*}(\omega)\, e^{-j\omega t_1}} \tag{5.3-11}$$

where $S(-\omega) = S^{*}(\omega)$ is the complex conjugate of the Fourier transform of the input signal $s(t)$.

The matched-filter impulse response is the inverse Fourier transform of $H^{\dagger}(\omega)$ and is given by Equation 4.3-16:

$$h^{\dagger}(t) = \frac{1}{2\pi}\int_{-\infty}^{\infty} S(-\omega)\, e^{-j\omega(t_1 - t)}\, d\omega \tag{5.3-12}$$

or

$$h^{\dagger}(t) = s(t_1 - t). \tag{5.3-13}$$

Early investigators in this field [1, 2, 3] considered only the case of a white noise background. The term "matched filter" arose from the form of Equation 5.3-13 which shows that the impulse response of the optimum linear system must be "matched" to the input signal. The solution is not so simple for a more general noise background and will be discussed in Section 5.4.

Thus far we have ignored the problem of the physical realizability of the filter $h^{\dagger}(t)$. The filter is physically realizable (see Equation 4.3-17) if its impulse response vanishes for negative time; that is, if

$$h^{\dagger}(t) = 0, \qquad t < 0. \tag{5.3-14}$$

In terms of Equation 5.3-13, this condition becomes

$$h^{\dagger}(t) = \begin{cases} 0, & t < 0, \\ s(t_1 - t), & t \geq 0. \end{cases} \tag{5.3-15}$$

If all of the input signal $s(t)$ is to contribute to the output signal component $s_o(t_1)$, then it follows from Equation 5.3-15 that

$$s(t) = 0, \qquad t > t_1. \tag{5.3-16}$$

This last relationship simply states that all of the signal $s(t)$ must have entered the filter $h(t)$ by the time $t = t_1$ at which it is desired to have a maximum signal-to-noise power ratio at the output.

The signal output $s_o(t)$ is given by Equation 5.2-2 as

$$s_o(t) = \int_{-\infty}^{t} s(x) h^{\dagger}(t - x)\, dx \tag{5.3-17}$$

or, from Equation 5.3-13, as

$$s_o(t) = \int_{-\infty}^{t} s(x) s(x - t + t_1)\, dx. \tag{5.3-18}$$

This last expression looks very similar to the time autocorrelation function $\mathscr{R}(\tau)$ defined in Equation 3.4-2 as

$$\mathscr{R}(\tau) = \lim_{T \to \infty} \frac{1}{2T} \int_{-T}^{T} s(x)s(x + \tau)\, dx. \tag{5.3-19}$$

For signals $s(t)$ with finite energy, this last function will be zero in the limit due to the multiplier $1/2T$. It is convenient, therefore, to define a finite time autocorrelation function $\mathscr{R}_F(\tau)$, when it exists, by

$$\mathscr{R}_F(\tau) = \int_{-\infty}^{t_1 - \tau} s(x)s(x + \tau)\, dx, \qquad \tau \geq 0. \tag{5.3-20}$$

A comparison of this definition with Equation 5.3-18 shows that the signal component of the output of the matched filter is

$$s_o(t) = \mathscr{R}_F(t_1 - t). \tag{5.3-21}$$

This function $\mathscr{R}_F(\tau)$ has a maximum at the origin (just as does the ordinary correlation function) and this maximum is

$$s_o(t_1) = \mathscr{R}_F(0) = \int_{-\infty}^{t_1} s^2(x)\, dx = E, \tag{5.3-22}$$

the total input signal energy in the interval $(-\infty, t_1)$. For a normalized input noise of $N_o/2 = 1$, this last expression is just R_o, the output signal-to-noise power ratio.

Let $f(t)$ be all signals with energy E in the interval $(-\infty, t_1)$, that is, all signals for which

$$\int_{-\infty}^{t_1} f^2(t)\, dt = E. \tag{5.3-23}$$

Then it is easily shown that the matched-filter output at time t_1 due to the signal alone is greatest for $f(t) = s(t)$, where $s(t)$ is the signal to which the filter is matched. From Equation 5.2-2, the signal output at t_1 due to the signal $f(t)$ is

$$s_o(t_1) = \int_{-\infty}^{t_1} f(\tau)h^{\dagger}(t_1 - \tau)\, d\tau. \tag{5.3-24}$$

If the filter is matched to $s(t)$, then it follows from Equation 5.3-15 that

$$s_o(t_1) = \int_{-\infty}^{t_1} f(\tau)s(\tau)\, d\tau. \tag{5.3-25}$$

From the Schwarz Inequality of Appendix E, we have

$$\left| \int_{-\infty}^{t_1} f(\tau)s(\tau)\, d\tau \right|^2 \leq \int_{-\infty}^{t_1} f^2(\tau)\, d\tau \int_{-\infty}^{t_1} s^2(\tau)\, d\tau = E^2 \tag{5.3-26}$$

or

$$s_o(t_1) \leq E. \tag{5.3-27}$$

The equality is satisfied when $f(t) = s(t)$ and has already been given by Equation 5.3-22.

Example 5.1 We consider the following signal corrupted by additive white noise:

$$s(t) = \begin{cases} B\, e^{bt}, & t < 0, \\ 0, & t \geq 0. \end{cases}$$

The transform $S(\omega)$ of this signal is

$$S(\omega) = \int_{-\infty}^{0} B\, e^{bt}\, e^{-j\omega t}\, dt = \frac{B}{b - j\omega}.$$

The matched-filter transfer function $H^{\dagger}(\omega)$ is given by Equation 5.3-11 as

$$H^{\dagger}(\omega) = S(-\omega)\, e^{-j\omega t_1} = \frac{B}{b + j\omega}\, e^{-j\omega t_1}$$

and the impulse response by Equation 5.3-13:

$$h^{\dagger}(t) = s(t_1 - t) = \begin{cases} B\, e^{b(t_1 - t)}, & t > t_1, \\ 0, & t < t_1. \end{cases}$$

The physical-realizability requirement that $h^{\dagger}(t)$ vanish for negative time is satisfied by taking $t_1 \geq 0$. The simplest choice is $t_1 = 0$ so that

$$H^{\dagger}(\omega) = \frac{B}{b + j\omega}$$

$$h^{\dagger}(t) = \begin{cases} B\, e^{-bt}, & t > 0, \\ 0, & t < 0. \end{cases}$$

This is just an *RC* low-pass filter preceded by an ideal amplifier. The transform $S_o(\omega)$ of the signal output is

$$S_o(\omega) = S(\omega) H^{\dagger}(\omega) = \frac{B^2}{b^2 + \omega^2}$$

and

$$s_o(t) = \frac{1}{2\pi} \int_{-\infty}^{\infty} \frac{B^2}{b^2 + \omega^2}\, e^{j\omega t}\, d\omega$$

$$= \begin{cases} \dfrac{B^2}{2b}\, e^{bt}, & t < 0, \\[2ex] \dfrac{B^2}{2b}\, e^{-bt}, & t > 0 \end{cases}$$

by contour integration or by pair number 5 of Table 3.1. The same results

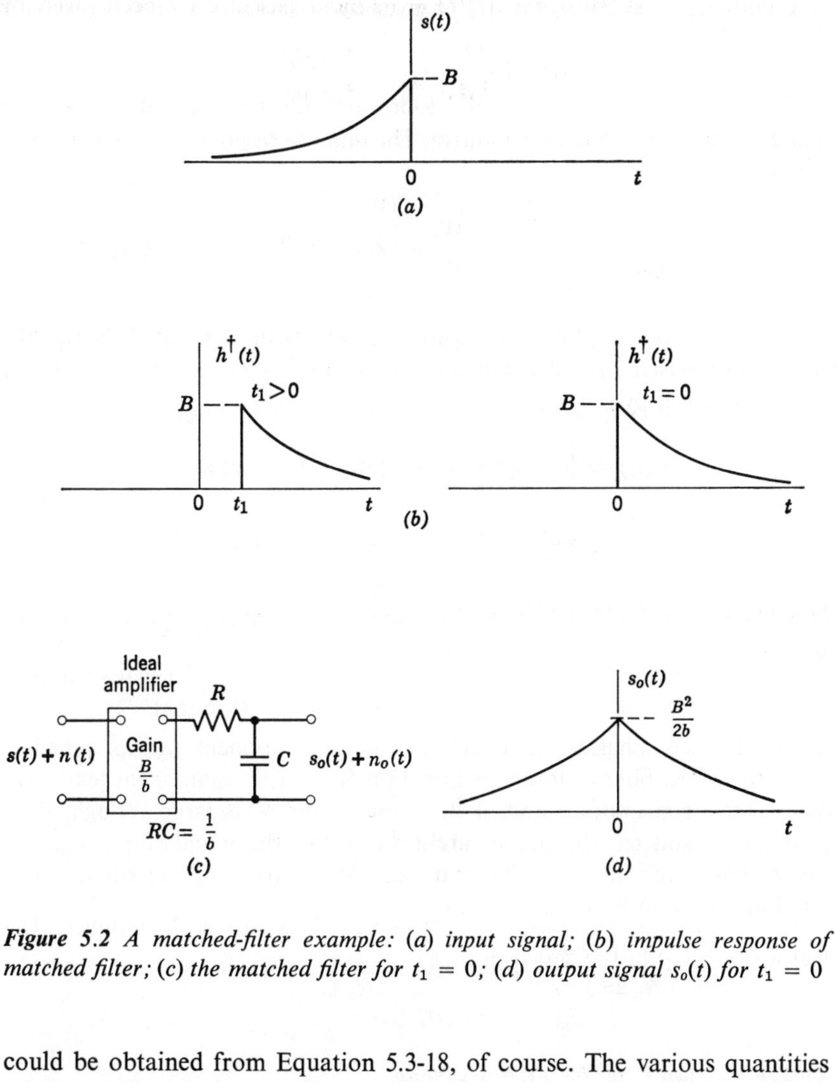

***Figure 5.2** A matched-filter example: (a) input signal; (b) impulse response of matched filter; (c) the matched filter for $t_1 = 0$; (d) output signal $s_o(t)$ for $t_1 = 0$*

could be obtained from Equation 5.3-18, of course. The various quantities relating to the problem are illustrated in Figure 5.2.

• •

The choice of signal has made the foregoing a rather trivial example. For arbitrary signals, however, the construction of the matched filter will not be simple and may involve complicated approximation techniques. Also it may not be possible to choose the time t_1 large enough so that all of the signal will have passed through the filter by t_1. In such cases, the output signal $s_o(t_1)$ will be less than the total signal energy E.

Example 5.2 Let the signal $s(t)$ be given by

$$s(t) = \begin{cases} 0, & t < 0, \\ B\,e^{-bt}, & t \geq 0, \end{cases}$$

and let the noise be white and additive. The impulse response of the matched filter is

$$h^{\dagger}(t) = \begin{cases} s(t_1 - t), & t \geq 0, \\ 0, & t < 0, \end{cases} = \begin{cases} 0, & t > t_1, \\ B\,e^{-b(t_1 - t)}, & 0 \leq t \leq t_1, \\ 0, & t < 0. \end{cases}$$

Both $s(t)$ and $h^{\dagger}(t)$ are plotted in Figure 5.3. For any finite t_1, all of the signal will not have passed through the filter; therefore the output signal is a function of t_1 and is given by

$$\begin{aligned} s_o(t_1) &= \int_{-\infty}^{t_1} s(\tau)h^{\dagger}(t_1 - \tau)\,d\tau = \int_0^{t_1} s^2(\tau)\,d\tau \\ &= B^2\left(\frac{1 - e^{-2bt_1}}{2b}\right). \end{aligned}$$

This function is zero when $t_1 = 0$ and a maximum when $t_1 \to \infty$, as shown in Figure 5.3.

• •

It should be emphasized that there is a noise component $n_o(t)$ present in the output of the filter as given by Equation 5.2-3. This component tends to obscure the signal output, and, if the noise power N_o is large enough, the signal may be undetectable in the output waveform. The actual output signal-to-noise power for the realizable matched filter is given by the equality in Equation 5.3-9 and by Equation 5.3-11 as

$$\frac{2}{N_o}\frac{1}{2\pi}\int_{-\infty}^{\infty} |H^{\dagger}(\omega)|^2\,d\omega = \frac{2}{N_o}\int_{-\infty}^{t_1} s^2(t)\,dt. \tag{5.3-28}$$

5.4 The Matched Filter for Arbitrary Noise

For the general case where the noise is not white, the signal-to-noise output power ratio R_o is given by Equation 5.2-5 and must be maximized by the proper choice of $h(t)$, the filter impulse response. It is apparent from the previous section that the optimum $h(t)$ (that is, the matched filter) will depend on the noise covariance $R_n(\tau, \sigma)$. We have already required that $h(t)$ be time invariant—that is, $h(t)$ instead of $h(t, \tau)$. This last requirement is not compatible in general with a nonstationary noise covariance $R_n(\tau, \sigma)$ and it will turn out that at least wide-sense stationarity $[R_n(\tau - \sigma)]$ must be

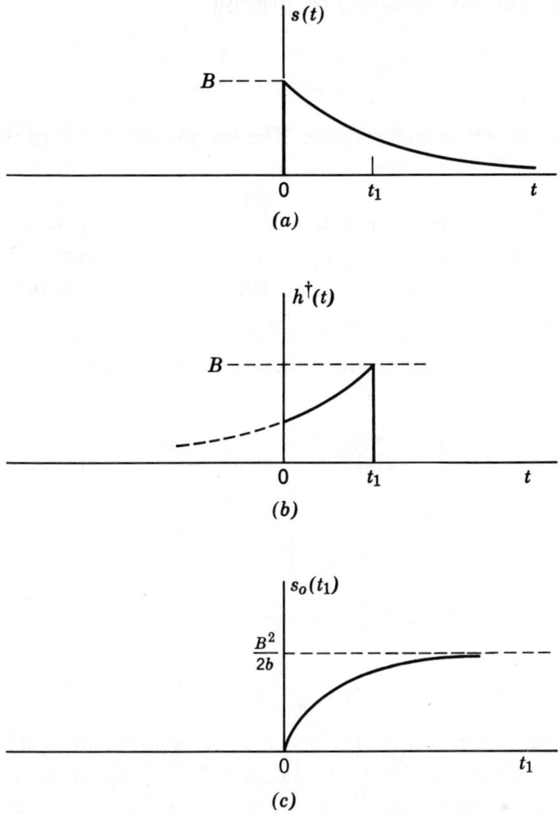

Figure 5.3 Another matched-filter example: (a) Input signal; (b) impulse response of matched filter; (c) output signal at time t_1

imposed on the noise for the matched filter to be realizable. However, we will meet this problem when it arises.

For a given signal $s(t)$ and noise covariance $R_n(\tau, \sigma)$, Equation 5.2-5 will yield a maximum value $R_{o\,\max} = 1/\alpha$ if the optimum filter $h^{\dagger}(t, \tau)$ is used Thus Equation 5.2-5 can be written as

$$E\{n_0^2(t_1)\} - \alpha s_0^2(t_1) \geq 0 \tag{5.4-1}$$

or

$$\int_{-\infty}^{t_1}\int_{-\infty}^{t_1} R_n(\tau, \sigma)h(t_1 - \tau)h(t_1 - \sigma)\,d\tau\,d\sigma - \alpha\left|\int_{-\infty}^{t_1} s(\tau)h(t_1 - \tau)\,d\tau\right|^2 = K \geq 0 \tag{5.4-2}$$

where the equality $K = 0$ holds only if the optimum filter $h^\dagger$ is used. From this point of view, we wish to minimize the first term (the output noise power) subject to a constraint on the second term (the output signal power). This is a conventional problem in the calculus of variations [4, 5, 6] with α a Lagrangian multiplier. A brief discussion of the techniques involved is given in Appendix F; however, the procedures to be used may be outlined and justified as follows.

Suppose $h^\dagger(\cdot)$, the optimum value of $h(\cdot)$, is replaced by $h^\dagger(\cdot) + \gamma\,\delta h(\cdot)$ where γ is a real variable and $\delta h(\cdot)$ is an arbitrary variation in $h^\dagger(\cdot)$. Then the right side of Equation 5.4-2 is a function of γ and can be written as $K(\gamma)$ where

$$\begin{aligned} K(\gamma) = \int_{-\infty}^{t_1}\int_{-\infty}^{t_1} & R_n(\tau,\sigma)[h^\dagger(t_1-\tau)+\gamma\,\delta h(t_1-\tau)] \\ & \times [h^\dagger(t_1-\sigma)+\gamma\,\delta h(t_1-\sigma)]\,d\tau\,d\sigma \\ & - \alpha\left|\int_{-\infty}^{t_1} s(\tau)[h^\dagger(t_1-\tau)+\gamma\,\delta h(t_1-\tau)]\,d\tau\right|^2 \end{aligned}$$

or

$$K(\gamma) = K(0) + 2\gamma A + \gamma^2 B \geq 0 \tag{5.4-3}$$

where $K(0)$ has been defined by Equation 5.4-2 and where

$$\begin{aligned} A = {} & \frac{1}{2}\int_{-\infty}^{t_1} \delta h(t_1-\tau)\int_{-\infty}^{t_1} R_n(\tau,\sigma)h^\dagger(t_1-\sigma)\,d\sigma\,d\tau \\ & + \frac{1}{2}\int_{-\infty}^{t_1} \delta h(t_1-\sigma)\int_{-\infty}^{t_1} R_n(\tau,\sigma)h^\dagger(t_1-\tau)\,d\tau\,d\sigma \\ & - \frac{\alpha}{2}\int_{-\infty}^{t_1} s(\sigma)h^\dagger(t_1-\sigma)\,d\sigma\int_{-\infty}^{t_1} s(\tau)\,\delta h(t_1-\tau)\,d\tau \\ & - \frac{\alpha}{2}\int_{-\infty}^{t_1} s(\tau)h^\dagger(t_1-\tau)\,d\tau\int_{-\infty}^{t_1} s(\sigma)\,\delta h(t_1-\sigma)\,d\sigma \end{aligned} \tag{5.4-4}$$

and

$$\begin{aligned} B = {} & \int_{-\infty}^{t_1}\int_{-\infty}^{t_1} R_n(\tau,\sigma)\,\delta h(t_1-\tau)\,\delta h(t_1-\sigma)\,d\tau\,d\sigma \\ & - \alpha\left|\int_{-\infty}^{t_1} s(\tau)\,\delta h(t_1-\tau)\,d\tau\right|^2. \end{aligned} \tag{5.4-5}$$

If the function $K(\gamma)$ has a minimum, it will occur when $h = h^\dagger$. A necessary condition for this minimum to occur is

$$\frac{\partial}{\partial\gamma}K(\gamma)|_{\gamma=0} = 0, \qquad \text{for all possible } \delta h \tag{5.4-6}$$

or, from Equation 5.4-3,

$$A = 0. \tag{5.4-7}$$

The expression for A given by Equation 5.4-3 can be simplified by noting that

$$R_n(\tau, \sigma) = R_n(\sigma, \tau) \tag{5.4-8}$$

and that

$$\int_{-\infty}^{t_1} s(\sigma)h^\dagger(t_1 - \sigma)\, d\sigma = \int_{-\infty}^{t_1} s(\tau)h^\dagger(t_1 - \tau)\, d\tau = s_o^\dagger(t_1) \tag{5.4-9}$$

is just the optimum output signal at time t_1. Now it is apparent from Equation 5.4-8 that the first two terms in Equation 5.4-4 are equal and from Equation 5.4-9 that the last two terms are equal. The condition $A = 0$ becomes

$$\int_{-\infty}^{t_1} \delta h(t_1 - \tau)\left[\int_{-\infty}^{t_1} R_n(\tau, \sigma)h^\dagger(t_1 - \sigma)\, d\sigma - \alpha s_o^\dagger(t_1)s(\tau)\right] d\tau = 0. \tag{5.4-10}$$

The function $\delta h(t_1 - \tau)$ in this expression was taken to be an *arbitrary* variation in $h^\dagger(t_1 - \tau)$; therefore it might have been taken, for example, to be positive when the term in brackets was positive and negative or zero when the term was negative. From this type of argument, it is clear that Equation 5.4-10 can be satisfied for arbitrary $\delta h(\cdot)$ and all σ in $(-\infty, t_1)$ only if the term in brackets is zero:

$$\int_{-\infty}^{t_1} R_n(\tau, \sigma)h^\dagger(t_1 - \sigma)\, d\sigma = \beta s(\tau), \qquad -\infty < \tau \leq t_1 \tag{5.4-11}$$

where β has been written for $\alpha s_o^\dagger(t_1)$. This *integral equation* must be solved to find $h^\dagger(\cdot)$, the matched filter. Note that $R_n(\tau, \sigma)$, the noise correlation function, and $s(t)$, the input signal, are presumed known. The unknown function $h^\dagger(\cdot)$ occurs in the integrand (hence the term *integral equation*).

We have so far ignored the quantity $\beta = \alpha s_o^\dagger(t_1)$. Suppose that $h^\dagger(t)$ is replaced by $bh^\dagger(t)$ where b is an arbitrary constant. Then the output signal-to-noise power ratio will be unchanged since b^2 will occur in both numerator and denominator of Equation 5.2-5 and will cancel. In other words, the signal-to-noise power ratio is unaffected by the system scale factor. Thus the system can be normalized insofar as absolute gain is concerned so that

$$\beta = \alpha s_o^\dagger(t_1) = 1 \tag{5.4-12}$$

or any other constant for that matter.

As mentioned previously, the restriction that the matched filter be time invariant will require that the input noise be at least wide-sense stationary or that

$$R_n(\tau, \sigma) = R_n(\tau - \sigma) = R_n(\sigma - \tau) = R_n(|\tau - \sigma|). \tag{5.4-13}$$

The derivation could have been carried out in terms of a time-varying filter $h(t, \tau)$. In that case, Equation 5.4-11 would have been obtained with $h^\dagger(t_1 - \sigma)$ replaced by $h^\dagger(t_1, \sigma)$. However, the solution of Equation 5.4-11 would then be prohibitively difficult for this treatment. Consequently, we are interested in solving Equation 5.4-11 with Equations 5.4-12 and 5.4-13 applying so that we have

$$\boxed{\int_{-\infty}^{t_1} R_n(\tau - \sigma)h^\dagger(t_1 - \sigma)\,d\sigma = s(\tau), \qquad -\infty < \tau \leq t_1} \tag{5.4-14}$$

as the equation specifying the matched filter.

For the white noise case, which was considered in Section 5.3, the input noise correlation function is

$$R_n(\tau - \sigma) = \frac{N_o}{2}\,\delta(\tau - \sigma) \tag{5.4-15}$$

and Equation 5.4-14 becomes

$$s(\tau) = \frac{N_o}{2}\,h^\dagger(t_1 - \tau) \tag{5.4-16}$$

or

$$h^\dagger(t) = \frac{2}{N_o}\,s(t_1 - t). \tag{5.4-17}$$

This result agrees with Equation 5.3-13 except for the constant multiplier $2/N_o$, the value of which has been shown to be irrelevant. Note that $h^\dagger(t)$ is physically realizable only if t_1 is large enough as mentioned previously in Section 5.3.

The output signal-to-noise power ratio R_o of Equation 5.2-5 can be found for the matched filter by replacing the impulse response $h(t)$ by $h^\dagger(t)$ obtained from a solution of Equation 5.4-14. Note that the expression for R_o can be simplified by rewriting the denominator with the help of Equation 5.4-14. The result is

$$R_o = \left|\int_{-\infty}^{t_1} s(\tau)h(t_1 - \tau)\,d\tau\right| = |s_o(t_1)| = \frac{1}{\alpha} \tag{5.4-18}$$

which is in agreement with Equation 5.3-22 for the normalized white noise case.

It is worth pointing out that Equation 5.4-14 is a sufficient, as well as necessary, condition for the function $K(\gamma)$ given by Equation 5.4-3 to be a minimum. Notice that, if Equation 5.4-14 is satisfied, then it follows from Equation 5.4-18 that $K(0)$ given by Equation 5.4-2 is zero,

$$K(0) = 0, \qquad \text{for} \quad h(t) = h^\dagger(t), \tag{5.4-19}$$

and that

$$A = 0, \qquad \text{for} \quad h(t) = h^\dagger(t), \tag{5.4-20}$$

from Equation 5.4-7. Then Equation 5.4-3 becomes

$$K(\gamma) = \gamma^2 B \geq 0, \qquad \gamma \neq 0, \tag{5.4-21}$$

and the filter $h^\dagger(t)$ yields a maximum output signal-to-noise power ratio. In other words, replacing $h^\dagger(t)$ by $h^\dagger(t) + \gamma\,\delta(t)$ can only decrease this ratio (or, equivalently, increase K). That $B \geq 0$ in Equation 5.4-21 follows from Equation 5.4-5 on noting that the first term in that equation can be written as

$$\int_{-\infty}^{t_1}\int_{-\infty}^{t_1} R_n(\tau, \sigma)\,\delta h(t_1 - \tau)\,\delta h(t_1 - \sigma)\,d\tau\,d\sigma$$

$$= E\left\{\left[\int_{-\infty}^{t_1} n(\tau)\,\delta h(t_1 - \tau)\,d\tau\right]^2\right\} \geq 0 \quad (5.4\text{-}22)$$

and that the second term must be zero

$$\int_{-\infty}^{t_1} s(\tau)\,\delta h(t_1 - \tau)\,d\tau = 0 \qquad (5.4\text{-}23)$$

since the constraint in Equation 5.4-2 was that the output signal power be held constant.

5.5 The Unrealizable Matched Filter

In this section we consider the solution of Equation 5.4-14 without regard to the realizability of the resulting filter $h^\dagger(t)$. This assumption is equivalent to rewriting Equation 5.4-14 as

$$\int_{-\infty}^{\infty} R_n(\tau - \sigma)h^\dagger(t_1 - \sigma)\,d\sigma = s(\tau), \qquad -\infty < \tau < \infty. \qquad (5.5\text{-}1)$$

Here we have replaced the upper limit t_1 on σ and on τ by ∞. In terms of the original convolutions of Equations 5.2-2 and 5.2-3, we are allowing contributions from the future, that is, we do not require that

$$h(t) = 0, \qquad t < 0.$$

Now the Fourier transform can be taken of both sides of Equation 5.5-1 to yield

$$\int_{-\infty}^{\infty}\int_{-\infty}^{\infty} R_n(\tau - \sigma)h^\dagger(t_1 - \sigma)\,d\sigma\,e^{-j\omega\tau}\,d\tau = \int_{-\infty}^{\infty} s(\tau)\,e^{-j\omega\tau}\,d\tau \qquad (5.5\text{-}2)$$

or, on rearranging,

$$\int_{-\infty}^{\infty} h^\dagger(t_1 - \sigma)\,e^{-j\omega\sigma}\,\varphi_n(\omega)\,d\sigma = S(\omega). \qquad (5.5\text{-}3)$$

After a change of variable $x = t_1 - \sigma$, this becomes

$$e^{-j\omega t_1}\varphi_n(\omega)\int_{-\infty}^{\infty} h^\dagger(x)\,e^{j\omega x}\,dx = S(\omega). \qquad (5.5\text{-}4)$$

Thus

$$H^{\dagger}(-\omega) = \frac{S(\omega)}{\varphi_n(\omega)} e^{j\omega t_1} \tag{5.5-5}$$

or, finally,

$$\boxed{H^{\dagger}(\omega) = \frac{S(-\omega)}{\varphi_n(\omega)} e^{-j\omega t_1}} \tag{5.5-6}$$

since $\varphi_n(\omega) = \varphi_n(-\omega)$. Equation 5.5-6 specifies the unrealizable matched filter for arbitrary (but time-invariant) noise spectrum $\varphi_n(\omega)$. For the white noise case where $\varphi_n(\omega)$ is a constant, this equation reduces to Equation 5.3-11 as expected.

It should be emphasized again that the filter given by Equation 5.5-6 is not necessarily realizable as the following two examples show.

Example 5.3 Let the signal $s(t)$ be given by

$$s(t) = u(-t)(e^{t/2} - e^{3t/2})$$

where $u(t)$ is the unit-step function, and let the noise be wide-sense stationary with power spectral density

$$\varphi_n(\omega) = \frac{4}{1 + 4\omega^2}$$

and autocorrelation function

$$R_n(\tau) = e^{-|\tau|/2}.$$

The transform $S(\omega)$ of the signal is

$$S(\omega) = \frac{2}{1 - j2\omega} - \frac{2}{3 - j2\omega} = \frac{4}{(1 - j2\omega)(3 - j2\omega)}.$$

It follows from Equation 5.5-6 that the matched filter has a transfer function

$$H^{\dagger}(\omega) = \frac{1 - j2\omega}{3 + j2\omega} e^{-j\omega t_1} = \left[-1 + \frac{2}{(\frac{3}{2}) + j\omega}\right] e^{-j\omega t_1}$$

and an impulse response

$$h^{\dagger}(t) = -\delta(t - t_1) + 2\, e^{-(3/2)(t - t_1)} u(t - t_1).$$

This filter is realizable for all $t_1 \geq 0$ as shown in Figure 5.4(*b*) and (*c*). The transform $S_o(\omega)$ of the output signal $s_o(t)$ for $t_1 = 0$ is given by

$$\begin{aligned} S_o(\omega) &= H^{\dagger}(\omega) S(\omega) \\ &= \frac{1 - j2\omega}{3 + j2\omega} \cdot \frac{4}{(1 - j2\omega)(3 - j2\omega)} = \frac{1}{(\frac{9}{4}) + \omega^2}. \end{aligned}$$

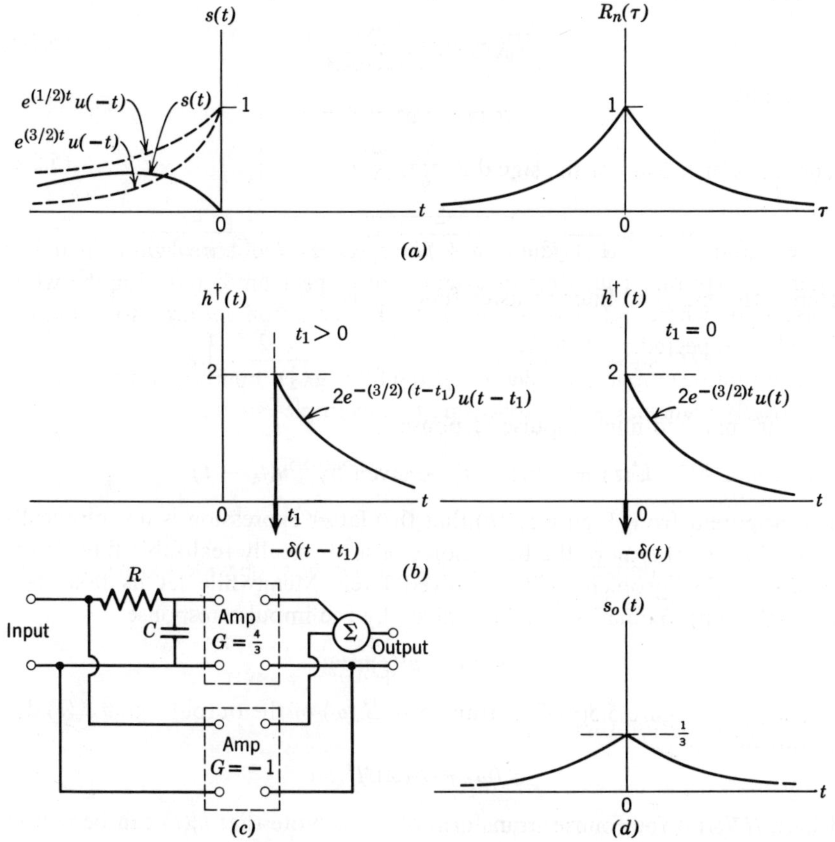

Figure 5.4 *A matched-filter example for nonwhite noise: (a) input signal and input noise correlation function; (b) impulse response of matched filter; (c) matched filter for $t_1 = 0$; (d) output signal $s_o(t)$*

The output signal is then

$$s_o(t) = (\tfrac{1}{3})\, e^{-(3/2)|t|}$$

as shown in Figure 5.4(*d*). For this example the generalization to nonwhite noise has produced little additional complication in filter realization.

• •

Example 5.4 We consider the input signal $s(t)$ given by

$$s(t) = u(t)[e^{-t/2} - e^{-3t/2}].$$

The input noise is the same as in the previous example; that is,

$$\varphi_n(\omega) = \frac{4}{1 + 4\omega^2}$$

$$R_n(\tau) = e^{-|\tau|/2}.$$

The transform $S(\omega)$ of the signal is now

$$S(\omega) = \frac{2}{1 + j2\omega} - \frac{2}{3 + j2\omega} = \frac{4}{(1 + j2\omega)(3 + j2\omega)}.$$

Hence the matched-filter transfer function is

$$H^\dagger(\omega) = \frac{1 + j2\omega}{3 - j2\omega} e^{-j\omega t_1} = \left[-1 + \frac{2}{(\frac{3}{2}) - j\omega}\right] e^{-j\omega t_1}$$

and the matched-filter impulse response is

$$h^\dagger(t) = -\delta(t - t_1) + 2\, e^{(3/2)(t - t_1)} u(t_1 - t).$$

It is apparent from Figure 5.5(*b*) that this latter expression is not physically realizable; furthermore the best choice of a physically realizable filter is not obvious. This problem will be solved later. Meanwhile let us take as a realizable approximation the filter given by the impulse response

$$h_r^\dagger(t) = h^\dagger(t) u(t)$$

as shown in Figure 5.5(*c*). The transform $S_o(\omega)$ of the output signal $s_o(t)$ due to this filter is

$$S_o(\omega) = S(\omega) H_r^\dagger(\omega)$$

where $H_r^\dagger(\omega)$ is the Fourier transform of $h_r^\dagger(t)$. Note that $h_r^\dagger(t)$ can be written as

$$h_r^\dagger(t) = -\delta(t - t_1) + 2\, e^{(3/2)(t - t_1)}[u(t_1 - t) - u(-t)].$$

The transform of this expression is

$$H_r^\dagger(\omega) = \int_0^{t_1 + 0} h_r^\dagger(t)\, e^{-j\omega t}\, dt$$

or

$$H_r^\dagger(\omega) = -e^{-j\omega t_1} + \frac{2}{(\frac{3}{2}) - j\omega} [e^{-j\omega t_1} - e^{-(3/2)t_1}].$$

Since $S(\omega)$ is already given, we have, after some rearranging,

$$S_o(\omega) = e^{-(3/2)t_1}\left[\frac{-1}{(\frac{1}{2}) + j\omega} + \frac{\frac{2}{3}}{(\frac{3}{2}) + j\omega} + \frac{-(\frac{1}{3})}{(\frac{3}{2}) - j\omega}\right]$$

$$+ e^{-j\omega t_1}\left[\frac{\frac{1}{3}}{(\frac{3}{2}) + j\omega} + \frac{\frac{1}{3}}{(\frac{3}{2}) - j\omega}\right].$$

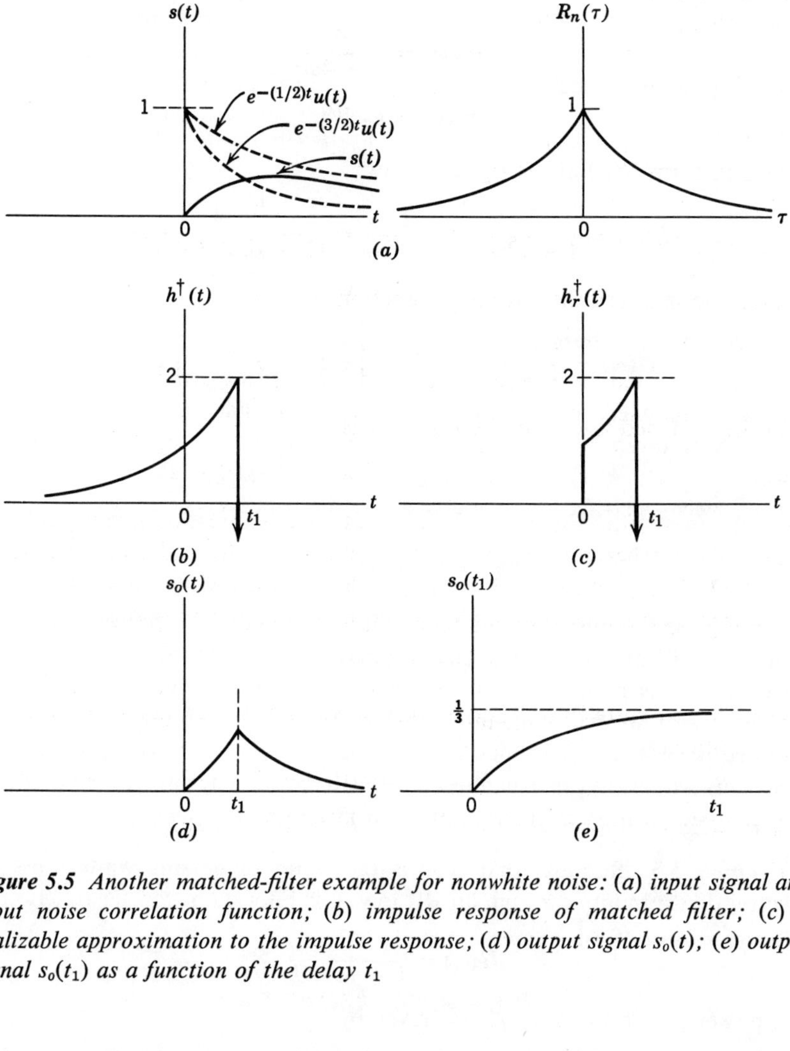

Figure 5.5 *Another matched-filter example for nonwhite noise: (a) input signal and input noise correlation function; (b) impulse response of matched filter; (c) a realizable approximation to the impulse response; (d) output signal $s_o(t)$; (e) output signal $s_o(t_1)$ as a function of the delay t_1*

This expression may be transformed term by term to yield the output signal

$$s_o(t) = -e^{-(3/2)t_1} e^{-(1/2)t}u(t) + (\tfrac{2}{3}) e^{-(3/2)(t_1+t)}u(t) - (\tfrac{1}{3}) e^{-(3/2)(t_1-t)}u(-t) + (\tfrac{1}{3}) e^{-(3/2)|t_1-t|}.$$

This signal is plotted in Figure 5.5(*d*) and does have a maximum value at $t = t_1$; nevertheless there is no guarantee that the procedure which has been used to find a realizable filter is optimum. Suppose we take $t = t_1 \geq 0$; then we have

$$s_o(t_1) = -e^{-2t_1} + (\tfrac{2}{3}) e^{-3t_1} + \tfrac{1}{3}.$$

This function is plotted in Figure 5.5(*e*) versus the delay t_1. As the delay approaches infinity, we would expect the filter to become optimum since, in this case, the impulse response $h^{\dagger}(t)$ of Figure 5.5(*b*) can be realized exactly. For this case the output signal $s_o(t_1)$ is given by Equation 5.3-22 as

$$\lim_{t_1 \to \infty} s_o(t_1) = \int_{-\infty}^{\infty} s^2(x)\,dx = \int_0^{\infty} [e^{-(1/2)x} - e^{-(3/2)x}]^2\,dx$$

$$= [-e^{-x} - (\tfrac{1}{3})\,e^{-3x} + e^{-2x}]_0^{\infty} = \tfrac{1}{3}$$

as expected from Figure 5.5(*e*).

• •

5.6 Spectral Factorization

We are interested ultimately in the solution of Equation 5.4-14 for the realizable matched filter. In this section we consider a topic which is a necessary preliminary to such a solution, namely the factorization of spectral densities.

The power spectral density $\varphi(\omega)$ of a random process is given by

$$\varphi(\omega) = \int_{-\infty}^{\infty} R(\tau)\,e^{-j\omega\tau}\,d\tau \tag{5.6-1}$$

where $R(\tau)$ is the autocorrelation function of the random process (which is assumed to be at least wide-sense stationary). It was pointed out in Section 3.8 that $\varphi(\omega)$ is real, even, and nonnegative for real values of ω. In most problems that we will encounter, this power spectral density $\varphi(\omega)$ can be assumed to be a rational function of ω; that is, the ratio of two polynomials in ω. This will always be true if $\varphi(\omega)$ arises from the filtering of a white noise by a lumped-parameter, linear, time-invariant filter.

Example 5.5 Suppose that $\varphi(\omega)$ is the result of passing white noise of spectral density unity through an *RC* low-pass filter with transfer function

$$H(\omega) = \frac{1}{1 + j\omega RC}.$$

The power spectral density $\varphi(\omega)$ is given by

$$\varphi(\omega) = |H(\omega)|^2 = H(\omega)H(-\omega) = \frac{1}{1 + \omega^2 R^2 C^2}$$

which is a rational function of ω.

• •

Example 5.6 Suppose that $\varphi(\omega)$ is the result of passing white noise of spectral density unity through the lumped-parameter, linear, time-invariant filter with transfer function

$$H(\omega) = K\frac{\omega^n + a_{n-1}\omega^{n-1} + \cdots + a_1\omega + a_0}{\omega^d + b_{d-1}\omega^{d-1} + \cdots + b_1\omega + b_0}.$$

Since the variable ω occurs only as powers of $j\omega$ in the determination of this transfer function, we can write

$$H(\omega) = A(\omega^2) + j\omega B(\omega^2)$$

where $A(\omega^2)$ and $B(\omega^2)$ are real polynomials in ω^2. Thus

$$\varphi(\omega) = |H(\omega)|^2 = H(\omega)H(-\omega) = A^2(\omega^2) + \omega^2 B^2(\omega^2).$$

The output spectral density $\varphi(\omega)$ can now be written as

$$\varphi(\omega) = K^2 \frac{N_n(\omega^2)}{D_d(\omega^2)}$$

where $N_n(\omega^2)$ and $D_d(\omega^2)$ are polynomials of degree n and d, respectively, in ω^2.

• •

In many cases not covered by the previous examples, the correlation function $R(\tau)$ can be closely approximated by a sum of damped exponentials or damped sinusoids

$$R(\tau) \simeq \sum_n A_n \, e^{-a_n|\tau|} \cos b_n\tau \qquad (5.6\text{-}2)$$

where some or all of the b_n may be zero. The corresponding power spectral density is given approximately by the transform of the right side of Equation 5.6-2:

$$\varphi(\omega) \simeq \sum_n a_n A_n \frac{(a_n^2 + b_n^2) + \omega^2}{\omega^4 + 2(a_n^2 - b_n^2)\omega^2 + (a_n^2 + b_n^2)^2} \qquad (5.6\text{-}3)$$

which is of the form

$$\varphi(\omega) = K^2 \frac{N(\omega^2)}{D(\omega^2)} \qquad (5.6\text{-}4)$$

and is a rational function of ω. For the remainder of this discussion it will be assumed that power spectral densities can be written in the form of Equation 5.6-4.

Since $\varphi(\omega)$ is real, then the polynomials $N(\omega^2)$ and $D(\omega^2)$ must have real coefficients. (It has been assumed that common factors have been deleted.) The roots of polynomials with real coefficients either are real or occur in complex conjugate pairs. At this point it is convenient to introduce the complex variable λ where

$$\lambda = \omega + j\sigma \qquad (5.6\text{-}5)$$

so that we can consider the complex λ plane as shown in Figure 5.6. In this plane roots of $N(\lambda^2)$ and $D(\lambda^2)$ must occur in pairs (λ_k, λ_k^*) where

$$\lambda_k = \omega_k + j\sigma_k, \qquad (5.6\text{-}6)$$

$$\lambda_k^* = \omega_k - j\sigma_k, \qquad (5.6\text{-}7)$$

as illustrated in Figure 5.6. Furthermore, since the polynomials $N(\lambda^2)$ and $D(\lambda^2)$ are even, for each root

$$\lambda_k = \omega_k + j\sigma_k \tag{5.6-8}$$

there is a root

$$-\lambda_k = -\omega_k - j\sigma_k \tag{5.6-9}$$

as shown in Figure 5.6. Thus purely real and purely imaginary roots occur in pairs $(\omega_k, -\omega_k)$ and $(j\sigma_k, -j\sigma_k)$, whereas complex roots occur in sets of four $(\lambda_k, \lambda_k^*, -\lambda_k, -\lambda_k^*)$ placed symmetrically with respect to both real and imaginary axes.

If we restrict our attention to random processes $x(t)$ with finite energy, then

$$E\{x^2(t)\} = R(0) = \frac{1}{2\pi}\int_{-\infty}^{\infty} \varphi(\omega)\, d\omega < \infty. \tag{5.6-10}$$

In this case $D(\lambda^2)$ can have no real roots since then $\varphi(\omega)$ would possess a pole on the real axis in violation of Equation 5.6-10. Also, since $\varphi(\omega)$ is nonnegative for all real ω, $N(\omega^2)$ cannot change sign and any real root of $N(\lambda^2)$ must be a multiple root of even order arising from a set of factors of the form

$$(\lambda - \omega_k)^{2r}(\lambda + \omega_k)^{2r}. \tag{5.6-11}$$

From the previous discussion it is apparent that the denominator polynomial $D(\lambda^2)$ can be factored into the form

$$D(\lambda^2) = \prod_k (\lambda - \lambda_k)(\lambda - \lambda_k^*)(\lambda + \lambda_k)(\lambda + \lambda_k^*) \times \prod_m (\lambda - j\sigma_m)(\lambda + j\sigma_m) \tag{5.6-12}$$

while the numerator polynomial can be written as

$$N(\lambda^2) = \prod_p (\lambda - \lambda_p)(\lambda - \lambda_p^*)(\lambda + \lambda_p)(\lambda + \lambda_p^*) \times \prod_q (\lambda - j\sigma_q)(\lambda + j\sigma_q) \times \prod_r (\lambda - \omega_r)^{n_r}(\lambda - \omega_r)^{n_r}(\lambda + \omega_r)^{n_r}(\lambda + \omega_r)^{n_r} \tag{5.6-13}$$

where n_r is an integer whose value depends on the index r. For purposes of notation, let us define λ_k, λ_p, $j\sigma_q$, $j\sigma_m$ to be in the upper-half λ plane (UHP).

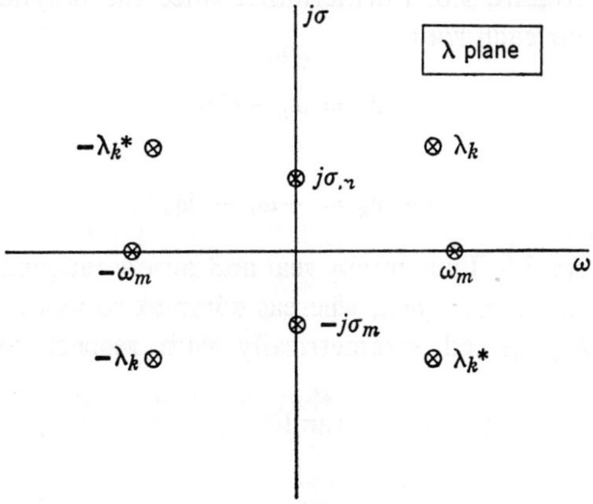

Figure 5.6 The complex λ plane and possible root locations

Note that $-\lambda_k^*$ and $-\lambda_p^*$ will also be in the UHP. Now we factor $D(\lambda^2)$ into two terms $D^+(\lambda)$ and $D^-(\lambda)$ such that

$$D(\lambda^2) = D^+(\lambda)D^-(\lambda) \tag{5.6-14}$$

where $D^+(\lambda)$ has only UHP roots and $D^-(\lambda)$ has only lower-half λ plane (LHP) roots. In the same way,

$$N(\lambda^2) = N^+(\lambda)N^-(\lambda). \tag{5.6-15}$$

If $N(\lambda^2)$ has real roots, they occur in even multiples and half are assigned to each plane. It is now apparent that

$$D^+(\lambda) = \prod_k (\lambda - \lambda_k)(\lambda + \lambda_k^*) \prod_m (\lambda - j\sigma_m) \tag{5.6-16}$$

$$D^-(\lambda) = \prod_k (\lambda - \lambda_k^*)(\lambda + \lambda_k) \prod_m (\lambda + j\sigma_m) \tag{5.6-17}$$

$$N^+(\lambda) = \prod_p (\lambda - \lambda_p)(\lambda + \lambda_p^*) \prod_q (\lambda - j\sigma_q)$$

$$\times \prod_r (\lambda - \omega_r)^{n_r}(\lambda + \omega_r)^{n_r} \tag{5.6-18}$$

and

$$N^-(\lambda) = \prod_p (\lambda - \lambda_p^*)(\lambda + \lambda_p) \prod_q (\lambda + j\sigma_q)$$

$$\times \prod_r (\lambda - \omega_r)^{n_r}(\lambda + \omega_r)^{n_r}. \tag{5.6-19}$$

We have now factored the power spectral density $\varphi(\lambda)$ into the form

$$\varphi(\lambda) = \varphi^+(\lambda)\varphi^-(\lambda) \tag{5.6-20}$$

where

$$\varphi^+(\lambda) = \frac{N^+(\lambda)}{D^+(\lambda)} \tag{5.6-21}$$

has all the zeros and poles of $\varphi(\lambda)$ that are in the UHP and where

$$\varphi^-(\lambda) = \frac{N^-(\lambda)}{D^-(\lambda)} \tag{5.6-22}$$

has all the zeros and poles of $\varphi(\lambda)$ that are in the LHP. Furthermore, $\varphi^+(\lambda)$ and $\varphi^-(\lambda)$ are conjugates so that

$$\varphi^+(\lambda) = [\varphi^-(\lambda)]^*, \tag{5.6-23}$$

$$\varphi^-(\lambda) = [\varphi^+(\lambda)]^*, \tag{5.6-24}$$

and

$$\varphi(\lambda) = \varphi^+(\lambda)\varphi^-(\lambda) = |\varphi^\pm(\lambda)|^2. \tag{5.6-25}$$

Example 5.7 Let the random process $x(t)$ be given as the sum of the two independent random processes $x_1(t)$ and $x_2(t)$, each with zero mean and spectral densities respectively of

$$\varphi_1(\omega) = 2$$

and

$$\varphi_2(\omega) = \frac{1}{1 + \omega^2}.$$

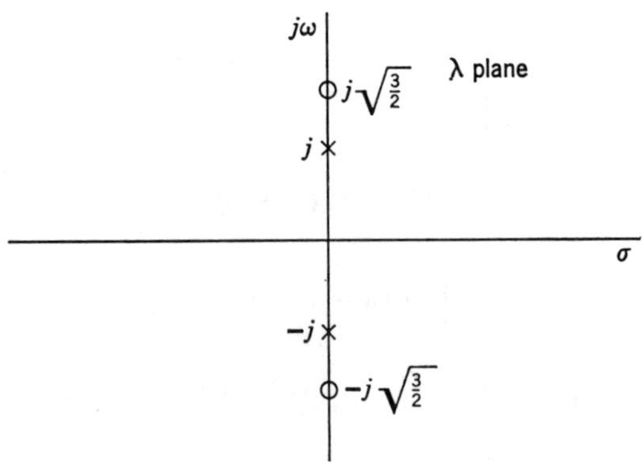

***Figure 5.7** Pole and zero locations of a power spectral density*

In this case the power spectral density of the process $x(t)$ is

$$\varphi(\omega) = \varphi_1(\omega) + \varphi_2(\omega) = \frac{2\omega^2 + 3}{\omega^2 + 1}$$

which may be written as

$$\varphi(\lambda) = \varphi^+(\lambda)\varphi^-(\lambda)$$

where

$$\varphi^+(\lambda) = \sqrt{2}\,\frac{\lambda - j\sqrt{\frac{3}{2}}}{\lambda - j},$$

$$\varphi^-(\lambda) = \sqrt{2}\,\frac{\lambda + j\sqrt{\frac{3}{2}}}{\lambda + j}.$$

Thus φ^+ has a zero at $\lambda = j\sqrt{\frac{3}{2}}$ and a pole at $\lambda = j$ while φ^- has a zero at $\lambda = -j\sqrt{\frac{3}{2}}$ and a pole at $\lambda = -j$ as shown in Figure 5.7.

• •

5.7 Prewhitening or Spectral Shaping

In the previous section we succeeded in factoring the power spectral density in the form

$$\varphi(\omega) = \varphi^+(\omega)\varphi^-(\omega).$$

We now wish to relate the location of the poles of the functions $\varphi^+(\omega)$ and $\varphi^-(\omega)$ to the time-domain behavior of their inverse Fourier transforms $R^-(t)$ and $R^+(t)$ where

$$R^+(t) = \frac{1}{2\pi}\int_{-\infty}^{\infty} \varphi^+(\omega)\, e^{j\omega t}\, d\omega \tag{5.7-1}$$

and

$$R^-(t) = \frac{1}{2\pi}\int_{-\infty}^{\infty} \varphi^-(\omega)\, e^{j\omega t}\, d\omega. \tag{5.7-2}$$

Suppose $f(t)$ is a function that vanishes on the negative half-line, that is,

$$f(t) = 0, \qquad t < 0. \tag{5.7-3}$$

If $f(t)$ is absolutely integrable; that is, if

$$\int_{-\infty}^{\infty} |f(t)|\, dt = \int_0^{\infty} |f(t)|\, dt < \infty, \tag{5.7-4}$$

then the Fourier transform

$$F(\omega) = \int_{-\infty}^{\infty} f(t)\, e^{-j\omega t}\, dt = \int_0^{\infty} f(t)\, e^{-j\omega t}\, dt \tag{5.7-5}$$

exists for all ω in the range $(-\infty, \infty)$. Now $F(\omega)$ may be written in complex form as

$$F(\lambda) = \int_0^\infty f(t)\, e^{-j\lambda t}\, dt = \int_0^\infty [f(t)\, e^{\sigma t}]\, e^{-j\omega t}\, dt \tag{5.7-6}$$

where λ is the complex variable

$$\lambda = \omega + j\sigma. \tag{5.7-7}$$

Clearly, if Equation 5.7-4 holds, then

$$\int_0^\infty |f(t)\, e^{\sigma t}|\, dt < \infty \tag{5.7-8}$$

for all $\sigma \leq 0$; that is, for all values of λ in the LHP. Thus $F(\lambda)$ exists for all λ in the LHP and, hence, has no singularities in the LHP including the ω axis. Since any function $F(\lambda)$ of a complex variable must have at least one singularity, unless the function is a constant or zero, the singularities of $F(\lambda)$ are in the UHP.

A function of $f(t)$, which vanishes on the negative half-line, may not be absolutely integrable but it may happen that

$$\int_0^\infty |f(t)\, e^{\sigma t}|\, dt < \infty, \qquad \sigma < 0. \tag{5.7-9}$$

In this case $F(\lambda)$ may have some of its singularities on the ω axis, although it will have none in the LHP. An example is the time function

$$f(t) = t^n u(t), \qquad n = 1, 2, \ldots, \tag{5.7-10}$$

where $u(t)$ is the unit-step function. The transform $F(\lambda)$ of this function is readily calculated to be

$$F(\lambda) = \int_0^\infty t^n\, e^{\sigma t}\, e^{-j\omega t}\, dt = \frac{n!}{\lambda^{n+1}} \tag{5.7-11}$$

for $\sigma < 0$. Thus $F(\lambda)$ has a pole of order $n + 1$ at the origin.

Suppose $g(t)$ is a function that vanishes on the positive half-line, that is,

$$g(t) = 0, \qquad t > 0. \tag{5.7-12}$$

If $g(t)$ is absolutely integrable so that

$$\int_{-\infty}^0 |g(t)|\, dt < \infty, \tag{5.7-13}$$

then the Fourier transform

$$G(\omega) = \int_{-\infty}^0 g(t)\, e^{-j\omega t}\, dt \tag{5.7-14}$$

exists for all ω in the range $(-\infty, \infty)$. This transform may be written in complex form as

$$G(\lambda) = \int_{-\infty}^{0} [g(t)\, e^{\sigma t}]\, e^{-j\omega t}\, dt. \tag{5.7-15}$$

Again it is clear from Equation 5.7-13 that

$$\int_{-\infty}^{0} |g(t)\, e^{\sigma t}|\, dt < \infty \tag{5.7-16}$$

for all $\sigma \geq 0$; that is, for all values of λ in the UHP including the ω axis. The function $G(\lambda)$ has no singularities in the UHP including the ω axis, and its singularities must be in the LHP. As before, if $g(t)$ is not absolutely integrable but

$$\int_{-\infty}^{0} |g(t)\, e^{\sigma t}|\, dt < \infty \tag{5.7-17}$$

for $\sigma > 0$, then $G(\lambda)$ may have singularities on the ω axis as well as in the LHP.

Thus we have shown that time-domain behavior is related to the location of frequency-domain singularities in the following two ways:

1. A function $\varphi^{+}(\lambda)$ of the complex variable $\lambda = \omega + j\sigma$ having all of its singularities in the upper half λ plane (UHP) has an inverse Fourier transform $R^{+}(t)$ which vanishes for negative time.
2. A function $\varphi^{-}(\lambda)$ of the complex variable $\lambda = \omega + j\sigma$ having all of its singularities in the lower-half λ plane (LHP) has an inverse Fourier transform $R^{-}(t)$ that vanishes for positive time.

We note in passing that, since $\varphi^{+}(\lambda)$ and $\varphi^{-}(\lambda)$ were defined previously to have their zeros as well as their poles in the UHP and LHP, respectively, the reciprocal functions $1/\varphi^{+}(\lambda)$ and $1/\varphi^{-}(\lambda)$ also have transforms that vanish for negative time and positive time respectively. Thus both $\varphi^{+}(\omega)$ and $1/\varphi^{+}(\omega)$ can be considered as the complex transfer functions of physically realizable filters since the corresponding impulse responses vanish for negative time.

The process of *prewhitening* or *spectral shaping* consists of filtering a random process so that the filter output is a normalized white noise; that is, has a constant power spectral density of unity. If the input process $x(t)$ is at least wide-sense stationary with power spectral density $\varphi(\omega)$, then the prewhitening filter $H(\omega)$ is given by

$$1 = \varphi(\omega)|H(\omega)|^2. \tag{5.7-18}$$

The situation is illustrated in Figure 5.8. Equation 5.7-18 may be rewritten as

$$1 = [\varphi^{+}(\omega)H(\omega)][\varphi^{-}(\omega)H(-\omega)]. \tag{5.7-19}$$

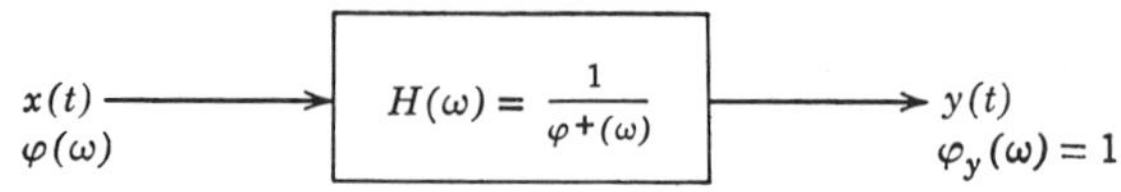

Figure 5.8 *The prewhitening filter*

This equation is satisfied if

$$1 = \varphi^+(\omega)H(\omega) \tag{5.7-20}$$

since, in this case,

$$\varphi^+(\omega)H(\omega) = \varphi^-(\omega)H(-\omega).$$

The prewhitening filter $H(\omega)$ is given by Equation 5.7-20 as

$$H(\omega) = \frac{1}{\varphi^+(\omega)} \tag{5.7-21}$$

and is physically realizable.

5.8 Solution for Matched Filter by Prewhitening Techniques

A solution was found in Section 5.3 for the matched filter when the input was white noise. In the general case we could prewhiten the input noise, changing the input signal in the process of doing this, and then find the matched filter for the white noise input and the new signal. The matched filter for the general case would then be obtained by combining the two filters. It is not clear at the moment that this procedure will yield the true matched filter, that is, the solution of Equation 5.4-14. It will be shown later, however, that this is indeed a valid technique and that the results correspond to the direct solution of Equation 5.4-14.

The approach to be used here can be stated more precisely in the notation of Figure 5.9. A given signal $s(t)$ in additive wide-sense stationary noise $n(t)$ will be passed through a linear filter $H_1(\omega)$ given by

$$H_1(\omega) = \frac{1}{\varphi_n^+(\omega)} \tag{5.8-1}$$

where $\varphi_n^+(\omega)$ is obtained by factoring the power spectral density $\varphi_n(\omega)$ of $n(t)$:

$$\varphi_n(\omega) = \varphi_n^+(\omega)\varphi_n^-(\omega). \tag{5.8-2}$$

The output of this filter consists of two components. One is the signal $s_1(t)$ with transform $S_1(\omega)$ given by

$$S_1(\omega) = H_1(\omega)S(\omega) = \frac{S(\omega)}{\varphi_n^+(\omega)} \tag{5.8-3}$$

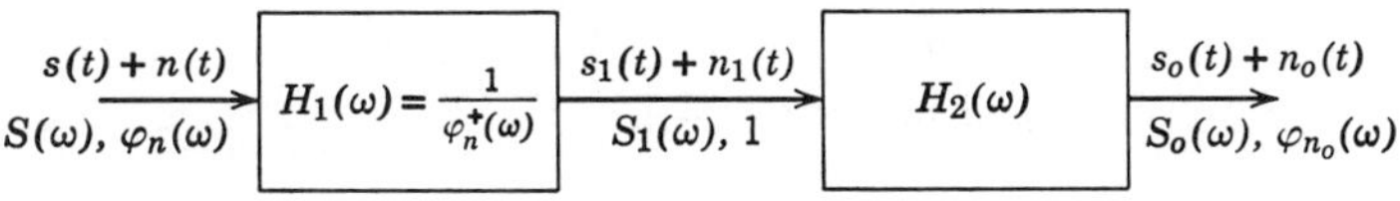

Figure 5.9 *Solution for the matched filter by prewhitening techniques:*

$$H(\omega) = H_1(\omega)H_2(\omega)$$

$$H_2(\omega) = \frac{1}{2\pi}\int_0^{\infty} e^{-j\omega t}\int_{-\infty}^{\infty}\frac{S(-\nu)}{\varphi_n^-(\nu)}\, e^{j\nu(t-t_1)}\, d\nu\, dt$$

where $S(\omega)$ is the transform of the input signal $s(t)$. The other component is the white noise $n_1(t)$ with a power spectral density of unity. The filter $H_2(\omega)$ is constructed from Equation 5.3-11 as follows. Let $H_3(\omega)$ be given by Equation 5.3-11; that is,

$$H_3(\omega) = S_1(-\omega)\, e^{-j\omega t_1} \tag{5.8-4}$$

where t_1 is the desired output observation instant. This filter may or may not be physically realizable, although it generally can be made so if t_1 is taken large enough. We are saying that the impulse response

$$h_3(t) = \frac{1}{2\pi}\int_{-\infty}^{\infty} S_1(-\omega)\, e^{-j\omega t_1}\, e^{j\omega t}\, d\omega \tag{5.8-5}$$

may not vanish for negative time. We define

$$h_2(t) = h_3(t)u(t). \tag{5.8-6}$$

Then the filter with transfer function

$$H_2(\omega) = \int_0^{\infty} h_3(t)\, e^{-j\omega t}\, dt \tag{5.8-7}$$

is physically realizable and the matched filter for the problem is given by

$$H(\omega) = H_1(\omega)H_2(\omega). \tag{5.8-8}$$

Suppose that we ignore the physical realizability of the second filter and take the general matched filter to be

$$H_u(\omega) = H_1(\omega)H_3(\omega). \tag{5.8-9}$$

It follows from Equations 5.8-1, 5.8-3, and 5.8-4 that

$$H_u(\omega) = \frac{S(-\omega)}{\varphi_n^+(\omega)\varphi_n^-(\omega)}\, e^{-j\omega t_1} = \frac{S(-\omega)}{\varphi_n(\omega)}\, e^{-j\omega t_1}. \tag{5.8-10}$$

This last expression is identical to Equation 5.5-6 which was obtained by direct solution of the integral equation (5.4-14) when physical realizability was ignored.

In the interests of completeness, we recapitulate the results of this section. For the input signal $s(t)$ with Fourier transform $S(\omega)$ and additive input noise $n(t)$ with power spectral density $\varphi_n(\omega)$, the realizable matched filter is

$$H(\omega) = \frac{1}{2\pi\varphi_n^+(\omega)} \int_0^\infty e^{-j\omega t} \int_{-\infty}^{\infty} \frac{S(-\nu)}{\varphi_n^-(\nu)} e^{j\nu(t-t_1)}\, d\nu\, dt. \qquad (5.8\text{-}11)$$

This result has been obtained by combining Equations 5.8-1, 5.8-3, 5.8-5, and 5.8-7 into Equation 5.8-8. The symbol ν is used as a variable of integration instead of ω to avoid confusion in evaluating the integrals. In the next section we will show that Equation 5.8-11 is obtained when Equation 5.4-14 is solved directly. It will become clear that the technique used for the solution is simply a slightly disguised version of the technique used in the present section.

Example 5.8 Let us refer again to Example 5.4 and consider the same input signal

$$s(t) = u(t)(e^{-t/2} - e^{-3t/2})$$

and the same input noise $n(t)$ with spectral density

$$\varphi_n(\omega) = \frac{1}{(\frac{1}{4}) + \omega^2}.$$

Both of these are shown in Figure 5.10(a). The noise spectral density may be factored as

$$\varphi_n(\omega) = \varphi_n^+(\omega)\varphi_n^-(\omega) = \frac{1}{(\frac{1}{2}) + j\omega} \frac{1}{(\frac{1}{2}) - j\omega}.$$

Now the prewhitening filter has a complex transfer function

$$H_1(\omega) = \frac{1}{\varphi_n^+(\omega)} = (\tfrac{1}{2}) + j\omega.$$

The signal $s_1(t)$ after this filter has a transform

$$S_1(\omega) = S(\omega)H_1(\omega) = \left[\frac{2}{1 + j2\omega} - \frac{2}{3 + j2\omega}\right] \frac{1 + j2\omega}{2}$$

or

$$S_1(\omega) = 1 - \frac{1 + j2\omega}{3 + j2\omega} = \frac{2}{3 + j2\omega}.$$

Thus, at the output of the prewhitening filter, the signal $s_1(t)$ is given by

$$s_1(t) = u(t)\, e^{-3t/2}$$

and exists in white noise of unity spectral density as shown in Figure 5.10(b). The unrealizable matched filter for this input has a transfer function

$$H_3(\omega) = S_1(-\omega)\, e^{-j\omega t_1} = \frac{2}{3 - j2\omega} e^{-j\omega t_1}$$

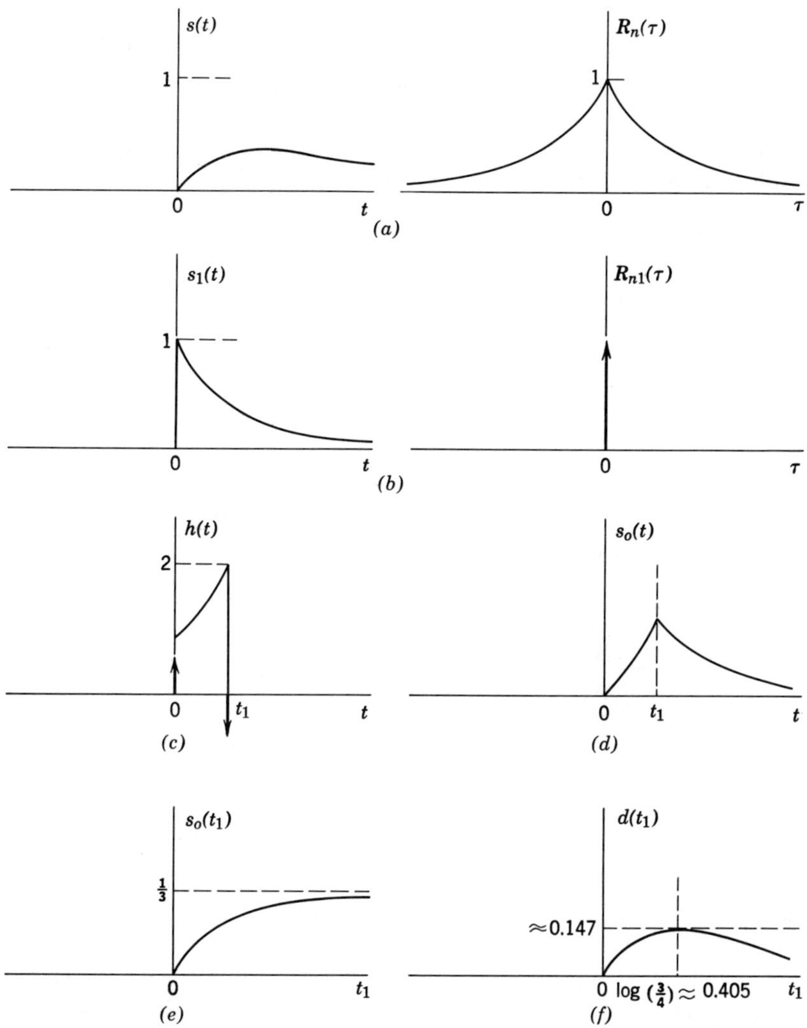

Figure 5.10 *A matched-filter example solved by prewhitening techniques:* (*a*) *input signal and input noise correlation function;* (*b*) *signal and noise correlation function after prewhitening;* (*c*) *impulse response of matched filter;* (*d*) *output signal* $s_o(t)$; (*e*) *output signal* $s_o(t_1)$ *as a function of the delay* t_1; (*f*) *difference between* $s_o(t_1)$ *of this example and of Example* 5.4

and an impulse response

$$h_3(t) = u(t_1 - t)\, e^{(3/2)(t-t_1)}.$$

We take as the realizable filter

$$h_2(t) = u(t)h_3(t)$$

with corresponding transfer function

$$H_2(\omega) = \int_0^{t_1} e^{(3/2)(t-t_1)}\, e^{-j\omega t}\, dt$$

$$= \frac{2}{3 - j2\omega}\,(e^{-j\omega t_1} - e^{-3t_1/2}).$$

Now we can combine $H_1(\omega)$ and $H_2(\omega)$ to form the overall matched filter

$$H(\omega) = H_1(\omega)H_2(\omega) = \frac{1 + j2\omega}{3 - j2\omega}\,(e^{-j\omega t_1} - e^{-3t_1/2})$$

$$= \left(-1 + \frac{4}{3 - j2\omega}\right)(e^{-j\omega t_1} - e^{-3t_1/2}).$$

We may take the inverse transform of $H(\omega)$ term by term to obtain the impulse response

$$h(t) = -\delta(t - t_1) + e^{-3t_1/2}\,\delta(t) + 2\,e^{(3/2)(t-t_1)}[u(t_1 - t) - u(-t)]$$

as shown in Figure 5.10(c).

The output signal $s_o(t)$ has a Fourier transform given by

$$S_o(\omega) = H(\omega)S(\omega) = \frac{4}{(3 - j2\omega)(3 + j2\omega)}\,[e^{-j\omega t_1} - e^{-3t_1/2}]$$

with inverse transform

$$s_o(t) = \tfrac{1}{3}[e^{-(3/2)|t-t_1|} - e^{-3t_1/2}\,e^{-(3/2)|t|}]$$

as shown in Figure 5.10(d). For $t = t_1 \geq 0$ this becomes

$$s_o(t_1) = \tfrac{1}{3}(1 - e^{-3t_1})$$

and is plotted in Figure 5.10(e) versus t_1. It is interesting to compare the quantity $s_o(t_1)$ with the same quantity obtained in Example 5.4. Call them $s_{o1}(t_1)$ and $s_{o2}(t_1)$ respectively. Recall, from Equation 5.4-18, that each of these is the actual signal-to-noise output power ratio at time t_1 as given by Equation 5.2-1 or 5.2-5. This quantity is a maximum for the matched filter. On forming the difference

$$d(t_1) = s_{o1}(t_1) - s_{o2}(t_1) = e^{-2t_1} - e^{-3t_1},$$

we see that this difference is nonnegative, or

$$d(t_1) \geq 0, \qquad \text{all} \quad t_1 \geq 0,$$

as shown in Figure 5.10(f). Note that this difference has a maximum of approximately 0.147 at $t_1 = \log(\frac{3}{2}) = 0.40547\ldots$. It is apparent that the prewhitening realization procedure used in this example yields a larger output signal-to-noise power ratio than the procedure used in Example 5.4 where the unrealizable impulse response corresponding to Equation 5.5-6 was simply truncated to yield a realizable filter.

The solution of this example has corresponded exactly to a step-by-step solution of Equation 5.8-11.

• •

5.9 Direct Solution of the Integral Equation for the Matched Filter

It was pointed out earlier that the difficulty in the solution of integral Equation 5.4-14 for the matched filter arises because the upper limits on σ and τ in this equation are both t_1 instead of ∞. The equation is not valid for $\sigma, \tau > t_1$. We begin the solution of Equation 5.4-14 by requiring that $h^\dagger(t)$ be physically realizable so that

$$h^\dagger(t) = 0, \qquad t < 0. \tag{5.9-1}$$

Then $h(t_1 - \sigma)$ is zero for $\sigma > t_1$ and Equation 5.4-14 becomes

$$\int_{-\infty}^{\infty} R_n(\tau - \sigma)h^\dagger(t_1 - \sigma)\, d\sigma = s(\tau), \qquad -\infty < \tau \le t_1. \tag{5.9-2}$$

After a linear change of variable $t_1 - \sigma = u$ and $t_1 - \tau = t$, this equation becomes

$$\int_{-\infty}^{\infty} R_n(u - t)h^\dagger(u)\, du = s(t_1 - t), \qquad 0 \le t < \infty, \tag{5.9-3}$$

which is a slightly more convenient form for our purposes.

Let us now define the function $g(t)$ from Equation 5.9-3 as

$$g(t) = s(t_1 - t) - \int_{-\infty}^{\infty} R_n(u - t)h^\dagger(u)\, du, \qquad -\infty < t < \infty. \tag{5.9-4}$$

It is apparent that $g(t)$ vanishes for positive t; that is,

$$g(t) = 0, \qquad t \ge 0. \tag{5.9-5}$$

Also the complex Fourier transform $G(\lambda)$ of $g(t)$ has no singularities in the upper-half λ plane (UHP). We take the complex Fourier transform of Equation 5.9-4 and obtain

$$G(\lambda) = S(-\lambda)\, e^{-j\lambda t_1} - \varphi_n(\lambda)H^\dagger(\lambda) \tag{5.9-6}$$

where $\lambda = \omega + j\sigma$. It follows from Equation 5.6-20 that the noise power spectral density $\varphi_n(\lambda)$ may be factored and the equation rearranged to yield

$$\frac{G(\lambda)}{\varphi_n^-(\lambda)} = \frac{S(-\lambda)}{\varphi_n^-(\lambda)}\, e^{-j\lambda t_1} - \varphi_n^+(\lambda)H^\dagger(\lambda). \tag{5.9-7}$$

Let us consider the inverse transform of this equation term by term and write

$$a(t) = b(t) - c(t) \tag{5.9-8}$$

where

$$a(t) = \frac{1}{2\pi}\int_{-\infty}^{\infty} \frac{G(\omega)}{\varphi_n^-(\omega)} e^{j\omega t}\, d\omega, \tag{5.9-9}$$

$$b(t) = \frac{1}{2\pi}\int_{-\infty}^{\infty} \frac{S(-\omega)}{\varphi_n^-(\omega)} e^{-j\omega t_1} e^{j\omega t}\, d\omega, \tag{5.9-10}$$

and

$$c(t) = \frac{1}{2\pi}\int_{-\infty}^{\infty} \varphi_n^+(\omega) H^\dagger(\omega)\, e^{j\omega t}\, d\omega. \tag{5.9-11}$$

It is clear that the term $G(\lambda)/\varphi_n^-(\lambda)$ has no UHP singularities and, hence, that its inverse Fourier transform $a(t)$ must vanish for $t \geq 0$. Therefore, for $t \geq 0$, it follows from Equation 5.9-8 that

$$c(t) = b(t), \qquad t \geq 0. \tag{5.9-12}$$

Furthermore the function $\varphi_n^+(\lambda)H^\dagger(\lambda)$ is free of singularities in the lower-half λ plane (LHP) so that

$$c(t) = 0, \qquad t \leq 0, \tag{5.9-13}$$

or

$$\varphi_n^+(\omega)H^\dagger(\omega) = \int_0^{\infty} c(t)\, e^{-j\omega t}\, dt. \tag{5.9-14}$$

On replacing $c(t)$ by $b(t)$ as given by Equation 5.9-10, we obtain

$$\boxed{H^\dagger(\omega) = \frac{1}{2\pi\varphi_n^+(\omega)}\int_0^{\infty} e^{-j\omega t}\int_{-\infty}^{\infty} \frac{S(-\nu)}{\varphi_n^-(\nu)} e^{j\nu(t-t_1)}\, d\nu\, dt} \tag{5.9-15}$$

which is identical to Equation 5.8-11 which was obtained by prewhitening techniques.

PART 2—OPTIMUM LINEAR FILTERING AND PREDICTION (WIENER FILTERS)

5.10 Formulation of the Least Mean-Squared-Error Filtering and Prediction Problem

Many of the fundamental ideas of statistical optimization theory must be credited to Norbert Wiener and to his book *Extrapolation, Interpolation, and Smoothing of Stationary Time Series* [7]. The material of this book was first published in 1942 as a classified defense report and has had a far-reaching impact on communications and control. Specifically, he considered least mean-squared-error linear filtering and prediction, although the influence of his work has extended over a much wider area.

We formulate the least mean-squared-error linear filtering and prediction problem in terms of the system of Figure 5.11(*a*). Here a random signal $s(t)$ in an additive random noise $n(t)$ is to be sent through a linear filter designed such that its output $s^*(t)$ is an approximation to some desired linear operation $L[s(t)]$ on the input signal $s(t)$. The filter is to be designed so that the least mean-squared-error

$$E\{\epsilon^2(t)\} = E\{|L[s(t)] - s^*(t)|^2\} \tag{5.10-1}$$

is as small as possible. We say that $s^*(t)$ is a least mean-squared-error estimate (l.m.s.e.e.) of $L[s(t)]$. In this treatment, we will restrict the problem somewhat further as shown in Figure 5.11(*b*) although these restrictions are not necessary [8]. We shall require that the desired linear operator $L[s(t)]$ have a complex transfer function

$$L(\omega) = e^{j\omega\alpha} \tag{5.10-2}$$

so that the desired filter output is $s(t + \alpha)$. It is convenient to distinguish three cases:

1. $\alpha > 0$ and $n(t) = 0$; the problem is one of the *pure prediction* of $s(t)$.
2. $\alpha > 0$ and $n(t) \neq 0$; the problem is one of the *prediction and estimation* of $s(t)$.

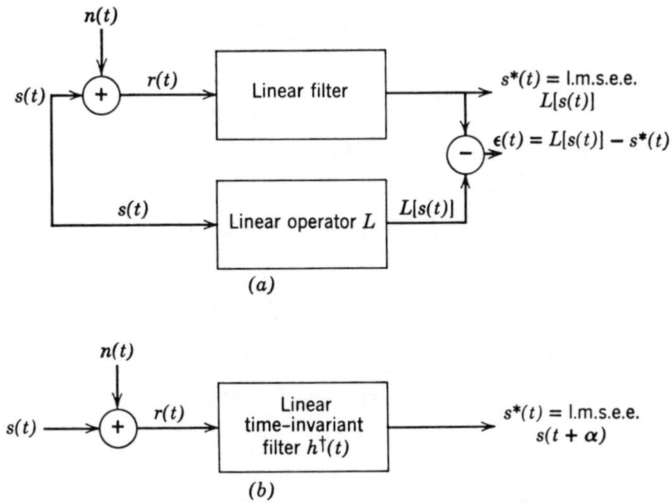

Figure 5.11 *Least mean-squared-error filtering and prediction:* (*a*) *the general problem;* (*b*) *filtering and/or prediction for the time-invariant case*

3. $\alpha \leq 0$ and $n(t) \neq 0$; the problem is one of the *estimation* of $s(t)$ with or without a time delay α in the estimate.

The specific restrictions on the input

$$r(t) = s(t) + n(t) \tag{5.10-3}$$

and on the filter $h^\dagger(t)$ are:

1. Both the signal $s(t)$ and the noise $n(t)$ are sample functions from random processes that are at least wide-sense stationary.
2. The filter, with impulse response $h^\dagger(t)$, is to be physically realizable.
3. The mean-squared error

$$E\{\epsilon^2\} = E\{|s(t + \alpha) - s^*(t)|^2\} \tag{5.10-4}$$

is to be a minimum.

We proceed now to find the filter $h^\dagger(t)$ that satisfies the condition of Equation 5.10-4.

For any linear filter $h(t)$, Equation 5.10-4 can be written as

$$E\{\epsilon^2\} = E\left\{\left| s(t + \alpha) - \int_0^\infty h(\tau)r(t - \tau)\, d\tau \right|^2\right\}. \tag{5.10-5}$$

This expression may be expanded and the expectation operator taken inside the integral to yield

$$E\{\epsilon^2\} = R_{ss}(0) - 2\int_0^\infty R_{rs}(\tau + \alpha)h(\tau)\, d\tau + \int_0^\infty h(\sigma)\int_0^\infty R_{rr}(\sigma - \tau)h(\tau)\, d\tau\, d\sigma \tag{5.10-6}$$

where $R_{ss}(\tau)$ is the autocorrelation function of the signal $s(t)$ and is given by

$$R_{ss}(\tau) = E\{s(t)s(t + \tau)\}; \tag{5.10-7}$$

$R_{rs}(\tau)$ is the cross-correlation function of the total input $r(t)$ and the input signal $s(t)$ and is given by

$$\begin{aligned} R_{rs}(\tau) &= E\{r(t)s(t + \tau)\} = E\{[s(t) + n(t)]s(t + \tau)\} \\ &= R_{ss}(\tau) + R_{ns}(\tau); \end{aligned} \tag{5.10-8}$$

and $R_{rr}(\tau)$ is the autocorrelation function of the total input $r(t)$:

$$\begin{aligned} R_{rr}(\tau) &= E\{r(t)r(t + \tau)\} = E\{[s(t) + n(t)][s(t + \tau) + n(t + \tau)]\} \\ &= R_{ss}(\tau) + R_{ns}(\tau) + R_{sn}(\tau) + R_{nn}(\tau). \end{aligned} \tag{5.10-9}$$

The problem is to choose $h^\dagger(t) = h(t)$ such that Equation 5.10-6 is a minimum.

We follow the procedure previously outlined in Section 5.4 and discussed in Appendix F. We assume that the filter $h^\dagger(t)$ to make Equation 5.10-6 a minimum is known. We replace $h^\dagger(\cdot)$ by $h^\dagger(\cdot) + \gamma\, \delta h(\cdot)$ where γ is a real variable and $\delta h(\cdot)$ is an arbitrary variation in $h^\dagger(\cdot)$. With this substitution the mean-squared error given by Equation 5.10-6 becomes a function of γ and can be written as

$$E\{\epsilon^2(\gamma)\} = J(\gamma) \tag{5.10-10}$$

where

$$J(0) = E\{\epsilon^2(0)\} = E\{\epsilon^2\}. \tag{5.10-11}$$

If the function $J(\gamma)$ has a minimum, it will occur when $h(\cdot) = h^\dagger(\cdot)$. A necessary condition for this minimum to occur is that

$$\frac{\partial}{\partial \gamma} J(\gamma)\bigg|_{\gamma=0} = 0. \tag{5.10-12}$$

Let us use Equation 5.10-6 to write

$$J(\gamma) = R_{ss}(0) - 2\int_0^\infty R_{rs}(\tau + \alpha)[h(\tau) + \gamma\, \delta h(\tau)]\, d\tau$$
$$+ \int_0^\infty [h(\sigma) + \gamma\, \delta h(\sigma)] \int_0^\infty R_{rr}(\sigma - \tau)[h(\tau) + \gamma\, \delta h(\tau)]\, d\tau\, d\sigma. \tag{5.10-13}$$

On noting that $R(\sigma - \tau) = R(\tau - \sigma)$, we can rearrange this expression to give

$$J(\gamma) = J(0) + 2\gamma C + \gamma^2 D \tag{5.10-14}$$

where $J(0)$ has been given by Equation 5.10-6 and where

$$C = \int_0^\infty \delta h(\tau)\left[-R_{rs}(\tau + \alpha) + \int_0^\infty R_{rr}(\tau - \sigma)h(\sigma)\, d\sigma\right] d\tau \tag{5.10-15}$$

and

$$D = \int_0^\infty \delta h(\tau) \int_0^\infty R_{rr}(\sigma - \tau)\, \delta h(\sigma)\, d\sigma\, d\tau. \tag{5.10-16}$$

On applying Equation 5.10-12 to Equation 5.10-14, we have that

$$C = 0 = \int_0^\infty \delta h(\tau)\left[-R_{rs}(\tau + \alpha) + \int_0^\infty R_{rr}(\tau - \sigma)h(\sigma)\, d\sigma\right] d\tau. \tag{5.10-17}$$

As before we argue that the term in brackets must be zero for this relation to be satisfied for arbitrary variation $\delta h(\tau)$; that is,

$$\boxed{\int_0^\infty R_{rr}(\tau - \sigma)h^\dagger(\sigma)\, d\sigma = R_{rs}(\tau + \alpha), \qquad 0 \le \tau < \infty} \tag{5.10-18}$$

is the integral equation that must be solved to obtain the least mean-squared-error filter $h^\dagger(t)$.

We have shown that Equation 5.10-18 is a necessary condition for the mean-squared error $J(\gamma)$ to be a minimum. It is easily shown that it is also sufficient. Note that the quantity D is nonnegative since Equation 5.10-16 can be written as

$$D = E\left\{\left|\int_0^{\infty} r(\tau)\, \delta h(\tau)\, d\tau\right|^2\right\} \geq 0. \tag{5.10-19}$$

Since $C = 0$ if Equation 5.10-18 is satisfied, it follows that

$$J(\gamma) \geq J(0). \tag{5.10-20}$$

In other words the mean-squared error cannot be decreased when $h^{\dagger}(t)$ is allowed to vary from its value determined from Equation 5.10-18.

It is clear from Equation 5.10-18 that the optimum filter will depend only on the correlation functions or power spectral densities of the input signal and noise. Two sets of inputs with the same power spectral densities will require the same optimum filters. This fact is a result of the mean-squared-error criterion and the restriction to a linear filter.

5.11 The Unrealizable Least Mean-Squared-Error Filter

Let us temporarily ignore the physical-realizability requirement and replace the bottom limit of Equation 5.10-18 by $-\infty$ without, however, ensuring that $h^{\dagger}(t) = 0,\ t < 0$. This is the same procedure as that followed in Section 5.5 for the matched filter. We may now transform both sides of Equation 5.10-18 and solve for $H^{\dagger}(\omega)$ to obtain

$$\boxed{H_u^{\dagger}(\omega) = \frac{\varphi_{rs}(\omega)}{\varphi_{rr}(\omega)}\, e^{j\omega\alpha}} \tag{5.11-1}$$

where the subscript u is used to denote the fact that $H_u^{\dagger}(\omega)$ is optimum but unrealizable; that is, it possesses, in general, singularities in the lower-half λ plane and, hence, its inverse Fourier transform does not vanish for negative time. This equation is the equivalent of Equation 5.5-6 for the matched filter.

A special case which occurs frequently is that where the input signal $s(t)$ and the input noise $n(t)$ are uncorrelated so that

$$R_{rr}(\tau) = R_{ss}(\tau) + R_{nn}(\tau), \tag{5.11-2}$$

$$\varphi_{rr}(\omega) = \varphi_{ss}(\omega) + \varphi_{nn}(\omega), \tag{5.11-3}$$

$$R_{rs}(\tau) = R_{ss}(\tau), \tag{5.11-4}$$

$$\varphi_{rs}(\omega) = \varphi_{ss}(\omega). \tag{5.11-5}$$

In this case, Equation 5.11-1 reduces to

$$H_u^{\dagger}(\omega) = \frac{\varphi_{ss}(\omega)}{\varphi_{ss}(\omega) + \varphi_{nn}(\omega)}\, e^{j\omega\alpha}. \tag{5.11-6}$$

For the noiseless case $[\varphi_{nn}(\omega) = 0]$, both Equations 5.11-1 and 5.11-6 become

$$H_u^\dagger(\omega) = e^{j\omega\alpha}, \tag{5.11-7}$$

a pure advance or delay, depending on whether α is positive or negative. This result is certainly to be expected.

The expression for the mean-squared error for the optimum unrealizable filter is given by Equation 5.10-6 with $h(\cdot) = h_u^\dagger(\cdot)$. This equation may be simplified by the use of Equation 5.10-18 to yield

$$\min E\{\epsilon^2\}_u = R_{ss}(0) - \int_{-\infty}^{\infty} R_{rs}(\tau + \alpha) h_u^\dagger(\tau)\, d\tau \tag{5.11-8}$$

or, on transforming,

$$\min E\{\epsilon^2\}_u = \frac{1}{2\pi} \int_{-\infty}^{\infty} [\varphi_{ss}(\omega) - \varphi_{rs}(\omega) H_u^\dagger(-\omega)\, e^{j\omega\alpha}]\, d\omega. \tag{5.11-9}$$

If the expression for $H_u^\dagger(\omega)$, as given by Equation 5.11-1, is substituted into this last expression, the result is

$$\min E\{\epsilon^2\}_u = \frac{1}{2\pi} \int_{-\infty}^{\infty} \frac{\varphi_{ss}(\omega)\varphi_{rr}(\omega) - |\varphi_{rs}(\omega)|^2}{\varphi_{rr}(\omega)}\, d\omega. \tag{5.11-10}$$

For the special case where $s(t)$ and $n(t)$ are uncorrelated, we have

$$\min E\{\epsilon^2\}_u = \frac{1}{2\pi} \int_{-\infty}^{\infty} \frac{\varphi_{ss}(\omega)\varphi_{nn}(\omega)}{\varphi_{ss}(\omega) + \varphi_{nn}(\omega)}\, d\omega. \tag{5.11-11}$$

Example 5.9 Let the signal input $s(t)$ be a wide-sense stationary process with autocorrelation function

$$R_{ss}(\tau) = \frac{a^2}{2b} e^{-|\tau|/b}$$

and power spectral density

$$\varphi_{ss}(\omega) = \frac{a^2}{b^2\omega^2 + 1}$$

where a and b are positive numbers. Let the input noise be independent and white with autocorrelation function

$$R_{nn}(\tau) = c^2\, \delta(\tau)$$

and power spectral density

$$\varphi_{nn}(\omega) = c^2.$$

Then

$$\begin{aligned} \varphi_{rr}(\omega) &= \varphi_{ss}(\omega) + \varphi_{nn}(\omega) \\ &= \frac{(a^2 + c^2) + b^2c^2\omega^2}{b^2\omega^2 + 1}. \end{aligned}$$

The optimum unrealizable filter has a complex transfer function given by Equation 5.11-6 or

$$H_u^\dagger(\omega) = \frac{a^2}{b^2c^2}\frac{e^{j\omega\alpha}}{[(a^2+c^2)/b^2c^2]+\omega^2}$$

which is easily inverted to yield

$$h_u^\dagger(t) = \frac{a^2}{2bc\sqrt{a^2+c^2}}\,e^{-(\sqrt{a^2+c^2}/bc)|t+\alpha|}$$

the impulse response for the least mean-squared-error unrealizable filter. This function is plotted in Figure 5.12 for various values of the prediction time α. It is apparent that the filter is unrealizable unless $\alpha \to -\infty$ (the infinite lag filter).

The mean-squared error can be found from Equation 5.11-11 to be

$$\min E\{\epsilon^2\}_u = \frac{1}{2\pi}\int_{-\infty}^{\infty}\frac{a^2c^2}{(a^2+c^2)+b^2c^2\omega^2}\,d\omega$$

$$= \frac{a^2c}{2b\sqrt{a^2+c^2}}$$

since

$$\int_{-\infty}^{\infty}\frac{a}{a^2+x^2}\,dx = \begin{cases}\pi, & a>0,\\ 0, & a=0,\\ -\pi, & a<0.\end{cases}$$

• •

5.12 Solution of the Integral Equation

The solution of Equation 5.10-18 is readily apparent from the work already done with the matched filter. Notice that Equation 5.10-18 is of the same form as Equation 5.4-14 which specified the matched filter. This similarity

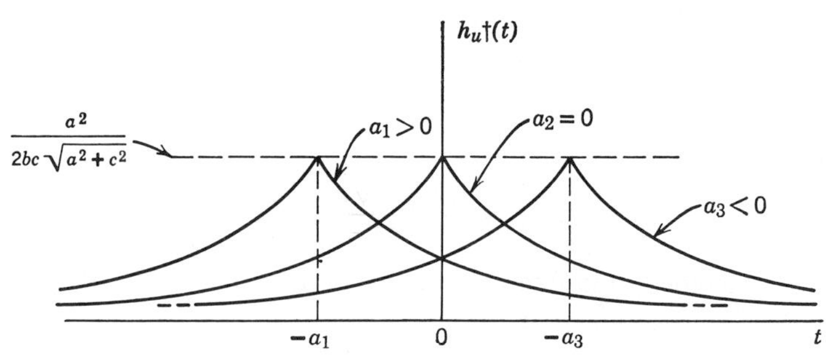

Figure 5.12 An unrealizable impulse response

may be more obvious if we examine Equation 5.9-3 which results after a linear change of variables in Equation 5.4-14. The form of this latter expression is identical to that of Equation 5.10-18 where $R_{rr}(\tau - \sigma)$ corresponds to $R_n(t - u)$, and $R_{rs}(\tau + \alpha)$ corresponds to $s(t_1 - t)$. In Equation 5.9-3 it has already been required that $h^\dagger(\cdot)$ be physically realizable so that the lower limit on the integral can be made $-\infty$ instead of zero, but this assumption must also be made for Equation 5.10-18. It follows, then, that the solution for the least mean-squared-error filter is given by Equation 5.9-15 or, after changing notation,

$$\boxed{H^\dagger(\omega) = \frac{1}{2\pi\varphi_{rr}^{+}(\omega)} \int_0^\infty e^{-j\omega t} \int_{-\infty}^{\infty} \frac{\varphi_{rs}(\nu)}{\varphi_{rr}^{-}(\nu)} e^{j\nu(\alpha+t)}\, d\nu\, dt} \tag{5.12-1}$$

where the input power spectral density $\varphi_{rr}(\omega)$ is given by

$$\varphi_{rr}(\omega) = \mathscr{F}\{R_{rr}(\tau)\} = \varphi_{rr}^{+}(\omega)\varphi_{rr}^{-}(\omega) \tag{5.12-2}$$

$\varphi_{rr}^{+}(\omega)$ has no LHP singularities

$\varphi_{rr}^{-}(\omega)$ has no UHP singularities

and where the cross-spectral density $\varphi_{rs}(\omega)$ is given by

$$\varphi_{rs}(\omega) = \mathscr{F}\{R_{rs}(\tau)\}. \tag{5.12-3}$$

Example 5.10 We consider the same inputs as for Example 5.9 except that we now find the least mean-squared-error *realizable* filter. Recall that the input spectral density was

$$\varphi_{rr}(\omega) = \frac{b^2c^2\omega^2 + (a^2 + c^2)}{b^2\omega^2 + 1}.$$

This spectral density may be factored as

$$\varphi_{rr}(\omega) = \varphi_{rr}^{+}(\omega)\varphi_{rr}^{-}(\omega)$$

where

$$\varphi_{rr}^{+}(\lambda) = \frac{bc\lambda - j\sqrt{a^2 + c^2}}{b\lambda - j}$$

is analytic, bounded, and free of poles and zeros in the lower-half λ plane (LHP), and where

$$\varphi_{rr}^{-}(\lambda) = \frac{bc\lambda + j\sqrt{a^2 + c^2}}{b\lambda + j}$$

is analytic, bounded, and free of poles and zeros in the upper-half λ plane (UHP). We form the function

$$\frac{\varphi_{rs}(\nu)}{\varphi_{rr}^{-}(\nu)} = \frac{\varphi_{ss}(\nu)}{\varphi_{rr}^{-}(\nu)} = \frac{K}{k_1 + j\nu} + \frac{K}{k_2 - j\nu}$$

where

$$K = \frac{a^2}{b(c + \sqrt{a^2 + c^2})}$$

$$k_1 = \frac{1}{b},$$

$$k_2 = \frac{\sqrt{a^2 + c^2}}{bc}.$$

We are now in a position to evaluate the inner integral of Equation 5.12-1. We have

$$\frac{1}{2\pi}\int_{-\infty}^{\infty} \frac{1}{k_1 + j\nu}\, e^{j\nu(t+\alpha)}\, d\nu = u(t + \alpha)\, e^{-k_1(t+\alpha)}$$

and

$$\frac{1}{2\pi}\int_{-\infty}^{\infty} \frac{1}{k_2 - j\nu}\, e^{j\nu(t+\alpha)}\, d\nu = u(-t - \alpha)\, e^{k_2(t+\alpha)}$$

from pair number 3 of Table 3.1 or from contour integration techniques. Hence

$$\frac{1}{2\pi}\int_{-\infty}^{\infty} \frac{\varphi_{rs}(\nu)}{\varphi_{rr}^{-}(\nu)}\, e^{j\nu(t+\alpha)}\, d\nu = u(-t - \alpha)\, e^{k_2(t+\alpha)} + u(t + \alpha)\, e^{-k_1(t+\alpha)}. \quad \text{(A)}$$

This expression is similar to Figure 5.12 except that the two sides of the exponential have different time constants $1/k_1$ and $1/k_2$. We must now perform the last integration in Equation 5.12-1. It is clear from Figure 5.12 that we may distinguish three cases, namely those where $\alpha = 0$, $\alpha < 0$, $\alpha > 0$.

Case I—Zero-Lag Filter ($\alpha = 0$): We need only consider the term

$$u(t)\, e^{-k_1 t}$$

on the right side of Equation A since the other term

$$u(-t)\, e^{k_2 t}$$

vanishes for positive time. The integration is easily carried out to give

$$\int_0^{\infty} e^{-j\omega t}\, e^{-k_1 t}\, dt = \frac{1}{k_1 + j\omega}$$

or, finally,

$$H^{\dagger}(\omega) = \frac{1}{\varphi_{rr}^{+}(\omega)} \frac{K}{k_1 + j\omega} = \frac{k_1 + j\omega}{c(k_2 + j\omega)} \frac{K}{k_1 + j\omega} = \frac{K}{(k_2 + j\omega)c}$$

$$= \frac{a^2}{c + \sqrt{a^2 + c^2}} \frac{1}{\sqrt{a^2 + c^2} + j\omega bc}.$$

This last expression is the complex transfer function of an *RC* low-pass filter preceded by an ideal attenuator as shown in Figure 5.13.

Case II—Lag Filter ($\alpha < 0$): On examining Equation A, we conclude that the t integral in Equation 5.12-1 can be written as

$$\int_0^{-\alpha} e^{k_2(t+\alpha)}\, e^{-j\omega t}\, dt + \int_{-\alpha}^{\infty} e^{-k_1(t+\alpha)}\, e^{-j\omega t}\, dt. \tag{B}$$

It should be kept in mind that $\alpha < 0$. Perhaps it would be clearer to define a positive number β by

$$\beta = -\alpha > 0.$$

Now expression B becomes

$$\int_0^{\beta} e^{(k_2 - j\omega)t}\, e^{-k_2\beta}\, dt + \int_{\beta}^{\infty} e^{-(k_1 + j\omega)t}\, e^{k_1\beta}\, dt$$

$$= \frac{(k_1 + k_2)\, e^{-j\omega\beta} - (k_1 + j\omega)\, e^{-k_2\beta}}{(k_2 - j\omega)(k_1 + j\omega)}.$$

Finally, the optimum filter is obtained by multiplying this last expression by $K/\varphi_{rr}^{+}(\omega)$. After simplification, we obtain

$$H^{\dagger}(\omega) = \frac{a^2}{c^2}\frac{k_1^2}{k_1 + k_2}\frac{(k_1 + k_2)\, e^{-j\omega\beta} - (k_1 + j\omega)\, e^{-k_2\beta}}{k_2^2 + \omega^2}$$

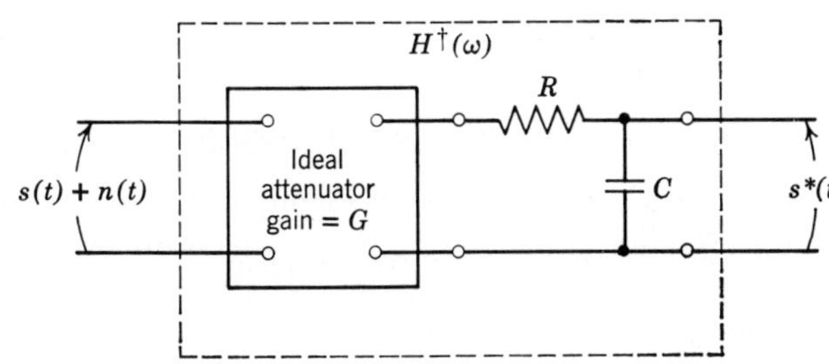

Figure 5.13 *A least mean-squared-error zero-lag filter:*

$$H^{\dagger}(\omega) = \frac{G}{1 + j\omega RC}$$

$$G = \frac{a^2}{a^2 + c^2 + c\sqrt{a^2 + c^2}} \leq 1$$

$$RC = \frac{bc}{\sqrt{a^2 + c^2}}.$$

where β is a positive number and is the desired delay. This filter is not a rational function of ω and cannot by synthesized exactly with a finite number of lumped, linear elements, although it may be approximated arbitrarily closely. For $\beta = \alpha = 0$, the expression reduces to that obtained in Case I.

Case III—Prediction Filter ($\alpha > 0$): It is clear from Equation A that only the term

$$e^{-k_1(t+\alpha)}$$

exists for positive time. Consequently we have

$$\int_0^\infty e^{-k_1(t+\alpha)}\, e^{-j\omega t}\, dt = \frac{e^{-k_1\alpha}}{k_1 + j\omega}.$$

The transfer function of the least mean-squared-error prediction filter is

$$H^\dagger(\omega) = \frac{K}{\varphi_{rr}^+(\omega)}\frac{e^{-k_1\alpha}}{k_1 + j\omega}$$

or

$$\begin{aligned} H^\dagger(\omega) &= \frac{K}{c}\frac{k_1 + j\omega}{k_2 + j\omega}\frac{e^{-k_1\alpha}}{k_1 + j\omega} \\ &= e^{-k_1\alpha}\frac{a^2}{a^2 + c^2 + c\sqrt{a^2 + c^2}}\frac{1}{1 + j\omega(bc/\sqrt{a^2 + c^2})}. \end{aligned}$$

It is convenient to rewrite this last expression as

$$H^\dagger(\omega) = G_1(\alpha)\frac{G}{1 + j\omega RC}$$

where $G_1(\alpha)$ is an attenuation factor given by

$$G_1(\alpha) = e^{-k_1\alpha}$$

and where G and RC have been defined in Figure 5.13. Notice that $G_1(0)$ is unity and that the optimum prediction filter is just the zero-lag filter of Case I and Figure 5.13 with another series attenuator $G_1(\alpha)$. As $\alpha \to \infty$, the transfer function approaches zero.

• •

5.13 The Mean-Squared Error

A general expression for the mean-squared error for the least mean-squared-error filter is given by Equation 5.10-6 when $h(\cdot)$ is replaced by $h^\dagger(\cdot)$ as given by Equation 5.10-18. The result is

$$\min E\{\epsilon^2\} = R_{ss}(0) - \int_0^\infty R_{rs}(\tau + \alpha)h^\dagger(\tau)\, d\tau \qquad (5.13\text{-}1)$$

corresponding to Equation 5.11-8 for the unrealizable filter. After transforming, Equation 5.13-1 becomes

$$\min E\{\epsilon^2\} = \frac{1}{2\pi}\int_{-\infty}^{\infty} [\varphi_{ss}(\omega) - \varphi_{rs}(\omega)H^{\dagger}(-\omega)\, e^{j\omega\alpha}]\, d\omega \qquad (5.13\text{-}2)$$

corresponding to Equation 5.11-9. For the case where the signal $s(t)$ and the noise $n(t)$ are uncorrelated so that Equations 5.10-25 and 5.10-27 apply, the last expression may be written as

$$\min E\{\epsilon^2\} = \frac{1}{2\pi}\int_{-\infty}^{\infty} \varphi_{ss}(\omega)[1 - H^{\dagger}(-\omega)\, e^{j\omega\alpha}]\, d\omega. \qquad (5.13\text{-}3)$$

Actually, in some ways, it is more instructive to transform Equation 5.10-6 term by term without using Equation 5.10-18. The result, in the general case, is

$$\min E\{\epsilon^2\} = \frac{1}{2\pi}\int_{-\infty}^{\infty} [\varphi_{ss}(\omega) - 2\varphi_{rs}(\omega)H^{\dagger}(-\omega)\, e^{j\omega\alpha} + \varphi_{rr}(\omega)|H^{\dagger}(\omega)|^2]\, d\omega \qquad (5.13\text{-}4)$$

and, for uncorrelated signal and noise,

$$\min E\{\epsilon^2\} = \frac{1}{2\pi}\int_{-\infty}^{\infty} [\varphi_{ss}(\omega)|1 - H^{\dagger}(-\omega)\, e^{j\omega\alpha}|^2 + \varphi_{nn}(\omega)|H^{\dagger}(\omega)|^2]\, d\omega. \qquad (5.13\text{-}5)$$

These last two equations are just slightly disguised forms of Equations 5.13-2 and 5.13-3, respectively. However, in Equation 5.13-5, two distinct error terms are readily identified. The first part of the integral, involving the integrand

$$\varphi_{ss}(\omega)|1 - H^{\dagger}(-\omega)\, e^{j\omega\alpha}|^2,$$

gives the error that results from the filter's distortion of the signal. The second part, involving the integrand

$$\varphi_{nn}(\omega)|H^{\dagger}(\omega)|^2,$$

gives the error that results from the input noise.

Let us return to Equation 5.13-1 for the minimum mean-squared error. We will call the last term $G(\alpha)$ and show that it is nonnegative, that is,

$$G(\alpha) = \int_{0}^{\infty} R_{rs}(\tau + \alpha)h^{\dagger}(\tau)\, d\tau \geq 0 \qquad (5.13\text{-}6)$$

for all α. This is easily seen by comparing Equations 5.13-1 and 5.10-6. From these equations, it is clear that

$$G(\alpha) = \int_0^\infty R_{rs}(\tau + \alpha)h^\dagger(\tau)\, d\tau$$

$$= \int_0^\infty h^\dagger(\sigma) \int_0^\infty R_{rr}(\sigma - \tau)h^\dagger(\tau)\, d\tau\, d\sigma$$

$$= E\left\{\left[\int_0^\infty r(\tau)h^\dagger(\tau)\, d\tau\right]^2\right\} \geq 0, \tag{5.13-7}$$

which proves Equation 5.13-6. Suppose we now rewrite Equation 5.13-1 with a linear change of variable $x = \tau + \alpha$ and note that min $E\{\epsilon^2\}$ must be nonnegative by definition:

$$\min E\{\epsilon^2\} = R_{ss}(0) - \int_\alpha^\infty R_{rs}(x)h(x - \alpha)\, dx \geq 0. \tag{5.13-8}$$

Now this expression is nonnegative and is the difference of two nonnegative terms, $R_{ss}(0)$ and $G(\alpha)$; therefore it will be a minimum when the term $G(\alpha)$ is a maximum and a maximum when $G(\alpha)$ is a minimum. It will be an absolute maximum when $G(\alpha)$ is zero:

$$\text{abs max}\,[\min E\{\epsilon^2\}] = R_{ss}(0). \tag{5.13-9}$$

Note that $R_{ss}(0)$ is just the normalized signal power. It is clear from Equation 5.13-6 that this absolute maximum can always be obtained by making $h(t)$ identically zero for all $t \in (0, \infty)$ or by allowing $\alpha \to \infty$.

Although it is not obvious, the function $G(\alpha)$ is monotone nonincreasing in α, that is,

$$G(\alpha_2) \leq G(\alpha_1), \qquad \alpha_2 > \alpha_1. \tag{5.13-10}$$

It follows then from Equation 5.13-1 that the minimum mean-squared error min $E\{\epsilon^2\}$ is monotone nondecreasing in α since the term $R_{ss}(0)$ is a constant. In other words, increasing α (the prediction time) increases the error, and the minimum value of the minimum mean-squared error is obtained when $\alpha \to -\infty$ (the infinite lag filter):

$$\lim_{\alpha \to -\infty} \min E\{\epsilon^2\} = \min\,[\min E\{\epsilon^2\}], \tag{5.13-11}$$

$$\lim_{\alpha \to \infty} \min E\{\epsilon^2\} = \max\,[\min E\{\epsilon^2\}] = R_{ss}(0). \tag{5.13-12}$$

The relationship of Equation 5.13-10 may be proved as follows. We write $G(\alpha)$ from Equation 5.13-6 in the form

$$G(\alpha) = \int_0^\infty R_{rs}(\tau + \alpha)h^\dagger(\tau)\, d\tau$$

$$= \int_0^\infty R_{rs}(\tau + \alpha)\frac{1}{2\pi}\int_{-\infty}^\infty H^\dagger(\omega)\, e^{j\omega\tau}\, d\omega\, d\tau. \tag{5.13-13}$$

Now we substitute Equation 5.12-1 for $H^{\dagger}(\omega)$:

$$G(\alpha) = \int_0^\infty R_{rs}(\tau + \alpha) \frac{1}{2\pi} \int_{-\infty}^\infty e^{j\omega\tau} \frac{1}{\varphi_{rr}^{+}(\omega)} \int_0^\infty e^{-j\omega t} \times \frac{1}{2\pi} \int_{-\infty}^\infty \frac{\varphi_{rs}(\nu)}{\varphi_{rr}^{-}(\nu)} e^{j\nu(t+\alpha)}\, d\nu\, dt\, d\omega\, d\tau. \tag{5.13-14}$$

Let us now define the function $K(\nu)$ by

$$K(\nu) = \frac{\varphi_{rs}(\nu)}{\varphi_{rr}^{-}(\nu)} \tag{5.13-15}$$

with inverse Fourier transform

$$k(t) = \frac{1}{2\pi} \int_{-\infty}^\infty K(\nu)\, e^{j\nu t}\, d\nu. \tag{5.13-16}$$

We see now that the integral with respect to ν becomes $k(t + \alpha)$. We may interchange the order of integration to obtain

$$G(\alpha) = \int_0^\infty k(t + \alpha) \frac{1}{2\pi} \int_{-\infty}^\infty \frac{e^{-j\omega t}}{\varphi_{rr}^{+}(\omega)} \int_{-\infty}^\infty R_{rs}(\tau + \alpha)\, e^{j\omega\tau}\, d\tau\, d\omega\, dt \tag{5.13-17}$$

or, since the inner integral is $\varphi_{rs}(-\omega)\, e^{-j\omega\alpha}$,

$$G(\alpha) = \int_0^\infty k(t + \alpha) \frac{1}{2\pi} \int_{-\infty}^\infty K(-\omega)\, e^{-j\omega(t+\alpha)} d\omega\, dt. \tag{5.13-18}$$

Finally, we have

$$G(\alpha) = \int_0^\infty [k(t + \alpha)]^2\, dt \tag{5.13-19}$$

or

$$G(\alpha) = \int_\alpha^\infty [k(t)]^2\, dt. \tag{5.13-20}$$

Since the integrand in Equation 5.13-20 is nonnegative for all t, it follows that $G(\alpha)$ is monotone nonincreasing in α as previously stated.

Let us now consider more closely the minimum value of the mean-squared error. This minimum is obtained when $\alpha \to -\infty$, where the function $G(\alpha)$ becomes

$$G(-\infty) = \int_{-\infty}^\infty [k(t)]^2\, dt = \frac{1}{2\pi} \int_{-\infty}^\infty |K(\omega)|^2\, d\omega. \tag{5.13-21}$$

The last relationship follows from Parseval's Theorem or from interchanging the order of integration in Equation 5.13-18. Since $K(\omega)$ is defined by Equation 5.13-15, we have

$$\begin{aligned}\min\,[\min E\{\epsilon^2\}] &= R_{ss}(0) - G(-\infty)\\ &= \frac{1}{2\pi}\int_{-\infty}^{\infty}\left[\varphi_{ss}(\omega) - \frac{|\varphi_{rs}(\omega)|^2}{\varphi_{rr}(\omega)}\right]d\omega\\ &= \frac{1}{2\pi}\int_{-\infty}^{\infty}\frac{\varphi_{ss}(\omega)\varphi_{rr}(\omega) - |\varphi_{rs}(\omega)|^2}{\varphi_{rr}(\omega)}\,d\omega. \qquad (5.13\text{-}22)\end{aligned}$$

This last expression is identical to Equation 5.11-10 for the minimum mean-squared error in the unrealizable case. The result is to be expected since ignoring realizability and allowing the lag to become infinite are equivalent.

Example 5.11 We return to the filter problem of Examples 5.9 and 5.10, where the function $K(\omega)$ is

$$K(\omega) = \frac{K}{k_1 + j\omega} + \frac{K}{k_2 - j\omega}$$

and where K, k_1, and k_2 have been defined in Example 5.10. The inverse transform of $K(\omega)$ is

$$k(t) = K\,e^{k_2 t}u(-t) + K\,e^{-k_1 t}u(t)$$

and its square is

$$[k(t)]^2 = K^2\,e^{2k_2 t}u(-t) + K^2\,e^{-2k_1 t}u(t)$$

as shown in Figure 5.14(*a*). Thus the function $G(\alpha)$ is given by

$$G(\alpha) = \int_{\alpha}^{\infty}[k(t)]^2\,dt$$

$$= \begin{cases}\dfrac{K^2}{2k_2}(1 - e^{2k_2\alpha}) + \dfrac{K^2}{2k_1}, & \alpha < 0,\\[2ex] \dfrac{K^2}{2k_1}, & \alpha = 0,\\[2ex] \dfrac{K^2}{2k_1}\,e^{-2k_1\alpha}, & \alpha > 0.\end{cases}$$

The minimum mean-squared error is

$$\min E\{\epsilon^2\} = R_{ss}(0) - G(\alpha) = \frac{a^2}{2b} - G(\alpha).$$

This expression is rather complicated. Note that, for $\alpha \to -\infty$, it becomes

$$\min E\{\epsilon^2\}_{\alpha\to-\infty} = \frac{a^2}{2b} - \frac{K^2(k_1 + k_2)}{2k_1k_2} = \frac{a^2c}{2b\sqrt{a^2 + c^2}}$$

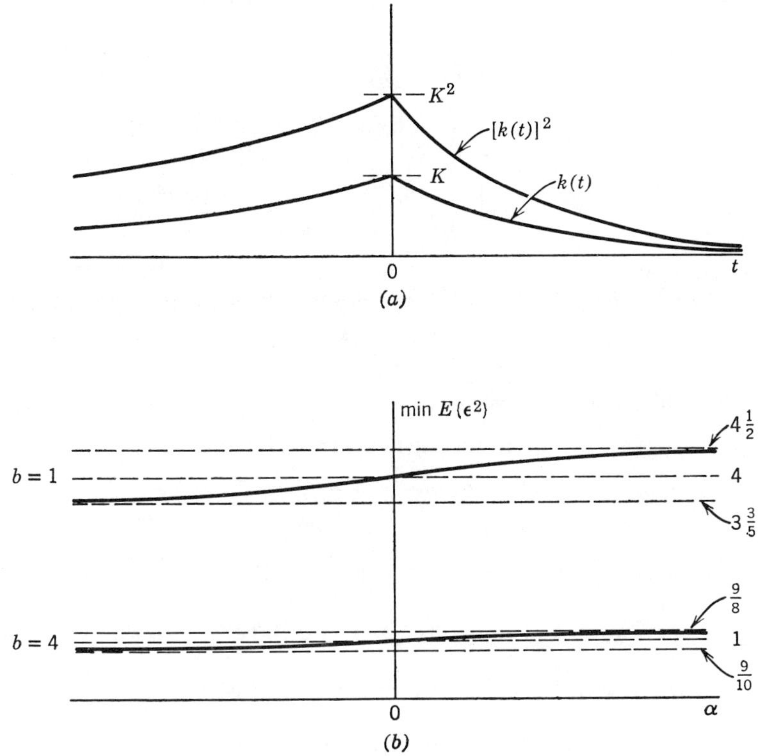

***Figure 5.14** A least mean-squared-error calculation: (a) plot of $k(t)$ and $k^2(t)$; (b) minimum mean-squared error as a function of prediction time α: $a = 3$, $b = 1$, $c = 4$, and $a = 3$, $b = 4$, $c = 4$*

in agreement with the results obtained in Example 5.9. Also, for $\alpha = 0$, we have

$$\min E\{\epsilon^2\}_{\alpha=0} = \frac{a^2}{2b} - \frac{K^2}{2k_1} = \frac{a^2 c}{bc + b\sqrt{a^2 + c^2}}.$$

For convenience in plotting, let us assume that

$$a = 3, \qquad c = 4$$

so that

$$K = k_1 = \frac{1}{b}, \qquad k_2 = \frac{5}{4b}.$$

Then the error is

$$\min E\{\epsilon^2\} = \begin{cases} \dfrac{8}{2b} - \dfrac{2}{5b}(1 - e^{5\alpha/2b}), & \alpha < 0, \\ \dfrac{8}{2b}, & \alpha = 0, \\ \dfrac{1}{2b}(9 - e^{-2\alpha/b}), & \alpha > 0, \end{cases}$$

and is plotted in Figure 5.14(b) for $b = 1$ and $b = 4$.

• •

5.14 The Pure Prediction Problem

In pure prediction, the input noise $n(t)$ is identically zero and

$$\varphi_{rs}(\omega) = \varphi_{ss}(\omega),$$
$$\varphi_{rr}(\omega) = \varphi_{ss}(\omega).$$

Hence Equation 5.12-1 for the least mean-squared-error filter becomes

$$H^{\dagger}(\omega) = \frac{1}{2\pi\varphi_{ss}^{+}(\omega)} \int_0^{\infty} e^{-j\omega t} \int_{-\infty}^{\infty} \varphi_{ss}^{+}(\nu)\, e^{j\nu(t+\alpha)}\, d\nu\, dt. \qquad (5.14\text{-}1)$$

The mean-squared error is given by Equations 5.13-1 and 5.13-20 as

$$\min E\{\epsilon^2\} = R_{ss}(0) - \int_{\alpha}^{\infty} [k(t)]^2\, dt \qquad (5.14\text{-}2)$$

where $k(t)$ is given by Equation 5.13-16, and, for the noiseless case, is just

$$k(t) = \frac{1}{2\pi} \int_{-\infty}^{\infty} \varphi_{ss}^{+}(\omega)\, e^{j\omega t}\, d\omega. \qquad (5.14\text{-}3)$$

Notice that $k(t) = 0$ for $t < 0$ since $\varphi_{ss}^{+}(\omega)$ has no LHP singularities. Also $R_{ss}(0)$ can be written as

$$R_{ss}(0) = \frac{1}{2\pi} \int_{-\infty}^{\infty} \varphi_{ss}^{+}(\omega)\varphi_{ss}^{-}(\omega)\, d\omega = \int_0^{\infty} [k(t)]^2\, dt. \qquad (5.14\text{-}4)$$

Now Equation 5.14-2 becomes

$$\min E\{\epsilon^2\} = \begin{cases} \displaystyle\int_0^{\alpha} [k(t)]^2\, dt, & \alpha > 0, \\ 0, & \alpha \le 0. \end{cases} \qquad (5.14\text{-}5)$$

Example 5.12 We consider Example 5.11 for the noiseless case. The signal power spectral density $\varphi_{ss}(\omega)$ may be factored as

$$\varphi_{ss}(\omega) = \varphi_{ss}^{+}(\omega)\varphi_{ss}^{-}(\omega) = \frac{a/b}{k_1 + j\omega}\, \frac{a/b}{k_1 - j\omega}$$

and

$$\frac{1}{2\pi}\int_{-\infty}^{\infty} \frac{a/b}{k_1 + j\nu} e^{j\nu(t+\alpha)}\, d\nu = \frac{a}{b} e^{-k_1(t+\alpha)}u(t+\alpha)$$

where $k_1 = 1/b$ has been defined previously. Now the optimum filter is

$$H^{\dagger}(\omega) = (k_1 + j\omega)\int_0^{\infty} e^{-j\omega t}\, e^{-k_1(t+\alpha)}u(t+\alpha)\, dt.$$

For the case where $\alpha = -\beta \leq 0$, this becomes

$$H^{\dagger}(\omega) = (k_1 + j\omega)\, e^{k_1\beta}\int_{\beta}^{\infty} e^{-(k_1+j\omega)t}\, dt = e^{-j\omega\beta},$$

a pure delay, as expected. For the prediction case where $\alpha > 0$, we have

$$H^{\dagger}(\omega) = (k_1 + j\omega)\, e^{-k_1\alpha}\int_0^{\infty} e^{-(k_1+j\omega)t}\, dt = e^{-k_1\alpha},$$

a constant attenuator. As $\alpha \to \infty$, the filter becomes an open circuit. Of course the same results could be obtained by allowing $c \to 0$ in Example 5.10.

The mean-squared error is zero for $\alpha \leq 0$ and, for $\alpha > 0$, it is given by

$$\begin{aligned}\min E\{\epsilon^2\} &= \int_0^{\alpha} [k(t)]^2\, dt = \frac{a^2}{b^2}\int_0^{\alpha} e^{-2k_1 t}\, dt \\ &= \frac{a^2}{2b}(1 - e^{-2k_1\alpha})\end{aligned}$$

since

$$k(t) = \frac{a}{b} e^{-k_1 t}u(t).$$

At $\alpha = 0$, the error is zero; as $\alpha \to \infty$, the error becomes

$$\frac{a^2}{2b} = R_{ss}(0),$$

the total signal energy.

• •

5.15 Solution of the Least Mean-Squared-Error Filtering and Prediction Problem by the Method of Bode and Shannon

In this section we solve the least mean-squared-error filtering and prediction problem by a technique due to Bode and Shannon [9]. Although this approach is not as elegant as that considered in Sections 5.10–5.14, it is motivated by physical considerations and has the advantages of greater simplicity and a direct physical interpretation.

The problem has already been formulated in the first three paragraphs of Section 5.10 in terms of Figure 5.11. It will be somewhat more pertinent for this discussion, however, to repeat the formulation in frequency-domain terms rather than time-domain terms. With reference to Figure 5.15, we assume:

1. Both the signal $s(t)$, with Fourier transform $S(\omega)$ and power spectral density $\varphi_{ss}(\omega)$, and the noise $n(t)$, with Fourier transform $N(\omega)$ and power spectral density $\varphi_{nn}(\omega)$, are sample functions from random processes which are at least wide-sense stationary. The input $r(t) = s(t) + n(t)$ has a power spectral density $\varphi_{rr}(\omega)$.
2. The linear, time-invariant filter with complex transfer function $H(\omega)$ is to be physically realizable.

These restrictions have been given previously in Section 5.10. For reasons of clarity in physical interpretation it will sometimes be convenient to add the *additional* condition:

3. The signal $s(t)$ and the noise $n(t)$ are uncorrelated so that the input power spectral density $\varphi_{rr}(\omega)$ is

$$\varphi_{rr}(\omega) = \varphi_{ss}(\omega) + \varphi_{nn}(\omega). \tag{5.15-1}$$

As before we want the output $s^*(t)$, with Fourier transform $S^*(\omega)$, to be a least mean-squared-error estimate of $s(t + \alpha)$, with Fourier transform $S(\omega)\, e^{j\omega\alpha}$, that is, we want

$$E\{\epsilon^2\} = E\{|s(t + \alpha) - s^*(t)|^2\} = \min. \tag{5.15-2}$$

This last equation may be expanded to give

$$E\{\epsilon^2\} = R_{ss}(0) - 2R_{s*s}(\alpha) + R_{s*s*}(0) \tag{5.15-3}$$

where $R_{s*s}(\tau)$ is the cross-correlation function of $s^*(t)$ and $s(t)$, and $R_{s*s*}(\tau)$ is the autocorrelation function of the output $s^*(t)$. This expression may be written as

$$E\{\epsilon^2\} = \frac{1}{2\pi}\int_{-\infty}^{\infty} [\varphi_{ss}(\omega) - 2\varphi_{rs}(\omega)H(-\omega)\, e^{j\omega\alpha} + \varphi_{rr}(\omega)|H(\omega)|^2]\, d\omega. \tag{5.15-4}$$

These relationships follow directly from Equations 4.4-7 and 4.4-14, and

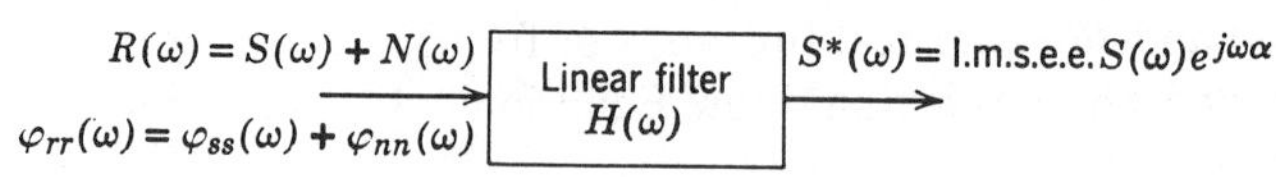

Figure 5.15 *Least mean-squared-error filtering and prediction as a frequency-domain problem*

Equation 5.15-4 is identical to Equation 5.13-4 as expected. The mean-squared error $E\{\epsilon^2\}$ of Equation 5.15-4 is to be minimized by the proper choice of the complex transfer function $H(\omega)$. For the case where the signal $s(t)$ and the noise $n(t)$ are uncorrelated, this last equation reduces to

$$E\{\epsilon^2\} = \frac{1}{2\pi}\int_{-\infty}^{\infty} \{\varphi_{ss}(\omega) - 2\varphi_{ss}(\omega)H(-\omega)\, e^{j\omega\alpha} + [\varphi_{ss}(\omega) + \varphi_{nn}(\omega)]|H(\omega)|^2\}\, d\omega. \quad (5.15\text{-}5)$$

The integrand may be rearranged to give

$$E\{\epsilon^2\} = \frac{1}{2\pi}\int_{-\infty}^{\infty} \{\varphi_{ss}(\omega)|e^{j\omega\alpha} - H(\omega)|^2 + \varphi_{nn}(\omega)|H(\omega)|^2\}\, d\omega \quad (5.15\text{-}6)$$

which is equivalent to Equation 5.13-5. As stated in Section 5.13, this latter form clearly shows the two error terms, first

$$E\{\epsilon_s^2\} = \frac{1}{2\pi}\int_{-\infty}^{\infty} \varphi_{ss}(\omega)|e^{j\omega\alpha} - H(\omega)|^2\, d\omega, \quad (5.15\text{-}7)$$

which is a result of the transfer function being $H(\omega)$ rather than the desired $e^{j\omega\alpha}$, and second,

$$E\{\epsilon_n^2\} = \frac{1}{2\pi}\int_{-\infty}^{\infty} \varphi_{nn}(\omega)|H(\omega)|^2\, d\omega, \quad (5.15\text{-}8)$$

which is the result of the input noise. Of course, the signal error $E\{\epsilon_s^2\}$ could be made zero by making $H(\omega)$ equal to $e^{j\omega\alpha}$ (provided $\alpha \leq 0$ so that $H(\omega)$ was physically realizable). The procedure would not necessarily minimize the sum $E\{\epsilon_s^2\} + E\{\epsilon_n^2\}$, however. Indeed, if the noise $n(t)$ were very large compared to the signal $s(t)$, it is apparent that such a procedure would not be reasonable.

We proceed now to find the choice of $H(\omega)$ which makes the error given by Equation 5.15-4 a minimum. The minimization has somewhat more physical significance if done for the special case where $s(t)$ and $n(t)$ are uncorrelated so that the error is given by Equation 5.15-6. The interested reader can parallel the development given here. Since $H(\omega)$, the complex transfer function of the system, is a complex number, it can be written as

$$H(\omega) = C(\omega)\, e^{jB(\omega)}, \quad (5.15\text{-}9a)$$

$$H(-\omega) = C(\omega)\, e^{-jB(\omega)}, \quad (5.15\text{-}9b)$$

where $C(\omega)$ and $B(\omega)$ are real functions of frequency, called respectively the amplitude and phase function. Now Equation 5.15-4 becomes

$$E\{\epsilon^2\} = \frac{1}{2\pi}\int_{-\infty}^{\infty} [\varphi_{ss}(\omega) - 2\varphi_{rs}(\omega)C(\omega)\, e^{j[\omega\alpha - B(\omega)]} + \varphi_{rr}(\omega)C^2(\omega)]\, d\omega. \quad (5.15\text{-}10)$$

Recall that the cross-spectral density $\varphi_{rs}(\omega)$ is complex and that we can introduce

$$\varphi_{rs}(\omega) = A(\omega)\, e^{j\alpha(\omega)} \tag{5.15-11}$$

where $A(\omega)$ and $\alpha(\omega)$ are real functions of frequency. Now Equation 5.15-10 becomes

$$E\{\epsilon^2\} = \frac{1}{2\pi}\int_{-\infty}^{\infty} [\varphi_{ss}(\omega) + \varphi_{rr}(\omega)C^2(\omega)]\, d\omega - \frac{2}{2\pi}\int_{-\infty}^{\infty} A(\omega)C(\omega)\, e^{j[\omega\alpha + \alpha(\omega) - B(\omega)]}\, d\omega. \tag{5.15-12}$$

Since e^{jx} has a maximum value of unity when $x = 0$, it is clear that the best choice of $B(\omega)$ is

$$\boxed{B(\omega) = \omega\alpha + \alpha(\omega).} \tag{5.15-13}$$

This choice makes the last integral in Equation 5.15-12 as large as possible and, hence, makes $E\{\epsilon^2\}$ as small as possible. Note that $B(\omega)$ does not appear in the first integral.

Having made the indicated choice for $B(\omega)$, we complete the square in $C(\omega)$ to obtain

$$E\{\epsilon^2\} = \frac{1}{2\pi}\int_{-\infty}^{\infty} \left\{\varphi_{ss}(\omega) - \frac{A^2(\omega)}{\varphi_{rr}(\omega)} + \varphi_{rr}(\omega)\left[C(\omega) - \frac{A(\omega)}{\varphi_{rr}(\omega)}\right]^2\right\} d\omega. \tag{5.15-14}$$

The error is obviously least when

$$\boxed{C(\omega) = \frac{A(\omega)}{\varphi_{rr}(\omega)}.} \tag{5.15-15}$$

It follows from Equations 5.15-9a, 5.15-13, and 5.15-16 that the optimum transfer function is

$$\boxed{H_u^\dagger(\omega) = \frac{\varphi_{rs}(\omega)}{\varphi_{rr}(\omega)}\, e^{j\omega\alpha}} \tag{5.15-16}$$

where the subscript u has been used to emphasize that this system is not necessarily physically realizable since we have done nothing to ensure that $h_u^\dagger(t) = 0,\ t < 0$. Note that this expression is identical to Equation 5.11-1 which was obtained by other means.

The problem of constructing the best physically realizable filter from Equation 5.15-16 has already been considered extensively in this chapter and has been solved in Section 5.12 by a different approach. Nevertheless we will continue here with the approach of Bode and Shannon [9]. This turns out to be just the prewhitening technique that was used in Section 5.8 for the solution of the matched-filter problem.

We consider the system shown in Figure 5.16 where two *realizable* filters $H_1(\omega)$ and $H_2(\omega)$ are to be constructed so that

$$H^{\dagger}(\omega) = H_1(\omega)H_2(\omega) \tag{5.15-17}$$

is the least mean-squared-error *realizable* filter. The first filter is chosen so that its output $m(t)$ has a flat spectral density of unity. As discussed in Section 5.6, this is accomplished by factoring the input spectral density $\varphi_{rr}(\omega)$ into

$$\varphi_{rr}(\omega) = \varphi_{rr}^{+}(\omega)\varphi_{rr}^{-}(\omega) \tag{5.15-18}$$

and by forming the realizable filter

$$H_1(\omega) = \frac{1}{\varphi_{rr}^{+}(\omega)}. \tag{5.15-19}$$

Suppose we attempt to fit the unrealizable filter of Equation 5.15-16 into this framework so that

$$H_u^{\dagger}(\omega) \stackrel{?}{=} H_1(\omega)H_2(\omega). \tag{5.15-20}$$

It is apparent that we cannot accomplish this and still require that $H_2(\omega)$ be realizable. Indeed, if we define a filter $H_3(\omega)$ by

$$H_u^{\dagger}(\omega) = H_1(\omega)H_3(\omega) \tag{5.15-21}$$

or

$$H_3(\omega) = \frac{\varphi_{rs}(\omega)}{\varphi_{rr}^{-}(\omega)}\, e^{j\omega\alpha}, \tag{5.15-22}$$

then $H_3(\omega)$ will not be realizable in general.

Let us consider the impulse response $h_3(t)$ corresponding to $H_3(\omega)$. In general

$$h_3(t) \neq 0, \qquad t < 0, \tag{5.15-23}$$

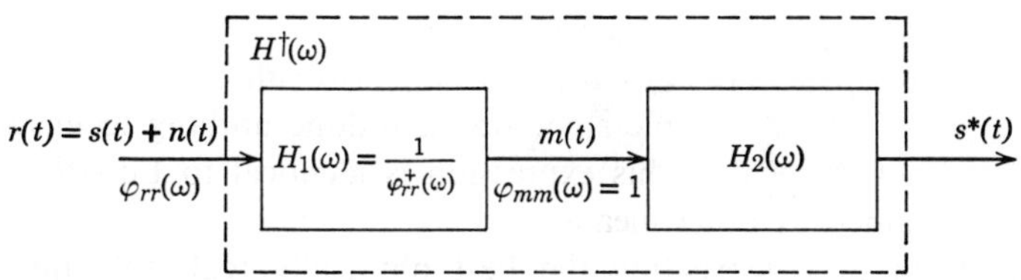

Figure 5.16 *Solution for the least mean-squared-error filter by prewhitening techniques*

and we can write, from Figure 5.16,

$$s^*(t) = \int_{-\infty}^{t} m(\tau)h_3(t - \tau)\, d\tau + \int_{t}^{t+\gamma} m(\tau)h_3(t - \tau)\, d\tau \quad (5.15\text{-}24)$$

where $\gamma > t$. The first integral represents operations on the past of the signal $m(t)$ while the second represents operations on the future and corresponds to the unrealizable part of $h_3(t)$. The situation with respect to $h_3(t)$ is diagrammed in Figure 5.17 for $t < 0 < \gamma$. Let us rewrite Equation 5.15-24 as

$$s^*(t) = I_p + I_f \quad (5.15\text{-}25)$$

where I_p represents the first integral, corresponding to contributions from $h_3(t)$ for $t \geq 0$, and I_f represents the second integral, corresponding to contributions from $h_3(t)$ for $t < 0$. The filter corresponding to I_p is realizable while the filter corresponding to I_f is not.

From physical intuition, we might argue that the contribution to $s^*(t)$ due to I_f should be ignored since it represents operations in the future on white noise, which is unpredictable in a mean-squared sense (being uncorrelated). More precisely, we show first that I_p and I_f are uncorrelated. We have

$$E\{I_p I_f\} = \int_{-\infty}^{t} h_3(t - \tau) \int_{t}^{t+\gamma} E\{m(\tau)m(\sigma)\}h_3(t - \sigma)\, d\sigma\, d\tau. \quad (5.15\text{-}26)$$

But $m(t)$ is white noise with unity spectral density so that

$$E\{m(\tau)m(\sigma)\} = \delta(\sigma - \tau). \quad (5.15\text{-}27)$$

Thus Equation 5.15-26 becomes

$$E\{I_p I_f\} = \int_{-\infty}^{t} h_3(t - \tau) \int_{t}^{t+\gamma} \delta(\sigma - \tau)h_3(t - \sigma)\, d\sigma\, d\tau = 0 \quad (5.15\text{-}28)$$

since the ranges of integration do not overlap.

Notice also that I and I_f are uncorrelated where I is the convolution integral giving any linear operation on the past of $m(t)$. In particular let I be given by

$$I = \int_{-\infty}^{t} m(\tau)h_2(t - \tau)\, d\tau \quad (5.15\text{-}29)$$

where $h_2(t)$ is the impulse response of the optimum physically realizable filter which approximates $s(t + \alpha)$ and which is given by Equation 5.15-17. The mean-squared error can now be written as

$$\begin{aligned} E\{\epsilon^2\} &= E\{|I - s^*(t)|^2\} = E\{|I - I_p - I_f|^2\} \\ &= E\{|(I - I_p) - I_f|^2\} \\ &= E\{|I - I_p|^2\} - 2E\{II_f\} + 2E\{I_p I_f\} + E\{|I_f|^2\} \\ &= E\{|I - I_p|^2\} + E\{|I_f|^2\} \end{aligned} \quad (5.15\text{-}30)$$

since $E\{II_f\} = E\{I_p I_f\} = 0$. It is clear that the best choice for I is that which makes

$$I = I_p \quad (5.15\text{-}31)$$

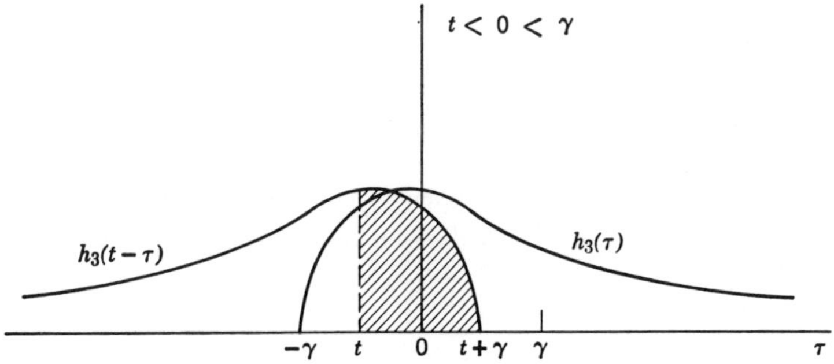

Figure 5.17 *An unrealizable impulse response*

or, in other words,

$$h_2(t) = h_3(t)u(t), \tag{5.15-32}$$

where $u(t)$ is the unit-step function and $h_2(t)$ is the impulse response of the filter which is placed in series with $h_1(t)$ to give the least mean-squared-error *realizable* filter $h^\dagger(t)$ with transfer function $H^\dagger(\omega)$.

We summarize the method of Bode and Shannon as follows:

1. Find the realizable filter $H_1(\omega)$ given by Equation 5.15-19:

$$H_1(\omega) = \frac{1}{\varphi_{rr}^+(\omega)}.$$

2. Form the unrealizable filter $H_3(\omega)$ given by Equation 5.15-22:

$$H_3(\omega) = \frac{\varphi_{rs}(\omega)}{\varphi_{rr}^-(\omega)}\, e^{j\omega\alpha}.$$

3. Find the impulse response $h_3(t)$ corresponding to $H_3(\omega)$ and let

$$h_2(t) = h_3(t)u(t).$$

4. Then the least mean-squared-error realizable filter is

$$H^\dagger(\omega) = H_1(\omega)H_2(\omega).$$

Note that $h_3(t)$ is given by

$$h_3(t) = \frac{1}{2\pi}\int_{-\infty}^{\infty} \frac{\varphi_{rs}(\nu)}{\varphi_{rr}^-(\nu)}\, e^{j\nu(\alpha+t)}\, d\nu. \tag{5.15-33}$$

Now $h_2(t)$ vanishes for negative time and its transform $H_2(\omega)$ is

$$H_2(\omega) = \int_0^{\infty} h_3(t)\, e^{-j\omega t}\, dt. \tag{5.15-34}$$

Combining these last two equations with Equations 5.15-17 and 5.15-19, we can write in closed form

$$H^{\dagger}(\omega) = \frac{1}{2\pi\varphi_{rr}^{+}(\omega)} \int_{0}^{\infty} e^{-j\omega t} \int_{-\infty}^{\infty} \frac{\varphi_{rs}(\nu)}{\varphi_{rr}^{-}(\nu)} e^{j\nu(\alpha+t)}\, d\nu\, dt. \tag{5.15-35}$$

This is just Equation 5.12-1 which was obtained by direct solution of an integral equation.

PART 3—THE PARAMETER OPTIMIZATION OF LINEAR SYSTEMS

5.16 Introduction and Formulation

The approach of Part 2 of this chapter provides an elegant solution to the least mean-squared-error filter problem. However, the realization of the resulting optimum filter may be prohibitively difficult in a given practical situation. A much more straightforward procedure is to assume a fixed-filter structure and to allow a finite (usually small) number of filter parameters to vary so as to minimize the mean-squared error [10, 11]. The problem can be quite easy to solve if the assumed filter structure is simple. Of course the resulting system will yield a larger mean-squared error, which is the price that must be paid for constraining the system form.

We consider the filter of Figure 5.11(*b*) or Figure 5.15 and rewrite the mean-squared error from Equation 5.15-4 as

$$E\{\epsilon^2\} = \frac{1}{2\pi} \int_{-\infty}^{\infty} \varphi_{\epsilon}(\omega)\, d\omega \tag{5.16-1}$$

where $\varphi_{\epsilon}(\omega)$ is the error power spectral density given by

$$\varphi_{\epsilon}(\omega) = \varphi_{ss}(\omega) - 2\varphi_{rs}(\omega)H(-\omega)\, e^{j\omega\alpha} + \varphi_{rr}(\omega)|H(\omega)|^2. \tag{5.16-2}$$

We simply write $\varphi_{\epsilon}(\omega)$ as a function of the N variable parameters of the filter $H(\omega)$ and evaluate Equation 5.16-1 for the mean-squared error. This mean-squared error, a function of the N parameters, is then minimized with respect to these parameters.

Example 5.13 We return to the example given in Part 2 where

$$R_{ss}(\tau) = \frac{a^2}{2b} e^{-|\tau|/b},$$

$$\varphi_{ss}(\omega) = \frac{a^2}{b^2\omega^2 + 1},$$

$$R_{nn}(\tau) = c^2\, \delta(\tau),$$

$$\varphi_{nn}(\omega) = c^2.$$

We take the prediction time α to be zero and assume that $s(t)$ and $n(t)$ are uncorrelated. We will assume that the filter is low pass with a single time constant $1/T$ so that

$$H(\omega) = \frac{1}{1 + j\omega T}.$$

The problem is to choose T so as to minimize the error $E\{\epsilon^2\}$.

The error power spectral density is obtained from Equation 5.16-2 or

$$\begin{aligned}\varphi_\epsilon(\omega) &= \left|1 - \frac{1}{1 + j\omega T}\right|^2 \frac{a^2}{1 + b^2\omega^2} + \left|\frac{1}{1 + j\omega T}\right|^2 c^2 \\ &= \frac{a^2T^2\omega^2}{(1 + \omega^2T^2)(1 + b^2\omega^2)} + \frac{c^2}{1 + T^2\omega^2} \\ &= -\frac{a^2T^2/(T^2 - b^2)}{1 + T^2\omega^2} + \frac{a^2T^2/(T^2 - b^2)}{1 + b^2\omega^2} + \frac{c^2}{1 + T^2\omega^2}.\end{aligned}$$

Since

$$\frac{1}{2\pi}\int_{-\infty}^{\infty} \frac{d}{1 + d^2x^2}\, dx = \begin{cases} \frac{1}{2}, & d > 0, \\ 0, & d = 0, \\ -\frac{1}{2}, & d < 0, \end{cases}$$

the mean-squared error is

$$\begin{aligned}E\{\epsilon^2\} &= \frac{1}{2\pi}\int_{-\infty}^{\infty} \varphi_\epsilon(\omega)\, d\omega \\ &= -\frac{a^2T}{2(T^2 - b^2)} + \frac{a^2T^2}{2b(T^2 - b^2)} + \frac{c^2}{2T}\end{aligned}$$

$$\boxed{E\{\epsilon^2\} = \frac{a^2}{2b}\frac{T}{T + b} + \frac{c^2}{2T}.} \qquad \text{(A)}$$

This last expression is easily minimized with respect to T

$$\frac{\partial E\{\epsilon^2\}}{\partial T} = 0 = \frac{(a^2 - c^2)T^2 - 2c^2bT - c^2b^2}{(b + T)^2 2T^2}$$

or

$$T^2 - \frac{2c^2b}{a^2 - c^2}T - \frac{c^2b^2}{a^2 - c^2} = 0$$

with solution

$$T = \frac{bc}{\pm a - c}.$$

Since T must be nonnegative, we allow

$$T = \frac{bc}{a - c}, \qquad a \geq c, \tag{B}$$

and note that if $a < c$, there is no minimum. On substituting Equation B into Equation A, we obtain the minimum mean-squared error

$$\min E\{\epsilon^2\} = \frac{2ac - c^2}{2b}. \tag{C}$$

Let us normalize Equation C by placing

$$\frac{\min E\{\epsilon^2\}}{b} = f(x, y) \geq 0,$$

$$x = \frac{c}{b} \geq 0,$$

$$y = \frac{a}{c} \geq 1,$$

so that the normalized mean-squared error is

$$f(x, y) = x^2(y - \tfrac{1}{2}), \qquad y \geq 1. \tag{D}$$

This error is plotted in Figure 5.18 versus y for various values of the parameter x. Note that $y = a/c$ must be greater than or equal to unity for T to be nonnegative and hence realizable; consequently the procedures are valid only for $a/c \geq 1$.

Let us compare the results obtained here to those of Examples 5.10 and 5.11 where the same inputs existed but no restrictions were placed on the form of the filter. For $\alpha = 0$, the least mean-squared-error filter is shown in Figure 5.13. The form is the same as the one fixed in the present example except that the gain G is variable in Example 5.11. Notice that the value of $T = RC = bc/\sqrt{a^2 + c^2}$ for Example 5.11 is not the same as the value $T = bc/(a - c)$ obtained in the present case. The error for the problem of Example 5.10 is calculated in Example 5.11 and is

$$E\{\epsilon^2\}_2 = \frac{a^2c}{bc + b\sqrt{a^2 + c^2}}$$

which should be compared to Equation C. It is easy to show that

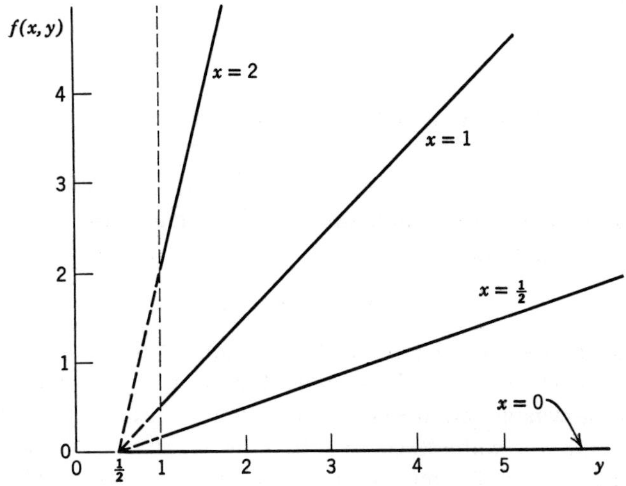

Figure 5.18 *Minimum mean-squared error for a fixed-form filter:*

$$f(x, y) = \frac{\min E[\epsilon^2]}{b}$$

$$x = \frac{c}{b} \geq 0$$

$$y = \frac{a}{c} \geq 1.$$

$$E\{\epsilon^2\}_2 \leq E\{\epsilon^2\}_1, \qquad \frac{a}{c} \geq 1. \tag{E}$$

We rewrite this last expression after canceling common terms as

$$\frac{y^2}{1 + \sqrt{y^2 + 1}} \leq y - \tfrac{1}{2}, \qquad y \geq 1, \tag{F}$$

where, as before, $y = a/c$. At $y = 1$, we have

$$\frac{1}{1 + \sqrt{2}} \leq \frac{1}{2}$$

which is certainly true. For $y > 1$, we have

$$\frac{y^2}{1 + \sqrt{y^2 + 1}} < \frac{y^2}{1 + y}$$

so that Equation F can be rewritten as

$$\frac{y^2}{1+y} < y - \tfrac{1}{2}, \qquad y > 1,$$

or, after cross-multiplication and simplification,

$$1 < y, \qquad y > 1.$$

Thus Equation E is correct and the fixed-form filter has a larger mean-squared error as expected.

• •

The larger mean-squared error in this example is the result of allowing too few parameters to vary. It is apparent that this problem could have been avoided by allowing a variable gain G as well as a variable time constant T and then maximizing with respect to both of these parameters. There are two difficulties that arise in the general case. First, there is no clear-cut procedure for determining *a priori* the form of the filter or the minimum number of parameters required. Second, even for a small number of parameters, the resulting computations may be prohibitively tedious. We will discuss some aspects of this latter problem in the next section. First, however, we return to the previous example and allow two parameters to vary.

Example 5.14 We consider the previous example where the filter transfer function is

$$H(\omega) = \frac{G}{1 + j\omega T}$$

and choose both G and T to make the mean-squared error a minimum.

Now the error power spectral density of Equation 5.16-2 is

$$\varphi_\epsilon(\omega) = \left|1 - \frac{G}{1 + j\omega T}\right|^2 \frac{a^2}{1 + b^2\omega^2} + \left|\frac{G}{1 + j\omega T}\right|^2 c^2$$

$$= \frac{(1 - G)^2 + \omega^2 T^2}{1 + \omega^2 T^2} \frac{a^2}{1 + b^2\omega^2} + \frac{c^2 G^2}{1 + \omega^2 T^2}$$

or

$$\varphi_\epsilon(\omega) = \frac{A_1}{1 + \omega^2 T^2} + \frac{A_2}{1 + b^2\omega^2} + \frac{c^2 G^2}{1 + \omega^2 T^2}$$

where

$$A_1 = -\frac{a^2 T^2[1 - (1 - G)^2]}{T^2 - b^2},$$

$$A_2 = \frac{a^2 T^2 + a^2 b^2 (1 - G)^2}{T^2 - b^2}.$$

We may now find the mean-squared error

$$E\{\epsilon^2\}_3 = \frac{1}{2\pi}\int_{-\infty}^{\infty} \varphi_\epsilon(\omega)\, d\omega$$

$$= \frac{A_1}{2T} + \frac{A_2}{2b} + \frac{c^2G^2}{2T}$$

$$= \frac{a^2}{2b}\,\frac{T + b(1-G)^2}{T+b} + \frac{c^2G^2}{2T}$$

$$= \frac{a^2T^2 + a^2bT(1-G)^2 + c^2G^2b(T+b)}{2bT(T+b)}. \tag{A}$$

This error is to be minimized with respect to both G and T. We begin with G:

$$\frac{\partial E\{\epsilon^2\}_3}{\partial G} = 0 = \frac{-2a^2bT(1-G) + c^2 2Gb(T+b)}{2bT(T+b)}$$

or

$$\boxed{G = \frac{a^2T}{(a^2+c^2)T + bc^2}.} \tag{B}$$

This value may be substituted into Equation A to obtain an expression in T only and the resulting error minimized with respect to T. Alternatively we may repeat the procedure we have just gone through with respect to T instead of G. The result is

$$\frac{\partial E\{\epsilon^2\}_3}{\partial T} = 0 = \frac{a^2T^2(T+b) - a^2T^2[T + b(1-G)^2] - c^2G^2b(T+b)^2}{2bT^2(T+b)^2}$$

which may be solved for T to yield

$$\boxed{T = \frac{bcG}{\pm a\sqrt{1-(1-G)^2} - Gc}.} \tag{C}$$

The plus sign must be used and we must impose the restriction

$$a\sqrt{1-(1-G)^2} \geq Gc$$

or

$$\frac{a}{c} = y \geq \frac{G}{\sqrt{1-(1-G)^2}}$$

to ensure that T is nonnegative and hence realizable. We now have two equations in the two unknowns T and G. We may substitute Equation C into Equation B to obtain

$$G = 1 - \frac{c}{\sqrt{a^2+c^2}} = \frac{a^2}{(a^2+c^2) + c\sqrt{a^2+c^2}} \tag{D}$$

and Equation B into Equation C to obtain

$$T = \frac{bc}{\sqrt{a^2 + c^2}}. \tag{E}$$

Both of these results agree with Example 5.10 and Figure 5.13. With these values of T and G, Equation A becomes

$$E\{\epsilon^2\}_3 = \frac{a^2 c}{bc + b\sqrt{a^2 + c^2}}$$

in agreement with Example 5.11.

• •

5.17 Evaluation of the Mean-Squared-Error Integral

From the last example it was quite apparent that the evaluation of the error integral

$$E\{\epsilon^2\} = \frac{1}{2\pi}\int_{-\infty}^{\infty} \varphi_\epsilon(\omega)\, d\omega \tag{5.17-1}$$

may become quite tedious. However, a useful tabulation is available [10] of the commonly encountered integrals of this type when $\varphi_\epsilon(\omega)$ is a rational function of ω.

It was pointed out in Section 3.8 that any power spectral density, say $\varphi_\epsilon(\omega)$, is real, nonnegative, and even. If $\varphi_\epsilon(\omega)$ is rational in ω, we can always write

$$\varphi_\epsilon(\omega) = \varphi(\omega)G(j\omega)G(-j\omega) \tag{5.17-2}$$

where $\varphi(\omega)$ is a power spectral density and $G(j\omega)$ is the complex transfer function of a realizable filter. This is just a form of the spectral factorization discussed in Section 5.6. In a different notation Equation 5.17-2 becomes

$$\varphi_\epsilon(\omega) = \frac{F(j\omega)}{H(j\omega)H(-j\omega)} \tag{5.17-3}$$

where $H(\lambda)$, with $\lambda = \omega + j\sigma$, has no lower half-plane roots and where $F(j\omega)$ is even in ω. Thus we may represent $H(j\omega)$ and $F(j\omega)$ as

$$H(j\omega) = a_0(j\omega)^n + a_1(j\omega)^{n-1} + \cdots + a_n, \tag{5.17-4}$$

$$F(j\omega) = b_0(j\omega)^{2n-2} + b_1(j\omega)^{2n-4} + \cdots + b_{n-1}. \tag{5.17-5}$$

It is clear that the order of $F(j\omega)$ must be at least two less than that of $H(j\omega)H(-j\omega)$ if the integral of Equation 5.17-2 is to converge; that is, if the error power is to be finite.

Now the integral

$$I_n = \frac{1}{2\pi}\int_{-\infty}^{\infty} \frac{F(j\omega)}{H(j\omega)H(-j\omega)}\, d\omega, \qquad n = 1, 2, \ldots, \tag{5.17-6}$$

has been tabulated [10] for $n = 1, 2, \ldots, 7$. The first three are

$$I_1 = \frac{b_0}{2a_0a_1}, \tag{5.17-7}$$

$$I_2 = \frac{-b_0 + (a_0b_1/a_2)}{2a_0a_1}, \tag{5.17-8}$$

$$I_3 = \frac{-a_2b_0 + a_0b_1 - (a_0a_1b_2/a_3)}{2a_0(a_0a_3 - a_1a_2)}. \tag{5.17-9}$$

Values of I_n, $n = 1, 2, 3, \ldots, 7$, are given in Appendix G.

Example 5.15 Let us return to Example 5.14 where the error power spectral density could have been written as

$$\varphi_\epsilon(\omega) = \frac{(a^2T^2 + c^2b^2G^2)\omega^2 + [a^2(1 - G)^2 + c^2G^2]}{[(1 - bT\omega^2) + j\omega(T + b)][(1 - bT\omega^2) - j\omega(T + b)]}$$

or, in terms of Equations 5.17-4 and 5.17-5,

$$H(j\omega) = bT(j\omega)^2 + (T + b)j\omega + 1$$

and

$$G(j\omega) = -(a^2T^2 + c^2b^2G^2)(j\omega)^2 + [a^2(1 - G)^2 + c^2G^2).$$

Hence $n = 2$ and

$$a_0 = bT, \qquad b_0 = -(a^2T^2 + c^2G^2b^2),$$

$$a_1 = T + b, \qquad b_1 = a^2(1 - G)^2 + c^2G^2,$$

$$a_2 = 1.$$

The integral I_2 is now given by

$$E\{\epsilon^2\} = I_2 = \frac{-b_0 + (a_0b_1/a_2)}{2a_0a_1}$$

or

$$E\{\epsilon^2\} = \frac{a^2T^2 + a^2bT(1 - G)^2 + c^2G^2b(T + b)}{2bT(T + b)}$$

in agreement with the results of Example 5.14.

• •

5.18 Conclusion

In this chapter we have investigated several types of optimum systems. In each case, the restriction to linear systems and to random processes which were at least wide-sense stationary led to a tractable formulation of a mean-squared-error expression. This expression then could be minimized with respect to the linear system to produce a least-error linear system.

Although it would be desirable to consider more general systems and criteria other than mean-squared error (or signal-to-noise power ratio), the resulting complexity in minimization has been great enough thus far to preclude significant results. A somewhat more general approach will be discussed in Part 3 of Chapter 6.

PROBLEMS

1. A signal $s(t) = A \sin \omega_0 t$ and independent additive white noise $n(t)$ with spectral density N_o is applied to a low-pass RC network. Find the ratio of the average signal power to expected noise power at the output. What value of RC maximizes this ratio and what is this maximum?
2. Consider the signal $s(t)$ as shown and additive white noise $n(t)$ with spectral density N_o. Find the output signal-to-noise power ratio for the appropriate matched filter with $t_1 = a$.

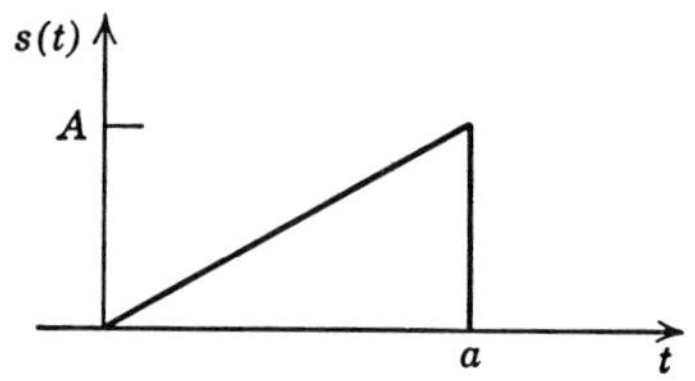

3. Consider the signal and noise in Problem 2 passed through the following linear system.

$$y(t) = \frac{1}{a}\int_{t-a}^{t} x(\lambda)\, d\lambda.$$

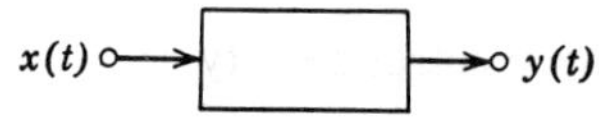

Express the output signal-to-noise power ratio S/N as a function of time. Find the value of S/N at $t = a$ and compare with the optimal value in Problem 2.

4. Consider the signal $s(t)$ in additive white noise $n(t)$ with spectral density N_o applied to an RC low-pass filter where

$$s(t) = \begin{cases} 1, & 0 \leq t \leq T, \\ 0, & \text{otherwise.} \end{cases}$$

Find the output signal-to-noise power ratio S/N as a function of t. Determine the value of RC which maximizes S/N at $t = T$ and find this maximum.

5. Find the matched filter for the signal and noise in Problem 4 with $t_1 = T$. Plot S/N as a function of t. Compare the optimal S/N to the result of Problem 4.

6. Find the matched filter for the signal

$$s(t) = \begin{cases} A \cos \omega_0 t, & 0 \leq t \leq T, \\ 0, & \text{otherwise,} \end{cases}$$

in white noise $n(t)$ with spectral density N_o and for $t_1 = T$. What is the output signal-to-noise ratio? For the case where the signal pulse contains many cycles ($\omega_0 T \gg 1$), to what does S/N reduce?

7. To what signal $s(t)$ in white noise $n(t)$ is the following system matched for $t_1 = b$?

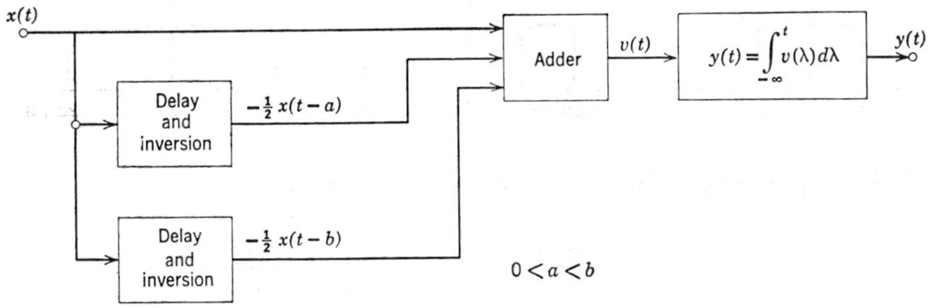

8. Consider the signal $s(t) = e^{-a(t/T)^2}$ in white noise $n(t)$ with spectral density N_o. Ignoring physical realizability, show that the matched filter with $t_1 = 0$ has the transfer function

$$H(\omega) = \sqrt{\frac{\pi}{a}}\, T\, e^{-(\omega T)^2/4a}.$$

Is $H(\omega)$ realizable?

9. Consider noise with the spectral density

$$\varphi_n(\omega) = N_o \frac{\omega^2 + 2}{\omega^4 + 4}.$$

What is the transfer function of the filter which has white noise of unity spectral density as its output with the given noise as input?

10. Consider the following linear system.

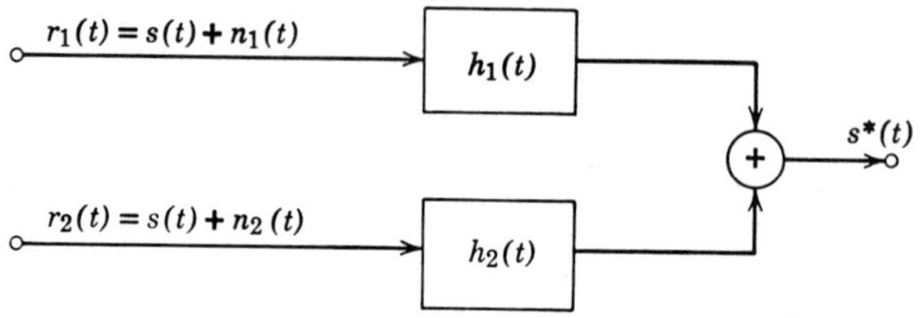

Find the pair of integral equations that specify the filters $h_1(t)$ and $h_2(t)$ such that the mean-squared error

$$E|s\{(t + \alpha) - s^*(t)|^2\}$$

is a minimum.

11. Find the *unrealizable* least mean-squared-error filters for the system of Problem 10. Assume that the input signal $s(t)$ and noises $n_1(t)$, $n_2(t)$ are mutually uncorrelated and have zero means. For the spectral densities shown, find $H_1(\omega)$ and $H_2(\omega)$ when the prediction time $\alpha = 0$.

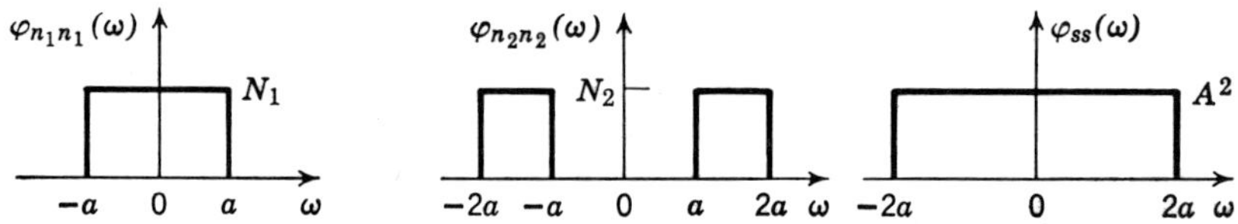

12. Solve Problem 11 for the following spectral densities:

$$\varphi_{n_1n_1}(\omega) = \frac{N_o\omega^2}{a^2 + \omega^2}, \qquad \varphi_{n_2n_2}(\omega) = \frac{N_o a^2}{a^2 + \omega^2},$$

$$\varphi_{ss}(\omega) = \begin{cases} A^2, & -2a \le \omega \le 2a, \\ 0, & \text{otherwise.} \end{cases}$$

13. Find the integral equation that specifies the filter $h(t)$ in the following linear system such that the mean-squared error

$$E\{|s(t + \alpha) - s^*(t)|^2\}$$

is a minimum.

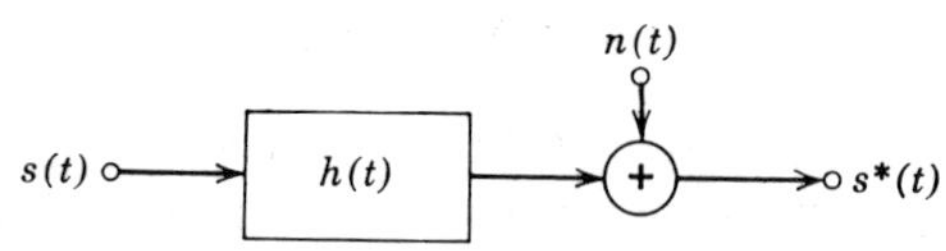

14. Find the least mean-squared-error predictor for the noise-free signal having spectral density

$$\varphi_{ss}(\omega) = \frac{1}{(1 + \omega^2)^2}.$$

15. A low-pass filter having transfer function

$$H(\omega) = \frac{1}{1 + j\omega T}$$

has uncorrelated signal and noise as its input, where

$$\varphi_{ss}(\omega) = \frac{A^2}{\omega^2 + 2a^2}, \qquad \varphi_{nn}(\omega) = \frac{2A^2}{5(\omega^2 + a^2)}.$$

Find the value of T that minimizes the mean-squared error

$$E\{\epsilon^2\} = E\{|s(t) - s^*(t)|^2\}.$$

Compare the value of $E\{\epsilon^2\}$ using the optimal value of T to the unrealizable minimum mean-squared error.

16. Consider additive uncorrelated signal and noise having spectral densities

$$\varphi_{ss}(\omega) = \frac{1}{1 + \omega^2}, \qquad \varphi_{nn}(\omega) = \frac{2\omega^2}{1 + \omega^4}.$$

Find the minimum mean-squared-error realizable linear prediction and estimation filter.

17. Find the realizable, linear, minimum mean-squared-error predictor for the noise-free signal having spectral density

$$\varphi_{ss}(\omega) = \frac{1}{(\omega^2 + a^2)(\omega^2 + b^2)}, \qquad 0 < a < b.$$

18. Find the realizable, linear, minimum mean-squared-error filter for the uncorrelated signal and noise having spectral densities

$$\varphi_{ss}(\omega) = \frac{\omega^2}{(1 + \omega^2)^2}, \qquad \varphi_{nn}(\omega) = \frac{1}{16}.$$

19. Find the transfer function for the realizable, linear, minimum mean-squared-error filter for uncorrelated signal and noise having the spectral densities

$$\varphi_{ss}(\omega) = \frac{1}{\omega^2 + 8}, \qquad \varphi_{nn}(\omega) = \frac{\omega^2 + 1}{\omega^2 + 16}.$$

REFERENCES

1. D. O. North, "Analysis of Factors Which Determine Signal-Noise Discrimination in Pulsed Carrier Systems," *RCA Tech. Rept.*, PTR-6C, June 1943.
2. R. S. Phillips and P. R. Weiss, "Theoretical Calculations on Best Smoothing of Position Data for Gunnery Predictions," *MIT Rad. Lab. Rept.*, 532, February 1944.
3. J. H. Van Vleck and D. Middleton, "A Theoretical Comparison of the Visual, Aural and Meter Reception of Pulsed Signals in the Presence of Noise," *J. Appl. Phys.*, **17**, 1946, 940.
4. I. S. Sokolnikoff and R. M. Redheffer, *Mathematics of Physics and Modern Engineering*, McGraw-Hill, New York, 1958.
5. I. M. Gelfand and S. V. Fomin, *Calculus of Variations* (trans. from Russian by R. A. Silverman), Prentice-Hall, Englewood Cliffs, N. J., 1963.
6. R. P. Weinstock, *Calculus of Variations*, McGraw-Hill, New York, 1952.
7. N. Wiener, *Extrapolation, Interpolation, and Smoothing of Stationary Time Series*, M.I.T. Press, Cambridge, Mass., 1949.
8. L. A. Zadeh and J. R. Ragazzini, "An Extension of Wiener's Theory of Prediction," *J. Appl. Phys.*, **21**, 1950, 645–655.
9. H. W. Bode and C. E. Shannon, "A Simplified Derivation of Linear Least Square Smoothing and Prediction Theory," *Proc. IRE*, **38**, 1950; 417–425.
10. H. M. James, N. B. Nichols, and R. S. Phillips, *Theory of Servomechanisms*, MIT Rad. Lab. Series, Vol. 25, McGraw-Hill, New York, 1947.
11. G. C. Newton, L. A. Gould, and J. F. Kaiser, *Analytical Design of Linear Feedback Controls*, Wiley, New York, 1957.

ADDITIONAL READING

1. Matched Filters

Books

(1953) Fano, R. M. "Communication in the Presence of Additive Gaussian Noise," in W. Jackson (Ed.), *Communication Theory*, Butterworth, London.

Woodward, P. M. *Probability and Information Theory, with Applications to Radar*, McGraw-Hill, New York.

(1958) Davenport, W. B., Jr., and W. L. Root. *An Introduction to the Theory of Random Signals and Noise*, McGraw-Hill, New York.

(1960) Helstrom, C. W. *Statistical Theory of Signal Detection*, Pergamon, New York.

Middleton, D. *An Introduction to Statistical Communication Theory*, McGraw-Hill, New York.

(1962) Wainstein, L. A., and V. D. Zubakov. *Extraction of Signals from Noise* (trans. from Russian by R. A. Silverman), Prentice-Hall, Englewood Cliffs, N. J.

Periodicals

(1946) Gabor, D. "Theory of Communication," *J. IEE*, **93**, Pt. III, 429–457.

(1947) Marcum, J. I. "A Statistical Theory of Target Detection by Pulsed Radar," *Rand Corp. Rept.*, RM-753 and RM-754, December 1947 and July 1948.

(1950) Dwork, B. M. "Detection of a Pulse Superimposed on Fluctuation Noise," *Proc. IRE*, **38**, 771–774.
(1952) Zadeh, L. A., and J. R. Ragazzini. "Optimum Filters for the Detection of Signals in Noise," *Proc. IRE*, **40**, 1123–1131.
(1953) Muller, F. A. "Communication in the Presence of Additive Gaussian Noise," *MIT Res. Lab. of Electronics Tech. Rept.*, 244, May 27.
Reich, E., and P. Swerling. "The Detection of a Sine Wave in Gaussian Noise," *J. Appl. Phys.*, **24**, 289–296.
(1954) Peterson, W. W., T. G. Birdsall, and W. C. Fox. "The Theory of Signal Detectability," *IRE Trans. on Information Theory*, PGIT-3, September, 171–212.
Rochefort, J. S. "Matched Filters for Detecting Pulsed Signals in Noise," *IRE Convention Record*, Pt. 4, 30–34.
(1955) Brennan, D. G. "On the Maximum Signal-to-Noise Ratio Realizable from Several Noisy Signals," *Proc. IRE*, **43**, 1530.
Helstrom, C. W. "The Resolution of Signals in White Gaussian Noise," *Proc. IRE*, **43**, 1111–1118.
(1956) Lytle, D. W. "Experimental Study of Tapped Delay-Line Filters," *Stanford Electronics Lab. Tech. Rept.*, 361–3, July 30.
Middleton, D., and D. Van Meter. "Detection and Extraction of Signals in Noise from the Point of View of Statistical Decision Theory," *J. Soc. Indus. Appl. Math.*, **3**, 1955, 192–253; **4**, 1956, 86–119.
Price, R. "Optimum Detection of Random Signals in Noise, with Application to Scatter-Multipath Communication," *IRE Trans. on Information Theory*, IT-2, December, 125–135.
Siebert, W. M. "A Radar Detection Philosophy," *IRE Trans. on Information Theory*, IT-2, September, 204–221.
(1957) Lytle, D. W. "On the Properties of Matched Filters," *Stanford Electronics Lab. Tech. Rept.*, 17, June 10.
(1958) Galejs, J. "Enhancement of Pulse Train Signals by Comb Filters," *IRE Trans. on Information Theory*, IT-4, September, 114–125.
(1960) Matched Filter Issue, *IRE Trans. on Information Theory*, IT-6, June.

2. Optimum Linear Filtering and Prediction

Books

(1956) Laning, J. H., and R. H. Battin. *Random Processes in Automatic Control*, McGraw-Hill, New York.
(1958) Bendat, J. S. *Principles and Applications of Random Noise Theory*, Wiley, New York.
Davenport, W. B., and W. L. Root. *An Introduction to the Theory of Random Signals and Noise*, McGraw-Hill, New York.
(1960) Lee, Y. W. *Statistical Theory of Communication*, Wiley, New York.
Middleton, D. *An Introduction to Statistical Communication Theory*, McGraw-Hill, New York.
Solodovnikov, V. V. *Introduction to the Statistical Dynamics of Automatic Control Systems* (trans. ed. by J. B. Thomas and L. A. Zadeh), Dover, New York.
(1962) Wainstein, L. A., and V. D. Zubakov. *Extraction of Signals from Noise* (trans. from Russian by R. A. Silverman), Prentice-Hall, Englewood Cliffs, N. J.
(1964) Merriam, C. W. *Optimization Theory and the Design of Feedback Control Systems*, McGraw-Hill, New York.

Periodicals

(1947) Levinson, N. "A Heuristic Exposition of Wiener's Mathematical Theory of Prediction and Filtering," *J. Math. Phys.*, **26**, 110–119; also App. C of Wiener [7] under References herein.

(1952) Booton, R. C., Jr. "An Optimization Theory for Time-Varying Linear Systems with Nonstationary Statistical Inputs," *Proc. IRE*, **40**, 977–981.

Davis, R. C. "On the Theory of Prediction of Nonstationary Stochastic Processes," *J. Appl. Phys.*, **23**, 1047–1053.

Zadeh, L. A., and J. R. Ragazzini. "Optimum Filters for the Detection of Signals in Noise," *Proc. IRE*, **40**, 1123–1131.

6

Nonlinear Systems with Random Inputs

6.1 Introduction

Linear systems were introduced and defined in Chapter 4. As pointed out in Section 4.3, a *linear* system is completely specified by its impulse response $h(t, \tau)$, relating the output $y(t)$ to the input $x(t)$ through the integral

$$y(t) = \int_{-\infty}^{\infty} x(\tau)h(t, \tau)\, d\tau. \tag{6.1-1}$$

For the *time-invariant* linear system, this expression becomes

$$y(t) = \int_{-\infty}^{\infty} x(\tau)h(t - \tau)\, d\tau, \tag{6.1-2}$$

the convolution of $x(t)$ and $h(t)$. We have seen in Section 4.5 that the statistical properties of the outputs of linear systems are easily described in terms of the statistical properties of the inputs. Thus the nth joint output moment is just

$$\begin{aligned} E\{y(t_1)y(t_2)\cdots y(t_n)\} &= \int_{-\infty}^{\infty}\cdots\int_{-\infty}^{\infty} E\{x(\tau_1)x(\tau_2)\cdots x(\tau_n)\} \\ &\quad \times h(t_1, \tau_1)h(t_2, \tau_2)\cdots h(t_n, \tau_n)\, d\tau_1\, d\tau_2\cdots d\tau_n. \end{aligned} \tag{6.1-3}$$

Of even more fundamental importance is the fact (discussed in Section 2.23) that linear operations on gaussian processes result in gaussian processes. Thus, under the common assumption of a gaussian input, the linear problem reduces to finding the first two moments, $E\{y(t_1)\}$ and $E\{y(t_1)y(t_2)\}$, since these completely specify the gaussian output.

The difficulties that arise in problems involving nonlinear systems with random inputs are myriad. Not the least of these is the vagueness of the term "nonlinear." We define a nonlinear system in a negative way—it is a system that is not linear. A completely general approach to the nonlinear problem does not appear possible. It seems to be necessary to classify nonlinear systems in some logical fashion and to develop specific techniques of analysis for each of the resulting classes.

6.2 Classification of Nonlinear Systems

L. A. Zadeh [1] has suggested a classification of nonlinear filters into a sequence $\kappa_1, \kappa_2, \kappa_3, \ldots$, such that each class in the sequence includes all of the preceding classes. Furthermore, the class of linear filters is a subclass of every class in the sequence. Since the class κ_1 includes all of the systems that we will treat and since the classification is interesting in its own right, we outline his approach.

Consider Figure 6.1 where N represents a nonlinear system with input $x(t)$ and output $y(t)$. We shall say that N belongs to class κ_n if the output $y(t)$ can be represented by

$$y(t) = \int_0^\infty \cdots \int_0^\infty K[x(t - \tau_1), \ldots, x(t - \tau_n), \tau_1, \ldots, \tau_n] \, d\tau_1 \cdots d\tau_n \tag{6.2-1}$$

where K is a real function of its $2n$ arguments and is called the *characteristic function* of the corresponding filter. Thus

$$N \in \kappa_1 : y(t) = \int_0^\infty K[x(t - \tau), \tau] \, d\tau \tag{6.2-2}$$

$$N \in \kappa_2 : y(t) = \int_0^\infty \int_0^\infty K[x(t - \tau_1), x(t - \tau_2), \tau_1, \tau_2] \, d\tau_1 \, d\tau_2. \tag{6.2-3}$$

$$\vdots \qquad \vdots \qquad\qquad \vdots$$

The class κ_1 given by Equation 6.2-2 includes three special subclasses of considerable importance. They are:

1. *The time-invariant, physically realizable linear system (Type I)*:

$$K[x(t - \tau), \tau] = h(\tau)x(t - \tau) \tag{6.2-4}$$

where $h(\tau) = 0$ for $\tau < 0$, and Equation 6.2-2 becomes

$$y(t) = \int_0^\infty h(\tau)x(t - \tau) \, d\tau. \tag{6.2-5}$$

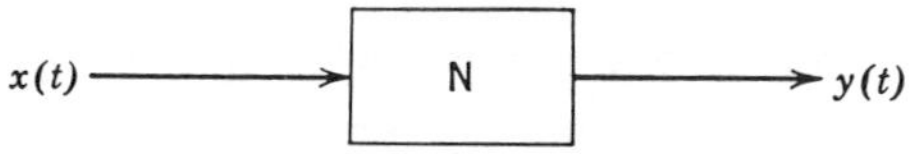

Figure 6.1 A nonlinear system

2. *The instantaneous (or zero-memory) nonlinear system (Type II):*
$$K[x(t-\tau), \tau] = \delta(\tau)g[x(t-\tau)] \tag{6.2-6}$$
and Equation 6.2-2 becomes
$$y(t) = g[x(t)]. \tag{6.2-7}$$
Thus the output at a time t is a function of the input at the same time t only. The system has no memory.

3. *The instantaneous nonlinearity followed by a noninteracting, physically realizable linear system (Type III):*
$$K[x(t-\tau), \tau] = h(\tau)g[x(t-\tau)] \tag{6.2-8}$$
and Equation 6.2-2 becomes
$$y(t) = \int_0^\infty h(\tau)g[x(t-\tau)]\,d\tau. \tag{6.2-9}$$

Each of these systems is shown schematically in Figure 6.2.

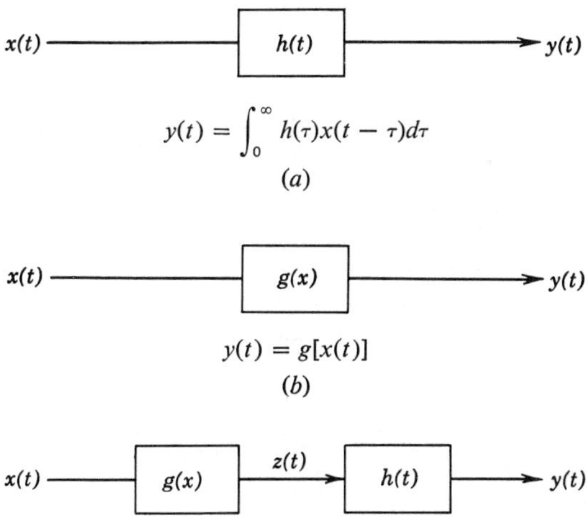

$$y(t) = \int_0^\infty h(\tau)x(t-\tau)d\tau$$
(*a*)

$$y(t) = g[x(t)]$$
(*b*)

$$y(t) = \int_0^\infty h(\tau)g[x(t-\tau)d\tau$$
(c)

Figure 6.2 *Some subclasses of the class κ_1 of nonlinear filters: (a) the linear, time-invariant, physically realizable filter (Type I); (b) the instantaneous (or zero-memory), nonlinear filter (Type II); (c) the instantaneous nonlinearity followed by a noninteracting, physically realizable linear filter (Type III)*

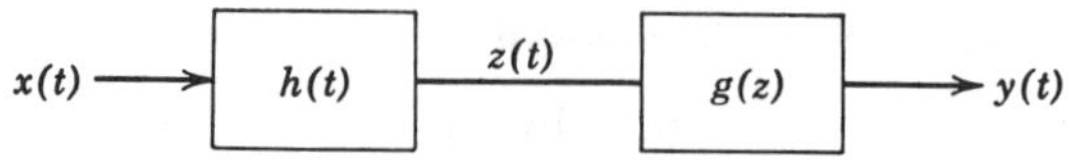

Figure 6.3 *A time-invariant, physically realizable filter followed by a noninteracting, instantaneous nonlinearity:*

$$y(t) = g\left[\int_0^\infty x(t - \tau)h(\tau)\, d\tau\right]$$

Notice that relatively simple systems are not necessarily representable by a class κ_m where m is of low order. For example, consider the system consisting of a linear filter followed by an instantaneous nonlinearity as shown in Figure 6.3. The output y is given by

$$y(t) = g\left[\int_0^\infty x(t - \tau)h(\tau)\, d\tau\right]. \qquad (6.2\text{-}10)$$

In general this expression cannot be equated to a nonlinear class κ_m for finite m.

The linear filter of Figure 6.2(*a*) has been considered extensively in Chapter 4. Most of the remainder of this chapter will be taken up with a treatment of the nonlinear systems shown in Figure 6.2(*b*) and (*c*). Since the instantaneous (or zero-memory) nonlinearity is relatively easy to handle and since it occurs frequently in problems of communication and control, it will be considered in some detail in the following sections.

PART 1—ZERO-MEMORY NONLINEAR SYSTEMS (CLASS κ_1—TYPE II NONLINEARITIES)

6.3 Direct Computation of Moments (The Density Function Method)

We consider the system represented by Figure 6.2(*b*) and require that $y = g(x)$ be a single-valued transformation from x to y, although the converse may not be true. For example, the relationship may be

$$y = x^2, \qquad -\infty < x < \infty. \qquad (6.3\text{-}1)$$

Thus the output y will be in the range $0 \le y < \infty$, and to each value of y there will correspond two values of x, $+x$ and $-x$.

The nth output moment is easily written as

$$E\{y_1 y_2 \cdots y_n\} = E\{g(x_1)g(x_2)\cdots g(x_n)\} \qquad (6.3\text{-}2)$$

or

$$\int_{-\infty}^{\infty} \cdots \int_{-\infty}^{\infty} y_1 \cdots y_n W_y(y_1, \ldots, y_n)\, dy_1 \cdots dy_n$$

$$= \int_{-\infty}^{\infty} \cdots \int_{-\infty}^{\infty} g(x_1)\cdots g(x_n) W_x(x_1, \ldots, x_n)\, dx_1 \cdots dx_n \qquad (6.3\text{-}3)$$

where, for convenience, we have replaced

$$y(t_i) \text{ by } y_i$$

and

$$x(t_i) \text{ by } x_i$$

and where $W_y(y_1, \ldots, y_n)$ and $W_x(x_1, \ldots, x_n)$ are the nth joint densities of the y's and the x's respectively. Thus the mean of y is given by

$$E\{y(t)\} = E\{g[x(t)]\} = \int_{-\infty}^{\infty} yW_y(y)\,dy = \int_{-\infty}^{\infty} g(x)W_x(x)\,dx \qquad (6.3\text{-}4)$$

and the autocorrelation function by

$$R_y(t_1, t_2) = E\{y_1 y_2\} = \int_{-\infty}^{\infty}\int_{-\infty}^{\infty} y_1 y_2 W_y(y_1, y_2)\,dy_1\,dy_2 \qquad (6.3\text{-}5a)$$

or

$$\boxed{R_y(t_1, t_2) = \int_{-\infty}^{\infty}\int_{-\infty}^{\infty} g(x_1)g(x_2)W_x(x_1, x_2)\,dx_1\,dx_2.} \qquad (6.3\text{-}5b)$$

It is apparent that the two definite integrals occurring in Equations 6.3-3 through 6.3-5 represent the same expression in two different coordinate systems. As discussed in Sections 2.11 and 2.20, the nth joint cumulative distribution function of the y's is given by

$$F_y(b_1, b_2, \ldots, b_n) = \int\int_{R_n}\cdots\int W_x(x_1, x_2, \ldots, x_n)\,dx_1\,dx_2 \cdots dx_n \qquad (6.3\text{-}6)$$

where the region of integration R_n is the set

$$R_n = \{(x_1, x_2, \ldots, x_n) \mid (y_1 \le b_1, y_2 \le b_2, \ldots, y_n \le b_n)\}. \qquad (6.3\text{-}7)$$

Furthermore, if F_y has a derivative, then the nth joint density of the y's is

$$W_y(y_1, y_2, \ldots, y_n) = \frac{\partial^n F_y(y_1, y_2, \ldots, y_n)}{\partial y_1\,\partial y_2 \cdots \partial y_n}. \qquad (6.3\text{-}8)$$

We proceed to illustrate these techniques by considering several of the nonlinearities that commonly are encountered in communications and controls. The first of these is the square-law nonlinearity which often is approximated at small signal levels by diodes and other electronic devices.

6.4 The Square-Law Device

In Section 2.11, the square-law transformation

$$y = x^2 \qquad (6.4\text{-}1)$$

was discussed. It was shown that the density function of y was given by

$$W_y(y) = \begin{cases} \dfrac{W_x(\sqrt{y}) + W_x(-\sqrt{y})}{2\sqrt{y}}, & y \geq 0, \\ 0, & y < 0. \end{cases} \tag{6.4-2}$$

If the random variable x was normally distributed with a mean of zero and a variance of σ^2 so that

$$W_x(x) = \frac{1}{\sqrt{2\pi}\,\sigma} e^{-x^2/2\sigma^2}, \qquad -\infty < x < \infty, \tag{6.4-3}$$

then

$$W_y(y) = \begin{cases} \dfrac{1}{\sqrt{2\pi}\,\sigma} y^{-1/2} e^{-y/2\sigma^2}, & y \geq 0, \\ 0, & y < 0. \end{cases} \tag{6.4-4}$$

Both of these last two densities were plotted in Figure 2.10.

Let us now find the autocorrelation function of y, given that x is a wide-sense (and, hence, strict-sense) stationary gaussian process with a second-order density given by Equation 2.21-8:

$$W_x(x_1, x_2; \tau) = \frac{1}{2\pi\sigma^2\sqrt{1-\rho^2}} \exp\left[-\frac{x_1^2 - 2\rho x_1 x_2 + x_2^2}{2\sigma^2(1-\rho^2)}\right] \tag{6.4-5}$$

where

$$x_1 = x(t_1),$$
$$x_2 = x(t_2) = x(t_1 + \tau).$$

Note that the normalized autocorrelation function $\rho(\tau) = \rho$ is given by

$$\rho = \rho(\tau) = \frac{E\{x(t)x(t+\tau)\}}{E\{x^2(t)\}} = \frac{R_x(\tau)}{R_x(0)} \tag{6.4-6}$$

and the variance σ^2 is

$$\sigma^2 = E\{x^2(t)\} = R_x(0). \tag{6.4-7}$$

It follows from Equation 6.3-5 that the autocorrelation function of y may be written as

$$\begin{aligned} R_y(\tau) &= E\{y(t_1)y(t_1+\tau)\} = E\{x^2(t_1)x^2(t_1+\tau)\} \\ &= \frac{1}{2\pi\sigma^2\sqrt{1-\rho^2}} \int_{-\infty}^{\infty}\int_{-\infty}^{\infty} x_1^2 x_2^2 \exp\left[-\frac{x_1^2 - 2\rho x_1 x_2 + x_2^2}{2\sigma^2(1-\rho^2)}\right] dx_1\, dx_2. \end{aligned} \tag{6.4-8}$$

It is clear from this last integral that $E\{y(t_1)y(t_1 + \tau)\}$ will be a function only of τ and hence that $y(t)$ will be a wide-sense stationary process. This is a consequence of the fact that the input process is at least wide-sense stationary and that the nonlinearity $y = x^2$ is time invariant.

The integral in Equation 6.4-8 may be evaluated by completing the square in one variable and making a linear change of variable. After completing the square in x_1, we have

$$R_y(\tau) = \frac{1}{2\pi\sigma^2\sqrt{1-\rho^2}} \int_{-\infty}^{\infty} x_2^2\, e^{-x_2^2/2\sigma^2} \int_{-\infty}^{\infty} x_1^2 \exp\left[-\frac{(x_1-\rho x_2)^2}{2\sigma^2(1-\rho^2)}\right] dx_1\, dx_2. \tag{6.4-9}$$

We now make the change of variable

$$z = \frac{x_1 - \rho x_2}{\sigma\sqrt{1-\rho^2}} \tag{6.4-10}$$

and evaluate the inner integral. We have

$$dx_1 = \sigma\sqrt{1-\rho^2}\, dz \tag{6.4-11}$$

and

$$x_1^2 = \rho^2 x_2^2 + 2\rho\sigma\sqrt{1-\rho^2}\, x_2 z + \sigma^2(1-\rho^2)z^2. \tag{6.4-12}$$

The inner integral of Equation 6.4-9 can now be written as

$$\sigma\sqrt{1-\rho^2}\; \rho^2 x_2^2 \sqrt{2\pi}\left(\frac{1}{\sqrt{2\pi}}\int_{-\infty}^{\infty} e^{-z^2/2}\, dz\right)$$
$$+ 2\rho\sigma^2(1-\rho^2)x_2\sqrt{2\pi}\left(\frac{1}{\sqrt{2\pi}}\int_{-\infty}^{\infty} z\, e^{-z^2/2}\, dz\right)$$
$$+ \sigma^3(1-\rho^2)^{3/2}\sqrt{2\pi}\left(\frac{1}{\sqrt{2\pi}}\int_{-\infty}^{\infty} z^2\, e^{-z^2/2}\, dz\right).$$

From the properties of the unit normal density function we have

$$\frac{1}{\sqrt{2\pi}}\int_{-\infty}^{\infty} e^{-z^2/2}\, dz = 1, \tag{6.4-13}$$

$$\frac{1}{\sqrt{2\pi}}\int_{-\infty}^{\infty} z\, e^{-z^2/2}\, dz = 0, \tag{6.4-14}$$

and

$$\frac{1}{\sqrt{2\pi}}\int_{-\infty}^{\infty} z^2\, e^{-z^2/2}\, dz = 1. \tag{6.4-15}$$

Now Equation 6.4-9 can be written as

$$R_y(\tau) = \frac{1}{\sqrt{2\pi}\,\sigma}\int_{-\infty}^{\infty} [\rho^2 x^4 + \sigma^2(1-\rho^2)x^2]\, e^{-x^2/2\sigma^2}\, dx. \tag{6.4-16}$$

For convenience, we have replaced x_2 by x. This last expression could be written as

$$\begin{aligned} R_y(\tau) &= \rho^2 E_x\{x^4\} + \sigma^2(1-\rho^2)E_x\{x^2\} \\ &= \rho^2 E_x\{x^4\} + \sigma^4(1-\rho^2) \end{aligned} \tag{6.4-17}$$

where $E_x\{\ \}$ denotes expectation with respect to the original input random variable x which was normally distributed with zero mean and variance of σ^2. The fourth moment $E_x\{x^4\}$ can be evaluated from Equation 2.22-21 which gives the fourth moment $E\{x_1x_2x_3x_4\}$ where the x_i are correlated normal random variables with means u_i. Setting $x = x_1 = x_2 = x_3 = x_4$, and $u_1 = u_2 = u_3 = u_4 = 0$, we have

$$E_x\{x^4\} = 3\sigma^4 = 3R_x^2(0). \tag{6.4-18}$$

Now Equation 6.4-17 becomes

$$\boxed{R_y(\tau) = R_x^2(0) + 2R_x^2(\tau),} \tag{6.4-19}$$

the autocorrelation function of a squarer when the input is gaussian with zero mean and correlation function $R_x(\tau)$.

We might note that all of the previous calculations could be avoided and $R_y(\tau)$ could be found directly from Equation 2.22-21. This equation states that, if x_1, x_2, x_3, and x_4 are correlated normal random variables with means u_1, u_2, u_3, and u_4, respectively, then the fourth moment $E\{x_1x_2x_3x_4\}$ is given by

$$\begin{aligned} E\{x_1x_2x_3x_4\} = E\{x_1x_2\}E\{x_3x_4\} + E\{x_1x_3\}E\{x_2x_4\} \\ + E\{x_1x_4\}E\{x_2x_3\} - 2u_1u_2u_3u_4. \end{aligned} \tag{2.22-21}$$

In the case that we are considering here, we can let

$$x_1 = x_2 = x(t), \qquad x_3 = x_4 = x(t + \tau);$$

then Equation 2.22-21 becomes

$$E\{x^2(t)x^2(t + \tau)\} = R_y(\tau) = R_x^2(0) + 2R_x^2(\tau)$$

since $x(t)$ was assumed to have zero mean.

The power spectral density $\varphi_y(\omega)$ of the squarer output may be found by transforming Equation 6.4-19

$$\begin{aligned} \varphi_y(\omega) &= R_x^2(0)\int_{-\infty}^{\infty} e^{-j\omega\tau}\, d\tau + 2\int_{-\infty}^{\infty} R_x^2(\tau)\, e^{-j\omega\tau}\, d\tau \\ &= 2\pi R_x^2(0)\delta(\omega) + 2\int_{-\infty}^{\infty} R_x(\tau)\, e^{-j\omega\tau}\, \frac{1}{2\pi}\int_{-\infty}^{\infty} \varphi_x(\nu)\, e^{j\nu\tau}\, d\nu\, d\tau \\ &= 2\pi R_x^2(0)\delta(\omega) + \frac{1}{\pi}\int_{-\infty}^{\infty} \varphi_x(\nu)\int_{-\infty}^{\infty} R_x(\tau)\, e^{-j(\omega-\nu)\tau}\, d\tau\, d\nu \end{aligned}$$

or, finally,

$$\boxed{\varphi_y(\omega) = 2\pi R_x^2(0)\delta(\omega) + \frac{1}{\pi}\int_{-\infty}^{\infty} \varphi_x(\nu)\varphi_x(\omega - \nu)\, d\nu.} \tag{6.4-20}$$

The first term in this expression is the d-c component that has resulted from the squaring operation. The second term is a convolution of the input spectrum with itself.

The square-law device is encountered frequently in AM demodulation systems. It is instructive, therefore, to consider a bandpass input spectral density such as might arise in a typical AM problem. The exact shape of the spectrum is not important; similar results will be obtained for any spectrum of this general type.

Example 6.1 We consider the input spectral density $\varphi_x(\omega)$ shown in Figure 6.4(*a*). The corresponding autocorrelation function $R_x(\tau)$ is

$$R_x(\tau) = \frac{1}{2\pi}\int_{-\infty}^{\infty} \varphi_x(\omega)\, e^{j\omega\tau}\, d\omega = \frac{1}{2\pi}\int_{-\omega_0 - B/2}^{-\omega_0 + B/2} \frac{N}{2}\, e^{j\omega\tau}\, d\omega$$

$$+ \frac{1}{2\pi}\int_{\omega_0 - B/2}^{\omega_0 + B/2} \frac{N}{2}\, e^{j\omega\tau}\, d\omega,$$

$$R_x(\tau) = \frac{NB}{2\pi}\frac{\sin B\tau/2}{B\tau/2}\cos \omega_0\tau$$

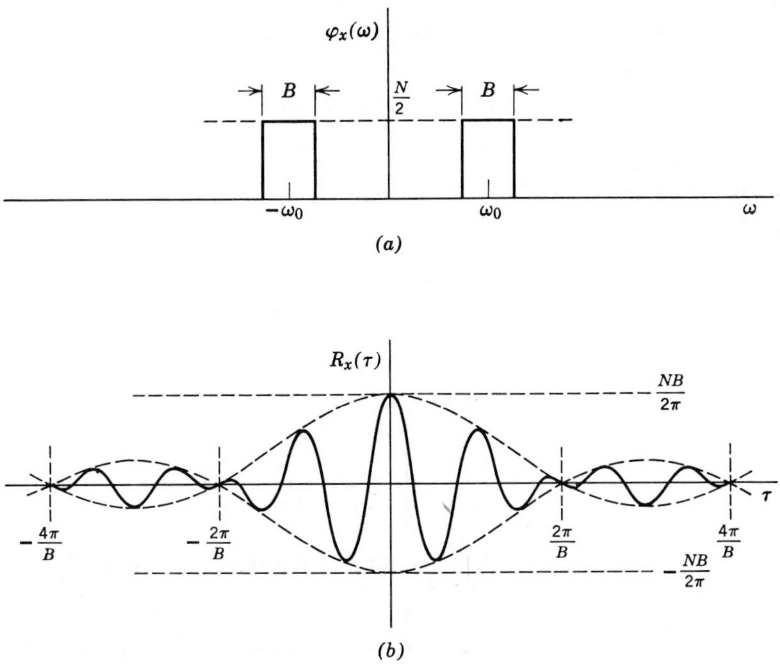

Figure 6.4 *A bandpass Fourier-transform pair: (a) a bandpass power spectral density, $\omega_0 = 2B$; (b) The corresponding autocorrelation function, $\omega_0 = 2B$*

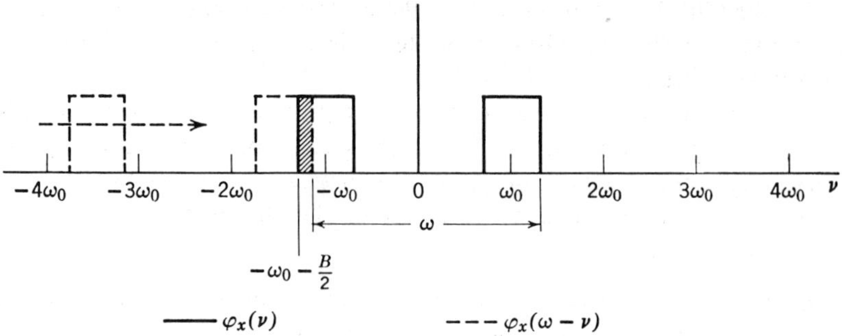

Figure 6.5 *The integrand $\phi_x(\nu)\phi_x(\omega - \nu)$ for $-B - 2\omega_0 \leq \omega \leq -2\omega_0$*

and a typical plot is shown in Figure 6.4(*b*). It is clear that

$$R_x(0) = \frac{NB}{2\pi}.$$

The integral of Equation 6.4-20 may be evaluated as follows. Consider the interval $-B - 2\omega_0 \leq \omega \leq -2\omega_0$. For ω in this interval, the integrand is shown in Figure 6.5 and the integral becomes

$$\frac{1}{\pi}\int_{-\omega_0 - B/2}^{\omega + \omega_0 + B/2} \left(\frac{N}{2}\right)^2 d\nu = \frac{N^2}{4\pi}(\omega + 2\omega_0 + B).$$

In a similar way, the integral may be evaluated for all ω. The result is shown in Figure 6.6. The part of the spectrum that was just calculated is shown shaded.

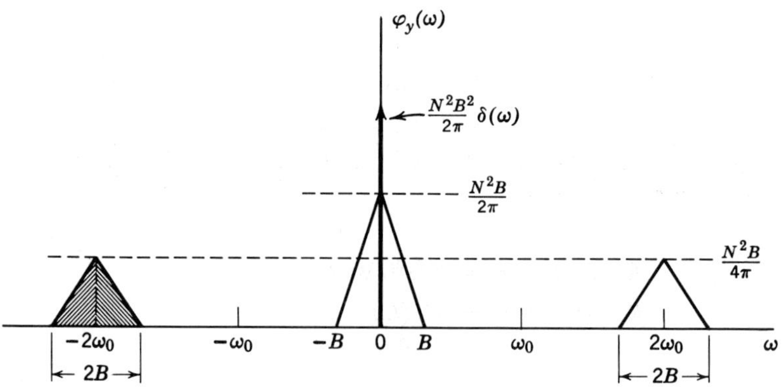

Figure 6.6 *The output power spectral density of the square-law device*

It is apparent that the squaring operation has translated the original spectrum by amount ω_0 in both directions. In the process, it has doubled the basic bandwidth from B to $2B$ and has distorted the shape of the original spectrum.

● ●

6.5 The Square-Law Device—Signal Plus Noise Input

Let us continue the investigation of the square-law device, considering now that the input $x(t)$ consists of two parts, a signal $s(t)$ and an independent additive gaussian noise $n(t)$.

$$x(t) = s(t) + n(t). \tag{6.5-1}$$

It will be convenient to assume that the noise is at least wide-sense stationary but to place no restrictions on the stationarity of the signal. Consequently we must write the output autocorrelation function as

$$\begin{aligned} R_y(t, t + \tau) &= E_y\{y(t)y(t + \tau)\} = E_x\{x^2(t)x^2(t + \tau)\} \\ &= E_x\{[s(t) + n(t)]^2[s(t + \tau) + n(t + \tau)]^2\}. \end{aligned} \tag{6.5-2}$$

This last expression may be expanded and the expectation taken term by term since the expectation of a sum is the sum of the expectations. It is also true that the expectation of the product of two random variables is the product of the expectations provided the two random variables are independent. We obtain

$$\begin{aligned} R_y(t, t + \tau) = {} & E_s\{s^2(t)s^2(t + \tau)\} + E_n\{n^2(t)n^2(t + \tau)\} \\ & + 2E_s\{s(t)s^2(t + \tau)\}E_n\{n(t)\} \\ & + E_s\{s^2(t + \tau)\}E_n\{n^2(t)\} \\ & + 2E_s\{s^2(t)s(t + \tau)\}E_n\{n(t + \tau)\} \\ & + E_s\{s^2(t)\}E_n\{n^2(t + \tau)\} \\ & + 2E_s\{s(t + \tau)\}E_n\{n^2(t)n(t + \tau)\} \\ & + 2E_s\{s(t)\}E_n\{n(t)n^2(t + \tau)\} \\ & + 4E_s\{s(t)s(t + \tau)\}E_n\{n(t)n(t + \tau)\}. \end{aligned} \tag{6.5-3}$$

This expression may be simplified considerably by assuming that both $n(t)$ and $s(t)$ have zero means. Then all first moments are zero; furthermore, it follows from the gaussian assumption and Equation 2.22-20 that all third moments of $n(t)$ are zero. The result is

$$R_y(t, t + \tau) = R_{s \times s}(t, t + \tau) + R_{s \times n}(t, t + \tau) + R_{n \times n}(t, t + \tau) \tag{6.5-4}$$

where

$$R_{s \times s}(t, t + \tau) = E_s\{s^2(t)s^2(t + \tau)\}, \tag{6.5-5}$$

$$\begin{aligned} R_{s \times n}(t, t + \tau) = {} & 4R_s(t, t + \tau)R_n(t, t + \tau) \\ & + R_s(t, t)R_n(t + \tau, t + \tau) \\ & + R_s(t + \tau, t + \tau)R_n(t, t), \end{aligned} \tag{6.5-6}$$

$$R_{n \times n}(t, t + \tau) = E_n\{n^2(t)n^2(t + \tau)\}. \tag{6.5-7}$$

On the assumption already made that $n(t)$ is a wide-sense stationary gaussian process with zero mean and autocorrelation function $R_n(\tau)$, the term $R_{n\times n}$ can be written from Equation 6.4-19 as

$$R_{n\times n}(t, t+\tau) = R_{n\times n}(\tau) = R_n^2(0) + 2R_n^2(\tau), \tag{6.5-8}$$

and the term $R_{s\times n}(t, t+\tau)$ as

$$\begin{aligned} R_{s\times n}(t, t+\tau) = 4R_s(t, t+\tau)R_n(\tau) + R_s(t, t)R_n(0) \\ + R_s(t+\tau, t+\tau)R_n(0). \end{aligned} \tag{6.5-9}$$

6.6 The Square-Law Device—AM Signal Plus Noise Input

Let us assume that the signal $s(t)$ is given by

$$s(t) = C[1 + a(t)] \cos [\omega_0 t + \theta] \tag{6.6-1}$$

where C and ω_0 are constants, θ is a random variable uniformly distributed in $(0, 2\pi)$, and $a(t)$ is a wide-sense stationary gaussian process with zero mean and autocorrelation function $R_a(\tau)$. We will require that the noise $n(t)$, the modulating signal $a(t)$, and the random phase θ be statistically independent. Notice that $s(t)$ has a correlation function given by

$$\begin{aligned} R_s(t, t+\tau) &= E\{s(t)s(t+\tau)\} \\ &= C^2E_\theta\{\cos [\omega_0 t + \theta] \cos [\omega_0(t+\tau) + \theta]\} \\ &\quad \times E_a\{[1 + a(t)][1 + a(t+\tau)]\} \end{aligned}$$

or, finally, from Example 3.6 of Chapter 3,

$$R_s(t, t+\tau) = R_s(\tau) = \frac{C^2}{2} \cos \omega_0\tau[1 + R_a(\tau)]. \tag{6.6-2}$$

The corresponding power spectral density is

$$\begin{aligned} \varphi_s(\omega) &= \int_{-\infty}^{\infty} R_s(\tau)\, e^{-j\omega\tau}\, d\tau \\ &= C^2\frac{\pi}{2}[\delta(\omega - \omega_0) + \delta(\omega + \omega_0)] \\ &\quad + \frac{C^2}{4}[\varphi_a(\omega - \omega_0) + \varphi_a(\omega + \omega_0)]. \end{aligned} \tag{6.6-3}$$

Typical plots of $R_s(\tau)$ and $\varphi_s(\omega)$ are shown in Figure 6.7 for a random process $a(t)$ with a given power spectral density as shown.

We may now find each of the terms of $R_y(t, t+\tau)$ given by Equation 6.5-4:

1. *The noise × noise term* $R_{n\times n}(t, t+\tau)$:
   This term has already been given by Equation 6.5-8 as

$$R_{n\times n}(t, t+\tau) = R_{n\times n}(\tau) = R_n^2(0) + 2R_n^2(\tau) \tag{6.6-4}$$

modulating signal a(t)

$$R_a(\tau) = \frac{AB}{2\pi} \frac{\sin B\tau/2}{B\tau/2}$$

$$\varphi_a(\omega) = \begin{cases} A, & |\omega| \le \dfrac{B}{2} \\ 0, & |\omega| > \dfrac{B}{2} \end{cases}$$

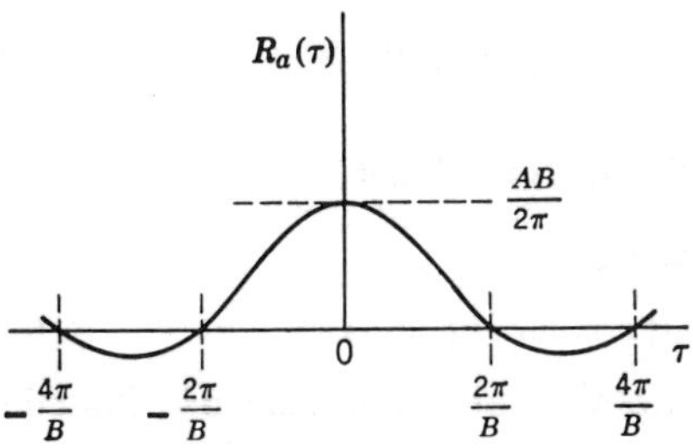

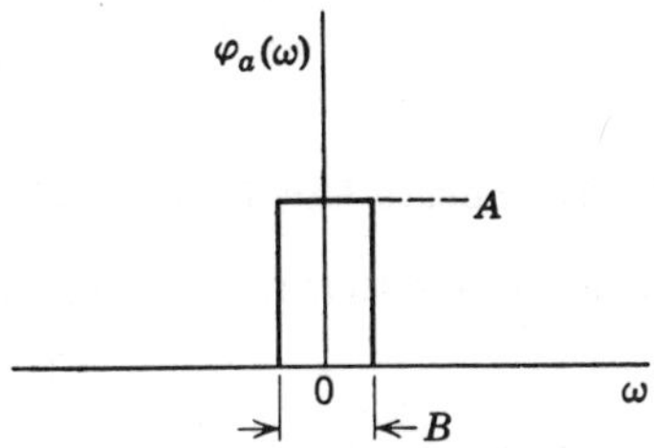

modulated signal $s(t) = C[1 + a(t)] \cos (\omega_0 t + \theta)$

$$R_s(\tau) = \frac{C^2}{2}[1 + R_a(\tau)] \cos \omega_0 \tau$$

$$\varphi_s(\omega) = \frac{C^2\pi}{2}[\delta(\omega - \omega_0) + \delta(\omega + \omega_0)] + \frac{C^2}{4}[\varphi_a(\omega - \omega_0) + \varphi_a(\omega + \omega_0)]$$

$$\omega_0 = 2B$$

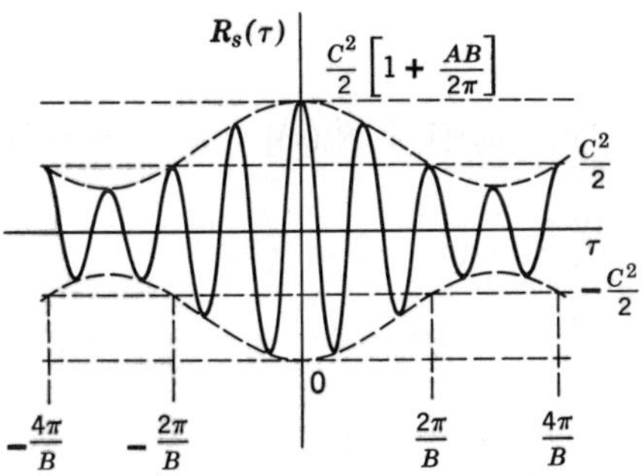

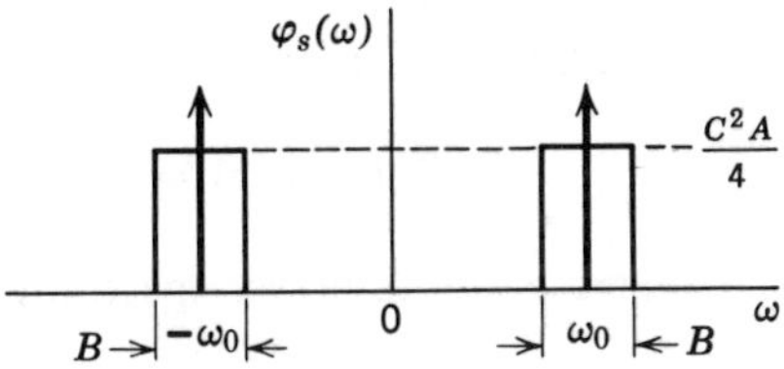

Figure 6.7 *Correlation function and power spectral density of an AM signal*

with power spectral density obtained from Equation 6.4-20 as

$$\varphi_{n \times n}(\omega) = R_n^2(0) 2\pi\delta(\omega) + \frac{1}{\pi}\int_{-\infty}^{\infty} \varphi_n(\nu)\varphi_n(\omega - \nu)\, d\nu. \tag{6.6-5}$$

2. *The signal* $\times$ *signal term* $R_{s \times s}(t, t + \tau)$:
   On substituting for $s(t)$ in Equation 6.5-5, we obtain

$$R_{s \times s}(t, t + \tau) = C^4 E_\theta\{\cos^2 [\omega_0 t + \theta] \cos^2 [\omega_0(t + \tau) + \theta]\} \times E_a\{[1 + a(t)]^2[1 + a(t + \tau)]^2\}. \tag{6.6-6}$$

We have

$$\begin{aligned}E_\theta\{\cos^2 [\omega_0 t + \theta] \cos^2 [\omega_0(t + \tau) + \theta]\} &= E_\theta\{\tfrac{1}{2} + \tfrac{1}{2}\cos 2[\omega_0 t + \theta]\}\{\tfrac{1}{2} + \tfrac{1}{2}\cos 2[\omega_0(t + \tau) + \theta]\} \\ &= \tfrac{1}{4} + \tfrac{1}{4}E_\theta\{\cos 2[\omega_0 t + \theta] \cos 2[\omega_0(t + \tau) + \theta]\} \\ &= \tfrac{1}{4} + \tfrac{1}{8}\cos 2\omega_0\tau.\end{aligned}$$

The last result follows from Example 3.6 of Chapter 3. Since $a(t)$ is assumed to have zero mean and autocorrelation function $R_a(\tau)$, it follows from Equation 6.4-19 that

$$E_a\{[1 + a(t)]^2[1 + a(t + \tau)]^2\} = 1 + 2R_a(0) + 4R_a(\tau) + R_a^2(0) + 2R_a^2(\tau).$$

Again we have made use of the fact that the third moments of a gaussian process with zero mean are zero (see Equation 2.22-20). Thus Equation 6.6-6 becomes

$$R_{s\times s}(t, t + \tau) = R_{s\times s}(\tau) = C^4\{\tfrac{1}{4} + \tfrac{1}{8}\cos 2\omega_0\tau\}\{[1 + R_a(0)]^2 + 4R_a(\tau) + 2R_a^2(\tau)\}. \quad (6.6\text{-}7)$$

The corresponding power spectral density is

$$\begin{aligned}\varphi_{s\times s}(\omega) &= C^4\frac{\pi}{8}[1 + R_a(0)]^2[4\delta(\omega) + \delta(\omega - 2\omega_0) + \delta(\omega + 2\omega_0)] \\ &\quad + \frac{C^4}{4}[4\varphi_a(\omega) + \varphi_a(\omega - 2\omega_0) + \varphi_a(\omega + 2\omega_0)] \\ &\quad + \frac{C^4}{16\pi}\int_{-\infty}^{\infty} \varphi_a(\nu)[4\varphi_a(\omega - \nu) + \varphi_a(\omega - 2\omega_0 - \nu) \\ &\qquad + \varphi_a(\omega + 2\omega_0 - \nu)]\, d\nu. \quad (6.6\text{-}8)\end{aligned}$$

3. *The signal × noise term* $R_{s\times n}(t, t + \tau)$:
   This term is given by Equation 6.5-9 after substitution of Equation 6.6-2 for $R_s(\tau)$:

$$R_{s\times n}(\tau) = C^2R_n(0)[1 + R_a(0)] + 2C^2R_n(\tau)[1 + R_a(\tau)]\cos\omega_0\tau. \quad (6.6\text{-}9)$$

The power spectral density is

$$\begin{aligned}\varphi_{s\times n}(\omega) &= 2\pi C^2R_n(0)[1 + R_a(0)]\delta(\omega) + C^2[\varphi_n(\omega - \omega_0) + \varphi_n(\omega + \omega_0)] \\ &\quad + \frac{C^2}{2\pi}\int_{-\infty}^{\infty} \varphi_n(\nu)[\varphi_a(\omega - \omega_0 - \nu) + \varphi_a(\omega + \omega_0 - \nu)]\, d\nu.\end{aligned}$$

(6.6-10)

These expressions for the autocorrelation function and the power spectral density are somewhat complicated functions of $\varphi_a(\omega)$ and $\varphi_n(\omega)$. In the following example, we consider the power spectral density of $y(t)$ for a particular choice of these spectral densities.

Example 6.2 Let us choose the noise spectral density $\varphi_n(\omega)$ shown in Figure 6.4 and the modulating signal spectral density $\varphi_a(\omega)$ shown in Figure 6.7. This choice could be regarded as the worst choice since the noise and modulated signal have spectra that are indistinguishable in shape and which completely overlap. We have

$$R_n(\tau) = \frac{NB}{2\pi}\frac{\sin B\tau/2}{B\tau/2}\cos\omega_0\tau, \tag{A}$$

$$R_n(0) = \frac{NB}{2\pi}, \tag{B}$$

$$R_a(\tau) = \frac{AB}{2\pi}\frac{\sin B\tau/2}{B\tau/2}, \tag{C}$$

$$R_a(0) = \frac{AB}{2\pi}. \tag{D}$$

We now proceed to evaluate the various terms in $\varphi_y(\omega)$.

1. *The noise $\times$ noise term* $\varphi_{n\times n}(\omega)$:
   The pertinent equation here is Equation 6.6-5, which becomes

$$\varphi_{n\times n}(\omega) = \frac{N^2B^2}{2\pi}\,\delta(\omega) + \frac{1}{\pi}\int_{-\infty}^{\infty}\varphi_n(\nu)\varphi_n(\omega-\nu)\,d\nu. \tag{E}$$

   This integral was already evaluated in Example 6.1 and shown in Figure 6.6. It is repeated in Figure 6.8(*a*).

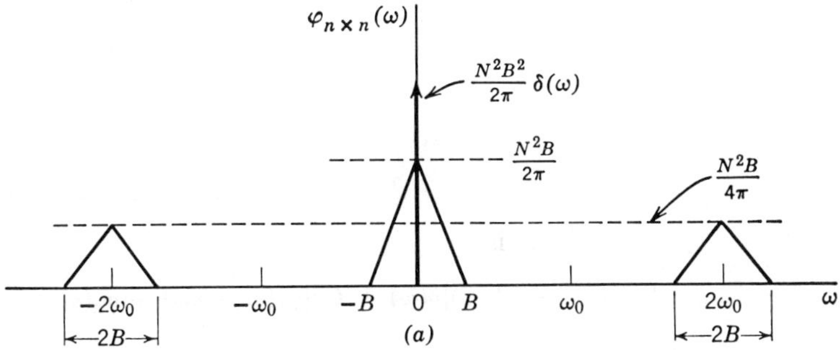

Figure 6.8 *The power spectral density for an AM signal at the output of a square-law device: (a) the noise $\times$ noise term*

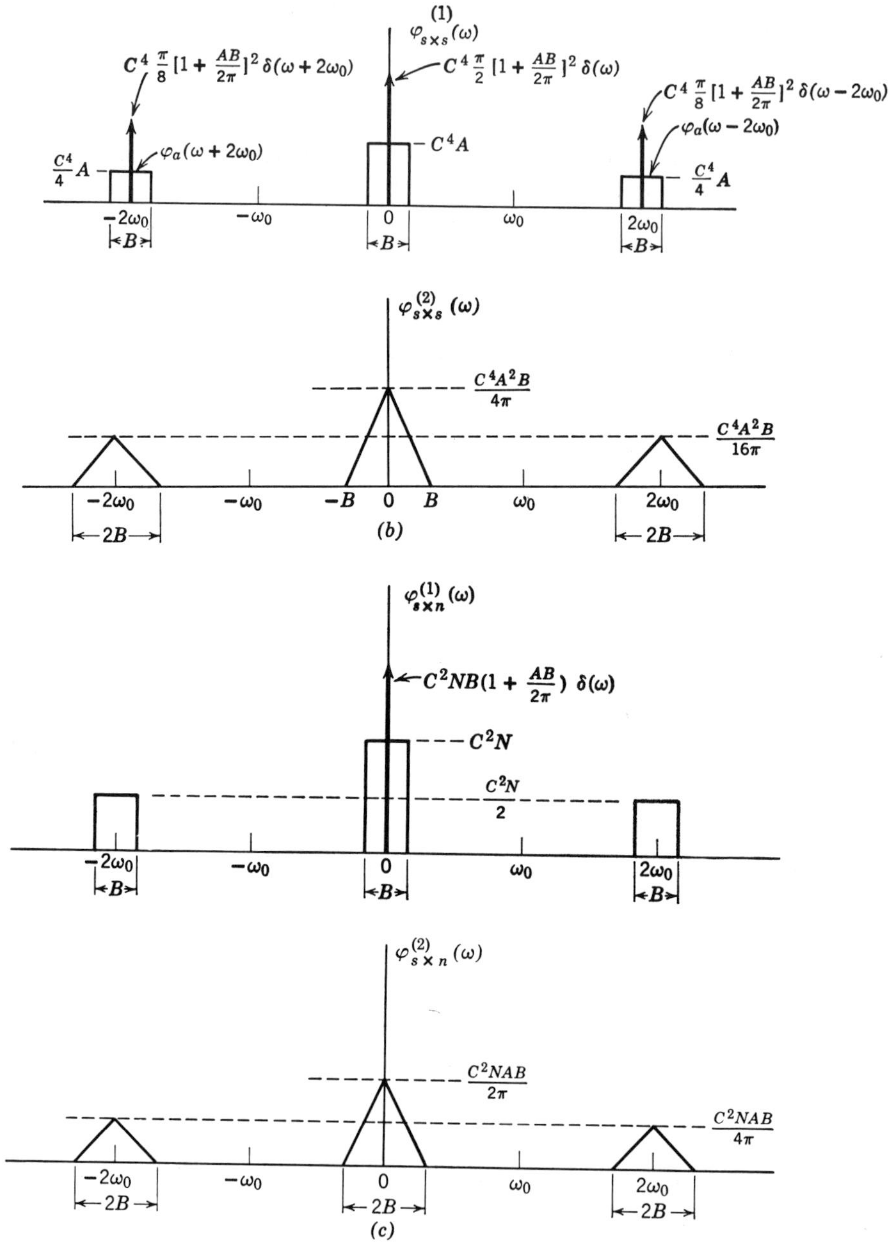

Figure 6.8 *(b) the signal × signal term; (c) the signal × noise term*

2. *The signal × signal term* $\varphi_{s\times s}(\omega)$:
This term is given by Equation 6.6-8. The substitution is straightforward and the results are shown in Figure 6.8(*b*). It is convenient to divide this spectrum into two parts

$$\varphi_{s\times s}(\omega) = \varphi^{(1)}_{s\times s}(\omega) + \varphi^{(2)}_{s\times s}(\omega) \tag{F}$$

where the first part $\varphi^{(1)}_{s\times s}(\omega)$ consists of delta functions (sinusoids and d-c terms) and the translated spectrum of $a(t)$, while the second part $\varphi^{(2)}_{s\times s}(\omega)$ arises from the squaring operation on $a(t)$ and is given by the convolution integral of Equation 6.6-8. The two parts are shown separately in Figure 6.8(*b*).
3. *The signal × noise term* $\varphi_{s\times n}(\omega)$:
This term is given by Equation 6.6-10. As in the previous case, it is convenient to consider two parts

$$\varphi_{s\times n}(\omega) = \varphi^{(1)}_{s\times n}(\omega) + \varphi^{(2)}_{s\times n}(\omega).$$

The first part contains the delta functions and the translated noise power spectral density, while the second part is the convolution of noise and modulating signal spectrum. These are shown in Figure 6.8(*c*).

• •

6.7 The Square-Law Receiver

If a low-pass filter is added after the square-law device as shown in Figure 6.9(*a*), the result is a square-law AM receiver. For simplicity, we will assume that the low-pass filter has the ideal transfer function shown in Figure 6.9(*b*). The results will not be appreciably different for reasonable approximations to this transfer function.

For the input $x(t)$ discussed in Section 6.6, the power spectral density $\varphi_y(\omega)$ after the squarer can be written as the sum of three terms

$$\varphi_y(\omega) = \varphi_{s\times s}(\omega) + \varphi_{s\times n}(\omega) + \varphi_{n\times n}(\omega) \tag{6.7-1}$$

given by Equations 6.6-9, 6.6-10, and 6.6-5, respectively. The output $\varphi_o(\omega)$ from the low-pass filter of Figure 6.9 is just

$$\varphi_o(\omega) = \varphi_y(\omega)|H(\omega)|^2 \tag{6.7-2}$$

or

$$\varphi_o(\omega) = \begin{cases} \dfrac{\varphi_y(\omega)}{C^4}, & |\omega| \le \dfrac{B}{2}, \\ 0, & |\omega| > \dfrac{B}{2}. \end{cases} \tag{6.7-3}$$

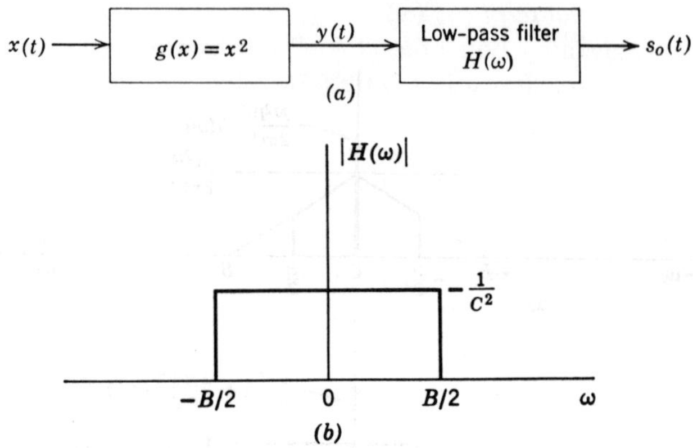

Figure 6.9 The square-law receiver: (a) system block diagram:

$$x(t) = s(t) + n(t) = C[1 + a(t)] \cos \omega_0 t + n(t)$$
$$y(t) = x^2(t);$$

(b) an ideal low-pass filter transfer function

Thus the output $s_o(t)$ of the square-law receiver will consist only of those terms with frequency components below $B/2$ appropriately modified by the scale factor $1/C^4$. Note that, in general, this output will contain components from each of the three classes of terms (signal × signal), (signal × noise), and (noise × noise).

Example 6.3 We continue our example using the inputs described in Example 6.2 and the square-law receiver of Figure 6.9. The ideal low-pass filter of Figure 6.9(*b*) will modify the spectra shown in Figure 6.8 in two ways:

1. The spectral components in the frequency range $|\omega| \leq B/2$ will be multiplied by the scale factor $1/C^4$.
2. The spectral components in the frequency range $|\omega| > B/2$ will be eliminated.

The resulting spectrum of $s_o(t)$, the square-law receiver output, is shown in Figure 6.10 and can be written as

$$\varphi_o(\omega) = \begin{cases} \left(\dfrac{A^2}{4\pi} + \dfrac{AN}{2\pi C^2} + \dfrac{N^2}{2\pi C^4}\right)(B - |\omega|) + A + \dfrac{N}{C^2}, & |\omega| \leq \dfrac{B}{2}, \\ 0, & |\omega| > \dfrac{B}{2}, \end{cases}$$

+ d-c terms represented by delta functions at the origin.

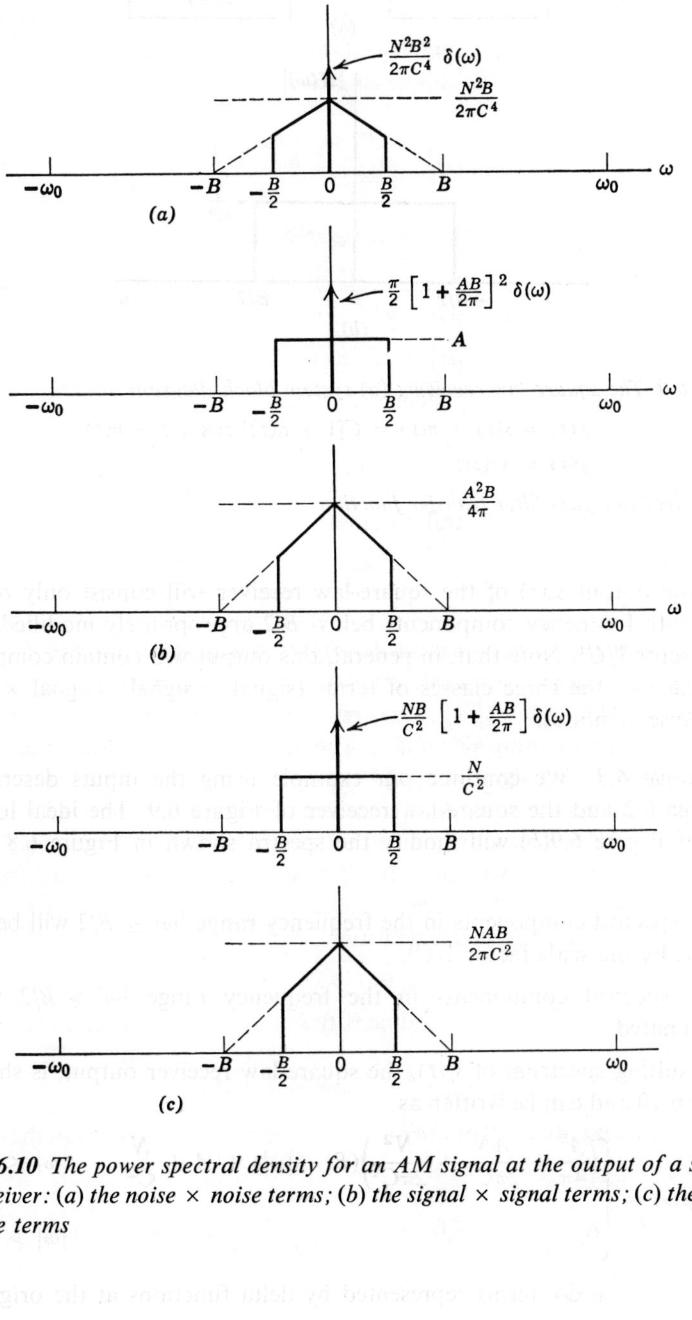

***Figure 6.10** The power spectral density for an AM signal at the output of a square-law receiver: (a) the noise × noise terms; (b) the signal × signal terms; (c) the signal × noise terms*

In the usual application where it is desired to reproduce the modulating signal $a(t)$, all of the spectral terms except one represent undesirable output signal components. This one term is the first term (linear term) in the signal × signal component and is just the power spectral density $\varphi_a(\omega)$ of the modulating signal $a(t)$. In this example it is the constant A in the interval $(-B/2, B/2)$.

• •

6.8 Receiver Evaluation—The Mean-Squared Error

Let us now consider some measures of the receiver's merit. We begin by finding the mean-squared error, which may be defined as

$$E\{\epsilon_t^2\} = E\{[a(t) - s_o(t)]^2\}. \tag{6.8-1}$$

This expression may be expanded to give

$$E\{\epsilon_t^2\} = R_a(0) - 2R_{as_o}(0) + R_{s_o}(0) \tag{6.8-2}$$

where the various correlation functions have been defined in the usual way. In terms of the power spectral densities

$$E\{\epsilon_t^2\} = \frac{1}{2\pi}\int_{-\infty}^{\infty} [\varphi_a(\omega) - 2\varphi_{as_o}(\omega) + \varphi_{s_o}(\omega)]\, d\omega. \tag{6.8-3}$$

Here $\varphi_a(\omega)$ is the power spectral density of the modulating signal $a(t)$ and $\varphi_{s_o}(\omega)$ is the power spectral density of the total output $s_o(t)$ as shown in Figure 6.10. The term $\varphi_{as_o}(\omega)$ is the cross-spectral density between the output $s_o(t)$ and the modulating signal $a(t)$. The assumption has already been made that the modulating signal $a(t)$ and the noise $n(t)$ are independent gaussian processes with zero means. As a consequence $a(t)$ is uncorrelated with all of the (noise × noise) components and all of the (signal × noise) components. It is also uncorrelated with all of the (signal × signal) components (the third moments of $a(t)$ are zero) except that component which is $a(t)$ itself. The result is, for arbitrary spectral densities,

$$\varphi_{as_o}(\omega) = \varphi_a(\omega). \tag{6.8-4}$$

Furthermore, the output spectral density $\varphi_{s_o}(\omega)$ can be written as the sum of two terms

$$\varphi_{s_o}(\omega) = \varphi_u(\omega) + \varphi_a(\omega) \tag{6.8-5}$$

where $\varphi_u(\omega)$ consists of all of the undesired spectral components in the output. Now Equation 6.8-3 becomes

$$E\{\epsilon_t^2\} = \frac{1}{2\pi}\int_{-\infty}^{\infty} \varphi_u(\omega)\, d\omega = \frac{1}{2\pi}\int_{-\infty}^{\infty} [\varphi_{s_o}(\omega) - \varphi_a(\omega)]\, d\omega \tag{6.8-6}$$

as expected.

Example 6.4 For the square-law receiver outputs shown in Example 6.3, the mean-squared error is easily evaluated. In the interest of simplicity we will assume that the d-c terms can be filtered out without appreciable effect on the rest of the spectrum of $\varphi_{s_0}(\omega)$. This will certainly be true if $B \gg 0$. Thus we will drop all delta functions from the spectrum of $\varphi_{s_0}(\omega)$. The result is

$$E\{\epsilon_t^2\} = \frac{1}{2\pi}\int_{-\infty}^{\infty} \varphi_u(\omega)\, d\omega$$

$$= \frac{1}{2\pi}\text{ [area under curves of Figure 6.10 except for that of the first (signal} \times \text{signal) term]}$$

$$= \frac{1}{2\pi}\left(\frac{3N^2B^2}{8\pi C^4} + \frac{3A^2B^2}{16\pi} + \frac{NB}{C^2} + \frac{3NAB^2}{8\pi C^2}\right).$$

This expression may be rearranged as

$$E\{\epsilon_t^2\} = \frac{3}{8}\frac{A^2B^2}{4\pi^2}\left(\frac{2N^2}{A^2C^4} + 1 + \frac{2N}{AC^2}\right) + \frac{AB}{2\pi}\frac{N}{AC^2}.$$

We note that the variance (mean-squared normalized power) of the modulating signal $a(t)$ is given by

$$\sigma_a^2 = R_a(0) = \frac{1}{2\pi}\int_{-B/2}^{B/2} A\, d\omega = \frac{AB}{2\pi}.$$

Let us also define a parameter P by

$$P = \frac{N}{AC^2}.$$

This parameter is a measure of the inverse signal-to-noise power ratio at the receiver input. Now the mean-squared error is simply

$$E\{\epsilon_t^2\} = \tfrac{3}{8}\sigma_a^4(2P^2 + 2P + 1) + \sigma_a^2 P$$

which is a function of the two parameters σ_a^2, the modulating signal power, and P, the input inverse S/N power ratio.

It is instructive to plot the mean-squared error as a function of P for various values of σ_a^2 as shown in Figure 6.11. Reasonable values for σ_a^2 have been chosen to prevent overmodulation by the gaussian modulating signal $a(t)$. For overmodulation it is necessary that $a(t) < -1$; hence a variance σ_a^2 which is much larger than unity will give rise to an appreciable amount of overmodulation. An examination of Figure 6.11 shows that the mean-squared error increases with increasing P (i.e., with increasing noise) as would be expected, since, for large values of P, the (noise $\times$ noise) error terms predominate. Also as σ_a^2 (the signal power) increases, the mean-squared error increases due to the (signal $\times$ signal) error terms. Thus we might reasonably expect some optimum condition at intermediate signal and noise levels.

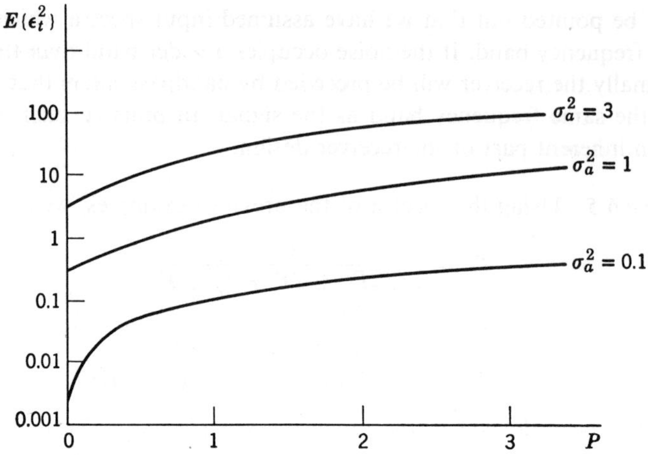

Figure 6.11 *The mean-squared error for the square-law receiver:*

$$E\{\epsilon_t^2\} = \tfrac{3}{8}\sigma_a^4(2P^2 + 2P + 1) + \sigma_a^2 P$$

To show that such is indeed the case, we consider another measure of the receiver's merit.

• •

6.9 Receiver Evaluation—Improvement Factor

The expression $E\{\epsilon_t^2\}$ for the mean-squared error is actually the noise power (normalized) out of the receiver, if we consider all undesired output components to be noise. The output signal power (normalized) is just σ_a^2. Thus the output signal-to-noise power ratio is

$$R_o = \frac{\int_{-\infty}^{\infty} \varphi_a(\omega)\, d\omega}{\int_{-\infty}^{\infty} [\varphi_{s_o}(\omega) - \varphi_a(\omega)]\, d\omega} \tag{6.9-1}$$

where $\varphi_a(\omega)$ is the modulating signal power spectral density and $\varphi_{s_o}(\omega)$ is that of the total receiver output. The input signal-to-noise power ratio R_i can be written as

$$R_i = \frac{(C^2/2)\int_{-\infty}^{\infty} \varphi_a(\omega)\, d\omega}{\int_{-\infty}^{\infty} \varphi_n(\omega)\, d\omega} = \frac{\sigma_a^2(C^2/2)}{\sigma_n^2} \tag{6.9-2}$$

where $\varphi_n(\omega)$ is the input noise power spectral density and σ_n^2 is the input noise power.

At this point, it is convenient to define an improvement ratio I as

$$I = \frac{R_o}{R_i} = \frac{\int_{-\infty}^{\infty} \varphi_n(\omega)\, d\omega}{(C^2/2)\int_{-\infty}^{\infty} [\varphi_{s_o}(\omega) - \varphi_a(\omega)]\, d\omega}. \tag{6.9-3}$$

It should be pointed out that we have assumed input spectra which occupy the same frequency band. If the noise occupies a wider band than the signal, then normally the receiver will be preceded by bandpass filters that limit the noise to the same frequency band as the signal. In practice such filters are usually an inherent part of the receiver design.

Example 6.5 Using the spectra of the previous examples, we have

$$R_o = \frac{1}{\frac{3}{8}\sigma_a^2(2P^2 + 2P + 1) + P}$$

and

$$R_i = \frac{\sigma_a^2(C^2/2)}{\sigma_n^2} = \frac{ABC^2/4\pi}{NB/2\pi} = \frac{1}{2}\frac{AC^2}{N} = \frac{1}{2P}$$

where P has been defined in the previous example. Now the improvement factor becomes

$$I = \frac{R_o}{R_i} = \frac{2}{1 + \frac{3}{8}\sigma_a^2(2P + 2 + 1/P)}.$$

This expression is plotted versus P in Figure 6.12 for the same values of σ_a^2 as was used in Figure 6.11. The maxima occur at the value $P = 1/\sqrt{2}$ and, in a certain sense, define the signal and noise conditions for which the square-law receiver performs best. It is clear that, for either error criterion

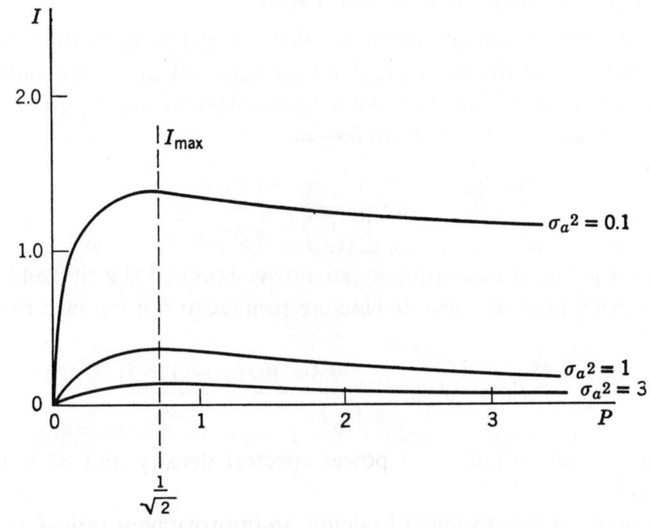

Figure 6.12 *The improvement factor for the square-law receiver:*

$$I = \frac{R_o}{R_i} = \frac{2}{1 + \frac{3}{8}\sigma_a^2(2P + 2 + 1/P)}$$

considered here, the square-law receiver is not a high-performance device.

For low-noise situations ($P \to 0$) the improvement factor I is given approximately by

$$I \approx \frac{2P}{\frac{3}{8}\sigma_a^2}.$$

Thus the curves of Figure 6.12 are linear functions of P in the vicinity of the origin. On the other hand, for small-signal conditions ($P \to \infty$), I is approximated by

$$I \approx \frac{1}{\frac{3}{8}\sigma_a^2 P}.$$

This last relationship illustrates the small-signal suppression effect which is typical of squarers and, indeed, of nonlinear demodulators in general.

• •

6.10 The Hard Limiter

As a further example of the direct method of calculation, we consider the *hard limiter* with zero-memory nonlinear characteristic shown in Figure 6.13(*a*). A typical input and resulting output are shown in Figure 6.13(*b*). This device may be described by

$$y = g(x) = \begin{cases} 1, & x > 0, \\ 0, & x = 0, \\ -1, & x < 0, \end{cases} \tag{6.10-1}$$

or by

$$y = g(x) = 2u(x) - 1 \tag{6.10-2}$$

where $u(x)$ is the unit-step function.

It is clear that, if x is a continuous random variable, then y will be a discrete random variable which takes on the value -1 with a probability given by $P(x < 0)$ and the value $+1$ with probability $P(x > 0)$. In terms of the Dirac delta function, the density function of y can be written as

$$W_y(y) = \delta(y + 1)\int_{-\infty}^{0} W_x(x)\,dx + \delta(y - 1)\int_{0}^{\infty} W_x(x)\,dx \tag{6.10-3}$$

where $W_x(x)$ is the density function of x. The cumulative distribution function of y is then given by

$$F_y(y) = u(y + 1)\int_{-\infty}^{0} W_x(x)\,dx + u(y - 1)\int_{0}^{\infty} W_x(x)\,dx. \tag{6.10-4}$$

Both of these functions are illustrated in Figure 6.14. For any input distribution which has zero median [so that $F_x(0) = \frac{1}{2}$], it is clear that $A = B = \frac{1}{2}$.

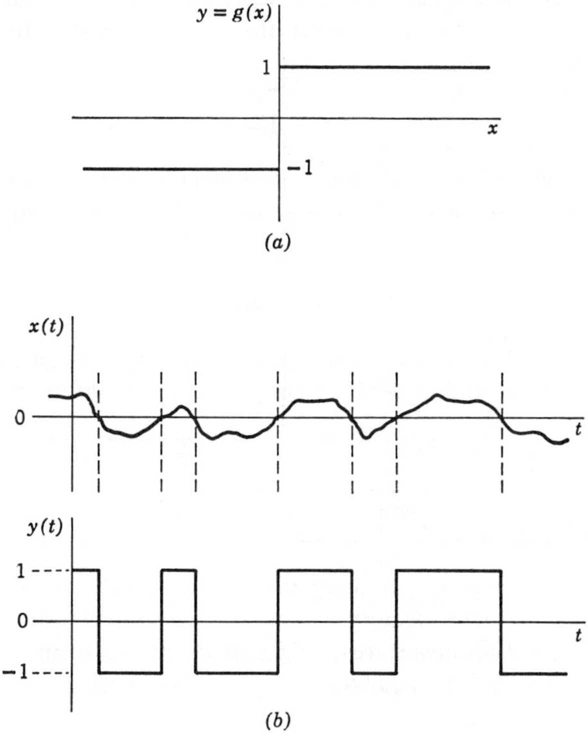

Figure 6.13 *The hard limiter: (a) the zero-memory characteristic: (b) a typical input and output*

We proceed now to consider the output autocorrelation function $R_y(\tau)$ on the assumption that the input $x(t)$ is a wide-sense stationary random process with zero mean, autocorrelation function $R(\tau)$, and normalized autocorrelation function $\rho(\tau)$ given by

$$\rho(\tau) = \frac{R(\tau)}{R(0)}. \tag{6.10-5}$$

The output autocorrelation function is

$$\begin{aligned} R_y(\tau) &= E\{y(t)y(t+\tau)\} \\ &= \int_{-\infty}^{\infty}\int_{-\infty}^{\infty} g(x_1)g(x_2)W_x(x_1, x_2; \tau)\, dx_1\, dx_2. \end{aligned} \tag{6.10-6}$$

This expression may be written somewhat more explicitly by noting that the product $g(x_1)g(x_2)$ is $+1$ in the first and third quadrants of the x_1x_2 plane

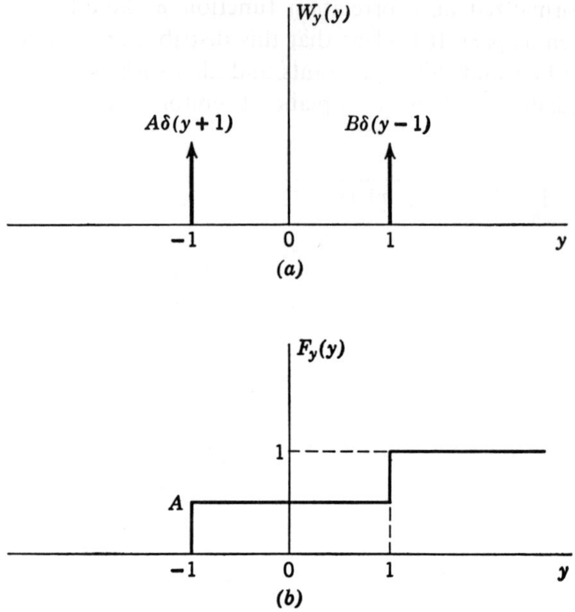

Figure 6.14 *The hard limiter output distributions: (a) the output density function for a hard limiter; (b) the output cumulative distribution function for a hard limiter:*

$$A = \int_{-\infty}^{0} W_x(x)\,dx; \qquad B = \int_{0}^{\infty} W_x(x)\,dx; \qquad A + B = 1$$

and -1 in the second and fourth quadrants. Consequently Equation 6.10-6 becomes

$$\begin{aligned} R_y(\tau) = & \int_0^\infty \int_0^\infty W_x(x_1, x_2; \tau)\,dx_1\,dx_2 \\ & + \int_{-\infty}^{0} \int_{-\infty}^{0} W_x(x_1, x_2; \tau)\,dx_1\,dx_2 \\ & - \int_0^\infty \int_{-\infty}^{0} W_x(x_1, x_2; \tau)\,dx_1\,dx_2 \\ & - \int_{-\infty}^{0} \int_0^\infty W_x(x_1, x_2; \tau)\,dx_1\,dx_2. \end{aligned} \tag{6.10-7}$$

At this point we will make the specific assumption that the input process $x(t)$ is gaussian so that the joint distribution can be written from Equation 6.4-5 as

$$W_x(x_1, x_2; \tau) = \frac{1}{2\pi\sigma^2\sqrt{1-\rho^2}} \exp\left[-\frac{x_1^2 + x_2^2 - 2\rho x_1 x_2}{2\sigma^2(1-\rho^2)}\right] \tag{6.10-8}$$

where the normalized autocorrelation function ρ should be more appropriately written as $\rho(\tau)$. It is clear that this distribution is symmetrical with respect to the first and third quadrants and also with respect to the second and fourth quadrants of the x_1x_2 plane. Therefore Equation 6.10-7 can be written as

$$R_y(\tau) = 2\int_0^\infty \int_0^\infty W_x(x_1, x_2; \tau)\, dx_1\, dx_2 - 2\int_0^\infty \int_{-\infty}^0 W_x(x_1, x_2; \tau)\, dx_1\, dx_2 \qquad (6.10\text{-}9)$$

or, after replacing x_1 by $-x_1$ in the second integral, as

$$R_y(\tau) = 2\int_0^\infty \int_0^\infty W_x(x_1, x_2; \tau) - W_x(-x_1, x_2; \tau)\, dx_1\, dx_2. \qquad (6.10\text{-}10)$$

Let us now substitute Equation 6.10-8 into this last expression to obtain

$$R_y(\tau) = \frac{1}{\pi\sigma^2\sqrt{1-\rho^2}}\int_0^\infty \int_0^\infty \left\{\exp\left[-\frac{x_1^2 + x_2^2 - 2\rho x_1 x_2}{2\sigma^2(1-\rho^2)}\right] - \exp\left[-\frac{x_1^2 + x_2^2 + 2\rho x_1 x_2}{2\sigma^2(1-\rho^2)}\right]\right\} dx_1\, dx_2. \qquad (6.10\text{-}11)$$

At this point it is convenient to transform to polar coordinates by the substitutions

$$x_1 = r\cos\theta, \qquad x_2 = r\sin\theta,$$
$$dx_1\, dx_2 = r\, dr\, d\theta. \qquad (6.10\text{-}12)$$

We obtain

$$R_y(\tau) = \frac{1}{\pi\sigma^2\sqrt{1-\rho^2}}\int_0^{\pi/2} \int_0^\infty \left(\exp\left\{-r^2\left[\frac{1-\rho\sin 2\theta}{2\sigma^2(1-\rho^2)}\right]\right\} - \exp\left\{-r^2\left[\frac{1+\rho\sin 2\theta}{2\sigma^2(1-\rho^2)}\right]\right\}\right) r\, dr\, d\theta \qquad (6.10\text{-}13)$$

where we have used the identity

$$2\sin\theta\cos\theta = \sin 2\theta. \qquad (6.10\text{-}14)$$

Now we may integrate with respect to r by making use of the relationship

$$\int_0^\infty e^{-r^2k} r\, dr = -\frac{1}{2k} e^{-r^2k}\bigg|_0^\infty = \frac{1}{2k}. \qquad (6.10\text{-}15)$$

The result is that Equation 6.10-13 becomes

$$R_y(\tau) = \frac{\sqrt{1-\rho^2}}{\pi}\int_0^{\pi/2} \left(\frac{1}{1-\rho\sin 2\theta} - \frac{1}{1+\rho\sin 2\theta}\right) d\theta. \qquad (6.10\text{-}16)$$

If we now make the change in variable [2]

$$z = \tan\theta \tag{6.10-17}$$

this equation becomes

$$R_y(\tau) = \frac{\sqrt{1-\rho^2}}{\pi}\int_0^\infty \left(\frac{1}{1-2\rho z+z^2} - \frac{1}{1+2\rho z+z^2}\right) dz. \tag{6.10-18}$$

This last expression may be integrated to give [3]

$$R_y(\tau) = \frac{1}{\pi}\left(\arctan\frac{z-\rho}{\sqrt{1-\rho^2}} - \arctan\frac{z+\rho}{\sqrt{1-\rho^2}}\right)_0^\infty$$

$$= \frac{1}{\pi}\left(\frac{\pi}{2} - \arctan\frac{-\rho}{\sqrt{1-\rho^2}} - \frac{\pi}{2} + \arctan\frac{\rho}{\sqrt{1-\rho^2}}\right)$$

$$= \frac{2}{\pi}\arctan\frac{\rho}{\sqrt{1-\rho^2}} \tag{6.10-19}$$

or, finally [4],

$$\boxed{R_y(\tau) = \frac{2}{\pi}\arcsin\rho(\tau).} \tag{6.10-20}$$

The function arc sin ρ may be expanded in a power series [5] to yield

$$\frac{2}{\pi}\arcsin\rho = \frac{2}{\pi}\left(\rho + \frac{1}{6}\rho^3 + \frac{1\cdot 3}{2\cdot 4}\frac{1}{5}\rho^5 + \frac{1\cdot 3\cdot 5}{2\cdot 4\cdot 6}\frac{1}{7}\rho^7 + \cdots\right) \tag{6.10-21}$$

which is valid for $\rho^2 < 1$. Thus, for values of τ near zero so that the normalized input autocorrelation function $\rho(\tau)$ is nearly unity, we have the approximation that

$$R_y(\tau) \approx \rho(\tau). \tag{6.10-22}$$

On the other extreme, for values of $\rho(\tau)$ that are nearly zero,

$$R_y(\tau) \approx \frac{2}{\pi}\rho(\tau). \tag{6.10-23}$$

The effect of clipping has been to make the correlation function fall off faster and to broaden the original spectrum.

6.11 The Characteristic Function Method

It is clear from the previous discussion that the second-moment analysis of instantaneous nonlinearities by the direct method reduces to the evaluation of the definite integral of Equation 6.3-5; that is,

$$R_y(t_1, t_2) = E\{y_1 y_2\}$$

$$= \int_{-\infty}^{\infty}\int_{-\infty}^{\infty} g(x_1)g(x_2)W_x(x_1, x_2)\, dx_1\, dx_2. \tag{6.11-1}$$

It will frequently happen that this integral is easier to evaluate if rewritten in terms of the joint characteristic function $F_x(v_1, v_2)$ given by Equation 2.20-41:

$$F_x(v_1, v_2) = E\{e^{j(v_1x_1+v_2x_2)}\}$$
$$= \int_{-\infty}^{\infty}\int_{-\infty}^{\infty} W_x(x_1, x_2)\, e^{j(v_1x_1+v_2x_2)}\, dx_1\, dx_2. \tag{6.11-2}$$

As discussed in Chapter 2, this equation may be inverted to give

$$W_x(x_1, x_2) = \frac{1}{(2\pi)^2}\int_{-\infty}^{\infty}\int_{-\infty}^{\infty} F_x(v_1, v_2)\, e^{-j(v_1x_1+v_2x_2)}\, dv_1\, dv_2. \tag{6.11-3}$$

If we substitute this last expression into Equation 6.11-1 and interchange the order of integration, we obtain

$$R_y(t_1, t_2) = \frac{1}{(2\pi)^2}\int_{-\infty}^{\infty}\int_{-\infty}^{\infty} F_x(v_1, v_2)\left[\int_{-\infty}^{\infty} g(x_1)\, e^{-jv_1x_1}\, dx_1\right]$$
$$\times \left[\int_{-\infty}^{\infty} g(x_2)\, e^{-jv_2x_2}\, dx_2\right] dv_1\, dv_2. \tag{6.11-4}$$

Each of the expressions in brackets in this last equation is the Fourier transform of the instantaneous nonlinearity $g(x)$. Let us denote this transform by $G(\omega)$ so that

$$G(\omega) = \int_{-\infty}^{\infty} g(x)\, e^{-j\omega x}\, dx \tag{6.11-5}$$

is the Fourier transform of $g(x)$. Now Equation 6.11-4 may be written formally as

$$R_y(t_1, t_2) = \frac{1}{(2\pi)^2}\int_{-\infty}^{\infty}\int_{-\infty}^{\infty} F_x(v_1, v_2)G(v_1)G(v_2)\, dv_1\, dv_2. \tag{6.11-6}$$

For nonlinearities $g(x)$ that have simple transforms $G(\omega)$, the last equation may be more convenient than Equation 6.11-1 for the evaluation of the output autocorrelation function $R_y(t_1, t_2)$. Also for wide classes of nonlinearities where the transform of Equation 6.11-5 does not properly exist, this approach is still preferable to the direct method and may be implemented by considering the complex variable s where

$$s = \sigma + j\omega \tag{6.11-7}$$

replaces $j\omega$ in Equation 6.11-5. Now the integral of Equation 6.11-5 may be evaluated over appropriate regions of convergence in the complex s plane, these regions depending on the values of σ. The results may then be used for $\sigma = 0$.

Suppose that we define the functions $g_+(x)$ and $g_-(x)$ by

$$g_+(x) = g(x)u(x) \tag{6.11-8}$$

and

$$g_-(x) = g(x)u(-x) \tag{6.11-9}$$

where $u(x)$ is the unit-step function; then

$$g(x) = g_+(x) + g_-(x). \tag{6.11-10}$$

Now we define the *unilateral Laplace transforms* of $g_+(x)$ and $g_-(x)$ by

$$G_+(s) = \int_0^\infty g(x)\, e^{-sx}\, dx \tag{6.11-11}$$

and

$$G_-(s) = \int_{-\infty}^0 g(x)\, e^{-sx}\, dx. \tag{6.11-12}$$

Laplace transforms are discussed in Appendix D where references to the literature also are given. It follows from Equation 6.11-10 and the inversion theorem for Laplace transforms that the nonlinearity $g(x)$ may be written as

$$g(x) = \frac{1}{2\pi j}\int_{C_+} G_+(s)\, e^{sx}\, ds + \frac{1}{2\pi j}\int_{C_-} G_-(s)\, e^{sx}\, ds \tag{6.11-13}$$

where C_+ and C_- are the appropriate Bromwich contours in the complex s plane. For any region $\sigma_1 < \text{Re}\,(s) < \sigma_2$ in the s plane where the contours C_+ and C_- overlap, the nonlinearity $g(x)$ has a *bilateral Laplace transform*

$$G(s) = \int_{-\infty}^\infty g(x)\, e^{-sx}\, dx. \tag{6.11-14}$$

In subsequent sections we will use the foregoing approach to solve a number of classes of instantaneous nonlinearities which are encountered extensively in system analysis. Now Equation 6.11-6 becomes

$$R_y(t_1, t_2) = \left(\frac{1}{2\pi j}\right)^2 \int_C \int_C F_x(s_1, s_2) G(s_1) G(s_2)\, ds_1\, ds_2 \tag{6.11-15}$$

where C is the appropriate contour in the s plane.

6.12 νth-Law Devices

An important class of instantaneous nonlinear devices are the νth-law devices whose output characteristics are represented by powers of the input. For our purposes, it will be convenient to categorize these as follows:

1. *Half-wave νth-law devices*, $g_1(x)$:

$$y = g_1(x) = \begin{cases} cx^\nu, & x > 0, \quad \nu \geq 0 \\ 0, & x \leq 0 \end{cases} \tag{6.12-1}$$

where c is an arbitrary positive constant. These include the ideal half-wave rectifier when $\nu = 1$ and the rectifier-hard-limiter when $\nu = 0$. A plot of this nonlinearity is given in Figure 6.15(*a*). It will sometimes be convenient to write this characteristic as

$$g_1(x, \nu) = cx^\nu u(x) \tag{6.12-2}$$

where $u(x)$ is the unit-step function.

2. *The full-wave even νth-law device,* $g_2(x)$:

$$y = g_2(x) = \begin{cases} cx^\nu, & x > 0, \quad \nu \geq 0, \\ 0, & x = 0, \\ c(-x)^\nu, & x < 0, \quad \nu \geq 0. \end{cases} \tag{6.12-3}$$

Note that this nonlinearity can be written in terms of the half-wave νth-law device as

$$y = g_2(x) = g_1(x) + g_1(-x) = g_1(x, \nu) + g_1(-x, \nu). \tag{6.12-4}$$

It is illustrated in Figure 6.15(*b*).

3. *The full-wave odd νth-law device,* $g_3(x)$:

$$y = g_3(x) = \begin{cases} cx^\nu, & x > 0, \quad \nu \geq 0, \\ 0, & x = 0, \\ -c(-x)^\nu, & x < 0, \quad \nu \geq 0. \end{cases} \tag{6.12-5}$$

Again we can write this nonlinearity as

$$y = g_3(x) = g_1(x) - g_1(-x) = g_1(x, \nu) - g_1(-x, \nu) \tag{6.12-6}$$

as shown in Figure 6.15(*c*).

We consider first the problem of finding the density and cumulative distribution function of the random variable y, given the density of the random variable x. It follows from Section 2.11 that we may write

$$W_y(y) = W_x(x)\left|\frac{dx}{dy}\right| \tag{6.12-7}$$

and

$$F_y(b) = P(y \leq b) = \int_{I=\{x|y\leq b\}} W_x(x)\, dx \tag{6.12-8}$$

where $y = g(x)$, $b = g(a)$, $W_y(y)$ and $W_x(x)$ are the densities of y and x respectively, and $F_y(b)$ is the cumulative distribution function of y. For the νth-law devices it will be more convenient to work with Equation 6.12-8 and then to determine the density of y from the relationship

$$W_y(y) = \frac{d}{dy} F_y(y). \tag{6.12-9}$$

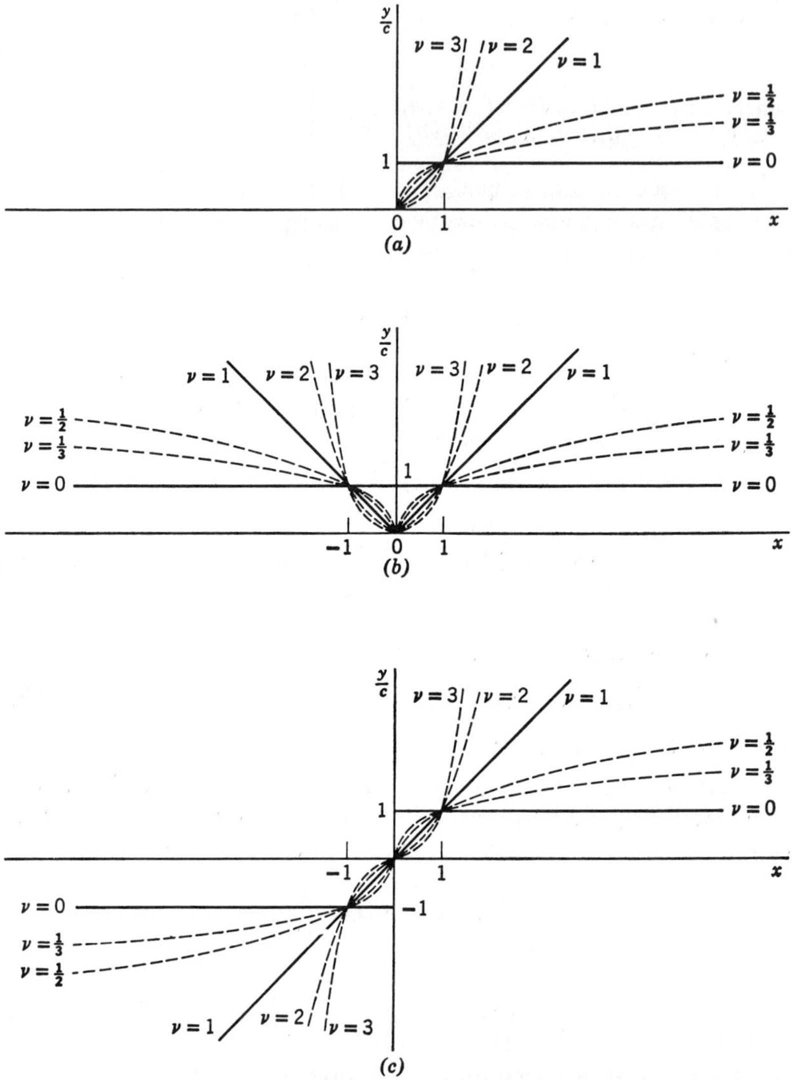

Figure 6.15 The νth-law device: (a) the half-wave νth-law device, $y = g_1(x)$; (b) the full-wave even νth-law device, $y = g_2(x)$; (c) the full-wave odd νth-law device, $y = g_3(x)$

This derivative may be obtained by noting that, if

$$G(u) = \int_{a(u)}^{b(u)} H(x, u)\, du,$$

then

$$\frac{dG}{du} = \int_a^b \frac{\partial H}{\partial u}\, dx + H(b, u)\frac{db}{du} - H(a, u)\frac{da}{du} \tag{6.12-10}$$

if $H(x, u)$ is a continuous function of x and u.

Let us illustrate the procedure by considering the half-wave νth-law device of Figure 6.15(a). For this case Equation 6.12-8 can be written as

$$F_y(b) = \begin{cases} F_x(0) + \displaystyle\int_0^{(b/c)^{1/\nu}} W_x(x)\, dx, & b \geq 0, \\ 0, & b < 0, \end{cases} \tag{6.12-11}$$

where $F_x(0)$ is given by

$$F_x(0) = \int_{-\infty}^{0} W_x(x)\, dx. \tag{6.12-12}$$

Now we make use of Equations 6.12-9 and 6.12-10 to obtain

$$W_y(y) = \begin{cases} F_x(0)\, \delta(y) + W_x\left[\left(\dfrac{y}{c}\right)^{1/\nu}\right]\left(\dfrac{y^{1-\nu}}{c}\right)^{1/\nu}\dfrac{1}{\nu}, & y \geq 0, \\ 0, & y < 0. \end{cases} \tag{6.12-13}$$

For the case where x is normally distributed with density

$$W_x(x) = \frac{1}{\sqrt{2\pi}\,\sigma} e^{-x^2/2\sigma^2}, \qquad -\infty < x < \infty, \tag{6.12-14}$$

then the density of y becomes

$$W_y(y) = \begin{cases} \dfrac{1}{2}\,\delta(y) + \dfrac{1}{\sqrt{2\pi}\,\sigma\nu} e^{-(1/2\sigma^2)(y/c)^{2/\nu}}\left(\dfrac{y^{1-\nu}}{c}\right)^{1/\nu}, & y \geq 0, \\ 0, & y < 0. \end{cases} \tag{6.12-15}$$

In the same way, the cumulative distribution function and the density function of y may be found for the full-wave even and full-wave odd νth-law devices and also for the particular case where the input x is normally distributed with mean zero and variance σ^2. As a matter of convenience the results are summarized in Table 6.1.

Having found the density and cumulative distribution function for the output of the νth-law device, we now consider the output autocorrelation function given by Equation 6.11-15. To use this expression we must find the

Table 6.1. Distributions and Densities for νth-Law Devices[a]

Device	Cumulative Distribution Function $F_y(y)$	Density Function $W_y(y)$	Density Function for Normal Input, Zero Mean, Variance σ^2
Half-wave	$\left[F_x(0) + \int_0^{(y/c)^{1/\nu}} W_x(x)\,dx\right]u(y)$	$F_x(0)\delta(y) + \frac{1}{\nu} W_x\left[\left(\frac{y}{c}\right)^{1/\nu}\right] \times \left(\frac{y^{1-\nu}}{c}\right)^{1/\nu} u(y)$	$\frac{1}{2}\delta(y) + \frac{1}{\sqrt{2\pi}\,\sigma\nu} e^{-(1/2\sigma^2)(y/c)^{2\nu}} \times \left(\frac{y^{1-\nu}}{c}\right)^{1/\nu} u(y)$
Full-wave, even	$\int_{-(y/c)^{1/\nu}}^{(y/c)^{1/\nu}} W_x(x)\,dx u(y)$	$\frac{1}{\nu}\left\{W_x\left[\left(\frac{y}{c}\right)^{1/\nu}\right] + W_x\left[-\left(\frac{y}{c}\right)^{1/\nu}\right]\right\} \times \left[\left(\frac{y^{1-\nu}}{c}\right)^{1/\nu}\right]u(y)$	$\frac{2}{\sqrt{2\pi}\,\sigma\nu} e^{-(1/2\sigma^2)(y/c)^{2/\nu}} \left(\frac{y^{1-\nu}}{c}\right)^{1/\nu} u(y)$
Full-wave, odd	$\int_{-\infty}^{-(y/c)^{1/\nu}} W_x(x)\,dx u(-y)$ + $\int_{-\infty}^{(y/c)^{1/\nu}} W_x(x)\,dx u(y)$	$\frac{1}{\nu} W_x\left[-\left(\frac{-y}{c}\right)^{1/\nu}\right]\left[\frac{(-y)^{1-\nu}}{c}\right]^{1/\nu} u(-y)$ + $\frac{1}{\nu} W_x\left[\left(\frac{y}{c}\right)^{1/\nu}\right]\left(\frac{y^{1-\nu}}{c}\right)u(y)$	$\frac{1}{\sqrt{2\pi}\,\sigma\nu} e^{-(1/2\sigma^2)(-y/c)^{2/\nu}} \times \left[\frac{(-y)^{1-\nu}}{c}\right]^{1/\nu} u(-y)$ + $\frac{1}{\sqrt{2\pi}\,\sigma\nu} e^{-(1/2\sigma^2)(y/c)^{2/\nu}} \left(\frac{y^{1-\nu}}{c}\right)^{1/\nu} u(y)$

[a] $u(y)$ = unit-step function.

Laplace transform $G(s)$ of the nonlinearity. This transform for the half-wave νth-law device $g_1(x)$ of Equation 6.12-1 is given by

$$G_1(s) = \mathscr{L}[g_1(x)] = c \int_0^{\infty} x^{\nu} e^{-sx}\, dx \tag{6.12-16}$$

which converges for all Re $(s) > 0$. After a change of variable $t = sx$, this last expression becomes

$$G_1(s) = \frac{c}{s^{\nu+1}} \int_0^{\infty} t^{\nu} e^{-t}\, dt = \frac{c\Gamma(\nu + 1)}{s^{\nu+1}} \tag{6.12-17}$$

where $\Gamma(\nu)$ is the gamma function [6, 7] defined by

$$\Gamma(\nu) = \int_0^{\infty} u^{\nu-1} e^{-u}\, du. \tag{6.12-18}$$

Since it follows directly from Equation 6.12-16 that

$$\mathscr{L}[g_1(-x)] = c \int_{-\infty}^{0} (-x)^{\nu} e^{-sx}\, dx = G_1(-s) \tag{6.12-19}$$

for Re $s > 0$, we have explicit expressions for the Laplace transforms of the various νth-law devices of Equations 6.12-1, 6.12-3, and 6.12-5. These transforms may be substituted into Equation 6.11-15 and, in principle, the autocorrelation function of $y = g(x)$ can be found if the input bivariate characteristic function $F_x(s_1, s_2)$ is known.

As a practical matter, Equation 6.11-15 may become somewhat complicated. Let us consider the case where the input $x(t)$ is gaussian and stationary with autocorrelation function $R_x(\tau)$ and bivariate density given by Equation 6.10-8. The corresponding characteristic function has been found in Chapter 2 and is given by Equation 2.21-5:

$$F_x(v_1, v_2) = e^{-[R_x(0)/2][v_1^2 + 2\rho(\tau)v_1v_2 + v_2^2]} \tag{6.12-20}$$

where the normalized autocorrelation function $\rho(\tau)$ is defined by

$$\rho(\tau) = \frac{R_x(\tau)}{R_x(0)}. \tag{6.12-21}$$

On replacing jv by s so that we deal with the complex s plane of Equation 6.11-7, this characteristic function becomes

$$F_x(s_1, s_2) = e^{[R_x(0)/2][s_1^2 + 2\rho(\tau)s_1s_2 + s_2^2]}. \tag{6.12-22}$$

Let us expand the cross-term in this expression in a power series so that

$$e^{R_x(0)\rho(\tau)s_1s_2} = e^{R_x(\tau)s_1s_2} = \sum_{k=0}^{\infty} \frac{R_x^k(\tau)s_1^k s_2^k}{k!}. \tag{6.12-23}$$

Now if we substitute Equations 6.12-17, 6.12-22, and 6.12-23 into Equation 6.11-15, we have an expression for the output autocorrelation function for the half-wave νth-law device with a stationary gaussian input. Furthermore, the double integral can be factored into the product of two identical integrals to obtain

$$\boxed{R_y(\tau) = \sum_{k=0}^{\infty} a_{k\nu}^2 \frac{R_x^k(\tau)}{k!}} \tag{6.12-24}$$

where the constant $a_{k\nu}$ is given by

$$a_{k\nu} = \frac{c\Gamma(\nu + 1)}{2\pi j} \int_{C+} s^{k-\nu-1} e^{[R_x(0)/2]s^2} ds \tag{6.12-25}$$

and can be evaluated by contour integration. It should be kept in mind that the region of convergence for the half-wave νth-law device in the complex s plane was that half of the s plane where Re $s > 0$; therefore C^+ must be taken along the vertical line $s = \epsilon + j\omega$ where $\epsilon > 0$. The result [8] is that

$$\begin{aligned} a_{k\nu} &= \frac{c}{2} \frac{\Gamma(1 + \nu)}{\Gamma[(2 + \nu - k)/2]} \left(\frac{R_x(0)}{2}\right)^{(\nu-k)/2} \\ &= \frac{c}{2\pi} \Gamma(1 + \nu)\Gamma\left(\frac{k - \nu}{2}\right)\left[\frac{R_x(0)}{2}\right]^{(\nu-k)/2} \sin \pi\left(\frac{k - \nu}{2}\right) \end{aligned} \tag{6.12-26}$$

where use has been made of the reflection formula [7]

$$\Gamma(z)\Gamma(1 - z) = \pi \csc \pi z. \tag{6.12-27}$$

Note that $a_{k\nu}$ is zero if the quantity $(k - \nu)/2$ is an integer since, in that case, $\sin \pi[(k - \nu)/2] = 0$. We now consider specifically each of the three νth-law devices.

Half-Wave νth-Law Device

The output autocorrelation function for a stationary gaussian input is given by Equation 6.12-24 where the coefficients are determined from Equation 6.12-26. Let us consider some special cases corresponding to various values of ν as shown in Figure 6.15(a).

Case I—$\nu = 0$, Half-Wave Limiter: Here the coefficient a_{k0} becomes

$$\begin{aligned} a_{k0} &= \frac{c}{2\pi} \Gamma\left(\frac{k}{2}\right)\left[\frac{2}{R_x(0)}\right]^{k/2} \sin \pi \frac{k}{2} \\ &= \frac{c}{2} \frac{1}{\Gamma[(2 - k)/2]} \left[\frac{2}{R_x(0)}\right]^{k/2} \end{aligned} \tag{6.12-28}$$

or, for $k = 0, 1, 2, \ldots$, we have

$$a_{00} = \frac{c}{2}, \tag{6.12-29}$$

$$a_{10} = \frac{c}{\sqrt{2\pi R_x(0)}}, \tag{6.12-30}$$

$$a_{20} = 0, \tag{6.12-31}$$

$$a_{30} = -\frac{c}{\sqrt{2\pi R_x^3(0)}}, \tag{6.12-32}$$

$$\vdots$$

$$|a_{k0}| = \begin{cases} 0, & k \neq 0, \quad k \text{ even}, \\ \dfrac{c}{\sqrt{2\pi R_x^k(0)}} 1\cdot 3\cdot 5\cdot 7\cdots(k-2), & k \text{ odd}, \end{cases} \tag{6.12-33}$$

where use has been made of the fact that [7]

$$\begin{aligned} \Gamma(\tfrac{1}{2}) &= \sqrt{\pi}, \qquad & \Gamma(-\tfrac{1}{2}) &= -2\sqrt{\pi}, \\ \Gamma(\tfrac{3}{2}) &= \tfrac{1}{2}\sqrt{\pi}, \qquad & \Gamma(-\tfrac{3}{2}) &= -\tfrac{1}{3}\sqrt{\pi}, \end{aligned} \tag{6.12-34}$$

and, more generally,

$$\begin{aligned} \Gamma(n + \tfrac{1}{2}) &= \frac{1\cdot 3\cdot 5\cdot 7\cdots(2n-1)}{2^n}\sqrt{\pi} \\ \Gamma(-z) &= -\frac{\Gamma(z)}{z}. \end{aligned} \tag{6.12-35}$$

These values of a_{k0} may be substituted into Equation 6.12-24, and, with the aid of Equation 6.12-21, we obtain

$$\begin{aligned} R_y(\tau) &= \frac{c^2}{2\pi}\left[\frac{\pi}{2} + \rho(\tau) + \frac{\rho^3(\tau)}{2\cdot 3} + \cdots + \frac{[1\cdot 3\cdot 5\cdot 7\cdots(k-2)]^2}{k!}\rho^k(\tau) + \cdots\right] \\ &= \frac{c^2}{2\pi}\left[\frac{\pi}{2} + \rho(\tau) + \frac{\rho^3(\tau)}{2\cdot 3} + \cdots \right. \\ &\qquad \left. + \frac{1\cdot 3\cdot 5\cdot 7\cdots(k-2)}{2\cdot 4\cdot 6\cdots(k-1)}\frac{\rho^k(\tau)}{k} + \cdots\right] \end{aligned} \tag{6.12-36}$$

or

$$\boxed{R_y(\tau) = \frac{c^2}{2\pi}\left[\frac{\pi}{2} + \arcsin \rho(\tau)\right].} \tag{6.12-37}$$

The last relation follows from Equation 6.10-21. This result could have been obtained directly from Section 6.10 by using only the first term in Equation

6.10-7 or, equivalently, one-half of the first term in Equation 6.10-19. Note that the half-wave rectification has added a d-c component responsible for the constant $c^2/4$ in Equation 6.12-36.

Case II—ν = 1, Half-Wave Linear Rectifier: For this case, Equation 6.12-26 becomes

$$a_{k1} = \frac{c}{2}\frac{1}{\Gamma[(3-k)/2]}\left[\frac{R_x(0)}{2}\right]^{(1-k)/2}$$

$$= \frac{c}{2\pi}\Gamma\left(\frac{k-1}{2}\right)\left[\frac{R_x(0)}{2}\right]^{(1-k)/2}\sin\pi\left(\frac{k-1}{2}\right) \tag{6.12-38}$$

or

$$a_{01} = c\sqrt{\frac{R_x(0)}{2\pi}}, \tag{6.12-39}$$

$$a_{11} = \frac{c}{2}, \tag{6.12-40}$$

$$a_{21} = -\frac{c}{\sqrt{2\pi R_x(0)}}, \tag{6.12-41}$$

$$a_{31} = 0, \tag{6.12-42}$$

$$a_{41} = -\frac{c}{\sqrt{2\pi R_x^3(0)}}, \tag{6.12-43}$$

$$\vdots$$

$$|a_{k1}| = \begin{cases} 0, & k \text{ odd}, \\ \dfrac{c}{\sqrt{2\pi R_x^{k-1}(0)}}\,1\cdot3\cdot5\cdot7\cdots(k-3), & k \text{ even}. \end{cases} \tag{6.12-44}$$

As before we may substitute these values of $a_{k\nu}$ into Equation 6.12-24 and rearrange to obtain

$$R_y(\tau) = \frac{c^2}{4}R_x(\tau) + \frac{c^2}{2\pi}R_x(0)\left\{1 + \frac{1}{2!}\rho^2(\tau) + \frac{1}{4!}\rho^4(\tau) + \cdots + \frac{1}{k!}[1\cdot3\cdot5\cdot7\cdots(k-3)]^2\rho^k(\tau) + \cdots\right\} \tag{6.12-45}$$

for k even. If we make use of the series expansion [9]

$$\sqrt{1-x^2} = 1 - \frac{x^2}{2} - \frac{x^4}{2\cdot4} - \frac{1\cdot3x^6}{2\cdot4\cdot6} - \frac{1\cdot3\cdot5x^8}{2\cdot4\cdot6\cdot8} - \cdots \tag{6.12-46}$$

and Equation 6.10-21, then this autocorrelation function can be written as

$$R_y(\tau) = \frac{c^2}{4} R_x(\tau) + \frac{c^2}{2\pi} R_x(0)[\rho(\tau) \arcsin \rho(\tau) + \sqrt{1 - \rho^2(\tau)}]. \tag{6.12-47}$$

Case III—ν = 2, Half-Wave Square-Law Device: The constant $a_{k\nu}$ is

$$a_{k2} = c \frac{2^{(k-2)/2}}{\Gamma[(4 - k)/2][R_x(0)]^{(k-2)/2}}$$

$$= \frac{c}{\pi} \Gamma\left(\frac{k-2}{2}\right) \frac{2^{(k-2)/2}}{[R_x(0)]^{(k-2)/2}} \sin \pi\left(\frac{k-2}{2}\right) \tag{6.12-48}$$

and

$$a_{02} = \frac{c}{2} R_x(0), \tag{6.12-49}$$

$$a_{12} = c\sqrt{\frac{2R_x(0)}{\pi}}, \tag{6.12-50}$$

$$a_{22} = c, \tag{6.12-51}$$

$$a_{32} = c\sqrt{\frac{2}{\pi R_x(0)}}, \tag{6.12-52}$$

$$a_{42} = 0, \tag{6.12-53}$$

$$\vdots$$

$$|a_{k2}| = \begin{cases} 0, & k \neq 0, \quad k \text{ even}, \\ c\sqrt{\dfrac{2}{\pi R_x^{k-2}(0)}}\, 1\cdot 3\cdot 5\cdot 7\cdots(k-4), & k \text{ odd}. \end{cases} \tag{6.12-54}$$

Finally, the autocorrelation function may be written from Equation 6.12-24 as

$$R_y(\tau) = \frac{2c^2}{\pi} R_x^2(0)\left[\frac{\pi}{8} + \rho(\tau) + \frac{\pi}{4}\rho^2(\tau) + \rho^3(\tau)\frac{1}{2\cdot 3} + \rho^5(\tau)\frac{1}{2\cdot 3\cdot 4\cdot 5} + \cdots + \rho^k(\tau)\frac{1\cdot 3\cdot 5\cdot 7\cdots(k-4)}{(k-3)(k-2)(k-1)k}\right] \tag{6.12-55}$$

for k odd. Let us consider now the second of the νth-law devices, the full-wave even device.

Full-Wave ν th-Law Device

If we examine Equations 6.12-14 and 6.12-19, we see that the transform $G_2(\omega)$ for this nonlinearity $g_2(x)$ is just

$$G_2(s) = G_1(s) + G_1(-s) = G_+(s) + G_-(s) \tag{6.12-56}$$

where $G_1(s)$ is given by Equation 6.12-17. It is clear from Equation 6.11-13 that two different contours C_+ and its mirror image C_- must be used in Equation 6.11-15 for the evaluation of the integrands containing $G_1(s)$ and $G_1(-s)$, respectively. Thus the new output autocorrelation function corresponding to Equation 6.12-24 is

$$\boxed{R_y(\tau) = \sum_{k=0}^{\infty} b_{k\nu}^2 \frac{R_x^k(\tau)}{k!}} \tag{6.12-57}$$

where the constant $b_{k\nu}$ is given by

$$b_{k\nu} = \frac{c\Gamma(\nu + 1)}{2\pi j} \int_{C_+} s^{k-\nu-1} e^{[R_x(0)/2]s^2}\, ds + \frac{c\Gamma(\nu + 1)}{2\pi j} \int_{C_-} s^k(-s)^{-\nu-1} e^{[R_x(0)/2]s^2}\, ds \tag{6.12-58}$$

or, after replacing $-s$ by s in the second integral,

$$b_{k\nu} = \frac{c\Gamma(\nu + 1)}{2\pi j} \int_{C_+} e^{[R_x(0)/2]s^2} s^{-\nu-1}[s^k + (-s)^k]\, ds \tag{6.12-59}$$

so that

$$b_{k\nu} = \begin{cases} 0, & k \text{ odd}, \\ 2a_{k\nu}, & k \text{ even}, \end{cases} \tag{6.12-60}$$

and Equation 6.12-57 becomes

$$\boxed{R_y(\tau) = 4 \sum_{k \text{ even}} a_{k\nu}^2 \frac{R_x^k(\tau)}{k!}.} \tag{6.12-61}$$

We now take up the examples where $\nu = 0$, 1, and 2 that were considered previously in the half-wave case.

Case IV—ν = 0, Full-Wave Even Limiter: It follows directly from Equations 6.12-37 and 6.12-60 that all of the coefficients b_{k0} are zero except for b_{00} and that

$$b_{00} = 2a_{00} = c. \tag{6.12-62}$$

Therefore we have

$$R_y(\tau) = c^2. \tag{6.12-63}$$

This result is scarcely surprising if we consider Figure 6.15(*b*) for $\nu = 0$; the output is the constant c with autocorrelation function c^2.

Case V—ν = 1, Full-Wave Linear Rectifier: We see from Equations 6.12-39 through 6.12-44 that

$$R_y(\tau) = \frac{2}{\pi} c^2 R_x(0)[\rho(\tau) \arcsin \rho(\tau) + \sqrt{1 - \rho^2(\tau)}]. \tag{6.12-64}$$

Note that the linear term $R_x(\tau)$ has disappeared.

Case VI—ν = 2, Full-Wave Even Squarer: We notice from Equations 6.12-49 through 6.12-54 that only b_{02} and b_{22} are nonzero and that

$$R_y(\tau) = c^2 R_x^2(0)[1 + 2\rho^2(\tau)]. \tag{6.12-65}$$

This is exactly the result obtained in Equation 6.4-19 by direct calculation. We look now briefly at the full-wave odd νth-law devices.

Full-Wave Odd νth-Law Devices

This device has a characteristic given by Equation 6.12-6 and a transform

$$G_3(s) = G_1(s) - G_1(-s) = G_+(s) - G_-(s). \tag{6.12-66}$$

It follows from Equations 6.12-57 and 6.12-59 that the autocorrelation function is

$$R_y(\tau) = \sum_{k=0}^{\infty} d_{k\nu}^2 \frac{R_x^k(\tau)}{k!} \tag{6.12-67}$$

where

$$d_{k\nu} = \begin{cases} 0, & k = 0, \\ 2a_{k\nu}, & k \text{ odd}, \\ 0, & k \text{ even}. \end{cases} \tag{6.12-68}$$

Case VII—ν = 0, Full-Wave Odd Limiter: It is clear from Case I and Equation 6.12-37 that

$$\boxed{R_y(\tau) = c^2 \frac{2}{\pi} \arc \sin \rho(\tau).} \tag{6.12-69}$$

This case is the hard limiter discussed in Section 6.10 with autocorrelation function given by Equation 6.10-20.

Case VIII—ν = 1, Linear Amplifier: Here we are dealing with a linear amplifier with a constant gain of c. We examine Equations 6.12-39 through 6.12-44 and see that, according to Equation 6.12-68, only d_{11} is nonzero and

$$d_{11} = 2a_{11} = c \tag{6.12-70}$$

so that

$$\boxed{R_y(\tau) = c^2 R_x(\tau)} \tag{6.12-71}$$

as expected.

Case IX—ν = 2, Full-Wave Odd Squarer: On omitting the terms a_{02} and a_{22} in Equation 6.12-55, we obtain

$$R_y(\tau) = \frac{2c^2}{\pi} R_x^2(0)\left[\rho(\tau) + \rho^3(\tau)\frac{1}{2\cdot 3} + \rho^5(\tau)\frac{1}{2\cdot 3\cdot 4\cdot 5} + \cdots \right.$$
$$\left. + \rho^k(\tau)\frac{1\cdot 3\cdot 5\cdot 7\cdots(k-4)}{(k-3)(k-2)(k-1)k} + \cdots\right] \tag{6.12-72}$$

for k odd.

6.13 The Smooth Limiter; Price's Theorem

In this section we consider an interesting zero-memory nonlinearity which is amenable to explicit solution [10] and which may be a reasonable approximation to systems subject to saturation. We consider the *smooth limiter* with characteristic given by

$$y = g(x) = \frac{1}{K}\frac{1}{\sqrt{2\pi\sigma^2}}\int_0^x e^{-z^2/2\sigma^2}\,dz \tag{6.13-1}$$

as shown in Figure 6.16(a).

We begin by considering the density $W_y(y)$ given by

$$W_y(y) = W_x(x)\left|\frac{dx}{dy}\right| \tag{6.13-2}$$

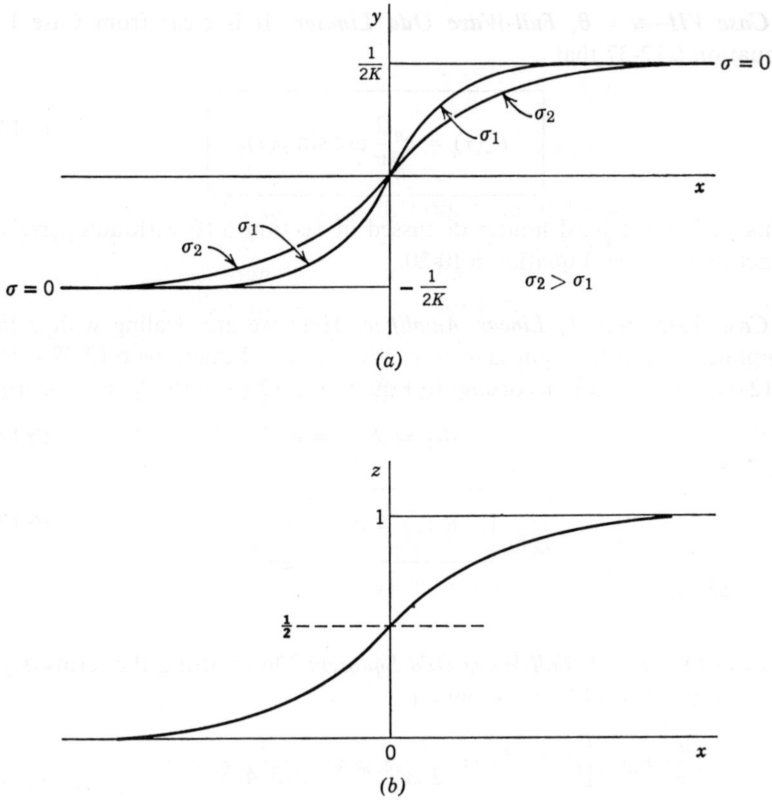

Figure 6.16 The smooth limiter: (a) $y = (1/K)(1/\sqrt{2\pi\sigma^2}) \int_0^x e^{-z^2/2\sigma^2}\,dz$; (b) $z = Ky + \frac{1}{2}$.

where the derivative dy/dx is easily obtained from Equation 6.13-1 as

$$\frac{dy}{dx} = \frac{1}{K}\frac{1}{\sqrt{2\pi\sigma^2}}\,e^{-x^2/2\sigma^2}. \tag{6.13-3}$$

We will now assume, as in the previous section, that the input $x(t)$ is a stationary zero-mean gaussian process with autocorrelation function $R_x(\tau)$ and variance

$$R_x(0) = \sigma_x^2. \tag{6.13-4}$$

On this assumption, the density of x becomes

$$W_x(x) = \frac{1}{\sqrt{2\pi\sigma_x^2}}\,e^{-x^2/2\sigma_x^2}. \tag{6.13-5}$$

Now Equation 6.13-2 becomes

$$W_y(y) = K\sqrt{\alpha}\, e^{-(x^2/2\sigma^2)(\alpha-1)} \tag{6.13-6}$$

where α is given by

$$\alpha = \frac{\sigma^2}{\sigma_x^2} = \frac{\sigma^2}{R_x(0)} \tag{6.13-7}$$

and y is determined from x through Equation 6.13-1 which is not easily inverted. We may use Equations 6.13-1 and 6.13-6 to plot $W_y(y)$ versus y as shown in Figure 6.17. For $\alpha = 0$, we have the hard limiter of Section 6.10. It is interesting to note, that for $\alpha = 1$, the output distribution is uniform. Higher values of α give distributions which are peaked about the origin while lower values give distributions peaked at $\pm 1/2K$.

Let us now proceed to find the autocorrelation function $R_y(\tau)$ using the characteristic function method of Equation 6.11-6. It will be convenient to define a new characteristic $z(x)$ given by

$$z(x) = Ky(x) + \tfrac{1}{2} \tag{6.13-8}$$

and shown in Figure 6.16(*b*). This new function is more tractable and the d-c term can be dropped later. Now $z(x)$ has a Fourier transform $G(\omega)$ given by

$$G(\omega) = \frac{1}{j\omega}\, e^{-\sigma^2\omega^2/2} \tag{6.13-9}$$

since, from Equation 2.18-21,

$$\mathscr{F}\left(\frac{1}{\sqrt{2\pi\sigma^2}}\, e^{-t^2/2\sigma^2}\right) = e^{-\sigma^2\omega^2/2} \tag{6.13-10}$$

and the factor $1/j\omega$ accounts for the integration in Equation 6.13-1.

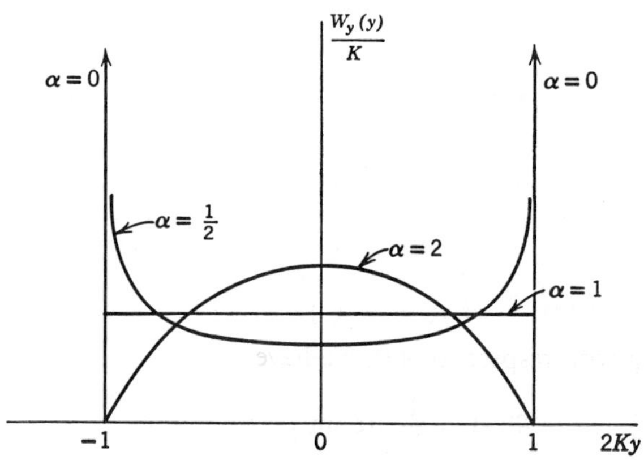

Figure 6.17 The output density of the smooth limiter for gaussian input

We have assumed that the input $x(t)$ is a stationary, zero-mean gaussian process with characteristic function given by Equation 6.12-20. Now the autocorrelation function of z becomes, with the aid of Equations 6.11-6 and 6.13-9,

$$R_z(\tau) = -\frac{1}{4\pi^2}\int_{-\infty}^{\infty}\int_{-\infty}^{\infty}\frac{1}{v_1 v_2}\, e^{-[(\sigma_x^2+\sigma^2)/2](v_1^2+v_2^2)-\rho\sigma_x^2 v_1 v_2}\, dv_1\, dv_2 \qquad (6.13\text{-}11)$$

where $\rho(\tau)$ has been defined by Equation 6.12-21 and σ_x^2 has been written, as a matter of convenience, for $R_x(0)$. Suppose now that we differentiate this last expression with respect to ρ. The result will be the cancellation of the term $1/v_1v_2$ in the integrand; therefore

$$\frac{dR_z(\tau)}{d\rho(\tau)} = \frac{\sigma_x^2}{4\pi^2}\int_{-\infty}^{\infty}\int_{-\infty}^{\infty} e^{-[(\sigma_x^2+\sigma^2)/2](v_1^2+v_2^2)-\rho\sigma_x^2 v_1 v_2}\, dv_1\, dv_2. \qquad (6.13\text{-}12)$$

This last expression may be integrated after the proper change in variables

$$v_1 = \sqrt{\frac{2}{\sigma_x^2+\sigma^2}}\left(y_1 - \frac{a}{\sqrt{1-a^2}}\, y_2\right), \qquad (6.13\text{-}13)$$

$$v_2 = \sqrt{\frac{2}{\sigma_x^2+\sigma^2}}\,\frac{1}{\sqrt{1-a^2}}\, y_2 \qquad (6.13\text{-}14)$$

where a is defined by

$$a = \frac{\rho\sigma_x^2}{\sigma_x^2+\sigma^2}. \qquad (6.13\text{-}15)$$

The result, after some straightforward algebra, is that Equation 6.13-12 becomes

$$\frac{dR_z(\tau)}{d\rho(\tau)} = \frac{1}{2\pi^2}\,\frac{\sigma_x^2}{\sigma_x^2+\sigma^2}\,\frac{1}{\sqrt{1-a^2}}\int_{-\infty}^{\infty}\int_{-\infty}^{\infty} e^{-(y_1^2+y_2^2)}\, dy_1\, dy_2. \qquad (6.13\text{-}16)$$

After the change in variable

$$y_1 = r\cos\theta, \qquad y_2 = r\sin\theta, \qquad (6.13\text{-}17)$$

the double integral becomes

$$\int_0^{2\pi}\int_0^{\infty} r\, e^{-r^2}\, dr\, d\theta = -\pi\, e^{-r^2}\Big|_0^{\infty} = \pi \qquad (6.13\text{-}18)$$

and Equation 6.13-16 can be written as

$$\frac{dR_z(\tau)}{d\rho(\tau)} = \frac{1}{2\pi}\,\frac{\sigma_x^2}{\sigma_x^2+\sigma^2}\,\frac{1}{\sqrt{1-[\rho\sigma_x^2/(\sigma_x^2+\sigma^2)]^2}}. \qquad (6.13\text{-}19)$$

On integrating with respect to $\rho(\tau)$, we have

$$R_z(\tau) = \frac{1}{2\pi}\arcsin\left[\frac{\rho(\tau)}{1+\alpha}\right] + C \qquad (6.13\text{-}20)$$

where C is the constant of integration and α has been defined by Equation 6.13-7.

We would now like to evaluate the constant C, which corresponds to the d-c term in z. Note that the original $y = g(x)$ of Equation 6.13-1 is symmetrical about $y = 0$ and therefore the output $y(t)$ for a zero-mean gaussian input will have zero mean. Consequently the constant C is zero for $R_y(\tau)$ and it follows from Equations 6.13-1 and 6.13-8 that

$$R_y(\tau) = \frac{1}{2\pi K^2} \arcsin \left[\frac{\rho(\tau)}{1 + \alpha}\right]. \tag{6.13-21}$$

The constant C also may be evaluated by considering the case $\alpha = 1$ which is the uniform distribution shown in Figure 6.17. Since $\rho(0) = 1$, it follows from Equation 6.13-20 that

$$R_z(0)|_{\alpha=1} = \frac{1}{2\pi}\cdot\frac{\pi}{6} + C = \frac{1}{12} + C. \tag{6.13-22}$$

By direct calculation, however,

$$R_z(0)\Big|_{\alpha=1} = E\{z^2\}\Big|_{\alpha=1} = \int_{-1/2}^{1/2} z^2\, dz = \frac{z^3}{3}\Big|_{-1/2}^{1/2} = \frac{1}{12}. \tag{6.13-23}$$

A comparison of these last two equations shows that C must be zero. Now the normalized correlation coefficient $\rho_y(\tau)$ is given by

$$\boxed{\rho_y(\tau) = \frac{R_y(\tau)}{R_y(0)} = \frac{\arcsin [\rho(\tau)/(1 + \alpha)]}{\arcsin [1/(1 + \alpha)]}} \tag{6.13-24}$$

since $\rho_y(0) = 1$.

A number of special cases are immediately obvious from an examination of Equation 6.13-24.

$\alpha = 0$; Hard Limiter: This is the case where $\sigma^2 = 0$ as shown in Figure 6.16. Since arc sin 1 $= \pi/2$, Equation 6.13-24 becomes

$$\boxed{\rho_y(\tau) = \frac{2}{\pi} \arcsin \rho(\tau)} \tag{6.13-25}$$

which agrees with Equation 6.10-20 obtained by direct calculation.

$\alpha = 1$; Uniform Density: In this case $\sigma^2 = \sigma_x^2$ and

$$\boxed{\rho_y(\tau) = \frac{6}{\pi} \arcsin \frac{\rho(\tau)}{2}.} \tag{6.13-26}$$

$\alpha \to \infty$; Linear Amplification: This is the case where the input power σ_x^2 is vanishingly small compared to σ^2 so that the operating point is in the vicinity of the origin of the characteristic shown in Figure 6.16. For small argument, it follows from Equation 6.10-21 that

$$\text{arc sin } x \approx x$$

and that

$$\boxed{\rho_y(\tau) \approx \rho(\tau).} \tag{6.13-27}$$

A general expression can be obtained for the output autocorrelation function of a zero-memory nonlinearity when the input is a stationary gaussian process. Consider two random variables x_1 and x_2 which are jointly gaussian with zero means, equal variances σ^2, and second-order density $W_x(x_1, x_2)$ given by Equation 6.10-8. The correlation coefficient ρ is

$$\rho = \frac{E\{x_1 x_2\}}{\sigma^2}. \tag{6.13-28}$$

If $x_1 = x(t)$ and $x_2 = x(t + \tau)$, this last expression reduces to Equation 6.10-5.

Suppose that x_1 is passed through the zero-memory nonlinearity $g_1(x_1)$ and x_2 is passed through the zero-memory nonlinearity $g_2(x_2)$. The cross-correlation between the two outputs is given by

$$R = E\{g_1(x_1)g_2(x_2)\} \tag{6.13-29}$$

or, from Equation 6.11-6, by

$$R = \frac{1}{(2\pi)^2} \int_{-\infty}^{\infty} \int_{-\infty}^{\infty} G_1(v_1)G_2(v_2)F_x(v_1, v_2)\, dv_1\, dv_2 \tag{6.13-30}$$

where $F_x(v_1, v_2)$ is the joint characteristic function of x_1 and x_2 and is given by Equation 2.21-5:

$$F_x(v_1, v_2) = e^{-(\sigma^2/2)(v_1^2 + v_2^2 + 2\rho v_1 v_2)}; \tag{6.13-31}$$

and $G_1(v_1)$ and $G_2(v_2)$ follow from Equation 6.11-5, that is,

$$G_i(v_i) = \int_{-\infty}^{\infty} g_i(x)\, e^{-jv_i x}\, dx, \qquad i = 1, 2, \tag{6.13-32}$$

and

$$g_i(x_i) = \frac{1}{2\pi} \int_{-\infty}^{\infty} G_i(v)\, e^{jvx_i}\, dv, \qquad i = 1, 2. \tag{6.13-33}$$

Equation 6.13-30 may be differentiated k times with respect to ρ to yield

$$\frac{\partial^k R}{\partial \rho^k} = \frac{(-1)^k}{(2\pi)^2} (\sigma^2)^k \int_{-\infty}^{\infty} \int_{-\infty}^{\infty} G_1(v_1)G_2(v_2)v_1^k v_2^k F_x(v_1, v_2)\, dv_1\, dv_2. \tag{6.13-34}$$

The characteristic function $F_x(v_1, v_2)$ may be represented as the Fourier transform of $W_x(x_1, x_2)$ as given by Equation 6.11-2. Thus Equation 6.13-34 become

$$\frac{\partial^k R}{\partial \rho^k} = \frac{(-1)^k}{(2\pi)^2}(\sigma^2)^k \int_{-\infty}^{\infty}\int_{-\infty}^{\infty} G_1(v_1)G_2(v_2)v_1^k v_2^k$$

$$\times \int_{-\infty}^{\infty}\int_{-\infty}^{\infty} W_x(x_1, x_2)\, e^{j(v_1x_1+v_2x_2)}\, dx_1\, dx_2\, dv_1\, dv_2. \quad (6.13\text{-}35)$$

The order of integration may be interchanged and the expression rearranged to yield

$$\frac{\partial^k R}{\partial \rho^k} = (\sigma^2)^k \int_{-\infty}^{\infty}\int_{-\infty}^{\infty} W_x(x_1, x_2)\, \frac{\partial^k g_1(x_1)}{\partial x_1^k}\, \frac{\partial^k g_2(x_2)}{\partial x_2^k}\, dx_1\, dx_2 \quad (6.13\text{-}36)$$

since it follows from Equation 6.13-33 that

$$\frac{\partial^k g_i(x_i)}{\partial x_i^k} = \frac{(j)^k}{2\pi}\int_{-\infty}^{\infty} v^k G_i(v)\, e^{jvx_i}\, dv, \qquad i = 1, 2. \quad (6.13\text{-}37)$$

As a matter of notational convenience, Equation 6.13-36 often is rewritten as

$$\boxed{\frac{\partial^k R}{\partial \rho^k} = (\sigma^2)^k E\left\{\frac{\partial^k g_1(x_1)}{\partial x_1^k}\, \frac{\partial^k g_2(x_2)}{\partial x_2^k}\right\}.} \quad (6.13\text{-}38)$$

This last expression is a simplified form of *Price's Theorem.** It is subject to some integrability conditions on $g_1(x_1)$ and $g_2(x_2)$ which have been ignored. A more careful proof also would have involved writing $g_1(x_1)$ and $g_2(x_2)$ as in Equation 6.11-13 to ensure that the transforms exist. The theorem has further significance in that, if Equation 6.13-38 is satisfied, x_1 and x_2 *must be* jointly gaussian, but we shall not prove this fact here.

This approach may be applied to the various systems already considered, for example, the hard limiter of Section 6.10 or the smooth limiter of this section. In some cases the amount of manipulation will be reduced but in others it will not be changed appreciably. The technique involved is to differentiate $g_1(x_1)$ and $g_2(x_2)$ often enough to simplify the evaluation of the right side of Equation 6.13-36. Then R is obtained by integration.

As a particularly simple example, consider the hard limiter of Section 6.10. In this case $x_1 = x(t)$, $x_2 = x(t + \tau)$, and $g_1(\cdot) = g_2(\cdot) = g(x)$, the hard-limiter characteristic shown in Figure 6.13(*a*) and given by Equation 6.10-1. The first derivative of $g(x)$ is given by

$$\frac{\partial}{\partial x} g(x) = 2\delta(x) \quad (6.13\text{-}39)$$

* See article by R. Price (1958) in Additional Reading list for a more general treatment.

where $\delta(x)$ is the Dirac delta function. Thus Equation 6.13-36 is easily evaluated to yield

$$\frac{\partial R}{\partial \rho} = 4\sigma^2 W_x(0, 0) = \frac{2}{\pi} \frac{1}{\sqrt{1 - \rho^2}} \tag{6.13-40}$$

or

$$R = R(\tau) = \frac{2}{\pi} \int_0^{\rho(\tau)} \frac{1}{\sqrt{1 - \rho^2}} \, d\rho = \frac{2}{\pi} \text{ arc sin } \rho(\tau) \tag{6.13-41}$$

in agreement with Equation 6.10-20 as expected. On the other hand, no amount of differentiation will reduce the smooth-limiter characteristic to a delta function so that either Price's method or the approach used earlier in this section involves essentially the same amount of work.

6.14 General Techniques—Series Expansions of Input Densities

Thus far we have confined our attention to specific classes of zero-memory nonlinearities. In this and the following section we will attempt to develop approaches that are somewhat more general. In particular we shall consider series expansions—first, for the input densities, and later, for the nonlinearities themselves.

It was pointed out in Sections 2.24 and 2.25 that an arbitrary density $W(y)$ could be expanded in a *Gram-Charlier* series of the form

$$\begin{aligned} W(y) &= \sum_{n=0}^{\infty} a_n \alpha(y) H_n(y) \\ &= \alpha(y)[1 + \tfrac{1}{6}u_3 H_3(y) + \tfrac{1}{24}(u_4 - 3)H_4(y) + \cdots] \end{aligned} \tag{6.14-1}$$

where $\alpha(y)$ is the unit normal distribution, u_n is the nth moment of $W(y)$ about the mean, and $H_n(y)$ is a *Hermite polynomial* of degree n defined by

$$H_n(y) = (-1)^n \, e^{y^2/2} \frac{d^n}{dy^n} (e^{-y^2/2}). \tag{6.14-2}$$

Also the coefficient a_n of Equation 6.14-1 is given by

$$a_n = \frac{1}{n!} \int_{-\infty}^{\infty} W(y) H_n(y) \, dy \tag{6.14-3}$$

since the Hermite polynomials are orthogonal over the interval $(-\infty, \infty)$ with respect to the weighting function $\alpha(y)$; that is,

$$\int_{-\infty}^{\infty} \alpha(y) H_m(y) H_n(y) \, dy = n! \, \delta_{mn} \tag{6.14-4}$$

where δ_{mn} is the Kronecker delta. Since $\alpha(y)$ is the *unit* normal distribution it is desirable to treat y as a normalized random variable with mean of zero and variance of unity. If the actual random variable is z with mean u_z and variance σ_z^2, then y is given by

$$y = \frac{z - u_z}{\sigma_z}.$$

The series as given by Equation 6.14-1 will be most useful in representing a density function $W(y)$ which is approximately normal. Indeed the first term in the series is just the unit normal distribution and subsequent terms may be looked upon as correction terms for nearly normal distributions. On the other hand the nonlinearities that we have examined thus far in this chapter produce output distributions that, for normally distributed inputs, are themselves quite different from the normal. Thus we might expect that the Gram-Charlier series is not particularly appropriate for many of the nonlinear problems encountered.

A very wide class of zero-memory nonlinearities, including those treated previously in this chapter, can be represented exactly or approximated closely by a Taylor series of the form

$$y = g(x) = g(0) + xg^{(1)}(0) + \frac{x^2}{2} g^{(2)}(0) + \cdots + \frac{x^n}{n!} g^{(n)}(0) + \cdots \tag{6.14-5}$$

where

$$g^{(n)}(0) = \frac{d^n}{dx^n} [g(x)]\bigg|_{x=0}. \tag{6.14-6}$$

If we assume that the input $x(t)$ to this nonlinearity $g(x)$ is a gaussian process, then we want to develop a sequence of polynomials appropriate for the representation of random variables which are powers of the normally distributed random variable x. For this purpose we will define a class of polynomials [11] which will be called the Q polynomials.

The first set of polynomials will be those appropriate for the representation of the density $W_y(y)$ when the first power of x predominates in Equation 6.14-5 and hence should be appropriate for the normal distribution. We therefore turn to the Hermite polynomials $H_n(y)$ of the Gram-Charlier series. As a matter of convenience, however, we change the normalization slightly and define $Q_n^{(1)}(y)$ as

$$Q_n^{(1)}(y) = \frac{H_n(y)}{\sqrt{n!}} = \frac{(-1)^n}{\sqrt{n!}} e^{y^2/2} \frac{d^n}{dy^n} (e^{-y^2/2}). \tag{6.14-7}$$

Now the orthogonality relationship of Equation 6.14-4 becomes

$$\frac{1}{\sqrt{2\pi}} \int_{-\infty}^{\infty} Q_m^{(1)}(y) Q_n^{(1)}(y)\, e^{-y^2/2}\, dy = \delta_{mn} \tag{6.14-8}$$

since the unit normal distribution $\alpha(y)$ is given by

$$\alpha(y) = \frac{1}{\sqrt{2\pi}} e^{-y^2/2}, \qquad -\infty < y < \infty. \tag{6.14-9}$$

If the second (or linear) term of Equation 6.14-5 predominates in the Taylor series expansion of $g(x)$, then, for a normally distributed input x, we would expect the first-order density of y to be appropriately represented by the series

$$W_y(y) = \sum_{n=0}^{\infty} c_n Q_n^{(1)}(y)\, e^{-y^2/2} \tag{6.14-10}$$

where c_n is obtained from Equation 6.14-8 as

$$c_n = \frac{1}{\sqrt{2\pi}} \int_{-\infty}^{\infty} W_y(y) Q_n^{(1)}(y)\, dy. \tag{6.14-11}$$

Thus far we have done no more than renormalize the Gram-Charlier series of Equation 2.25-1, while recognizing that such a series is useful if the density $W_y(y)$ is nearly normal; that is, if it is the result of a nearly linear operation on a normal input density.

Now we consider ways to take account of the term in x^k in Equation 6.14-5. The functions $Q_n^{(1)}(y)$ are polynomials of degree n as shown by Equation 2.24-10. They can be written, therefore, as

$$Q_n^{(1)}(y) = a_0 + a_1 y + a_2 y^2 + \cdots + a_n y^n. \tag{6.14-12}$$

Let us now define a polynomial of order nk by

$$Q_n^{(k)}(y) = b_0 + b_1 y^k + b_2 y^{2k} + \cdots + b_n y^{nk}. \tag{6.14-13}$$

If we make the change in variable

$$w = y^k \tag{6.14-14}$$

then this new polynomial becomes

$$Q_n^{(k)}(w) = b_0 + b_1 w + b_2 w^2 + \cdots + b_n w^n \tag{6.14-15}$$

and the orthogonality relationship of Equation 6.14-8 becomes

$$\frac{1}{k\sqrt{2\pi}} \int_{-\infty}^{\infty} Q_m^{(k)}(w) Q_n^{(k)}(w)\, e^{-w^{2/k}/2} w^{(1-k)/k}\, dw = \delta_{mn}, \qquad k \text{ odd}, \tag{6.14-16a}$$

and

$$\frac{2}{k\sqrt{\pi}} \int_{0}^{\infty} Q_m^{(k)}(w) Q_n^{(k)}(w)\, e^{-w^{2/k}/2} w^{(1-k)/k}\, dw = \delta_{mn}, \qquad k \text{ even}. \tag{6.14-16b}$$

In analogy with Equation 6.14-10, we attempt to represent the output density $W_y(y)$ by the series

$$W_y(w) = \sum_{n=0}^{\infty} c_{nk} Q_n^{(k)}(w)\, e^{-(w^{2/k})/2} w^{(1-k)/k}. \tag{6.14-17}$$

Here the coefficients c_{nk} are found by the use of Equation 6.14-16 to be

$$c_{nk} = \frac{1}{k}\frac{1}{\sqrt{2\pi}}\int_{-\infty}^{\infty} W_y(w)Q_n^{(k)}(w)\,dw, \qquad n \text{ odd}, \tag{6.14-18a}$$

$$c_{nk} = \frac{2}{k}\frac{1}{\sqrt{2\pi}}\int_{0}^{\infty} W_y(w)Q_n^{(k)}(w)\,dw, \qquad n \text{ even}. \tag{6.14-18b}$$

The idea now is to use Equation 6.14-17 to represent the output density. The proper choice of the integer k will depend on which term predominates in the Taylor series expansion of Equation 6.14-5. Thus, if the term in x^m predominates, we might hope that a few terms of the right side of Equation 6.14-17 with $k = m$ would provide a reasonable approximation to the output density $W_y(y)$.

The various polynomials $Q_n^{(k)}(y)$ can be found from the orthogonality requirement of Equation 6.14-18. As an example, let us take $k = 1$ and find the first few $Q_n^{(1)}(y)$. If the input x has a unit normal distribution and if the nonlinearity is $g(x) = x$, then we must have

$$W_y(y) = \frac{1}{\sqrt{2\pi}}e^{-y^2/2}. \tag{6.14-19}$$

Hence we should require that

$$Q_0^{(1)}(y) = 1 \tag{6.14-20}$$

so that

$$c_{01} = \frac{1}{\sqrt{2\pi}}\int_{-\infty}^{\infty} W_y(y)\,dy = \frac{1}{\sqrt{2\pi}}. \tag{6.14-21}$$

Let us now write $Q_1^{(1)}(y)$ in the known form

$$Q_1^{(1)}(y) = b_0 + b_1 y. \tag{6.14-22}$$

It follows from Equation 6.14-16 that

$$\frac{1}{\sqrt{2\pi}}\int_{-\infty}^{\infty} Q_0^{(1)}(y)Q_1^{(1)}(y)\,e^{-y^2/2}\,dy = \frac{1}{\sqrt{2\pi}}\int_{-\infty}^{\infty}(b_0 + b_1 y)\,e^{-y^2/2}\,dy = 0 \tag{6.14-23}$$

and

$$\frac{1}{\sqrt{2\pi}}\int_{-\infty}^{\infty} [Q_1^{(1)}(y)]^2\,e^{-y^2/2}\,dy = \frac{1}{\sqrt{2\pi}}\int_{-\infty}^{\infty}(b_0^2 + 2b_0b_1 y + b_1^2 y^2)\,e^{-y^2/2}\,dy = 1 \tag{6.14-24}$$

or

$$b_0 = 0, \qquad b_1^2 = 1, \tag{6.14-25}$$

from the known properties of the moments of the unit normal distribution. Hence we have

$$Q_1^{(1)}(y) = y \tag{6.14-26}$$

and, from Equation 6.14-18,

$$c_{11} = \frac{1}{\sqrt{2\pi}} \int_{-\infty}^{\infty} y W_y(y)\, dy = \frac{E\{y\}}{\sqrt{2\pi}}. \tag{6.14-27}$$

In the same way we find that

$$Q_2^{(1)}(y) = \frac{y^2 - 1}{\sqrt{2}}, \qquad c_{21} = \frac{E\{y^2\} - 1}{\sqrt{4\pi}}, \tag{6.14-28}$$

$$Q_3^{(1)}(y) = \frac{y^3 - 3y}{\sqrt{6}}, \qquad c_{31} = \frac{E\{y^3\} - 3E\{y\}}{\sqrt{12\pi}}. \tag{6.14-29}$$

$$\vdots \qquad\qquad \vdots$$

Since the $Q_n^{(1)}(y)$ are just normalized Hermite polynomials, these results may be checked against Equations 2.24-11 through 2.24-14.

For the case $k = 2$, we begin as before by assuming that

$$Q_0^{(2)}(x) = 1 \tag{6.14-30}$$

and, from Equation 6.14-18b,

$$c_{02} = \frac{1}{\sqrt{2\pi}} \int_0^{\infty} W_y(w)\, dw = \frac{1}{\sqrt{2\pi}} \tag{6.14-31}$$

so that, for $g(x) = x^2$, we have

$$W_y(w) = \frac{1}{\sqrt{2\pi}} e^{-w/2} w^{-1/2}, \qquad 0 \le w < \infty. \tag{6.14-32}$$

This is the chi-squared distribution of Section 2.11 as expected. Now we place

$$Q_1^{(2)}(w) = b_0 + b_1 w \tag{6.14-33}$$

and use the orthogonality relationship to obtain

$$Q_1^{(2)}(w) = \frac{w - 1}{\sqrt{2}} \tag{6.14-34}$$

and, in the same way,

$$Q_2^{(2)}(w) = \frac{w^2 - 6w + 3}{\sqrt{4!}}, \tag{6.14-35}$$

$$Q_3^{(2)}(w) = \frac{w^3 - 15w^2 + 45w - 15}{\sqrt{6!}}. \tag{6.14-36}$$

Similar procedures may be followed for higher values of k. It should be apparent that these procedures will be practical only if a few terms of Equation 6.14-17 give a reasonable approximation to $W_y(y)$. Further details may be found in the literature [11].

Thus far in this section, we have considered series expansions of first-order densities. For the calculation of autocorrelation functions and power spectral densities it may be desirable to have similar representations for second-order densities. We consider the second-order density $W(x_1, x_2)$ with associated first-order densities

$$W_1(x_1) = \int_{-\infty}^{\infty} W(x_1, x_2)\, dx_2 \tag{6.14-37}$$

and

$$W_2(x_2) = \int_{-\infty}^{\infty} W(x_1, x_2)\, dx_1. \tag{6.14-38}$$

We will be interested in series expansions of the form

$$W(x_1, x_2) = W_1(x_1)W_2(x_2) \sum_{m=0}^{\infty} \sum_{n=0}^{\infty} a_{mn}\theta_m(x_1)\psi_n(x_2) \tag{6.14-39}$$

where the $\theta_m(x_1)$ and $\psi_n(x_2)$ are two sets of polynomials of order m and n respectively. Each set will be restricted to be orthonormal with respect to the corresponding weighting function $W_1(x_1)$ or $W_2(x_2)$ so that

$$\int_{-\infty}^{\infty} W_1(x_1)\theta_m(x_1)\theta_n(x_1)\, dx_1 = \delta_{mn} \tag{6.14-40a}$$

and

$$\int_{-\infty}^{\infty} W_2(x_2)\psi_m(x_2)\psi_n(x_2)\, dx_2 = \delta_{mn}. \tag{6.14-40b}$$

A direct application of these last two relations to Equation 6.14-39 yields the coefficient a_{mn}:

$$a_{mn} = \int_{-\infty}^{\infty}\int_{-\infty}^{\infty} W(x_1, x_2)\theta_m(x_1)\psi_n(x_2)\, dx_1\, dx_2. \tag{6.14-41}$$

We have already seen that it is possible to use either Equation 6.14-40a or 6.14-40b to construct a set of polynomials orthonormal with respect to a given weighting function. In fact, using the unit normal distribution as the weighting function, we found the first few polynomials $Q_n^{(1)}(y)$ and noted that they were simply the normalized Hermite polynomials of Section 2.24. Let us look briefly at the general situation where the form of the weighting function is not specified. We consider Equation 6.14-40a, for example, and set $m = n = 0$ so that

$$\int_{-\infty}^{\infty} W_1(x_1)\theta_0^2(x_1)\,dx_1 = 1. \tag{6.14-42}$$

This relation is satisfied, obviously, for

$$\theta_0(x_1) = 1. \tag{6.14-43}$$

Now take the two cases $m = 0, n = 1$ and $m = 1, n = 1$ so that

$$\int_{-\infty}^{\infty} W_1(x_1)\theta_1(x_1)\,dx_1 = 0, \tag{6.14-44}$$

$$\int_{-\infty}^{\infty} W_1(x_1)\theta_1^2(x_1)\,dx_1 = 1 \tag{6.14-45}$$

where $\theta_1(x)$ must be a polynomial of first degree; that is,

$$\theta_1(x_1) = ax_1 + b. \tag{6.14-46}$$

We have, from Equations 6.14-44 and 6.14-45,

$$E\{ax_1 + b\} = au_1 + b = 0 \tag{6.14-47}$$

and

$$E\{(ax + b)^2\} = a^2(\sigma_1^2 + u_1^2) + 2abu_1 + b^2 = 1. \tag{6.14-48}$$

We solve these last two equations simultaneously for a and b to obtain

$$a = \frac{1}{\sigma_1}, \qquad b = -\frac{u_1}{\sigma_1}, \tag{6.14-49}$$

or

$$\theta_1(x_1) = \frac{x_1 - u_1}{\sigma_1} \tag{6.14-50}$$

where u_1 and σ_1^2 are the mean and variance respectively of the random variable x_1. It is clear that we can continue this procedure by assuming that

$$\theta_2(x_1) = a_1x^2 + b_1x + c_1 \tag{6.14-51}$$

and then using the three orthogonality relationships involving $\theta_0(x_1)$, $\theta_1(x_1)$, and $\theta_2(x_1)$. The resulting three algebraic equations may be used to find the constants a_1, b_1, and c_1 in terms of the moments of x_1. This general technique for generating orthogonal polynomials is called the *Gram-Schmidt* procedure [12].

We have now set up a procedure whereby, for given first-order densities $W_1(x_1)$ and $W_2(x_2)$, the sets of polynomials $\theta_m(x_1)$ and $\psi_n(x_2)$ can be generated and, at least formally, the representation of Equation 6.14-39 can be applied to the second-order density $W(x_1, x_2)$. The procedure is greatly simplified and the results more easily applied if we restrict our attention to those second-order densities $W(x_1, x_2)$ where the matrix $[a_{mn}]$ is *diagonal*; that is, where only the elements on the main diagonal are different from zero:

$$[a_{mn}] = \begin{bmatrix} a_{11} & a_{12} & a_{13} & \cdots \\ a_{21} & a_{22} & a_{23} & \cdots \\ a_{31} & a_{32} & a_{33} & \cdots \\ \vdots & \vdots & \vdots & \end{bmatrix} = \begin{bmatrix} a_{11} & 0 & 0 & \cdots \\ 0 & a_{22} & 0 & \cdots \\ 0 & 0 & a_{33} & \cdots \\ \vdots & \vdots & \vdots & \end{bmatrix}. \tag{6.14-52}$$

In this case, Equation 6.14-39 becomes

$$W(x_1, x_2) = W_1(x_1)W_2(x_2)\sum_{n=0}^{\infty} a_n\theta_n(x_1)\psi_n(x_2) \tag{6.14-53}$$

where a_n has been written for a_{nn} and is given by Equation 6.14-41 as

$$a_n = a_{nn} = \int_{-\infty}^{\infty}\int_{-\infty}^{\infty} W(x_1, x_2)\theta_n(x_1)\psi_n(x_2)\, dx_1\, dx_2. \tag{6.14-54}$$

Note that the first two a_n are

$$a_0 = \int_{-\infty}^{\infty}\int_{-\infty}^{\infty} W(x_1, x_2)\, dx_1\, dx_2 = 1, \tag{6.14-55}$$

$$a_1 = \int_{-\infty}^{\infty}\int_{-\infty}^{\infty} W(x_1, x_2)\left(\frac{x_1 - u_1}{\sigma_1}\right)\left(\frac{x_2 - u_2}{\sigma_2}\right) dx_1\, dx_2 = \rho \tag{6.14-56}$$

where ρ is the normalized covariance and is called the correlation coefficient. This special expansion where the matrix $[a_{mn}]$ is diagonal has been treated in some detail in the literature [13].

For the special case where the second-order density is symmetric in x_1 and x_2, further simplifications are possible. If we assume that

$$W(x_1, x_2) = W(x_2, x_1) \tag{6.14-57}$$

then Equation 6.14-53 becomes

$$W(x_1, x_2) = W_1(x_1)W_1(x_2)\sum_{n=0}^{\infty} a_n\theta_n(x_1)\theta_n(x_2) \tag{6.14-58}$$

where a_n is given by

$$a_n = \int_{-\infty}^{\infty}\int_{-\infty}^{\infty} W(x_1, x_2)\theta_n(x_1)\theta_n(x_2)\, dx_1\, dx_2. \tag{6.14-59}$$

This case is of particular interest since many second-order densities encountered in practice have the symmetric property of Equation 6.14-57.

Example 6.6 Let us assume as before that x_1 is $x(t_1)$ and x_2 is $x(t_2)$ where $x(t)$ is a gaussian random process with zero mean and autocorrelation function $R_x(\tau)$. Then $W(x_1, x_2)$ is given by Equation 6.10-8 as

$$W(x_1, x_2) = \frac{1}{2\pi\sigma^2\sqrt{1-\rho^2}} \exp\left[-\frac{x_1^2 + x_2^2 - 2\rho x_1 x_2}{2\sigma^2(1-\rho^2)}\right] \tag{A}$$

where

$$\rho = \rho(\tau) = \frac{R_x(\tau)}{R_x(0)}, \qquad \tau = t_2 - t_1, \tag{B}$$

and

$$\sigma^2 = R_x(0). \tag{C}$$

For this case, the two corresponding first-order densities are identical and we have, for example,

$$W_1(x_1) = \frac{1}{\sqrt{2\pi}\,\sigma} e^{-x_1^2/2\sigma^2}, \qquad -\infty < x_1 < \infty. \tag{D}$$

We have already developed the set of polynomials which are orthogonal with respect to this density function of Equation D. These are the Hermite polynomials $H_n(x)$ of Section 2.24. It follows, therefore, from Equations 2.24-22 and 6.14-40 that

$$\int_{-\infty}^{\infty} \frac{1}{\sqrt{2\pi}\,\sigma} e^{-x^2/2\sigma^2} H_m\left(\frac{x}{\sigma}\right) H_n\left(\frac{x}{\sigma}\right) dx = n!\delta_{mn} \tag{E}$$

or

$$\theta_n(x_1) = \frac{1}{\sqrt{n!}} H_n\left(\frac{x_1}{\sigma}\right) \tag{F}$$

and

$$\psi_n(x_2) = \theta_n(x_2) = \frac{1}{\sqrt{n!}} H_n\left(\frac{x_2}{\sigma}\right). \tag{G}$$

Also, from Equation 6.14-59,

$$a_n = \frac{1}{n!2\pi\sigma^2\sqrt{1-\rho^2}} \int_{-\infty}^{\infty}\int_{-\infty}^{\infty} \exp\left[-\frac{x_1^2 + x_2^2 - 2\rho x_1 x_2}{2\sigma^2(1-\rho^2)}\right] \times H_n\left(\frac{x_1}{\sigma}\right) H_n\left(\frac{x_2}{\sigma}\right) dx_1\, dx_2. \tag{H}$$

It is clear from Equations 6.14-55 and 6.14-56 or by direct integration that

$$a_0 = 1, \tag{I}$$

$$a_1 = \rho. \tag{J}$$

We can continue in this fashion and use Equation 2.24-13 to write

$$a_2 = \frac{1}{2} E\left\{\left(\frac{x_1^2}{\sigma^2} - 1\right)\left(\frac{x_2^2}{\sigma^2} - 1\right)\right\} \tag{K}$$

where the expectation is with respect to the joint normal distribution of Equation A. On expanding we have

$$a_2 = \frac{1}{2\sigma^4} E\{x_1^2 x_2^2 - \sigma^4\}. \tag{L}$$

We can now use either Equation 2.22-21 or 6.4-19 to obtain

$$E\{x_1^2 x_2^2\} = 2R_x^2(\tau) + R_x^2(0) \tag{M}$$

or

$$a_2 = \frac{R_x^2(\tau)}{R_x^2(0)} = \rho^2 \tag{N}$$

since $R_x(0) = \sigma^2$. If we continue in the same way, we obtain

$$a_n = \rho^n. \tag{O}$$

Now Equation 6.14-58 becomes

$$W(x_1, x_2) = \frac{1}{2\pi\sigma^2} e^{-(x_1^2 + x_2^2)/2\sigma^2} \sum_{n=0}^{\infty} \frac{\rho^n}{n!} H_n\left(\frac{x_1}{\sigma}\right) H_n\left(\frac{x_2}{\sigma}\right). \tag{P}$$

This last result is well known and frequently appears in the literature [14] in slightly different forms.

• •

6.15 General Techniques—Series Expansions of Nonlinearities

In the previous section we have treated zero-memory nonlinear problems by the use of series representation of input densities. We might expect that similar representations of the nonlinearities themselves would be useful. In terms of Figure 6.2(*b*) we consider the series

$$y = g(x) = \sum_{n=0}^{\infty} c_n \Psi_n(x) \tag{6.15-1}$$

where, as before, the $\Psi_n(x)$ are required to be polynomials of degree n which are orthonormal in the interval (a, b) with respect to some weighting function $w(x)$; that is,

$$\int_a^b w(x)\Psi_m(x)\Psi_n(x)\, dx = \delta_{mn}. \tag{6.15-2}$$

The coefficient c_n is

$$c_n = \int_a^b g(x)w(x)\Psi_n(x)\, dx. \tag{6.15-3}$$

Suppose we choose the $\Psi_n(x)$ to be the $\psi_n(x)$ of Equation 6.14-40b so that the weighting function $w(x)$ is the first-order density $W_2(x)$. Since $\psi_0(x)$ is equal to unity, Equation 6.15-3 becomes, for $n = 0$,

$$c_0 = \int_{-\infty}^{\infty} g(x_2)W_2(x_2)\, dx_2 = E\{g(x_2)\} \tag{6.15-4}$$

and we can rewrite Equation 6.15-1 as

$$g(x_2) - E\{g(x_2)\} = \sum_{n=1}^{\infty} c_n \psi_n(x_2). \tag{6.15-5}$$

Now let us define a cross-correlation function $C_{xy}(t_1, t_2)$ by

$$C_{xy}(t_1, t_2) = E\{(x_1 - E\{x_1\})(y_2 - E\{y_2\})\} \tag{6.15-6}$$

where, as before, x_1 corresponds to $x(t_1)$ and y_2 to $y(t_2)$. Since $y_2 = g(x_2)$, we may use Equation 6.15-5 to obtain

$$\begin{aligned} C_{xy}(t_1, t_2) &= E\left\{(x_1 - E\{x_1\}) \sum_{n=1}^{\infty} c_n \psi_n(x_2)\right\} \\ &= \int_{-\infty}^{\infty}\int_{-\infty}^{\infty} W(x_1, x_2)(x_1 - E\{x_1\}) \sum_{n=1}^{\infty} c_n \psi_n(x_2)\, dx_1\, dx_2 \end{aligned} \tag{6.15-7}$$

where $W(x_1, x_2)$ is the joint density of x_1 and x_2. Let us now make use of Equation 6.14-50 to replace

$$x_1 - E\{x_1\} = \sigma_1 \theta_1(x_1). \tag{6.15-8}$$

After this substitution,

$$C_{xy}(t_1, t_2) = \sigma_1 \sum_{n=1}^{\infty} c_n \int_{-\infty}^{\infty}\int_{-\infty}^{\infty} W(x_1, x_2)\theta_1(x_1)\psi_n(x_2)\, dx_1\, dx_2. \tag{6.15-9}$$

If the joint density $W(x_1, x_2)$ is representable by Equation 6.14-53, then it follows immediately from the orthogonality relations of Equations 6.14-40a and 6.14-40b that

$$C_{xy}(t_1, t_2) = \sigma_1 \sum_{n=1}^{\infty} c_n \sum_{m=0}^{\infty} a_m \delta_{1m} \delta_{mn} = \sigma_1 c_1 a_1. \tag{6.15-10}$$

However, the coefficient a_1 has already been evaluated in Equation 6.14-56 and is just the correlation coefficient ρ. Thus we have

$$C_{xy}(t_1, t_2) = c_1 \sigma_1 \rho(t_1, t_2) = c_1 \frac{R_x(t_1, t_2)}{\sigma_2}. \tag{6.15-11}$$

Now the coefficient c_1 is given by Equation 6.15-3 and $\psi_1(x_2)$ by Equation 6.14-50 so that, finally,

$$\boxed{C_{xy}(t_1, t_2) = cR_x(t_1, t_2)} \tag{6.15-12}$$

where c is a constant determined by

$$c = \frac{c_1}{\sigma_2} = \int_{-\infty}^{\infty} g(x_2) W_2(x_2)\left(\frac{x_2 - u_2}{\sigma_2^2}\right) dx_2. \tag{6.15-13}$$

If the input process $x(t)$ is at least wide-sense stationary so that

$$R_x(t_1, t_2) = R_x(t_2 - t_1) = R_x(\tau), \tag{6.15-14}$$

then

$$\boxed{C_{xy}(\tau) = cR_x(\tau).} \tag{6.15-15}$$

This result was originally due to Bussgang [15]. If the input process $x(t)$ has a second-order density which can be represented by the diagonal series of Equation 6.14-53, then the cross-correlation function between the input and output of $g(x)$ is proportional to the autocorrelation function of $x(t)$.

We now illustrate the use of the series expansion of Equation 6.15-1 by solving the smooth limiter problem of Section 6.13. Referring to Equation 6.13-1 and Figure 6.17(a), we write the zero-memory nonlinearity as

$$y = g(x) = M\phi\left(\frac{x}{\sigma}\right) \tag{6.15-16}$$

where M is any constant and $\phi(x)$ is the function defined by

$$\phi(x) = \frac{1}{\sqrt{2\pi}} \int_0^x e^{-z^2/2}\, dz. \tag{6.15-17}$$

As in Section 6.13, the input $x(t)$ is assumed to be a wide-sense stationary, zero-mean random process with autocorrelation function $R_x(\tau)$ and normalized correlation function $\rho(\tau)$. It follows from our previous discussions that the nonlinearity $g(x)$ should be represented by the series

$$g(x) = M\phi\left(\frac{x}{\sigma}\right) = \sum_{n=0}^{\infty} a_n \frac{H_n(x)}{\sqrt{n!}} = \sum_{n=0}^{\infty} a_n Q_n^{(1)}(x) \tag{6.15-18}$$

where $H_n(x)$ are the Hermite polynomials and $Q_n^{(1)}(x)$ was defined by Equation 6.14-7. From Equation 6.14-8, we have

$$a_n = \frac{1}{\sqrt{2\pi n!}} \int_{-\infty}^{\infty} g(x)\, e^{-x^2/2} H_n(x)\, dx. \tag{6.15-19}$$

Let us now refer to Equation 2.24-6 and substitute

$$e^{-x^2/2} H_n(x) = (-1)^n \frac{d^n}{dx^n}(e^{-x^2/2}) \tag{6.15-20}$$

into the integrand of this last expression to obtain

$$a_n = \frac{(-1)^n}{\sqrt{2\pi n!}} \int_{-\infty}^{\infty} g(x) \frac{d^n}{dx^n}(e^{-x^2/2})\, dx. \tag{6.15-21}$$

We may integrate by parts to find that

$$a_n = \frac{(-1)^{n-1}}{\sqrt{n!}} \int_{-\infty}^{\infty} \frac{d}{dx}[g(x)] \frac{d^{n-1}}{dx^{n-1}}\left(\frac{e^{-x^2/2}}{\sqrt{2\pi}}\right) dx. \qquad (6.15\text{-}22)$$

But $g(x)$, as given by Equation 6.15-16, is easily differentiated so that

$$a_n = \frac{M}{\sqrt{n!}} \int_{-\infty}^{\infty} \frac{e^{-x^2/2\sigma^2}}{\sqrt{2\pi}\sigma} (-1)^{n-1} \frac{d^{n-1}}{dx^{n-1}}\left(\frac{e^{-x^2/2}}{\sqrt{2\pi}}\right) dx \qquad (6.15\text{-}23)$$

or, from *Parseval's Theorem**

$$a_n = \frac{M}{2\pi\sqrt{n!}} \int_{-\infty}^{\infty} e^{-\omega^2\sigma^2/2}(j\omega)^{n-1} e^{-\omega^2/2}\, d\omega. \qquad (6.15\text{-}24)$$

It is clear that the integrand of this last expression is an odd function of ω for n even; therefore a_n is zero for n even and, for n odd, we can write

$$a_{2k+1} = \frac{(-1)^k M}{\sqrt{2\pi(2k+1)!}} \int_{-\infty}^{\infty} \frac{e^{-(\omega^2/2)(1+\sigma^2)}\omega^{2k}}{\sqrt{2\pi}}\, d\omega. \qquad (6.15\text{-}25)$$

We may make a linear change in variable in the definite integral to obtain

$$a_{2k+1} = \frac{(-1)^k M}{\sqrt{2\pi(2k+1)!}(\sqrt{1+\sigma^2})^{2k+1}} \int_{-\infty}^{\infty} \frac{\omega^{2k} e^{-\omega^2/2}}{\sqrt{2\pi}}\, d\omega. \qquad (6.15\text{-}26)$$

This integral is just the $2k$th moment of the unit normal distribution and is given by Equation 2.18-24. Consequently we have

$$a_{2k+1} = \frac{(-1)^k M (2k)!}{k!2^k\sqrt{2\pi(2k+1)!}(\sqrt{1+\sigma^2})^{2k+1}}. \qquad (6.15\text{-}27)$$

We now substitute this expression into the series expansion of Equation 6.15-18 and have

$$\boxed{g(x) = \frac{M}{\sqrt{2\pi}} \sum_{k=0}^{\infty} \left(-\tfrac{1}{2}\right)^k \frac{H_{2k+1}(x)}{k!(2k+1)(\sqrt{1+\sigma^2})^{2k+1}}.} \qquad (6.15\text{-}28)$$

* We consider the two functions $h(x)$ and $k(x)$ with Fourier transforms $H(\omega)$ and $K(\omega)$, respectively. We have

$$\int_{-\infty}^{\infty} h(x)k(x)\, dx = \int_{-\infty}^{\infty} \frac{1}{2\pi}\int_{-\infty}^{\infty} H(\omega)\, e^{j\omega x}\, d\omega k(x)\, dx.$$

On interchanging the order of integration, we obtain one form of Parseval's Theorem:

$$\int_{-\infty}^{\infty} h(x)k(x)\, dx = \frac{1}{2\pi}\int_{-\infty}^{\infty} H(\omega)K(-\omega)\, d\omega.$$

This theorem is discussed in Appendix D in connection with Fourier series. Now Equation 6.15-24 follows on using pair numbers 8 and 9 of Table 3.1.

As previously mentioned, the representation of $g(x)$ as given by Equation 6.15-28 is most appropriate for those problems where the input $x(t)$ is gaussian. Let us use this equation to find the autocorrelation function $R_y(\tau)$ of the output $y = g(x)$ of the smooth limiter for such a gaussian input. We have

$$R_y(\tau) = E\{y_1 y_2\} = \int_{-\infty}^{\infty}\int_{-\infty}^{\infty} W(x_1, x_2) g(x_1) g(x_2)\, dx_1\, dx_2 \quad (6.15\text{-}29)$$

where $\tau = t_2 - t_1$ and where $W(x_1, x_2)$ is the bivariate normal distribution of Equation A of Example 6.6. Let us keep in mind, however, that the σ^2 of Equation A is the variance of the random variable x while the σ^2 of Equations 6.15-16 and 6.15-28 is the parameter in the nonlinear characteristic of the smooth limiter $g(x)$ as shown in Figure 6.17(a). To avoid confusion, we will call the former σ_x^2 for the remainder of this discussion. This notation is then consistent with that of Section 6.13. Furthermore, as in Equation 6.13-7, we define the convenient parameter α by

$$\alpha = \frac{\sigma^2}{\sigma_x^2} = \frac{\sigma^2}{R_x(0)}. \quad (6.15\text{-}30)$$

Now Equation P of Example 6.6 becomes

$$W(x_1, x_2) = \frac{1}{2\pi\sigma_x^2} e^{-(x_1^2 + x_2^2)/2\sigma_x^2} \sum_{n=0}^{\infty} \frac{\rho^n}{n!} H_n\left(\frac{x_1}{\sigma_x}\right) H_n\left(\frac{x_2}{\sigma_x}\right). \quad (6.15\text{-}31)$$

Note that the smooth limiter $g(x)$ was represented in Equation 6.15-28 in terms of Hermite polynomials with argument x. We want to replace these with Hermite polynomials with argument x/σ_x as in Equation 6.15-31 so that orthogonality properties can be used later. To make this change, we examine Equations 6.15-18 and 6.15-19 and note that we want to replace Equation 6.15-18 by

$$g(x) = M\phi\left(\frac{x}{\sigma}\right) = \sum_{n=0}^{\infty} b_n \frac{H_n(x/\sigma_x)}{\sqrt{n!}} \quad (6.15\text{-}32)$$

where the coefficient b_n is given by

$$b_n = \frac{M}{\sqrt{2\pi n!}\,\sigma_x} \int_{-\infty}^{\infty} \phi\left(\frac{x}{\sigma}\right) e^{-x^2/2\sigma_x^2} H_n\left(\frac{x}{\sigma_x}\right) dx. \quad (6.15\text{-}33)$$

We now replace x/σ_x by x in this definite integral to obtain

$$b_n = \frac{M}{\sqrt{2\pi n!}} \int_{-\infty}^{\infty} \phi\left(\frac{\sigma_x}{\sigma} x\right) e^{-x^2/2} H_n(x)\, dx. \quad (6.15\text{-}34)$$

A comparison of this last equation with Equation 6.15-19 shows that $g(x)$ may be expanded in a series of Hermite polynomials with argument x/σ_x if, in Equation 6.15-28, the parameter σ is replaced by $\sqrt{\alpha} = \sigma/\sigma_x$ and the argument x by x/σ_x. Thus Equation 6.15-28 becomes

$$g(x) = \frac{M}{\sqrt{2\pi}} \sum_{k=0}^{\infty} (-\tfrac{1}{2})^k \frac{H_{2k+1}(x/\sigma_x)}{k!(2k+1)(\sqrt{1+\alpha})^{2k+1}}. \tag{6.15-35}$$

At this point we are ready to find the output autocorrelation function $R_y(\tau)$ by substituting Equations 6.15-31 and 6.15-35 into Equation 6.15-29. We make use of the orthogonality relationship

$$\frac{1}{\sqrt{2\pi}\sigma_x} \int_{-\infty}^{\infty} e^{-x^2/2\sigma_x^2} H_n\left(\frac{x}{\sigma_x}\right) H_{2k+1}\left(\frac{x}{\sigma_x}\right) dx = n!\delta_{n(2k+1)} \tag{6.15-36}$$

to write the resulting expression

$$R_y(\tau) = \frac{M^2}{2\pi} \sum_{k=0}^{\infty} \left(\frac{\rho}{1+\alpha}\right)^{2k+1} \frac{(2k)!}{2^{2k}(k!)^2(2k+1)} \tag{6.15-37}$$

or

$$R_y(\tau) = \frac{M^2}{2\pi} z \sum_{k=0}^{\infty} \frac{(z^2)^k}{k!} \frac{(2k-1)(2k-3)\cdots 3\cdot 1}{2^k(2k+1)} \tag{6.15-38}$$

where z has been used to denote

$$z = \frac{\rho}{1+\alpha}. \tag{6.15-39}$$

The sum given in Equation 6.15-38 is a special case of the hypergeometric function $F(a, b; c; z)$ defined by [16]

$$F(a, b; c; z) = \sum_{k=0}^{\infty} \frac{z^k}{k!} \frac{(a)_k(b)_k}{(c)_k} \tag{6.15-40}$$

where, for example,

$$(a)_k = \frac{\Gamma(a+k)}{\Gamma(a)} \tag{6.15-41}$$

and the gamma function $\Gamma(x)$ has been defined previously by Equation 6.12-18. We may rewrite Equation 6.15-38 as

$$R_y(\tau) = \frac{M^2}{2\pi} z \sum_{k=0}^{\infty} \frac{(z^2)^k}{k!} \frac{(\frac{1}{2})_k(\frac{1}{2})_k}{(\frac{3}{2})_k} = \frac{M^2}{2\pi} zF(\tfrac{1}{2}; \tfrac{1}{2}; \tfrac{3}{2}; z^2) \tag{6.15-42}$$

and make use of the further relationship [17] that

$$F(\tfrac{1}{2}; \tfrac{1}{2}; \tfrac{3}{2}; z^2) = \frac{1}{z} \text{arc sin } z \tag{6.15-43}$$

to obtain, finally,

$$R_y(\tau) = \frac{M^2}{2\pi} \arcsin \frac{\rho(\tau)}{1 + \alpha} \tag{6.15-44}$$

which is identical to Equation 6.13-21 after proper adjustment of the scale factors M and K.

Another type of zero-memory nonlinearity which is easily treated by these series techniques is that where $g(x)$ is piecewise constant with jumps of amount Δ_i at N points x_i where $i = 1, 2, \ldots, N$. Such a characteristic is shown in Figure 6.18(a). As in Equation 6.15-18, we use the representation

$$g(x) = \sum_{n=0}^{\infty} a_n \frac{H_n(x)}{\sqrt{n!}} \tag{6.15-45}$$

where the coefficients are given by Equation 6.15-19 as

$$a_0 = \frac{1}{\sqrt{2\pi}} \int_{-\infty}^{\infty} g(x)\, e^{-x^2/2}\, dx \tag{6.15-46}$$

and, for $n \geq 1$,

$$a_n = \frac{1}{\sqrt{2\pi n!}} \int_{-\infty}^{\infty} g(x)\, e^{-x^2/2} H_n(x)\, dx. \tag{6.15-47}$$

An equivalent, and more convenient, form is given by Equation 6.15-22 so that

$$a_n = \frac{1}{\sqrt{2\pi n!}} \int_{-\infty}^{\infty} \sum_{i=1}^{N} \Delta_i \delta(x - x_i) H_{n-1}(x)\, e^{-x^2/2}\, dx \tag{6.15-48}$$

or, after rearranging and using the sampling property of the delta function,

$$a_n = \frac{1}{\sqrt{2\pi n!}} \sum_{i=1}^{N} \Delta_i H_{n-1}(x_i)\, e^{-x_i^2/2}. \tag{6.15-49}$$

With this substitution, Equation 6.15-45 becomes

$$g(x) = a_0 + \frac{1}{\sqrt{2\pi}} \sum_{n=1}^{\infty} \sum_{i=1}^{N} \Delta_i H_{n-1}(x_i)\, e^{-x_i^2/2} \frac{H_n(x)}{n!}. \tag{6.15-50}$$

We illustrate this type of problem by a simple example of a nonlinearity encountered in control problems.

Example 6.7 Suppose that the specific nonlinearity $g_1(x)$ is the limiter with dead zone shown in Figure 6.19. For this case, Equation 6.15-46 becomes

$$a_0 = 0$$

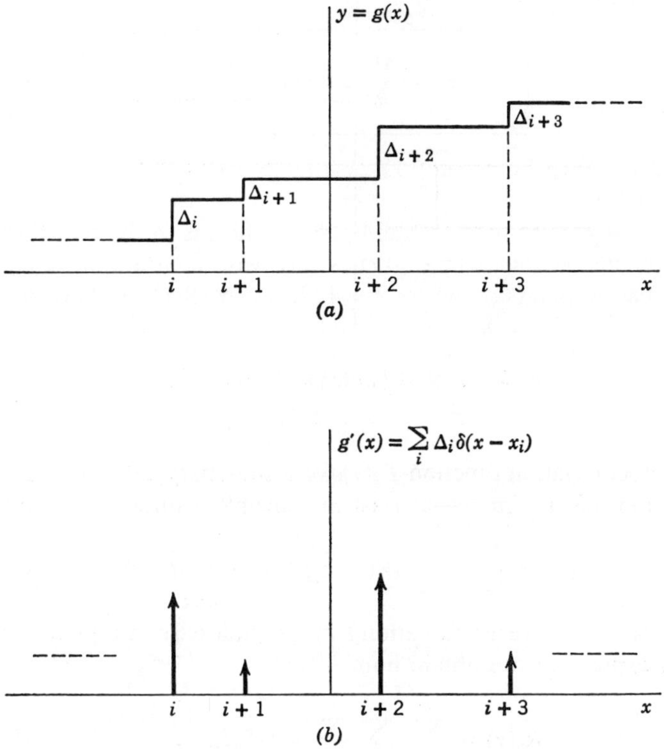

Figure 6.18 A piecewise constant instantaneous nonlinearity: (a) the nonlinearity; (b) the derivative of the nonlinearity

and Equation 6.15-49 is

$$a_n = \frac{M}{\sqrt{2\pi} n!}[H_{n-1}(-a)\, e^{-a^2/2} + H_{n-1}(a)\, e^{-a^2/2}]$$

or

$$a_n = \begin{cases} 0, & n \text{ even}, \\ \dfrac{M\, e^{-a^2/2}}{\sqrt{2\pi} n!}\, 2H_{n-1}(a), & n \text{ odd}. \end{cases}$$

• •

It should be apparent from the previous discussion that, if an arbitrary instantaneous nonlinearity $g(x)$ can be expanded in a series of Hermite polynomials so that

$$g(x) = \sum_{n=0}^{\infty} b_n H_n\left(\frac{x}{\sigma}\right), \tag{6.15-51}$$

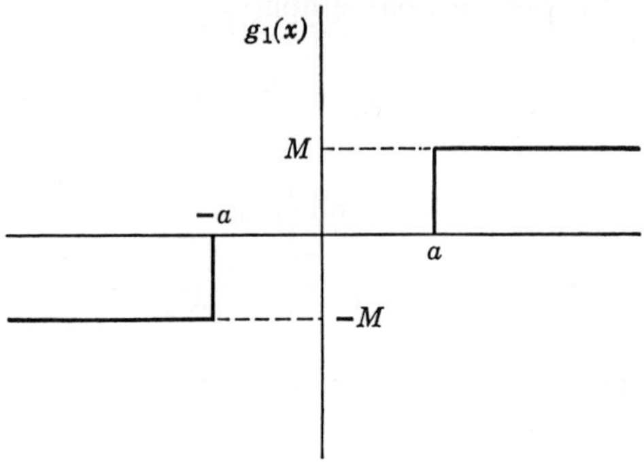

Figure 6.19 *A limiter with dead zone*

then the autocorrelation function $R_y(\tau)$ for a stationary gaussian input has a representation that is simple—at least in principle. Consider the expression

$$R_y(\tau) = \int_{-\infty}^{\infty}\int_{-\infty}^{\infty} W(x_1, x_2)g(x_1)g(x_2)\,dx_1\,dx_2 \tag{6.15-52}$$

where $\tau = t_2 - t_1$. If we use Equation P of Example 6.6 and Equation 6.15-51 in this last expression, we obtain immediately

$$R_y(\tau) = \sum_{m=0}^{\infty}\sum_{k=0}^{\infty}\sum_{n=0}^{\infty} \rho^m b_k b_n \delta_{mk}\delta_{mn} \tag{6.15-53}$$

or

$$R_y(\tau) = \sum_{m=0}^{\infty} \rho^m b_m^2 \tag{6.15-54}$$

where

$$\rho = \rho(\tau) = \rho(t_2 - t_1) = \frac{R_x(t_2 - t_1)}{\sigma^2} \tag{6.15-55}$$

and where the coefficient b_m is determined by

$$b_m = \frac{1}{\sqrt{2\pi}\sigma n!}\int_{-\infty}^{\infty} g(x)\,e^{-x^2/2\sigma^2}H_n\!\left(\frac{x}{\sigma}\right) dx. \tag{6.15-56}$$

More generally, suppose the input joint density $W(x_1, x_2)$ is symmetric and admits the diagonal expansion of Equation 6.14-58 with orthogonal polynomials $\theta_n(x)$. If the nonlinearity can be written in terms of these same polynomials as

$$g(x) = \sum_{n=0}^{\infty} c_n\theta_n(x), \tag{6.15-57}$$

then, exactly as in the previous paragraph,

$$R_y(t_1, t_2) = \sum_{n=0}^{\infty} c_n^2 a_n \tag{6.15-58}$$

with a_n given by Equation 6.14-59 and c_n by

$$c_n = \int_{-\infty}^{\infty} g(x) W_1(x) \theta_n(x)\, dx. \tag{6.15-59}$$

Whether or not these approaches are useful will depend on the tractability of the resulting sums such as that given by Equation 6.15-58.

PART 2—ZERO-MEMORY NONLINEAR DEVICES FOLLOWED BY LINEAR FILTERS (CLASS κ_1—TYPE III NONLINEARITIES)

In the following three sections we consider some of the nonlinear systems illustrated in Figure 6.2(*c*). It will soon become apparent that such systems are much harder to analyze than the linear systems of Chapter 4 and the zero memory nonlinear systems of Part 1 of this chapter. Reasonably simple procedures, however, can be applied to some of those most commonly encountered. We begin in the next section with a specific system of considerable importance.

6.16 Squaring and Filtering a Gaussian Input

The situation illustrated in Figure 6.20 is commonly encountered in problems involving the detection and demodulation of weak signals and almost deserves the appellation *classical.* A special case was encountered in Section 6.7 and illustrated in Figure 6.9(*a*) where the square law receiver was discussed. As shown in Figure 6.20, we have a stationary gaussian input $x(t)$ which is squared and then passed through a noninteracting linear filter with impulse

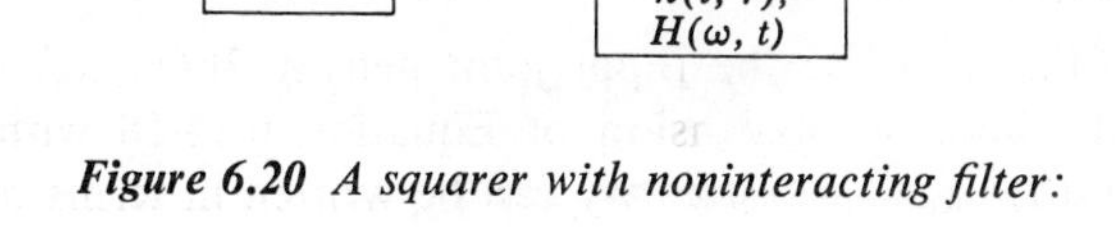

Figure 6.20 *A squarer with noninteracting filter:*

$$y = x^2$$

$$z(t) = \int_{-\infty}^{t} h(t, \tau) y(\tau)\, d\tau$$

response $h(t, \tau)$ and complex transfer function $H(\omega, t)$. The output $z(t)$ is given by

$$z(t) = \int_{-\infty}^{t} h(t, \tau)x^2(\tau)\, d\tau. \tag{6.16-1}$$

As in previous problems we are interested in the first order probability distribution $W_z(z)$ of the output $z(t)$ as well as its autocorrelation function and power spectral density, given that the gaussian input $x(t)$ has zero mean and an autocorrelation function $R_x(\tau)$ and corresponding power spectral density $\varphi_x(\omega)$.

We consider now the first order distribution $W_z(z)$ of the output. It turns out that results are easier to obtain if we work with the characteristic function $F_z(v)$ as given by Equation 2.15-8

$$F_z(v) = E\{e^{jvz}\} = \int_{-\infty}^{\infty} W_z(z)\, e^{jvz}\, dz. \tag{6.16-2}$$

Then the density $W_z(z)$ can be obtained from Equation 2.15-11 as the Fourier transform

$$W_z(z) = \frac{1}{2\pi}\int_{-\infty}^{\infty} F_z(v)\, e^{-jvz}\, dv. \tag{6.16-3}$$

It is clear from Equations 6.16-1 and 6.16-2 that the characteristic function $F_z(v)$ can be written as

$$F_z(v) = E\left\{\exp\left[jv\int_{-\infty}^{t} h(t, \tau)x^2(\tau)\, d\tau\right]\right\}. \tag{6.16-4}$$

In order to rewrite this last expression in a more tractable form, it is necessary for us to refer to Section 4.6 and Equation 4.6-38. There we developed an orthogonal representation for the random process $x(t)$ of the form

$$x(t) = \sum_{n=1}^{\infty} X_i\varphi_i(t) \tag{6.16-5}$$

where the X_i are random variables and the $\varphi_i(t)$ are given functions that satisfy an orthogonality relationship and are the solutions of a certain homogeneous integral equation. We shall now generalize those results somewhat in order to make them more useful for our particular problem. We will assume that the $\varphi_i(t)$ obey a more general orthogonality with respect to the weight function $h(t, \tau)$ in the interval $(-\infty, t)$ so that

$$\int_{-\infty}^{t} h(t, \tau)\varphi_i(\tau)\varphi_j(\tau)\, d\tau = \delta_{ij} \tag{6.16-6}$$

where δ_{ij} is the Kronecker delta. It then follows from this last expression that the X_i are random variables given by

$$X_i = \int_{-\infty}^{t} h(t, \tau)x(\tau)\varphi_i(\tau)\, d\tau. \tag{6.16-7}$$

Let us now suppose that the $\varphi_i(t)$ are the solutions of the homogeneous integral equation

$$\varphi_i(\sigma) = \lambda_i \int_{-\infty}^{t} R_x(\tau - \sigma)h(t, \tau)\varphi_i(\tau)\, d\tau \tag{6.16-8}$$

with kernel $R_x(\tau - s)h(t, \tau)$, eigenvalues λ_i, and eigenfunctions $\varphi_i(t)$. We form the product

$$E\{X_i X_j\} = \int_{-\infty}^{t}\int_{-\infty}^{t} R_x(\tau - \sigma)h(t, \tau)h(t, \sigma)\varphi_i(\tau)\varphi_j(\sigma)\, d\tau\, d\sigma \tag{6.16-9}$$

from Equation 6.16-7. However, if Equation 6.16-8 is satisfied, then this last expression may be written as

$$E\{X_i X_j\} = \frac{1}{\lambda_i}\int_{-\infty}^{t} h(t, \sigma)\varphi_i(\sigma)\varphi_j(\sigma)\, d\sigma = \frac{\delta_{ij}}{\lambda_i} \tag{6.16-10}$$

if the orthogonality of Equation 6.16-6 is valid. Thus the X_i are uncorrelated random variables with zero means and variances $1/\lambda_i$ provided Equations 6.16-6 and 6.16-8 are consistent. In addition it follows from Equation 6.16-7 that the X_i (being linear operations on a gaussian process) are normally distributed and hence (being uncorrelated) are independent. It remains only to show that Equation 6.16-6 is a direct result of Equation 6.16-8.

We use Equation 6.16-8 to form

$$\int_{-\infty}^{t} h(t, \sigma)\varphi_i(\sigma)\varphi_j(\sigma)\, d\sigma$$
$$= \lambda_i \int_{-\infty}^{t}\int_{-\infty}^{t} R_x(\tau - \sigma)h(t, \tau)h(t, \sigma)\varphi_i(\tau)\varphi_j(\sigma)\, d\tau\, d\sigma \tag{6.16-11}$$

and

$$\int_{-\infty}^{t} h(t, \sigma)\varphi_i(\sigma)\varphi_j(\sigma)\, d\sigma$$
$$= \lambda_j \int_{-\infty}^{t}\int_{-\infty}^{t} R_x(\tau - \sigma)h(t, \tau)h(t, \sigma)\varphi_j(\tau)\varphi_i(\sigma)\, d\tau\, d\sigma. \tag{6.16-12}$$

After noting that $R_x(\tau - \sigma) = R_x(\sigma - \tau)$, we may equate the double integrals in both of these last two equations to obtain

$$\frac{1}{\lambda_i}\int_{-\infty}^{t} h(t, \sigma)\varphi_i(\sigma)\varphi_j(\sigma)\, d\sigma = \frac{1}{\lambda_j}\int_{-\infty}^{t} h(t, \sigma)\varphi_i(\sigma)\varphi_j(\sigma)\, d\sigma. \tag{6.16-13}$$

For distinct eigenvalues $\lambda_i \neq \lambda_j$, the orthogonality relationship of Equation 6.16-6 must follow. Other properties of the $\varphi_i(t)$ such as their existence and completeness are easily proved [18].

We now proceed to make use of this representation for the gaussian process $x(t)$ by substituting Equation 6.16-5 into Equation 6.16-1 to obtain, for time invariant linear filters $h(t, \tau) = h(t - \tau)$,

$$z(t) = \int_{-\infty}^{t} h(t-\tau) \sum_{i=1}^{\infty} \sum_{j=1}^{\infty} X_i \varphi_i(\tau) X_j \varphi_j(\tau)\, d\tau$$

$$= \sum_{i=1}^{\infty} \sum_{j=1}^{\infty} X_i X_j \int_{-\infty}^{t} h(t-\tau) \varphi_i(\tau) \varphi_j(\tau)\, d\tau. \qquad (6.16\text{-}14)$$

The orthogonality relationship of Equation 6.16-6 makes this last expression become

$$z(t) = \sum_{i=1}^{\infty} X_i^2. \qquad (6.16\text{-}15)$$

The characteristic function $F_z(v)$ of Equation 6.16-2 can now be written as

$$F_z(v) = E\{e^{jvz}\} = E\left\{\exp\left(jv \sum_{i=1}^{\infty} X_i^2\right)\right\}. \qquad (6.16\text{-}16)$$

However, the X_i are independent random variables so that the expectation of the product can be written as the product of the expectations to obtain

$$F_z(v) = \prod_{i=1}^{\infty} E\{e^{jvX_i^2}\}. \qquad (6.16\text{-}17)$$

After expanding $e^{jvX_i^2}$ in a power series and taking the expectation term by term, we have

$$F_z(v) = \prod_{i=1}^{\infty} \left[1 + jvE\{X_i^2\} + \frac{(jv)^2}{2!} E\{X_i^4\} + \cdots + \frac{(jv)^n}{n!} E\{X_i^{2n}\} + \cdots\right]. \qquad (6.16\text{-}18)$$

It follows from Equation 6.16-10 that

$$E\{X_i^2\} = \frac{1}{\lambda_i} \qquad (6.16\text{-}19)$$

and from Equation 2.18-24 that

$$E\{X_i^{2n}\} = \frac{2n(2n-1)\cdots(n+1)}{2^n} \frac{1}{\lambda_i}. \qquad (6.16\text{-}20)$$

Thus Equation 6.16-18 can be written as

$$F_z(v) = \prod_{i=1}^{\infty} \left[1 + \frac{1}{2}\frac{2jv}{\lambda_i} + \frac{1\cdot 3}{2\cdot 4}\left(\frac{2jv}{\lambda_i}\right)^2 + \frac{1\cdot 3\cdot 5}{2\cdot 4\cdot 6}\left(\frac{2jv}{\lambda_i}\right)^3 + \cdots\right]. \quad (6.16\text{-}21)$$

On comparing this last expression with the series [19]

$$(1 - x)^{-1/2} = 1 + \frac{1}{2}x + \frac{1\cdot 3}{2\cdot 4}x^2 + \frac{1\cdot 3\cdot 5}{2\cdot 4\cdot 6}x^3 + \cdots, \quad (6.16\text{-}22)$$

we conclude that $F_z(v)$ can be expressed as

$$\boxed{F_z(v) = \prod_{i=1}^{\infty}\left(1 - \frac{2jv}{\lambda_i}\right)^{-1/2}.} \quad (6.16\text{-}23)$$

This last equation is an explicit expression for the characteristic function of z. It remains only to solve the homogeneous integral equation given by Equation 6.16-8 and substitute the resulting eigenvalues λ_i into Equation 6.16-23.

An alternative procedure, which avoids the evaluation of the infinite product of Equation 6.16-23, is as follows. Let us note first that the autocorrelation function $R_x(\tau - \sigma)$ possesses an expansion in terms of the eigenfunctions $\varphi_i(\sigma)$; suppose that

$$R_x(\tau - \sigma) = \sum_{i=1}^{\infty} m_i(\tau)\varphi_i(\sigma). \quad (6.16\text{-}24)$$

Then, from Equation 6.16-6, it follows that

$$m_i(\tau) = \int_{-\infty}^{t} R_x(\tau - \sigma)h(t - \sigma)\varphi_i(\sigma)\, d\sigma = \frac{\varphi_i(\tau)}{\lambda_i} \quad (6.16\text{-}25)$$

from Equation 6.16-8. In other words, Equation 6.16-24 becomes

$$R_x(\tau - \sigma) = \sum_{i=1}^{\infty} \frac{\varphi_i(\tau)\varphi_i(\sigma)}{\lambda_i}. \quad (6.16\text{-}26)$$

This is a special case of Mercer's Theorem and it can be shown that the right side of Equation 6.16-26 converges absolutely and uniformly to $R_x(\tau - \sigma)$ under very general conditions.

Now let us define the function $g(\tau, \sigma)$ by

$$g(\tau, \sigma) = \sum_{i=1}^{\infty} \frac{1}{1 - (2jv/\lambda_i)}\, \frac{\varphi_i(\tau)\varphi_i(\sigma)}{\lambda_i} \quad (6.16\text{-}27)$$

and form the expression

$$g(\tau, u) - 2jv \int_{-\infty}^{t} R_x(\sigma - \tau)h(t - \sigma)g(\sigma, u)\, d\sigma$$

$$= \sum_{i=1}^{\infty} \frac{1}{1 - (2jv/\lambda_i)} \left[\frac{\varphi_i(\tau)\varphi_i(u)}{\lambda_i} - \frac{2jv}{\lambda_i}\varphi_i(u) \int_{-\infty}^{t} R_x(\sigma - \tau)h(t - \sigma)\varphi_i(\sigma)\, d\sigma\right]$$

$$= \sum_{i=1}^{\infty} \frac{1}{1 - (2jv/\lambda_i)} \left[\frac{\varphi_i(\tau)\varphi_i(u)}{\lambda_i} - \frac{2jv}{\lambda_i}\frac{\varphi_i(u)\varphi_i(\tau)}{\lambda_i}\right]$$

$$= \sum_{i=1}^{\infty} \frac{\varphi_i(\tau)\varphi_i(u)}{\lambda_i} = R_x(\tau - u). \tag{6.16-28}$$

Thus we have that the function $g(\tau, \sigma)$ is the solution of the inhomogeneous integral equation

$$g(\tau, u) - 2jv \int_{-\infty}^{t} R_x(\sigma - \tau)h(t - \sigma)g(\sigma, u)\, d\sigma = R_x(\tau - u). \tag{6.16-29}$$

We return now to Equation 6.16-23 and write

$$\ln F_z(v) = -\frac{1}{2}\sum_{i=1}^{\infty} \ln\left(1 - \frac{2jv}{\lambda_i}\right) \tag{6.16-30}$$

and

$$\frac{1}{j}\frac{d}{dv}\ln F_z(v) = \sum_{i=1}^{\infty} \frac{1}{\lambda_i}\left[\frac{1}{1 - (2jv/\lambda_i)}\right]. \tag{6.16-31}$$

This last expression may be identified readily by using Equation 6.16-27 to form the integral

$$j\int_{-\infty}^{t} h(t - \sigma)g(\sigma, \sigma)\, d\sigma = j\sum_{i=1}^{\infty} \frac{1}{1 - (2jv/\lambda_i)}\frac{1}{\lambda_i}\int_{-\infty}^{t} h(t - \sigma)\varphi_k(\sigma)\varphi_k(\sigma)\, d\sigma$$

$$= j\sum_{i=1}^{\infty} \frac{1}{\lambda_i}\left[\frac{1}{1 - (2jv/\lambda_i)}\right] = \frac{d}{dv}\ln F_z(v). \tag{6.16-32}$$

On integrating this last expression with respect to v we obtain

$$\ln F_z(v) = j\int_{0}^{v}\int_{-\infty}^{t} h(t - \sigma)g(\sigma, \sigma)\, d\sigma\, dv \tag{6.16-33}$$

or, finally,

$$\boxed{F_z(v) = \exp\left[j\int_{0}^{v}\int_{-\infty}^{t} h(t - \sigma)g(\sigma, \sigma)\, d\sigma\, dv\right].} \tag{6.16-34}$$

The solution of the homogeneous integral equation of Equation 6.16-8 and the evaluation of the infinite product of Equation 6.16-23 has been replaced by the solution of the inhomogeneous integral equation of Equation 6.16-29 and the evaluation of the integral of Equation 6.16-34. In a given situation one of these two approaches may be preferable to the other. However, it will develop in general that neither yields results that are easy to obtain. It was pointed out in Section 5.6 of Chapter 5 that it is frequently reasonable to assume that the power spectral density $\varphi_x(\omega)$ is a rational function of ω; that is, the ratio of two polynomials in ω^2. Such will be the case, for example, if the random process $x(t)$ arises from the filtering of white noise by a lumped-parameter, linear, time-invariant filter. In such cases, the foregoing problem is more readily attacked as will now be shown.

Let us assume that the spectral density $\varphi_x(\omega)$ can be written as

$$\varphi_x(\omega) = K^2 \frac{N(\omega^2)}{D(\omega^2)} \tag{6.16-35}$$

where K^2 is a constant and where $N(\omega^2)$ and $D(\omega^2)$ are polynomials of degree n and d, respectively, in ω^2. The autocorrelation function $R_x(\sigma - \tau)$ can now be written as

$$\begin{aligned} R_x(\sigma - \tau) &= \frac{1}{2\pi} \int_{-\infty}^{\infty} \varphi_x(\omega)\, e^{j\omega(\sigma-\tau)}\, d\omega \\ &= \frac{1}{2\pi} \int_{-\infty}^{\infty} K^2 \frac{N(\omega^2)}{D(\omega^2)}\, e^{j\omega(\sigma-\tau)}\, d\omega. \end{aligned} \tag{6.16-36}$$

Suppose we define the differential operator $p = d/d\tau$ and differentiate both sides of this equation with respect to τ. The result is

$$-pR_x(\sigma - \tau) = \frac{K^2}{2\pi} \int_{-\infty}^{\infty} \frac{N(\omega^2)}{D(\omega^2)} (j\omega)\, e^{j\omega(\sigma-\tau)}\, d\omega. \tag{6.16-37}$$

If we integrate, on the other hand, we have formally

$$-\frac{1}{p} R_x(\sigma - \tau) = \frac{K^2}{2\pi} \int_{-\infty}^{\infty} \frac{N(\omega^2)}{D(\omega^2)} \left(\frac{1}{j\omega}\right) e^{j\omega(\sigma-\tau)}\, d\omega. \tag{6.16-38}$$

More generally,

$$\frac{D(-p^2)}{N(-p^2)} R_x(\sigma - \tau) = \frac{K^2}{2\pi} \int_{-\infty}^{\infty} e^{j\omega(\sigma-\tau)}\, d\omega = K^2 \delta(\sigma - \tau) \tag{6.16-39}$$

since the effect of the operators $D(-p^2)$ and $1/N(-p^2)$ is to build up $D(\omega^2)$ and $1/N(\omega^2)$ respectively in the integrand on the right side. Thus we have the operator relationship

$$D(-p^2) R_x(\sigma - \tau) = K^2 N(-p^2) \delta(\sigma - \tau). \tag{6.16-40}$$

Let us now operate on both sides of Equation 6.16-29 with $D(-p^2)$ and make use of this last relationship to obtain

$$[D(-p^2) - 2jvK^2 N(-p^2) h(t, \tau)] g(\tau, u) = K^2 N(-p^2) \delta(\tau - u). \tag{6.16-41}$$

Now this expression is just a differential equation in the independent variable τ. It may be solved for the unknown function $g(\tau, u)$ over the two regions $\tau < u$ and $\tau > u$, and the two sets of arbitrary constants related by matching the solutions at $\tau = u$. Then the resulting one set of arbitrary constants (equal in number to the order of the differential equation) may be determined by substituting the function $g(\tau, u)$ into the integral Equation 6.16-29, evaluating the integral for $\tau < u$ and $\tau > u$, and equating coefficients on each side of the resulting equalities. Then the characteristic function $F_z(v)$ may be found from Equation 6.16-34. We now illustrate this rather complicated procedure with several examples.

Example 6.8 We begin with the degenerate case where the filter $h(t, \tau)$ is replaced by a direct connection so that

$$h(t, \tau) = h(t - \tau) = \delta(t - \tau)$$

and Equation 6.16-1 becomes

$$z(t) = x^2(t).$$

We already know from Section 2.11 that, in this case, the random variable z has a chi-squared distribution with one degree of freedom and that the density of z can be written as

$$W_z(z) = \frac{1}{\sqrt{2\pi}\sigma} z^{-1/2} e^{-z/2\sigma^2}, \qquad 0 \leq z < \infty,$$

where

$$\sigma^2 = R_x(0) = E\{x^2(t)\}.$$

Also the characteristic function $F_z(v)$ is easily found. We have

$$F_z(v) = E\{e^{jvz}\} = \frac{1}{\sqrt{2\pi}\sigma} \int_0^\infty z^{-1/2} e^{-z[(1/2\sigma^2) - jv]} dz.$$

Let us make the linear change in variable

$$\eta = z\left(\frac{1}{2\sigma^2} - jv\right);$$

then we have

$$F_z(v) = \frac{1}{(1 - 2jv\sigma^2)^{1/2}} \frac{1}{\sqrt{\pi}} \int_0^\infty \eta^{-1/2} e^{-\eta} d\eta.$$

However, the gamma function $\Gamma(\mu)$ is defined by [6, 7]

$$\Gamma(\mu) = \int_0^\infty \eta^{\mu - 1} e^{-\eta} d\eta$$

and, in particular (see Equation 6.12-34),

$$\Gamma(\tfrac{1}{2}) = \sqrt{\pi}$$

so that

$$F_z(v) = \frac{1}{(1 - 2jv\sigma^2)^{1/2}}.$$

Let us now see if this result can be obtained using the procedures that have been developed in this section. From Equation 6.16-29, we have

$$g(\tau, u) = \frac{R_x(\tau - u)}{1 - 2jvR_x(t - \tau)}$$

and, hence,

$$g(\sigma, \sigma) = \frac{R_x(0)}{1 - 2jvR_x(t - \sigma)}.$$

Now the inner integral of Equation 6.16-34 becomes

$$\int_{-\infty}^{t} \delta(t - \sigma)g(\sigma, \sigma)\, d\sigma = g(t, t) = \frac{R_x(0)}{1 - 2jvR_x(0)}$$

and $F_z(v)$ can be written as

$$\begin{aligned} F_z(v) &= \exp\left\{j\int_0^v \frac{R_x(0)}{1 - 2jvR_x(0)}\, dv\right\} \\ &= \exp\{-\tfrac{1}{2}\ln[1 - 2jvR_x(0)]_0^v\} \\ &= \exp\{-\tfrac{1}{2}\ln[1 - 2jvR_x(0)]\} \\ &= [1 - 2jvR_x(0)]^{-1/2} = \frac{1}{(1 - 2jv\sigma^2)^{1/2}}. \end{aligned}$$

Thus, for this simple example, the methods developed in this section yield the same results as a direct calculation.

• •

We now consider a less trivial example. Unfortunately, the amount of labor involved will now become considerable and will serve as a good illustration of the difficulty of obtaining results for all but the simplest nonlinear problems.

Example 6.9 Let us suppose that the linear filter of Figure 6.20 is given by

$$h(t, \tau) = u(\tau),$$

where $u(\tau)$ is the unit-step function, so that Equation 6.16-1 becomes

$$z(t) = \int_0^t x^2(\tau)\, d\tau. \tag{A}$$

We will also assume that the autocorrelation function of the gaussian input $x(t)$ is

$$R_x(\tau) = e^{-\alpha|\tau|} \tag{B}$$

with the corresponding power spectral density

$$\varphi_x(\omega) = \frac{2\alpha}{\alpha^2 + \omega^2}. \tag{C}$$

For this case the spectral density numerator polynomial $N(-p^2)$ is

$$N(-p^2) = 1,$$

the denominator polynomial is

$$D(-p^2) = \alpha^2 - p^2,$$

and the constant K^2 becomes

$$K^2 = 2\alpha.$$

The differential equation of Equation 6.16-41 reduces to

$$[\alpha^2 - p^2 - 4jv\alpha u(\tau)]g(\tau, u) = 2\alpha\delta(\tau - u), \qquad u \geq 0,$$

or

$$[\alpha^2 - p^2 - 4jv\alpha]g(\tau, u) = 2\alpha\delta(\tau - u), \qquad \tau, u \geq 0.$$

If we define the constant β by

$$\beta^2 = \alpha^2 - 4jv\alpha, \tag{D}$$

then this last expression becomes

$$\frac{d^2g}{d\tau^2} - \beta^2 g = -2\alpha\delta(\tau - u), \qquad \tau, u \geq 0. \tag{E}$$

Thus we have an ordinary second-order differential equation with constant coefficients. For $\tau \neq u$, this reduces to the homogeneous expression

$$\frac{d^2g}{d\tau^2} - \beta^2 g = 0, \qquad \tau, u \geq 0, \tag{F}$$

with solutions

$$g_1 = g_1(\tau, u) = A_1(u)\, e^{-\beta\tau} + A_2(u)\, e^{\beta\tau}, \qquad \tau > u \geq 0,$$

and

$$g_2 = g_2(\tau, u) = B_1(u)\, e^{-\beta\tau} + B_2(u)\, e^{\beta\tau}, \qquad u > \tau \geq 0.$$

At the point $\tau = u \geq 0$, the two solutions must coincide so that

$$g_1(u, u) = g_2(u, u), \qquad u \geq 0,$$

or

$$[A_1(u) - B_1(u)]\, e^{-\beta u} + [A_2(u) - B_2(u)]\, e^{\beta u} = 0. \tag{G}$$

Also it is clear from Equation E that, at $\tau = u$, the second derivative of g must contain a delta function of strength -2α; consequently the first derivative of g must contain a discontinuity of amount -2α at this point. We have

$$\frac{d}{d\tau}[g_1(\tau, u) - g_2(\tau, u)]_{\tau=u} = -2\alpha$$

or

$$[B_1(u) - A_1(u)]\,e^{-\beta u} - [B_2(u) - A_2(u)]\,e^{\beta u} = -\frac{2\alpha}{\beta}. \qquad \text{(H)}$$

On adding and subtracting Equations G and H we obtain immediately that

$$B_1(u) = A_1(u) - \frac{\alpha}{\beta}\,e^{\beta u},$$

$$B_2(u) = A_2(u) + \frac{\alpha}{\beta}\,e^{-\beta u}.$$

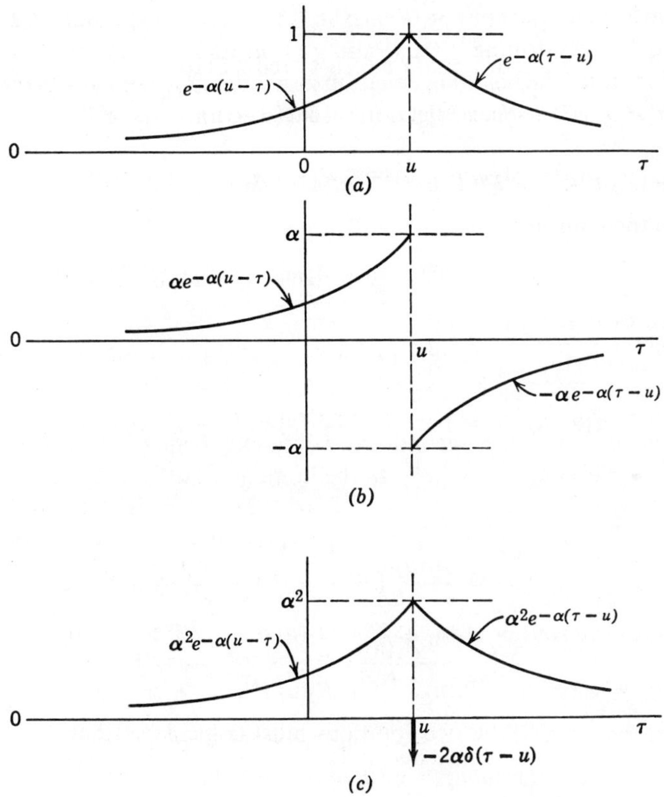

Figure 6.21 *The function* $e^{-\alpha|\tau-u|}$ *and its first two derivatives: (a) the function* $e^{-\alpha|\tau-u|}$ *(b) the function*

$$\frac{d}{d\tau}\left(e^{-\alpha|\tau-u|}\right) = \begin{cases} \alpha e^{-\alpha(u-\tau)}; & 0 \le \tau \le u, \\ -\alpha e^{-\alpha(\tau-u)}, & u \le \tau < \infty; \end{cases}$$

(c) The function

$$\frac{d^2}{d\tau^2}\left(e^{-\alpha|\tau-u|}\right) = \alpha^2\,e^{-\alpha|\tau-u|} - 2\alpha\delta(\tau - u)$$

Now the general solution to Equation E becomes

$$g_1(\tau, u) = A_1(u)\, e^{-\beta\tau} + A_2(u)\, e^{\beta\tau}, \qquad \tau > u \geq 0, \tag{I}$$

$$g_2(\tau, u) = g_1(\tau, u) - \frac{\alpha}{\beta}\, e^{-\beta(\tau-u)} + \frac{\alpha}{\beta}\, e^{\beta(\tau-u)}, \qquad u > \tau \geq 0. \tag{J}$$

As a matter of interest, the function $e^{-\alpha|\tau-u|}$ and its first two derivatives are shown in Figure 6.21. This is one of the simplest examples where the first derivative contains a discontinuity at $\tau = u$ and, consequently, the second derivative contains a delta function at this point. We will see later that this function $e^{-\alpha|\tau-u|}$ has a prominent part in the solution to our problem.

The two arbitrary constants $A_1(u)$ and $A_2(u)$ of Equations I and J can now be evaluated by substituting $g_1(\tau, u)$ and $g_2(\tau, u)$ into the integral Equation 6.16-29 and treating the resulting equation as an identity. Let us take the case where $t \geq \tau \geq u \geq 0$ so that Equation 6.16-29 becomes

$$A_1(u)\, e^{-\beta\tau} + A_2(u)\, e^{\beta\tau} - 2jv \int_0^t e^{-\alpha|\sigma-\tau|} g_1(\sigma, u)\, d\sigma$$

$$- 2jv \int_0^u e^{-\alpha(\tau-\sigma)}[g_2(\sigma, u) - g_1(\sigma, u)]\, d\sigma = e^{-\alpha(\tau-u)}. \tag{K}$$

We consider the two integrals first and find

$$\begin{aligned}
\int_0^t e^{-\alpha|\sigma-\tau|} g_1(\sigma, u)\, d\sigma &= \int_0^\tau e^{-\alpha(\tau-\sigma)}[A_1(u)\, e^{-\beta\sigma} + A_2(u)\, e^{\beta\sigma}]\, d\sigma \\
&\quad + \int_\tau^t e^{-\alpha(\sigma-\tau)}[A_1(u)\, e^{-\beta\sigma} + A_2(u)\, e^{\beta\sigma}]\, d\sigma \\
&= \frac{A_1(u)}{\alpha - \beta}(e^{\beta\tau} - e^{-\alpha\tau}) + \frac{A_2(u)}{\alpha + \beta}(e^{\beta\tau} - e^{-\alpha\tau}) \\
&\quad - \frac{A_1(u)}{\alpha + \beta}[e^{\alpha\tau}\, e^{-(\alpha+\beta)t} - e^{-\beta\tau}] \\
&\quad - \frac{A_2(u)}{\alpha - \beta}[e^{\alpha\tau}\, e^{-(\alpha-\beta)t} - e^{\beta\tau}],
\end{aligned}$$

$$\begin{aligned}
\int_0^u e^{-\alpha(\tau-\sigma)}[g_2(\sigma, u) - g_1(\sigma, u)]\, d\sigma &= \int_0^u e^{-\alpha(\tau-\sigma)} \frac{\alpha}{\beta}[e^{\beta(\sigma-u)} - e^{-\beta(\sigma-u)}]\, d\sigma \\
&= \frac{\alpha}{\beta} e^{-\alpha\tau}\left[\frac{e^{\alpha u} - e^{-\beta u}}{\alpha + \beta} - \frac{e^{\alpha u} - e^{\beta u}}{\alpha - \beta}\right].
\end{aligned}$$

Now we make these substitutions into Equation K and equate like terms. We have four sets of such terms:

1. *Coefficient of* $e^{\beta\tau}$:

$$-2jv\left[\frac{A_2(u)}{\alpha+\beta}+\frac{A_2(u)}{\alpha-\beta}\right]+A_2(u)=0.$$

For $A_2(u) \neq 0$, this becomes

$$-\frac{4jv\alpha}{\alpha^2-\beta^2}+1=0.$$

Since we have defined β by Equation D, this is just an identity.

2. *Coefficient of* $e^{-\beta\tau}$:

$$-2jv\left[\frac{A_1(u)}{\alpha-\beta}+\frac{A_1(u)}{\alpha+\beta}\right]+A_1(u)=0.$$

Again, for $A_1(u) \neq 0$, this is an identity.

3. *Coefficient of* $e^{\alpha\tau}$:

$$2jv\left[\frac{A_1(u)}{\alpha+\beta}\,e^{-(\alpha+\beta)t}+\frac{A_2(u)}{\alpha-\beta}\,e^{-(\alpha-\beta)t}\right]=0.$$

We conclude from this expression that

$$A_2(u)=-\frac{\alpha-\beta}{\alpha+\beta}\,e^{-2\beta t}A_1(u). \tag{L}$$

4. *Coefficient of* $e^{-\alpha\tau}$:

$$2jv\left[\frac{A_2(u)}{\alpha+\beta}+\frac{A_1(u)}{\alpha-\beta}-\frac{\alpha}{\beta}\,\frac{e^{\alpha u}-e^{-\beta u}}{\alpha+\beta}+\frac{\alpha}{\beta}\,\frac{e^{\alpha u}-e^{\beta u}}{\alpha-\beta}\right]=e^{\alpha u}.$$

We use Equation L to replace $A_2(u)$ in this expression and solve for $A_1(u)$ to obtain

$$A_1(u)=B(u)(\alpha+\beta)\,e^{\beta t}, \tag{M}$$

$$A_2(u)=-B(u)(\alpha-\beta)\,e^{-\beta t} \tag{N}$$

where, for convenience, we have defined the function $B(u)$ by

$$B(u)=B(u,t)=\frac{\alpha}{\beta}\,\frac{(\alpha-\beta)\,e^{-\beta u}-(\alpha+\beta)\,e^{\beta u}}{(\alpha-\beta)^2\,e^{-\beta t}-(\alpha+\beta)^2\,e^{\beta t}}. \tag{O}$$

It follows immediately from Equations I and J that

$$g_1(\tau,u)=B(u)[(\alpha+\beta)\,e^{\beta(t-\tau)}-(\alpha-\beta)\,e^{-\beta(t-\tau)}] \tag{P}$$

and

$$g_2(\tau,u)=g_1(u,\tau). \tag{Q}$$

We return to Equation 6.16-34 and note that we want to evaluate the integral

$$K(v)=\int_{-\infty}^{t}h(t,\sigma)g(\sigma,\sigma)\,d\sigma=\int_0^t g(\sigma,\sigma)\,d\sigma$$

where $g(\sigma, \sigma) = g_1(\sigma, \sigma) = g_2(\sigma, \sigma)$. We obtain

$$K(v) = \frac{\alpha}{\beta^2} \frac{(\alpha^2 - \beta^2)(e^{\beta t} - e^{-\beta t}) - \beta t[(\alpha - \beta)^2 e^{-\beta t} + (\alpha + \beta)^2 e^{\beta t}]}{(\alpha - \beta)^2 e^{-\beta t} - (\alpha + \beta)^2 e^{\beta t}}.$$

Again we refer to Equation 6.16-34 and evaluate

$$j \int_0^v \int_{-\infty}^t h(t, \sigma) g(\sigma, \sigma)\, d\sigma\, dv = j \int_0^v K(v)\, dv. \tag{R}$$

It is convenient at this point to rewrite this last definite integral in terms of the two variables β and y. The variable $\beta = \beta(v)$ has been defined by Equation D and y is given by

$$y = y(v) = (\alpha + \beta)^2 e^{\beta t} - (\alpha - \beta)^2 e^{-\beta t}.$$

Now Equation R becomes

$$j \int_0^v K(v)\, dv = \frac{1}{2} \int_{\beta(0)}^{\beta(v)} \frac{d\beta}{\beta} - \frac{1}{2} \int_{y(0)}^{y(v)} \frac{dy}{y}$$

$$= \ln \left[\frac{\beta(v) y(0)}{\beta(0) y(v)}\right]^{1/2}$$

and, finally,

$$F_z(v) = \exp\left[j \int_0^v K(v)\, dv\right] = \left[\frac{\beta(v) y(0)}{\beta(0) y(v)}\right]^{1/2}. \tag{S}$$

We have

$$\beta(0) = \alpha, \qquad \beta(v) = \alpha a,$$
$$y(0) = 4\alpha^2 e^{\alpha t}, \qquad y(v) = e^{\beta t}(\alpha + \beta)^2 - e^{-\beta t}(\alpha - \beta)^2,$$

where a has been used to denote

$$a = a(v) = \left(1 - \frac{4jv}{\alpha}\right)^{1/2}. \tag{T}$$

Now Equation S becomes

$$F_z(v) = e^{\alpha t/2} \left[\frac{(\alpha + \beta)^2}{4\alpha^2 a} e^{\alpha a t} - \frac{(\alpha - \beta)^2}{4\alpha^2 a} e^{-\alpha a t}\right]^{-1/2} \tag{U}$$

or, more conveniently,

$$\boxed{F_z(v) = e^{\alpha t/2}\left(\cosh \alpha a t + \frac{1 + a^2}{2a} \sinh \alpha a t\right)^{-1/2}} \tag{V}$$

where $a(v)$ is given by Equation T. The final expression for the first-order characteristic function of z is relatively complicated and the procedures used to find this expression were somewhat involved, to say the least.

• •

We turn now to the problem of finding the autocorrelation function and power spectral density of the output $z(t)$ of the system of Figure 6.20. We will restrict ourselves to the case where the low-pass filter is time invariant with impulse response $h(t)$ and complex transfer function $H(\omega)$. In that case, if the input process $x(t)$ is at least wide-sense stationary, then the output $z(t)$ is also at least wide-sense stationary and we may write $R_z(\tau)$ for the output autocorrelation function and $\varphi_z(\omega)$ for the output power spectral density. As before, we shall assume the input $x(t)$ is a zero-mean gaussian process with autocorrelation function $R_x(\tau)$ and power spectral density $\varphi_x(\omega)$.

It follows from Equations 4.4-3 and 4.4-7 that

$$R_z(\tau) = \int_{-\infty}^{\infty}\int_{-\infty}^{\infty} R_y(\tau - \beta + \alpha)h(\alpha)h(\beta)\, d\alpha\, d\beta \tag{6.16-42}$$

and

$$\varphi_z(\omega) = \varphi_y(\omega)|H(\omega)|^2 \tag{6.16-43}$$

where $y(t)$ is the output of the squarer. In Section 6.4, however, we found the autocorrelation function and power spectral density of the output of a squarer for a gaussian input. The results are given by Equations 6.4-19 and 6.4-20 and are

$$R_y(\tau) = R_x^2(0) + 2R_x^2(\tau) \tag{6.16-44}$$

and

$$\varphi_y(\omega) = 2\pi R_x^2(0)\delta(\omega) + \frac{1}{\pi}\int_{-\infty}^{\infty} \varphi_x(\nu)\varphi_x(\omega - \nu)\, d\nu. \tag{6.16-45}$$

Thus Equations 6.16-42 and 6.16-43 become

$$R_z(\tau) = R_x^2(0)\left|\int_{-\infty}^{\infty} h(\alpha)\, d\alpha\right|^2 + 2\int_{-\infty}^{\infty}\int_{-\infty}^{\infty} R_x^2(\tau - \beta + \alpha)h(\alpha)h(\beta)\, d\alpha\, d\beta \tag{6.16-46}$$

and

$$\varphi_z(\omega) = 2\pi R_x^2(0)\left|\int_{-\infty}^{\infty} h(\alpha)\, d\alpha\right|^2\delta(\omega) + \frac{1}{\pi}|H(\omega)|^2\int_{-\infty}^{\infty} \varphi_x(\nu)\varphi_x(\omega - \nu)\, d\nu. \tag{6.16-47}$$

These two equations can be evaluated, at least in principle, for a given filter and input autocorrelation function $R_x(\tau)$. We illustrate with several simple examples.

Example 6.10 Let $x(t)$ be a gaussian process with the bandpass power spectral density shown in Figure 6.4(a). Then, as discussed in Example 6.1, the corresponding autocorrelation function $R_x(\tau)$ is

$$R_x(\tau) = \frac{NB}{2\pi}\frac{\sin B\tau/2}{B\tau/2}\cos \omega_0\tau.$$

Suppose that the filter is an ideal (unrealizable) low-pass filter with the complex transfer function

$$H(\omega) = \begin{cases} 1, & |\omega| \leq B, \\ 0, & \text{elsewhere,} \end{cases}$$

as shown in Figure 6.22(*a*) with corresponding (unrealizable) impulse response

$$h(t) = \frac{1}{2\pi}\int_{-B}^{B} e^{j\omega t}\, d\omega = \frac{B}{\pi}\frac{\sin \omega B}{\omega B}.$$

The power spectral density $\varphi_y(\omega)$ of $y = x^2$ is shown in Figure 6.6. It is clear that the effect of the ideal low-pass filter will be to retain all of the low-frequency components and remove all of those centered about $\pm 2\omega_0$. The resulting $\varphi_z(\omega)$ is shown in Figure 6.22(*b*). It is also obvious that it is easier to find $R_z(\tau)$ by transforming $\varphi_z(\omega)$ than by using Equation 6.16-46. We have

$$\begin{aligned} R_z(\tau) &= \frac{1}{2\pi}\int_{-\infty}^{\infty} \varphi_z(\omega)\, e^{j\omega t}\, d\omega \\ &= \frac{N^2B^2}{4\pi^2}\int_{-\infty}^{\infty} \delta(\omega)\, e^{j\omega t}\, d\omega + \frac{N^2B}{4\pi^2}\int_{-B}^{B}\left(1 - \frac{|\omega|}{B}\right) e^{j\omega t}\, d\omega \\ &= \frac{N^2B^2}{4\pi^2} + \frac{N^2B}{4\pi^2}\int_{-B}^{0}\left(1 + \frac{\omega}{B}\right) e^{j\omega t}\, d\omega + \frac{N^2B}{4\pi^2}\int_{0}^{B}\left(1 - \frac{\omega}{B}\right) e^{j\omega t}\, d\omega \\ &= \frac{N^2B^2}{4\pi^2}\left[1 + \frac{2}{B}\int_0^B \left(1 - \frac{\omega}{B}\right)\cos \omega t\, d\omega\right]. \end{aligned}$$

But we have

$$\frac{2}{B}\int_0^B \cos \omega t\, d\omega = 2\frac{\sin Bt}{Bt}$$

and

$$-\frac{2}{B^2}\int_0^B \omega \cos \omega t\, d\omega = -2\frac{\sin Bt}{Bt} + \frac{\sin^2 Bt/2}{(Bt/2)^2}$$

or, finally,

$$R_z(\tau) = \frac{N^2B^2}{4\pi^2}\left[1 + \frac{\sin^2 Bt/2}{(Bt/2)^2}\right].$$

• •

Example 6.11 Let $x(t)$ have an autocorrelation function and power spectral density given by

$$R_x(\tau) = e^{-a|\tau|},$$

$$\varphi_x(\omega) = \frac{2a}{a^2 + \omega^2}.$$

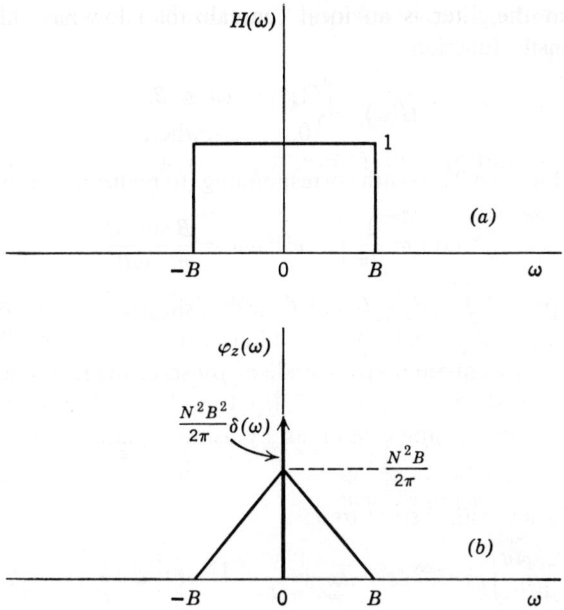

Figure 6.22 Low-pass filtering of a squared random process: (a) the complex transfer function of the time-invariant filter: (b) the power spectral density of z(t)

Then, for the squarer output,

$$R_y(\tau) = R_x^2(0) + 2R_x^2(\tau) = 1 + 2\,e^{-2a|\tau|},$$

and, by direct transformation,

$$\varphi_y(\omega) = 2\pi\delta(\omega) + \frac{8a}{4a^2 + \omega^2}.$$

Suppose that the low-pass filter is a simple *RC* filter with transfer function

$$H(\omega) = \frac{1}{1 + j(\omega/b)}$$

and impulse response

$$h(t) = b\,e^{-bt}u(t).$$

The output power spectral density $\varphi_z(\omega)$ is

$$\begin{aligned}\varphi_z(\omega) &= \varphi_y(\omega)|H(\omega)|^2\\ &= 2\pi\delta(\omega) + \frac{8a}{4a^2 + \omega^2}\frac{b^2}{b^2 + \omega^2}.\end{aligned}$$

The corresponding autocorrelation function is most easily found from Equation 6.16-46:

$$R_z(\tau) = 1 + 2b^2 \int_0^\infty \int_0^\infty e^{-2a|\tau+\alpha-\beta|}\, e^{-b\alpha}\, e^{-b\beta}\, d\alpha\, d\beta.$$

We consider the integral with respect to β for $\tau \geq 0$ and we have

$$\int_0^\infty e^{-2a|\tau+\alpha-\beta|}\, e^{-b\beta}\, d\beta$$

$$= \int_0^{\tau+\alpha} e^{-2a(\tau+\alpha-\beta)}\, e^{-b\beta}\, d\beta + \int_{\tau+\alpha}^\infty e^{-2a(\beta-\tau-\alpha)}\, e^{-b\beta}\, d\beta$$

$$= e^{-2a(\tau+\alpha)} \int_0^{\tau+\alpha} e^{(2a-b)\beta}\, d\beta + e^{2a(\tau+\alpha)} \int_{\tau+\alpha}^\infty e^{-(2a-b)\beta}\, d\beta$$

$$= \frac{4a}{4a^2 - b^2}\, e^{-b(\tau+\alpha)} - \frac{1}{2a-b}\, e^{-2a(\tau+\alpha)}.$$

Now we integrate with respect to α:

$$\int_0^\infty e^{-b\alpha}\left[\frac{4a}{4a^2 - b^2}\, e^{-b(\tau+\alpha)} - \frac{1}{2a-b}\, e^{-2a(\tau+\alpha)}\right] d\alpha = \frac{4a\, e^{-b\tau} - 2b\, e^{-2a\tau}}{2b(4a^2 - b^2)}.$$

Finally, since the autocorrelation function is even, we have

$$R_z(\tau) = 1 + \frac{4ab\, e^{-b|\tau|} - 2b^2\, e^{-2a|\tau|}}{4a^2 - b^2}.$$

• •

6.17 Narrow-Band Inputs to νth-Law Systems

In the previous section, we considered the situation where a square-law device was followed by a linear filter. A more general problem is that where a νth-law device is followed by a linear filter to form a *νth-law system.* If the filter is bandpass, the resulting system might be called a *νth-law nonlinear amplifier* as illustrated in Figure 6.23(a). Such systems occur, for example, in FM receivers where the νth-law device may be a hard limiter. On the other hand, if the filter is low pass, the system could be called a *νth-law detector* as illustrated in Figure 6.23(b). In many cases, the input to the νth-law system will be a narrow-band random process or a sum of such processes with or without added narrow-band deterministic signals. In this section we consider such narrow-band inputs to the νth-law systems of Figure 6.23.

Let us refer to Section 4.8 where a narrow-band process $x(t)$ was represented as

$$x(t) = R(t) \cos\,[\omega_m t + \theta(t)] \tag{6.17-1}$$

where $\omega_m/2\pi$ is the "center frequency" of the process and where the *envelope function* $R(t)$ and the *phase function* $\theta(t)$ are random processes with $R(t) \geq 0$.

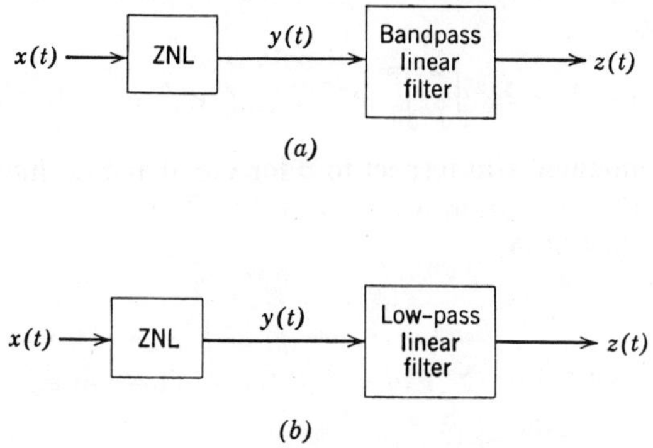

Figure 6.23 *νth-law systems: (a) νth-law nonlinear amplifier; (b) νth-law detector*

We shall use this representation to express the output $y = g(x)$ of the various νth-law devices already studied in Section 6.12, namely the half-wave, full-wave even, and full-wave odd devices.

Half-Wave νth-Law Device with Narrow-Band Input

The characteristic $y = g_1(x)$ of this device is shown in Figure 6.15(*a*) and given by Equation 6.12-1. The Laplace transform $G_1(s)$ of $g_1(x)$ has been found in Equation 6.12-17 to be

$$\mathscr{L}[g_1(x)] = G_1(s) = c\,\frac{\Gamma(\nu + 1)}{s^{\nu+1}}. \tag{6.17-2}$$

On inverting $G_1(s)$ we have

$$y = g_1(x) = \frac{1}{2\pi j}\int_C G_1(s)\, e^{sx}\, ds \tag{6.17-3}$$

where C is the appropriate contour in the complex s plane. On replacing x in this last expression by Equation 6.17-1, we obtain

$$y = g_1(x) = \frac{1}{2\pi j}\int_C G_1(s)\, e^{sR\cos[\omega_m t + \theta]}\, ds \tag{6.17-4}$$

or

$$y = g_1(x) = \frac{c\Gamma(\nu + 1)}{2\pi j}\int_C \frac{e^{sR\cos[\omega_m t + \theta]}}{s^{\nu+1}}\, ds. \tag{6.17-5}$$

We now make use of a well-known relationship [20] called the *Jacobi-Anger formula*

$$e^{z\cos\phi} = \sum_{n=0}^{\infty} \mathscr{E}_n I_n(z)\cos n\phi \tag{6.17-6}$$

where $\mathscr{E}_n$ is the *Neumann factor*

$$\mathscr{E}_n = \begin{cases} 1, & n = 0, \\ 2, & n = 1, 2, 3, \ldots, \end{cases} \tag{6.17-7}$$

and $I_n(z)$ is the modified Bessel function which we first encountered in Section 4.8. On substituting Equation 6.17-6 into 6.17-5 and making the change in variable $z = sR$, we obtain

$$\boxed{y = g_1(x) = \sum_{n=0}^{\infty} F(n, \nu)\mathscr{E}_n R^\nu(t) \cos n[\omega_m t + \theta(t)]} \tag{6.17-8}$$

where, for convenience, the notation $F(n, \nu)$ has been used to denote the integral

$$F(n, \nu) = c\,\frac{\Gamma(\nu + 1)}{2\pi j} \int_{C'} \frac{I_n(z)}{z^{\nu+1}}\, dz \tag{6.17-9}$$

and where C' is an appropriate new contour. For this case we can consider as our contour the line with endpoints $\sigma_1 - j\infty$ and $\sigma_1 + j\infty$, where $\sigma_1 > 0$, and then allow $\sigma_1 \to 0$. Thus we evaluate formally the integral

$$F(n, \nu) = c\,\frac{\Gamma(\nu + 1)}{2\pi j} \int_{-j\infty}^{j\infty} \frac{I_n(z)}{z^{\nu+1}}\, dz. \tag{6.17-10}$$

Let us rewrite the relationship of Equation 4.8-16 as

$$I_n(jx) = j^n J_n(x) \tag{6.17-11}$$

and use this in Equation 6.17-10 to obtain

$$F(n, \nu) = c\,\frac{\Gamma(\nu + 1)}{2\pi j}\, j^{n-\nu} \int_{-\infty}^{\infty} \frac{J_n(z)}{z^{\nu+1}}\, dz \tag{6.17-12}$$

or, since

$$J_n(-z) = (-1)^z J_n(z), \tag{6.17-13}$$

we can write

$$F(n, \nu) = c\,\frac{\Gamma(\nu + 1)}{2\pi j} \int_{0}^{\infty} \frac{J_n(z)}{z^{\nu+1}}\, dz j^{n-\nu}[1 - (-1)^{n-\nu}]. \tag{6.17-14}$$

We see that $[1 - (-1)^{n-\nu}]$ has the value zero if $n - \nu$ is even and the value two if $n - \nu$ is odd so that

$$\begin{aligned} j^{n-\nu}[1 - (-1)^{n-\nu}] &= 2j \sin\,(n - \nu)\frac{\pi}{2} \\ &= \frac{2\pi j}{\Gamma[(n - \nu)/2]\Gamma[1 - (n - \nu)/2]}. \end{aligned} \tag{6.17-15}$$

The last equality follows [7] from Equation 6.12-27. Another well-known relationship [21] is

$$\int_0^\infty \frac{J_n(z)}{z^{\nu+1}}\,dz = \frac{\Gamma[(n-\nu)/2]}{2^{\nu+1}\Gamma[1+(n+\nu)/2]}. \tag{6.17-16}$$

On putting together the last three equations, we have finally

$$F(n,\nu) = c\,\frac{\Gamma(\nu+1)}{2^{\nu+1}\Gamma[1-(n-\nu)/2]\Gamma[1+(n+\nu)/2]}. \tag{6.17-17}$$

Thus Equation 6.17-8 provides a representation for the output of a half-wave, νth-law device when the input is a narrow-band process $x(t)$ with envelope $R(t)$ and phase $\theta(t)$.

Full-Wave, Even, νth-Law Device with Narrow-Band Input

This device is shown in Figure 6.15(*b*) and can be represented by Equation 6.12-4 as

$$y = g_2(x) = g_1(x) + g_1(-x). \tag{6.17-18}$$

Since we have from Equation 6.12-19 that

$$\mathscr{L}[g_1(-x)] = G_1(-s), \tag{6.17-19}$$

we can write immediately

$$y = g_2(x) = \frac{c\Gamma(\nu+1)}{2\pi j}\int_C e^{sR\cos[\omega_m t+\theta]}\left[\frac{1}{s^{\nu+1}} + \frac{1}{(-s)^{\nu+1}}\right] ds. \tag{6.17-20}$$

After replacing s by $-s$ in the second integral and making use of the fact that

$$I_n(-z) = (-1)^n I_n(z), \tag{6.17-21}$$

we obtain, with the use of Equation 6.17-6,

$$y = g_2(x) = \sum_{n=0}^{\infty} [1+(-1)^n]F(n,\nu)\mathscr{E}_n R^\nu(t)\cos n[\omega_m t + \theta(t)] \tag{6.17-22}$$

or, finally,

$$y = g_2(x) = \sum_{\substack{n=0\\ n\ \text{even}}}^{\infty} 2F(n,\nu)\mathscr{E}_n R^\nu(t)\cos n[\omega_m t + \theta(t)]. \tag{6.17-23}$$

Full-Wave, Odd, νth-Law Device with Narrow-Band Input

This device is shown in Figure 6.15(*c*) and can be represented by Equation 6.12-6 as

$$y = g_3(x) = g_1(x) - g_1(-x). \tag{6.17-24}$$

It follows directly from Equation 6.17-22 that

$$y = g_3(x) = \sum_{n=0}^{\infty} [1 - (-1)^n]F(n, \nu)\mathscr{E}_n R^\nu(t) \cos n[\omega_m t + \theta(t)] \quad (6.17\text{-}25)$$

or

$$\boxed{y = g_3(x) = \sum_{\substack{n=1 \\ n \text{ odd}}}^{\infty} 2F(n, \nu)\mathscr{E}_n R^\nu(t) \cos n[\omega_m t + \theta(t)].} \quad (6.17\text{-}26)$$

An examination of Equations 6.17-8, 6.17-23, and 6.17-26 shows, as expected, that the output of the full-wave, even, νth-law device contains only even harmonics of the input while the output of the full-wave, odd, νth-law device contains only odd harmonics.

Example 6.12 The *half-wave linear device* is illustrated in Figure 6.15(a) for $\nu = 1$. In this case Equation 6.17-8 becomes

$$y = g_1(x) = \sum_{n=0}^{\infty} F(n, 1)\mathscr{E}_n R(t) \cos n[\omega_m t + \theta(t)] \quad \text{(A)}$$

with

$$F(n, 1) = \frac{c}{4} \frac{1}{\Gamma[(3/2) - (n/2)]\Gamma[(3/2) + (n/2)]}. \quad \text{(B)}$$

The *full-wave even device* is illustrated in Figure 6.15(b) for $\nu = 2$. Now Equation 6.17-23 is

$$y = g_2(x) = \sum_{\substack{n=0 \\ n \text{ even}}}^{\infty} 2F(n, 2)\mathscr{E}_n R^2(t) \cos n[\omega_m t + \theta(t)] \quad \text{(C)}$$

where

$$F(n, 2) = \frac{c}{4} \frac{1}{\Gamma[2 - (n/2)]\Gamma(2 + (n/2)]}. \quad \text{(D)}$$

Finally we consider the *hard limiter* of Section 6.10 which is the full-wave odd device of Figure 6.15(c) for $\nu = 0$. The output y is given by Equation 6.17-26 as

$$y = g_3(x) = \sum_{\substack{n=1 \\ n \text{ odd}}}^{\infty} 2F(n, 0)\mathscr{E}_n \cos n[\omega_m t + \theta(t)] \quad \text{(E)}$$

where

$$F(n, 0) = \frac{c}{2} \frac{1}{\Gamma[1 - (n/2)]\Gamma[1 + (n/2)]} = \frac{c}{2} \frac{\sin \pi/2}{n\pi/2}. \quad \text{(F)}$$

The last equality in Equation F follows from Equation 6.12-27 and from the relationship [22]

$$\Gamma(1 + z) = z\Gamma(z). \tag{G}$$

It is apparent from Equation E that the hard limiter has removed the envelope function $R(t)$.

• •

The output autocorrelation function $R_y(\tau)$ for the νth-law devices and for a stationary gaussian input has been considered in some detail in Section 6.12. If the input is narrow band as discussed in this section, it will frequently be convenient to assume that this input is the sum of a narrow-band stationary gaussian noise $n(t)$ and an independent modulated sinusoidal signal $s(t)$ so that

$$x(t) = s(t) + n(t). \tag{6.17-27}$$

We will restrict $n(t)$ to have zero mean and autocorrelation function $R_n(\tau)$ and require that $s(t)$ be of the form

$$s(t) = m(t) \cos \phi = m(t) \cos [\omega_m t + \psi] \tag{6.17-28}$$

where $m(t)$ is a stationary random process with autocorrelation function $R_m(\tau)$ and where ψ is a random variable independent of $m(t)$ and uniformly distributed on the interval $(0, 2\pi)$ with density

$$W_\psi(\psi) = \begin{cases} \dfrac{1}{2\pi}, & 0 \leq \psi \leq 2\pi, \\ 0, & \text{elsewhere.} \end{cases} \tag{6.17-29}$$

Note that we have not required the process $m(t)$ to have zero mean: thus we can accept

$$m_1(t) = C[1 + a(t)] \tag{6.17-30}$$

where C is a constant and $a(t)$ is a random process. This gives an ordinary double-sideband AM signal for $s(t)$. We can also accept

$$m_2(t) = C \tag{6.17-31}$$

so that $s(t)$ is a simple sinusoid of random phase.

The joint characteristic function of the input $x(t)$ may be written as

$$F_x(v_1, v_2) = E\{e^{j(v_1 x_1 + v_2 x_2)}\} \tag{6.17-32}$$

where, as usual, x_1 is used to denote $x(t_1)$ and x_2 to denote $x(t_2)$. It follows from Equation 6.17-27 and from the independence of $s(t)$ and $n(t)$ that

$$\begin{aligned} F_x(v_1, v_2) &= E\{e^{j(v_1 s_1 + v_2 s_2)}\, e^{j(v_1 n_1 + v_2 n_2)}\} \\ &= E\{e^{j(v_1 s_1 + v_2 s_2)}\}E\{e^{j(v_1 n_1 + v_2 n_2)}\} \\ &= F_s(v_1, v_2)F_n(v_1, v_2) \end{aligned} \tag{6.17-33}$$

where F_s and F_n are the joint characteristic functions of $s(t)$ and $n(t)$ respectively. We now use Equation 6.11-6 to write the output autocorrelation function

$$R_y(t_1, t_2) = \frac{1}{(2\pi)^2} \int_{-\infty}^{\infty} \int_{-\infty}^{\infty} F_s(v_1, v_2) F_n(v_1, v_2) G(v_1) G(v_2)\, dv_1\, dv_2 \quad (6.17\text{-}34)$$

where $G(\omega)$ is the Fourier transform of the nonlinearity $g(x)$ as given by Equation 6.11-5. We have already found the joint characteristic function for a stationary gaussian process. It is given, for example, by Equation 6.12-20 so that

$$F_n(v_1, v_2) = e^{-[R_n(0)/2][v_1^2 + 2\rho_n(\tau) v_1 v_2 + v_2^2]} \quad (6.17\text{-}35)$$

where $\rho_n(\tau)$ is

$$\rho_n(\tau) = \frac{R_n(\tau)}{R_n(0)}. \quad (6.17\text{-}36)$$

The joint characteristic function of the signal $s(t)$ is

$$F_s(v_1, v_2) = E\{e^{j(v_1 m_1 \cos \phi_1)}\, e^{j(v_2 m_2 \cos \phi_2)}\}. \quad (6.17\text{-}37)$$

We may now use Equation 6.17-6 to write

$$F_s(v_1, v_2) = \sum_{m=0}^{\infty} \sum_{n=0}^{\infty} \mathscr{E}_m \mathscr{E}_n E\{I_m(v_1 m_1) I_n(v_2 m_2) \cos m\phi_1 \cos n\phi_2\} \quad (6.17\text{-}38)$$

or, since $m(t)$ and $\phi(t)$ are statistically independent,

$$F_s(v_1, v_2) = \sum_{m=0}^{\infty} \sum_{n=0}^{\infty} \mathscr{E}_m \mathscr{E}_n E\{I_m(v_1 m_1) I_n(v_2 m_2)\} \times E\{\cos m\phi_1 \cos n\phi_2\}. \quad (6.17\text{-}39)$$

We note that

$$\begin{aligned} E\{\cos m\phi_1 \cos n\phi_2\} &= \tfrac{1}{2} E\{\cos [\omega_m(mt_1 - nt_2) + (m - n)\psi] \\ &\qquad + \cos [\omega_m(mt_1 + nt_2) + (m + n)\psi]\} \\ &= \begin{cases} \dfrac{1}{\mathscr{E}_n} \cos n\omega_m t, & m = n, \\ 0, & m \neq n, \end{cases} \end{aligned} \quad (6.17\text{-}40)$$

since ψ is uniformly distributed on $(0, 2\pi)$.

At this point let us replace jv by the complex s and define the contour integral

$$h_{mk}(t_i) = \frac{1}{2\pi j} \int_C G(s) s^k I_m(sm_i)\, e^{[R_n(0)/2]s^2}\, ds \quad (6.17\text{-}41)$$

where t_i is either t_1 or t_2, m_i is either m_1 or m_2, and the contour C is the appropriate Bromwich contour to ensure existence of the integral. We now define the autocorrelation function of this integral by

$$R_{mk}(t_1, t_2) = E\{h_{mk}(t_1) h_{mk}(t_2)\}. \quad (6.17\text{-}42)$$

On substituting Equations 6.17-35, 6.17-39, and 6.17-40 into Equation 6.17-34 and using the last two relationships as well as Equation 6.12-23, we obtain finally

$$R_y(t_1, t_2) = \sum_{m=0}^{\infty} \sum_{n=0}^{\infty} \frac{\mathscr{E}_m}{k!} R_{mk}(t_1, t_2) R_n^k(\tau) \cos m\omega_m \tau. \tag{6.17-43}$$

We have already assumed that the input noise $n(t)$ is at least wide-sense stationary. If, in addition, we place the same restriction on the modulation $m(t)$, then $R_{mk}(t_1, t_2)$ becomes a function only of the time difference $\tau = t_2 - t_1$ and Equation 6.17-43 can be written as

$$R_y(\tau) = \sum_{m=0}^{\infty} \sum_{n=0}^{\infty} \frac{\mathscr{E}_m}{k!} R_{mk}(\tau) R_n^k(\tau) \cos m\omega_m \tau. \tag{6.17-44}$$

At this point let us refer to Section 6.5 where we examined the output of the square-law device for an input of the form $x(t) = s(t) + n(t)$. In that case we found that both the output autocorrelation function and power spectral density could be expressed as the sum of three terms, a (signal × signal) term, a (noise × noise) term, and a (signal × noise) term. In the same way, the output of the νth-law detector can be so written. If we examine Equations 6.17-41 and 6.17-43, we see that the (signal × signal) terms correspond to those where the index k is zero; the (noise × noise) terms are those where the index m is zero; and the (signal × noise) terms are those where neither m nor k is zero. Thus we may rewrite Equation 6.17-43 as

$$R_y(t_1, t_2) = R_{s\times s}(t_1, t_2) + R_{n\times n}(t_1, t_2) + R_{s\times n}(t_1, t_2) \tag{6.17-45}$$

where

$$R_{s\times s}(t_1, t_2) = \sum_{m=0}^{\infty} \mathscr{E}_m R_{m0}(t_1, t_2) \cos m\omega_m \tau, \tag{6.17-46}$$

$$R_{n\times n}(t_1, t_2) = \sum_{k=0}^{\infty} \frac{1}{k!} R_{0k}(t_1, t_2) R_n^k(\tau), \tag{6.17-47}$$

$$R_{s\times n}(t_1, t_2) = 2 \sum_{m=1}^{\infty} \sum_{k=1}^{\infty} \frac{1}{k!} R_{mk}(t_1, t_2) R_n^k(\tau) \cos m\omega_m \tau. \tag{6.17-48}$$

As in Section 6.5, if the modulation $m(t)$ is stationary, we may invert Equation 6.17-45 term by term to obtain the output power spectral density

$$\varphi_y(\omega) = \varphi_{s\times s}(\omega) + \varphi_{n\times n}(\omega) + \varphi_{s\times n}(\omega). \tag{6.17-49}$$

The term $\varphi_{s\times s}(\omega)$ is the Fourier transform of Equation 6.17-46

$$\varphi_{s\times s}(\omega) = \sum_{m=0}^{\infty} \mathscr{E}_m \int_{-\infty}^{\infty} R_{m0}(\tau) \cos m\omega_m\tau\, e^{-j\omega\tau}\, d\tau. \qquad (6.17\text{-}50)$$

Let us define the Fourier transform of $R_{mk}(\tau)$ by

$$\varphi_{mk}(\omega) = \int_{-\infty}^{\infty} R_{mk}(\tau)\, e^{-j\omega\tau}\, d\tau. \qquad (6.17\text{-}51)$$

Now Equation 6.17-50 may be written as

$$\varphi_{s\times s}(\omega) = \sum_{m=0}^{\infty} \frac{\mathscr{E}_m}{2} [\varphi_{m0}(\omega - m\omega_m) + \varphi_{m0}(\omega + m\omega_m)]. \qquad (6.17\text{-}52)$$

In the same way, the transform of Equation 6.17-47 becomes

$$\varphi_{n\times n}(\omega) = \sum_{k=0}^{\infty} \frac{1}{k!} \int_{-\infty}^{\infty} R_{0k}(\tau) R_n^k(\tau)\, e^{-j\omega\tau}\, d\tau. \qquad (6.17\text{-}53)$$

For convenience, we define the transform

$$_k\varphi_n(\omega) = \int_{-\infty}^{\infty} R_n^k(\tau)\, e^{-j\omega\tau}\, d\tau = \int_{-\infty}^{\infty} R_n^{k-1}(\tau) R_n(\tau)\, e^{-j\omega\tau}\, d\tau. \qquad (6.17\text{-}54)$$

This expression may be written as

$$_k\varphi_n(\omega) = \frac{1}{2\pi} \int_{-\infty}^{\infty} {}_{k-1}\varphi_n(\nu) \varphi_n(\omega - \nu)\, d\nu \qquad (6.17\text{-}55)$$

or

$$_k\varphi_n(\omega) = \frac{1}{(2\pi)^{k-1}} \int_{-\infty}^{\infty} \cdots \int_{-\infty}^{\infty} \varphi_n(\nu_{k-1}) \varphi_n(\nu_{k-2} - \nu_{k-1}) \cdots \varphi_n(\omega - \nu_1)\, d\nu_{k-1} \cdots d\nu_1. \qquad (6.17\text{-}56)$$

Now Equation 6.17-53 becomes

$$\varphi_{n\times n}(\omega) = \sum_{k=0}^{\infty} \frac{1}{k!(2\pi)^k} \int_{-\infty}^{\infty} \varphi_{0k}(\nu)\, {}_k\varphi_n(\omega - \nu)\, d\nu. \qquad (6.17\text{-}57)$$

The (signal × noise) term in the output power spectral density is the transform of Equation 6.17-48 and can be written as

$$\varphi_{s\times n}(\omega) = 2 \sum_{m=1}^{\infty} \sum_{k=1}^{\infty} \frac{1}{k!} \int_{-\infty}^{\infty} R_{mk}(\tau) R_n^k(\tau) \cos m\omega_m\tau\, e^{-j\omega\tau}\, d\tau \qquad (6.17\text{-}58)$$

or

$$\varphi_{s\times n}(\omega) = \sum_{m=1}^{\infty} \sum_{k=1}^{\infty} \frac{1}{k!(2\pi)^k} \int_{-\infty}^{\infty} \varphi_{mk}(\nu) [{}_k\varphi_n(\omega - m\omega_m - \nu) + {}_k\varphi_n(\omega + m\omega_m - \nu)]\, d\nu. \qquad (6.17\text{-}59)$$

The coefficient $h_{mk}(t_i)$ is given by Equation 6.17-41 for an arbitrary zero-memory nonlinearity $g(x)$ with transform $G(s)$. We consider now the evaluation of this coefficient for the specific νth-law devices that we have been treating. For the *half-wave* ν*th-law device* it is clear from Equation 6.12-17 that the transform $G(s)$ is

$$G(s) = G_1(s) = \frac{c\Gamma(\nu + 1)}{s^{\nu+1}} \tag{6.17-60}$$

and, hence, that Equation 6.17-41 can be written as

$$h_{mk}(t_i) = \frac{c\Gamma(\nu + 1)}{2\pi j} \int_{C_+} s^{k-\nu-1} I_m(sm_i)\, e^{[R_n(0)/2]s^2}\, ds. \tag{6.17-61}$$

Note that this last expression is identical to Equation 6.12-25 except for the presence of the term $I_m(sm_i)$ in the integrand. Since

$$I_m(0) = \begin{cases} 1, & m = 0, \\ 0, & m = 1, 2, \ldots, \end{cases} \tag{6.17-62}$$

it follows that, if the input signal is absent, that is, $m(t) \equiv 0$, then

$$h_{mk}(t_i) = h_{mk} = \begin{cases} a_{k\nu}, & m = 0, \\ 0, & m = 1, 2, \ldots. \end{cases} \tag{6.17-63}$$

In this case we are simply considering the situation discussed in Section 6.12 where the input was assumed to be a stationary gaussian process.

For the general case where $m(t) \neq 0$, we may use Equation 6.17-11 and replace s by jv to rewrite Equation 6.17-61 as

$$h_{mk}(t_i) = \frac{c\Gamma(\nu + 1)}{2\pi j} j^{k+m-\nu} \int_C v^{k-\nu-1} J_m(vm_i)\, e^{-[R_n(0)/2]v^2}\, dv. \tag{6.17-64}$$

In this form the contour integral has been evaluated [23] and the result is

$$h_{mk}(t_i) = \frac{c\Gamma(\nu + 1)(m_i^2/2\sigma_n^2)^{m/2}{}_1F_1[(k + m - \nu)/2; m + 1; -(m_i^2/2\sigma_n^2)]}{2m!(\sigma_n^2/2)^{(k-\nu)/2}\Gamma\{1 - [(k + m - \nu)/2]\}} \tag{6.17-65}$$

where σ_n^2 has been written for $R_n(0)$ and where ${}_1F_1(\alpha; \beta; \pm y)$ is the confluent hypergeometric function first encountered in Equation 4.8-27 of Section 4.8. Note that, if $k + m - \nu$ is an even integer greater than zero, then $h_{mk}(t_i)$ vanishes since $|\Gamma\{1 - [(k + m - \nu)/2]\}|$ approaches infinity.

The coefficient $h_{mk}(t_i)$ is now easily found for the full-wave devices. For the *full-wave, even,* ν*th-law device* it follows as in Equation 6.12-60 that

$$h_{mk}(t_i)_e = \begin{cases} 0, & m + k \text{ odd}, \\ 2h_{mk}(t_i), & m + k \text{ even}. \end{cases} \tag{6.17-66}$$

Also, as in Equation 6.12-68, for the *full-wave, odd, νth-law device* we have

$$h_{mk}(t_i)_o = \begin{cases} 0, & m = k = 0, \\ 2h_{mk}(t_i), & m + k \text{ odd}, \\ 0, & m + k \text{ even}. \end{cases} \tag{6.17-67}$$

Let us now illustrate this rather complicated procedure by an example in which we consider the square-law device that was treated extensively in Section 6.6.

Example 6.13 We take $\nu = 2$ and use Equations 6.17-64 and 6.17-65. Since $m + k$ must be even and since $\Gamma\{1 - [(k + m - 2)/2]\}$ is infinite for this case if $k + m - 2$ is greater than zero, we have the restrictions

$$k + m \leq 2, \qquad k + m \text{ even}.$$

Hence only h_{00}, h_{11}, h_{02}, and h_{20} are nonzero. We use Equation 6.17-64 to find

$$h_{00}(t_i) = \frac{c\sigma_n^2}{2}\,{}_1F_1\left(-1; 1; -\frac{m_i^2}{2\sigma_n^2}\right),$$

$$h_{11}(t_i) = \frac{cm_i}{2}\,{}_1F_1\left(0; 2; -\frac{m_i^2}{2\sigma_n^2}\right),$$

$$h_{02}(t_i) = c\,{}_1F_1\left(0; 1; -\frac{m_i^2}{2\sigma_n^2}\right),$$

$$h_{20}(t_i) = \frac{cm_i^2}{8}\,{}_1F_1\left(0; 3; -\frac{m_i^2}{2\sigma_n^2}\right).$$

It follows from the series expansion of ${}_1F_1(\alpha; \beta; \pm y)$ as given by Equation 4.8-27 that

$${}_1F_1(-1; 1; \pm y) = 1 - (\pm y)$$

and

$${}_1F_1(0; \beta; \pm y) = 1.$$

Thus the coefficients h_{mk} are

$$h_{00}(t_i) = \frac{c\sigma_n^2}{2}\left(1 + \frac{m_i^2}{2\sigma_n^2}\right),$$

$$h_{11}(t_i) = \frac{cm_i}{2},$$

$$h_{02}(t_i) = c,$$

$$h_{20}(t_i) = \frac{cm_i^2}{8}.$$

We now use Equation 6.17-65 in Equation 6.17-43 to obtain

$$R_{mk}(t_1t_2)_e = 4R_{mk}(t_1, t_2) \tag{A}$$

for $m + k$ even and less than or equal to two. We have

$$R_{00}(t_1, t_2) = E\left\{\frac{c\sigma_n^2}{2}\left(1 + \frac{m_1^2}{2\sigma_n^2}\right)\frac{c\sigma_n^2}{2}\left(1 + \frac{m_2^2}{2\sigma_n^2}\right)\right\}$$

$$= \frac{c^2\sigma_n^4}{4}\left(1 + \frac{\sigma_m^2}{\sigma_n^2} + \frac{E\{m_1^2m_2^2\}}{4\sigma_n^4}\right)$$

where we have made the assumption

$$\sigma_m^2 = E\{m_1^2\} = E\{m_2^2\}.$$

In the same way

$$R_{11}(t_1, t_2) = E\left\{\frac{cm_1}{2}\frac{cm_2}{2}\right\} = \frac{c^2}{4}R_m(\tau)$$

where we have assumed that $m(t)$ is at least wide-sense stationary with autocorrelation function $R_m(\tau)$ so that

$$R_m(0) = \sigma_m^2.$$

We proceed to find

$$R_{02}(t_1, t_2) = c^2$$

and

$$R_{20}(t_1, t_2) = E\left\{\frac{cm_1^2}{8}\frac{cm_2^2}{8}\right\} = \frac{c^2}{64}E\{m_1^2m_2^2\}.$$

We now use Equation A in Equation 6.17-43 to obtain

$$R_y(t_1, t_2) = 4R_{00}(t_1, t_2) + 8R_{11}(t_1, t_2)R_n(\tau)\cos\omega_m\tau + 2R_{02}(t_1, t_2)R_n^2(\tau) + 8R_{20}(t_1, t_2)\cos 2\omega_m\tau$$

or, after substituting for the R_{mk} and rearranging,

$$R_y(t_1, t_2) = c^2[R_n^2(0) + 2R_n^2(\tau)] + c^2[R_n(0)R_m(0) + 2R_n(\tau)R_m(\tau)\cos\omega_m\tau] + c^2E\{m_1^2m_2^2\}(\tfrac{1}{4} + \tfrac{1}{8}\cos 2\omega_m\tau)$$

where we have written $R_n(0)$ for σ_n^2. That these results are identical to those of Section 6.6 is easily seen by making the substitution

$$m(t) = C[1 + a(t)]$$

and noting that

$$R_m(\tau) = C^2[1 + R_a(\tau)]$$

and that

$$E\{m_1^2m_2^2\} = C^4\{[1 + R_a(0)]^2 + 4R_a(\tau) + 2R_a^2(\tau)\}$$

since $a(t)$ was assumed in Section 6.6 to be a stationary gaussian process with zero mean.

• •

Example 6.14 Let us now consider an example where the output autocorrelation function R_y is not so simply expressed. We treat the full-wave, even, rectifier so that $\nu = 1$. We let $m(t)$ be the constant A so that the signal $s(t)$ is an unmodulated sinusoid of random phase. Then the coefficients $h_{mk}(t_i)$ are found from Equation 6.17-65 as

$$h_{mk}(t_i) = h_{mk} = \frac{(A^2/2\sigma_n^2)^{m/2} {}_1F_1[(k + m - 1)/2; m + 1; -(A^2/2\sigma_n^2)]}{2m!(\sigma_n^2/2)^{(k-1)/2}\Gamma\{1 - [(k + m - 1)/2]\}} \tag{A}$$

where, for convenience, we have placed $c = 1$. We note from Equation 6.17-66 that $m + k$ must be even if h_{mk} is to be nonzero; therefore we may write Equation 6.17-42 as

$$R_{mk}(t_1, t_2)_e = \underset{(m+k \text{ even})}{4R_{mk}(t_1, t_2)} = \underset{(m+k \text{ even})}{4R_{mk}} = \underset{(m+k \text{ even})}{4h_{mk}^2}.$$

The output autocorrelation function given by Equation 6.17-44 is now conveniently written as the sum of four sets of terms:

$$R_y(\tau) = 4h_{00}^2 + 8 \sum_{\substack{m=2 \\ (m \text{ even})}}^{\infty} h_{m0}^2 \cos m\omega_m\tau + 4 \sum_{\substack{k=2 \\ (k \text{ even})}}^{\infty} h_{0k}^2 \frac{R_n^k(\tau)}{k!}$$

$$+ 8 \sum_{m=1}^{\infty} \sum_{\substack{k=1 \\ (m+k \text{ even})}}^{\infty} h_{mk}^2 \frac{R_n^k(\tau)}{k!} \cos m\omega_m\tau. \tag{B}$$

In this form it is easy to see that the first term corresponds to a constant (d-c) output term and that the second set of terms corresponds to a set of sinusoids of random phase and radian frequency $2\omega_m, 4\omega_m, 6\omega_m, \ldots$. If $R_n(\tau) \to 0$ as $\tau \to \infty$, then the last two sets of terms go to zero as $\tau \to \infty$.

The term h_{00} is given by Equation A as

$$h_{00} = \frac{\sigma_n}{\sqrt{2\pi}} {}_1F_1\left(-\tfrac{1}{2}; 1; -\frac{A^2}{2\sigma_n^2}\right)$$

or, on using Equation 4.8-30,

$$h_{00} = \frac{\sigma_n}{\sqrt{2\pi}} e^{-A^2/4\sigma_n^2}\left[\left(1 + \frac{A^2}{2\sigma_n^2}\right)I_0\left(\frac{A^2}{4\sigma_n^2}\right) + \frac{A^2}{2\sigma_n^2} I_1\left(\frac{A^2}{4\sigma_n^2}\right)\right].$$

If the output $y(t)$ could be sent through an ideal low-pass filter of infinitely narrow bandwidth, then only the d-c term

$$[z(t)]_{\text{d-c}} = 2h_{00}$$

would be obtained. For the small signal-to-noise case, we have $A/\sigma_n \to 0$ and

$$[z(t)]_{\text{d-c}} \to \sqrt{\frac{2}{\pi}}\, \sigma_n.$$

For large signal-to-noise, the ratio $A/\sigma_n \to \infty$ and

$$[z(t)]_{\text{d-c}} \to \frac{2A}{\pi}$$

as expected. These results should be compared with Equation 4.8-23 and with Figure 4.8 where we found the same thing, that is, the expected value (normalized) of the envelope of a sinusoid in additive narrow-band noise.

Let us now find the output autocorrelation function $R_y(\tau)$ for the case where the noise $n(t)$ is identically zero so that the input is just

$$s(t) = A \sin(\omega_m t + \theta).$$

What we are finding in this case is the autocorrelation function of $|s(t)|$. We note that only the first two sets of terms of Equation B are nonzero since $R_n(\tau) \equiv 0$. We wish to find h_{m0} as $A/\sigma_n \to 0$. We make use of the asymptotic relationship [24]

$$\lim_{y\to\infty} {}_1F_1(\alpha; \beta; -y) \to \frac{\Gamma(\beta)}{\Gamma(\beta-\alpha)} \frac{1}{y^\alpha}$$

to write, for m even,

$$\lim_{A^2/2\sigma_n^2\to\infty} h_{m0} = \frac{A}{4} \frac{1}{\Gamma[(3-m)/2]\Gamma[(3+m)/2]} = \frac{A}{\pi} \frac{1}{1-m^2} \sin\left(\frac{m+1}{2}\right)\pi$$

where we have made use of the formulas [25]

$$\Gamma(1 \pm z) = \pm z\Gamma(\pm z),$$

$$\Gamma(z)\Gamma(1-z) = \frac{\pi}{\sin \pi z},$$

and

$$\Gamma(n + \tfrac{1}{2}) = \frac{1\cdot3\cdot5\cdot7\cdots(2n-1)}{2^n}\sqrt{\pi}.$$

Now Equation B becomes

$$\lim_{A^2/2\sigma_n^2\to\infty} R_y(\tau) = \frac{4A^2}{\pi^2} + \frac{8A^2}{\pi^2} \sum_{\substack{m=2\\ m \text{ even}}}^{\infty} \frac{\cos m\omega_m\tau}{(1-m^2)^2}.$$

This result should be compared with problem 5 of Chapter 3.

• •

6.18 νth-Law Nonlinear Amplifiers and Detectors

In this section we consider briefly the systems of Figure 6.23 when the input $x(t)$ is a stationary narrow-band process (or processes) and when the time-invariant linear filter $h(t)$ following the νth-law device is also narrow band; that is, the filter has a complex transfer function $H(\omega)$ which is substantially different from zero only over a limited frequency range. As mentioned at the beginning of the previous section, if the filter is bandpass, we will call the resulting system a νth-law *nonlinear amplifier*; if the filter is low pass, we will call the system a *νth-law detector.*

The power spectral density $\varphi_z(\omega)$ for the output $z(t)$ can be found in each case from the relationship first given by Equation 4.4-7:

$$\varphi_z(\omega) = \varphi_y(\omega)|H(\omega)|^2 \tag{6.18-1}$$

where $H(\omega)$ is the complex transfer function of the linear filter and $\varphi_y(\omega)$ is given by Equations 6.17-49, 6.17-52, 6.17-57, and 6.17-59. In the same way, the output autocorrelation function $R_z(\tau)$ is given by Equation 4.4-3:

$$R_z(\tau) = \int_{-\infty}^{\infty}\int_{-\infty}^{\infty} R_y(\tau - \beta + \alpha)h(\alpha)h(\beta)\,d\alpha\,d\beta. \tag{6.18-2}$$

In the absence of signal $[s(t) \equiv 0]$, the input as given by Equation 6.17-27 is simply a narrow-band stationary gaussian noise $n(t)$ with zero mean and autocorrelation function $R_n(\tau)$. For this case the output $z(t)$ is easily written for each of the νth-law devices previously studied:

νth-Law Half-Wave Device (Noise Only)

Detector: Since the filter $h(t)$ of Figure 6.24(b) is low pass, it follows from Equation 6.17-8 that

$$z(t) = F(0, \nu)R^{\nu}(t) \tag{6.18-3}$$

where $F(0, \nu)$ is obtained from Equation 6.17-17 as

$$F(0, \nu) = c\,\frac{\Gamma(\nu + 1)}{2^{\nu+1}\{\Gamma[1 + (\nu/2)]\}^2}. \tag{6.18-4}$$

Example 6.15 Let $\nu = 1$ so that we have the half-wave linear rectifier; then

$$F(0, \nu) = \frac{c}{\pi}$$

and

$$z(t) = \frac{c}{\pi}R(t)$$

where $R(t)$ is the envelope of the input narrow-band process $x(t)$. Thus, as expected, the half-wave linear detector is an envelope detector.

• •

Nonlinear Amplifier: If the bandpass filter is centered on ω_m, then Equation 6.17-8 becomes

$$z(t) = 2F(1, \nu)R^{\nu}(t)\cos[\omega_m t + \theta(t)] \tag{6.18-5}$$

where

$$F(1, \nu) = c \frac{\Gamma(\nu + 1)}{2^{\nu+1}\Gamma[(1 + \nu)/2]\Gamma[(3 + \nu)/2]}. \tag{6.18-6}$$

Example 6.16 For $\nu = 1$, we have the simple result that

$$F(1, 1) = c$$

and

$$z(t) = 2cR(t) \cos [\omega_m t + \theta(t)].$$

• •

Example 6.17 For $\nu = 0$, we have a *bandpass half-wave limiter* where

$$F(1, 0) = \frac{c}{\pi}$$

and

$$z(t) = \frac{2c}{\pi} \cos [\omega_m t + \theta(t)].$$

Thus the envelope $R(t)$ has been removed and the angle modulation has remained unchanged.

• •

νth-Law, Full-Wave, Even, Device (Noise Only)

Detector: We make use of Equation 6.17-23 to write

$$z(t) = 2F(0, \nu)R^{\nu}(t) \tag{6.18-7}$$

where $F(0, \nu)$ has already been given by Equation 6.18-4. It is clear that, except for the factor of two, the output is identical to that of the νth-law half-wave detector.

Example 6.18 Let $\nu = 2$ to obtain the full-wave, even, detector (receiver) studied in Section 6.7. We have

$$F(0, 2) = \frac{c}{4}$$

and

$$z(t) = \frac{c}{2} R^2(t).$$

• •

Nonlinear Amplifier: In this case, Equation 6.17-23 becomes

$$z(t) = 0 \tag{6.18-8}$$

as expected.

νth-Law, Full-Wave, Odd, Device (Noise Only)

Detector: We make use of Equation 6.17-26 to write

$$z(t) = 0.$$

Nonlinear Amplifier: Again we use Equation 6.17-26 to obtain

$$z(t) = 4F(1, \nu)R^{\nu}(t) \cos [\omega_m t + \theta(t)]$$

where $F(1, \nu)$ has already been given by Equation 6.18-6.

Example 6.19 Let $\nu = 1$ so that the νth-law device becomes a linear change of scale; then

$$F(1, 1) = \frac{c}{4}$$

and

$$z(t) = cR(t) \cos [\omega_m t + \theta(t)].$$

• •

For the more general case, where the narrow-band input is the sum of a signal and a noise, the situation is not so simple. In principle the output autocorrelation function $R_z(\tau)$ and the output power spectral density $\varphi_z(\omega)$ are obtained by applying Equations 6.17-44 and 6.17-49, respectively, to Equations 6.18-2 and 6.18-1. As a practical matter, the expressions become quite complicated. One of the simplest examples is the square-law detector (receiver) which has been treated in some detail in Section 6.7. Other examples are given in the literature [26].

PART 3—GENERAL (NONLINEAR) LEAST MEAN-SQUARE-ERROR FILTERING AND PREDICTION

6.19 Formulation and General Solution

A short discussion of the philosophy of optimum system design was given in Section 5.1 while the remainder of Chapter 5 was devoted to a treatment of optimum *linear* systems using the least mean-squared-error criterion. In this section we consider least mean-squared-error filtering from a more general point of view in that we do not place any restrictions of linearity on the filters.

We shall consider the problem illustrated in Figure 6.24. Here the input will be a random process called $x(t)$ and the output will be denoted by $y(t)$. They will be presumed to be related through the general (nonlinear) operation N. We shall not attempt to restrict the system N further except to require that it be *nonrandom* or *deterministic*; that is, that it contain no random

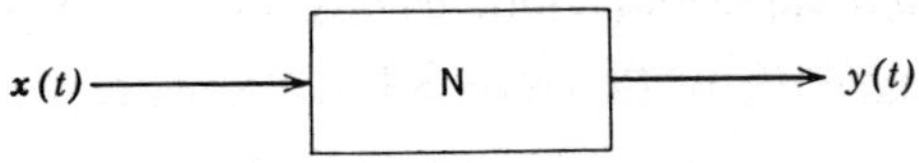

Figure 6.24 *Least mean-squared-error filtering*

variables. We require $y(t)$ to be a least mean-squared-error estimate of some desired output $z(t)$. In other words, the nonnegative function

$$L = E\{(y - z)^2\} \tag{6.19-1}$$

is to be minimized by the proper choice of the filter N. For the case analogous to the Wiener filter covered in Part 2 of Chapter 5, typical choices of x and z might be

$$x(t) = s(t) + n(t), \tag{6.19-2}$$

$$z(t) = s(t + \alpha) \tag{6.19-3}$$

where $s(t)$ is a desired signal and $n(t)$ is a noise. Other choices, however, also are possible; for example, we might let

$$x(t) = m(t)s(t), \tag{6.19-4}$$

$$z(t) = s(t) \tag{6.19-5}$$

where $m(t)$ may or may not be a random process. Such a choice could correspond to an optimum demodulation problem.

We may proceed to minimize Equation 6.19-1 as discussed in Sections 5.4 and F.4. We assume that the system N is optimum and then allow it to vary from this optimum. Thus Equation 6.19-1 becomes

$$\begin{aligned} L + \delta L &= E\{(y + \epsilon\delta y - z)^2\} \\ &= E\{[(y - z) + \epsilon\delta y]^2\} \end{aligned} \tag{6.19-6}$$

where δL is the variation in L produced by the arbitrary variation $\epsilon\delta y$ and ϵ is a real variable. This last expression can be written as

$$L + \delta L = L + \epsilon E\{\delta y(y - z)\} + \epsilon^2 E\{(\delta y)^2\}. \tag{6.19-7}$$

As discussed in Section F.4, a *necessary* condition for the original expression of Equation 6.19-1 to be a minimum is that

$$\left.\frac{\partial(L + \delta L)}{\partial\epsilon}\right|_{\epsilon=0} = 0 \tag{6.19-8}$$

or

$$\boxed{E\{\delta y(y - z)\} = 0.} \tag{6.19-9}$$

We note immediately that this condition is also *sufficient*. Suppose Equation 6.19-9 is satisfied; then it follows from Equation 6.19-7 that

$$\delta L = \epsilon^2 E\{(\delta y)^2\} \geq 0. \tag{6.19-10}$$

In other words, varying y from its value given by Equation 6.19-9 cannot decrease the mean-squared error. Thus Equation 6.19-9 is a necessary and sufficient condition that must be satisfied by the system N for y to be a least mean-squared-error estimate of z.

A number of consequences follow immediately from the fact that the variation δy in Equation 6.19-9 is *arbitrary*. Suppose we allow

$$\delta y = 1. \tag{6.19-11}$$

Then we obtain directly that

$$\boxed{E\{y\} = E\{z\}.} \tag{6.19-12}$$

The mean value of the estimate of z is equal to the mean value of z. Any estimate y which has this property of reproducing the mean of the quantity to be estimated is called an *unbiased* estimate. Now suppose we require that

$$\delta y = y. \tag{6.19-13}$$

Then Equation 6.19-9 becomes

$$E\{y^2\} = E\{yz\}. \tag{6.19-14}$$

Equation 6.19-1 may be expanded to give

$$L_{\min} = E\{y^2\} + E\{z^2\} - 2E\{yz\} \tag{6.19-15}$$

or, after making use of Equation 6.19-14 we have

$$\boxed{L_{\min} = E\{z^2\} - E\{y^2\}} \tag{6.19-16}$$

as a convenient expression for the least mean-squared error. Since z is the desired estimate and since $E\{z^2\}$ is independent of y, the largest value for $L_{\min}$ is

$$\max L_{\min} = E\{z^2\} \tag{6.19-17}$$

and can be obtained, for example, by having a system N that transmits nothing. The reader will recall that this was the result found in Section 5.13 for linear filtering in overwhelming noise. Also, since $L_{\min}$, $E\{z^2\}$, and $E\{y^2\}$ are all nonnegative, we must have the inequality

$$E\{y^2\} \leq E\{z^2\}. \tag{6.19-18}$$

Thus the normalized power in the estimate of z is less than or equal to the normalized power in z, the quantity to be estimated. Since the means of y and z are equal, Equation 6.19-16 may be rewritten as

$$L_{\min} = \sigma_z^2 - \sigma_y^2 \tag{6.19-19}$$

where the variances σ_z^2 and σ_y^2 are defined by

$$\sigma_z^2 = E\{[z - E\{z\}]^2\} = E\{z^2\} - [E\{z\}]^2 \tag{6.19-20}$$

and

$$\sigma_y^2 = E\{[y - E\{y\}]^2\} = E\{y^2\} - [E\{y\}]^2. \tag{6.19-21}$$

Although Equation 6.19-9 specifies the optimum choice of y, it is more instructive to obtain an explicit expression for y. Let us rewrite Equation 6.19-9 as

$$\int_{-\infty}^{\infty} \int_{-\infty}^{\infty} \delta y(y - z) f(x, z)\, dz\, dx = 0 \tag{6.19-22}$$

where $f(x, z)$ is the joint density of x and z. Now we can make use of the conditional probability relationship

$$f(x, z) = f_x(x) f_{z/x}(z/x) \tag{6.19-23}$$

to rewrite this last expression as

$$\int_{-\infty}^{\infty} \delta y f_x(x) \left[\int_{-\infty}^{\infty} (y - z) f_{z/x}(z/x)\, dz \right] dx = 0. \tag{6.19-24}$$

Since δy is arbitrary it is necessary that the term in brackets be zero. (Note that $f_x(x)$ is nonnegative; then we could choose $\delta y = 1$, for example):

$$\int_{-\infty}^{\infty} (y - z) f_{z/x}(z/x)\, dz = 0. \tag{6.19-25}$$

However, when x is given, y is also determined so that

$$\int_{-\infty}^{\infty} y f_{z/x}(z/x)\, dz = y \int_{-\infty}^{\infty} f_{z/x}(z/x)\, dz = y \tag{6.19-26}$$

or, finally,

$$\boxed{y = E\{z/x\}.} \tag{6.19-27}$$

Thus the least mean-squared-error estimate of z is the conditional expectation of z given x. For the special case given by Equations 6.19-2 and 6.19-3, this expression becomes

$$y = E\{s(t + \alpha)/[s(t) + n(t)]\}. \tag{6.19-28}$$

6.20 The Linear Case

Let us assume that the system N is restricted to be linear and time invariant so that the output y can be written as

$$y(t) = \int_0^{\infty} h(\tau) x(t - \tau)\, d\tau \tag{6.20-1}$$

or

$$y = h * x \tag{6.20-2}$$

where the asterisk has been used to denote convolution. Now the variation δy becomes

$$\delta y = \delta h * x \tag{6.20-3}$$

and Equation 6.20-9 can be written as

$$E\{\delta h * x(h * x - z)\} = 0 \tag{6.20-4}$$

or

$$\delta h * E\{x h * x - x \delta * z\} = 0 \tag{6.20-5}$$

where $\delta * z$ is the identity convolution, that is,

$$z = \delta * z \qquad \text{or} \qquad z(t) = \int_0^\infty \delta(\tau) z(t - \tau)\, d\tau. \tag{6.20-6}$$

Again, since δh is arbitrary, we have

$$\boxed{E\{x h * x - x \delta * z\} = 0} \tag{6.20-7}$$

or

$$E\{x(t - \sigma) z(t)\} = \int_0^\infty h(\tau) R_x(t - \sigma, t - \tau)\, d\tau, \qquad 0 \le \sigma < \infty, \tag{6.20-8}$$

where $R_x(t - \sigma, t - \tau)$ is the autocorrelation function of $x(t)$ and is given by

$$R_x(t - \sigma, t - \tau) = E\{x(t - \sigma) x(t - \tau)\}. \tag{6.20-9}$$

If it is assumed that $x(t)$ is at least wide-sense stationary and that $x(t)$ and $z(t)$ are given by Equations 6.20-2 and 6.20-3, then Equation 6.20-8 becomes

$$E\{x(t - \sigma) s(t + \alpha)\} = \int_0^\infty h(\tau) R_x(\sigma - \tau)\, d\tau, \qquad 0 \le \sigma < \infty. \tag{6.20-10}$$

This expression is identical to Equation 5.10-18 for the Wiener filter. Such *linear* least mean-squared-error filters as specified by Equation 6.20-10 were treated extensively in Chapter 5.

The least mean-squared-error filter is given by Equation 6.19-27. In general the system determined by this expression will not be linear and, consequently, will not coincide with the (linear) system given by Equation 6.20-10. It is interesting to consider the conditions under which Equation 6.19-27 specifies a linear system and hence coincides with Equation 6.20-10. A *sufficient* (but *not* necessary) condition is that the random processes x and z of Equation 6.19-27 be jointly gaussian. This will be the case, for example, if we are dealing with the problem of predicting a gaussian signal in additive gaussian noise so that x and z are given by Equations 6.19-2 and 6.19-3 with $s(t)$ and $n(t)$ gaussian processes. We now prove that Equation 6.19-27 specifies a linear system if x and z are jointly gaussian processes.

We refer to Sections 2.22 and 3.10 and to Equation 2.22-14 which gives the n-variate normal moment-generating function $M(\mathbf{t})$. For convenience we will assume that $x(t)$ and $z(t)$ each have zero mean. We define the column vector $\mathbf{x}$ by

$$\mathbf{x} = \begin{bmatrix} x(t_1) \\ x(t_2) \\ \vdots \\ x(t_n) \end{bmatrix} = \begin{bmatrix} x_1 \\ x_2 \\ \vdots \\ x_n \end{bmatrix} \tag{6.20-11}$$

and the covariance matrix $\mathbf{R}_{xx}$ by

$$\mathbf{R}_{xx} = [\sigma_{ij}] = [E\{(x_i - u_i)(x_j - u_j)\}] = [E\{x_i x_j\}] \tag{6.20-12}$$

since the means u_i have been taken to be zero. We denote $z(t)$ by z and let

$$E\{z^2\} = A. \tag{6.20-13}$$

Now the joint characteristic function of z and x may be written from Equations 2.20-41 and 2.22-14 as

$$\begin{aligned} F_{zx}(v_0, \mathbf{v}) &= \int_{\mathbf{x}} \int_{-\infty}^{\infty} f(\mathbf{x}, z)\, e^{jv_0 z}\, e^{j\mathbf{v}'\mathbf{x}}\, dz\, d\mathbf{x} \\ &= \exp\left\{-\tfrac{1}{2}[v_0, \mathbf{v}'] \begin{bmatrix} A & \mathbf{R}_{zx} \\ \mathbf{R}'_{zx} & R_{xx} \end{bmatrix} \begin{bmatrix} v_0 \\ \mathbf{v} \end{bmatrix}\right\} \\ &= \exp\left[-\tfrac{1}{2}(v_0^2 A + 2v_0 \mathbf{R}_{zx}\mathbf{v} + \mathbf{v}' R_{xx}\mathbf{v})\right] \end{aligned} \tag{6.20-14}$$

where $\mathbf{R}_{zx}$ is the row matrix

$$\mathbf{R}_{zx} = E\{z\mathbf{x}'\}, \tag{6.20-15}$$

$\mathbf{v}$ is the column matrix

$$\mathbf{v} = \begin{bmatrix} v_1 \\ v_2 \\ \vdots \\ v_n \end{bmatrix} \tag{6.20-16}$$

with transpose $\mathbf{v}'$, and where we have used the notation

$$\int_{\mathbf{x}} (\ )\, d\mathbf{x} = \int_{-\infty}^{\infty} \cdots \int_{-\infty}^{\infty} (\ )\, dx_1 \cdots dx_n. \tag{6.20-17}$$

The joint characteristic function of $\mathbf{x}$ will be denoted by $F_x(\mathbf{v})$ and is

$$F_x(\mathbf{v}) = \int_{\mathbf{x}} f_x(\mathbf{x})\, e^{j\mathbf{v}'\mathbf{x}}\, d\mathbf{x} \tag{6.20-18}$$

where $f_x(\mathbf{x})$ is the n-variant normal distribution. We may also write

$$F_x(\mathbf{v}) = F_{zx}(v_0, \mathbf{v})\big|_{v_0 = 0} = \exp\left(-\tfrac{1}{2}\mathbf{v}' R_{xx}\mathbf{v}\right). \tag{6.20-19}$$

We now return to Equation 6.19-27 which specified the least mean-squared-error system and write

$$y = E\{z/x\} = \frac{\int_{-\infty}^{\infty} zf(\mathbf{x}, z)\, dz}{f_x(\mathbf{x})} \tag{6.20-20}$$

where we have made use of Equations 6.19-23 and 6.19-26. From Equation 6.20-14 we find that

$$\begin{aligned}\left.\frac{\partial F_{zx}(v_0, \mathbf{v})}{\partial j v_0}\right|_{v_0=0} &= \int_{\mathbf{x}} \left[\int_{-\infty}^{\infty} zf(\mathbf{x}, z)\, dz\right] e^{j\mathbf{v}'\mathbf{x}}\, d\mathbf{x} \\ &= \mathscr{F}\left[\int_{-\infty}^{\infty} zf(\mathbf{x}, z)\, dz\right] \end{aligned} \tag{6.20-21}$$

or that

$$\int_{-\infty}^{\infty} zf(\mathbf{x}, z)\, dz = \mathscr{F}^{-1}\left[\left.\frac{\partial F_{zx}(v_0, \mathbf{v})}{\partial j v_0}\right|_{v_0=0}\right] \tag{6.20-22}$$

where $\mathscr{F}[\ ]$ and $\mathscr{F}^{-1}[\ ]$ denote the Fourier transform and inverse Fourier transform, respectively. It also follows from Equation 6.20-14 that

$$\left.\frac{\partial F_{zx}(v_0, \mathbf{v})}{\partial v_0}\right|_{v_0=0} = -\mathbf{R}_{zx}\mathbf{v}F_x(\mathbf{v}) \tag{6.20-23}$$

and, from Equation 6.20-19, that

$$\frac{\partial F_x(\mathbf{v})}{\partial \mathbf{v}'} = -\boldsymbol{R}_{xx}\mathbf{v}F_x(\mathbf{v}). \tag{6.20-24}$$

Thus we may write Equation 6.20-23 as

$$\begin{aligned}\left.\frac{\partial F_{zx}(v_0, \mathbf{v})}{\partial v_0}\right|_{v_0=0} &= \mathbf{R}_{zx}\boldsymbol{R}_{xx}^{-1}\frac{\partial F_x(\mathbf{v})}{\partial \mathbf{v}'} \\ &= \mathbf{R}_{zx}\boldsymbol{R}_{xx}^{-1}\int_{\mathbf{x}} \mathbf{x}f_x(\mathbf{x})\, e^{j\mathbf{v}'\mathbf{x}}\, d\mathbf{x} \\ &= \mathbf{R}_{zx}\boldsymbol{R}_{xx}^{-1}\mathscr{F}[\mathbf{x}f_x(\mathbf{x})]. \end{aligned} \tag{6.20-25}$$

The last equality follows from Equation 6.20-18. Thus we conclude that

$$\mathscr{F}^{-1}\left[\left.\frac{\partial F_{zx}(v_0, \mathbf{v})}{\partial v_0}\right|_{v_0=0}\right] = \mathbf{R}_{zx}\boldsymbol{R}_{xx}^{-1}\mathbf{x}f_x(\mathbf{x}) \tag{6.20-26}$$

and Equation 6.20-20 becomes

$$y = E\{z/x\} = \mathbf{R}_{zx}\boldsymbol{R}_{xx}^{-1}\mathbf{x}. \tag{6.20-27}$$

In other words, the output $y = y(t)$ is a linear operation on the input $\mathbf{x}$. Let us define the linear vector operator

$$\mathbf{h}(t, \mathbf{t}) = \mathbf{R}_{zx}\boldsymbol{R}_{xx}^{-1} = \mathbf{R}_{zx}(t, \mathbf{t})\mathrm{w}_{xx}(\mathbf{t}, \mathbf{t}) \tag{6.20-28}$$

or, after postmultiplication of both sides by $\boldsymbol{R}_{xx}$,

$$R_{zx}(t, t_i) = \sum_{k=1}^{n} h(t, t_k) R_{xx}(t_k, t_i). \tag{6.20-29}$$

This is just the discrete analog of the linear Wiener filter of Equation 6.20-10. On passing to the limit we have

$$R_{zx}(t, \sigma) = \int_0^\infty h(t, \tau) R_{xx}(\tau, \sigma)\, d\tau. \tag{6.20-30}$$

On assuming stationary statistics and $z(t) = s(t + \alpha)$, we obtain Equation 6.20-10.

We have shown that, if z and x are jointly gaussian, then the least mean-squared-error filter, given by Equation 6.19-27, is linear.

6.21 The ZNL Least Mean-Squared-Error Filter

In this section we consider briefly Equation 6.19-27 on the condition that

$$y = E\{z/x\} = g(x) \tag{6.21-1}$$

so that the filter is constrained to be a zero-memory nonlinearity. For simplicity we limit ourselves to the case where

$$x(t) = s(t) + n(t), \tag{6.21-2}$$

$$z(t) = s(t) \tag{6.21-3}$$

so that we are attempting to recover a signal in additive noise.

Suppose that $s(t)$ and $n(t)$ are jointly gaussian so that the least mean-squared-error filter is linear. If, in addition, $s(t)$ and $n(t)$ are uncorrelated and have the same power spectral density shapes $K_1\varphi(\omega)$ and $K_2\varphi(\omega)$, it follows from Equation 5.11-6 that the filter is the constant $K_1/(K_1 + K_2)$ which is a ZNL, although a rather simple one. For processes $s(t)$ and $n(t)$ with continuous parameter t, there are no simple conditions for which Equation 6.21-1 holds since, in general, the conditional expectation in that equation depends on the past of $x(t)$. On the other hand suppose that $s(t_i)$ and $n(t_i)$ are random processes with discrete parameter t_i, $i = 1, 2, \ldots,$ and suppose that all $s(t_i)$ and $n(t_i)$ are mutually independent. In other words the signal and noise are discrete series of independent random variables. Then it is clear that Equation 6.21-1 will hold for each $x_i = s_i + n_i$. If the processes are jointly stationary to at least order one, then $g(x)$ will be fixed and can be written from Equations 6.19-23 and 6.20-20 as

$$g(x) = E\{z/x\} = \frac{\int_{-\infty}^{\infty} s f_n(x - s) f_s(s)\, ds}{\int_{-\infty}^{\infty} f_n(x - s) f_s(s)\, ds} \tag{6.21-4}$$

where $f_n(\cdot)$ is the density function for each n_i and $f_s(\cdot)$ is the density function for each s_i. We proceed to illustrate with two examples.

Example 6.20 Let each s_i be uniformly distributed on the interval $(-a, a)$ so that

$$f_s(s) = \begin{cases} \dfrac{1}{2a}, & -a \le s \le a, \\ 0, & \text{elsewhere,} \end{cases}$$

and let the n_i have an arbitrary density function $f_n(\cdot)$. The denominator of Equation 6.21-4 becomes

$$\int_{-\infty}^{\infty} f_n(x - s)f_s(s)\,ds = \frac{1}{2a}\int_{-a}^{a} f_n(x - s)\,ds = \frac{1}{2a}\int_{x-a}^{x+a} f_n(y)\,dy. \tag{A}$$

The numerator becomes

$$\begin{aligned}\int_{-\infty}^{\infty} sf_n(x - s)f_s(s)\,ds &= \frac{1}{2a}\int_{-a}^{a} sf_n(x - s)\,ds \\ &= \frac{1}{2a}\int_{x-a}^{x+a} (x - y)f_n(y)\,dy \\ &= \frac{x}{2a}\int_{x-a}^{x+a} f_n(y)\,dy - \frac{1}{2a}\int_{x-a}^{x+a} yf_n(y)\,dy. \end{aligned} \tag{B}$$

From the last two equations, it follows that Equation 6.21-4 can be written as

$$g(x) = x - \frac{\int_{x-a}^{x+a} yf_n(y)\,dy}{\int_{x-a}^{x+a} f_n(y)\,dy}. \tag{C}$$

Suppose now that the noise distribution is gaussian so that

$$f_n(n) = \frac{1}{\sqrt{2\pi}\sigma} e^{-n^2/2\sigma^2}, \qquad -\infty < n < \infty. \tag{D}$$

The two integrals in Equation C are easily evaluated to give

$$\begin{aligned}\frac{1}{\sqrt{2\pi}\sigma}\int_{x-a}^{x+a} y\,e^{-y^2/2\sigma^2}\,dy &= -\frac{\sigma}{\sqrt{2\pi}} e^{-y^2/2\sigma^2}\Big|_{x-a}^{x+a} \\ &= -\frac{\sigma}{\sqrt{2\pi}}[e^{-(x+a)^2/2\sigma^2} - e^{-(x-a)^2/2\sigma^2}] \end{aligned} \tag{E}$$

and

$$\frac{1}{\sqrt{2\pi}\sigma}\int_{x-a}^{x+a} e^{-y^2/2\sigma^2}\,dy = \Phi\left(\frac{x + a}{\sigma}\right) - \Phi\left(\frac{x - a}{\sigma}\right) \tag{F}$$

where $\Phi(\cdot)$ is the unit normal cumulative distribution function. On combining Equations E and F with C, we obtain the least mean-squared error filter

$$g(x) = x + \frac{\sigma}{\sqrt{2\pi}} \frac{e^{-(x+a)^2/2\sigma^2} - e^{-(x-a)^2/2\sigma^2}}{\Phi[(x+a)/\sigma] - \Phi[(x-a)/\sigma]}. \tag{G}$$

• •

Example 6.21 Let each s_i be uniformly distributed as in Example 6.20. Let each n_i have a two-sided exponential distribution

$$f_n(n) = \frac{1}{\sqrt{2}\sigma} e^{-(\sqrt{2}/\sigma)|n|}, \qquad -\infty < n < \infty. \tag{A}$$

Again we consider Equation 6.21-4 and find

$$\begin{aligned}\int_{-\infty}^{\infty} f_n(x-s)f_s(s)\,ds &= \frac{\sqrt{2}}{4a\sigma}\int_{-a}^{a} e^{-(\sqrt{2}/\sigma)|x-s|}\,ds \\ &= \begin{cases} \dfrac{1}{2a}(1 - e^{-\alpha}\cosh\xi), & |\xi| \le \alpha, \\ \dfrac{1}{2a} e^{-|\xi|}\,\text{sine}\,\alpha, & |\xi| \ge \alpha, \end{cases}\end{aligned} \tag{B}$$

where

$$\alpha = \frac{a\sqrt{2}}{\sigma}, \qquad \xi = \frac{x\sqrt{2}}{\sigma}. \tag{C}$$

In the same way we find

$$\begin{aligned}&\int_{-\infty}^{\infty} s f_n(x-s)f_s(s)\,ds \\ &\quad = \frac{\sqrt{2}}{4a\sigma}\int_{-a}^{a} s\, e^{-(\sqrt{2}/\sigma)|x-s|}\,ds \\ &\quad = \begin{cases} \dfrac{1}{2\alpha}[\xi - e^{-\alpha}(\alpha+1)\sinh\xi], & |\xi| \le \alpha, \\ \dfrac{\text{sgn}\,\xi\, e^{-|\xi|}}{4\alpha}[e^{\alpha}(\alpha-1) + e^{-\alpha}(\alpha+1)], & |\xi| \ge \alpha, \end{cases}\end{aligned} \tag{D}$$

where sgn $(\cdot)$ is the signum function

$$\text{sgn}\, x = \begin{cases} 1, & x > 0, \\ 0, & x = 0, \\ -1, & x < 0. \end{cases} \tag{E}$$

Thus the nonlinearity becomes

$$g(x) = \begin{cases} \dfrac{\sigma}{\sqrt{2}} \dfrac{\xi - e^{-\alpha}(\alpha + 1)\sinh \xi}{1 - e^{-\alpha}\cosh \xi}, & |\xi| \le \alpha, \\ \dfrac{\sigma}{\sqrt{2}}(\alpha \coth \alpha - 1)\operatorname{sgn} \xi, & |\xi| \ge \alpha. \end{cases} \tag{F}$$

• •

PROBLEMS

1. Consider a square-law device with an input which is a normally distributed random variable x with mean u and variance σ^2. Find:
   (a) the probability density function of the output y;
   (b) the mean and variance of y.
2. A square-law device has as its input a normally distributed random process $x(t)$ with zero mean. If $x(t)$ is wide-sense stationary with the spectral density shown, find the autocorrelation function and spectral density of the output $y(t)$.

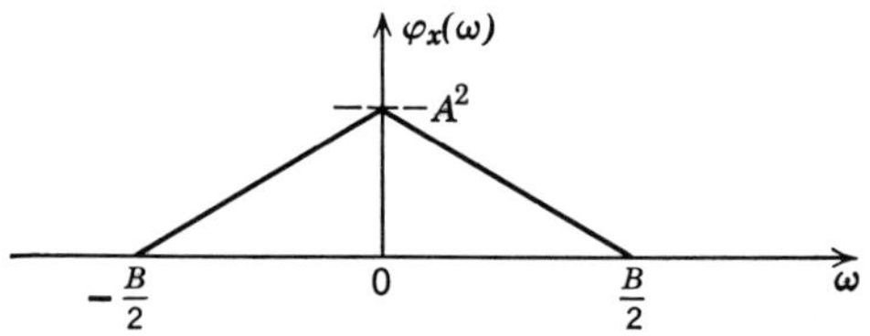

3. Consider the synchronous receiver shown.

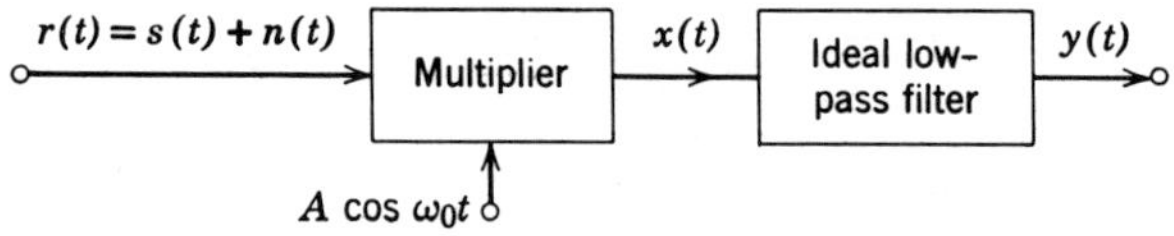

The signal $s(t)$ is a randomly modulated sine wave given by $s(t) = q(t) \cos \omega_0 t$ where $q(t)$ is a wide-sense stationary process with a spectral density given by

$$\varphi_q(\omega) = \begin{cases} Q, & |\omega| \le \dfrac{B}{2}, \\ 0, & |\omega| > \dfrac{B}{2}. \end{cases}$$

The noise $n(t)$ is a wide-sense stationary random process with spectral density

$$\varphi_n(\omega) = \begin{cases} N, & |\omega \pm \omega_0| \le \dfrac{B}{2}, \\[2ex] 0, & |\omega \pm \omega_0| > \dfrac{B}{2}. \end{cases}$$

The low-pass filter has a transfer function as shown.

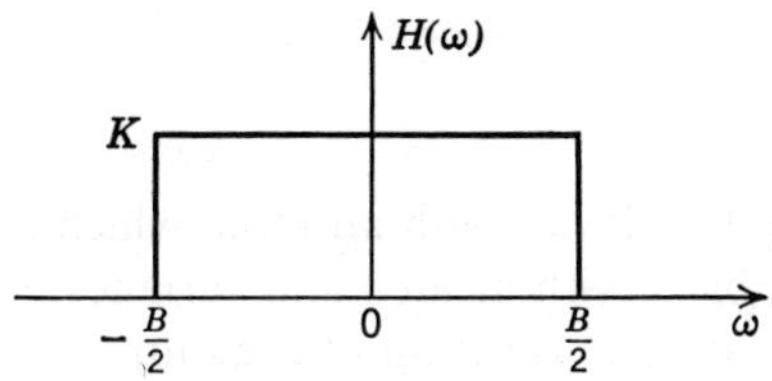

Find:

(a) the time average of $R_x(\tau)$ and $\varphi_x(\omega)$; that is, $\langle R_x(\tau)\rangle$ and $\langle \varphi_x(\omega)\rangle$;

(b) $\langle R_y(\tau)\rangle$ and $\langle \varphi_y(\omega)\rangle$;

(c) the mean-squared error $E\{[y(t) - q(t)]^2\}$ and the value of K that minimizes the error.

4. Consider a limiter having the transfer characteristic as shown. Find the probability density function of y in terms of the density of x. Find the mean and variance of y if x is normally distributed with zero mean and variance σ^2.

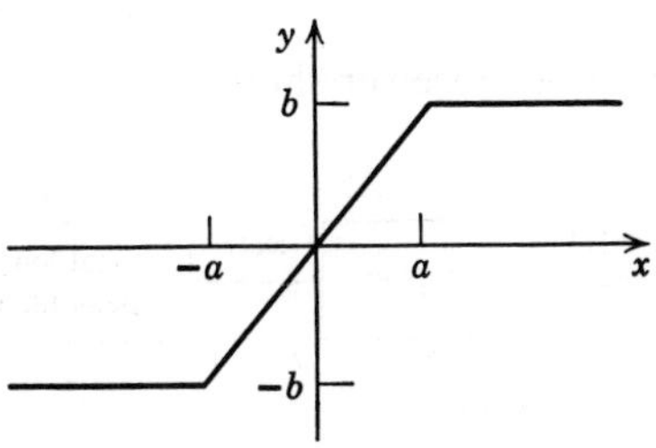

5. Consider the following system:

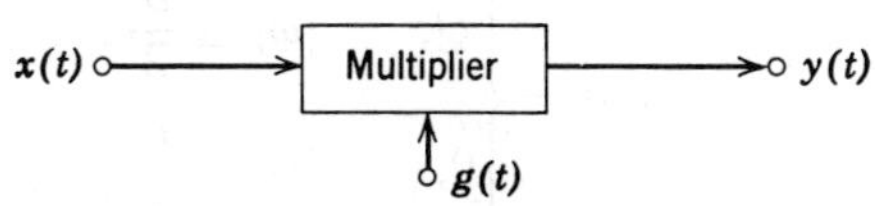

where $g(t)$ is the square wave shown. The phase θ is a uniformly distributed random variable on $(0, 2\pi)$:

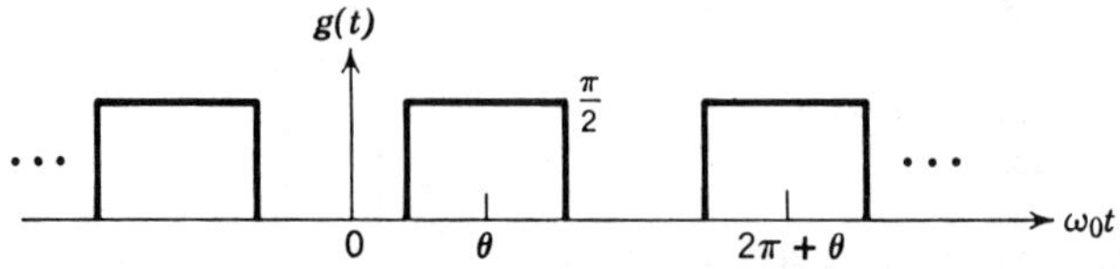

The wide-sense stationary random process $x(t)$ has the spectral density

$$\varphi_x(\omega) = \begin{cases} A^2, & |\omega| < \dfrac{\omega_0}{4}, \\ 0, & \text{otherwise.} \end{cases}$$

The input $x(t)$ and θ are independent. Find the spectra density of $y(t)$.

6. Using Price's Theorem, find the output autocorrelation function of the smooth limiter of Section 6.13 when the input is the same as in Section 6.13.
7. Repeat problem 6 for the half-wave linear rectifier with characteristic

$$y = \begin{cases} cx, & x > 0, \\ 0, & x \le 0. \end{cases}$$

8. Consider an arbitrary zero-memory nonlinearity with a narrow-band input. Show that the odd part of the nonlinearity does not yield a low-frequency output.
9. Consider an arbitrary zero-memory nonlinearity with an input consisting of a gaussian signal and an additive gaussian noise. Let the signal and noise be independent with zero means. Define the output signal as the total output when the input is signal only, and define the output noise similarly. Find the cross-correlation between the output signal and the output noise.
10. The density functions of two independent random variables x and y are given by

$$W_x(x) = \begin{cases} \frac{1}{3}, & 1 \le x \le 4, \\ 0, & \text{elsewhere,} \end{cases} \qquad W_y(y) = \begin{cases} e^{-(y-2)}, & 2 \le y < \infty, \\ 0, & \text{elsewhere.} \end{cases}$$

Find the density function of the new random variable $z = x/y$.

11. The density functions for two independent random variables x and y are given by

$$W_x(x) = k_1\, e^{-\alpha|x-1|}, \qquad -\infty < x < \infty,$$
$$W_y(y) = k_2\, e^{-\beta|x-2|}, \qquad -\infty < y < \infty,$$

where $\alpha < \beta$ and k_1 and k_2 are normalizing constants. Find k_1 and k_2 and the density function of the new random variable $z = x + y$.

REFERENCES

1. L. A. Zadeh, "Optimum Nonlinear Filters," *J. Appl. Phys.*, **24**, 1953, 396–404.
2. B. O. Pierce, *A Short Table of Integrals*, Blaisdell, New York, 1956, p. 44, no. 302.
3. *Ibid.*, p. 12, no. 70.
4. *Ibid.*, p. 84, no. 667.
5. *Ibid.*, p. 101, no. 852.
6. *Ibid.*, p. 136.
7. M. Abramowitz and I. Stegun (Eds.), *Handbook of Mathematical Functions*, National Bureau of Standards, Applied Mathematics Series 55, U.S. Govt. Printing Office, Washington, D. C., 1964, pp. 255–258.
8. D. Middleton, *An Introduction to Statistical Communication Theory*, McGraw-Hill, New York, 1960, Sect. 2.3, Chap. 5, and App. 1.
9. Pierce, *op. cit.*, p. 98, no. 821.
10. R. F. Baum, "The Correlation Function of Smoothly Limited Gaussian Noise," *IRE Trans. on Information Theory*, IT-3, September 1957, 193–197.
11. G. O. Young, "Random Function Probability Distributions after a Nonlinear Filter," *IRE WESCON Convention Record*, Pt. 4, August 1958, 164–172.
12. E. T. Whittaker and G. N. Watson, *A Course of Modern Analysis*, Cambridge Univ. Press, New York, 1952, p. 224.
13. J. F. Barrett and D. G. Lampard, "An Expansion of Some Second-Order Probability Distributions and Its Application to Noise Problems," *IRE Trans. on Information Theory*, IT-1, March 1955, 10–15.
14. P. M. Morse and H. Feshbach, *Methods of Theoretical Physics*, McGraw-Hill, New York, 1953, p. 786.
15. J. J. Bussgang, "Cross Correlation Functions of Amplitude Distorted Gaussian Signals," *MIT Res. Lab. of Electronics Tech. Rept.*, 216, March 1952.
16. Abramowitz and Stegun, *op. cit.*, p. 556, Eq. 15.1.1.
17. *Ibid.*, p. 556, Eq. 15.1.6.
18. Whittaker and Watson, *op. cit.*, Chap. 11.
19. Pierce, *op. cit.*, p. 98, no. 818.
20. Abramowitz and Stegun, *op. cit.*, p. 376, no. 9.6.34.
21. *Ibid.*, p. 486, no. 11.4.16.
22. *Ibid.*, p. 256, no. 6.1.15.
23. Middleton, *op. cit.*, p. 1079, Eq. A.1.54.
24. Abramowitz and Stegun, *op. cit.*, p. 508, Eq. 13.5.1.
25. *Ibid.*, pp. 255–256.
26. Middleton, *op. cit.*, Chaps. 5, 9, 12–15.

ADDITIONAL READING

Books

(1950) Lawson, J. L., and G. E. Uhlenbeck. *Threshold Signals*, MIT Rad. Lab. Series, Vol. 24, McGraw-Hill, New York.

(1953) Johnson, R. A., and D. Middleton. "Measurement of Correlation Functions of Modulated Carriers and Noise following a Nonlinear Device," in W. Jackson (Ed.), *Communication Theory*, Butterworth, London.

(1954) Wax, N. (Ed.). *Selected Papers on Noise and Stochastic Processes*, Dover, New York.

(1958) Davenport, W. B., and W. L. Root. *An Introduction to the Theory of Random Signals and Noise*, McGraw-Hill, New York, Chaps. 12, 13.

Wiener, N. *Nonlinear Problems in Random Theory*, M.I.T. Press, Cambridge, Mass.

(1962) Deutsch, R. *Nonlinear Transformations of Random Processes*, Prentice-Hall, Englewood Cliffs, N. J.

Van Trees, H. L. *Synthesis of Optimum Nonlinear Control Systems*, M.I.T. Press, Cambridge, Mass., Chap. 2.

(1963) Strantonovich, R. L. *Topics in the Theory of Random Noise*, Vols. I and II (trans. from Russian by R. A. Silverman), Gordon and Breach, New York.

(1964) Krasnosel'skii, M. A. *Topological Methods in the Theory of Nonlinear Integral Equations*, Macmillan, New York.

(1965) Kuznetsov, P. I., R. L. Strantonovich, and V. I. Tikhonov. *Nonlinear Transformations of Stochastic Processes* (trans. from Russian), Pergamon, London.

Pervozvanskii, A. A. *Random Processes in Nonlinear Control Systems* (trans. from Russian), Academic, New York.

Periodicals

(1936) Craig, C. C. "On the Frequency Function of *xy*," *Ann. Math. Stat.*, **1**, 1–15.

(1942) Ragazzini, J. R. "The Effect of Fluctuation Voltages on the Linear Detector," *Proc. IRE*, **30**, 277.

Wiener, N. "Response of a Non-Linear Device to Noise," *MIT Rad. Lab. Rept.*, **168**, April.

(1944) Bennett, W. R. "Response of a Linear Rectifier to Signal and Noise," *J. Acoust. Soc. Am.*, **15**, 165–172.

(1945) Rice, S. O. "Mathematical Analysis of Random Noise," *Bell Syst. Tech. J.*, **24**.

Young, G. O., and B. Gold. "Effect of Limiting on the Information Content of Noisy Signals," *IRE Convention Record*, Pt. 4, p. 76.

(1946) Kac, M. "On the Average of a Certain Wiener Functional and a Related Theorem in the Calculus of Probability," *Am. Math. Soc. Trans.*, **59**, 401–414.

Middleton, D. "The Response of Biased, Saturated Linear and Quadratic Rectifiers to Random Noise," *J. Appl. Phys.*, **17**, 778–801.

(1947) Aroian, L. A. "The Probability Function of the Product of Two Normally Distributed Variables," *Ann. Math. Stat.*, **18**, 265–271.

Bennett, W. R. "The Biased Ideal Rectifier," *Bell Syst. Tech. J.*, **26**, 139–169.

Cameron, R. H., and W. T. Martin. "The Orthogonal Development of Nonlinear Functionals in Series of Fourier-Hermite Functionals," *Ann. Math. Stat.*, **48**, 385–392.

Jastram, P. "The Effect of Nonlinearity and Frequency Distortions on the Amplitude Distribution for Stationary Random Processes," doctoral dissertation, Univ. of Michigan, Ann Arbor.

Kac, M., and A. J. F. Seigert. "On the Theory of Noise in Radio Receivers with Square Law Detectors," *J. Appl. Phys.*, **18**, 383.

(1948) Bennett, W. R. "Spectra of Quantized Signals," *Bell Syst. Tech. J.*, **27**, 446–472.

Fubini, E. G., and D. C. Johnson. "Signal-to-Noise Ratio in AM Modulated Receivers," *Proc. IRE*, **36**, 146.

Middleton, D. "Rectification of a Sinusoidally Modulated Carrier in the Presence of Noise," *Proc. IRE*, **36**, 1467.

Middleton, D. "Some General Remarks in the Theory of Noise through Nonlinear Devices," *Quart. Appl. Math.*, **5**, 445–498.

Rice, S. O. "Properties of a Sine Wave Plus Random Noise," *Bell Syst. Tech. J.* **27**, 109–157.

(1950) Middleton, D. "The Spectrum of Frequency-Modulated Waves after Reception in Random Noise, II," *Quart. Appl. Math.*, **8**, 59–80.

(1951) Burgess, R. E. "The Rectification and Observation of Signals in the Presence of Noise," *Phil. Mag.* Series 7, **42**, 475–503.

Ikehara, S. "A Method of Wiener in a Non-Linear Circuit," *MIT Rad. Lab. Rept.*, **217**, December.

Middleton, D. "The Distribution of Energy in Randomly Modulated Waves," *Phil. Mag.*, Series 7, **42**, 689–707.

(1952) Arthur, G. R. "A Note on the Approach of Narrow-Band Noise after a Nonlinear Device to a Normal Probability Density," *J. Appl. Phys.*, **23**, October.

Johnson, R. A. "Rectification of Signals and Random Noise by a Linear Detector," *Cruft Lab. Tech. Rept.*, **144**, March; **145**, May.

Middleton, D. "On the Distribution of Energy in Noise-and-Signal Modulated Waves, II: Simultaneous Amplitude and Noise Modulation," *Quart. Appl. Math.*, **10**, 37–56.

Slattery, T. G. "The Detection of a Sine Wave in the Presence of Noise by the Use of Nonlinear Filter," *Proc. IRE*, **40**, 1232–1236.

Tucker, G. D. "Linear Rectifiers and Limiters," *Wireless Engr.* (London), **29**, 128.

(1953) Blachman, N. M. "The Output Signal-to-Noise Ratio of a Power-Law Device," *J. Appl. Phys.*, **24**, 783.

Davenport, W. B., Jr. "Signal-to-Noise Ratios in Band-Pass Limiters," *J. Appl. Phys.*, **24**, 720.

Emerson, R. C. "First Probability Densities for Receivers with Square-Law Detectors," *J. Appl. Phys.*, **24**, 1168.

Stone, N., and D. Middleton. "Full-wave Detection of Signals in Noise," *Cruft Lab. Tech. Rept.*, **182**, August.

Weinberg, L., and L. G. Kraft. "Measurement of Detector Output Spectra by Correlation Methods," *Proc. IRE*, **41**, 1157.

Zadeh, L. A. "Optimum Nonlinear Filters for the Extraction and Detection of Signals," *IRE Convention Record*, Pt. 8, 57–65.

Zadeh, L. A. "A Contribution to the Theory of Nonlinear Systems," *J. Franklin Inst.*, **255**, 387–408.

(1954) Arthur, G. R. "The Statistical Properties of the Output of a Frequency Sensitive Device," *J. Appl. Phys.*, **25**, 4.

Deutsch, R. "Detection of Modulated Noise-like Signals," *IRE Trans. on Information Theory*, PGIT-3, March, 107–122.

Hoffman, W. C. "The Joint Distribution of *n* Successive Outputs of a Linear Detector," *J. Appl. Phys.*, **25**, 1006–1007.

Lampard, D. G. "A New Method of Determining Correlation Functions of Stationary Time Series," *IEE Monogr.*, No. 4, August 16.

Meyer, M. A., and D. Middleton. "On the Distributions of Signals and Noise after Rectification and Filtering," *J. Appl. Phys.*, **25**, 1037.

Siegert, A. J. F. "Passage of Stationary Processes through Linear and Non-Linear Devices," *IRE Trans. on Information Theory*, PGIT-3, March, 4–25.

(1955) Barrett, J. F. "Application of the Theory of Functionals to Communication Problems," *Eng. Lab. Rept.*, Cambridge Univ., England.

Deutsch, R. "On a Method of Wiener for Noise through Non-Linear Devices," *IRE Convention Record*, Pt. 4, 186–192.

Thomson, W. E. "The Response of a Nonlinear System to Random Noise," *Proc. IEE*, **102**, Pt. C, 46–48.

(1956) Balakrishnan, A. V., and R. Drenick. "On Optimum Non-Linear Extraction and Coding Filters," *IRE Trans. on Information Theory*, IT-2, September, 166–171.

Bennett, W. R. "Methods of Solving Noise Problems," *Proc. IRE*, **44**, 609–638.

Bose, A. "A Theory of Nonlinear Systems," *MIT Res. Lab. of Electronics Tech. Rept.*, **309**, May 15.

Campbell, L. L. "Rectification of Two Signals in Random Noise," *IRE Trans. on Information Theory*, IT-2, December, 119.

Deutsch, R. "On the Distribution of Nonlinear Circuit Transformations of a Markov Process," *Proc. of the Symposium on Nonlinear Circuit Analysis*, Polytechnic Institute of Brooklyn, N.Y., Vol. 6, 243–253.

Deutsch, R. "Piecewise Quadratic Detector," *IRE Convention Record*, Pt. 4, 15.

Fellows, G. E., and D. Middleton. "An Experimental Study of Intensity Spectra after Half-Wave Rectification of Signals in Noise," *Proc. IEE* (London), Mon.160R, January.

Horton, C. W. "On the Use of Gram-Charlier Series to Represent Noise," *J. Appl. Phys.*, **27**, 350–355.

Lampard, D. G. "The Probability Distribution for the Filtered Output of a Multiplier Whose Inputs Are Correlated, Stationary, Gaussian, Time-Series," *IRE Trans. on Information Theory*, IT-2, March, 4.

McFadden, J. A. "The Correlation Function of a Sine Wave Plus Noise after Extreme Clipping," *IRE Trans. on Information Theory*, IT-2, June, 82–83.

(1957) Brown, J. L. "On a Cross-Correlation Property for Stationary Random Processes," *IRE Trans. on Information Theory*, IT-3, March, 28–31.

Darling, D. A., and A. J. F. Siegert. "A Systematic Approach to a Class of Problems in the Theory of Noise and Other Random Phenomena—Part I," *IRE Trans. on Information Theory*, IT-3, March, 32–37.

Nuttall, A. H. "Shaping Correlation Functions with Nonlinear-No-Memory Networks," *Quart. Prog. Rept.*, MIT Res. Lab. of Electronics, July 15, 68.

Siegert, A. J. F. "A Systematic Approach to a Class of Problems in the Theory of Noise and Other Random Phenomena—Part II, Examples," *IRE Trans. on Information Theory*, IT-3, March, 38–43.

Zadeh, L. A. "On the Representation of Nonlinear Operators," *IRE WESCON Convention Record*, Pt. 2, 105–113.

(1958) Brilliant, M. B. "Theory of Analysis of Nonlinear Systems," *MIT Res. Lab. of Electronics Tech. Rept.*, **345**, March, 3.

Dugundji, J. "Envelopes and Pre-Envelopes of Real Waveforms," *IRE Trans. on Information Theory*, IT-4, March, 53.

Leipnik, R. "The Effect of Instantaneous Nonlinear Devices on Cross-Correlation," *IRE Trans. on Information Theory*, IT-4, January, 73–76.

Nuttall, A. H. "Theory and Application of the Separable Class of Random Processes," *MIT Res. Lab. of Electronics Tech. Rept.*, 343, May.

Price, R. "A Useful Theorem for Nonlinear Devices Having Gaussian Inputs," *IRE Trans. on Information Theory*, IT-4, June, 69–72.

Sherman, S. "Non-Mean-Square Error Criteria," *IRE Trans. on Information Theory*, IT-4, September, 125–126.

Siegert, A. J. F. "A Systematic Approach to a Class of Problems in the Theory of Noise and Other Random Phenomena—Part III, Examples," *IRE Trans. on Information Theory*, IT-4, March, 4–14.

(1959) Brown, J. L. "On the Cross Correlation between Two Noisy Channels," *IRE Trans. on Information Theory*, IT-5, June, 77.

Brown, W. M. "Some Results on Noise through Circuits," *IRE Trans. on Information Theory*, IT-5, May, 217.

Chuang, K., and L. F. Kazda. "A Study of Nonlinear Systems with Random Inputs," *AIEE Trans.*, **78**, Pt. II, *Appl. and Ind.*, 100–105.

George, D. A. "Continuous Nonlinear Systems," *MIT Res. Lab. of Electronics Tech. Rept.*, 355, July 24.

Keilson, J., N. D. Mermin, and P. Bello. "A Theorem on Cross Correlation between Noisy Channels," *IRE Trans. on Information Theory*, IT-5, June, 77.

(1960) Balakrishnan, A. V. "On a Characterization of Processes for Which Optimal Mean-Square Systems are of Specified Form," *IRE Trans. on Information Theory*, IT-6, September, 490–500.

Blachman, N. M. "The Effect of a Limiter upon Signals in the Presence of Noise," *IRE Trans. on Information Theory*, IT-6, March, 52.

Chesler, D. A. "Nonlinear Systems with Gaussian Inputs," *MIT Res. Lab. of Electronics Tech. Rept.*, 366, February 15.

Hause, A. D. "Nonlinear Least-Squares Filtering and Frequency Modulation," *MIT Res. Lab. of Electronics Tech. Rept.*, 371, August 25.

Leland, H. E. "Input-Output Cross-Correlation Functions for Some Memory Type Nonlinear Systems with Gaussian Inputs," *AIEE Trans.*, **79**, Pt. II, *Appl. and Ind.*, 219–223.

Lubbock, J. F. "The Optimization of a Class of Nonlinear Filters," *Proc. IEE* (London), **107**, Pt. C, 60–74.

Marcum, J. I. "A Statistical Theory of Target Detection by Pulsed Radar," *IRE Trans. on Information Theory*, IT-6, April.

Max, J. "Quantizing for Minimum Distortion," *IRE Trans. on Information Theory*, IT-6, March, 7.

Pugachev, V. S. "Statistical Theory of Systems Reducible to Linear," *IRE Trans. on Circuit Theory*, CT-7, December, 506–513.

Smets, H. B. "Analysis and Synthesis of Nonlinear Systems," *IRE Trans. on Circuit Theory*, CT-7, December, 459–469.

Zadeh, L. A. "A Contribution to the Theory of Nonlinear Systems," *J. Franklin Inst.*, **255**, 387–408.

Zakai, M. "Second-Order Properties of the Pre-Envelope and Envelope Processes," *IRE Trans. on Information Theory*, IT-6, December, 556.

Zames, G. "Nonlinear Operators for System Analysis," *MIT Res. Lab. of Electronics Tech. Rept.*, 370, August 25.

(1961) Cahn, C. R. "A Note on Signal-to-Noise Ratio in Band-Pass Limiters," *IRE Trans. on Information Theory*, IT-7, January.

Clavier, P. A. "Construction of Probability Densities from Their Moments," *Proc. IRE*, **49**, 1580.

Landau, H. J., and W. L. Miranker. "The Recovery of Distorted Band-Limited Signals," *J. Math. Anal. and Appl.*, **2**, 97–104.

Roe, G. M., and G. M. White. "Probability Density Functions for Correlators with Noisy Reference Signals," *IRE Trans. on Information Theory*, IT-7, January, 13.

Thomas, J. B., and E. Wong. "The Distribution of Quadratic Functionals," *Quart. Appl. Math.*, **19**, 150–153.

Tung, F., and R. S. Schwarz. "Optimum Nonlinear Filters for Quantized Inputs," *IRE Trans. on Information Theory*, IT-7, October, 257–265.

Wolf, A. A. "Some Recent Advances in the Analysis and Synthesis of Nonlinear Systems," *AIEE Trans.*, **80**, Pt. II, *Appl. and Ind.*, 289–300.

(1962) Bluestein, L. I., and R. J. Schwarz. "Optimum Zero-Memory Filters," *IRE Trans. on Information Theory*, IT-8, October, 337–341.

Grumet, A. "Demodulation Effect of an Envelope Detector at Low Signal-to-Noise Ratios," *Proc. IRE*, **50**, October.

Katznelson, J., and L. A. Gould. "The Design of Nonlinear Filters and Control Systems, Part I, Information and Control," **5**, 108–143.

Schetzen, M. "Some Problems in Nonlinear Theory," *MIT Res. Lab. of Electronics Tech. Rept.*, 390, July 6.

Sinha, N. K. "Optimum Nonlinear Filters for Random Signals," *IRE International Convention Record*, Vol. 10, Pt. 2, 88–91.

Tung, F. "Optimal Message for Zero-Memory Systems with Additive Gaussian Noise," *Proc. IRE*, **50**, 481.

Wong, E., and J. B. Thomas. "On Polynomial Expansions of Second-Order Distributions," *J. Soc. Ind. Appl. Math.*, **10**, 507–576.

(1963) Barrett, J. F. "The Use of Functionals in the Analysis of Nonlinear Physical Systems," *J. Elec. Control*, **15**, 567–615.

Jones, J. J. "Hard-Limiting of Two Signals in Random Noise," *IEEE Trans. on Information Theory*, IT-9, January, 34–42.

Richard, R. H., and W. C. Gore. "A Nonlinear Filter for Non-Gaussian Interference," *IEEE Trans. on Communication Systems*, CS-11, December, 436–443.

Sandberg, I. W. "On the Properties of Some Systems That Distort Signals—I," *Bell Syst. Tech. J.*, **42**, 2033–2046.

Shutterly, H. B. "General Results in the Mathematical Theory of Random Signals and Noise in Nonlinear Devices," *IEEE Trans. on Information Theory*, IT-9, April, 74–84.

(1964) Algazi, V. R. "A Study of the Performance of Linear and Nonlinear Filters," *MIT Res. Lab. of Electronics Tech. Rept.*, 420, June 22.

Bonet, G. "Nonlinear, Zero-Memory Transformations of Random Signals," *Ann. Telecom.*, **19**, 203–220.

Kushner, H. J. "On the Differential Equations Satisfied by Conditional Probability Densities of Markov Processes with Applications," *RIAS Tech. Rept.*, 64-4, March.

Sandberg, I. W. "On the Properties of Some Systems That Distort Signals—II," *Bell Syst. Tech. J.*, **43**, 91–112.

Wonham, W. M. "Some Applications of Stochastic Differential Equations to Optimal Nonlinear Filtering," *RIAS Tech. Rept.*, 64-3, February.

(1965) Bello, P. A. "On the rms Bandwidth of Nonlinearly Envelope Detected Narrow-Band Gaussian Noise," *IEEE Trans. on Information Theory*, IT-11, April, 236–239.

Fine, T. "Optimum Mean-Square Quantization of a Noisy Input," *IEEE Trans. on Information Theory*, IT-11, April, 293–294.

Katznelson, J., and L. A. Gould. "A Spectrum Factorization Method for the Calculation of Nonlinear Filters of the Volterra Type," *Information and Control*, **8**, 239–250.

Lechleider, J. W. "Summary of Some Results on Optimum Nonlinear Operations in PAM," *Proc. IEEE*, **53**, 744–745.

Lee, Y. W. "Some Aspects of the Wiener Theory of Nonlinear Systems," *Proc. NEC*, **21**, 759–764.

Lindsey, W. C. "Optimum Coherent Linear Demodulation," *IEEE Trans. on Communication Technology*, COM-13, June, 183–190.

Schetzen, M. "Determination of Optimum Nonlinear Systems for Generalized Error Criteria Based on the Use of Gate Functions," *IEEE Trans. on Information Theory*, IT-11, January, 117–125.

Wierwille, W. W. "A Theory and Method for Correlation Analysis of Nonstationary Signals," *IEEE Trans. on Electronic Computers*, EC-4, December.

(1966) Lal, M. "A Note on Optimization of a Class of Nonlinear Systems," *IEEE Trans. on Automatic Control*, AC-11, January, 137–138.

Van Vleck, J. H., and D. Middleton. "The Spectrum of Clipped Noise," *Proc. IEEE*, **54**, 2–19.

(1967) Brown, J. L. Jr. "A Generalized Form of Price's Theorem and Its Converse," *IEEE Trans. on Information Theory*, IT-13, January, 27–30.

Pawula, R. F. "A Modified Version of Price's Theorem," *IEEE Trans. on Information Theory*, IT-13, April, 285–288.

Abramson, N. "Nonlinear Transformations of Random Processes," *IEEE Trans. on Information Theory*, IT-13, July, 502–505.

Fisher, J. R., and E. B. Stear. "Optimal Nonlinear Filtering for Independent Increment Processes," *IEEE Trans. on Information Theory*, IT-13, October, 558–578.

(1968) Haddad, A. H., and J. B. Thomas. "On Optimal and Suboptimal Nonlinear Filters for Discrete Inputs," *IEEE Trans. on Information Theory*, IT-14, January, 16–21.

Blachman, N. M. "The Signal × Signal, Noise × Noise, and Signal × Noise Output of a Nonlinearity," *IEEE Trans. on Information Theory*, IT-14, January, 21–27.

Blachman, N. M. "The Uncorrelated Output Components of a Nonlinearity," *IEEE Trans. on Information Theory*, IT-14, March, 250–255.

Gorman, C. D., and J. Zaborozky. "Functional Calculus in the Theory of Nonlinear System with Stochastic Signals," *IEEE Trans. on Information Theory*, IT-14, July, 528–531.

Haddad, A. H. "On a Nonlinear Feedback Filtering System," *IEEE Trans. on Information Theory*, IT-14, July, 532–535.

7

Modulation Theory and Systems

7.1 Introduction

A central problem in communication and control is the transfer of information originating in a source to some destination by means of electrical signals of various kinds. The typical situation is illustrated in Figure 7.1(a) where the *source signal* (sometimes called the *intelligence* or *message*) is denoted by $a(t)$. This signal is transmitted from an information source S through a communication channel C to the destination D. In the general case what is received is not exactly the original signal $a(t)$ but a distorted version or estimate $a^*(t)$. This distortion may be caused by imperfections in the channel transmission characteristics or may be the result of noises or other disturbances.

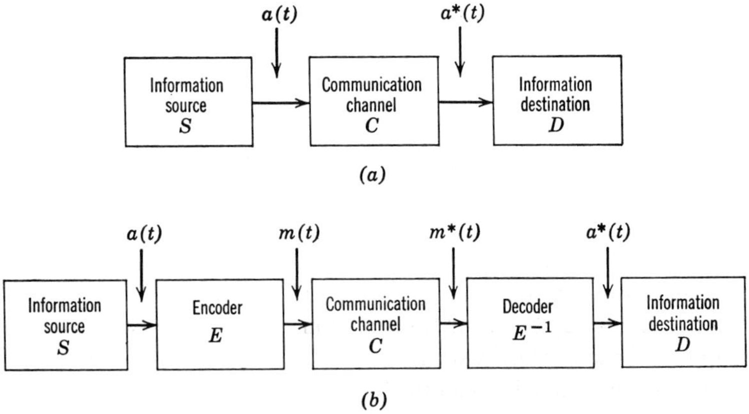

Figure 7.1 *Block diagrams of elementary communication systems:* (a) *the basic communication system;* (b) *a communication system with coding and decoding*

In many situations the message $a(t)$ may not be in suitable form for transmission over the channel C and a system such as that shown in Figure 7.1(b) must be used. Here an encoder E is provided to produce a coded signal $m(t)$ for transmission. A decoder E^{-1} is then required at the receiving end of the channel to accept the distorted signal $m^*(t)$ and yield the estimate $a^*(t)$. As before only estimates of $m(t)$ and $a(t)$ can be obtained due to channel limitations. Of course, distortion may take place in the encoder and decoder also. Sometimes, however, it will be convenient to assume that these are ideal and to consider that all undesirable signal changes occur in the channel.

The message signals used in communication and control systems are usually limited in frequency range to some maximum frequency $f_m = \omega_m/2\pi$ cycles per second. This frequency is typically in the range of a few cycles per second for control systems and moves upward to a few megacycles per second for television video signals. In addition the bandwidth of the signal is often of the order of this maximum frequency so that the signal spectrum is approximately low pass in character. It frequently happens that the transmission of such a spectrum through a given communication channel is inefficient or impossible.

Modulation may be defined as the modification of one signal, called the *carrier*, by another, called the *modulation, modulating signal,* or *message signal.* The result of the modulation process is a *modulated wave* or signal. In most cases a frequency shift is one of the results. There are a number of reasons for producing modulated waves. The following list gives some of the major ones:

1. *Frequency translation for efficient antenna design.* It may be necessary to transmit the modulating signal through space as electromagnetic radiation. If the antenna used is to radiate an appreciable amount of power, it must be large compared to the signal wavelength. Thus translation to higher frequencies (and hence to smaller wavelengths) will permit antenna structures of reasonable size and cost at both the transmitter and receiver.
2. *Frequency translation for ease of signal processing.* It may be easier to amplify and/or shape a signal in one frequency range than in another. For example, a d-c signal may be converted to a-c, amplified, and converted back again.
3. *Frequency translation to assigned location.* A signal may be translated to an assigned frequency band for transmission or radiation as, for example, in commercial radio broadcasting.
4. *Changing bandwidth.* The bandwidth of the original message signal may be increased or decreased by the modulation process. In general, decreased bandwidth will result in channel economies at the cost of fidelity. On the other hand, increased bandwidth will be accompanied by increased

immunity to channel disturbances as, for example, in wideband frequency modulation.

5. *Multiplexing*. It may be necessary or desirable to transmit several signals occupying the same frequency range or the same time range over a single channel. Various modulation techniques allow the signals to share the same channel and yet be recovered separately. Such techniques are given the generic name of *multiplexing*. As will be discussed later, multiplexing is possible in either the frequency domain (frequency-domain multiplexing, FDM) or in the time domain (time-domain multiplexing, TDM). As a simple example, the signals may be translated in frequency so that they occupy separate and distinct frequency ranges as mentioned in item 2 above.

Thus it can be seen that, in a general way, the process of modulation can be considered as a form of encoding used to match the message signal arising from the information source to the communication channel. At the same time it is generally true that the channel itself has certain undesirable characteristics resulting in distortion of the signal during transmission. A part of such distortion can be accounted for frequently by postulating noise disturbances in the channel. These noises may be additive and, in addition, may affect the modulated wave in a more complicated fashion, although it is usually sufficient (and much simpler) to assume additive noise only. Also, the received signal must be decoded (demodulated) to recover the original signal.

In view of the discussion of this last paragraph, it is convenient to change the block diagram of Figure 7.1(b) to that shown in Figure 7.2 so that the modulation and demodulation processes are shown explicitly. The notations used in this figure will be retained for the remainder of this chapter. Specifically the waveform received at the demodulator (receiver) will be denoted by $r(t)$ where

$$r(t) = m^*[t, a(t), p(t)] + n(t) \tag{7.1-1}$$

where $a(t)$ is the original message signal, $m[t, a(t)]$ is the modulated wave, $m^*[t, a(t), p(t)]$ is a corrupted version of $m[t, a(t)]$, and $p(t)$ and $n(t)$ are noises whose characteristics depend on the channel. Unless absolutely necessary for an accurate characterization of the channel, we will assume that $p(t) \equiv 0$ to avoid the otherwise complicated analysis that results.

As usual, the aim is to find modulators M and demodulators M^{-1} which make $a^*(t)$ a "good" estimate of the message signal $a(t)$. It should be emphasized that M^{-1} is not uniquely specified by M; for example, it is not intended to imply that $MM^{-1} = 1$. The form of the demodulator, for a given modulator, will depend on the characteristics of the message signal $a(t)$ and the channel as well as on the criterion of "goodness of estimation" used.

The remainder of this chapter will be taken up with a study of various forms of modulation and demodulation, their principal characteristics, their

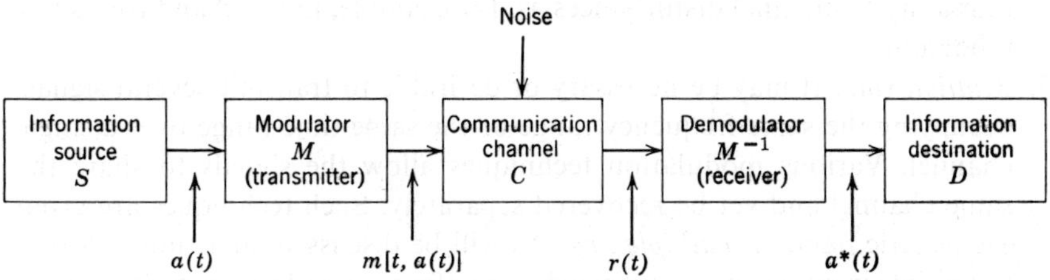

Figure 7.2 *A communication system involving modulation and demodulation:*

$a(t)$ = *message signal (intelligence)*
$m[t, a(t)]$ = *modulated wave*
$r(t) = m^*[t, a(t), p(t)] + n(t)$ = *received wave*
$p(t)$ = *noise affecting the modulated wave*
$n(t)$ = *additive noise*
$a^*(t)$ = *the estimate of the original message signal*

behavior in conjunction with noisy channels, and their advantages and disadvantages. Various modulation schemes will be considered for three types of *modulation signals* $a(t)$:

1. periodic,
2. pulsed,
3. random;

and for two types of *carriers*:

1. continuous,
2. pulsed.

PART 1—ELEMENTS OF CONTINUOUS SIGNAL THEORY

In the interest of making this chapter as self-contained as possible, we repeat here some of the elementary notions of the analysis of continuous signals.

7.2 Transforms and Convolutions

The Fourier transform and its properties are discussed in some detail in Section 6 of Chapter 3 and in the references given at the end of that chapter. If a signal $f(t)$ has finite energy and is reasonably well behaved, then its (direct) Fourier transform $F(\omega)$ exists and is given by

$$F(\omega) = \int_{-\infty}^{\infty} f(t)e^{-j\omega t}\,dt. \tag{7.2-1}$$

Also the original signal $f(t)$ may be obtained from the (inverse) Fourier transform

$$f(t) = \frac{1}{2\pi}\int_{-\infty}^{\infty} F(\omega)\, e^{j\omega t}\, d\omega = \int_{-\infty}^{\infty} F(\omega)\, e^{j2\pi ft}\, df \tag{7.2-2}$$

where the frequency f in cycles per second and the (radian) frequency ω in radians per second are related by

$$f = \frac{\omega}{2\pi}. \tag{7.2-3}$$

The pair of functions $f(t)$ and $F(\omega)$ will be called a *Fourier-transform pair.* A short list of some of the commonest pairs is given in Table 3.1 in Chapter 3. A more complete table is given in Appendix I.

Let us consider the total energy E of the signal $f(t)$ where

$$E = \int_{-\infty}^{\infty} f^2(t)\, dt. \tag{7.2-4}$$

It follows from Equation 7.2-2 that this expression may be written as

$$E = \int_{-\infty}^{\infty} f(t)\left[\frac{1}{2\pi}\int_{-\infty}^{\infty} F(\omega)\, e^{j\omega t}\, d\omega\right] dt. \tag{7.2-5}$$

On interchanging the order of integration,† we have

$$E = \frac{1}{2\pi}\int_{-\infty}^{\infty} F(\omega)\left[\int_{-\infty}^{\infty} f(t)\, e^{j\omega t}\, dt\right] d\omega$$

or

$$E = \frac{1}{2\pi}\int_{-\infty}^{\infty} F(\omega)F^*(\omega)\, d\omega \tag{7.2-6}$$

where the asterisk indicates the complex conjugate. On combining Equations 7.2-4 and 7.2-6 we have a special case of Parseval's Theorem:

$$E = \int_{-\infty}^{\infty} f^2(t)\, dt = \frac{1}{2\pi}\int_{-\infty}^{\infty} |F(\omega)|^2\, d\omega. \tag{7.2-7}$$

† It will often be necessary to interchange the order of integration; usually such a procedure is justified although there are significant cases where it is not (see, for example, Section 8 of Chapter 3). Specifically, the order of integration in a double integral may be interchanged if (1) the range of integration is finite and the integrand is bounded or (2) the range of integration is infinite and the double integral is absolutely convergent (convergent when the integrand is replaced by its absolute value). For a further discussion see K. S. Miller, *Advanced Real Calculus*, Harper, New York, 1957, Chap. 7.

Suppose that we are given two functions $f(t)$ and $g(t)$ with Fourier transforms $F(\omega)$ and $G(\omega)$ respectively. Then the *convolution* of $f(t)$ and $g(t)$ is defined as

$$f(t)*g(t) = \int_{-\infty}^{\infty} f(\tau)g(t-\tau)\,d\tau = \int_{-\infty}^{\infty} f(t-\tau)g(\tau)\,d\tau. \qquad (7.2\text{-}8)$$

By direct substitution we can write

$$f(t)*g(t) = \int_{-\infty}^{\infty} f(\tau)\frac{1}{2\pi}\int_{-\infty}^{\infty} G(\omega)\,e^{j\omega(t-\tau)}\,d\omega\,d\tau$$

or, on interchanging the order of integration,

$$f(t)*g(t) = \frac{1}{2\pi}\int_{-\infty}^{\infty} G(\omega)\,e^{j\omega t}\int_{-\infty}^{\infty} f(\tau)\,e^{-j\omega\tau}\,d\tau\,d\omega$$

$$= \frac{1}{2\pi}\int_{-\infty}^{\infty} [F(\omega)G(\omega)]\,e^{j\omega t}\,d\omega. \qquad (7.2\text{-}9)$$

This last expression is the (inverse) Fourier transform of the product $F(\omega)G(\omega)$. In the same way we can show

$$F(\omega)*G(\omega) = \int_{-\infty}^{\infty} F(\nu)G(\omega-\nu)\,d\nu = 2\pi\int_{-\infty}^{\infty} [f(t)g(t)]\,e^{-j\omega t}\,dt. \qquad (7.2\text{-}10)$$

In other words, the convolution of two functions in one domain (time or frequency) is the transform of the product of the two functions in the other domain. These relationships are listed in Table 7.1.

Table 7.1. Fourier Transform and Convolution

Time Domain	Frequency Domain
$f(t)$	$F(\omega)$
$g(t)$	$G(\omega)$
$f(t)*g(t)$	$F(\omega)G(\omega)$
$f(t)g(t)$	$\dfrac{F(\omega)*G(\omega)}{2\pi}$

For a linear system with impulse response $g(t)$ and input signal $f(t)$, the convolution $f(t)*g(t)$ gives the output $h(t)$

$$h(t) = \int_{-\infty}^{\infty} f(\tau)g(t-\tau)\,d\tau = \int_{-\infty}^{\infty} f(t-\tau)g(\tau)\,d\tau. \qquad (7.2\text{-}11)$$

As discussed in more detail in Chapter 4, the system is said to be *causal* or *physically realizable* if the impulse response vanishes for negative time:

$$g(t) = 0, \qquad t < 0. \qquad (7.2\text{-}12)$$

In this case, Equation 7.2-11 can be written as

$$h(t) = \int_{-\infty}^{t} f(\tau)g(t-\tau)\,d\tau = \int_{0}^{\infty} f(t-\tau)g(\tau)\,d\tau. \tag{7.2-13}$$

Of course, either the input $f(t)$ or the impulse response $g(t)$ may exist only over a finite interval; then the limits on the convolution integral will both be finite. For example, if the impulse response $g(t)$ vanishes for $t > T$ or if the input $f(t)$ begins T seconds in the past, then

$$h(t) = \int_{t-T}^{t} f(\tau)g(t-\tau)\,d\tau = \int_{0}^{T} f(t-\tau)g(\tau)\,d\tau. \tag{7.2-14}$$

It is apparent from the last equation that, when $g(t)$ is nonzero (or significantly different from zero) only over $(0, T)$, the output $h(t)$ is a result of a weighting of the values of the input $f(t)$ only over an interval of the past equal to T. The time T is roughly what is meant by the *response time* of the system. We desire to develop some more convenient measures of this quantity and to relate these measures to the *bandwidth* of the system. At the same time we want to apply those concepts to input and output signals as well as to systems.

7.3 Time and Frequency Domain Relations

In this section, we are interested in relating general properties in one domain (time or frequency) to the corresponding properties in the other domain. In the next section we will use these correspondences to relate system response time to system bandwidth and signal duration to signal bandwidth.

Let us consider the *real time function* $f(t)$ with Fourier transform $F(\omega)$ given by Equation 7.2-1. This transform may be written as

$$F(\omega) = \int_{-\infty}^{\infty} f(t)\cos\omega t\,dt - j\int_{-\infty}^{\infty} f(t)\sin\omega t\,dt. \tag{7.3-1}$$

The real part of this function is even since

$$\operatorname{Re} F(\omega) = \int_{-\infty}^{\infty} f(t)\cos\omega t\,dt = \operatorname{Re} F(-\omega) \tag{7.3-2}$$

and the imaginary part is odd since

$$\operatorname{Im} F(\omega) = -\int_{-\infty}^{\infty} f(t)\sin\omega t\,dt = -\operatorname{Im} F(-\omega). \tag{7.3-3}$$

Thus *the transform of a real time function has a real part that is even and an imaginary part that is odd.*

Suppose the real time function $f(t)$ vanishes for negative time:

$$f(t) = 0, \qquad t < 0. \tag{7.3-4}$$

Then the real and imaginary parts of its transform satisfy the pair of equations

$$\text{Re } F(\omega) = \int_0^\infty f(t) \cos \omega t \, dt, \tag{7.3-5}$$

$$\text{Im } F(\omega) = -\int_0^\infty f(t) \sin \omega t \, dt. \tag{7.3-6}$$

It is shown in Appendix H that these last two relations constitute a *Hilbert-transform pair*. Note that the impulse response of any physically realizable linear system satisfies Equation 7.3-4; therefore the real and imaginary parts of the complex transfer function of the system form a Hilbert-transform pair. Hilbert transforms will be useful also in representing frequency-domain functions that vanish over half the frequency range of interest. Such functions arise, for example, in the study of single-sideband modulation systems.

If we assume, on the other hand, that the real time function $f(t)$ is even so that

$$f(t) = f(-t),$$

then it is clear from Equation 7.3-3 that $\text{Im } F(\omega) \equiv 0$ and from Equation 7.3-2 that $\text{Re } F(\omega)$ is even. In other words, *the transform of an even, real time function is real and even.*

Let us refer again to the Fourier-transform pair given by Equations 7.2-1 and 7.2-2. On setting the arguments of this pair equal to zero, we obtain

$$F(0) = \int_{-\infty}^{\infty} f(t) \, dt \tag{7.3-7}$$

and

$$f(0) = \frac{1}{2\pi} \int_{-\infty}^{\infty} F(\omega) \, d\omega. \tag{7.3-8}$$

Thus the value of the function at the origin in one domain determines the area under the corresponding transform in the other domain.

Example 7.1 Let $f_1(t)$ and $F_1(\omega)$ be the pair

$$f_1(t) = \cos \omega_0 t,$$
$$F_1(\omega) = \pi\delta(\omega - \omega_0) + \pi\delta(\omega + \omega_0).$$

Then

$$f_1(0) = 1 = \tfrac{1}{2} \int_{-\infty}^{\infty} [\delta(\omega - \omega_0) + \delta(\omega + \omega_0)] \, d\omega$$

and also

$$F_1(0) = 0 = \int_{-\infty}^{\infty} \cos \omega_0 t \, dt, \qquad \omega_0 \neq 0.$$

On the other hand, consider the pair

$$f_2(t) = \sin \omega_0 t,$$
$$F_2(\omega) = -j\pi\delta(\omega - \omega_0) + j\pi\delta(\omega + \omega_0).$$

Now we have

$$f_2(0) = 0 = \frac{j}{2}\int_{-\infty}^{\infty} [-\delta(\omega - \omega_0) + \delta(\omega + \omega_0)]\, d\omega$$

and

$$F_2(0) = 0 = \int_{-\infty}^{\infty} \sin \omega_0 t\, dt.$$

• •

We consider now the function $f(t)$ shown in Figure 7.3(a). This might be a pulsed signal or the impulse response of a linear system, for example. The time ΔT over which $f(t)$ is appreciably different from zero might be called the *duration* of $f(t)$, and some measure, such as τ_d, of the center of the pulse could be called the *delay* of $f(t)$. In system terms, the quantity ΔT is the system *response time* or *rise time* and τ_d is the system delay. The integral of $f(t)$ is

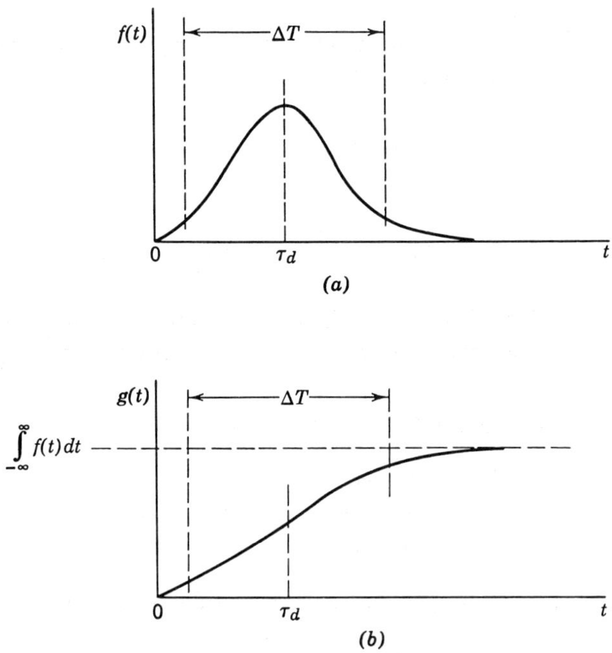

Figure 7.3 *Illustrations of duration and delay:* (a) *a typical pulse;* (b) *the integral of a typical pulse*

shown in Figure 7.3(*b*) and corresponds to the step function response of a system with impulse response $f(t)$. Note that thus far neither ΔT nor τ_d is well defined.

If the function $f(t)$ of Figure 7.3 is nonnegative, then the new function

$$\frac{f(t)}{\int_{-\infty}^{\infty} f(t)\,dt}$$

is nonnegative with unit area and can be considered to be a probability density. For the moment let us suppose that this is the case. We will seek measures of duration and delay that are meaningful both in terms of communication problems and mathematical tractability. It will be clear that some of the results that we obtain will not be universally applicable and, in particular, must be used with care when the function $f(t)$ can be negative for some values of t. The results, however, will be useful for wide classes of problems.

Suppose that we look at a frequency function $F(\omega)$ which we will assume to be real. If $F(\omega)$ is not real then we can consider either $|F(\omega)|^2 = F(\omega)F(-\omega)$ or $|F(\omega)|$. Such a function might be similar to that shown in Figure 7.4(*a*). The radian frequency range $\Delta\Omega$ (or the frequency range ΔF) over which $F(\omega)$ is appreciably different from zero might be called the *bandwidth* of the function. Of course, if the function is a *bandpass* function such as that shown in Figure 7.4(*b*), then the bandwidth usually will be taken to be some measure of the width of the positive-frequency (or negative-frequency) part of the function only.

As in the case of the time function previously discussed, we may normalize to unit area and consider

$$\frac{F(\omega)}{\int_{-\infty}^{\infty} F(\omega)\,d\omega}.$$

Again this new function, being nonnegative with unit area, has the properties of a probability density.

Let us consider the Fourier pair $f(t)$ and $F(\omega)$ and change the time scale by the factor a, replacing $f(t)$ by $af(at)$ so that both the old and the new signal have the same area, that is,

$$\int_{-\infty}^{\infty} f(t)\,dt = \int_{-\infty}^{\infty} af(at)\,dt. \tag{7.3-9}$$

For $a < 1$, the new signal $af(at)$ is stretched in time and reduced in height; its "duration" has been increased. For $a > 1$, $af(at)$ has been compressed in time and increased in height; its "duration" has been decreased. The transform of this new function is

$$\int_{-\infty}^{\infty} af(at)\,e^{-j\omega t}\,dt = \int_{-\infty}^{\infty} f(x)\,e^{-j(\omega/a)x}\,dx = F\left(\frac{\omega}{a}\right). \tag{7.3-10}$$

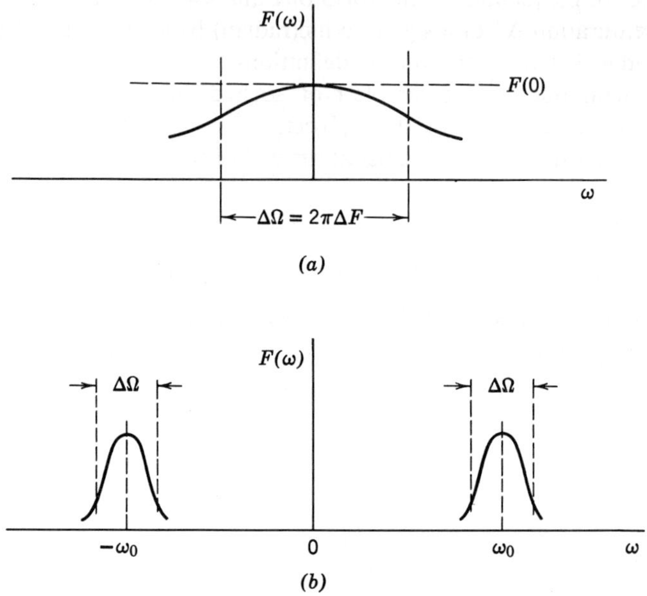

Figure 7.4 *Illustrations of bandwidth: (a) a typical low-pass frequency function; (b) a typical bandpass frequency function.*

The effect on the "bandwidth" of $F(\omega)$ has been the opposite of the effect on the "duration" of $f(t)$. When the signal duration is increased (decreased), then the bandwidth is decreased (increased) in the same proportion. From this discussion, we might suspect that more fundamental relationships hold between properly defined durations and bandwidths of signals.

7.4 Duration (Response Time) and Bandwidth—Uncertainty Relationships

It is apparent from the discussion of the previous section that treatments of duration and bandwidth are mathematically similar although one is defined in the time domain and the other in the frequency domain. In this section we proceed to find several specific measures of these two quantities and to show that they are intimately related to each other through various so-called *uncertainty relationships*. The term "uncertainty" arises from the *Heisenberg uncertainty principle* of quantum mechanics [1] which states that it is not possible to determine simultaneously and exactly the position and momentum coordinates of a particle. More specifically, if Δx and Δp are the uncertainties in position and momentum, then

$$(\Delta x)(\Delta p) \geq h \tag{7.4-1}$$

where h is a constant.

A number of inequalities of the form of Equation 7.4-1 can be developed relating the duration ΔT of a signal to its (radian) bandwidth $\Delta\Omega$. The value of the constant h will depend on the definitions of duration and bandwidth. We begin with one of the simplest of such definitions. For the reader interested in further study of this subject, a list of references is given in part 1 of the Additional Reading list at the end of this chapter.

Equivalent Rectangular Bandwidth $\Delta\Omega_1$ and Duration ΔT_1

In Section 4.11, we introduced the idea of the *equivalent noise bandwidth* β of a linear amplifier. In the same way we define the *equivalent rectangular bandwidth* $\Delta\Omega_1$ of a frequency function $F(\omega)$ as

$$\Delta\Omega_1 = \frac{\int_{-\infty}^{\infty} F(\omega)\, d\omega}{F(\omega_0)} \tag{7.4-2}$$

where ω_0 is some characteristic center frequency of the function $F(\omega)$. It is clear from this definition that we are replacing the original function $F(\omega)$ by a rectangular function of equal area, of width $\Delta\Omega_1$, and of height $F(\omega_0)$ as shown in Figure 7.5. For the case shown in Figure 7.4(*a*), the characteristic frequency would be taken to be $\omega_0 = 0$ while in Figure 7.4(*b*) it would be ω_0 as noted. If we restrict ourselves to the low-pass case ($\omega_0 \equiv 0$), then it follows from Equations 7.3-7 and 7.3-8 that Equation 7.4-2 can be rewritten as

$$\Delta\Omega_1 = \frac{2\pi f(0)}{\int_{-\infty}^{\infty} f(t)\, dt} \tag{7.4-3}$$

where $f(t)$ is the time function which is the inverse Fourier transform of $F(\omega)$.

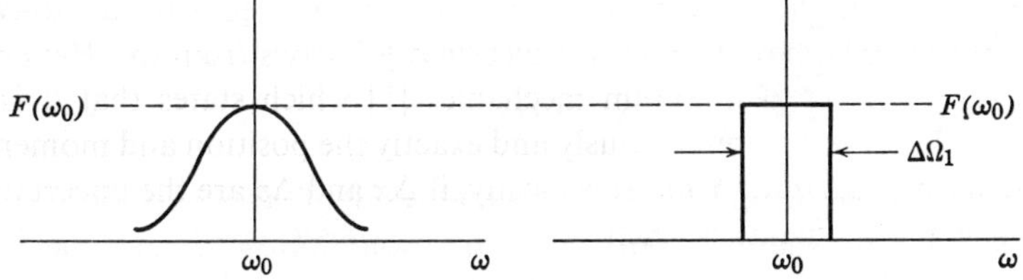

Figure 7.5 *Equivalent rectangular bandwidth* $\Delta\Omega_1$

Suppose now that we follow the same procedure in the time domain and define the *equivalent rectangular duration* ΔT_1 of the signal $f(t)$ by

$$\Delta T_1 = \frac{\int_{-\infty}^{\infty} f(t)\, dt}{f(t_0)} \tag{7.4-4}$$

where t_0 is some characteristic time denoting the center of the pulse. For the case where $t_0 \equiv 0$, it is clear, then, from Equations 7.4-3 and 7.4-4 that equivalent rectangular duration and bandwidth are connected by the uncertainty relationship

$$\boxed{(\Delta T_1)(\Delta \Omega_1) = 2\pi.} \tag{7.4-5}$$

Example 7.2 We consider the strictly bandlimited flat spectrum shown in Figure 7.6(*a*) with corresponding time function

$$f(t) = 2WA \frac{\sin 2\pi Wt}{2\pi Wt}.$$

It is immediately apparent from Equation 7.4-2 that

$$\Delta \Omega_1 = 4\pi W.$$

The expression for ΔT_1 follows from Equation 7.4-4:

$$\Delta T_1 = \frac{2WA \int_{-\infty}^{\infty} [(\sin 2\pi Wt)/2\pi Wt]\, dt}{2WA}$$

$$= \frac{1}{2\pi W} \int_{-\infty}^{\infty} \frac{\sin x}{x}\, dx = \frac{1}{2W}$$

since the definite integral

$$\int_{-\infty}^{\infty} \frac{\sin x}{x}\, dx = \pi$$

is well known [2]. Thus, as expected, we have the product

$$(\Delta T_1)(\Delta \Omega_1) = 2\pi.$$

It is clear that we would obtain the same results for the flat, strictly time-limited pulse of Figure 7.6(*b*) with duration $\Delta T_1 = 2b$ and bandwidth $\Delta \Omega_1 = \pi/b$.

• •

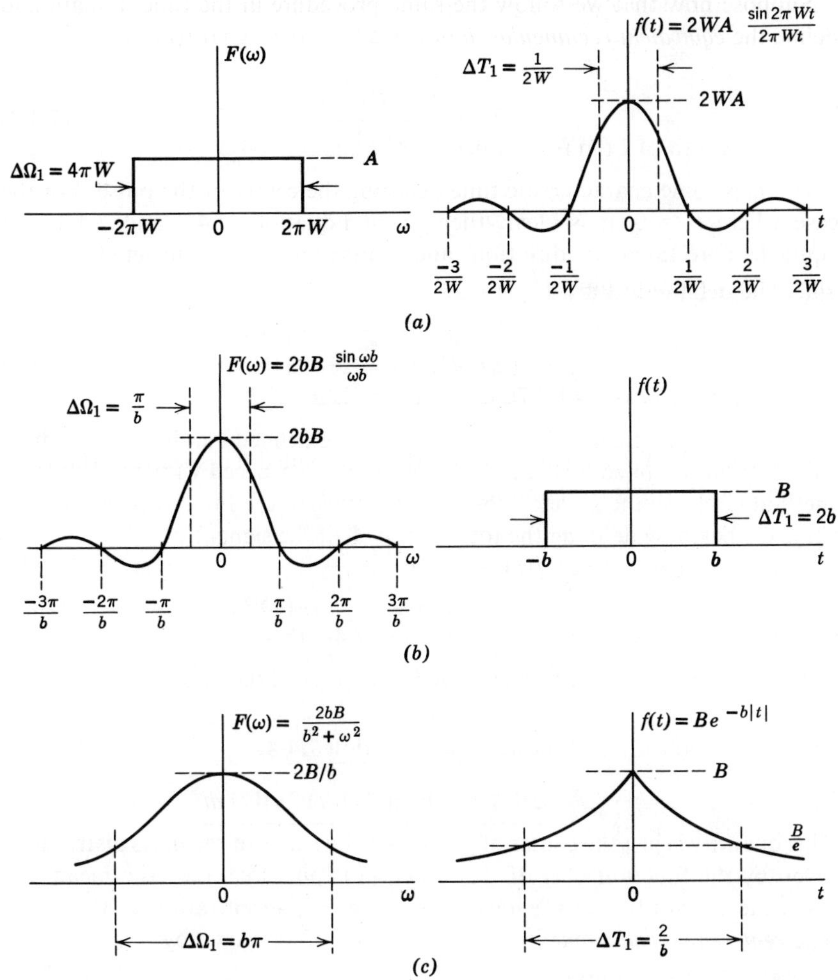

Figure 7.6 *Some bandwidth-duration relationships: (a) a flat, strictly bandlimited function; (b) a flat, strictly time-limited function; (c) the double exponential pulse*

Example 7.3 Let us consider the double exponential pulse

$$f(t) = B\,e^{-b|t|}, \qquad b > 0,$$

as shown in Figure 7.6(c) with transform

$$F(\omega) = \frac{2bB}{b^2 + \omega^2}.$$

Then the duration of $f(t)$ is given from Equation 7.4-4 as

$$\Delta T_1 = 2\int_0^{\infty} e^{-bt}\, dt = \frac{2}{b}.$$

The bandwidth of $F(\omega)$ is obtained from Equation 7.4-2

$$\Delta\Omega_1 = 2b\int_0^{\infty} \frac{b}{b^2 + \omega^2}\, d\omega = b\pi$$

since the definite integral [3] is $\pi/2$.

• •

Second-Moment Bandwidth $\Delta\Omega_2$ and Duration $T\Delta_2$

In the preceding part of this section, we have defined bandwidth and duration in terms of an equivalent rectangle. Now we proceed with an alternative method based on the second-moment properties of the Fourier pair $F(\omega)$ and $f(t)$. We begin by defining the total energy $\mathscr{E}$ of the signal by

$$\mathscr{E} = \int_{-\infty}^{\infty} f^2(t)\, dt = \frac{1}{2\pi}\int_{-\infty}^{\infty} |F(\omega)|^2\, d\omega. \tag{7.4-6}$$

For an arbitrary, real, nonnegative function $g(x)$, the *mean* $\bar{x}$ is given by

$$\bar{x} = \frac{\int_{-\infty}^{\infty} xg(x)\, dx}{\int_{-\infty}^{\infty} g(x)\, dx}. \tag{7.4-7}$$

This quantity is just the center of gravity on the x axis of the mass distribution given by the function $g(x)$. If this function is pulselike, then the mean $\bar{x}$ of x is a measure of the pulse's location or point of concentration on the x axis. The *centered second moment* or *variance* $(\Delta x)^2$ of x is given by*

$$(\Delta x)^2 = \frac{\int_{-\infty}^{\infty} (x - \bar{x})^2 g(x)\, dx}{\int_{-\infty}^{\infty} g(x)\, dx}. \tag{7.4-8}$$

This quantity is a measure of the dispersion of the function $g(x)$ about the mean $\bar{x}$. A small value for $(\Delta x)^2$ indicates that most of the area under the $g(x)$ curve is concentrated near the point $x = \bar{x}$. It is apparent that, at least for functions $g(x)$ which are nonnegative, the quantity $(\Delta x)^2$ can serve as a measure of the "bandwidth" or "duration" of $g(x)$.

* In our previous treatments involving probability densities, the variance has been denoted by σ^2. In this section we adopt the notation $(\Delta x)^2$ for this quantity in order to be consistent with most of the literature.

Example 7.4 Consider the double exponential pulse

$$g(x) = B\, e^{-b|x|}$$

with mean $\bar{x}$ given by

$$\bar{x} = \frac{\int_{-\infty}^{\infty} x\, e^{-b|x|}\, dx}{\int_{-\infty}^{\infty} e^{-b|x|}\, dx} = 0$$

and variance

$$(\Delta x)^2 = \frac{\int_{-\infty}^{\infty} x^2\, e^{-b|x|}\, dx}{\int_{-\infty}^{\infty} e^{-b|x|}\, dx} = \frac{\int_0^{\infty} x^2\, e^{-bx}\, dx}{\int_0^{\infty} e^{-bx}\, dx} = \frac{2}{b^2}$$

since [4]

$$\int_0^{\infty} x^n\, e^{-ax}\, dx = \frac{n!}{a^{n+1}}, \qquad \text{for} \quad a > 0.$$

• •

We now proceed to use this second moment as a measure of bandwidth and duration by replacing the function $g(\cdot)$ in the time domain by

$$g(\cdot) \sim |f(t)|^2 \tag{7.4-9}$$

and in the frequency domain by

$$g(\cdot) \sim \frac{1}{2\pi}\, |F(\omega)|^2. \tag{7.4-10}$$

Thus we have defined a *second-moment bandwidth* $\Delta\Omega_2$ by

$$(\mathscr{E})(\Delta\Omega_2)^2 = \frac{1}{2\pi} \int_{-\infty}^{\infty} (\omega - \bar{\omega})^2 |F(\omega)|^2\, d\omega \tag{7.4-11}$$

and a *second-moment duration* ΔT_2 by

$$(\mathscr{E})(\Delta T_2)^2 = \int_{-\infty}^{\infty} (t - \bar{t})^2 |f(t)|^2\, dt. \tag{7.4-12}$$

It is obvious that we can assume without loss of generality that

$$\mathscr{E} = \frac{1}{2\pi} \int_{-\infty}^{\infty} |F(\omega)|^2\, d\omega = \int_{-\infty}^{\infty} |f(t)|^2\, dt = 1 \tag{7.4-13}$$

provided we remember to divide all results by the actual value of $\mathscr{E}$. In other words we can replace $g(\cdot)$ by $g(\cdot)/\mathscr{E}$. We proceed now to find a relationship between $\Delta\Omega_2$ and ΔT_2.

In order to simplify Equations 7.4-12 and 7.4-13 let us note that any t translation of $f(t)$, a linear change of variable, will not affect the value of $|F(\omega)|^2$ in Equation 7.4-11. Similarly any ω translation in Equation 7.4-11 will not affect the value of $|f(t)|^2$ in Equation 7.4-12. In other words translation of either $f(t)$ or $F(\omega)$ will not affect the values of $\Delta\Omega_2$ and ΔT_2. Thus we can assume without loss of generality that $\bar{t}$ and $\bar{\omega}$ are zero. For example, we can replace $f(t)$ by $h(t - \bar{t})$, then $H(\omega)$ becomes $F(\omega)\, e^{j\omega\bar{t}}$, $|H(\omega)|^2 = |F(\omega)|^2$, and both $\Delta\Omega_2$ and ΔT_2 are unchanged. The derivative of $f(t)$ can be written from Equation 7.2-2 as

$$f'(t) = \frac{1}{2\pi}\int_{-\infty}^{\infty} j\omega F(\omega)\, e^{j\omega t}\, d\omega \tag{7.4-14}$$

and we can form

$$\begin{aligned}\int_{-\infty}^{\infty} f'(t)f^{*\prime}(t)\, dt &= \left(\frac{1}{2\pi}\right)^2 \int_{-\infty}^{\infty}\int_{-\infty}^{\infty} \omega\nu F(\omega)F^*(\nu)\int_{-\infty}^{\infty} e^{-jt(\nu-\omega)}\, dt\, d\nu\, d\omega \\ &= \frac{1}{2\pi}\int_{-\infty}^{\infty}\int_{-\infty}^{\infty} \omega\nu F(\omega)F^*(\nu)\delta(\omega - \nu)\, d\nu\, d\omega \end{aligned} \tag{7.4-15}$$

where $\delta(\omega - \nu)$ is the Dirac delta function which is discussed in more detail in Appendix B.* It follows then, from the sampling property of the delta function, that Equation 7.4-15 can be written as

$$\int_{-\infty}^{\infty} f'(t)f^{*\prime}(t)\, dt = \frac{1}{2\pi}\int_{-\infty}^{\infty} \omega^2|F(\omega)|^2\, d\omega. \tag{7.4-16}$$

From the results of the last paragraph and on the assumption that $\mathscr{E} = 1$, the bandwidth product $(\Delta\Omega_2)(\Delta T_2)$ can be written as

$$(\Delta\Omega_2)^2(\Delta T_2)^2 = \int_{-\infty}^{\infty} tf(t)tf^*(t)\, dt \int_{-\infty}^{\infty} f'(t)f^{*\prime}(t)\, dt. \tag{7.4-17}$$

We now apply a form of the Schwarz Inequality developed in Appendix E. Let us refer to Equation E.4-9 and replace $f(t)$ by $tf(t)$ and $g(t)$ by $f'(t)$. We obtain from Equation 7.4-17

$$\begin{aligned}(\Delta\Omega_2)^2(\Delta T_2)^2 &\geq \frac{1}{4}\left|\int_{-\infty}^{\infty} [tf^*(t)f'(t) + tf(t)f^{*\prime}(t)]\, dt\right|^2 \\ &= \frac{1}{4}\left|\int_{-\infty}^{\infty} t\frac{d}{dt}[f(t)f^*(t)]\, dt\right|^2. \end{aligned} \tag{7.4-18}$$

This expression may be integrated by parts to yield

$$(\Delta\Omega_2)^2(\Delta T_2)^2 \geq \frac{1}{4}\left[t|f(t)|^2\Big|_{-\infty}^{\infty} - \int_{-\infty}^{\infty} |f(t)|^2\, dt\right]\cdot \tag{7.4-19}$$

* We make use here of the Fourier pair $f(t) = 1$ and $F(\omega) = 2\pi\delta(\omega)$, that is,

$$f(t) = \frac{1}{2\pi}\int_{-\infty}^{\infty} 2\pi\delta(\omega)\, e^{j\omega t}\, d\omega = 1$$

and

$$F(\omega) = \int_{-\infty}^{\infty} e^{-j\omega t}\, dt = 2\pi\delta(\omega).$$

Let us require that $|f(t)|$ vanish at infinity at least as rapidly as $1/\sqrt{t}$ so that

$$\lim_{t \to \pm \infty} t|f(t)|^2 = 0. \tag{7.4-20}$$

Then Equation 7.4-19 becomes, as a result of the assumption that $\mathscr{E} = 1$,

$$\boxed{(\Delta\Omega_2)(\Delta T_2) \geq \tfrac{1}{2}.} \tag{7.4-21}$$

This expression may be regarded as a second uncertainty relationship connecting the bandwidth and duration of a signal.

We note in passing that the equality in Equation E.4-9 is attained if and only if $f(t)$ is linearly proportional to $g(t)$, that is, if

$$-kf(t) = g(t)$$

or, in terms of the present problem, if

$$-ktf(t) = f'(t). \tag{7.4-22}$$

This is an ordinary differential equation which may be solved by the separation of variables

$$\frac{f'(t)}{f(t)} = -kt,$$

$$\ln f(t) = \frac{-kt^2}{2} + C$$

or, finally,

$$f(t) = K e^{-kt^2/2}, \qquad -\infty < t < \infty. \tag{7.4-23}$$

If we normalize to unit energy, $\mathscr{E} = 1$, then $f(t)$ becomes

$$f(t) = \sqrt[4]{\frac{k}{\pi}}\, e^{-kt^2/2}, \qquad -\infty < t < \infty, \tag{7.4-24}$$

which is a gaussian pulse (a pulse having the shape of the normal or gaussian distribution). The second-moment duration ΔT_2 for this pulse is given by

$$(\Delta T_2)^2 = \int_{-\infty}^{\infty} t^2 f^2(t)\, dt = \sqrt{\frac{k}{\pi}} \int_{-\infty}^{\infty} t^2 e^{-kt^2}\, dt = \frac{1}{2k^2} \tag{7.4-25}$$

since [5]

$$\int_0^{\infty} x^2 e^{-ax^2}\, dx = \frac{1}{4a^2}\sqrt{\frac{\pi}{a}}.$$

The transform of $f(t)$ can be shown (see Equations 2.18-21 and 2.18-31 and accompanying development) to be

$$F(\omega) = \sqrt[4]{\frac{4\pi}{k}}\, e^{-\omega^2/2k}, \qquad -\infty < \omega < \infty. \tag{7.4-26}$$

Now the second-moment bandwidth $\Delta\Omega_2$ is found from

$$(\Delta\Omega_2)^2 = \frac{1}{2\pi}\int_{-\infty}^{\infty} \omega^2 |F(\omega)|^2 \, d\omega$$

$$= \frac{1}{2\pi}\sqrt{\frac{4\pi}{k}}\int_{-\infty}^{\infty} \omega^2 \, e^{-\omega^2/k} \, d\omega = \frac{k^2}{2}. \tag{7.4-27}$$

Thus, in this case of the gaussian pulse, we attain the equality

$$(\Delta\Omega_2)(\Delta T_2) = \tfrac{1}{2}. \tag{7.4-28}$$

Example 7.5 Consider the double exponential pulse of Example 7.3 normalized to unit energy:

$$f(t) = \sqrt{b}\, e^{-b|t|}, \qquad b > 0.$$

The duration ΔT_2 can be found as

$$(\Delta T_2)^2 = b\int_{-\infty}^{\infty} t^2 \, e^{-2b|t|} \, dt = 2b\int_{0}^{\infty} t^2 \, e^{-2bt} \, dt = \frac{1}{2b^2}$$

from [4]. The transform of $f(t)$ is given by pair number 4 of Table 3.1 or by

$$F(\omega) = \sqrt{b}\int_{-\infty}^{0} e^{bt} \, e^{-j\omega t} \, dt + \sqrt{b}\int_{0}^{\infty} e^{-bt} \, e^{-j\omega t} \, dt = \frac{2b\sqrt{b}}{b^2 + \omega^2}.$$

The bandwidth is obtained from

$$(\Delta\Omega_2)^2 = \frac{1}{2\pi}\int_{-\infty}^{\infty} \omega^2 |F(\omega)|^2 \, d\omega = \frac{4b^3}{2\pi}\int_{-\infty}^{\infty} \frac{\omega^2}{(b^2 + \omega^2)^2} \, d\omega.$$

We have the integral [6]

$$\int_{-\infty}^{\infty} \frac{\omega^2}{(b^2 + \omega^2)^2} \, d\omega = -\frac{\omega}{2(b^2 + \omega^2)}\bigg|_{-\infty}^{\infty} + \frac{1}{2b}\arctan\frac{\omega}{b}\bigg|_{-\infty}^{\infty} = \frac{\pi}{2b}.$$

Consequently,

$$(\Delta\Omega_2)(\Delta T_2) = (b)\left(\frac{1}{\sqrt{2}b}\right) = \frac{1}{\sqrt{2}},$$

a number considerably greater than the lower bound of $\frac{1}{2}$.

• •

Note that, for any real signal $f(t)$, the Fourier transform $F(\omega)$ satisfies

$$F(\omega) = \int_{-\infty}^{\infty} f(t)\, e^{-j\omega t} \, dt = F^*(-\omega) \tag{7.4-29}$$

or

$$|F(\omega)| = |F(-\omega)| \tag{7.4-30}$$

and

$$\text{ang } F(\omega) = -\text{ang } F(-\omega). \tag{7.4-31}$$

Thus, for real signals, the mean $\bar{\omega}$ given by Equations 7.4-7 and 7.4-10 is always zero. It is apparent, therefore, that the second-moment bandwidth $\Delta\Omega_2$ is not an appropriate measure for the bandpass function of Figure 7.4(*b*). In this case we would like a second moment defined about $\omega = \omega_0$ rather than $\omega = 0$.

The problem of the previous paragraph can be resolved by defining the *analytic signal* [7]

$$f_+(t) = \frac{1}{\sqrt{2\pi}} \int_0^\infty F(\omega)\, e^{j\omega t}\, d\omega. \tag{7.4-32}$$

This function is complex and is discussed in more detail in Appendix H. Its Fourier transform

$$F_+(\omega) = \int_{-\infty}^{\infty} f_+(t)\, e^{-j\omega t}\, dt \tag{7.4-33}$$

is called the *positive frequency spectrum* of $f(t)$ and has the property

$$F_+(\omega) = \sqrt{2}\, F(\omega) u(\omega) \tag{7.4-34}$$

where $u(\omega)$ is the unit-step function. Thus the function $F_+(\omega)$ exists only for positive frequencies. Note that

$$\int_{-\infty}^{\infty} |f_+(t)|^2\, dt = \frac{1}{2\pi} \int_{-\infty}^{\infty} |F_+(\omega)|^2\, d\omega = \mathscr{E} \tag{7.4-35}$$

where $\mathscr{E}$ is the total signal energy. Now we can define a mean $\bar{\omega}_+$ by

$$\mathscr{E}\bar{\omega}_+ = \int_{-\infty}^{\infty} \omega |F_+(\omega)|^2\, d\omega = 2 \int_0^\infty \omega |F(\omega)|^2\, d\omega. \tag{7.4-36}$$

For the bandpass case of Figure 7.4(*b*), the value of this new mean $\bar{\omega}_+$ will be ω_0. In the same way we define a second-moment bandwidth $\Delta\Omega_+$ by

$$\begin{aligned} \mathscr{E}(\Delta\Omega_+)^2 &= \int_{-\infty}^{\infty} (\omega - \bar{\omega}_+)^2 |F_+(\omega)|^2\, d\omega \\ &= 2 \int_0^\infty (\omega - \bar{\omega}_+)^2 |F(\omega)|^2\, d\omega. \end{aligned} \tag{7.4-37}$$

It can be shown [7, 8] that the product $(\Delta\Omega_+)(\Delta T_2)$ obeys the uncertainty relationship

$$(\Delta\Omega_+)(\Delta T_2) \geq \tfrac{1}{2}|1 - 2|F(0)|^2 \bar{\omega}_+|. \tag{7.4-38}$$

In this case, however, the equality cannot be achieved since all $F(\omega)$ corresponding to real $f(t)$ have negative frequency components. Note that, for the commonly encountered case where $F(0) \equiv 0$, we have

$$(\Delta\Omega_+)(\Delta T_2) \geq \tfrac{1}{2}. \tag{7.4-39}$$

We proceed now to consider another uncertainty relationship based on a third definition of duration and bandwidth.

Bandwidth $\Delta\Omega_3$ and Duration ΔT_3 for a Given Percentage of Energy Concentration —Essentially Band- and Time-Limited Signals

In the first two parts of this section we have defined bandwidth and duration first in terms of an equivalent rectangle and then by a second-moment property. In both cases we have considered integrals over all time or over all frequencies. We may obtain more insight into what is really happening in many cases if we consider the bandwidth (or duration) to be determined by the finite interval (if it exists) where most of the signal energy is concentrated [9, 10]. Specifically, we define a *bandwidth* $\Delta\Omega_3 = \Delta\Omega_3(\beta)$ of the signal pair $f(t)$ and $F(\omega)$ by the equation

$$\beta^2\mathscr{E} = \frac{1}{2\pi}\int_{-\Delta\Omega_3}^{\Delta\Omega_3} |F(\omega)|^2\, d\omega \tag{7.4-40}$$

where $\mathscr{E}$ is, as before, the total signal energy given by

$$\mathscr{E} = \frac{1}{2\pi}\int_{-\infty}^{\infty} |F(\omega)|^2\, d\omega = \int_{-\infty}^{\infty} f^2(t)\, dt$$

and β^2 must satisfy

$$0 < \beta^2 \leq 1. \tag{7.4-41}$$

If $\beta^2 = 1$, we will say that the signal is *exactly bandlimited*; if $\beta^2 = 1 - \epsilon_1 \approx 1$, we will call it *essentially bandlimited.* In the same way we define a duration $\Delta T_3 = \Delta T_3(\alpha)$ by

$$\alpha^2\mathscr{E} = \int_{-\Delta T_3}^{\Delta T_3} f^2(t)\, dt \tag{7.4-42}$$

where

$$0 < \alpha^2 \leq 1. \tag{7.4-43}$$

Again the signal will be called *exactly time limited* if $\alpha^2 = 1$ and *essentially time limited* if $\alpha^2 = 1 - \epsilon_2 \approx 1$.

It is apparent that, for a given signal, β^2 (or α^2) is a monotone nondecreasing function of $\Delta\Omega_3$ (or ΔT_3). Conversely, $\Delta\Omega_3$ (or ΔT_3) is a monotone increasing function of β^2 (or α^2). In other words, as the percentage of energy included in the basic definition of bandwidth (duration) increases, the defined bandwidth (duration) must increase for a given signal.

At this point it will be somewhat more convenient to normalize the signal energy to unity; that is, to replace $\alpha^2\mathscr{E}$ by α^2 and $\beta^2\mathscr{E}$ by β^2. Also for simplicity of notation in the development that follows we shall let

$$T = \Delta T_3 \tag{7.4-44}$$

and

$$\Omega = \Delta\Omega_3. \tag{7.4-45}$$

This change of notation also will be convenient when referring to the literature [9, 10]. We shall now proceed to develop the fact that, for fixed α^2 and β^2, the product ΩT must be bounded from below; that is,

$$\boxed{(\Delta\Omega_3)(\Delta T_3) = \Omega T \geq k = k(\alpha, \beta)} \tag{7.4-46}$$

where k is a function that depends on α and β (we shall always take α and β to be the positive square roots of α^2 and β^2). Stated in somewhat different terms, for fixed ΩT and α (or β), then β (or α) must remain at or below some maximum value which is a function of ΩT.

Equation 7.4-46 is another uncertainty relationship of the same general type as that given by Equations 7.4-5 and 7.4-21. Unfortunately its derivation and the determination of the bound k are not very simple. We shall content ourselves here with an outline of the approach and refer the interested reader to the literature [9, 10]. From this point we shall assume that $\mathscr{E} = 1$.

We begin by considering the integral equation

$$\lambda\phi(t) = \int_{-T}^{T} \phi(\tau) \frac{\sin \Omega(t - \tau)}{\pi(t - \tau)} \, d\tau \tag{7.4-47}$$

with *parameter* λ and *kernel* $[\sin \Omega(t - \tau)]/\pi(t - \tau)$. A solution to this equation is any function $\phi(t)$ which satisfies the equation. It is apparent that $\phi(t) \equiv 0$ is one such solution. As already mentioned in Section 4.6, however, there is a set of values of the parameter λ, called the *eigenvalues* λ_i of the equation, for which there are corresponding nonzero solutions $\phi_i(t)$, called the *eigenfunctions* of the equation. It can be shown that all λ_i are positive in this case. Let λ_0 denote the largest eigenvalue and $\phi_0(t)$ the corresponding eigenfunction.

Note that all of the functions $\phi(t)$ obtained from Equation 7.4-47 are band-limited. We can show this easily by defining a truncated function $\phi_T(t)$ by

$$\phi_T(t) = \begin{cases} \phi(t), & |t| \leq T, \\ 0, & |t| > T. \end{cases} \tag{7.4-48}$$

Then the right side of Equation 7.4-47 is just the convolution (see Equation 7.2-8) of $\phi_T(t)$ and the centered kernel of the integral equation; that is,

$$\lambda\phi(t) = \phi_T(t) * \frac{\sin \Omega t}{\pi t}. \tag{7.4-49}$$

The Fourier transform of $(\sin \Omega t)/\pi t$ is a pulse of unit height extending over the interval $(-\Omega, \Omega)$. Consequently Equation 7.4-49 indicates that $\lambda\phi(t)$ is the output of an ideal low-pass filter with input $\phi_T(t)$, or

$$\lambda\Phi(\omega) = \begin{cases} \Phi_T(\omega), & |\omega| \leq \Omega, \\ 0, & |\omega| > \Omega, \end{cases} \tag{7.4-50}$$

where $\Phi(\omega)$ and $\Phi_T(\omega)$ are the Fourier transforms of $\phi(t)$ and $\phi_T(t)$, respectively. Thus $\phi(t)$ is strictly bandlimited to $(-\Omega, \Omega)$ radians per second. It follows from Equations 7.2-10 and 7.4-50 that

$$\int_{-T}^{T} \phi^2(t)\, dt = \int_{-\infty}^{\infty} \phi(t)\phi_T(t)\, dt = \frac{1}{2\pi}\int_{-\infty}^{\infty} \Phi(\omega)\Phi_T(\omega)\, d\omega$$

$$= \frac{\lambda}{2\pi}\int_{-\Omega}^{\Omega} \Phi^2(\omega)\, d\omega = \lambda \tag{7.4-51}$$

if $\phi(t)$ has been normalized to unit energy.

It can be shown [10] that the function strictly bandlimited to $(-\Omega, \Omega)$ with the largest energy α^2 in the interval $(-T, T)$ is the function $\phi_0(t)$ corresponding to the largest eigenvalue λ_0 of Equation 7.4-47. It follows from Equation 7.4-51 that, for any function $f(t)$ strictly bandlimited $(-\Omega, \Omega)$, we have

$$\alpha^2 \le \lambda_0. \tag{7.4-52}$$

It can also be shown that λ_0 depends only on the product ΩT as does $\phi_0(t)$ (except for a normalization factor). The family of functions $\phi_0(t)$ with parameter ΩT are *prolate spheroidal wave functions* and are illustrated in [9, 10]. The function $\lambda_0(\Omega T)$ is plotted in Figure 7.7(*a*) and the inverse $\lambda_0^{-1}(\alpha^2)$ in Figure 7.7(*b*).

Thus for any function $f(t)$ strictly bandlimited $(-\Omega, \Omega)$, the bandwidth-duration product $(\Delta\Omega_3)(\Delta T_3)$ must obey the uncertainty relationship

$$\boxed{(\Delta\Omega_3)(\Delta T_3) = \Omega T \ge \lambda_0^{-1}(\alpha^2), \qquad \beta^2 = 1.} \tag{7.4-53}$$

This last relationship is obviously a special case of Equation 7.4-46. In the same way, if $f(t)$ is now assumed to be strictly time limited to the interval $(-T, T)$ so that $\alpha^2 = 1$, then it can be shown that

$$\beta^2 \le \lambda_0 \tag{7.4-54}$$

and the function $f(t)$ which makes $\beta^2 = \lambda_0$ is

$$f(t) = \frac{\phi_{0T}(t)}{\sqrt{\lambda_0}}, \tag{7.4-55}$$

the truncated version of $\phi_0(t)$ normalized by the factor $1/\sqrt{\lambda_0}$ so that, from Equation 7.4-51,

$$\int_{-T}^{T} f^2(t)\, dt = \frac{1}{\lambda_0}\int_{-\infty}^{\infty} \phi_{0T}^2(t)\, dt = \frac{1}{\lambda_0}\int_{-\infty}^{\infty} \phi_0^2(t)\, dt = 1 = \alpha^2. \tag{7.4-56}$$

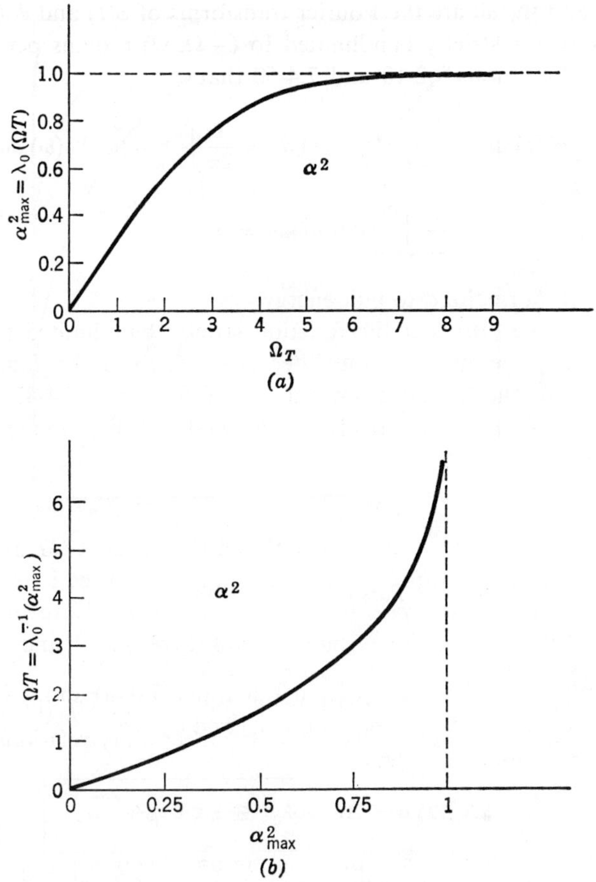

***Figure 7.7** Bounds on α^2 and ΩT for $\beta^2 = 1$*

In other words, Equation 7.4-53 holds also for the strictly time-limited case in the form

$$(\Delta\Omega_3)(\Delta T_3) = \Omega T \geq \lambda_0^{-1}(\beta^2), \qquad \alpha^2 = 1 \tag{7.4-57}$$

where the plot of Figure 7.7 holds with α and β replacing each other.

For the general problem where neither α^2 nor β^2 is unity, it can be shown that the function to maximize β^2 (or α^2) for fixed α^2 (or β^2) is a linear combination of the functions found previously, that is,

$$f(t) = a_1\phi_0(t) + a_2\phi_{0T}(t) \tag{7.4-58}$$

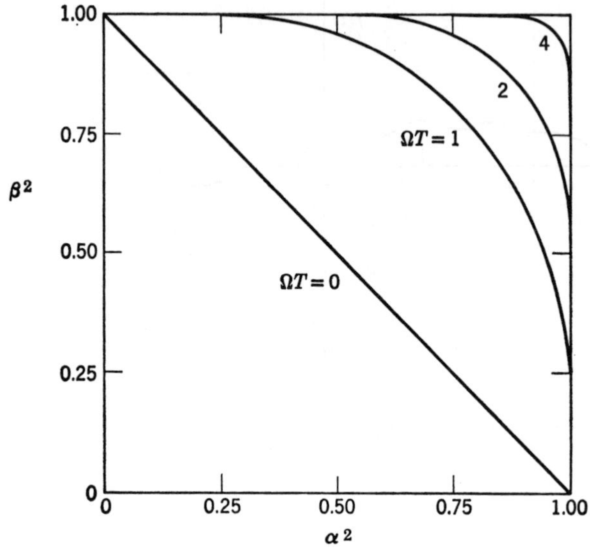

Figure 7.8 *Possible α^2 and β^2 for various ΩT*

where a_1 and a_2 are determined from Equations 7.4-40 and 7.4-42 and the unit energy requirement. On eliminating a_1 and a_2, we can obtain a relationship among α, β, and λ_0 in the form

$$\text{arc cos } \alpha + \text{arc cos } \beta \leq \text{arc cos } \sqrt{\lambda_0} \tag{7.4-59}$$

where the equality holds only for the optimum function given by Equation 7.4-58. The curve of Figure 7.7(*a*) may be combined with Equation 7.4-59 to eliminate λ_0 and give a plot of α^2 versus β^2 for various values of ΩT as shown in Figure 7.8. This figure shows the best α^2 (or β^2) obtainable for a given β^2 (or α^2) and a fixed product ΩT. It is apparent that the upper bound on the inequality of Equation 7.4-46 can be obtained from interpolation of this set of curves. Further discussion is beyond the scope of this treatment.

7.5 The Sampling Theorem

Suppose we consider a given arbitrary signal $f(t)$ available for all time as shown in Figure 7.9(*a*). We ask ourselves if it is necessary to know the amplitude of the signal for every value of time in order to characterize it uniquely. In other words, can $f(t)$ be represented (and reconstructed) from some set of *sample values* or *samples* $\ldots, f(t_{-1}), f(t_0), f(t_1), \ldots$? Surprisingly enough, it develops that, under certain fairly reasonable conditions, a signal

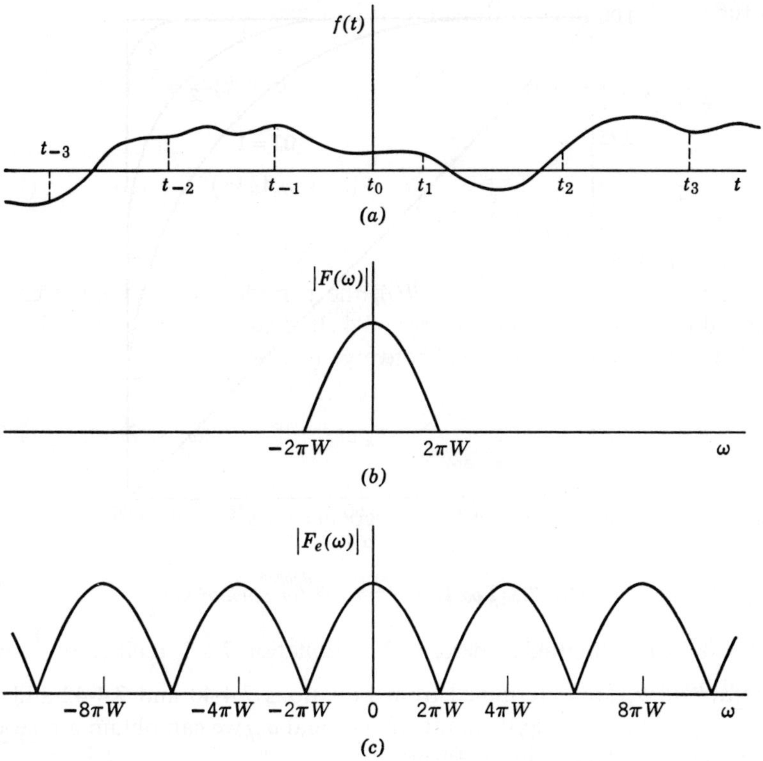

***Figure** 7.9 Sampling and bandlimited spectra:* (*a*) *a given signal with sample values* $\ldots, f(t_1),\ f(t_0),\ f(t_1), \ldots$; (*b*) *a strictly bandlimited spectrum;* (*c*) *the extension of a strictly bandlimited spectrum to a periodic function*

may be represented exactly by samples spaced relatively far apart. The reasonable conditions are that the signal be *strictly bandlimited.*

As defined previously, a (real) signal $f(t)$ will be called *strictly bandlimited* $(-2\pi W, 2\pi W)$ if its Fourier transform $F(\omega)$ has the property

$$F(\omega) = 0, \qquad |\omega| > 2\pi W. \tag{7.5-1}$$

The magnitude of such a spectrum is shown in Figure 7.9(*b*) where the shape is arbitrary. It is clear that this spectrum could be extended into a periodic frequency function with period $4\pi W$ as shown in Figure 7.9(*c*). In other words, we define a new function $F_e(\omega)$ by

$$F_e(\omega) = \sum_{n=-\infty}^{\infty} F(\omega + n4\pi W). \tag{7.5-2}$$

This function is periodic with period $4\pi W$ since

$$F_e(\omega + k4\pi W) = \sum_{n=-\infty}^{\infty} F[\omega + (n + k)4\pi W]$$

$$= \sum_{m=-\infty}^{\infty} F(\omega + m4\pi W) = F_e(\omega) \tag{7.5-3}$$

where $m = n + k$.

For reasonably well-behaved* $F(\omega)$, the periodic function $F_e(\omega)$ can be expanded in a Fourier series with period $4\pi W$, and, in the interval $-2\pi W \leq \omega \leq 2\pi W$, this Fourier series will converge to $F(\omega)$; that is,

$$F(\omega) = \sum_{k=-\infty}^{\infty} F_k\, e^{-jk2\pi\omega/4\pi W} = \sum_{k=-\infty}^{\infty} F_k\, e^{-jk\omega/2W}, \qquad |\omega| \leq 2\pi W, \tag{7.5-4}$$

where $j = \sqrt{-1}$ and F_k is the Fourier coefficient given by

$$F_k = \frac{1}{4\pi W}\int_{-2\pi W}^{2\pi W} F(\omega)\, e^{jk\omega/2W}\, d\omega. \tag{7.5-5}$$

Since $F(\omega)$ is bandlimited as described by Equation 7.5-1, its inverse Fourier transform is

$$f(t) = \frac{1}{2\pi}\int_{-2\pi W}^{2\pi W} F(\omega)\, e^{j\omega t}\, d\omega. \tag{7.5-6}$$

If we define the *Nyquist instants* as the set of times

$$\{t_n\} = \left\{t_n \middle| t_n = \frac{n}{2W}, \qquad n = \ldots, -1, 0, 1, \ldots\right\}, \tag{7.5-7}$$

then it is clear that $f(t_n)$ is given from Equation 7.5-6 as

$$\boxed{f(t_n) = f\left(\frac{n}{2W}\right) = \frac{1}{2\pi}\int_{-2\pi W}^{2\pi W} F(\omega)\, e^{jn\omega/2W}\, d\omega.} \tag{7.5-8}$$

A comparison of this last equation with Equation 7.5-5 shows that the Fourier coefficient F_k is related to the sample value $f(k/2W)$ by

$$\boxed{F_k = \frac{1}{2W} f\left(\frac{k}{2W}\right).} \tag{7.5-9}$$

* The conditions that a function must satisfy in order to be expanded in a Fourier series are discussed in Appendix D, particularly Section D.4.

If the sample values $f(n/2W)$ are given for all time, then the Fourier series

$$F(\omega) = \frac{1}{2W} \sum_{k=-\infty}^{\infty} f\left(\frac{k}{2W}\right) e^{-jk\omega/2W}, \qquad |\omega| \le 2\pi W, \tag{7.5-10}$$

determines $F(\omega)$ exactly and hence $f(t)$ through the inverse Fourier transform given by Equation 7.5-6. This completes the proof of the existence of a sampling theorem: *a function $f(t)$, strictly bandlimited $(-2\pi W, 2\pi W)$ radians per second, is uniquely and exactly determined by its sample values spaced $1/2W$ seconds apart throughout the time domain.* Of course, there are an infinite number of such samples.

We now consider the problem of the reconstruction of $f(t)$ from its sample values. On substituting Equation 7.5-10 into Equation 7.5-6, we obtain

$$f(t) = \frac{1}{2\pi} \int_{-2\pi W}^{2\pi W} \frac{1}{2W} \sum_{k=-\infty}^{\infty} f\left(\frac{k}{2W}\right) e^{-jk\omega/2W} e^{j\omega t}\, d\omega. \tag{7.5-11}$$

We may interchange the order of summation and integration:

$$f(t) = \sum_{k=-\infty}^{\infty} f\left(\frac{k}{2W}\right) \frac{1}{4\pi W} \int_{-2\pi W}^{2\pi W} e^{j\omega[t-(k/2W)]}\, d\omega. \tag{7.5-12}$$

The integral is the Fourier transform of the symmetrical pulse centered on $t = k/2W$. It may be evaluated easily as

$$\frac{1}{4\pi W} \int_{-2\pi W}^{2\pi W} e^{j\omega[t-(k/2W)]}\, d\omega = \frac{1}{2\pi Wt - k\pi} \left[\frac{e^{j(2\pi Wt - k\pi)} - e^{-j(2\pi Wt - k\pi)}}{2j}\right]$$

$$= \frac{\sin(2\pi Wt - k\pi)}{2\pi Wt - k\pi}. \tag{7.5-13}$$

This last result may be used in Equation 7.5-12 to give the *sampling representation*

$$f(t) = \sum_{k=-\infty}^{\infty} f\left(\frac{k}{2W}\right) \frac{\sin(2\pi Wt - k\pi)}{(2\pi Wt - k\pi)}. \tag{7.5-14}$$

This expression is sometimes called the *Cardinal Series.* Although it has been known to mathematicians at least since 1915 [11], its extensive use in engineering dates from some of the pioneer work of Shannon in 1949 in the field of information theory [12]. This expression, together with Equation 7.5-8, relates the discrete time domain $\{k/2W\}$ with sample values $f(k/2W\}$ to the continuous time domain $\{t\}$ of the function $f(t)$.

The interpolation function

$$k(t) = \frac{\sin 2\pi Wt}{2\pi Wt} \tag{7.5-15}$$

has a Fourier transform $K(\omega)$ given by

$$K(\omega) = \begin{cases} \dfrac{1}{4\pi W}, & |\omega| < 2\pi W, \\ 0, & |\omega| > 2\pi W. \end{cases} \tag{7.5-16}$$

Also the shifted function $k[t - (k/2W)]$ has the transform

$$\mathscr{F}\left\{k\left(t - \frac{k}{2W}\right)\right\} = K(\omega)\, e^{-j\omega(k/2W)}. \tag{7.5-17a}$$

Therefore, each term on the right side of Equation 7.5-14 is a time function which is strictly bandlimited $(-2\pi W, 2\pi W)$. Note also that

$$k\left(t - \frac{k}{2W}\right) = \frac{\sin(2\pi Wt - k\pi)}{2\pi Wt - k\pi} = \begin{cases} 1, & t = t_k = \dfrac{k}{2W}, \\ 0, & t = t_n,\ n \neq k. \end{cases} \tag{7.5-17b}$$

Thus this sampling function $k[t - (k/2W)]$ is zero at all Nyquist instants except t_k, where it equals unity.

The expansion of Equation 7.5-14 might be called a sampling theorem in the time domain. It is apparent from the symmetrical properties of the Fourier transform that a similar relationship could be written in the frequency domain for strictly time-limited signals. We can proceed as before and define a strictly time-limited signal $g(t)$ by

$$g(t) = 0, \qquad |t| > T. \tag{7.5-18}$$

Then the transform $G(\omega)$ is given by

$$G(\omega) = \int_{-T}^{T} g(t)\, e^{-j\omega t}\, dt. \tag{7.5-19}$$

Now let $g(t)$ be expanded in a Fourier series of period $2T$ in the interval $(-T, T)$

$$g(t) = \sum_{k=-\infty}^{\infty} G_k\, e^{jk\pi t/T} \tag{7.5-20}$$

with Fourier coefficient G_k given by

$$G_k = \frac{1}{2T}\int_{-T}^{T} g(t)\, e^{-jk\pi t/T}\, dt = \frac{1}{2T}\, G\left(\frac{k\pi}{T}\right) \tag{7.5-21}$$

from Equation 7.5-19. We use this last relation and substitute Equation 7.5-20 for $g(t)$ in Equation 7.5-19. On interchanging the order of summation and integration, we obtain

$$G(\omega) = \frac{1}{2T}\sum_{k=-\infty}^{\infty} G\left(\frac{k\pi}{T}\right)\int_{-T}^{T} e^{-jt[\omega - (k\pi/T)]}\, dt \tag{7.5-22}$$

or, as before,

$$G(\omega) = \sum_{k=-\infty}^{\infty} G\left(\frac{k\pi}{T}\right) \frac{\sin(\omega T - k\pi)}{\omega T - k\pi}. \tag{7.5-23}$$

Thus there is a sampling theorem in the frequency domain as follows: *a function $f(t)$, strictly time-limited $(-T, T)$ seconds, is uniquely and exactly determined by the sample values $G(k\pi/T)$ of its frequency spectrum $G(\omega)$ spaced π/T radians per second apart.* Note that each interpolation function in Equation 7.5-23 is exactly time limited $(-T, T)$.

For the strictly bandlimited signal $f(t)$, the spectrum $F(\omega)$ is nonzero only for $|\omega| \leq 2\pi W$, and the Fourier series of Equations 7.5-10 is considered only for this frequency interval. On the other hand, the Fourier transform of the sampling expression of Equation 7.5-14 is $F(\omega)$ only for $|\omega| \leq 2\pi W$, and, for all frequencies, is the function $F_e(\omega)$ given by Equation 7.5-2:

$$F_e(\omega) = \sum_{n=-\infty}^{\infty} F(\omega - n4\pi W). \tag{7.5-2}$$

We have already pointed out that $F_e(\omega)$ is periodic with period $4\pi W$. Let us call the set of frequencies ω_k, defined by

$$\omega_k = \omega - k4\pi W, \qquad k = \ldots, -1, 0, 1, \ldots, \tag{7.5-24}$$

the *aliases* of ω. It is clear that

$$F(\omega) = F_e(\omega_k), \qquad |\omega| < 2\pi W. \tag{7.5-25}$$

Suppose that a function $h(t)$ is not strictly bandlimited to at least $(-2\pi W, 2\pi W)$ radians per second and an attempt is made to reconstruct the function using Equation 7.5-14 with sample values space $1/2W$ seconds apart. It is apparent that the reconstructed signal [which is strictly bandlimited $(-2\pi W, 2\pi W)$ as already mentioned] will differ from the original. Moreover a given set of sample values $\{f(k/2W)\}$ could have been obtained from a whole class of different signals. Thus we emphasize that the reconstruction of Equation 7.5-14 is unambiguous only for signals strictly bandlimited to at least $(-2\pi W, 2\pi W)$ radians per second. The set of different possible signals with the same set of sample values $\{f(k/2W)\}$ are called the *aliases* of the bandlimited signal $f(t)$.

7.6 Bandpass Signals and Systems

The treatment of the previous sections is most applicable to low-pass frequency functions of the general form of Figure 7.4(a). For typical bandpass functions as shown in Figure 7.4(b) it is often more convenient to consider special techniques of analysis and methods of representation. Since most of the modulated signals to be studied later will have a bandpass frequency characteristic, we devote this section to such topics.

We shall say that a (real) function $f(t)$ is a *bandpass* type of function provided its Fourier transform $F(\omega)$ has the property that it is approximately zero outside of some interval which does not include the point $\omega = 0$ and its vicinity. In other words

$$F(\omega) \approx 0, \qquad \begin{cases} |\omega - \omega_0| > 2\pi W, & \omega > 0, \\ |\omega + \omega_0| > 2\pi W, & \omega < 0, \end{cases} \tag{7.6-1}$$

where $4\pi W < 2\omega_0$ is a measure of the (radian) bandwidth. Such a function is shown in Figure 7.4(b) with $4\pi W \approx \Delta\Omega$. If the approximation becomes equality in Equation 7.6-1, we shall say that $f(t)$ is a *strictly bandlimited, bandpass function.*

We shall begin by showing that every bandpass signal has a low-frequency equivalent signal of the same bandwidth. Let the bandpass signal $f(t)$ have the Fourier transform $F(\omega)$ given by Equation 7.6-1. We shall say that this signal has a bandwidth $4\pi W$ as shown in Figure 7.10(a). This figure has been drawn to emphasize that the spectrum $F(\omega)$ need not be symmetrical about ω_0 and $-\omega_0$. However, since $f(t)$ is real, we must have

$$F(\omega) = F^*(-\omega). \tag{7.6-2}$$

Let us assume that we wish to write the signal $f(t)$ as

$$f(t) = A(t)\cos[\omega_0 t + \Psi(t)] = \text{Re}\{A(t)\,e^{j[\omega_0 t + \Psi(t)]}\} \tag{7.6-3}$$

where ω_0 is the *center frequency* as shown in Figure 7.10(a) and $A(t)$ and $\Psi(t)$ are called the *envelope function* and *phase function*, respectively, for the bandpass case. A method for determining these functions is developed in Section H.5 of Appendix H for the low-pass case, that is, where $\omega_0 = 0$.

Let us follow the procedure of Appendix H and, as in Equation (H.4-3), define the *positive frequency spectrum* $F_+(\omega)$ by

$$F_+(\omega) = \sqrt{2}\,F(\omega)u(\omega) \tag{7.6-4}$$

and the *negative frequency spectrum* $F_-(\omega)$ by

$$F_-(\omega) = \sqrt{2}\,F(\omega)u(-\omega) \tag{7.6-5}$$

where $u(\omega)$ is the unit-step function. Now $F(\omega)$ may be written as

$$F(\omega) = \frac{1}{\sqrt{2}}[F_+(\omega) + F_-(\omega)]. \tag{7.6-6}$$

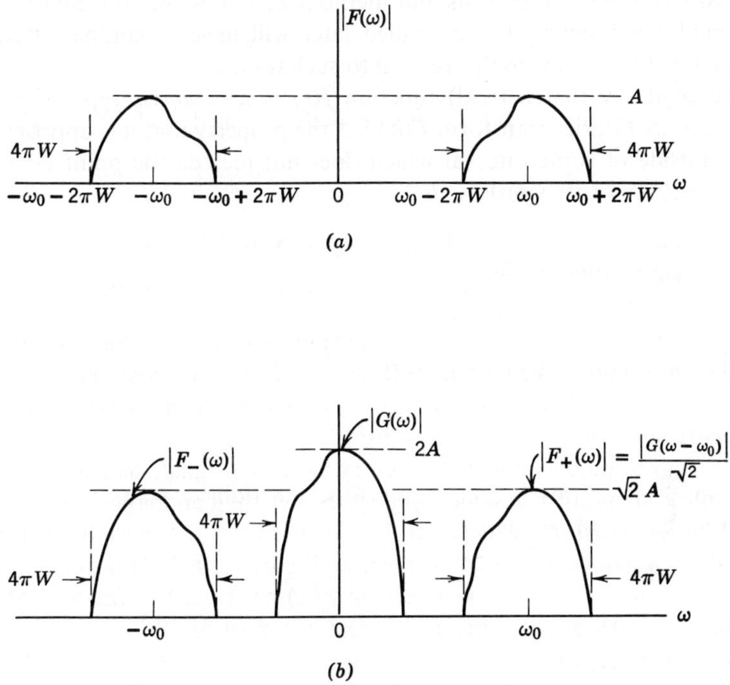

Figure 7.10 *Bandpass spectra and low-frequency equivalents: (a) a bandpass signal spectrum of bandwidth $4\pi W$; (b) the equivalent low-pass spectrum $G(\omega)$ of a bandpass signal*

On taking the inverse transform, we have

$$f(t) = \frac{1}{\sqrt{2}}\,[f_+(t) + f_-(t)] \tag{7.6-7}$$

where

$$f_+(t) = \frac{\sqrt{2}}{2\pi}\int_0^{\infty} F(\omega)\,e^{j\omega t}\,d\omega \tag{7.6-8}$$

and

$$f_-(t) = \frac{\sqrt{2}}{2\pi}\int_{-\infty}^{0} F(\omega)\,e^{j\omega t}\,d\omega = \frac{\sqrt{2}}{2\pi}\int_0^{\infty} F^*(\omega)\,e^{j\omega t}\,d\omega = f_+^*(t) \tag{7.6-9}$$

from Equation 7.6-2. Note that $f_+(t)$ is just the *analytic signal associated with* $f(t)$ and defined in Section H.4 of Appendix H. Thus Equation 7.6-7 becomes

$$f(t) = \frac{1}{\sqrt{2}}\,[f_+(t) + f_+^*(t)]. \tag{7.6-10}$$

If we let

$$f_+(t) = \frac{A(t)}{\sqrt{2}} e^{j\Psi(t)} e^{j\omega_0 t}, \tag{7.6-11}$$

then it is clear from a comparison of Equations 7.6-10 and 7.6-3 that $f(t)$ is indeed given by

$$f(t) = A(t) \cos [\omega_0 t + \Psi(t)]. \tag{7.6-12}$$

We shall not bother to prove the uniqueness of the result.

Let us define function $g(t)$ by

$$g(t) = A(t) e^{j\Psi(t)} \tag{7.6-13}$$

with Fourier transform $G(\omega)$, given by

$$G(\omega) = \int_{-\infty}^{\infty} A(t) e^{j\Psi(t)} e^{-j\omega t} dt. \tag{7.6-14}$$

A comparison of this last equation with the Fourier transform $F_+(\omega)$ of Equation 7.6-11 shows that

$$F_+(\omega) = \frac{G(\omega - \omega_0)}{\sqrt{2}} \tag{7.6-15}$$

where $G(\omega)$ is called the *equivalent low-pass spectrum* of $F(\omega)$ and $g(t)$, as given by Equation 7.6-13, the *equivalent low-pass signal.* It is obvious that $G(\omega)$ has the same bandwidth $4\pi W$ as the original spectrum $F(\omega)$ as illustrated in Figure 7.10(b). Thus, as we set out to show, every bandpass signal has a low-frequency equivalent signal of the same bandwidth.

By *bandpass systems,* we shall mean (real) linear filters with (real) impulse response $h(t)$ and complex transfer function $H(\omega)$ satisfying Equation 7.6-1. Thus the properties developed in the previous paragraphs apply equally to bandpass systems. In addition, however, we shall frequently require that the bandpass system be *causal* or physically realizable; that is, that

$$h(t) = 0, \qquad t < 0. \tag{7.6-16}$$

It is pointed out in Section H.2 of Appendix H that *causal functions* such as that of Equation 7.6-16 have transforms whose real and imaginary parts form a Hilbert-transform pair. Thus we have

$$H(\omega) = \text{Re } H(\omega) + j \text{ Im } H(\omega) \tag{7.6-17}$$

and

$$\text{Re } H(\omega) = \frac{1}{\pi} \fint_{-\infty}^{\infty} \frac{\text{Im } H(\nu)}{\omega - \nu} d\nu, \tag{7.6-18}$$

$$\text{Im } H(\omega) = -\frac{1}{\pi} \fint_{-\infty}^{\infty} \frac{\text{Re } H(\nu)}{\omega - \nu} d\nu \tag{7.6-19}$$

in agreement with Equations H.2-16 and H.2-17. Of course, since $h(t)$ is real, Equation 7.6-2 is satisfied in the form

$$H(\omega) = \int_0^\infty h(t)\, e^{-j\omega t}\, dt = H^*(-\omega). \tag{7.6-20}$$

7.7 The Sampling Theorem for Bandpass Signals

We shall consider a strictly bandlimited, bandpass signal $f(t)$ with Fourier transform $F(\omega)$ such that

$$F(\omega) = 0 \quad \text{except for} \quad \begin{cases} \omega_1 \le \omega \le \omega_2, & \omega > 0, \\ -\omega_2 \le \omega \le -\omega_1, & \omega < 0. \end{cases} \tag{7.7-1}$$

A typical spectrum of this type is illustrated in Figure 7.11(*a*). We wish to develop a sampling theorem for such signals which is similar to the sampling theorem already presented in Section 7.5. The procedure to be followed is straightforward. We can define the analytic signal $f_+(t)$ corresponding to $f(t)$ as discussed in Section H.4 of the Appendices. The spectrum $F_+(\omega)$ of this analytic signal is (except for a scale factor) simply the positive frequency components of $F(\omega)$ as shown in Figure 7.11(*b*). We then proceed as in Section 7.5 to develop a sampling expansion centered on ω_0 rather than zero

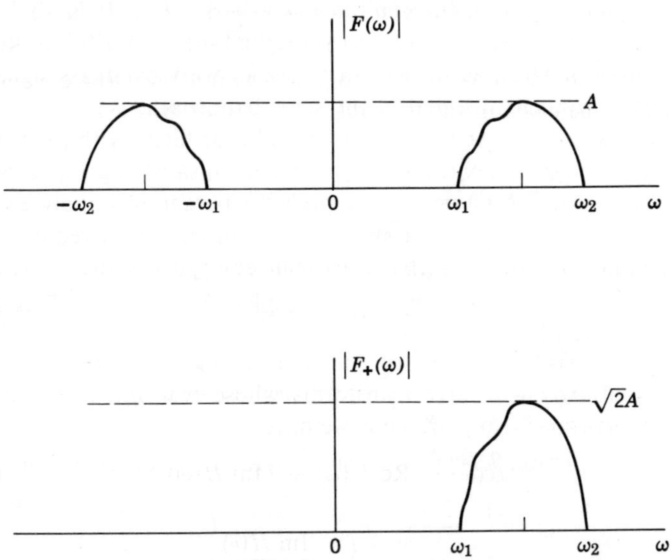

Figure 7.11 *The spectrum of a bandpass signal and its associated analytic signal; (a) a strictly bandlimited, bandpass signal spectrum; (b) the associated analytic signal spectrum*

and using samples of $g(t)$, the analytic signal, rather than of the original signal $f(t)$.

We begin by defining the analytic signal $f_+(t)$ in accordance with Equation H.4-1

$$f_+(t) = \frac{f(t) - jf_1(t)}{\sqrt{2}} \tag{7.7-2}$$

where $f_1(t)$ is the Hilbert transform of $f(t)$ and is given by

$$f_1(t) = -\frac{1}{\pi} \fint_{-\infty}^{\infty} \frac{f(x)}{t - x}\, dx \tag{7.7-3}$$

as discussed in Appendix H in Section H.1. Also, as shown in Equation H.1-17, the spectrum $F_1(\omega)$ is

$$F_1(\omega) = jF(\omega) \operatorname{sgn} \omega \tag{7.7-4}$$

where $\operatorname{sgn} \omega$ is $+1$ for $\omega > 0$ and -1 for $\omega < 0$. Thus the spectrum of $f_+(t)$ is

$$F_+(\omega) = \frac{F(\omega) - jF_1(\omega)}{\sqrt{2}} = \sqrt{2}\, F(\omega) u(\omega) \tag{7.7-5}$$

where $u(\omega)$ is the unit-step function. Since $F_+(\omega)$ is nonzero only in the interval $\omega_1 \leq \omega \leq \omega_2$, we can write its inverse Fourier transform as

$$f_+(t) = \frac{1}{2\pi} \int_{-\infty}^{\infty} F_+(\omega)\, e^{j\omega t}\, d\omega = \frac{1}{\sqrt{2}\pi} \int_{\omega_1}^{\omega_2} F(\omega)\, e^{j\omega t}\, d\omega. \tag{7.7-6}$$

We will now proceed as in Section 7.5 to find a sampling expansion for $f(t)$. Let $F(\omega)$ be given in the interval $\omega_1 \leq \omega \leq \omega_2$ by the complex Fourier series

$$F(\omega) = \sum_{n=-\infty}^{\infty} C_n\, e^{-j[2\pi n/(\omega_2 - \omega_1)]\omega}. \tag{7.7-7}$$

We use a complex series in order to be able to relate the Fourier coefficient to the analytic signal $f_+(t)$, which is complex for real $f(t)$. This Fourier coefficient is given by

$$C_n = \frac{1}{\omega_2 - \omega_1} \int_{\omega_1}^{\omega_2} F(\omega)\, e^{j[2\pi n/(\omega_2 - \omega_1)]\omega}\, d\omega. \tag{7.7-8}$$

If we rewrite Equation 7.7-6 for the case where t takes on the set of values

$$\{t_n\} = \left\{ \frac{2\pi n}{\omega_2 - \omega_1} \,\middle|\, n = \ldots, -1, 0, 1, \ldots \right\}, \tag{7.7-9}$$

we have

$$f_+(t_n) = f_+\left(\frac{2\pi n}{\omega_2 - \omega_1}\right) = \frac{1}{\sqrt{2}\pi} \int_{\omega_1}^{\omega_2} F(\omega)\, e^{j[2\pi n/(\omega_2 - \omega_1)]\omega}\, d\omega. \tag{7.7-10}$$

A comparison of Equations 7.7-8 and 7.7-10 shows that

$$C_n = \frac{\sqrt{2}\pi}{\omega_2 - \omega_1} f_+\left(\frac{2\pi n}{\omega_2 - \omega_1}\right). \tag{7.7-11}$$

Thus the coefficients in the Fourier series of Equation 7.7-7 are determined by the sample values of the analytic signal $f_+(t)$ spaced $2\pi/(\omega_2 - \omega_1)$ seconds apart throughout the time domain. Since the original signal $f(t)$ is assumed to be real, its Fourier transform given by

$$F(\omega) = \int_{-\infty}^{\infty} f(t)\, e^{-j\omega t}\, dt \tag{7.7-12}$$

must have the property

$$F(-\omega) = F^*(\omega); \tag{7.7-13}$$

consequently a knowledge of $F(\omega)$ for the positive frequency range $\omega_1 \leq \omega \leq \omega_2$ as given by Equation 7.7-7 is sufficient to determine $F(\omega)$ for all frequencies. Therefore we have a bandpass sampling theorem: *If the (real) signal $f(t)$ is a bandpass function strictly bandlimited (ω_1, ω_2), then it is determined uniquely and exactly by the sample values of its analytic signal $f_+(t)$ spaced $2\pi/(\omega_2 - \omega_1)$ seconds apart throughout the time domain.*

We now proceed to find a sampling expansion representation for $f(t)$. Substituting Equations 7.7-7 and 7.7-11 into Equation 7.7-6 and rearranging, we obtain

$$f_+(t) = \frac{1}{\omega_2 - \omega_1} \sum_{n=-\infty}^{\infty} f_+\left(\frac{2\pi n}{\omega_2 - \omega_1}\right) \int_{\omega_1}^{\omega_2} e^{j\omega\{t - [2\pi n/(\omega_2 - \omega_1)]\}}\, d\omega. \tag{7.7-14}$$

On performing the integration and noting from Equation 7.7-2 that $f(t)$ is just $\sqrt{2}$ times the real part of $f_+(t)$, we have

$$f(t) = \text{Re}\left\{\frac{\sqrt{2}}{\omega_2 - \omega_1} \sum_{n=-\infty}^{\infty} f_+\left(\frac{2\pi n}{\omega_2 - \omega_1}\right) \times \frac{e^{j\omega_2\{t - [2\pi n/(\omega_2 - \omega_1)]\}} - e^{j\omega_1\{t - [2\pi n/(\omega_2 - \omega_1)]\}}}{j\{t - [2\pi n/(\omega_2 - \omega_1)]\}}\right\}. \tag{7.7-15}$$

This rather unattractive equation is the equivalent for the bandpass case of Equation 7.5-14 for the low-pass case. It is clear that the original sampling expansion of Equation 7.5-14 could be used with upper frequency $2\pi W = \omega_2$. In this event, however, samples would be spaced $1/2W$ or π/ω_2 seconds apart. If the expansion of Eq. 7.7-15 is used, the sample spacing is $2\pi/(\omega_2 - \omega_1)$ seconds but two samples [that is, $f(t_n)$ and $f_1(t_n)$] must be taken at each point. Thus in the first case the sample rate is ω_2/π samples per second and in the second case only $(\omega_2 - \omega_1)/\pi$, which is always smaller than ω_2/π for all ω_1 such that $0 < \omega_1 < \omega_2$. In other words, for maximum economy in sampling, the bandpass representation must be used for bandpass functions since its sampling rate is determined by $(\omega_2 - \omega_1)$ rather than by ω_2 only.

We may use Equation 7.7-2 to replace $f_+[2\pi n/(\omega_2 - \omega_1)]$ in Equation 7.7-15; then we can take the real part of the equation. After some re-arranging and simplification, Equation 7.7-15 can be rewritten as

$$f(t) = \sqrt{2} \sum_{n=-\infty}^{\infty} \frac{1}{(\omega_2 - \omega_1)t - 2\pi n}$$
$$\times \left\{ \begin{array}{l} f\left(\dfrac{2\pi n}{\omega_2 - \omega_1}\right)\left[\sin \omega_2\left(t - \dfrac{2\pi n}{\omega_2 - \omega_1}\right) - \sin \omega_1\left(t - \dfrac{2\pi n}{\omega_2 - \omega_1}\right)\right] \\ - f_1\left(\dfrac{2\pi n}{\omega_2 - \omega_1}\right)\left[\cos \omega_2\left(t - \dfrac{2\pi n}{\omega_2 - \omega_1}\right) - \cos \omega_1\left(t - \dfrac{2\pi n}{\omega_2 - \omega_1}\right)\right] \end{array} \right\} \tag{7.7-16}$$

Although this expression may seem scarcely more attractive than Equation 7.7-15, it contains no complex functions. If we let the bandwidth $\omega_2 - \omega_1$ be

$$\omega_2 - \omega_1 = 2\pi W, \tag{7.7-17}$$

then sample values $f(n/W)$ and $f_1(n/W)$ are each taken $1/W$ seconds apart throughout the time domain. Other forms of the bandpass sampling theorem can be obtained [13].

7.8 Power Spectrum and Time Autocorrelation Functions

In Chapter 3 we considered in some detail the concepts of power spectral density and correlation function as related to random processes. Also, in Appendix D, we treated signal energies in terms of Fourier series and transforms. In this section we take up briefly the application of these concepts to deterministic signals.

We are given the deterministic signal $f(t)$ and define the truncated signal

$$f_T(t) = \begin{cases} f(t), & |t| \le T, \\ 0, & |t| > T. \end{cases} \tag{7.8-1}$$

This truncated signal has a Fourier transform

$$F_T(\omega) = \int_{-\infty}^{\infty} f_T(t)\, e^{-j\omega t}\, dt = \int_{-T}^{T} f(t)\, e^{-j\omega t}\, dt. \tag{7.8-2}$$

If $f(t)$ has a Fourier transform $F(\omega)$, then we have

$$\lim_{T\to\infty} F_T(\omega) = F(\omega). \tag{7.8-3}$$

The total energy $\mathscr{E}_T$ in $f_T(t)$ is given by Parseval's Theorem as

$$\mathscr{E}_T = \int_{-\infty}^{\infty} f_T^2(t)\, dt = \int_{-T}^{T} f^2(t)\, dt = \frac{1}{2\pi}\int_{-\infty}^{\infty} |F_T(\omega)|^2\, d\omega. \tag{7.8-4}$$

The average (normalized) power P_T can be defined as the average rate of flow of energy or

$$P_T = \frac{\mathscr{E}_T}{2T} = \frac{1}{2T}\frac{1}{2\pi}\int_{-\infty}^{\infty} |F_T(\omega)|^2\, d\omega. \tag{7.8-5}$$

In the limit the energy becomes

$$\mathscr{E} = \lim_{T\to\infty}\int_{-T}^{T} f^2(t)\, dt \tag{7.8-6}$$

and the power is

$$P = \lim_{T\to\infty}\frac{1}{2T}\int_{-T}^{T} f^2(t)\, dt = \lim_{T\to\infty}\frac{1}{2T}\frac{1}{2\pi}\int_{-T}^{T} |F_T(\omega)|^2\, d\omega. \tag{7.8-7}$$

In general, if $\mathscr{E}$ exists, then P will be zero. On the other hand, P may exist and be well defined while $\mathscr{E}$ becomes infinite. If P exists, then Equation 7.8-7 can be written as

$$P = \frac{1}{2\pi}\int_{-\infty}^{\infty}\lim_{T\to\infty}\frac{|F_T(\omega)|^2}{2T}\, d\omega. \tag{7.8-8}$$

Let us define the *power spectrum* $G(\omega)$ of the signal $f(t)$ by

$$G(\omega) = \lim_{T\to\infty}\frac{|F_T(\omega)|^2}{2T}. \tag{7.8-9}$$

When this quantity exists, we have from Equation 7.8-8

$$P = \frac{1}{2\pi}\int_{-\infty}^{\infty} G(\omega)\, d\omega. \tag{7.8-10}$$

It is apparent that the power spectrum $G(\omega)$ of the deterministic signal $f(t)$ is analogous to the power spectral density $\varphi(\omega)$ of the random process $v(t)$ as defined by Equation 3.8-13.

Example 7.6 Let us assume that the deterministic signal $f(t)$ is the constant A so that

$$f(t) = A.$$

It follows from Equation 7.8-9 that

$$\begin{aligned} G(\omega) &= \lim_{T\to\infty}\frac{1}{2T}\left(\int_{-T}^{T} A\, e^{-j\omega t}\, dt \int_{-T}^{T} A\, e^{j\omega t}\, dt\right) \\ &= A^2 \lim_{T\to\infty}\frac{1}{2T}\left(\frac{2\sin\omega T}{\omega}\frac{2\sin\omega T}{\omega}\right) \\ &= A^2 \lim_{T\to\infty}\frac{1}{T}\left(\frac{1-\cos 2\omega T}{\omega^2}\right). \end{aligned}$$

It is clear that, for $\omega \neq 0$, the power spectrum $G(\omega)$ is zero. On the other hand, for investigating $G(0)$ we rewrite $G(\omega)$ as

$$G(\omega) = 2\pi A^2 \lim_{T\to\infty} \frac{T}{\pi}\left(\frac{\sin \omega T}{\omega T}\right)^2.$$

As in Appendix B, it can be shown that a limiting expression for the Dirac delta function is

$$\delta(\omega) = \lim_{T\to\infty} \frac{T}{\pi}\left(\frac{\sin \omega T}{\omega T}\right)^2.$$

Thus the power spectrum becomes

$$G(\omega) = A^2 2\pi\delta(\omega)$$

as expected. We note in passing that the inverse Fourier transform of $G(\omega)$ is

$$\mathscr{F}^{-1}[G(\omega)] = A^2 \int_{-\infty}^{\infty} \delta(\omega)\, e^{j\omega t}\, d\omega = A^2.$$

• •

In Section 3.4 we defined the time average $A[v(t)]$ of a random process $v(t)$ by

$$A[v(t)] = \lim_{T\to\infty} \frac{1}{2T}\int_{-T}^{T} v(t)\, dt. \tag{7.8-11}$$

Since this equation represents an operation on each member of the ensemble, it can also be used as a definition of the time average of the deterministic signal $f(t)$. There will be cases where it will be somewhat more convenient to define the asymmetrical time average $A_a[f(t)]$ by

$$A_a[f(t)] = \lim_{T_1\to\infty} \frac{1}{2T_1}\int_{-T_1}^{0} f(t)\, dt + \lim_{T_2\to\infty} \frac{1}{2T_2}\int_{0}^{T_2} f(t)\, dt. \tag{7.8-12}$$

This time average has the desirable property of being invariant to time translation; that is,

$$A_a[f(t)] = A_a[f(t + b)] \tag{7.8-13}$$

and will be useful in treating signals which are analogous in a certain sense to nonstationary random processes. It is easy to show that, if $A_a[f(t)]$ exists, then $A[f(t)]$ exists also and is equal to $A_a[f(t)]$. However, the converse is not true.

Example 7.7 We consider the deterministic signal given by

$$f(t) = t.$$

The symmetrical time average is

$$A(t) = \lim_{T\to\infty} \frac{1}{2T}\int_{-T}^{T} t\, dt = \lim_{T\to\infty} \frac{1}{2T}\left(\frac{T^2}{2} - \frac{T^2}{2}\right) = 0.$$

For arbitrary time translation b, we have

$$A(t + b) = \lim_{T\to\infty} \frac{1}{2T} \int_{-T}^{T} (t + b)\, dt = \lim_{T\to\infty} \frac{1}{2T} (2bT) = b.$$

On the other hand, we find that the asymmetrical time average

$$A_a(t) = \lim_{T_1\to\infty} \frac{1}{2T_1} \int_{-T_1}^{0} t\, dt + \lim_{T_2\to\infty} \frac{1}{2T_2} \int_{0}^{T_2} t\, dt$$

does not exist in the sense that both of the terms on the right side become infinite in the limit.

• •

As in Section 3.4, we now define the time autocorrelation function of a signal $f(t)$ by

$$\mathscr{R}(\tau) = A[f(t)f(t + \tau)] = \lim_{T\to\infty} \frac{1}{2T} \int_{-T}^{T} f(t)f(t + \tau)\, dt. \tag{7.8-14}$$

Note that this expression can also be written as

$$\mathscr{R}(\tau) = \lim_{T\to\infty} \frac{1}{2T} \int_{-T}^{T} f_T(t)f(t + \tau)\, dt \tag{7.8-15}$$

where $f_T(t)$ is given by Equation 7.8-1. Suppose that we now find the Fourier transform of $\mathscr{R}(\tau)$ and write

$$\int_{-\infty}^{\infty} \mathscr{R}(\tau)\, e^{-j\omega\tau}\, d\tau = \int_{-\infty}^{\infty} \lim_{T\to\infty} \frac{1}{2T} \int_{-T}^{T} f_T(t)f(t + \tau)\, dt\, e^{-j\omega\tau}\, d\tau. \tag{7.8-16}$$

If the order of integration can be interchanged we obtain

$$\int_{-\infty}^{\infty} \mathscr{R}(\tau)\, e^{-j\omega\tau}\, d\tau = \lim_{T\to\infty} \frac{1}{2T} \int_{-T}^{T} f_T(t) \int_{-\infty}^{\infty} f(t + \tau)\, e^{-j\omega\tau}\, d\tau\, dt. \tag{7.8-17}$$

But the inner integral is just $e^{j\omega t}F(\omega)$ so that

$$\begin{aligned} \int_{-\infty}^{\infty} \mathscr{R}(\tau)\, e^{-j\omega\tau}\, d\tau &= \lim_{T\to\infty} \frac{1}{2T} F(\omega) \int_{-T}^{T} f(t)\, e^{j\omega t}\, dt \\ &= \lim_{T\to\infty} \frac{F(\omega)F_T(-\omega)}{2T} \\ &= \lim_{T\to\infty} \frac{F_T(\omega)F_T(-\omega)}{2T} \end{aligned} \tag{7.8-18}$$

whenever the limit exists. A comparison of this last expression with Equation 7.8-9 shows that the power spectrum $G(\omega)$ is given by

$$G(\omega) = \int_{-\infty}^{\infty} \mathscr{R}(\tau)\, e^{-j\omega\tau}\, d\tau. \tag{7.8-19}$$

The inverse transform is

$$\mathscr{R}(\tau) = \frac{1}{2\pi} \int_{-\infty}^{\infty} G(\omega)\, e^{j\omega\tau}\, d\omega. \tag{7.8-20}$$

Thus, when they exist, the time autocorrelation function and power spectrum form a Fourier-transform pair.

The time autocorrelation function $\mathscr{R}(\tau)$ has a set of properties exactly analogous to those of the autocorrelation function $R(\tau)$ of a wide-sense stationary random process. In particular we have:

1. $\mathscr{R}(0) = A[f^2(t)]$.
2. $\mathscr{R}(\tau) = \mathscr{R}(-\tau)$.
3. $\mathscr{R}(0) \geq |\mathscr{R}(\tau)|$.

These properties can be verified by the same procedure as used in Section 3.3 with the time average operator $A(\ )$ replacing the expectation operator $E\{\ \}$.

7.9 Periodic Signals and Discrete Spectra

In this section we consider the special case where the deterministic signal $f(t)$ is periodic with period P so that

$$f(t) = f(t + P). \tag{7.9-1}$$

As discussed in Appendix D, such a signal may be expanded in a Fourier series of the form

$$f(t) = \frac{a_0}{2} + \sum_{k=1}^{\infty} (a_k \cos k\omega_1 t + b_k \sin k\omega_1 t) \tag{7.9-2}$$

or, in complex notation,

$$f(t) = \sum_{k=-\infty}^{\infty} c_k\, e^{-jk\omega_1 t} \tag{7.9-3}$$

where $\omega_1 = 2\pi/P$ and where the Fourier coefficients a_k, b_k, and c_k are given by the integrals

$$a_k = \frac{2}{P} \int_{-P/2}^{P/2} f(t) \cos k\omega_1 t\, dt, \tag{7.9-4}$$

$$b_k = \frac{2}{P} \int_{-P/2}^{P/2} f(t) \sin k\omega_1 t\, dt, \tag{7.9-5}$$

and

$$c_k = \frac{1}{P} \int_{-P/2}^{P/2} f(t)\, e^{jk\omega_1 t}\, dt. \tag{7.9-6}$$

Let us consider a signal $k(t)$ of the form

$$k(t) = A\cos(\omega_0 t + \theta) \tag{7.9-7}$$

where A, ω_0, and θ are constants. The time autocorrelation function $\mathscr{R}(\tau)$ of this signal is given by Equation 7.8-14 and may be written as

$$\begin{aligned}
\mathscr{R}(\tau) &= \lim_{T\to\infty}\frac{1}{2T}\int_{-T}^{T} A^2\cos(\omega_0 t + \theta)\cos[\omega_0(t+\tau)+\theta]\,dt \\
&= \frac{A^2}{2}\lim_{T\to\infty}\frac{1}{2T}\int_{-T}^{T}\{\cos\omega_0\tau + \cos[\omega_0(2t+\tau)+2\theta]\}\,dt \\
&= \frac{A^2}{2}\cos\omega_0\tau + \frac{A^2}{2}\lim_{T\to\infty}\frac{1}{2T}\int_{-T}^{T}\cos[\omega_0(2t+\tau)+2\theta]\,dt \\
&= \frac{A^2}{2}\cos\omega_0\tau.
\end{aligned} \tag{7.9-8}$$

Thus the phase angle information has been lost in the calculation of the time autocorrelation function of a sinusoid although the amplitude and frequency information have been retained. By allowing θ to take on the value $\pm\pi/2 + \theta_1$, we note as a trivial consequence that Equation 7.9-8 also gives the time autocorrelation function of $A\sin(\omega_0 t + \theta)$. We now let $k(t)$ be given by

$$k(t) = A\cos k\omega_1 t + B\cos m\omega_1 t \tag{7.9-9}$$

where k and m are integers. The time autocorrelation function of this signal is obtained straightforwardly as

$$\mathscr{R}(\tau) = \begin{cases} \dfrac{(A+B)^2}{2}\cos k\omega_1\tau, & k = m, \\[2ex] \dfrac{A^2}{2}\cos k\omega_1\tau + \dfrac{B^2}{2}\cos m\omega_1\tau, & k \neq m. \end{cases} \tag{7.9-10}$$

On applying Equations 7.9-8 and 7.9-10 to the Fourier series of Equation 7.9-2, we obtain for the time autocorrelation function of the arbitrary periodic signal $f(t)$

$$\mathscr{R}(\tau) = \frac{a_0^2}{4} + \sum_{k=1}^{\infty}\frac{a_k^2 + b_k^2}{2}\cos k\omega_1\tau. \tag{7.9-11}$$

We transform this expression term by term to find the power spectrum $G(\omega)$ to be

$$G(\omega) = \frac{\pi}{2}a_0^2\delta(\omega) + \sum_{k=1}^{\infty}\pi\frac{a_k^2 + b_k^2}{2}[\delta(\omega - k\omega_1) + \delta(\omega + k\omega_1)] \tag{7.9-12}$$

as expected.

Example 7.8 Let us find the power spectrum of the sinusoid

$$f(t) = A \cos \omega_0 t$$

by direct calculation from Equation 7.8-9:

$$\begin{aligned} G(\omega) &= \lim_{T\to\infty} \frac{1}{2T} \left| \int_{-T}^{T} A \cos \omega_0 t \, e^{-j\omega t} \, dt \right|^2 \\ &= A^2 \lim_{T\to\infty} \frac{1}{2T} \left| \int_{-T}^{T} \tfrac{1}{2}[e^{jt(\omega_0-\omega)} + e^{-jt(\omega_0+\omega)}] \, dt \right|^2 \\ &= \frac{A^2}{2} \lim_{T\to\infty} \left[\sqrt{T} \frac{\sin (\omega - \omega_0)T}{(\omega - \omega_0)T} + \sqrt{T} \frac{\sin (\omega + \omega_0)T}{(\omega + \omega_0)T} \right]^2. \end{aligned}$$

When this last expression is squared, it is clear that the cross-terms will be nonintersecting since one will peak at the point $\omega = \omega_0$ and the other at $\omega = -\omega_0$. Thus

$$G(\omega) = \frac{A^2\pi}{2} \lim_{T\to\infty} \left\{ \frac{T}{\pi} \left[\frac{\sin (\omega - \omega_0)T}{(\omega - \omega_0)T} \right]^2 + \frac{T}{\pi} \left[\frac{\sin (\omega + \omega_0)T}{(\omega + \omega_0)T} \right]^2 \right\}.$$

As pointed out in Example 7.6, we have

$$\lim_{T\to\infty} \frac{T}{\pi} \left(\frac{\sin \omega T}{\omega T} \right)^2 = \delta(\omega)$$

so that

$$G(\omega) = \frac{A^2\pi}{2} [\delta(\omega - \omega_0) + \delta(\omega + \omega_0)].$$

But this is just the Fourier transform of Equation 7.9-8.

• •

7.10 Pulsed Signals—Signals with Finite Energy

For the case where the signal energy is finite, that is,

$$\mathscr{E} = \lim_{T\to\infty} \int_{-T}^{T} f^2(t) \, dt < \infty, \tag{7.10-1}$$

the time autocorrelation function $\mathscr{R}(\tau)$ will be zero for all τ. A deterministic signal $f(t)$ with the property given by Equation 7.10-1 will be called a *pulse* or a *pulsed signal*. Somewhat as in Section 5.3 it is convenient to define the energy autocorrelation function $\mathscr{R}_{\mathscr{E}}(\tau)$ by

$$\mathscr{R}_{\mathscr{E}}(\tau) = \lim_{T\to\infty} \int_{-T}^{T} f(t)f(t + \tau) \, dt \tag{7.10-2}$$

when this expression exists. The function defined by this last expression has the same properties as the time autocorrelation function of Section 7.8 except that

$$\mathscr{R}_{\mathscr{E}}(0) = \mathscr{E}. \tag{7.10-3}$$

Clearly, the corresponding energy spectrum is

$$G_{\mathscr{E}}(\omega) = \lim_{T \to \infty} F_T(\omega) F_T(-\omega). \tag{7.10-4}$$

Example 7.9 Let $f(t)$ be given by

$$f(t) = A\, e^{-\alpha t}\, u(t)$$

where $u(t)$ is the unit-step function. Then the total signal energy is

$$\mathscr{E} = \int_0^\infty A^2\, e^{-2\alpha t}\, dt = \frac{A^2}{2\alpha}$$

while the energy autocorrelation function is, for $\tau \geq 0$,

$$\mathscr{R}_{\mathscr{E}}(\tau) = \int_0^\infty A\, e^{-\alpha t} A\, e^{-\alpha(t+\tau)}\, dt$$

$$= \frac{A^2}{2\alpha} e^{-\alpha\tau}, \qquad \tau \geq 0,$$

and, for $\tau \leq 0$,

$$\mathscr{R}_{\mathscr{E}}(\tau) = \int_{-\tau}^\infty A\, e^{-\alpha t} A\, e^{-\alpha(t+\tau)}\, dt$$

$$= \frac{A^2}{2\alpha} e^{\alpha\tau}, \qquad \tau \leq 0,$$

or, finally,

$$\mathscr{R}_{\mathscr{E}}(\tau) = \frac{A^2}{2\alpha} e^{-\alpha|\tau|}.$$

The transform $F_T(\omega)$ is given by

$$F_T(\omega) = \int_0^T A\, e^{-\alpha t}\, e^{-j\omega t}\, dt = \frac{A}{\alpha + j\omega}[1 - e^{-(\alpha + j\omega)T}]$$

and the energy spectrum by

$$G_{\mathscr{E}}(\omega) = \lim_{T \to \infty} \frac{A^2}{\alpha^2 + \omega^2}[1 - e^{-(\alpha + j\omega)T}][1 - e^{-(\alpha - j\omega)T}]$$

$$= \frac{A^2}{\alpha^2 + \omega^2}.$$

• •

7.11 Random Signals

The subject of random signals and noises and their spectral analysis has been treated in some detail in Chapter 3. The results of that chapter will be used freely throughout the following treatment of modulation.

In most cases it will be assumed that the random processes involved are stationary and gaussian with zero means. When the modulating intelligence $a(t)$ is random, it will possess an autocorrelation function $R_a(\tau)$ and a corresponding power spectral density $\varphi_a(\omega)$. The noise $n(t)$ will always be additive with autocorrelation function $R_n(\tau)$ and power spectral density $\varphi_n(\omega)$.

PART 2—CONTINUOUS MODULATION

As discussed in Section 7.1, *modulation* may be defined as the modification of one signal, called the *carrier*, by another, called the *modulation*, *modulating signal*, or *message signal*. In the following sections we will be concerned with those situations in which both the carrier and the modulation are continuous functions of time. Later we will treat the cases in which the carrier and/or the modulation have the form of pulse trains.

For our analysis, Figure 7.2 may be modified to the system shown in Figure 7.12. As shown in the latter, the message is sent through a modulator (or transmitter) to produce the modulated continuous signal $m[t, a(t)]$. This waveform is corrupted by additive noise $n(t)$ in transmission so that the received (continuous) waveform $r(t)$ may be written as

$$r(t) = m[t, a(t)] + n(t).$$

The purpose of the demodulator (or receiver) is to produce some "best estimate" $a^*(t)$ of the original message signal $a(t)$. As pointed out in Section 7.1, a more general model of the transmission medium would allow corrup-

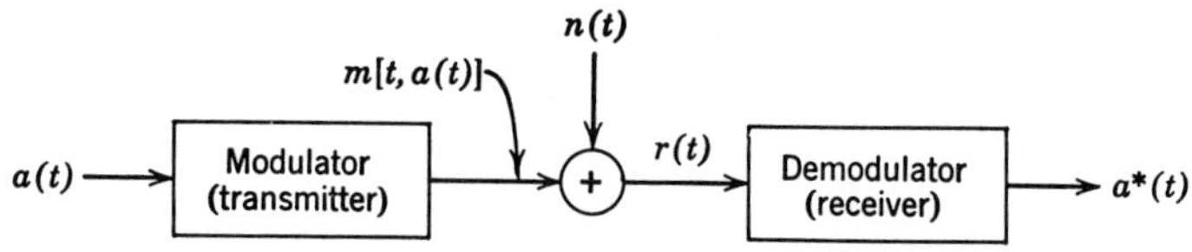

Figure 7.12 *A communication system model for continuous modulation and demodulation:*

$$\begin{aligned} a(t) &= \textit{message signal (intelligence)} \\ m[t, a(t)] &= \textit{modulated signal} \\ r(t) &= m[t, a(t)] + n(t) = \textit{received signal} \\ n(t) &= \textit{additive noise} \\ a^*(t) &= \textit{the estimate of the original message signal} \end{aligned}$$

tion of the modulated waveform itself so that the received signal was of the form of Equation 7.1-1. For example, in ionospheric radio-wave propagation, multiplicative disturbances might result from multipath transmission or fading so that the received signal was of the form

$$r_1(t) = p(t)m[t, a(t)] + n(t)$$

where both $p(t)$ and $n(t)$ are "noises." We shall not treat such systems, however, but shall confine ourselves to the simpler additive noise model of Figure 7.12.

7.12 Linear or Amplitude Modulation (AM)

In a general way, we can define *linear (or amplitude) modulation* as a system in which a *carrier wave* $c(t)$ has its amplitude varied linearly by some *message signal* $a(t)$. More precisely, we will say that a waveform is linearly modulated (or amplitude modulated) by a given message $a(t)$ if the partial derivative of that waveform with respect to $a(t)$ is independent of $a(t)$. In other words, we can write the modulated signal $m[t, a(t)]$ in the form

$$m[t, a(t)] = a(t)c(t) + d(t) \tag{7.12-1}$$

where $c(t)$ and $d(t)$ are independent of $a(t)$. Now we have

$$\frac{\partial m[t, a(t)]}{\partial a(t)} = c(t) \tag{7.12-2}$$

and $c(t)$ will be called the *carrier*. In most of the cases that we will treat, the waveform $d(t)$ either will be zero or will be linearly related to $c(t)$. It will be more convenient, therefore, to write Equation 7.12-1 as

$$m[t, a(t)] = b(t)c(t) \tag{7.12-3}$$

where $b(t)$ will be either

$$b_1(t) \equiv 1 + a(t) \tag{7.12-4}$$

or

$$b_2(t) \equiv a(t). \tag{7.12-5}$$

Also, at present it will be sufficient to allow the carrier $c(t)$ to be of the form

$$c(t) = C \cos(\omega_0 t + \theta) \tag{7.12-6}$$

where C and ω_0 are constants and where θ is either a constant or a random variable uniformly distributed on $(0, 2\pi)$.

Whenever Equation 7.12-4 applies it will be convenient to assume that $b_1(t)$ is nearly always nonnegative. This implies that, if $a(t)$ is a deterministic signal, then

$$a(t) \geq -1. \tag{7.12-7}$$

It also implies that, if $a(t)$ is a random process, the probability that $a(t)$ is less than -1 in any finite interval $(-T, T)$ is arbitrarily small; that is,

$$p[a(t) < -1] \leq \epsilon \ll 1, \qquad -T \leq t \leq T. \tag{7.12-8}$$

For example, if $a(t)$ is a stationary gaussian process with zero mean, then the variance σ^2 of the process must be small enough so that

$$\frac{1}{\sqrt{2\pi}\sigma} \int_{-\infty}^{-1} e^{-x^2/2\sigma^2}\, dx \leq \epsilon \ll 1. \tag{7.12-9}$$

It will become clear later that the purpose of these restrictions on $b_1(t)$ is to ensure that the message signal $a(t)$ is easily recovered by certain simple receivers.

We may take the Fourier transform of both sides of Equation 7.12-3 to obtain an expression for the frequency spectrum $M(\omega)$ of the general form of the linear modulated waveform $m[t, a(t)]$:

$$M(\omega) = \frac{1}{2\pi} \int_{-\infty}^{\infty} B(\omega - \nu) C(\nu)\, d\nu \tag{7.12-10}$$

where $B(\omega)$ and $C(\omega)$ are the Fourier transforms of $b(t)$ and $c(t)$ respectively. If the carrier $c(t)$ is the sinusoid of Equation 7.12-6, then

$$C(\omega) = \pi C\, e^{j\theta} \delta(\omega - \omega_0) + \pi C\, e^{-j\theta} \delta(\omega + \omega_0) \tag{7.12-11}$$

and Equation 7.12-10 becomes

$$M(\omega) = \frac{C}{2} e^{j\theta} B(\omega - \omega_0) + \frac{C}{2} e^{-j\theta} B(\omega + \omega_0). \tag{7.12-12}$$

Thus, for a sinusoidal carrier, linear modulation is essentially a symmetrical frequency translation of the message signal through an amount ω_0 rad/sec and no new signal components are generated. On the other hand, if $c(t)$ is not a sinusoid so that the spectrum $C(\omega)$ has nonzero width, then $M(\omega)$ will represent a spreading and shaping as well as a translation of $B(\omega)$.

Suppose that the message signal $a(t)$ [and hence $b(t)$] is low pass and strictly bandlimited $(0, \omega_s)$ rad/sec. Then the frequency spectrum of $b(t)$ obeys

$$B(\omega) = 0, \qquad |\omega| > \omega_s < \omega_0, \tag{7.12-13}$$

and $B(\omega - \omega_0)$ and $B(\omega + \omega_0)$ do not overlap. The spectrum $B(\omega - \omega_0)$ occupies only a range of positive frequencies while $B(\omega + \omega_0)$ occupies only negative frequencies.

As discussed in Section H.5, the *envelope of the modulated waveform* $m[t, a(t)]$ is the magnitude $|b(t)|$. If, as mentioned earlier, the function $b(t)$ is restricted to be nonnegative almost always, then this envelope becomes just $b(t)$. In such cases, an *envelope detector* will uniquely recover $b(t)$ and, hence, $a(t)$.

We now consider the common forms of simple amplitude or linear modulation.

Double-Sideband Amplitude Modulation (DSBAM)

In this case we allow the function $b(t)$ to be given by Equation 7.12-4 so that

$$m[t, a(t)] = C[1 + a(t)] \cos [\omega_0 t + \theta]. \qquad (7.12\text{-}14)$$

The transform of $b_1(t)$ is just

$$B_1(\omega) = \mathscr{F}[1 + a(t)] = 2\pi\delta(\omega) + A(\omega) \qquad (7.12\text{-}15)$$

where $A(\omega)$ is the Fourier transform of $a(t)$. It follows, therefore, from Equation 7.12-12 that the frequency spectrum of $m[t, a(t)]$ is given by

$$M(\omega) = \frac{C}{2} e^{j\theta}[2\pi\delta(\omega - \omega_0) + A(\omega - \omega_0)] + \frac{C}{2} e^{-j\theta}[2\pi\delta(\omega + \omega_0) + A(\omega + \omega_0)]. \qquad (7.12\text{-}16)$$

Depending on the form of $a(t)$, we can distinguish a number of special cases that are of interest.

Case I—a(t) Is Periodic: If we assume that $a(t)$ is the special periodic function given by

$$a(t) = \eta \cos \omega_s t, \qquad \begin{cases} \omega_s < \omega_0, \\ 0 \le \eta \le 1, \end{cases} \qquad (7.12\text{-}17)$$

then the frequency spectrum $A(\omega)$ is

$$A(\omega) = \pi\eta\delta(\omega - \omega_s) + \pi\eta\delta(\omega + \omega_s). \qquad (7.12\text{-}18)$$

For convenience, let us take the phase angle θ to be zero

$$\theta = 0 \qquad (7.12\text{-}19)$$

so that

$$M(\omega) = \pi C\Big[\delta(\omega - \omega_0) + \delta(\omega + \omega_0) + \frac{\eta}{2}\delta(\omega - \omega_0 - \omega_s) + \frac{\eta}{2}\delta(\omega - \omega_0 + \omega_s) + \frac{\eta}{2}\delta(\omega + \omega_0 - \omega_s) + \frac{\eta}{2}\delta(\omega + \omega_0 + \omega_s)\Big]\cdot \qquad (7.12\text{-}20)$$

This spectrum is illustrated in Figure 7.13 together with the corresponding modulated waveform

$$m[t, a(t)] = C(1 + \eta \cos \omega_s t) \cos \omega_0 t. \qquad (7.12\text{-}21)$$

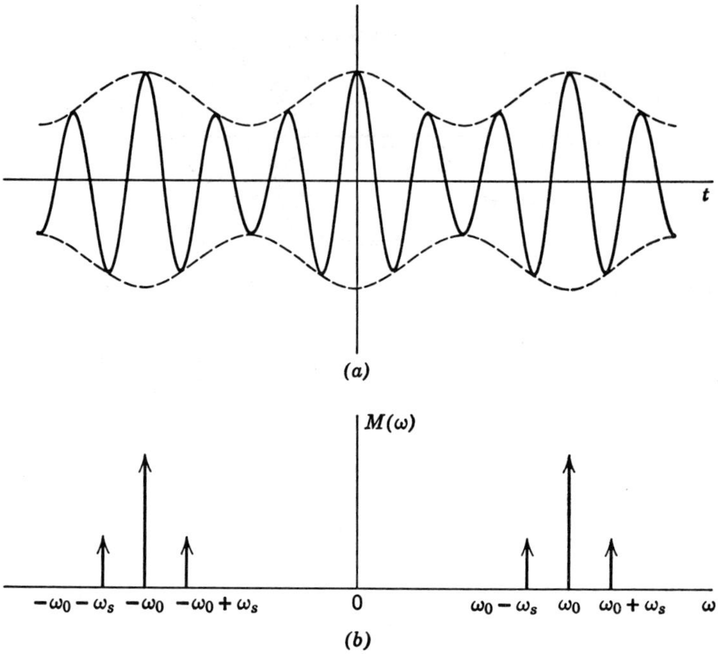

Figure 7.13 *A double-sideband, amplitude-modulated signal and frequency spectrum: (a) the signal* $m[t, a(t)] = C(1 - \eta \cos \omega_s t) \cos \omega_0 t$; *(b) the frequency spectrum* $M(\omega)$

Note that this last equation can also be written as

$$m[t, a(t)] = \underset{\text{(carrier)}}{C \cos \omega_0 t} + \underset{\text{(lower sideband)}}{\frac{C}{2} \eta \cos (\omega_0 - \omega_s)t} + \underset{\text{(upper sideband)}}{\frac{C}{2} \eta \cos (\omega_0 + \omega_s)t} \tag{7.12-22}$$

where we have used the identity

$$\cos x \cos y = \tfrac{1}{2} \cos (x - y) + \tfrac{1}{2} \cos (x + y). \tag{7.12-23}$$

In this form it is easy to distinguish the carrier, the *lower sideband*, and the *upper sideband.*

It is of some interest to consider the phasor diagram of Equation 7.12-22 using the carrier phase as reference. Such a diagram is shown in Figure 7.14. Note that the two sideband phasors rotate in opposite directions, have phases that are equal in magnitude but opposite in sign, and are equal in length. Thus their resultant is always in phase with the carrier and has a length $\eta C \cos \omega_s t$ as expected.

Suppose that $a(t)$ is not a single sinusoid but is periodic with period P. Then it may be expanded in a Fourier series as in Equation 7.12-20.

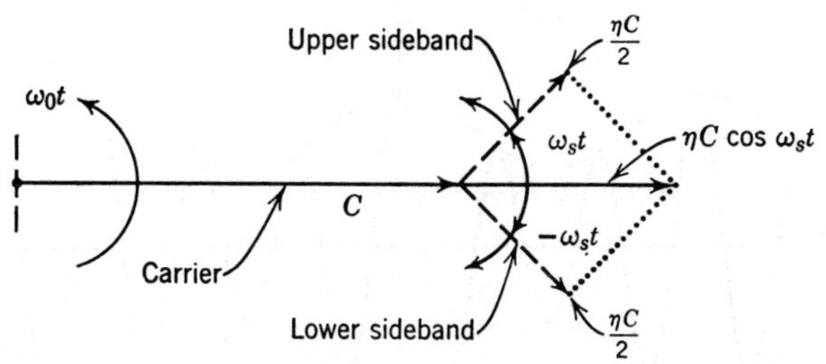

Figure 7.14 A phasor diagram of double-sideband amplitude modulation

Case II—a(t) Is Deterministic But Nonperiodic with Spectrum A(ω): In this case, Equation 7.12-16 applies directly and the magnitude of $M(\omega)$ is given by

$$|M(\omega)| = \pi C[\delta(\omega - \omega_0) + \delta(\omega + \omega_0)] + \frac{C}{2}[|A(\omega - \omega_0)| + |A(\omega + \omega_0)|] \quad (7.12\text{-}24)$$

if the initial phase angle θ is taken to be zero. For an arbitrary low-pass frequency spectrum $A(\omega)$, this spectrum is shown in Figure 7.15. Note that

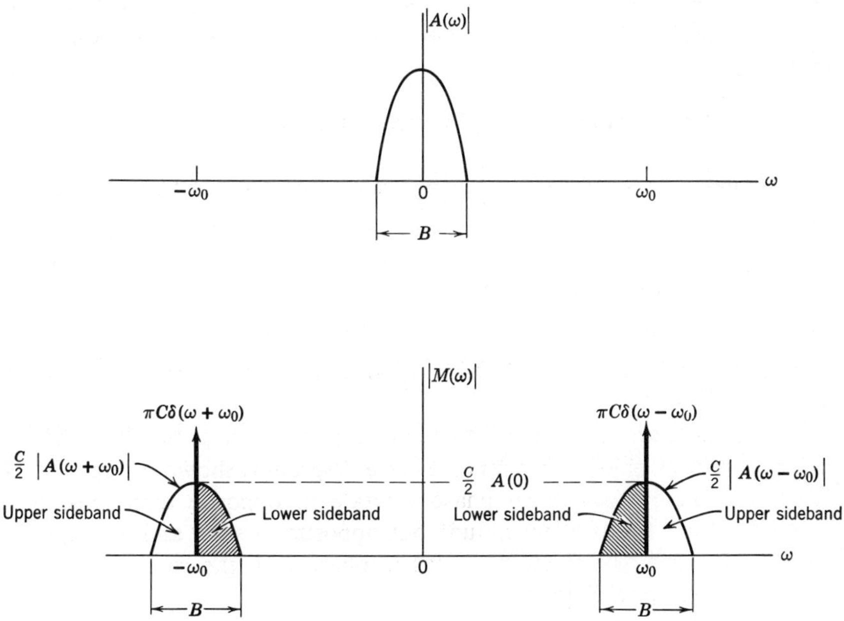

Figure 7.15 The frequency spectrum for double-sideband amplitude modulation

the real bandwidth has been doubled by the modulation process. As a matter for future attention, we have indicated the upper sidebands (USB) and lower sidebands (LSB).

Case III—a(t) Is a Random Process: We will assume that $a(t)$ is a random process which is at least wide-sense stationary with zero mean, autocorrelation function $R_a(\tau)$, and power spectral density $\varphi_a(\omega)$. In this case it is desirable to consider the power spectral density $\varphi_m(\omega)$ of $m[t, a(t)]$ rather than the frequency spectrum $M(\omega)$. We have

$$m[t, a(t)] = C[1 + a(t)] \cos(\omega_0 t + \theta) \tag{7.12-25}$$

where θ will be assumed to be a random variable, independent of $a(t)$, and uniformly distributed on $(0, 2\pi)$ so that its density function $p(\theta)$ is

$$p(\theta) = \begin{cases} \dfrac{1}{2\pi}, & 0 \leq \theta \leq 2\pi, \\ 0, & \text{elsewhere.} \end{cases} \tag{7.12-26}$$

The autocorrelation function of Equation 7.12-25 can be obtained in a straightforward fashion. We have

$$R_m(\tau) = C^2 E_a\{[1 + a(t)][1 + a(t + \tau)]\} \times E_\theta\{\cos[\omega_0 t + \theta] \cos[\omega_0(t + \tau) + \theta]\} \tag{7.12-27}$$

since $a(t)$ and θ have been assumed to be independent. Since $a(t)$ has zero mean we find

$$R_m(\tau) = \frac{C^2}{2}[1 + R_a(\tau)] \cos \omega_0 \tau \tag{7.12-28}$$

as in Example 3.6 or Section 6.6. On transforming this last expression we obtain the power spectral density $\varphi_m(\omega)$

$$\varphi_m(\omega) = \frac{C^2\pi}{2}[\delta(\omega - \omega_0) + \delta(\omega + \omega_0)] + \frac{C^2}{4}[\varphi_a(\omega - \omega_0) + \varphi_a(\omega + \omega_0)]. \tag{7.12-29}$$

Except for the multiplicative constants and the substitution of power spectral density for frequency spectrum, this form is the same as that of Equation 7.12-24.

Suppose now that θ is not a uniformly distributed random variable but is simply a constant but unknown phase angle in the interval $(0, 2\pi)$. Now Equation 7.12-27 becomes

$$R_m(t, t + \tau) = \frac{C^2}{2}[1 + R_a(\tau)][\cos \omega_0 \tau + \cos(2\omega_0 t + 2\theta + \omega_0 \tau)] \tag{7.12-30}$$

and the autocorrelation function of the modulated signal $m[t, a(t)]$ is no longer a function of τ only. One way out of this difficulty is to take the time average of Equation 7.12-30 as in Equation 7.9-8 to obtain

$$A[R_m(t, t + \tau)] = \lim_{T \to \infty} \frac{1}{2T} \int_{-T}^{T} R_m(t, t + \tau)\, dt$$

$$= \frac{C^2}{2} [1 + R_a(\tau)] \cos \omega_0 \tau \qquad (7.12\text{-}31)$$

as in Equation 7.12-28. It should be kept in mind, however, that the process $m[t, a(t)]$ with θ nonrandom is no longer wide-sense stationary.

Double-Sideband Amplitude Modulation—Suppressed Carrier (DSBAM-SC)

If $b(t)$ is given by Equation 7.12-5 so that

$$m[t, a(t)] = Ca(t) \cos (\omega_0 t + \theta), \qquad (7.12\text{-}32)$$

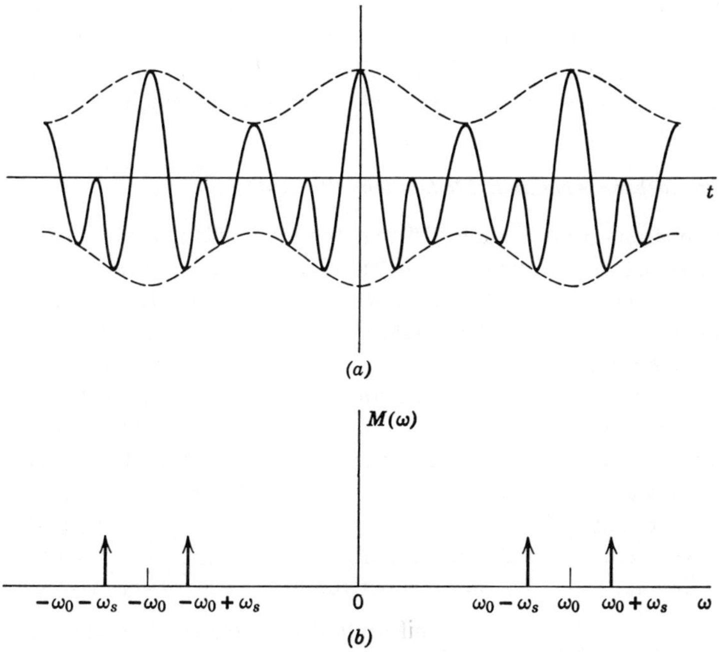

Figure 7.16 *A double-sideband, amplitude-modulated, suppressed-carrier signal and frequency spectrum: (a) the signal* $m[t, a(t)] = C\eta \cos \omega_s t \cos \omega_0 t$*; (b) the frequency spectrum* $M(\omega)$

then the carrier component has been suppressed and Equation 7.12-16 reduces to

$$M(\omega) = \frac{C}{2} e^{j\theta} A(\omega - \omega_0) + \frac{C}{2} e^{-j\theta} A(\omega + \omega_0). \tag{7.12-33}$$

For the three cases already considered, it is only necessary to remove the carrier component from Equations 7.12-22, 7.12-24, and 7.12-29, respectively. The principal advantage of DSBAM-SC over DSBAM is that the carrier is not transmitted in the former case with a consequent saving in transmitted power. The principal disadvantages relate to problems in generating and demodulating the suppressed-carrier waveform. These topics will be discussed later in more detail.

It is of some interest to consider the appearance of the DSBAM-SC waveform for a sinusoidal message signal. The wave

$$m[t, a(t)] = C\eta \cos \omega_s t \cos \omega_0 t \tag{7.12-34}$$

is plotted in Figure 7.16. Note the phase reversals at the zero crossings of the envelope. These did not occur for the DSBAM waveform given by Equation 7.12-21 and shown in Figure 7.13(*a*). When the problem of demodulation is considered, the phase reversal will have considerable significance.

Vestigial-Sideband Amplitude Modulation (VSBAM)

It is apparent from Figure 7.15 that the total information regarding the message signal $a(t)$ is contained in either the upper or lower sideband in conventional DSBAM. In principle it is only necessary to transmit one of these sidebands and the other could be eliminated, say by filtering, with a consequent reduction in bandwidth and transmitted power. Such a procedure actually is followed in single-sideband amplitude modulation (SSBAM), which will be discussed in a later section. Completely filtering out one sideband, however, requires an ideal bandpass filter with infinitely sharp cutoff or an equivalent technique. A policy intermediate between the production of DSBAM and SSBAM is vestigial-sideband amplitude modulation (VSBAM) where one sideband is attenuated much more than the other with some saving in bandwidth and transmitted power. The carrier may be present or suppressed.

Let us consider that the modulated signal

$$m_1(t) = m_1[t, a(t)] = Ca(t) \cos (\omega_0 t + \theta) \tag{7.12-35}$$

has been passed through a linear bandpass filter with impulse response $h(t)$

and complex transfer function $H(\omega)$. The output $m[t, a(t)]$ is given by the convolution

$$
\begin{aligned}
m[t, a(t)] &= \int_{-\infty}^{\infty} h(\tau) m_1(t - \tau)\, d\tau \\
&= C \int_{-\infty}^{\infty} h(\tau) a(t - \tau) \cos\left[\omega_0(t - \tau) + \theta\right] d\tau \qquad (7.12\text{-}36)
\end{aligned}
$$

or, after expanding $\cos(\alpha - \beta)$,

$$
\begin{aligned}
m[t, a(t)] = {} & C \cos(\omega_0 t + \theta) \int_{-\infty}^{\infty} h(\tau) a(t - \tau) \cos \omega_0 \tau \, d\tau \\
& + C \sin(\omega_0 t + \theta) \int_{-\infty}^{\infty} h(\tau) a(t - \tau) \sin \omega_0 \tau \, d\tau. \qquad (7.12\text{-}37)
\end{aligned}
$$

From a systems point of view, the waveform $m[t, a(t)]$ could be considered to have been produced either by the modulator of Figure 7.17(a) or that of

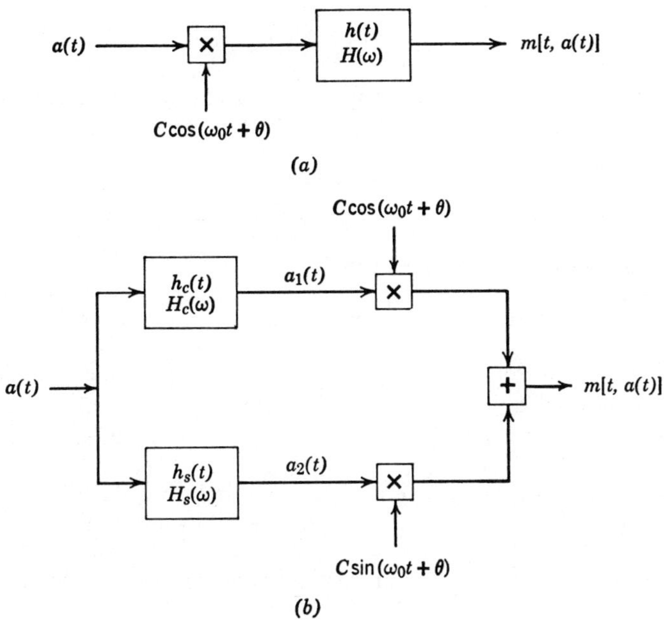

Figure 7.17 *A vestigial-sideband modulator: (a) original modulator; (b) equivalent circuit:*

$$
\begin{aligned}
h_c(t) &= h(t) \cos \omega_0 t \\
h_s(t) &= h(t) \sin \omega_0 t
\end{aligned}
$$

Figure 7.17(*b*). In the latter figure the filters $h_c(t)$ and $h_s(t)$ are low-pass equivalent filters to $h(t)$ with impulse responses given by

$$h_c(t) = h(t)\cos\omega_0 t \tag{7.12-38}$$

and

$$h_s(t) = h(t)\sin\omega_0 t, \tag{7.12-39}$$

and with complex transfer functions

$$H_c(\omega) = \int_{-\infty}^{\infty} h_c(t)\, e^{-j\omega t}\, dt = \tfrac{1}{2}[H(\omega - \omega_0) + H(\omega + \omega_0)] \tag{7.12-40}$$

and

$$H_s(\omega) = \int_{-\infty}^{\infty} h_s(t)\, e^{-j\omega t}\, dt = \frac{1}{2j}[H(\omega - \omega_0) - H(\omega + \omega_0)]. \tag{7.12-41}$$

It is clear that the filter $h(t)$ cannot be arbitrary if the message $a(t)$ is to be recovered. For example, if the filter has an ideal rectangular passband which is less than the bandwidth of $a(t)$, then some information will be lost. From an examination of Figure 7.17(*b*) we see that, if $a(t)$ is to be recovered from $m[t, a(t)]$ without further spectral shaping (but possibly with a scale change), it is necessary that

$$H_c(\omega) + H_s(\omega) = \mathscr{K}\, e^{j\psi} \tag{7.12-42}$$

over the passband of $a(t)$. Here the factors $\mathscr{K}$ and ψ are arbitrary constants. It follows from Equations 7.12-40 and 7.12-41 that this last expression can be rewritten as

$$(1 - j)H(\omega - \omega_0) + (1 + j)H(\omega + \omega_0) = 2\mathscr{K}\, e^{j\psi} \tag{7.12-43}$$

or, if only magnitudes are considered,

$$|H(\omega - \omega_0)| + |H(\omega + \omega_0)| = \sqrt{2}\,\mathscr{K}. \tag{7.12-44}$$

It should be kept in mind that $H(\omega)$ is a bandpass filter and that these last three relations need hold only for those frequencies where the spectrum of $a(t)$ is nonzero. A transfer function magnitude $|H(\omega)|$ satisfying Equation 7.12-44 is shown in Figure 7.18.

The principal use of VSBAM has been in commercial television. In this case the video (picture) signal is transmitted by VSBAM with a consequent reduction in the total transmitted signal bandwidth and in the frequency difference that must be allowed between adjacent channels.

7.13 Vector Linear Modulation

It was shown in the last section that single-input VSBAM could be regarded also as a two-input problem of the type illustrated in Figure 7.17.

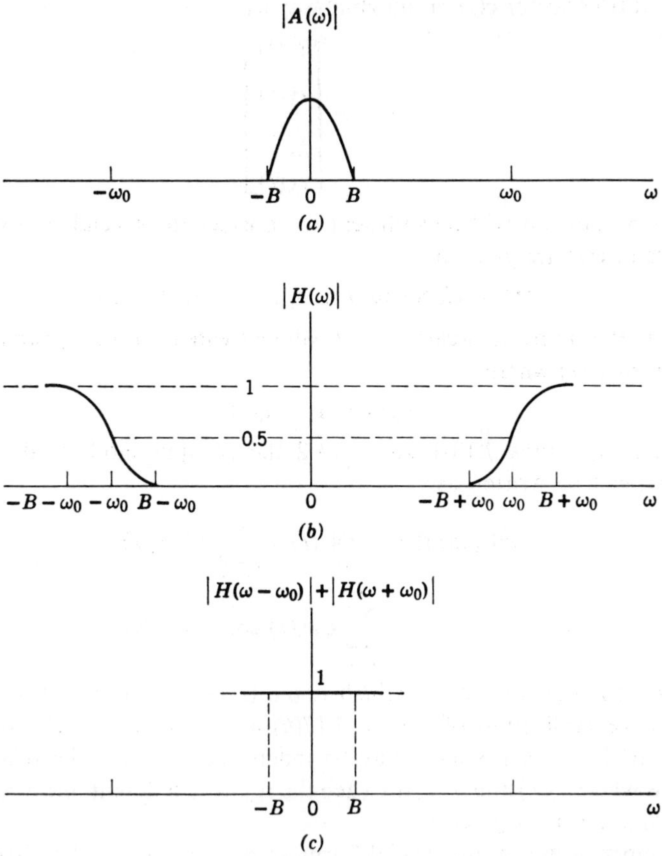

Figure 7.18 *Filtering for vestigial-sideband amplitude modulation:* (*a*) *the spectrum of the message signal;* (*b*) *the gain magnitude of the bandpass filter;* (*c*) *the gain magnitude of the equivalent low-pass filter*

This concept is easily generalized to account for a number of multi-input forms of AM. As a matter of convenience, all such forms of modulation will be called *vector linear modulation.*

Consider an N-input system with N separate message signals $a_i(t)$, $i = 1, 2, 3, \ldots, N$. The N-element message column vector $\mathbf{a}(t)$ will be defined by

$$\mathbf{a}(t) = \begin{bmatrix} a_1(t) \\ a_2(t) \\ \vdots \\ a_N(t) \end{bmatrix} \tag{7.13-1}$$

and the carrier vector $\mathbf{c}(t)$ in the same way by

$$\mathbf{c}(t) = \begin{bmatrix} c_1(t) \\ c_2(t) \\ \vdots \\ c_N(t) \end{bmatrix} \tag{7.13-2}$$

For our purposes it will be sufficient to consider the special case where the elements of $\mathbf{c}(t)$ are given by

$$c_i(t) = C_i \cos(\omega_0 t + \theta_i), \qquad 0 \leq \theta_i \leq 2\pi. \tag{7.13-3}$$

Also only the suppressed-carrier case will be treated since $a_i(t)$ can always be replaced by $b_i(t)$ where

$$b_i(t) = 1 + a_i(t). \tag{7.13-4}$$

On using Equations 7.13-1 and 7.13-2 the N-input modulated waveform $m[t, \mathbf{a}(t)]$ can be written as

$$\begin{aligned} m[t, a(t)] &= \mathbf{a}(t)\mathbf{c}'(t) = \sum_{i=1}^{N} a_i(t)c_i(t) \\ &= \sum_{i=1}^{N} C_i a_i(t) \cos(\omega_0 t + \theta_i) \end{aligned} \tag{7.13-5}$$

where $\mathbf{c}'(t)$ is the row vector which is the transpose of $\mathbf{c}(t)$. The system is simply a generalization of Figure 7.17(b) and is shown in Figure 7.19(a). Note that the messages $a_i(t)$ may be independent or may be related as in VSBAM where $a_1(t)$ and $a_2(t)$ are each the result of different linear operations on an original message $a(t)$.

A number of important special forms of modulation can be distinguished which correspond to Equation 7.13-5. Attention will be restricted to the case in which $N = 2$.

Quadrature Amplitude Modulation (QAM)

The general form for quadrature amplitude modulation is obtained by setting

$$\theta_1 = \theta \tag{7.13-6}$$

and

$$\theta_2 = -\frac{\pi}{2} + \theta \tag{7.13-7}$$

where θ may be an arbitrary constant or a uniformly distributed random variable with density given by Equation 7.12-26. Now Equation 7.13-5 becomes

$$m[t, \mathbf{a}(t)] = Ca_1(t) \cos(\omega_0 t + \theta) + Ca_2(t) \sin(\omega_0 t + \theta). \tag{7.13-8}$$

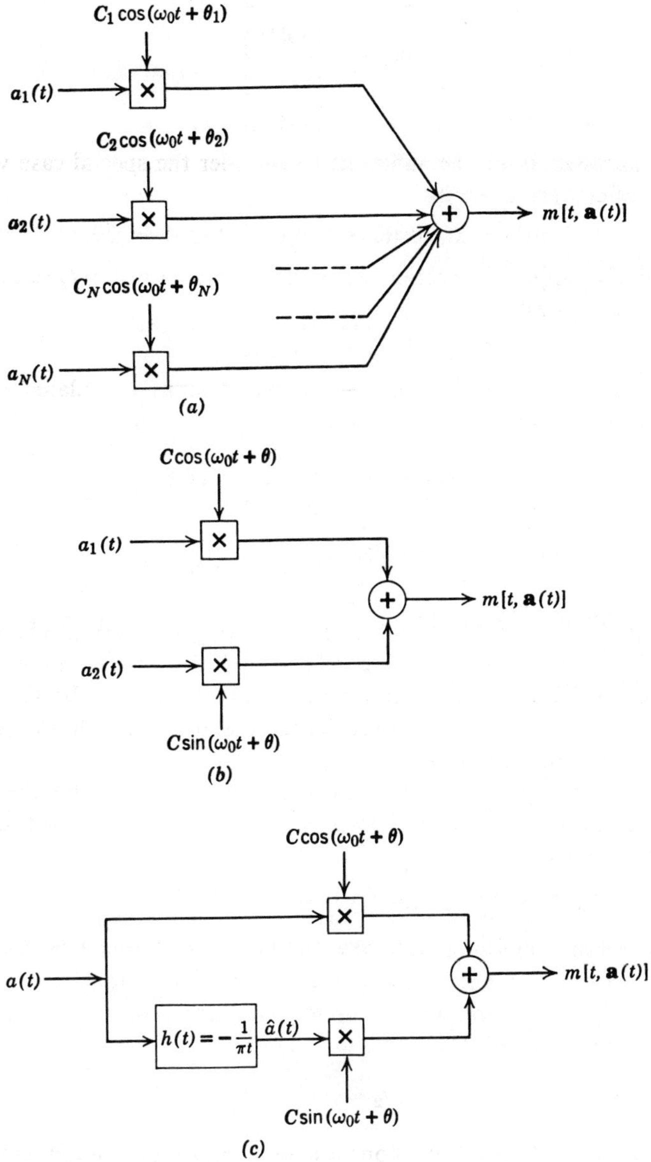

Figure 7.19 *Vector linear modulation: (a) a vector linear modulator; (b) a quadrature AM modulator; (c) a single-sideband modulator*

The name "quadrature modulation" comes from the fact that the two carriers are in phase quadrature. The modulator is shown schematically in Figure 7.19(b) and is obviously a special case of Figure 7.19(a). As previously mentioned, in general the messages $a_1(t)$ and $a_2(t)$ are arbitrary. Let us now consider two special cases in which $a_1(t)$ and $a_2(t)$ are uniquely related.

Vestigial-Sideband Amplitude Modulation (VSBAM)

As a matter of emphasis, it should be mentioned again that, for VSBAM, the messages $a_1(t)$ and $a_2(t)$ are both obtained by linear operations on an original message $a(t)$. The linear relations are obtained directly from Figure 7.17(b) as the convolutions

$$a_1(t) = \int_{-\infty}^{\infty} h(\tau)a(t - \tau) \cos \omega_0\tau \, d\tau \qquad (7.13\text{-}9)$$

and

$$a_2(t) = \int_{-\infty}^{\infty} h(\tau)a(t - \tau) \sin \omega_0\tau \, d\tau. \qquad (7.13\text{-}10)$$

Single-Sideband Amplitude Modulation (SSBAM)

This important type of amplitude modulation can be considered as a limiting form of VSBAM where the filter $h(t)$ of Figure 7.17(a) is an ideal filter with infinitely sharp cutoff so that one sideband (e.g., the lower) is completely eliminated. It is shown in Section H.3 that SSBAM is a special case of quadrature modulation where

$$a_1(t) = a(t) \qquad (7.13\text{-}11)$$

and

$$a_2(t) = -\frac{1}{\pi} \fint_{-\infty}^{\infty} \frac{a(x)}{t - x} \, dx = \hat{a}(t). \qquad (7.13\text{-}12)$$

Here $a_2(t)$ is a particular linear operation on $a(t)$ and is called the *Hilbert transform* of $a(t)$ as discussed in Appendix H. In the remainder of this section the notation $\hat{a}(t)$ will be used to denote the Hilbert transform* of $a(t)$. It is shown in Section H.2 that this (unrealizable) linear operation, or Hilbert "transformer," has impulse response $h(t)$ and corresponding complex transfer function $H(\omega)$ given, respectively, by

$$h(t) = -\frac{1}{\pi t}, \qquad -\infty < t < \infty, \qquad (7.13\text{-}13)$$

and

$$H(\omega) = j \operatorname{sgn} \omega \qquad (7.13\text{-}14)$$

* In the remainder of the book, the Hilbert transform of a function $f(t)$ is denoted by $f_1(t)$.

where sgn $(\cdot)$ is the signum function of Equation H.1-18. Thus a SSBAM modulator is a special case of the QAM modulator as shown in Figure 7.19(*c*).

Because of the importance of SSBAM, let us now consider the three special types of modulating signal $a(t)$ already treated for DSBAM in Section 7.12; that is, where $a(t)$ is (1) a periodic signal, (2) a general deterministic signal, and (3) a random process.

Case I—a(t) Is Periodic: As before it is sufficient to consider the special case where $a(t)$ is a single-frequency sinusoid given by

$$a(t) = a_1(t) = \eta \cos \omega_s t, \qquad \begin{cases} \omega_s < \omega_0, \\ 0 \leq \eta \leq 1, \end{cases} \tag{7.13-15}$$

with spectrum

$$A(\omega) = \pi\eta\delta(\omega - \omega_s) + \pi\eta\delta(\omega + \omega_s). \tag{7.13-16}$$

For convenience let us again take the phase angle θ to be zero. It is clear from Figure 7.19(*c*) and Equation 7.13-14 that the spectrum of $a_2(t) = \hat{a}(t)$ is

$$\begin{aligned} A_2(\omega) &= A(\omega)H(\omega) = A(\omega)j \operatorname{sgn} \omega \\ &= \pi\eta j\delta(\omega - \omega_s) - \pi\eta j\delta(\omega + \omega_s). \end{aligned} \tag{7.13-17}$$

But this last expression is just the Fourier transform of $-\eta \sin \omega_s t$ so that

$$a_2(t) = \hat{a}(t) = -\eta \sin \omega_s t. \tag{7.13-18}$$

In other words $a(t)$ of Equation 7.13-15 and $a_2(t)$ of this last equation are a Hilbert-transform pair as given by pair number 3 of Table H.1.

The modulated waveform $m[t, \mathbf{a}(t)]$ is obtained by substituting Equations 7.13-15 and 7.13-18 into Equation 7.13-8:

$$\begin{aligned} m[t, a(t)] &= \eta C(\cos \omega_s t \cos \omega_0 t - \sin \omega_s t \sin \omega_0 t) \\ &= \eta C \cos (\omega_0 + \omega_s)t \end{aligned} \tag{7.13-19}$$

with spectrum

$$M(\omega) = \pi\eta C\delta[\omega - (\omega_0 + \omega_s)] + \pi\eta C\delta[\omega + (\omega_0 + \omega_s)] \tag{7.13-20}$$

as expected. This spectrum is shown in Figure 7.20(*b*) and consists of the upper sideband only. For this special case, SSBAM modulation can be represented schematically as in Figure 7.20(*a*).

If $a(t)$ is a general periodic waveform, it can be expanded in a Fourier series and superposition can be applied.

Case II—a(t) Is Deterministic But Nonperiodic with Spectrum A(ω): This problem has been treated in detail in Section H.3. The modulated waveform

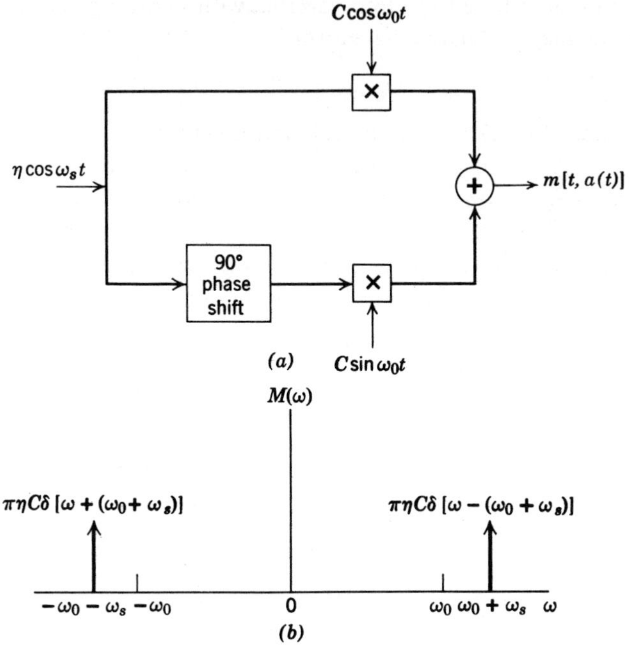

***Figure* 7.20** *Single-frequency, single-sideband amplitude modulation: (a) the SSB modulator; (b) the spectrum*

$m[t, \mathbf{a}(t)]$ is given by Equation H.3-1 with $\theta = 0$ and $C = 1$. In the notation of this section, the modulated signal is

$$m[t, \mathbf{a}(t)] = Ca(t) \cos \omega_0 t + C\hat{a}(t) \sin \omega_0 t \qquad (7.13\text{-}21)$$

where θ has been taken to be zero and $\hat{a}(t)$ is given by Equation 7.13-12. It follows from Equation H.3-3 or from Equation 7.13-14 that the spectrum $M(\omega)$ is given by

$$M(\omega) = \frac{C}{2}[A(\omega - \omega_0) + A(\omega - \omega_0) \operatorname{sgn}(\omega - \omega_0)]$$
$$+ \frac{C}{2}[A(\omega + \omega_0) - A(\omega + \omega_0) \operatorname{sgn}(\omega + \omega_0)]. \qquad (7.13\text{-}22)$$

As illustrated in Figure H.2, this is a single-sideband spectrum with the lower sideband missing.

Case III—a(t) Is a Random Process: As before let us assume that $a(t)$ is at least wide-sense stationary with zero mean, autocorrelation function $R_a(\tau)$, and power spectral density $\varphi_a(\omega)$. It will be assumed that the phase angle θ

is a uniformly distributed random variable with density given by Equation 7.12-26. Thus the modulated waveform is

$$m[t, \mathbf{a}(t)] = Ca(t)\cos(\omega_0 t + \theta) + C\hat{a}(t)\sin(\omega_0 t + \theta). \quad (7.13\text{-}23)$$

As in Equation 7.12-27 the autocorrelation function is

$$R_m(\tau) = \frac{C^2}{2}\cos\omega_0\tau[E\{a(t)a(t+\tau)\} + E\{\hat{a}(t)\hat{a}(t+\tau)\}]$$

$$+ \frac{C^2}{2}\sin\omega_0\tau[E\{a(t)\hat{a}(t+\tau)\} - E\{\hat{a}(t)a(t+\tau)\}]. \quad (7.13\text{-}24)$$

Equation 7.13-12 may now be used to evaluate the four expectations in this last expression. The first term is

$$E\{a(t)a(t+\tau)\} = R_a(\tau). \quad (7.13\text{-}25a)$$

The second term can be written as

$$E\{\hat{a}(t)\hat{a}(t+\tau)\} = \frac{1}{\pi^2}⨍_{-\infty}^{\infty}⨍_{-\infty}^{\infty}\frac{E\{a(x)a(y)\}}{(t-x)(t+\tau-y)}\,dx\,dy$$

$$= \frac{1}{\pi^2}⨍_{-\infty}^{\infty}⨍_{-\infty}^{\infty}\frac{R_a(y-x)}{(t-x)(t+\tau-y)}\,dx\,dy$$

$$= -\frac{1}{\pi}⨍_{-\infty}^{\infty}\frac{1}{t-x}\left[-\frac{1}{\pi}⨍_{-\infty}^{\infty}\frac{R_a(z)}{(t+\tau-x)-z}\,dz\right]dy.$$

(7.13-25b)

But the term in brackets is the Hilbert transform of $R_a(t+\tau-x)$ and will be denoted by $\hat{R}_a(t+\tau-x)$. Consequently, this last expression becomes

$$E\{\hat{a}(t)\hat{a}(t+\tau)\} = -\frac{1}{\pi}⨍_{-\infty}^{\infty}\frac{\hat{R}_a(t+\tau-x)}{t-x}\,dx$$

$$= \frac{1}{\pi}⨍_{-\infty}^{\infty}\frac{\hat{R}_a(z)}{t-z}\,dz. \quad (7.13\text{-}26)$$

The integral is the inverse Hilbert transform of $\hat{R}_a(\tau)$ so that

$$E\{\hat{a}(t)\hat{a}(t+\tau)\} = R_a(\tau). \quad (7.13\text{-}27)$$

In the same way we find

$$E\{a(t)\hat{a}(t+\tau)\} = \hat{R}_a(\tau) \quad (7.13\text{-}28)$$

and

$$E\{\hat{a}(t)a(t+\tau)\} = -\hat{R}_a(\tau). \quad (7.13\text{-}29)$$

Thus Equation 7.13-24 can be written as

$$R_m(\tau) = C^2R_a(\tau)\cos\omega_0\tau + C^2\hat{R}_a(\tau)\sin\omega_0\tau. \quad (7.13\text{-}30)$$

Note that this autocorrelation function is even (as it must be) since

$$\hat{R}_a(-\tau) = -\hat{R}_a(\tau). \tag{7.13-31}$$

The results obtained in Equation 7.13-30 are not very surprising and, in fact, are identical mathematically to Equation 7.13-21. It follows, therefore, from Equation 7.13-22 that the power spectral density of $m[t, \mathbf{a}(t)]$ is just

$$\varphi_m(\omega) = \frac{C^2}{2}[\varphi_a(\omega - \omega_0) + \varphi_a(\omega - \omega_0)\,\mathrm{sgn}\,(\omega - \omega_0)] + \frac{C^2}{2}[\varphi_a(\omega + \omega_0) - \varphi_a(\omega + \omega_0)\,\mathrm{sgn}\,(\omega + \omega_0)]. \tag{7.13-32}$$

If the frequency spectrum $A(\omega)$ is replaced by the power spectral density $\varphi_a(\omega)$, then the power spectral density $\varphi_m(\omega)$ is illustrated in Figure H.2.

7.14 Bandwidth and Power Relations for Amplitude Modulation

It is clear that the bandwidth of a given amplitude-modulated signal is related in a particularly simple fashion to the bandwidth of the modulating signal since amplitude modulation is essentially a frequency translation. If the modulating signal $a(t)$ is assumed to be low pass and to have a bandwidth of W cycles per second, then the bandwidth ΔF of the modulated signal $m[t, a(t)]$ cannot be greater than $2W$ or less than W; that is,

$$W \leq \Delta F \leq 2W. \tag{7.14-1}$$

From the discussions of the two previous sections, it is obvious that the upper limit of $2W$ cps holds for double-sideband modulation and the lower limit of W cps for single sideband. For the concept of bandwidth to be meaningful for a single-frequency sinusoidal-modulating signal, it is assumed, of course, that a low-frequency sinusoid of frequency W cps has a bandwidth W. Actually, what is really being assumed is that this sinusoid is the highest frequency component in the low-pass modulating signal.

The only case in which intermediate values of the inequality of Equation 7.14-1 are encountered is in vestigial-sideband modulation (VSBAM). In principle any bandwidth between W and $2W$ cps is possible so long as Equation 7.12-42 is satisfied. In this respect SSBAM can be considered as a limiting case of VSBAM. From a practical point of view, however, it is difficult (or expensive) to design a filter whose gain magnitude drops off too rapidly. As a matter of interest, Figure 7.21 shows the gain magnitude of the intermediate-frequency (IF) amplifier of a commercial television receiver. The magnitude of the equivalent DSBAM frequency spectrum has been given also for comparison. The bandwidth relationships for various types of AM modulation are summarized in Table 7.2.

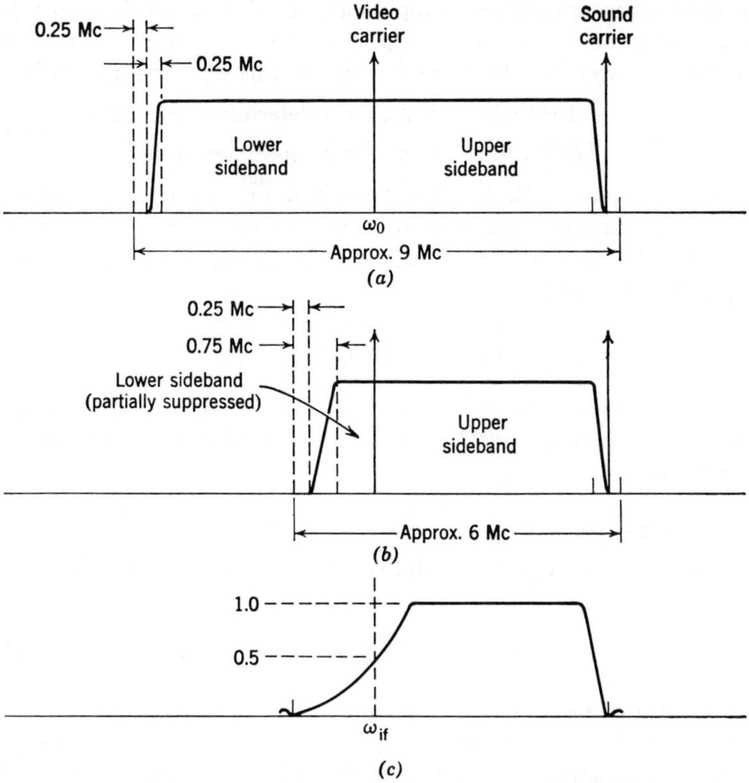

Figure 7.21 *Bandwidth relationships in commercial television: (a) typical DSBAM video signal before sideband suppression; (b) typical VSBAM video signal after sideband suppression; (c) typical IF response curve*

Table 7.2. Bandwidth Relations for Amplitude Modulation

Type of Modulation	Signal	Bandwidth ΔF (cps)
Modulating signal	$a(t)$	W
DSBAM	$C[1 + a(t)] \cos \omega_0 t$	$2W$
VSBAM	$\left\{ \begin{array}{l} C \cos \omega_0 t \int_{-\infty}^{\infty} h(\tau) a(t - \tau) \cos \omega_0 \tau \, d\tau \\ \quad + C \sin \omega_0 t \int_{-\infty}^{\infty} h(\tau) a(t - \tau) \sin \omega_0 \tau \, d\tau \end{array} \right\}$	$>W$ $<2W$
SSBAM	$Ca(t) \cos \omega_0 t + C\hat{a}(t) \sin \omega_0 t$	W

Power relationships are also simple and straightforward for AM signals. Consider Equation 7.12-14 which is a general expression for the modulated AM waveform $m(t)$. Let the average power in this signal be defined by

$$P_{\text{av}} = \begin{cases} A[m^2(t)], & \text{if } a(t) \text{ is a deterministic signal.} \\ E\{m^2(t)\}, & \text{if } a(t) \text{ is a random process.} \end{cases} \tag{7.14-2}$$

As before, the phase angle θ will be fixed if $a(t)$ is deterministic and will be taken to be a random variable uniformly distributed on $(0, 2\pi)$ if $a(t)$ is a random process. Also, it will be assumed that $A[a(t)]$ or $E\{a(t)]$ is zero. It is clear that P_{av} is given by

$$P_{\text{av}} = \frac{C^2}{2}\,[1 + a^2(t)_{\text{av}}] \tag{7.14-3}$$

where $a^2(t)_{\text{av}}$ is either $A[a^2(t)]$ or $E\{a^2(t)\}$. The first term $C^2/2$ is the carrier power and the second term $(C^2/2)a^2(t)_{\text{av}}$ is the signal power in the upper and lower sidebands.

Case I—a(t) Is Periodic: If $a(t)$ is the single-frequency sinusoid of Equation 7.13-15, then the time average $A[a^2(t)]$ is just $\eta^2/2$ and Equation 7.14-3 becomes

$$P_{\text{av}} = \frac{C^2}{2}\left(1 + \frac{\eta^2}{2}\right), \qquad 0 \le \eta \le 1. \tag{7.14-4}$$

Thus the ratio of modulating-signal power to carrier power is $\eta^2/2$, which cannot exceed 1/2. More significantly, the *fraction* of power bearing nonredundant information cannot exceed 1/6 since the upper and lower sidebands contain the same information.

Case II—a(t) Is Deterministic But Nonperiodic with Spectrum $A(\omega)$***:*** In this case $a^2(t)_{\text{av}}$ is given by

$$a^2(t)_{\text{av}} = \lim_{T\to\infty} \frac{1}{2T}\int_{-T}^{T} a^2(t)\,dt \le 1 \tag{7.14-5}$$

if $a(t)$ is constrained by $|a(t)| \le 1$ to prevent overmodulation. As in Case I, the fraction of power bearing nonredundant information cannot exceed 1/6.

Case III—a(t) Is a Random Process: If $a(t)$ is wide-sense stationary with autocorrelation function $R_a(\tau)$, then

$$P_{\text{av}} = \frac{C^2}{2}\,[1 + R_a(0)]. \tag{7.14-6}$$

The term $R_a(0)$ gives the mean-square value of $a(t)$. The magnitude of $R_a(0)$ will depend on the amount of overmodulation that can be tolerated and on the statistics of $a(t)$. For example, let $a(t)$ have a first-order amplitude distribution which is gaussian with zero mean. The condition necessary to prevent substantial overmodulation is given in Equation 7.12-9. Suppose ϵ is taken to be 0.0035. Then Equation 7.12-9 can be rewritten, after a change in variable, as

$$\frac{1}{\sqrt{2\pi}} \int_{-\infty}^{-1/\sigma} e^{-y^2/2}\, dy \leq 0.0035. \qquad (7.14\text{-}7)$$

It follows from Table I.1 of Appendix I that $1/\sigma$ is given by $1/\sigma = 2.7$ or

$$\sigma^2 = R_a(0) \approx 0.14. \qquad (7.14\text{-}8)$$

In this case, the ratio of modulating-signal power to carrier power is approximately 1/7 and the *fraction* of power that bears nonredundant information is approximately 1/16.

In double-sideband, suppressed-carrier operation, half the total power exists in each sideband. The fraction of information bearing power is $\frac{1}{2}$. For single-sideband, suppressed-carrier systems, all of the transmitted power is information bearing, and, in this sense, SSB-SC has maximum transmission efficiency.

The disadvantages of both suppressed-carrier and single-sideband operation lie in the difficulties of generating the signals for transmission and in the more complicated receivers required. Demodulation of suppressed-carrier AM signals involves the reinsertion of the carrier or an equivalent operation. The local generation of a sinusoid at the exact frequency and phase of the missing carrier is either difficult or impossible unless a pilot tone of reduced magnitude is transmitted with the modulated signal for synchronization purposes or unless some nonlinear operation is performed on the suppressed-carrier signal to regenerate the carrier term at the receiver. In SSBAM, not only is the receiver more complicated but transmission is considerably more difficult. It is usually necessary to generate the SSB signal at a low power level and then to amplify with a linear power amplifier to the proper level for transmission. On the other hand, DSB signals are easily generated at high power levels so that inefficient linear power amplifiers need not be used. Some of the difficulties encountered with suppressed-carrier and single-sideband receivers will become clear in the next section.

7.15 AM Receivers

In this section the performance of AM receivers most commonly encountered will be analyzed. For the analyses to be meaningful it will be assumed that an additive noise $n(t)$ exists at the receiver input so that the total received signal $r(t)$ is given by

$$r(t) = m(t) + n(t) \qquad (7.15\text{-}1)$$

where $m(t) = m[t, a(t)]$ is the modulated AM signal. The noise $n(t)$ may have been added in the transmission medium, or it may be present in the input stages of the receiver. It will be assumed that $n(t)$ is a wide-sense stationary random process with zero mean, autocorrelation function $R_n(\tau)$, and corresponding power spectral density $\varphi_n(\omega)$. When a given spectral density must be assumed in order to obtain explicit results, the one chosen will be that shown in Figure 6.4(a) with corresponding autocorrelation function

$$R_n(\tau) = \frac{NB}{2\pi}\frac{\sin B\tau/2}{B\tau/2}\cos\omega_0\tau \tag{7.15-2}$$

as shown in Figure 6.4(b). Note that

$$E\{n^2(t)\} = R_n(0) = \frac{1}{2\pi}\int_{-\infty}^{\infty}\varphi_n(\omega)\,d\omega = \frac{NB}{2\pi}. \tag{7.15-3}$$

It will be assumed that the modulating intelligence $a(t)$ is low pass and band-limited to $(-B/2, B/2)$ rad/sec. Thus the modulated signal $m(t)$ and the noise $n(t)$ have the same bandwidth in the DSBAM case. If the noise bandwidth were greater than B rad/sec, then prefiltering by bandpass filters could be used to reduce it without affecting the signal.

The Synchronous Receiver

From a conceptual point of view, the simplest AM demodulator or receiver is the *synchronous receiver* shown schematically in Figure 7.22. The incoming signal $r(t)$, consisting of the AM modulated waveform and additive noise, is multiplied by a locally generated sinusoid of the same frequency ω_0 as the carrier, resulting in a frequency translation of $\pm\omega_0$. The resulting waveform $v(t)$ is passed through a low-pass filter whose output $s(t)$ consists of a scaled

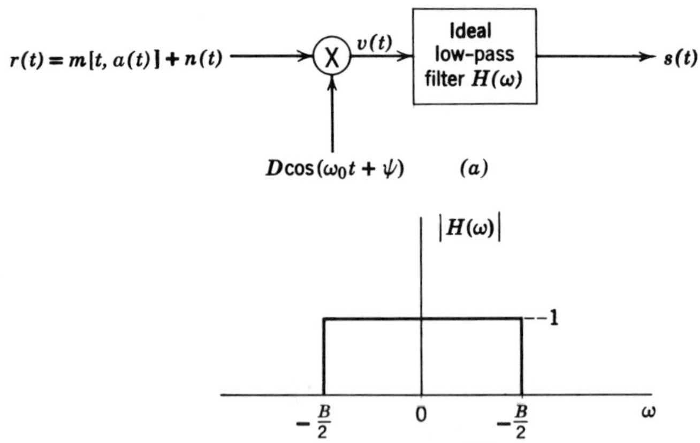

Figure 7.22 A synchronous receiver for amplitude modulation: (a) the form of the receiver; (b) the ideal low-pass filter

reproduction of the original modulation $a(t)$ as well as noise components resulting from the frequency translation of $n(t)$.

The analysis is straightforward. After the multiplier, the signal $v(t)$ can be written as

$$\begin{aligned} v(t) &= \{C[1 + a(t)] \cos(\omega_0 t + \theta)\} D \cos(\omega_0 t + \psi) + n(t) D \cos(\omega_0 t + \psi) \\ &= \frac{CD}{2}[1 + a(t)] \cos(\psi - \theta) + \frac{CD}{2}[1 + a(t)] \cos(2\omega_0 t + \theta + \psi) \\ &\quad + n(t) D \cos(\omega_0 t + \psi). \end{aligned} \tag{7.15-4}$$

Note that the first term in the last expression contains only low-frequency components, the second term is centered on $\pm 2\omega_0$ in the frequency domain, and the last term contains components centered both on zero frequency and on $\pm 2\omega_0$. The ideal low-pass filter $H(\omega)$ will pass the first term unchanged, will remove the second term entirely, and will pass only that part of the last term which is centered on zero frequency. Thus the output $s(t)$ of the synchronous receiver is

$$s(t) = \frac{CD}{2}[1 + a(t)] \cos(\psi - \theta) + [n(t) D \cos(\omega_0 t + \psi)]_{\text{lf}} \tag{7.15-5}$$

where the notation $[\ ]_{\text{lf}}$ indicates those frequency components of the shifted noise that are centered on zero frequency.

It is clear that the output depends on the difference $(\psi - \theta)$ between the phase angles of the incoming carrier and the locally generated sinusoid. For the moment, let us assume that this difference is zero. The average power in $s(t)$ when $\psi - \theta = 0$ is

$$P_{\text{sav}} = \frac{C^2 D^2}{4}[1 + a^2(t)_{\text{av}}] + \frac{D^2}{2}[n^2(t)_{\text{av}}]_{\text{lp}}. \tag{7.15-6}$$

Here the various terms are

$$\frac{C^2 D^2}{4} = \text{d-c power resulting from frequency translation of the carrier}$$

$$\frac{C^2 D^2}{4} a^2(t)_{\text{av}} = \text{useful output signal power}$$

$$\frac{D^2}{2}[n^2(t)_{\text{av}}]_{\text{lf}} = \text{output noise power.}$$

The last term $[n^2(t)_{\text{av}}]_{\text{lf}}$ is just one-half of $n^2(t)_{\text{av}}$ since half the input noise power is translated to the vicinity of zero frequency and half to the vicinity of $\pm 2\omega_0$. Thus Equation 7.15-6 becomes

$$P_{\text{sav}} = \frac{C^2 D^2}{4}[1 + a^2(t)_{\text{av}}] + \frac{D^2}{4} n^2(t)_{\text{av}}. \tag{7.15-7}$$

The output signal-to-noise power ratio is given by

$$\left(\frac{S}{N}\right)_o = C^2 \frac{a^2(t)_{\text{av}}}{n^2(t)_{\text{av}}}. \tag{7.15-8}$$

On the other hand, the input signal-to-noise power ratio is just

$$\left(\frac{S}{N}\right)_i = \frac{C^2}{2} \frac{a^2(t)_{\text{av}}}{n^2(t)_{\text{av}}}. \tag{7.15-9}$$

As in Section 6.9, it is convenient to define an *improvement factor* I as the ratio of Equation 7.15-8 to Equation 7.15-9; thus

$$I = \frac{(S/N)_o}{(S/N)_i} = 2 \tag{7.15-10}$$

and the improvement factor is a constant regardless of the form of the modulating signal.

It is clear that the synchronous receiver can be used for the reception of DSBAM-SC as well as for DSBAM. The only difference in the suppressed-carrier case will be the absence of the d-c term $C^2D^2/4$ in Equation 7.15-6. From a practical point of view, however, the presence of the carrier will aid materially in the production of the locally generated multiplying sinusoid. For example, the carrier may be removed from the DSBAM signal by a narrow-band filter and used for frequency and phase synchronization of the local sinusoid.

Let us now consider briefly the effect of the locally generated sinusoid's frequency and phase on the performance of the synchronous receiver. It is clear from Equation 7.15-5 that a phase difference $(\psi - \theta)$ produces a multiplicative factor $\cos(\psi - \theta)$ which reduces the magnitude of the output signal. Since the magnitude of $\cos(\psi - \theta)$ varies from unity to zero, it is clear that phase synchronization is very important in the performance of the synchronous receiver.

The effect of frequency deviations is somewhat different. Suppose the locally generated sinusoid has a frequency $\omega_1 = \omega_0 + \Delta\omega$. Assume that $\psi = \theta = 0$ for simplicity. Then the signal component after the multiplier is

$$\begin{aligned} &CDa(t)\cos\omega_0 t\cos(\omega_0 + \Delta\omega)t \\ &\quad = CDa(t)\cos^2\omega_0 t\cos\Delta\omega t - CDa(t)\cos\omega_0 t\sin\omega_0 t\sin\Delta\omega t. \end{aligned} \tag{7.15-11}$$

If $\Delta\omega$ is taken to be small so that $\cos\Delta\omega t$ and $\sin\Delta\omega t$ are low-frequency terms, then the signal output of the synchronous receiver after the ideal low-pass filter will be

$$\frac{CD}{2}a(t)\cos\Delta\omega t. \tag{7.15-12}$$

Thus the effect of a frequency error is an additional amplitude-variation factor.

The foregoing suggests that slight phase errors can be tolerated but frequency errors cannot. Thus carrier transmission (even at a greatly reduced magnitude) for synchronization purposes is desirable. Various types of automatic-frequency/phase-controlled loops are used in suppressed-carrier reception to minimize both frequency and phase errors.

We note in passing that SSB can be demodulated by a synchronous receiver. Refer, for example, to Equation 7.13-23. Multiplication of this modulated signal by $D \cos(\omega_0 t + \theta)$ will produce only one low-frequency component, namely $(CD/2)a(t)$, which may be recovered from the high-frequency components by the low-pass filter of Figure 7.22.

The Square-Law Receiver

This receiver is shown schematically in Figure 6.9. It has been analyzed in detail in Sections 6.4 through 6.9 for the case in which both the modulating intelligence $a(t)$ and the additive noise $n(t)$ are stationary gaussian processes with power spectral densities shown in Figure 6.7. The results are fairly complicated because of the large number of terms arising from the squaring operation. The total output power spectral density is shown in Figure 6.10. Only one of the "signal × signal" terms is the desired output, that is, only this term corresponds to the undistorted intelligence $a(t)$. The improvement factor I for the square-law receiver has been plotted in Figure 6.12 as a function of noise-to-signal power ratio P and for various values of $\sigma_a^2 = E\{a^2(t)\}$. As mentioned previously, the improvement factor is low for low noise $(P \to 0)$ because of the predominance of the signal-squared distortion terms and is low for small-signal conditions $(P \to \infty)$ as a result of the noise-squared terms. For small-signal conditions the improvement factor varies as $1/P$, illustrating a *small-signal suppression effect* typical of nonlinear demodulators. In contrast, the improvement factor for the synchronous receiver is the constant 2, independent of signal-to-noise ratio.

The square-law receiver has been analyzed extensively and often is used as a standard against which other receivers can be compared. Under small-signal conditions, many physical devices used as nonlinear demodulators have a response characteristic that is approximately square law.

The Envelope Receiver

This receiver is shown schematically in Figure 7.23(a). An ideal half-wave rectifier is followed by the ideal low-pass filter of Figure 7.22(b). A more practical implementation is shown in Figure 7.23(b). In this receiver, the rectifier passes the positive part of the modulated waveform while the low-pass filter transmits only the envelope of this positive part. In terms of the circuit of Figure 7.23(b), the product R_1C_1 is chosen to be large enough so that the voltage on capacitor C_1 follows the envelope of the carrier and not the carrier itself.

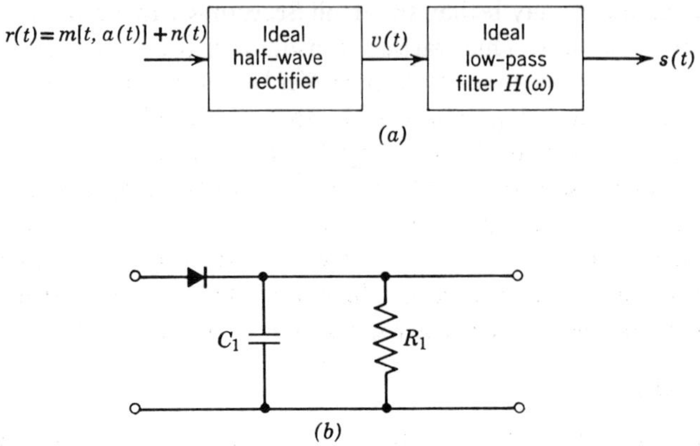

Figure 7.23 The envelope receiver: (a) the ideal envelope detector; (b) a practical implementation

In the analysis of this receiver it is convenient to represent the narrow-band input noise by the techniques developed in Section 4.7. Following Equation 4.7-1, we will write the additive gaussian noise $n(t)$ in the form

$$n(t) = E(t) \cos [\omega_0 t + \phi(t)] \tag{7.15-13}$$

where $E(t)$ is the Rayleigh-distributed envelope function and $\phi(t)$ is a phase function which is uniformly distributed on $(0, 2\pi)$. Thus the input signal $r(t)$ becomes

$$r(t) = C[1 + a(t)] \cos (\omega_0 t + \theta) + E(t) \cos (\omega_0 t + \phi). \tag{7.15-14}$$

The output $s(t)$ of the receiver is just the envelope of this last expression and is given by

$$s(t) = \sqrt{C^2[1 + a(t)]^2 + E^2(t) + 2CE(t)[1 + a(t)] \cos (\psi - \theta)}. \tag{7.15-15}$$

In general, it is difficult to find the average power in $s(t)$ even when the statistics of $a(t)$ and $n(t)$ are known. For the large-signal case, however, the term $E^2(t)$ may be ignored and Equation 7.15-15 may be approximated by

$$s(t) \approx \sqrt{x^2 + 2xy} \approx x + y$$

or

$$s(t) \approx C[1 + a(t)] + E(t) \cos (\psi - \theta). \tag{7.15-16}$$

The average output power is now

$$s^2(t)_{\text{av}} = C^2 + C^2 a^2(t)_{\text{av}} + \frac{E^2(t)_{\text{av}}}{2} \tag{7.15-17}$$

where the last term may be written from Section 4.7 as

$$\frac{E^2(t)_{\text{av}}}{2} = n^2(t)_{\text{av}}. \tag{7.15-18}$$

Thus both the input and output signal-to-noise power ratios have already been given by Equations 7.15-9 and 7.15-8, respectively. Hence the improvement ratio is

$$I = 2.$$

For large signal-to-noise conditions, the envelope receiver approaches the performance of the synchronous receiver.

For small signal-to-noise ratio, the term $C^2[1 + a(t)]^2$ in Equation 7.15-15 may be ignored and $s(t)$ is approximated by

$$s(t) \approx E(t) + C[1 + a(t)] \cos (\psi - \theta). \tag{7.15-19}$$

Now there is *no term* in the output corresponding to an uncorrupted version of the intelligence $a(t)$. Thus there is a *threshold* signal-to-noise ratio below which the signal is rapidly lost in corrupting noise and above which system performance approaches synchronous demodulation.

7.16 Angle Modulation

In angle modulation, the carrier $c(t)$ has either its phase angle or its frequency varied in accordance with the intelligence $a(t)$. The result is not a simple frequency translation as in the AM case but involves both translation and the production of entirely new frequency components. In general, the new spectrum is much wider than that of the intelligence $a(t)$. The greater bandwidth may be used to improve the signal-to-noise performance of the receiver. This ability to *exchange* bandwidth for signal-to-noise enhancement is one of the outstanding characteristics and advantages of angle modulation.

Let us restrict the modulated signal $m[t, a(t)]$ to be a sinusoid of the form

$$m(t) = m[t, a(t)] = C \cos \psi(t). \tag{7.16-1}$$

In the absence of modulation this signal reduces to the usual carrier $c(t)$ where

$$c(t) = m(t, 0) = C \cos (\omega_0 t + \theta) \tag{7.16-2}$$

and θ is either a constant or a random variable uniformly distributed on $(0, 2\pi)$.

In the form of angle modulation which will be called *phase modulation* (PM) the phase of the carrier is varied linearly with the intelligence $a(t)$. Thus the modulated signal is given by

$$m(t) = C \cos [\omega_0 t + \theta + k_p a(t)] \tag{7.16-3}$$

where k_p is a constant and the *modulation index* ϕ_m is defined by

$$\phi_m = \max |k_p a(t)| \text{ radians}. \tag{7.16-4}$$

The concept of *frequency modulation* (FM) is most conveniently introduced by defining an *instantaneous (radian) frequency* ω_i from Equation 7.16-1 by

$$\omega_i = \frac{d\psi(t)}{dt} \tag{7.16-5}$$

or, equivalently, by

$$\psi(t) = \int_{-\infty}^{t} \omega_i \, dt. \tag{7.16-6}$$

In the absence of modulation this frequency becomes

$$\omega_i = \omega_0$$

as expected. In frequency modulation the instantaneous frequency is made proportional to the intelligence $a(t)$:

$$\omega_i = \omega_i(t) = \omega_0 + k_f a(t) \tag{7.16-7}$$

where k_f is a constant. Now the argument $\psi(t)$ is given by

$$\psi(t) = \int_{-\infty}^{t} \omega_i \, dt = \omega_0 t + k_f \int_{-\infty}^{t} a(\tau) \, d\tau \tag{7.16-8}$$

and the modulated signal by

$$m(t) = C \cos \left[\omega_0 t + k_f \int_{-\infty}^{t} a(\tau) \, d\tau\right]. \tag{7.16-9}$$

It is convenient to define a *maximum deviation* $\Delta\omega$ by

$$\Delta\omega = \max |k_f a(t)| \text{ rad/sec} \tag{7.16-10}$$

and, as before, a *modulation index* ϕ_m by

$$\phi_m = \max \left| k_f \int_{-\infty}^{t} a(\tau) \, d\tau \right|. \tag{7.16-11}$$

In general, the analysis of angle-modulated signals is difficult even for simple modulating intelligence. However, the narrow-band case is relatively easy to treat although not of as general interest as wide-band angle modulation.

Narrow-Band Angle Modulation

It will become clear later that the narrow-band case corresponds to a small modulation index ϕ_m. Consider the PM case (the situation is essentially the same for FM) and restrict

$$|k_p a(t)| \ll 1 \tag{7.16-12}$$

in Equation 7.16-3. This equation may be expanded to yield

$$m(t) = C \cos [k_p a(t)] \cos (\omega_0 t + \theta) - C \sin [k_p a(t)] \sin (\omega_0 t + \theta). \tag{7.16-13}$$

On making use of the approximations

$$\cos [k_p a(t)] \approx 1, \qquad |k_p a(t)| \ll 1, \tag{7.16-14}$$

and

$$\sin [k_p a(t)] \approx k_p a(t), \qquad |k_p a(t)| \ll 1, \tag{7.16-15}$$

we can rewrite Equation 7.16-13 as

$$m(t) \approx C \cos (\omega_0 t + \theta) - C k_p a(t) \sin (\omega_0 t + \theta). \tag{7.16-16}$$

Thus narrow-band PM is very similar to AM with the carrier shifted 90 degrees in phase. It may be demodulated by a synchronous receiver using a locally generated sinusoid $D \sin (\omega_0 t + \theta)$. One way to implement such a receiver would be to recover the carrier by a narrow-band filter, pass this carrier through a 90-degree phase shifter, and use the output for the multiplying sinusoid.

Wide-Band Angle Modulation

We will consider only the case in which the modulating intelligence $a(t)$ is a sinusoid. Let $a(t)$ be given, as in Equation 7.13-15, by

$$a(t) = \eta \cos \omega_s t, \qquad \omega_s < \omega_0. \tag{7.16-17}$$

Then the corresponding PM signal is

$$m(t) = C \cos (\omega_0 t + k_p \eta \cos \omega_s t) \tag{7.16-18}$$

where the phase angle θ has been set equal to zero. In the same way, the FM signal is

$$m(t) = C \cos \left(\omega_0 t + \frac{k_f \eta}{\omega_s} \sin \omega_s t\right). \tag{7.16-19}$$

Note that, in the FM case, the instantaneous frequency ω_i is given by

$$\omega_i = k_f \eta \cos \omega_s t. \tag{7.16-20}$$

The constant $k_f \eta$ is often called the *(radian) frequency deviation.* It is the maximum difference between ω_i and ω_0 and is independent of modulating-signal frequency, depending only on modulating-signal amplitude and the construction of the modulator.

Let us now consider the FM signal of Equation 7.16-19. Essentially the same results will be obtained for PM. Equation 7.16-19 may be expanded to yield

$$m(t) = C \cos (\phi_m \sin \omega_s t) \cos \omega_0 t - C \sin (\phi_m \sin \omega_s t) \sin \omega_0 t \tag{7.16-21}$$

where the modulation index ϕ_m is given by

$$\phi_m = \frac{k_f \eta}{\omega_s}. \tag{7.16-22}$$

Sinusoids with sinusoidal arguments give rise to Bessel functions. Two well-known relationships are

$$\cos(\phi_m \sin \omega_s t) = J_0(\phi_m) + 2 \sum_{n=1}^{\infty} J_{2n}(\phi_m) \cos 2n\omega_s t \tag{7.16-23}$$

and

$$\sin(\phi_m \sin \omega_s t) = 2 \sum_{n=0}^{\infty} J_{2n+1}(\phi_m) \sin (2n+1)\omega_s t \tag{7.16-24}$$

where $J_m(x)$ is the Bessel function of the first kind, order m. These functions are given by the infinite series

$$J_m(x) = \sum_{k=0}^{\infty} \frac{(-1)^k (x/2)^{2k+m}}{k!(k+m)!} \tag{7.16-25}$$

and have been tabulated extensively. A few of the lower-order functions are plotted in Figure 7.24. Note that $J_m(0)$ is zero except for the single case where $m = 0$. Also, for large m, $J_m(\epsilon) \approx 0$ for small ϵ. In fact, $J_m(x)$ is quite small whenever $x < m$.

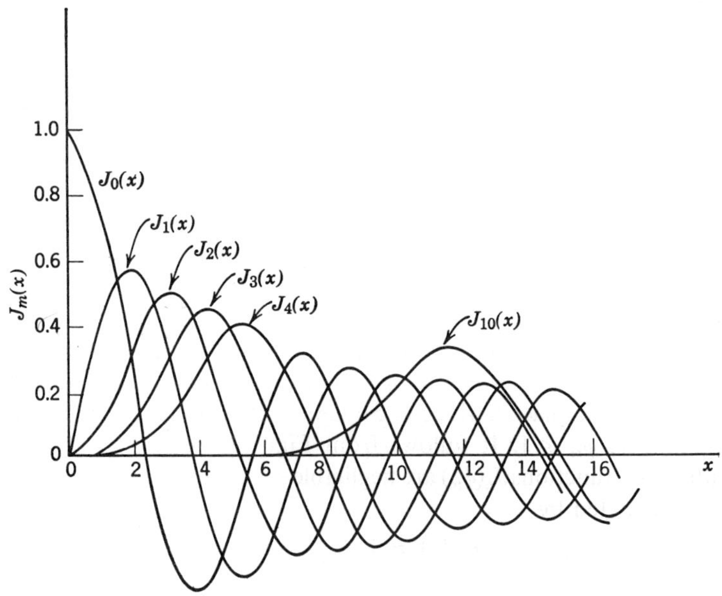

Figure 7.24 *Bessel functions of the first kind*

Now Equations 7.16-23 and 7.16-24 may be substituted in Equation 7.16-21 to obtain

$$m(t) = \underbrace{CJ_0(\phi_m)\cos\omega_0 t}_{\text{carrier}}$$
$$+ C\sum_{n=1}^{\infty} J_{2n}(\phi_m)[\underbrace{\cos(\omega_0 + 2n\omega_s)t}_{\text{USB}} + \underbrace{\cos(\omega_0 - 2n\omega_s)t}_{\text{LSB}}]$$
$$+ C\sum_{n=1}^{\infty} J_{2n-1}(\phi_m)\{\underbrace{\cos[\omega_0 + (2n-1)\omega_s]t}_{\text{USB}} - \underbrace{\cos[\omega_0 - (2n-1)\omega_s]t}_{\text{LSB}}\} \tag{7.16-26}$$

where use has been made of the trigonometric identities

$$2\cos\alpha\cos\beta = \cos(\alpha+\beta) + \cos(\alpha-\beta)$$

and

$$-2\sin\alpha\sin\beta = \cos(\alpha+\beta) - \cos(\alpha-\beta).$$

This expression for $m(t)$ is relatively complicated even though the modulating intelligence is a simple sinusoid. In addition to the carrier, there are an infinite number of upper and lower sidebands separated from the carrier (and from each other) by integral multiples of the modulating frequency ω_s. Each sideband, and the carrier, has an amplitude determined by the appropriate Bessel function. When there is more than one modulating sinusoid, the complexity increases rapidly. For example, when $a(t)$ is given by

$$a(t) = \eta_1\cos\omega_1 t + \eta_2\cos\omega_2 t,$$

then it follows directly from Equation 7.16-21 that products of the form

$$\cos(\phi_{m1}\sin\omega_1 t)\cos(\phi_{m2}\sin\omega_2 t)\cos\omega_0 t$$

exist. These terms give rise to sidebands not only at the infinite sets of frequencies $\omega_0 \pm n\omega_1$ and $\omega_0 \pm m\omega_2$ but also at the doubly infinite set $\omega_0 \pm (n\omega_1 \pm m\omega_2)$. On the other hand, the corresponding AM signal would contain only the sideband frequencies $\omega_0 \pm \omega_1$ and $\omega_0 \pm \omega_2$.

The condition under which the FM signal is narrow band is clear from an examination of Equation 7.16-26. Suppose the modulation index ϕ_m is very small, that is, $\phi_m \ll 1$. It follows from Figure 7.24 and Equation 7.16-25 that all Bessel functions $J_n(\phi_m)$ are negligible except for $n = 0, 1$ and that these are approximated by

$$J_0(\phi_m) \approx 1, \qquad \phi_m \ll 1, \tag{7.16-27}$$

$$J_1(\phi_m) \approx \frac{\phi_m}{2}, \qquad \phi_m \ll 1. \tag{7.16-28}$$

Thus Equation 7.16-26 reduces in the narrow-band case to

$$m(t) = C \cos \omega_0 t + \frac{C\phi_m}{2} [\cos (\omega_0 + \omega_s)t - \cos (\omega_0 - \omega_s)t]. \quad (7.16\text{-}29)$$

This last expression is equivalent to Equation 7.16-16 which was already obtained for narrow-band PM.

It is apparent from Equation 7.16-26 that, in the general case, an infinite number of sidebands exist dispersed throughout the whole frequency domain. In this sense, the bandwidth of an FM (or PM) signal is infinite. Outside of some interval centered on ω_0, however, the magnitude of the sidebands will be negligible; this interval may be taken as a practical measure of the bandwidth. An approximation may be obtained by noting that $J_n(\phi_m)$, considered as a function of n, decreases rapidly when $n < \phi_m$. Therefore only the first ϕ_m sidebands are significant and Equation 7.16-26 may be truncated. If the highest frequency sideband of significance is $\phi_m\omega_s$, then the bandwidth is given approximately by

$$BW \approx 2\omega_s\phi_m \text{ rad/sec.} \quad (7.16\text{-}30)$$

A more accurate "rule of thumb" is the slightly revised expression

$$BW \approx 2\omega_s(\phi_m + 1) \text{ rad/sec} \quad (7.16\text{-}31)$$

which may be considered a good approximation when $\phi_m \geq 5$.

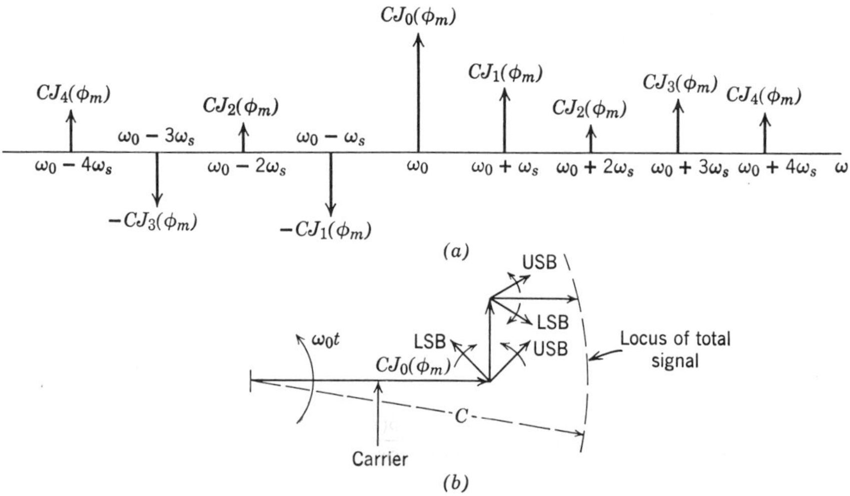

Figure 7.25 *The frequency spectrum and phasor diagram of an FM signal: (a) the frequency spectrum $M(\omega)$; (b) the phasor diagram:*

$$M(t) = C \cos (\omega_0 t + \phi_m \sin \omega_s t)$$

In the AM case, both the frequency spectrum $M(\omega)$ of $m(t)$ and the associated phasor diagram are relatively simple, as shown in Figures 7.13(*b*) and 7.14, respectively. The FM case is considerably more complicated. The spectrum of Equation 7.16-26 is shown in Figure 7.25(*a*) for the frequency interval near ω_0. Note that corresponding upper and lower sidebands are symmetrically placed about the carrier and have equal magnitude. Moreover, the even-order sidebands and the odd-order *upper* sidebands are all in the same phase relationship with each other and with the carrier as in the AM case. On the other hand, the odd-order *lower* sidebands are 180 degrees out of phase with their corresponding upper sidebands. The phasor diagram is shown in Figure 7.25(*b*). The odd- and even-order sidebands are in phase quadrature and, when added to the carrier, result in a total signal phasor of constant length C as must follow from Equation 7.16-19.

7.17 The Effect of Noise on FM Signals

The case will be considered in which the FM signal, as given by Equation 7.16-19, is received in stationary bandpass gaussian noise $n(t)$. The total received waveform $r(t)$ may be written as

$$\begin{aligned} r(t) &= m(t) + n(t) \\ &= C\cos(\omega_0 t + \phi_m \sin\omega_s t) + X(t)\cos\omega_0 t + Y(t)\sin\omega_0 t \end{aligned} \tag{7.17-1}$$

where $\phi_m = k_f\eta/\omega_s$ and where the noise $n(t)$ has been decomposed into its quadrature components $X(t)\cos\omega_0 t$ and $Y(t)\sin\omega_0 t$ as in Equation 4.7-6. The expression for $r(t)$ may be rearranged to read

$$\begin{aligned} r(t) = {} & [C + X(t)\cos\phi(t) - Y(t)\sin\phi(t)]\cos[\omega_0 t + \phi(t)] \\ & + [X(t)\sin\phi(t) + Y(t)\cos\phi(t)]\sin[\omega_0 t + \phi(t)] \end{aligned} \tag{7.17-2}$$

where $\phi(t)$ has been written for $\phi_m \sin\omega_s t$. In terms of envelope and phase this becomes

$$r(t) = E(t)\cos[\omega_0 t + \psi(t)] \tag{7.17-3}$$

where the envelope function $E(t)$ is

$$E(t) = \sqrt{C^2 + X^2(t) + Y^2(t) + 2CX(t)\cos\phi(t) - 2CY\sin\phi(t)} \tag{7.17-4}$$

and the phase function $\psi(t)$ is

$$\psi(t) = \phi(t) - \arctan\frac{X(t)\sin\phi(t) + Y(t)\cos\phi(t)}{C + X(t)\cos\phi(t) - Y(t)\sin\phi(t)}. \tag{7.17-5}$$

In the received signal of Equation 7.17-1 it is clear that the signal-to-noise power ratio (or, more correctly, the carrier-to-noise power ratio) is given by

$$\left(\frac{S}{N}\right)_i = \frac{C^2}{2\overline{n^2(t)}_{\text{av}}} \tag{7.17-6}$$

where $n^2(t)_{av}$ is the noise power in the effective FM bandwidth. The usual procedure in receiving an FM signal is to pass it through a *limiter* that removes all amplitude variations or, equivalently, to employ a detector that is insensitive to these variations. In either case, the signal of Equation 7.17-3 is effectively changed to

$$r_1(t) = \cos [\omega_0 t + \psi(t)]. \tag{7.17-7}$$

The problem is to recover the intelligence $\phi(t)$ from this last expression.

Let us consider first the case in which the signal is large so that there is a high probability that

$$C \gg \sqrt{X^2(t) + Y^2(t)}. \tag{7.17-8}$$

Then Equation 7.17-5 can be written approximately as

$$\psi(t) \approx \phi(t) - \frac{X(t) \sin \phi(t) + Y(t) \cos \phi(t)}{C}. \tag{7.17-9}$$

It is clear from this last expression that, when the signal is very large compared to the noise, the phase variations due to the noise are greatly attenuated. In this case it is sometimes said that the signal *captures* the demodulator. Consider now the instant of time at which $\phi(t) = 0$ and find the noise component; Equation 7.17-9 becomes

$$\psi(t)|_{\phi(t)=0} \approx -\frac{Y(t)}{C}. \tag{7.17-10}$$

As noted earlier and illustrated by Equations 7.16-5 and 7.16-7, the FM demodulator forms the derivative of the argument $\psi(t)$. Thus, in the large-signal case, the output from an FM demodulator can be approximated by

$$\frac{d\psi(t)}{dt} = \frac{d\phi(t)}{dt} - \frac{1}{C}\frac{dY(t)}{dt} \tag{7.17-11}$$

where $Y(t)$ is the low-frequency equivalent of one of the quadrature components of the noise and

$$\frac{d\phi(t)}{dt} = k_f \eta \cos \omega_s t. \tag{7.17-12}$$

It is clear that the output signal power arises from this last term and is given by $k_f^2\eta^2/2$. The output noise power arises from the term $-(1/C)[dY(t)/dt]$.

If the input noise power spectral density is constant with magnitude $N_0/2$ over the effective passband of the FM signal, as illustrated in Figure 6.4, then the effective noise power at the input to the FM demodulator is obtained simply from Equation 7.16-30 as

$$n^2(t)_{av} = \frac{2\omega_s \phi_m N_0}{2\pi}. \tag{7.17-13}$$

Thus the input signal-to-noise power ratio follows from Equation 7.17-6 as

$$\left(\frac{S}{N}\right)_i = \frac{\pi C^2}{2\omega_s \phi_m N_0} = \frac{\pi C^2}{2N_0 k_f \eta} \tag{7.17-14}$$

since $\phi_m = k_f\eta/\omega_s$. The expression given by Equation 7.17-11 can now be passed through an ideal low-pass filter with highest frequency ω_s rad/sec in order to recover the signal $k_f\eta \cos \omega_s t$. Let us consider the effect of this filter on the noise term $-(1/C)[dY(t)/dt]$ of Equation 7.17-11. We have assumed that the noise $n(t)$ is gaussian with a flat power spectral density over that frequency interval about ω_0 where appreciable sidebands of the FM signal exist. It was pointed out in Section 4.7 and in Equation 4.7-6 that, when a gaussian signal $n(t)$ is decomposed into quadrature components $X(t) \cos \omega_0 t$ and $Y(t) \sin \omega_0 t$, then both $X(t)$ and $Y(t)$ are gaussian processes. Furthermore, as given by Equation 4.7-18, we have

$$\begin{aligned} n^2(t)_{\text{av}} &= E\{n^2(t)\} = \tfrac{1}{2}E\{X^2(t)\} + \tfrac{1}{2}E\{Y^2(t)\} \\ &= E\{X^2(t)\} = E\{Y^2(t)\}. \end{aligned} \tag{7.17-15}$$

Thus the power spectral density of $Y(t)$ must be the constant N_0 over the low-frequency interval $(-\omega_s, \omega_s)$, as shown in Figure 7.26(a), with total power

$$\frac{1}{2\pi}\int_{-\omega_s}^{\omega_s} N_0 \, d\omega = \frac{N_0\omega_s}{\pi}. \tag{7.17-16}$$

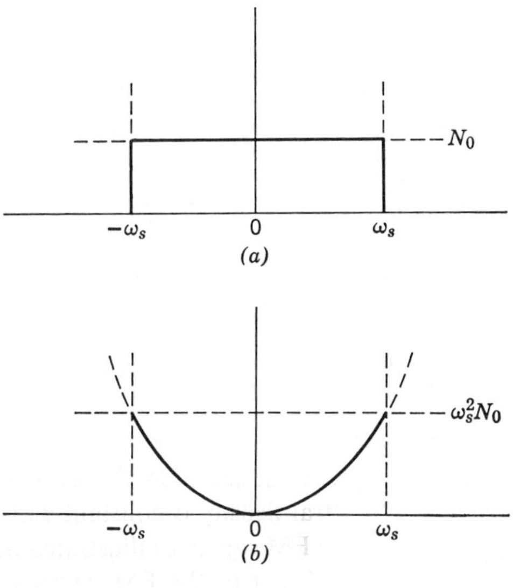

Figure 7.26 *The effect of an FM demodulator on wide-band noise: (a) the noise power spectral density before the FM demodulator; (b) the output noise power spectral density (large-signal case)*

The differentiator that performs the operation d/dt is a linear system with complex transfer function $H(\omega) = j\omega$. Thus the spectral density of $dY(t)/dt$ is equal to the spectral density of $Y(t)$ multiplied by $|H(\omega)|^2 = \omega^2$ and is shown in Figure 7.26(b) over the range $(-\omega_s, \omega_s)$. The total noise power after the low-pass filter is

$$\frac{1}{C^2}\frac{1}{2\pi}\int_{-\omega_s}^{\omega_s} \omega_s^2 N_0 \, d\omega = \frac{N_0\omega_s^3}{3\pi C^2}. \tag{7.17-17}$$

On the other hand, the signal output from the differentiator (and low-pass filter) is given by Equation 7.17-12 with power $k_f^2\eta^2/2$. Thus the output signal-to-noise power ratio is

$$\left(\frac{S}{N}\right)_o = \frac{k_f^2\eta^2}{2}\cdot\frac{3\pi C^2}{N_0\omega_s^3} = \frac{3\pi C^2 k_f^2\eta^2}{2N_0\omega_s^3}. \tag{7.17-18}$$

The improvement factor I follows immediately from this last expression and from Equation 7.17-14:

$$I = \frac{(S/N)_o}{(S/N)_i} = 3\left(\frac{k_f\eta}{\omega_s}\right)^3 = 3\phi_m^3. \tag{7.17-19}$$

Thus, in the large-signal case, the improvement factor varies as the third power of the modulation index. A more accurate analysis would use Equation 7.16-31 instead of Equation 7.16-30 for the bandwidth and would yield

$$I = 3\phi_m^2(\phi_m + 1). \tag{7.17-20}$$

In commercial FM broadcasting the maximum frequency $f_s = \omega_s/2\pi$ can be taken as 15 kc and the maximum frequency deviation $\Delta f = \Delta\omega/2\pi$ is usually fixed at ± 75 kc. Thus we have

$$\Delta\omega = 2\pi\,\Delta f = k_f\eta = 150\pi \times 10^3 \text{ rad/sec}$$

$$\phi_m = \frac{\Delta\omega}{\omega_s} = 5$$

and the improvement factor becomes

$$I = 3(5)^2(5 + 1) = 450.$$

Recall that the improvement factor for synchronous demodulation of DSBAM was 2. The idealized FM receiver for which this analysis has been performed is shown schematically in Figure 7.27.

It is sometimes convenient to compare the FM receiver with an AM system of the same carrier power. In this case the AM input signal would be $C \cos \omega_s t \cos \omega_0 t$ with power $C^2/2$. The associated input noise power would

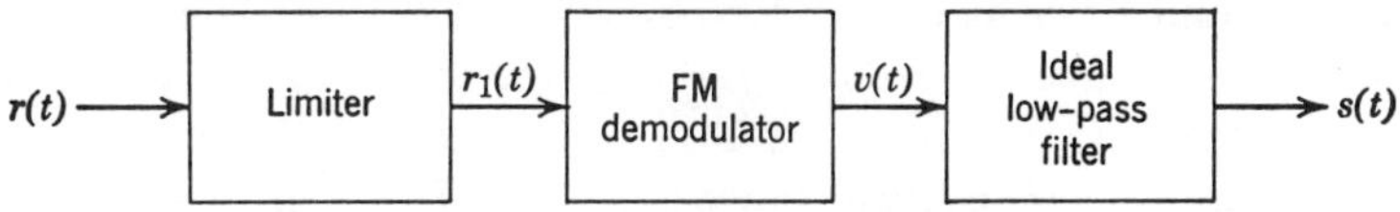

Figure 7.27 The block diagram of an idealized FM receiver:

$$\begin{aligned} r(t) &= C\cos(\omega_0 t + \phi_m \sin \omega_s t) + X(t)\cos\omega_0 t + Y(t)\sin\omega_0 t \\ &= E(t)\cos[\omega_0 t + \psi(t)] \\ r_1(t) &= \cos[\omega_0 t + \psi(t)] \\ v(t) &= \frac{d\psi(t)}{dt} \end{aligned}$$

be $\omega_s N_0/\pi$ in the frequency band $(\omega_0 - \omega_s, \omega_0 + \omega_s)$ and the output carrier-to-noise ratio would be

$$\left(\frac{C}{N}\right)_{o(\mathrm{AM})} = \frac{C^2\pi}{2\omega_s N_0}. \tag{7.17-21}$$

It follows from Equation 7.17-18 that

$$\left(\frac{S}{N}\right)_{o(\mathrm{FM})} = 3\phi_m^2\left(\frac{C}{N}\right)_{o(\mathrm{AM})}. \tag{7.17-22}$$

Thus it could be said that the FM system is $3\phi_m^2$ better in signal-to-noise performance but requires a bandwidth of $2\omega_s(\phi_m + 1)$ rad/sec which is $(\phi_m + 1)$ times the equivalent AM bandwidth. For the previous example where $\phi_m = 5$, the improvement in signal-to-noise is 75 in exchange for a bandwidth increase by a factor of 6.

The previous analysis has been based on the assumption of Equation 7.17-8, namely that the signal is large compared to the noise. For the high-noise case, it is more instructive to rewrite Equation 7.17-1 as

$$r(t) = C\cos[\omega_0 t + \phi(t)] + N(t)\cos[\omega_0 t + \theta(t)] \tag{7.17-23}$$

where the noise $n(t)$ has been expressed in the envelope-and-phase narrow-band representation of Equation 4.7-1. If $n(t)$ is gaussian, then the envelope function $N(t)$ is Rayleigh-distributed and the phase function $\theta(t)$ is uniformly distributed on $(0, 2\pi)$. This last equation can be rewritten as

$$r(t) = N(t)\cos[\omega_0 t + \theta(t)] + C\cos\phi(t)\cos\omega_0 t - C\sin\phi(t)\sin\omega_0 t. \tag{7.17-24}$$

A direct comparison of this equation with Equations 7.17-1 and 7.17-5 shows that the phase function $\psi(t)$ is given by

$$\psi(t) = \theta(t) - \arctan\frac{C\sin[\theta(t) - \phi(t)]}{N(t) + C\cos[\theta(t) - \phi(t)]}. \tag{7.17-25}$$

In the high-noise case, we assume that there is a high probability that

$$N(t) \gg C. \tag{7.17-26}$$

Since $\arctan x \approx x$ for small argument, then Equation 7.17-25 can be approximated by

$$\psi(t) \approx \theta(t) - \frac{C}{N(t)} \sin [\theta(t) - \phi(t)]. \tag{7.17-27}$$

It is clear that the random phase variations $\theta(t)$ predominate in the high-noise case and noise *captures* the instantaneous phase angle $\psi(t)$. The operation $d\psi(t)/dt$ will not recover the intelligence $d\phi(t)/dt$. Thus the FM receiver exhibits a *threshold* effect as did the nonlinear AM receivers. This effect is found to occur in the vicinity of 10 dB input signal-to-noise power ratio; that is, when the left side of Equation 7.17-6 obeys

$$10 = \log_{10} \left(\frac{S}{N}\right)_i.$$

Actually the threshold varies in the approximate range of 6–13 dB depending on the modulation index. A plot of output S/N versus input S/N given in Figure 7.28 clearly illustrates this threshold. For convenience, both $(S/N)_o$ and $(S/N)_i$ are given in decibels. Above the threshold, this figure is simply a plot of Equation 7.17-19.

It is clear from Figure 7.26(b) that the FM demodulator accentuates the higher-frequency noise components. In commercial FM transmitters the high-frequency components of the message are usually emphasized by passing the message signal through an *RC high-pass* filter. In this way, the effect of the high-frequency noise on this portion of the signal can be minimized. The process is known as *preemphasis*. At the FM receiver the demodulated message can then be *deemphasized* by passing it through an *RC low-pass* filter.

PART 3—PULSE MODULATION

In the previous sections, modulation schemes were considered which operated with sinusoidal carriers. In principle, any other continuous waveform could have been used, although analysis of the resulting modulated signal might have become very complicated. In the following sections, systems are considered in which the carrier is no longer continuous but consists of a pulse train, some parameter of which is suitably modified by the modulating intelligence. It will be seen that there are natural applications for such modulation schemes and that they may provide striking improvements in noise immunity.

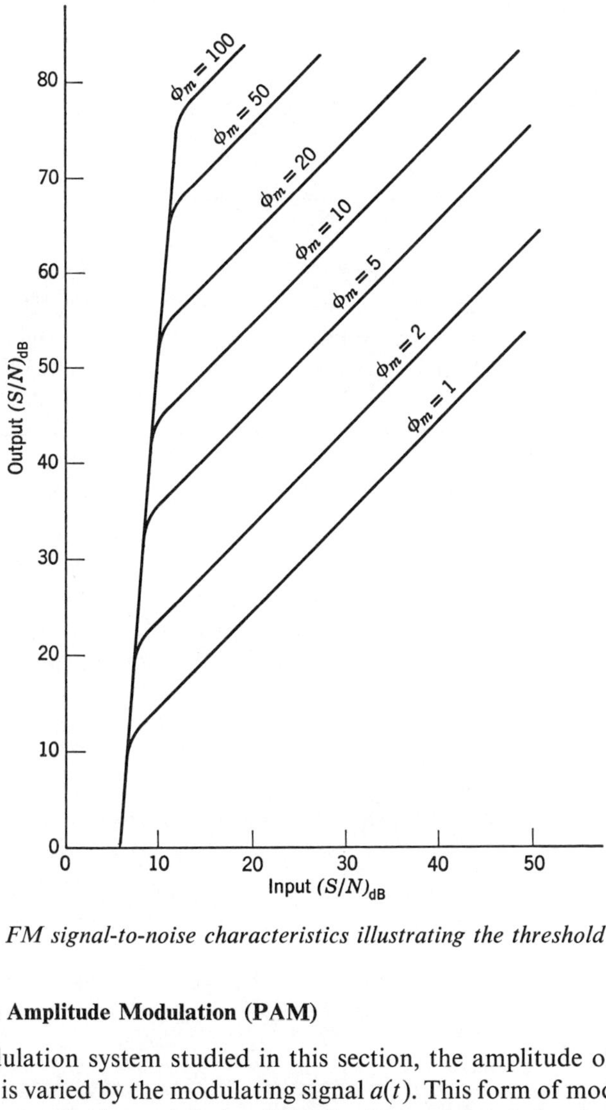

Figure 7.28 *FM signal-to-noise characteristics illustrating the threshold effect*

7.18 Pulse Amplitude Modulation (PAM)

In the modulation system studied in this section, the amplitude of a pulsed carrier $p(t)$ is varied by the modulating signal $a(t)$. This form of modulation is called pulse amplitude modulation (PAM) with modulated signal given by

$$m(t) = m[t, a(t)] = a(t)p(t). \tag{7.18-1}$$

Let us assume that the carrier $p(t)$ is the periodic pulse train shown in Figure 7.29(b) with basic pulse shape $f(t)$, as shown in Figure 7.29(a). Since the pulse train is periodic, it may be expanded in a Fourier series as

$$p(t) = \sum_{k=-\infty}^{\infty} P_k \, e^{jk\omega_0 t} \tag{7.18-2}$$

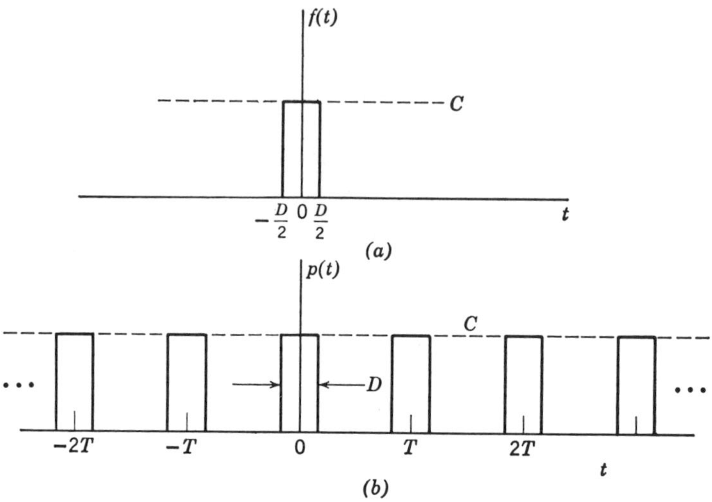

Figure 7.29 A pulsed carrier for PAM: (a) the basic pulse shape; (b) the pulsed carrier

where $\omega_0 = 2\pi/T$ and the Fourier coefficient P_k is given by

$$P_k = \frac{1}{T}\int_{-T/2}^{T/2} p(t)\, e^{-jk\omega_0 t}\, dt. \tag{7.18-3}$$

These last two relations hold for any periodic pulse train with period T. For the particular pulse train shown in Figure 7.29, Equation 7.18-3 may be evaluated to yield specifically

$$P_k = \frac{C}{T}\int_{-D/2}^{D/2} e^{-jk\omega_0 t}\, dt = \frac{CD}{T}\frac{\sin(k\omega_0 D/2)}{k\omega_0 D/2}. \tag{7.18-4}$$

The PAM signal of Equation 7.18-1 may be rewritten with the aid of Equation 7.18-2 as

$$m(t) = a(t)p(t) = \sum_{k=-\infty}^{\infty} P_k a(t)\, e^{jk\omega_0 t} \tag{7.18-5}$$

with Fourier transform $M(\omega)$ given by

$$M(\omega) = \sum_{k=-\infty}^{\infty} P_k \mathscr{F}[a(t)\, e^{jk\omega_0 t}] = \sum_{k=-\infty}^{\infty} P_k A(\omega - k\omega_0) \tag{7.18-6}$$

where $A(\omega)$ is the Fourier transform of $a(t)$. Suppose now that $a(t)$ is band-limited $(-B/2, B/2)$ rad/sec so that $A(\omega)$ has a shape approximately as shown in Figure 7.30(a). Then it is clear from Equation 7.18-6 that the PAM signal

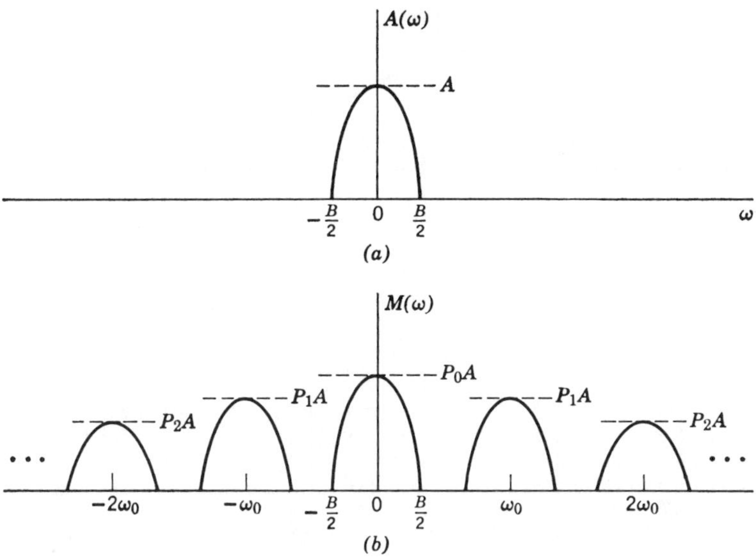

Figure 7.30 *Spectral relationships in PAM: (a) the spectrum of the modulating signal a(t); (b) the spectrum of the PAM signal m(t)*

$m(t)$ has a spectrum as shown in Figure 7.30(*b*). The basic spectral shape is repeated periodically throughout the frequency domain, but each spectral pulse is weighted by the appropriate Fourier coefficient P_k. The value of this coefficient depends on the amplitude, shape, and spacing of the carrier pulse train $p(t)$ as indicated by Equation 7.18-4.

A typical PAM signal is shown in Figure 7.31 together with the original modulating signal and pulse train. In practice, the pulse width D is small compared to the pulse spacing T so that the illustration is somewhat misleading in this respect. The effect of the modulating process is to sample the intelligence $a(t)$ at a fixed interval of T seconds throughout the time domain. If $a(t)$ is strictly bandlimited $(-B/2, B/2)$ rad/sec as shown in Figure 7.30(*a*), then the sampling theorem of Section 7.5 ensures that $a(t)$ may be reconstructed exactly from its samples spaced $2\pi/B$ seconds apart throughout the time domain. Thus a *necessary* condition for exact recovery of $a(t)$ from $a(t)p(t)$ is that

$$T \leq \frac{2\pi}{B} \qquad \text{or} \qquad BT \leq 2\pi. \tag{7.18-7}$$

It is clear from Figure 7.30(*b*) that this condition is the necessary restriction to prevent overlapping of the repeated spectra of the PAM signal $m(t)$. (Note that ω_0 has been defined as $2\pi/T$.)

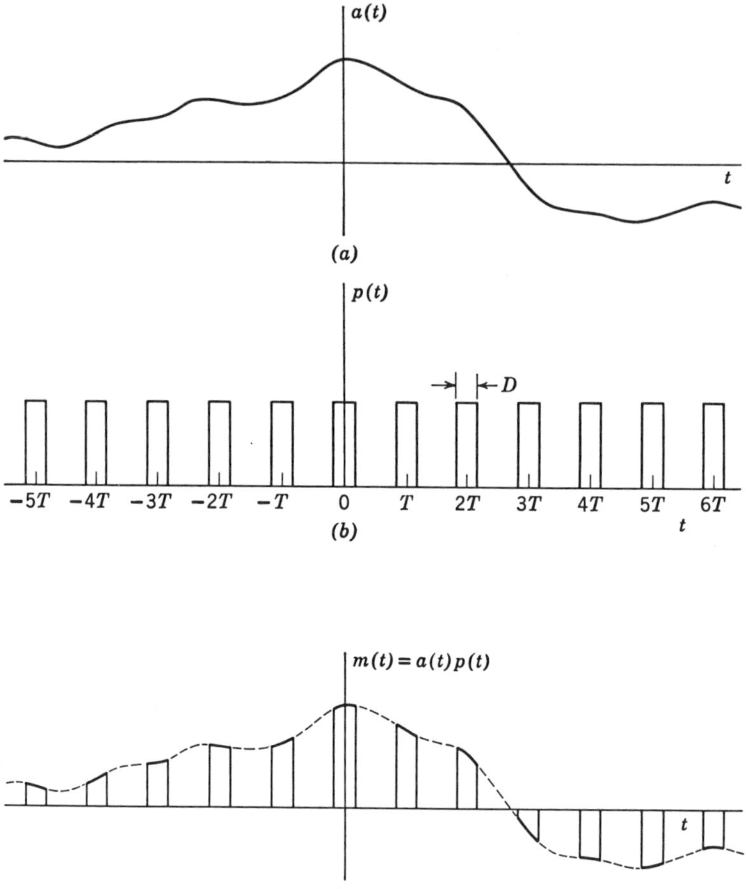

Figure 7.31 *Signal relationships in PAM: (a) the modulating signal $a(t)$; (b) the pulse train $p(t)$; (c) the PAM signal $m(t)$*

In the demodulation of PAM signals, two cases must be distinguished. If the Fourier coefficient P_0 is not zero, then it is clear from Figure 7.30(*b*) that the modulating intelligence $a(t)$ may be recovered by ideal low-pass filtering as shown in Figure 7.32(*a*). The resulting output is just the original intelligence $a(t)$ weighted by the constant P_0. In the case where P_0 is zero but P_1 is not, the intelligence $a(t)$ may be recovered by synchronous demodulation as shown in Figure 7.32(*b*). In this case, the signal $x(t)$ after the multiplier has a spectrum $X(\omega)$ given by

$$X(\omega) = \frac{1}{2} \sum_{k=-\infty}^{\infty} P_k\{A[\omega - (k - 1)\omega_0] + A[\omega - (k + 1)\omega_0]\}. \quad (7.18\text{-}8)$$

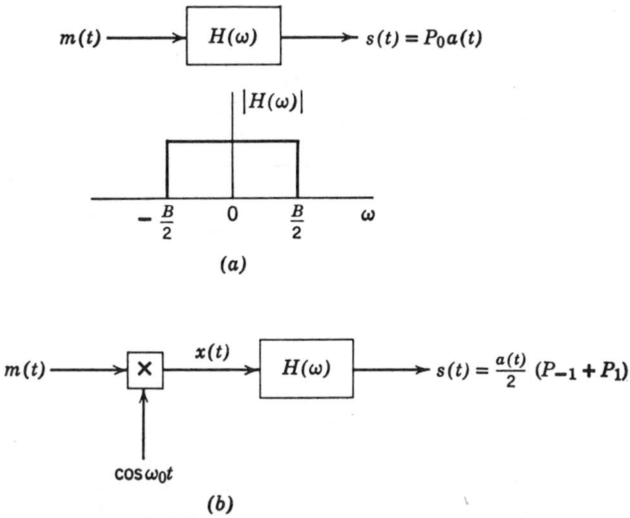

Figure 7.32 The demodulation of PAM signals: (a) the special case where $P_0 \neq 0$; (b) the special case where $P_0 = 0$ and $P_1 \neq 0$

The low-pass filter $H(\omega)$ will remove all high-frequency components (i.e., those where $k \neq 1$). The resulting output spectrum $S(\omega)$ is

$$S(\omega) = \tfrac{1}{2}A(\omega)(P_{-1} + P_1) \tag{7.18-9}$$

and the output signal $s(t)$ is

$$s(t) = \tfrac{1}{2}a(t)(P_{-1} + P_1). \tag{7.18-10}$$

As a practical matter, the exact waveform shown in Figure 7.31(c) is not usually employed in PAM. More specifically, the tops of the pulse train are not usually shaped by the modulating signal. Instead they are kept flat and the pulse height is determined by the value of $a(t)$ at some point in the pulse length interval D. Thus the PAM signal is not exactly $a(t)p(t)$ and the resulting spectrum $M(\omega)$ does not yield $A(\omega)$ with a constant scale factor. Instead there is some distortion of the spectrum of $A(\omega)$ depending on the exact shape of the pulse train $p(t)$. In many cases this distortion can be kept negligibly small and, in other cases, it can be substantially removed by a low-pass equalizing filter in the PAM receiver.

7.19 Quantizing and Quantizing Error

The PAM system studied in the previous section involved converting a time-continuous modulating intelligence $a(t)$ to a time-discrete form by *sampling*. The sampling theorem of Section 7.5 guaranteed that $a(t)$ could be reconstructed exactly from its samples $a(k/2W)$, $k = \ldots, -1, 0, 1, \ldots$, spaced throughout the time domain provided $a(t)$ had a spectrum $A(\omega)$ that was strictly bandlimited $(-2\pi W, 2\pi W)$ rad/sec. In a similar fashion, an amplitude-continuous signal is converted to an *amplitude-discrete* or *digital* signal by *quantizing* in the amplitude domain into a finite number of fixed distinguishable amplitude levels. An example of a signal that has been *sampled* and *quantized* is shown in Figure 7.33. Eight distinguishable quantization levels are possible, although only six have been used for the section of signal shown. Note that the particular quantization level chosen is closest to the signal amplitude at that time instant.

The process of quantizing is irreversible since, regardless of how small the quantization level Q is taken to be, an unresolvable uncertainty of $\pm Q/2$ is associated after quantizing with each amplitude value. Thus a *quantization noise* N_q is inevitably associated with all quantized signals. This noise may be made as small as desired by choosing enough quantization levels or, equivalently, making each quantization level small enough, but it cannot be eliminated. In the pulsed carrier systems to be discussed subsequently, it will develop that the noise added in transmission can be almost completely eliminated at the receiver. In other words, the type of noise interference that has been considered heretofore will be negligible and the principal source of contamination will be the quantization noise.

In Figure 7.33, the quantization levels are all shown to be equal, that is, the spacing between amplitude levels is uniform. Nonuniform quantizing can be used to favor small amplitudes where noise has a greater effect at the expense of large amplitudes. For this discussion, however, it will be assumed that the differences in amplitude levels are all equal to Q. Let us assume that the actual amplitude of the signal being sampled is equally likely to lie any-

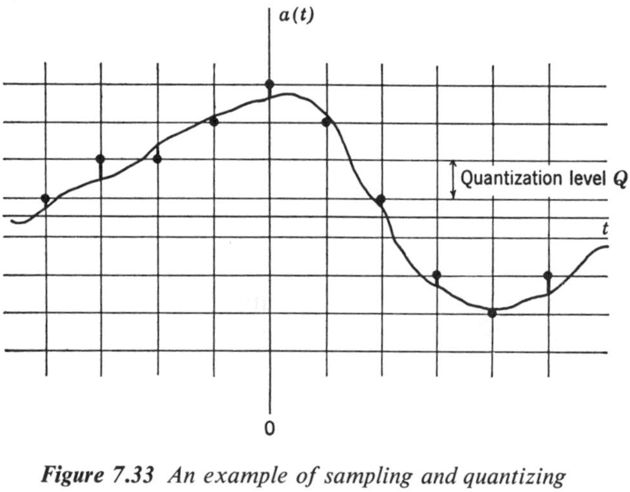

Figure 7.33 *An example of sampling and quantizing*

where within the particular quantization level. In other words, the quantizing error in an amplitude interval is taken to be a uniformly distributed random variable q with density $p_q(x)$ given by

$$p_q(x) = \begin{cases} 0, & |x| > \dfrac{Q}{2}, \\ \dfrac{1}{Q}, & |x| \leq \dfrac{Q}{2}. \end{cases} \tag{7.19-1}$$

The quantization noise N_q can be written as

$$N_q = \int_{-Q/2}^{Q/2} x^2 p_q(x)\, dx = \frac{Q^2}{12}. \tag{7.19-2}$$

The average value of the quantizing error is zero since the distribution of Equation 7.19-1 has zero mean. In terms of one-volt quantization levels the r.m.s. error voltage is $1/\sqrt{12}$ volts. Suppose that the peak-to-peak value of the signal to be quantized is limited to $\pm A$. Then the quantization level Q can be written as

$$Q = \frac{2A}{d} \tag{7.19-3}$$

where d is the number of quantization levels. The signal-to-quantization-noise power ratio for a sampled signal $a^*(t)$ can now be written as

$$\frac{S}{N_q} = \frac{12[a^*(t)]^2_{\text{av}}}{Q^2} = 3d^2 \frac{[a^*(t)]^2_{\text{av}}}{A^2} \tag{7.19-4}$$

where the quantity $[a^*(t)]^2_{\text{av}}/A^2$ is the normalized sampled signal power.

7.20 Signal Encoding

The result of amplitude quantizing and time sampling a message signal is a sequence of numbers. We now consider advantageous ways to represent this sequence. Each number could be represented by a pulse of height proportional to that number. The result would be a quantized PAM pulse train. However, there can be great advantages in representing *each* number by a *sequence* of numbers which are allowed to assume fewer values than the original set. Such a system of representation will be called a *code*. The principal advantages of *encoding* a message set are:

1. It is easier to discriminate in the presence of noise between a few possible numbers (pulse heights) than between many.
2. An error in the value of a given code number will affect only part of the information contained in the original sample value.

In general, an m-ary code will consist of m possible levels. Then a sequence of n symbols of this code can represent m^n possible message levels since there are $M = m^n$ distinguishable code groups of length n.

Example 7.10 Suppose the original message is quantized at eight levels. The message sequence is to be encoded into a *binary* code consisting of two levels or symbols "0" and "1." In this case $m = 2$ and, since M must equal 8, we have

$$M = m^n \qquad \text{or} \qquad 8 = 2^3.$$

Thus each binary code word, representing one of the possible quantization levels, must consist of three binary digits. A logical encoding is

Quantization Level	*Code Word*	*Quantization Level*	*Code Word*
0	000	4	100
1	001	5	101
2	010	6	110
3	011	7	111

• •

Of the possible m-ary codes that could be used, the binary code ($m = 2$) is encountered most frequently. There are at least two reasons for this choice:

1. The binary code offers the least number of choices, namely two, in decoding.
2. Binary systems are most easily implemented electronically since the two possible states can be made to correspond to *on-off*, *open-closed*, or *conducting-nonconducting* conditions in circuits and systems.

In the binary case, a code word $a_n a_{n-1} \cdots a_2 a_1 a_0$ is formed by the relationship

$$a_n a_{n-1} \cdots a_2 a_1 a_0 = (a_n \times 2^n) + (a_{n-1} \times 2^{n-1}) + \cdots + (a_2 \times 2^2) + (a_1 \times 2^1) + (a_0 \times 2^0)$$

where each a_i is either 0 or 1. Thus the number 21 in the decimal system becomes 10101 in the binary system since

$$21 = (1 \times 2^4) + (0 \times 2^3) + (1 \times 2^2) + (0 \times 2^1) + (1 \times 2^0).$$

The encoding of Example 7.10 further illustrates the procedure.

It is clear from the previous discussion that the process of amplitude quantization, time sampling, and encoding can be used to convert a continuous bandlimited message signal into a pulse train consisting of equally spaced pulses of two heights—for example, zero and unity. The steps by

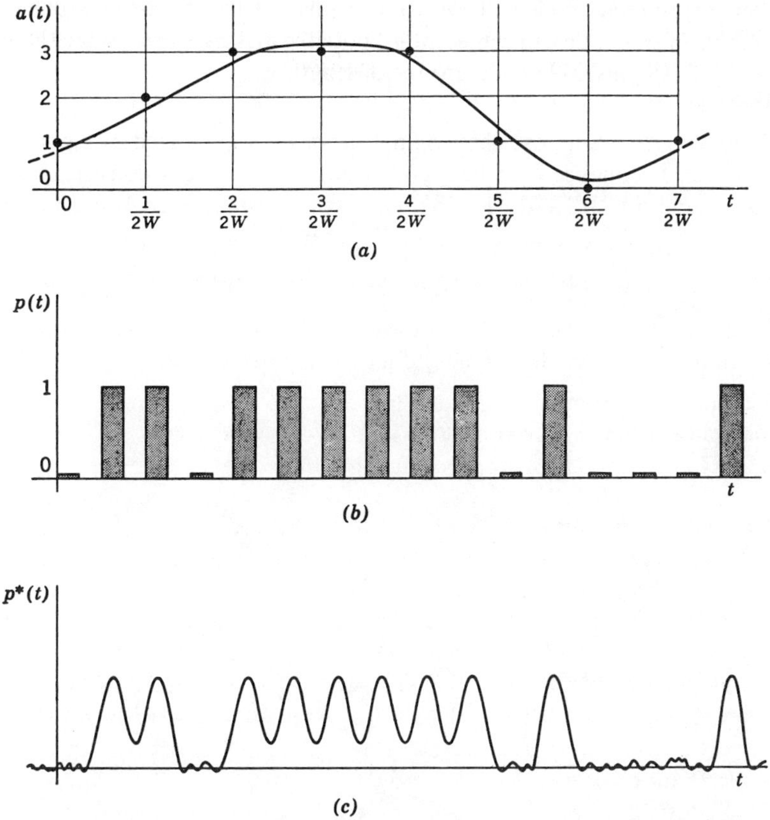

Figure 7.34 *Encoding a message signal:* (*a*) *message signal*—12333101; (*b*) *encoded pulse train*—0110111111010001; (*c*) *a recognizable but corrupted version of* (*b*)

which this is accomplished are shown graphically in Figure 7.34. A low-frequency signal $a(t)$ bandlimited $(-2\pi W, 2\pi W)$ has been quantized into four amplitude levels and sampled at an interval of $1/2W$ seconds. For the portion of $a(t)$ that is shown, the result is a sequence of numbers

$$...12333101....$$

Each of these numbers has been encoded in a binary code where

Quantization Level	*Code Word*
0	00
1	01
2	10
3	11

The corresponding binary sequence is ...0110111111010001... which is shown in Figure 7.34(*b*) as a pulse train consisting of pulses of height unity and zero to correspond to 1 and 0, respectively. In order to show their location, the pulses of zero height are shown as very short pulses.

7.21 Pulse Code Modulation (PCM)

All of the elements of pulse code modulation (PCM) have now been discussed. In fact, Figure 7.34(*b*) may be considered a PCM wave. The steps are seen to be:

1. A continuous signal $a(t)$ is sampled and quantized.
2. The quantized sample values are encoded into a pulse train $p(t)$.
3. The pulse train $p(t)$ may be transmitted directly or may be used as the modulating signal in any appropriate modulation scheme.

The great advantage of PCM is implied in Figure 7.34(*c*) which shows a corrupted version of $p(t)$ after transmission. It is clear that $p(t)$ can be recovered *exactly* from $p^*(t)$ since only a knowledge of the *presence* or *absence* of a pulse and not the pulse *shape* itself is needed at each pulse position to reproduce exactly the original pulse train. Thus, if the signal is large enough compared to corrupting noise, there will be *no error*. Furthermore, if it is desired to transmit the pulse train for a long distance, it can be regenerated exactly as often as desired at repeater stations along the way. In other words, no noise is added in transmission in the large-signal case. This situation is in striking contrast to the other modulation schemes already studied. Thus, in a properly designed large-signal PCM system, all error that results will arise from the original quantization noise discussed in Section 7.19.

The PCM system can be considered a coded wide-band system that trades complexity and bandwidth for noise immunity. That PCM is inherently wide band compared to the modulating signal can be seen from a comparison of Figure 7.34(*a*) and (*b*). Note that two pulses must occur in each sampling interval if four quantization levels are to be encoded. In general, if there are d quantization levels and if binary pulses are used, each code word must contain n pulses where

$$d = 2^n \qquad \text{or} \qquad n = \log_2 d. \tag{7.21-1}$$

Thus there must be $2nW$ pulses/sec in the PCM pulse train and any two adjacent pulses must be clearly distinguishable. If the bandwidth of the system transmitting the PCM pulse train is made too small, the pulses will be broadened and will interfere with each other so that errors in identification will result. In fact, the uncertainty relationships developed in Section 7.4

bear directly on this problem. Thus Equation 7.4-21 relates the bandwidth $\Delta\Omega$ and the duration ΔT of a pulse by

$$(\Delta\Omega)(\Delta T) \geq \tfrac{1}{2}. \tag{7.21-2}$$

Since there must be $2nW$ pulses/sec, each pulse cannot exceed a width $\Delta T \leq 1/2nW$ sec. It follows directly from Equation 7.21-2 that the system bandwidth $\Delta\Omega$ is bounded from below by

$$\Delta\Omega \geq nW \text{ cycles/sec.} \tag{7.21-3}$$

It is clear that the PCM signal requires a transmission bandwidth which is at least n times that of the original intelligence, where n is determined by Equation 7.21-1 and depends on the number of quantization levels desired. If the number of quantization levels is taken to be small (so that n is small), then the quantization noise will increase as shown by Equations 7.19-2 and 7.19-3.

Let us consider briefly the output signal-to-noise power ratio due to quantization noise. This ratio is given by Equation 7.19-4 where $[a^*(t)]^2_{av}$ is the power in the quantized signal and must be calculated. Assume that the quantizer is symmetric as shown in Figure 7.35, where $a(t)$ is the continuous-amplitude message and $a^*(t)$ is the resulting quantized signal. The quantization level is Q which is related to $2A$, the peak-to-peak value of $a(t)$, through

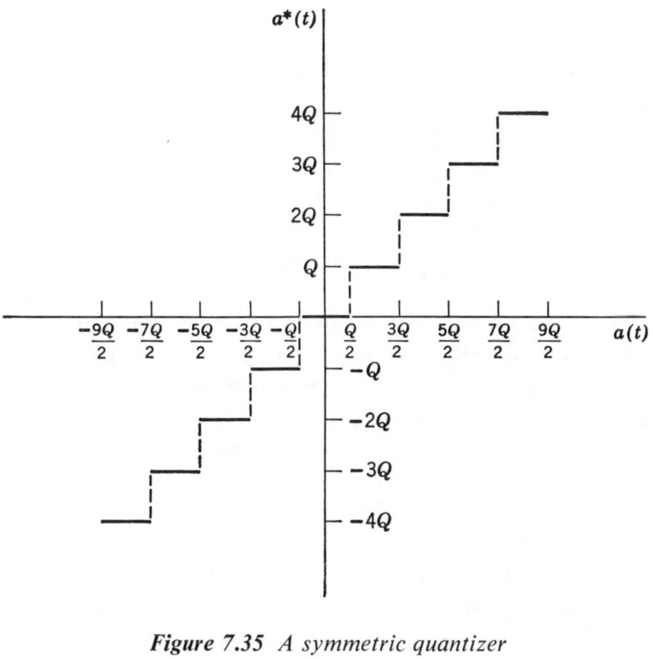

Figure 7.35 *A symmetric quantizer*

Equation 7.19-3 where d is the number of quantization levels. In the binary case, this number is $d = 2^n$ where n is the length of a code word.

In the quantized signal $a^*(t)$, the peak-to-peak signal is no longer $2A$ but has been reduced to $2A - Q$, as shown in Figure 7.35. On the assumption that the signal is uniformly distributed so that all possible amplitudes are equally likely, all quantization levels should occur on the average the same number of times. Thus, in an interval where 2^n levels occur, there should be one each of the levels

$$0, \pm Q, \pm 2Q, \ldots, \pm \frac{2^n - 2}{2} Q.$$

The average quantized signal power $[a^*(t)]^2_{\text{av}}$ should be given by

$$\begin{aligned}[a^*(t)]^2_{\text{av}} &= \frac{1}{2^n}\Bigg\{0 + Q^2 + (-Q)^2 + (2Q)^2 + (-2Q)^2 + \cdots \\ &\qquad + \left[\frac{(2^n - 2)Q}{2}\right]^2 + \left[-\frac{(2^n - 2)Q}{2}\right]^2\Bigg\} \\ &= \frac{2Q^2}{2^n} \sum_{k=0}^{(2^n-2)/2} k^2. \end{aligned} \tag{7.21-4}$$

A well-known [14] relationship is

$$\sum_{k=1}^{m} k^2 = \frac{m(m + 1)(2m + 1)}{6}. \tag{7.21-5}$$

Now Equation 7.21-4 can be rewritten as

$$[a^*(t)]^2_{\text{av}} = \frac{Q^2}{12}(2^n - 1)(2^n - 2). \tag{7.21-6}$$

This last expression may be substituted into Equation 7.19-4 to yield

$$\frac{S}{N_q} = (2^n - 1)(2^n - 2). \tag{7.21-7}$$

Thus the signal-to-noise ratio due to quantization noise in a quantized signal increases exponentially with n, the number of pulses per code word. For large n, the signal-to-noise ratio increases approximately as 2^{2n}. These results have been obtained on the assumption that the signal is uniformly distributed. For other signal distributions, essentially the same conclusions can be reached; that is, the signal-to-noise ratio increases approximately as 2^{2n}.

In many practical situations, equipment inadequacies and/or low-signal conditions will create situations in which transmission noise as well as quantization noise is added. In other words, some pulses in the coded pulse train will be incorrectly identified. This problem is most conveniently formulated in terms of a *probability of error* which gives the average rate at which incorrect pulse identification occurs. Space consideration, however, precludes further treatment here.

Thus far, no consideration has been given to the actual transmission of the PCM pulse train. It is clear that the pulse train of Figure 7.34(*b*) cannot be radiated from any convenient antenna. As in other forms of modulation, some frequency translation technique must be employed. A number of systems are used for such *digital* signaling. Some of the commonest are:

1. *Binary on-off keying* (*OOK*): This type of binary signaling is almost self-explanatory. A high-frequency sinusoidal signal is switched on and off so that "on" periods correspond to "one" and "off" periods to "zero" in the PCM wave. Figure 7.36(*a*) shows a PCM wave while Figure 7.36(*b*) shows the corresponding OOK signal. In practice, the pulsed sinusoids

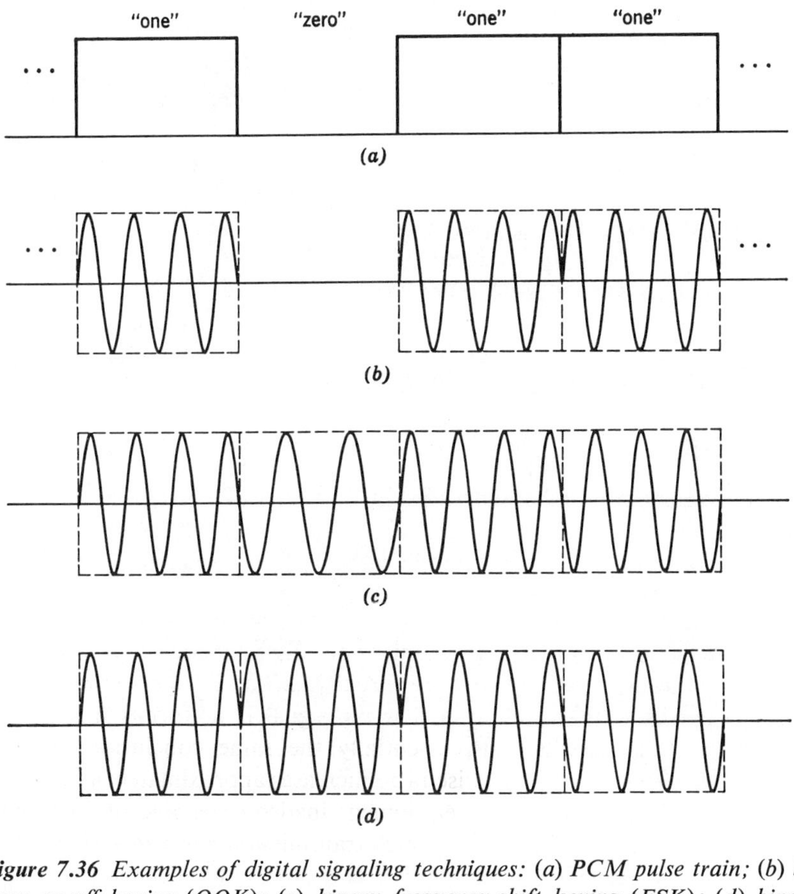

Figure 7.36 *Examples of digital signaling techniques:* (*a*) *PCM pulse train;* (*b*) *binary on-off keying* (*OOK*)*;* (*c*) *binary frequency-shift keying* (*FSK*)*;* (*d*) *binary phase-shift keying* (*PSK*)

will not start or stop as abruptly as shown but will exhibit a transient build-up and decay.

2. *Binary frequency-shift keying* (*FSK*): In this form of digital signaling, a continuous wave is sent which shifts between two frequencies, one representing the symbol "one" in the PCM wave and the other representing "zero." A typical realization is shown in Figure 7.36(*c*).
3. *Binary phase-shift keying* (*PSK*): Here the frequency is kept constant but the phase is shifted 180 degrees whenever the basic signal changes from "zero" to "one." A typical waveform is shown in Figure 7.36(*d*).

This chapter has been intended as a short treatment of some of the more fundamental aspects of modulation theory. More complete discussions will be found in the books in the Additional Reading list at the end of this chapter.

PROBLEMS

1. Consider a pulse train as shown in Figure 7.29(*b*) with pulse width D and repetition interval T. Assume that adjacent pulses can be resolved if the ratio of the least response between pulses to peak pulse output is 0.75 or less. Find the minimum required bandwidth of an ideal low-pass filter that will permit pulse resolution at its output. To what does this reduce when $D = 10^{-7}$ sec and $T = 10^{-6}$ sec?
2. Consider the product $a(t)b(t)$ of two time functions. Under what conditions on $a(t)$ and $b(t)$ is this product a periodic function with period P?
3. Find the output of the square-law receiver of Figure 6.9 when the input is the modulated signal

$$m(t) = C(1 + \cos \omega_s t) \cos \omega_0 t$$

where $\omega_s < B/2 = \frac{2}{3}\omega_0$.

4. Derive Equations 7.9-10 and 7.9-11.
5. Show that, if the asymmetrical time average $A_a[f(t)]$ exists, then the symmetrical time average $A[f(t)]$ exists also and is equal to $A_a[f(t)]$.
6. Consider the balanced modulator circuit shown below.

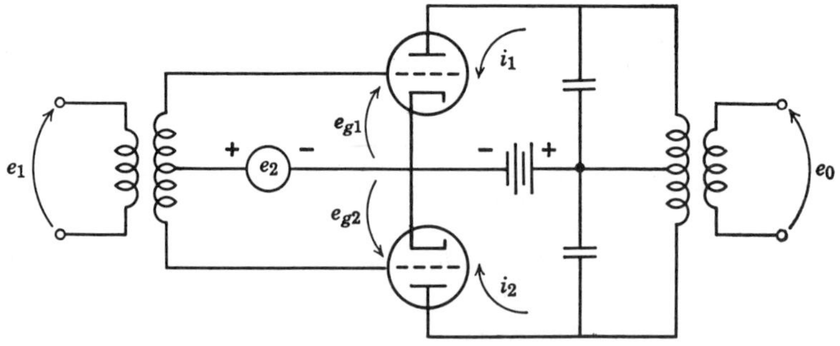

Assume that the tubes are identical and that plate currents satisfy $i_1 = a_1 e_{g1} + a_2 e_{g1}^2$, $i_2 = a_1 e_{g2} + a_2 e_{g2}^2$. The output circuit is tuned to ω_c. If $e_2 = s(t)$ and $e_1 = \cos \omega_c t$ show that e_0 is an amplitude modulated signal with carrier radian frequency ω_c.

7. Show that if e_2 and e_1 are reversed in problem 6, that is, let $e_1 = s(t)$ and $e_2 = \cos \omega_c t$, then the output is an amplitude-modulated signal with the carrier suppressed.

8. A suppressed carrier AM signal

$$r(t) = e_m(t) \cos \omega_c t$$

is demodulated using the following receiver:

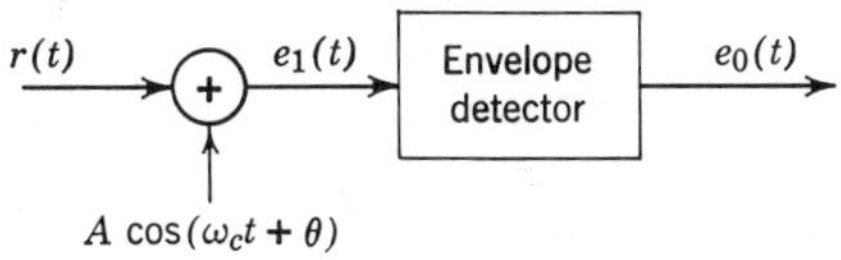

If $e_m(t)$ is slowly varying with respect to $\cos \omega_c t$, show that the output of the envelope detector is

$$e_0(t) = A\left[1 + \frac{e_m^2(t)}{A^2} + \frac{2e_m(t)}{A} \cos \theta\right]^{1/2}$$

and that for $A \cos \theta \gg e_m(t)$

$$e_0(t) = A + e_m(t) \cos \theta.$$

9. The circuit shown below can serve, in certain circumstances, as an amplitude modulator. The function $a(t)$ is the modulating intelligence and $C \cos \omega_0 t$ is the carrier. Assume that the diode is ideal and that the constant C is very large.
   (a) Find an approximate expression for $v(t)$.
   (b) If the circuit is followed by an ideal bandpass filter centered on ω_0, find the output of the filter.

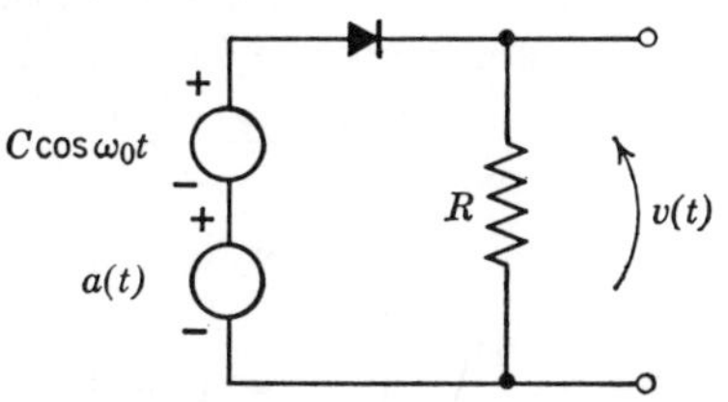

10. Repeat problem 9 when the ideal diode is replaced by a device with current-voltage relationship $i = av + bv^2$, where a and b are constants.

11. Sketch the following AM waveforms assuming a sinusoidal modulation and carrier:
    (a) DSBAM; (b) DSBAM-SC; (c) SSBAM; (d) SSBAM-SC.

12. Show that a SSBAM signal may be demodulated by the circuit shown below. Analyze the system in enough detail to indicate the disadvantages.

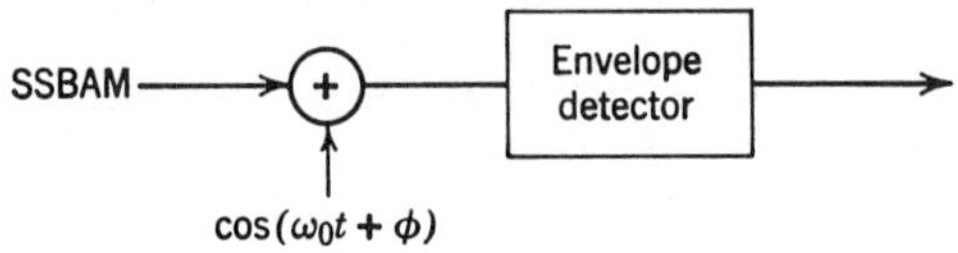

13. A modulated signal is given by

$$m(t) = C\,e^{-a_1(t)} \cos\,[\omega_0 t + a(t)]$$

where $a(t)$ is the modulating signal and $a_1(t)$ is its Hilbert transform given by Equation H.1-26. Show that the spectrum of $m(t)$ vanishes on one side of ω_0 and, hence, that $m(t)$ is a type of single-sideband FM signal.

14. Two modulated signals are given by

$$m_1(t) = a(t) \cos \omega_0 t$$

and

$$m_2(t) = \cos\,[\omega_0 t + b(t)].$$

Find a relationship between $a(t)$ and $b(t)$ so that $m_1(t) = m_2(t)$.

15. Sketch the waveform of an FM signal where the modulation is (a) a sinusoid; (b) a triangular wave; and (c) a square wave.

16. A modulated signal is given by

$$m(t) = C \cos\,(377t + 0.2 \sin t + 0.2 \sin 2t).$$

Find an *approximate* expression for the frequency spectrum $M(\omega)$ and for the bandwidth.

17. A modulated signal is given by

$$m(t) = C \cos\,(377t + 10 \sin t).$$

Find the modulation index, instantaneous frequency, and approximate bandwidth.

18. A signal with a bandwidth of 10 kc is to be transmitted using PCM. Find the minimum bandwidths required for transmission if the samples of the modulating signal are quantized in 8, 16, and 32 levels. What are the ratios of these bandwidths to that required for a standard DSBAM system? For a PAM system?

REFERENCES

1. P. M. Morse and H. Feshbach, *Methods of Theoretical Physics*, McGraw-Hill, New York, 1953, pp. 226–251.
2. B. O. Pierce, *A Short Table of Integrals*, Blaisdell, New York, 1956, p. 68, no. 499.
3. *Ibid.*, p. 67, no. 495.
4. *Ibid.*, p. 68, no. 508.

5. *Ibid.*, p. 68, no. 509.
6. G. Petit Bois, *Tables of Indefinite Integrals*, Dover, New York, 1961, p. 20.
7. D. Gabor, "Theory of Communication," *J. Inst. Elec. Engrs.* (London), **93**, 1946, 429–441.
8. I. Kay and R. A. Silverman, "On the Uncertainty Relation for Real Signals," *Information and Control*, **1**, 1957, 64–76.
9. D. Slepian and H. O. Pollak, "Prolate Spheroidal Wave Functions, Fourier Analysis and Uncertainty—I," *Bell Syst. Tech. J.*, **40**, 1961, 43–64.
10. H. J. Landau and H. O. Pollak, "Prolate Spheroidal Wave Functions, Fourier Analysis and Uncertainty—II," *Bell Syst. Tech. J.*, **40**, 1961, 65–84.
11. E. T. Whittaker, "On the Functions Which Are Represented by the Expansions of the Interpolation Theory," *Proc. Royal Soc.* (Edinburgh), **35**, 1915, 181–194.
12. C. E. Shannon, "Communication in the Presence of Noise," *Proc. IRE*, **37**, 1949, 10–21.
13. A. Kohlenberg, "Exact Interpolation of Band-Limited Functions," *J. Appl. Phys.*, **24**, 1953, 1432–1436.
14. I. S. Gradshteyn and I. M. Ryzhik, *Tables of Integrals, Series, and Products*, Academic, New York, 1965, p. 1.

ADDITIONAL READING

1. Duration, Response Time, and Bandwidth

Books

(1956) Brillouin, L. *Science and Information Theory*, Academic, New York, pp. 87–93.

(1960) Kharkevich, A. A. *Spectra and Analysis*, State Press for Theoretical and Technical Literature, Moscow, 1957; translated from the Russian and published by Consultants Bureau, New York, 1960, pp. 62–81.

Periodicals

(1934) Mayer, A. G., and E. A. Leontovich. "On an Inequality Related to the Fourier Integral," *Doklady Akad. Nauk SSSR*, **4**, 353–360 (English translation included).

(1956) Lampard, D. G. "Definition of 'Bandwidth' and 'Time Duration' of Signals Which are Connected by an Identity," *IRE Trans. on Circuit Theory*, CT-3, December, 286–288.

(1957) Lytle, D. W. "On the Properties of Matched Filters," *Stanford Electronics Res. Lab. Tech. Rept.*, 17, June.

(1958) Bourret, R. "A Note on an Information Theoretic Form of the Uncertainty Principle," *Information and Control*, **1**, 398–401.

Wolf, E. "Reciprocity Inequalities, Coherence Time and Bandwidth in Signal Analysis and Optics," *NYU Inst. Math. Sci. Div. Electromag. Res. Rept.*, EM-106; and *Proc. Phys. Soc.* (London), **71**, 257.

(1959) Kay, I., and R. A. Silverman. "On the Uncertainty Relation for Real Signals: Postscript," *Information and Control*, **2**, 396–397.

Leipnik, R. "Entropy and the Uncertainty Principle," *Information and Control*, **2**, 64–79.

(1960) Zakai, M. "A Class of Definitions of Duration and the Associated Uncertainty Relations," *Information and Control*, **3**, 110–115.

2. Modulation Theory and Systems.

Books

(1948) Goldman, S. *Frequency Analysis, Modulation, and Noise*, McGraw-Hill, New York.

(1953) Black, H. S. *Modulation Theory*, Van Nostrand, Princeton, N. J.

(1959) Schwartz, M. *Information Transmission, Modulation, and Noise*, McGraw-Hill, New York.

(1962) Russell, G. M. *Modulation and Coding in Information Systems*, Prentice-Hall, Englewood Cliffs, N. J.

(1964) Downing, J. J. *Modulation Systems and Noise*, Prentice-Hall, Englewood Cliffs, N. J.

(1965) Panter, P. F. *Modulation, Noise, and Spectral Analysis*, McGraw-Hill, New York.

Rowe, H. E. *Signals and Noise in Communication Systems*, Van Nostrand, Princeton, N. J.

(1966) Schwartz, M., W. R. Bennett, and S. Stein. *Communication Systems and Techniques*, McGraw-Hill, New York.

(1967) Stein, S., and J. J. Jones. *Modern Communication Principles*, McGraw-Hill, New York.

8

Information Theory

8.1 Introduction and Historical Background

The origins of information theory can be traced to ideas that are at least 100 years old [1, 2]. From a practical point of view, however, one of the first attempts to formulate a quantitative information measure was by Hartley [3] in 1928. His primary interest was in communication systems and the problems of signal transmission. He defined the information transmission rate of a communication system as the logarithm of the number of possible messages that could be sent through the system. He assumed that all messages were equally likely.

During World War II Norbert Wiener was largely responsible for the development of a general philosophy of communication and control called *cybernetics.* He formalized the concept that both desirable signals and undesirable signals (noises) could be defined in probabilistic terms as random processes (see Chapter 2). His work was oriented principally toward control problems (such as the military problem of fire control) and emphasized the recovery (interpolation) and prediction (extrapolation) of corrupted signals. His work was well known to initiates by the end of World War II but did not become readily available until 1948 [4] and 1949 [5]. Drawing on Wiener's concepts and taking into account the effect of noise and message probabilities, C. E. Shannon produced several classic papers in 1948 [6] and 1949 [7]. He introduced the concepts of entropy and channel capacity in communication systems and related them through the coding theorems. Shannon might be considered the father and creator of modern information theory.

The key to the development of information theory is the placing of severely simplifying restrictions on messages. The semantic content is ignored and only classes of messages characterized by probabilities are considered. The resulting system as developed in this chapter has the following desirable characteristics:

1. A logical mathematical structure defining quantitatively the concepts of the production, transmission, and reception of information.
2. A theory that provides upper bounds on system performance in terms of the information measure defined.

On the other hand, information theory suffers from some shortcomings, the major ones being:

1. The theory is not readily applied to practical situations except after extensive abstraction and description in statistical terms.
2. The theory does not yield synthesis procedures or constructive techniques whereby specific systems can be designed.

Despite these disadvantages, information theory has had a tremendous impact on the fields of communications and controls and has yielded insights into system performance which have had far-reaching consequences.

In a general way, the principal problem in many communication and control systems is the transmission of information in the form of messages or data from some originating *information source* S to some *destination* or *receiver* D. The method of transmission is frequently by means of electrical signals more or less under the control of the sender. These signals are transmitted via a *channel* C as shown in Figure 8.1(*a*). The set of messages sent by the source will be denoted by $\{U\}$. If, for arbitrary message set $\{U\}$, the channel were such that each member of U were received exactly, there would be no communication problem. Due to channel limitations and noise,

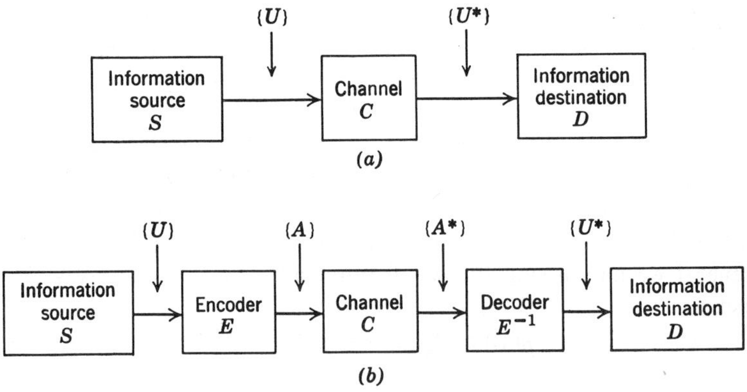

Figure 8.1 *Models of communication systems: (a) a basic communication system; (b) a communication system with encoding and decoding*

however, a corrupted version $\{U^*\}$ of $\{U\}$ is received at the information destination. Two results are generally desired:

1. The distorting effects of channel imperfections and noise should be minimized.
2. The number of messages sent over the channel in a given time should be maximized.

We will show later that these two requirements are interacting since, in general, increasing the rate of message transmission increases the distortion or error. It will develop, however, that some forms of messages are better suited for transmission over a given channel than others, in that they can be transmitted faster or with less error. This suggests that it may be desirable to modify the message set $\{U\}$ by a suitable *encoder* E to produce a new message set $\{A\}$ more suitable for a given channel. Then a decoder E^{-1} will be required at the destination to recover $\{U^*\}$ from the distorted set $\{A^*\}$. A typical block diagram of the resulting system is shown in Figure 8.1(*b*).

Thus far the discussion has been in terms of a quantity vaguely called information. We will now proceed to develop a precise definition of this term and to derive some of its properties.

8.2 Self-information and Entropy

A fundamental property of information theory is that it is concerned with the relative frequency of occurrence of messages and not with their meaning. In terms of the model of a communication system given in Figure 8.1, we consider the source message set $\{U\}$. We presume that each member of this set is expressible by means of some combination of a finite set of symbols called an *alphabet*. For example, all possible English text messages of arbitrary length are expressible as a combination of the twenty-six lower-case letters of the English alphabet plus the twenty-six upper-case letters plus certain punctuation symbols such as the period, the comma, the space, and so on. Similarly, all numbers of arbitrary size are expressible by means of the ten integers 0, 1, 2, 3, 4, 5, 6, 7, 8, 9, the plus sign, the minus sign, and the decimal point.

Let us denote the source alphabet by the set $\{X\}$ with elements (or symbols) $x_1, x_2, \ldots, x_M$ where M is the *size* of the alphabet. We will use the notation $p(x_i)$, $i = 1, 2, \ldots, M$, for the probability of occurrence of the ith symbol x_i. In general we may assign the set of numbers $\{p(x_i)\}$ arbitrarily so long, of course, as

$$p(x_i) \geq 0, \qquad i = 1, 2, \ldots, M \tag{8.2-1}$$

and

$$\sum_{i=1}^{M} p(x_i) = 1. \tag{8.2-2}$$

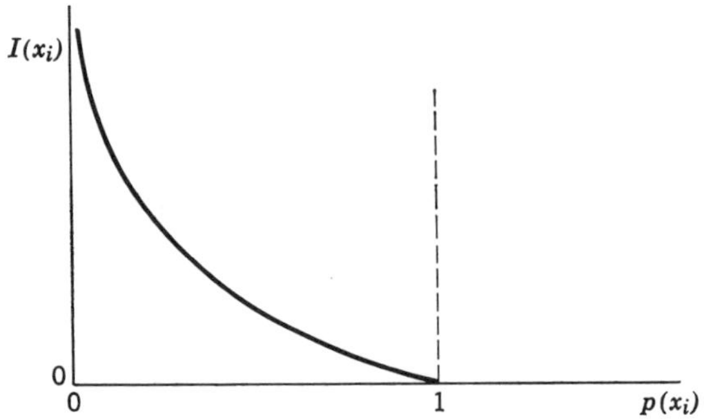

***Figure 8.2** Self-information as a function of probability:* $I(x_i) = -\log p(x_i)$

We now seek some measure of the amount of information contained in the ith symbol x_i based solely on the probability $p(x_i)$. Our intuitive notion of the concept of information suggests that the less likely symbols should contain more information since the occurrence of something improbable would appear to be more informative than the occurrence of something which is very likely to happen. Thus the information measure should be a decreasing function of $p(x_i)$. It is apparent that the reciprocal function $1/p(x_i)$ has the foregoing properties, as, indeed, do many other simple functions. It can be shown, however, and will be made plausible later that, if other very reasonable and desirable properties are required of the information measure, then the only simple function that will yield these properties is the logarithm of the reciprocal function $1/p(x_i)$. Thus, for reasons that will become clearer as we proceed, we define the *self-information* $I(x_i)$ of the ith symbol x_i as

$$\boxed{I(x_i) = \log \frac{1}{p(x_i)} = -\log p(x_i).} \tag{8.2-3}$$

This quantity is plotted in Figure 8.2 as a function of $p(x_i)$ over the allowable interval $0 \leq p(x_i) \leq 1$. As expected, it is a decreasing function of $p(x_i)$ with the endpoint values of infinity for the impossible event and zero for the certain event.

It follows directly from Equation 8.2-3 that $I(x_i)$ is a discrete random variable; that is, a real-valued function defined on the elements x_i of a probability space. Of the various statistical properties of this random variable $I(x_i)$, the most important is the expected value or mean given by

$$\boxed{E\{I(x_i)\} = H(X) = \sum_{i=1}^{M} p(x_i)I(x_i) = -\sum_{i=1}^{M} p(x_i) \log p(x_i).} \tag{8.2-4}$$

This quantity will be denoted by $H(X)$ and will be called the *entropy* (or sometimes the *comentropy*) of the distribution $p(x_i)$. If $p(x_i)$ is interpreted as the probability of the ith state of a system in phase space, then the expression of Equation 8.2-4 is identical to the entropy of statistical mechanics and thermodynamics. Furthermore, the relationship is more than a mathematical similarity. In statistical mechanics, entropy is a measure of the "disorder" of a system; in information theory, it is a measure of the "uncertainty" associated with a message source. Disorder is closely related to uncertainty since the greater the disorder, the more uncertain will it be that the system will be found in any particular microstate.

8.3 Units, Number Systems, and Binary Digits

In the definitions of self-information and entropy, the choice of the base for the logarithm is arbitrary, but of course each choice results in a different system of units for the information measures. The most common bases used are base 2, base e (the natural logarithm), and base 10. We shall use consistently the following notation:

1. $\log_2$ ( ) for base 2 logarithms;
2. ln ( ) for base e logarithms (natural logarithms);
3. log ( ) for base b logarithms when b is arbitrary.

When base 2 is used, the unit of $I(\ )$ is called the *binary digit* or *bit*. This base is most convenient when dealing with the binary case; that is, the case where $M = 2$ and the alphabet consists of only two symbols, which usually will be denoted by 0 and 1. When base e is used, the unit is the *nat*; this base is often used because of its convenient analytical properties in integration, differentiation, and so on. The base 10 is encountered only rarely and results in the *Hartley* or *decit*.

For convenience in our calculation, a table of logarithms is provided in the appendix (Table I.4, Appendix I) for base 2, base e, and base 10. The conversion between the various systems of units is given in Table 8.1.

We ordinarily use the *decimal system* or *base 10 system* of numbers. The basic symbols are 0, 1, 2, 3, 4, 5, 6, 7, 8, and 9. A given integer number $a_n a_{n-1} \cdots a_3 a_2 a_1 a_0$ (where $9 \leq a_i \leq 0$) is formed as follows:

$$a_n a_{n-1} \cdots a_3 a_2 a_1 a_0 = [(a_n \times 10^n) + (a_{n-1} \times 10^{n-1}) + \cdots + (a_3 \times 10^3) + (a_2 \times 10^2) + (a_1 \times 10^1) + (a_0 \times 10^0)]. \quad (8.3\text{-}1)$$

Thus, by the number $a_3 a_2 a_1 a_0 = 3769$, we mean

$$3769 = [(3 \times 10^3) + (7 \times 10^2) + (6 \times 10^1) + (9 \times 10^0)]. \quad (8.3\text{-}2)$$

Table 8.1. Conversion of Information Units

Unit:	Bits (base 2)	Nats (base e)	Decits (base 10)
Bits (base 2)	—	1 bit $= \dfrac{1}{\log_2 e}$ nats	1 bit $= \dfrac{1}{\log_2 10}$ decits
Nats (base e)	1 nat $= \dfrac{1}{\ln 2}$ bits	—	1 nat $= \dfrac{1}{\ln 10}$ decits
Decits (base 10)	1 decit $= \dfrac{1}{\log_{10} 2}$ bits	1 decit $= \dfrac{1}{\log_{10} e}$ nats	—

$\log_2 2 = 1;$	$\log_2 e = 1.4426;$	$\log_2 10 = 3.3219.$
$\ln 2 = 0.6932;$	$\ln e = 1;$	$\ln 10 = 2.3026.$
$\log_{10} 2 = 0.3010;$	$\log_{10} e = 0.4343;$	$\log_{10} 10 = 1.$

It is apparent that we could easily use any integer base b instead of 10. A number $a_n a_{n-1} \cdots a_3 a_2 a_1 a_0$ (where $(b - 1) \leq a_i \leq 0$) is formed in this number system by

$$a_n a_{n-1} \cdots a_3 a_2 a_1 a_0 = [(a_n \times b^n) + (a_{n-1} \times b^{n-1}) + \cdots + (a_3 \times b^3) + (a_2 \times b^2) + (a_1 \times b^1) + (a_0 \times b^0)]. \tag{8.3-3}$$

For example, let $b = 7$ and let the basic symbols be 0, 1, 2, 3, 4, 5, 6. Then the decimal number 173 would be written in the base 7 number system as 335; that is,

$$[173]_{\text{base } 10} = [(3 \times 7^2) + (3 \times 7^1) + (5 \times 7^0)] = [335]_{\text{base } 7}. \tag{8.3-4}$$

Similarly, the same decimal number 173 would become 20102 in the base 3 number system:

$$[173]_{\text{base } 10} = [(2 \times 3^4) + (0 \times 3^3) + (1 \times 3^2) + (0 \times 3^1) + (2 \times 3^0)] = [20102]_{\text{base } 3}. \tag{8.3-5}$$

As previously mentioned, the base 2 or binary system will be of particular importance in this study. A great many problems of data processing, transmission, and storage are conveniently reduced to a choice between two states, 0 and 1. Indeed, essentially the whole area of computer science uses the binary system exclusively. It follows from Equation 8.3-3 that our decimal number 173 becomes 10101101 in the binary system:

$$[173]_{\text{base } 10} = [(1 \times 2^7) + (0 \times 2^6) + (1 \times 2^5) + (0 \times 2^4) + (1 \times 2^3) + (1 \times 2^2) + (0 \times 2^1) + (1 \times 2^0)] = [10101101]_{\text{base } 2}. \tag{8.3-6}$$

A list of binary numbers and their decimal equivalent is given in Table 8.2.

Table 8.2. Binary Numbers

Decimal Number	Binary Equivalent	Decimal Number	Binary Equivalent	Decimal Number	Binary Equivalent
0	0				
1	1	17	10001	33	100001
2	10	18	10010	34	100010
3	11	19	10011	35	100011
4	100	20	10100	36	100100
5	101	21	10101	37	100101
6	110	22	10110	38	100110
7	111	23	10111	39	100111
8	1000	24	11000	40	101000
9	1001	25	11001	41	101001
10	1010	26	11010	42	101010
11	1011	27	11011	43	101011
12	1100	28	11100	44	101100
13	1101	29	11101	45	101101
14	1110	30	11110	46	101110
15	1111	31	11111	47	101111
16	10000	32	100000	48	110000

8.4 Properties of the Entropy of a Discrete Random Variable

The more elementary properties of the entropy of a discrete random variable may be inferred from several simple examples. Consider first the binary case where $M = 2$ so that the alphabet consists of the symbols 0 and 1 with probabilities p and $1 - p$ respectively. It follows from Equation 8.2-4 that

$$H_1(X) = -[p \log_2 p + (1 - p) \log_2 (1 - p)] \text{ bits.} \tag{8.4-1}$$

This equation may be plotted as a function of p as shown in Figure 8.3. (A Table of values of $H_1(X)$ is given in Appendix I.) This graph has a number of interesting properties:

1. $H_1(X)$ is nonnegative.
2. $H_1(X)$ is zero only for $p = 0$ or $p = 1$ [corresponding to $(1 - p) = 1$ or $(1 - p) = 0$].*
3. $H_1(X)$ is a maximum at $p = 1 - p = \frac{1}{2}$.
4. The maximum is not very sharp.

* The function $z \log z$ will be defined at $z = 0$ as the limit as z approaches zero through positive values; therefore we have

$$\lim_{\substack{z \to 0 \\ z > 0}} z \log z = \frac{1}{\ln b} \lim_{\substack{z \to 0 \\ z > 0}} z[(z - 1) - \tfrac{1}{2}(z - 1)^2 + \tfrac{1}{3}(z - 1)^3 - \cdots] = 0$$

for any base logarithm.

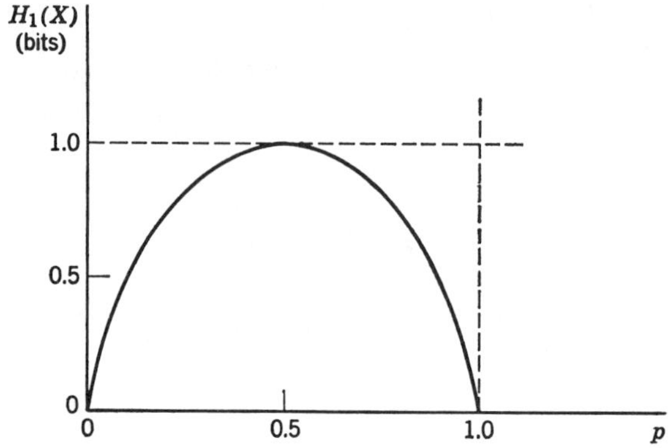

Figure 8.3 *Entropy in the binary case:* $p(0) = p$; $p(1) = 1 - p$

The maximum given in item 3 is easily found by forming

$$\frac{dH_1(X)}{dp} = 0 = -\ln 2 - \log_2 p + \ln 2 + \log_2 (1 - p)$$

$$\log_2 p = \log_2 (1 - p),$$

$$p = 1 - p \qquad \text{or} \qquad p = \tfrac{1}{2}.$$

Also, we have

$$\frac{d^2H_1(X)}{dp^2} = -\frac{1}{p} - \frac{1}{1 - p} < 0;$$

therefore $H_1(X)$ has a maximum at $p = \frac{1}{2}$ as stated.

Let us now consider the case where $M = 3$ so that the alphabet consists of three symbols x_1, x_2, and x_3 with probabilities p_1, p_2, and p_3 respectively. The entropy now is

$$H_2(X) = -(p_1 \log_2 p_1 + p_2 \log_2 p_2 + p_3 \log_2 p_3) \text{ bits} \qquad (8.4\text{-}2)$$

where

$$p_1 + p_2 + p_3 = 1. \qquad (8.4\text{-}3)$$

We may use this last relationship to eliminate one of the probabilities, say p_3, from Equation 8.4-2; now we have

$$H_2(X) = -p_1 \log_2 p_1 - p_2 \log_2 p_2 - (1 - p_1 - p_2) \log_2 (1 - p_1 - p_2). \qquad (8.4\text{-}4)$$

Suppose we now fix p_1 and plot $H_2(X)$ as a function of p_2. The result is shown in Figure 8.4 where p_1 has been fixed at the constant value k_1. Note that p_2 is

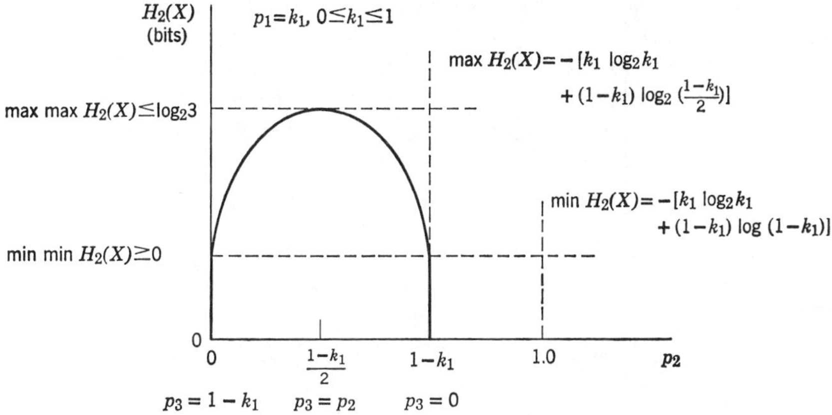

Figure 8.4 *Entropy in the ternary case, one probability fixed*

constrained by Equation 8.4-3 to be less than or equal to the quantity $(1 - k_1)$. As in the binary case, certain conclusions may be drawn:

1. $H_2(X)$ is nonnegative.
2. $H_2(X)$ is zero if, and only if, two of the probabilities, say p_1 and p_2, are zero and the other is equal to unity.
3. If any one of the three probabilities is fixed, then $H_2(X)$ is a maximum when the other two probabilities are equal; this suggests that $H_2(X)$ has an absolute maximum at $p_1 = p_2 = p_3 = \frac{1}{3}$.
4. The maximum is not very sharp.

Statement 3 is easily proved by maximizing the expression $\max H(X) = -\{k_1 \log_2 k_1 + (1 - k_1)\log_2 [(1 - k_1)/2]\}$ with respect to k_1. We have

$$\frac{d}{dk_1}[\max H(X)] = 0 = -\ln 2 - \log_2 k_1 + \ln 2 + \log_2 \left(\frac{1 - k_1}{2}\right),$$

$$k_1 = \frac{1 - k_1}{2}, \qquad k_1 = \tfrac{1}{3}.$$

Hence, as shown in Figure 8.4, for $p_1 = \frac{1}{3}$, then $p_2 = p_3 = (1 - k_1)/2 = \frac{1}{3}$.

We are now in a position to state and prove the two most important properties of the entropy for the general case of an alphabet of size M.

Property 1: (a) $H(X) \geq 0$. (8.4-5)

(b) $H(X) = 0$ *if, and only if, all of the probabilities are zero except for one, which must be unity.* (8.4-6)

Each of the terms on the right side of Equation 8.2-4 is nonnegative and, furthermore, is zero if, and only if, the probability occurring in that term is zero; therefore properties 1(a) and 1(b) follow immediately. When all of the probabilities but one are zero, then one symbol is certain to occur; there is no longer any uncertainty connected with the alphabet and $H(X)$ is zero.

Property 2: (a) $H(X) \leq \log_b M.$ (8.4-7)

(b) $H(X) = \log_b M$ *if, and only if, all of the probabilities are equal so that* $p(x_i) = p_i = 1/M$ *for all* $i.$ (8.4-8)

Here b is the particular logarithmic base used in defining $H(X)$. The proof of this property follows from the application of an inequality which may be derived as follows. Suppose we consider the real variable z and take first the region where $z \geq 1$; then

$$\ln z = \int_1^z \frac{dx}{x} \geq \frac{1}{z}\int_1^z dx = \frac{1}{z}(z - 1) = 1 - \frac{1}{z}, \qquad z \geq 1. \tag{8.4-9}$$

For the region $0 \leq z \leq 1$, we have

$$\ln z = -\int_z^1 \frac{dx}{x} \geq -\frac{1}{z}\int_z^1 dx = -\frac{1}{z}(1 - z) = 1 - \frac{1}{z}, \quad 0 \leq z < 1. \tag{8.4-10}$$

The situation is illustrated in Figure 8.5. It follows from Equations 8.4-9 and 8.4-10 that we can write

$$\log_b z = \frac{\ln z}{\ln b} \geq \frac{1}{\ln b}\left(1 - \frac{1}{z}\right), \qquad z \geq 0, \tag{8.4-11}$$

where the number b is any logarithm base and the equality occurs only when $z = 1$. Let us consider the probabilities p_i and q_i where $i = 1, 2, \ldots, M$ so that

$$\sum_{i=1}^{M} p_i = \sum_{i=1}^{M} q_i = 1. \tag{8.4-12}$$

Then, from Equations 8.4-11 and 8.4-12, we can write

$$\sum_{i=1}^{M} p_i \log_b \frac{p_i}{q_i} \geq \frac{1}{\ln b}\sum_{i=1}^{M} p_i\left(1 - \frac{q_i}{p_i}\right) = 0 \tag{8.4-13}$$

where the equality occurs only when $q_i = p_i$ for $i = 1, 2, \ldots, M$. Now, if we let

$$q_i = \frac{1}{M}, \qquad i = 1, 2, \ldots, M,$$

we obtain

$$\sum_{i=1}^{M} p_i \log_b (Mp_i) = \sum_{i=1}^{M} p_i \log_b p_i + \log_b M \sum_{i=1}^{M} p_i \geq 0$$

or

$$-\sum_{i=1}^{M} p_i \log_b p_i = H(X) \leq \log_b M \qquad (8.4\text{-}14)$$

with equality if, and only if, $p_i = 1/M$, as was to be proved. It is clear from Figure 8.5 that $\log_b M$ is a monotonically increasing function of M and,

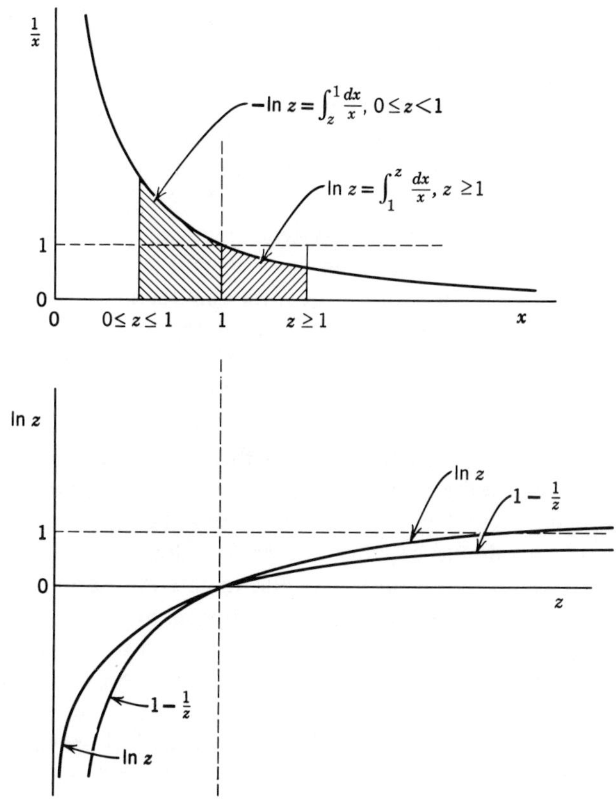

Figure 8.5 *An inequality for* ln *z:*

$$0 \leq z < 1, \qquad \ln z \geq 1 - \frac{1}{z}$$

$$z = 1, \qquad \ln z = 0 = 1 - \frac{1}{z}$$

$$z \geq 1, \qquad \ln z \geq 1 - \frac{1}{z}$$

hence, that the maximum value of the source entropy increases monotonically with the size of the source alphabet.

Actually, the inequality given by Equation 8.4-11 is a special case of an inequality that holds for all convex functions, the entropy $H(X)$ being a particular example of such functions.

8.5 Mutual Information

Thus far we have introduced the concept of self-information and applied it to an alphabet of size M. In the usual communication problem we are interested in the transfer of information from a source S through a channel C to a destination D as shown in Figure 8.1(a) and Figure 8.6. The source is assumed to have available for forming messages the alphabet X of size M. At this source a particular symbol x_i is selected out of the set of M possible symbols and is sent over the channel C. It is the limitations of the channel that produce the need for our study of information theory.

The information destination has available an alphabet Y of size N. For each symbol x_i sent from the source, a symbol y_j is selected at the destination. For the simplest communication systems involving perfect channels, the set Y would equal the set X and for each x_i sent, the same x_i would be received. We consider two probabilities to describe the "state of knowledge" at the destination. Prior to the reception of a communication, the "state of knowledge" of the destination about the symbol x_i is just the probability that x_i would be selected for transmission. We will call this the *a-priori probability* $p(x_i)$. However, after reception and the selection of the symbol y_j, the "state of knowledge" concerning x_i is the conditional probability $p(x_i/y_j)$ which will be called the *a-posteriori probability* of x_i. It is the probability that x_i was sent, given that y_j was received. Ideally this a-posteriori probability for each given y_j should be unity for one x_i and zero for all other x_i. In this case an observer at the destination is able to determine exactly which symbol x_i has been sent after the reception of each symbol y_j. Thus the uncertainty that existed previously and was expressed by the a-priori probability distribution of x_i has been removed completely by reception. In the general case it is not possible to remove all of the uncertainty and the best that can be hoped for is that it has been decreased. Thus the a-posteriori probability $p(x_i/y_j)$ is distributed over a number of x_i but should be different ("sharper" or

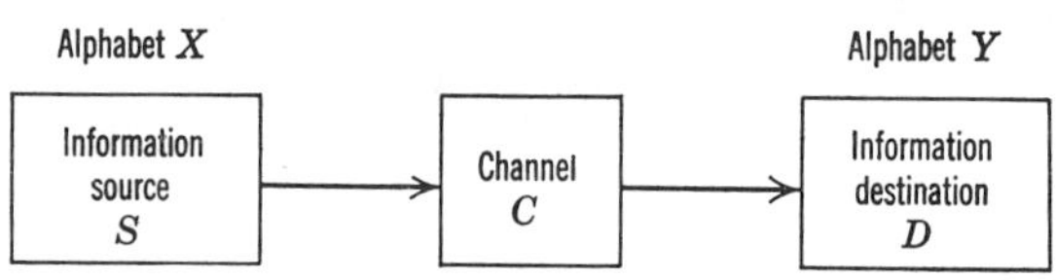

Figure 8.6 *Information transfer*

"more peaked" in some sense) from $p(x_i)$. If the two probabilities are the same, then no uncertainty has been removed by transmission, or we would say that no information has been transferred.

Based on this discussion and on other considerations that will become clearer as we proceed, we define $I(x_i; y_j)$ as the information gained about x_i by the reception of y_j:

$$\boxed{I(x_i; y_j) = \log_b \left[\frac{p(x_i/y_j)}{p(x_i)}\right]} \tag{8.5-1}$$

where, as before, the base b determines the system of units. This measure has a number of reasonable and desirable properties which we will now proceed to develop.

Property 1: *The information measure $I(x_i; y_j)$ is symmetric in x_i and y_j; that is,*

$$I(x_i; y_j) = I(y_j; x_i). \tag{8.5-2}$$

The proof of this property follows directly from the symmetry involved in the definition of conditional probability. The joint probability $p(x_i, y_j)$ is given by

$$\begin{aligned} p(x_i, y_j) &= p(x_i/y_j)p(y_j) \\ &= p(y_j/x_i)p(x_i) \end{aligned} \tag{8.5-3}$$

or the argument of Equation 8.5-1 admits the equality

$$\frac{p(x_i/y_j)}{p(x_i)} = \frac{p(y_j/x_i)}{p(y_j)} \tag{8.5-4}$$

which proves the desired result. The symmetry of our measure suggests that $I(x_i; y_j) = I(y_j; x_i)$ might be called the *mutual information* in x_i and y_j. We shall use this terminology frequently in the future.

Property 2: *The mutual information $I(x_i; y_j)$ is a maximum when $p(x_i/y_j) = 1$, that is, when the reception of y_j completely removes the uncertainty concerning x_i:*

$$I(x_i; y_j) \leq -\log p(x_i) = I(x_i). \tag{8.5-5}$$

The proof follows after writing Equation 8.5-1 as

$$I(x_i; y_j) = -\log p(x_i) + \log p(y_j/x_i). \tag{8.5-6}$$

Since $p(y_j/x_i)$ is a probability and must lie in the interval [0, 1], the last term $\log p(y_j/x_i)$ is always nonpositive and has a maximum value of zero when $p(y_j/x_i)$ is equal to unity. Note that this maximum value of $I(x_i; y_j)$ is just $I(x_i)$, the self-information of x_i.

Property 3: *If two communications* y_j *and* z_k *concerning the same message* x_i *are received successively, and if the observer at the destination takes the a-posteriori probability of the first as the a-priori probability of the second, then the total information gained about* x_i *is the sum of the gains from both communications:*

$$I(x_i; y_j, z_k) = I(x_i; y_j) + I(x_i; z_k/y_j). \tag{8.5-7}$$

It follows from Equation 8.5-1 that

$$I(x_i; y_j, z_k) = \log_b \frac{p[x_i/(y_j, z_k)]}{p(x_i)}. \tag{8.5-8}$$

The argument of the right side of this expression may be written as

$$\frac{p[x_i/(y_j, z_k)]}{p(x_i)} = \left[\frac{p(x_i/y_j)}{p(x_i)}\right]\left[\frac{p[x_i/(y_j, z_k)]}{p(x_i/y_j)}\right]. \tag{8.5-9}$$

On the substitution of this expression into Equation 8.5-8 we obtain

$$I(x_i; y_j, z_k) = \log_b \left[\frac{p(x_i/y_j)}{p(x_i)}\right] + \log_b \left[\frac{p[x_i/(y_j, z_k)]}{p(x_i/y_j)}\right] \tag{8.5-10}$$

which is identical to Equation 8.5-7.

Property 4: *If two communications* y_j *and* y_k *concerning two independent messages* x_i *and* x_m *are received, the total information gain is the sum of the two information gains considered separately:*

$$I(x_i, x_m; y_j, y_k) = I(x_i; y_j) + I(x_m; y_k). \tag{8.5-11}$$

This property is easily proved since, if x_i and x_m are independent,

$$p(x_i, x_m/y_j, y_k) = p(x_i/y_j)p(x_m/y_k) \tag{8.5-12}$$

and

$$p(x_i, x_m) = p(x_i)p(x_m). \tag{8.5-13}$$

Now if we apply Equation 8.5-1 to $I(x_i, x_m; y_j, y_k)$, we obtain

$$I(x_i, x_m; y_j, y_k) = \log_b \frac{p(x_i, x_m/y_j, y_k)}{p(x_i, x_m)}. \tag{8.5-14}$$

A direct substitution of Equations 8.5-12 and 8.5-13 into this last expression yields

$$I(x_i, x_m; y_j, y_k) = \log_b \frac{p(x_i/y_j)}{p(x_i)} + \log_b \frac{p(x_m/y_k)}{p(x_m)} \tag{8.5-15}$$

which was to be proved.

These four properties of mutual information are intuitively satisfying and desirable. Moreover, if one begins by requiring these properties, it is easily shown [8] that the logarithmic definition of Equation 8.5-1 is the simplest form that can be obtained. We now illustrate the concepts of the previous sections with some examples.

Example 8.1 (*Perfect Reception*) We consider a communication system with source alphabet of size $M = 4$ with the four possible symbols denoted by x_1, x_2, x_3, and x_4 with a-priori probabilities $\frac{1}{3}$, $\frac{1}{6}$, $\frac{1}{6}$, and $\frac{1}{3}$ respectively. We will presume that, for each symbol sent, two communications y_j and z_k are received. The destination alphabet will consist of the two symbols A and B. The situation is summarized in Table 8.3. The various joint and conditional

Table 8.3. A Simple Communication Problem

Source Symbol x_i	A-priori Probability $p(x_i)$	Received Symbols y_j	z_k
x_1	$\frac{1}{3}$	A	A
x_2	$\frac{1}{6}$	A	B
x_3	$\frac{1}{6}$	B	A
x_4	$\frac{1}{3}$	B	B

probabilities are easily found. For example, it is clear that each of the symbols y_A, y_B, z_A, and z_B occurs in one-half of the cases. Thus $p(y_A) = p(y_B) = p(z_A) = p(z_B) = \frac{1}{2}$ as shown in column 2 of Table 8.4. It is also apparent that the conditional probabilities of each of the y_j and z_k, given any x_i, is either zero or one as shown in column 3. In other words, for each source symbol x_i, we know exactly which received symbols will occur. Now we can construct the joint probabilities $p(x_i, y_j)$ and $p(x_i, z_k)$ of column 4 and the conditional probabilities $p(x_i/y_j)$ and $p(x_i/z_k)$ of column 5 from the relationships

$$p(x_i, y_j) = p(y_j/x_i)p(x_i) = p(x_i/y_j)p(y_j)$$

and

$$p(x_i, z_k) = p(z_k/x_i)p(x_i) = p(x_i/z_k)p(z_k).$$

We can find the conditional probabilities $p(y_j/z_k)$ and $p[x_i/(y_j, z_k)]$ by direct counting from Table 8.3 as shown in columns 6 and 7. Now the joint probability $p(x_i, y_j, z_k)$ is determined from

$$p(x_i, y_j, z_k) = p[x_i/(y_j, z_k)]p(y_j/z_k)p(z_k)$$

and is shown in column 8. The numbers in parentheses refer to Example 8.2.

It is clear from Table 8.3 that, if both y_j and z_k are considered, it is always possible to determine exactly which x_i was sent. Both column 7 and column 8

Table 8.4. Probabilities for Examples 8.1 and 8.2[a]

(1) $p(x_i)$	(2) $p(y_j)$, $p(z_k)$	(3) $p(y_j/x_i)$, $p(z_k/x_i)$	(4) $p(x_i, y_j)$, $p(x_i, z_k)$	(5) $p(x_i/y_j)$, $p(x_i/z_k)$	(6) $p(y_j/z_k)$	(7) $p[x_i/(y_j, z_k)]$	(8) $p(x_i, y_j, z_k)$
$p(x_1) = \frac{1}{3}$	$p(y_A) = \frac{1}{2}$	$p(y_A/x_1) = 1\ (\frac{2}{3})$	$p(x_1, y_A) = \frac{1}{3}\ (\frac{2}{9})$	$p(x_1/y_A) = \frac{2}{3}\ (\frac{4}{9})$	$p(y_A/z_A) = \frac{2}{3}\ (\frac{5}{9})$	$p[x_1/(y_A, z_A)] = 1\ (\frac{4}{5})$	$p(x_1, y_A, z_A) = \frac{1}{3}\ (\frac{2}{9})$
		$p(y_A/x_2) = 1\ (\frac{2}{3})$	$p(x_2, y_A) = \frac{1}{6}\ (\frac{1}{9})$	$p(x_2/y_A) = \frac{1}{3}\ (\frac{2}{9})$	$p(y_A/z_B) = \frac{1}{3}\ (\frac{4}{9})$	$p[x_1/(y_A, z_B)] = 0\ (0)$	$p(x_1, y_A, z_B) = 0\ (0)$
		$p(y_A/x_3) = 0\ (\frac{1}{3})$	$p(x_3, y_A) = 0\ (\frac{1}{18})$	$p(x_3/y_A) = 0\ (\frac{1}{9})$	$p(y_B/z_A) = \frac{1}{3}\ (\frac{4}{9})$	$p[x_1/(y_B, z_A)] = 0\ (\frac{1}{2})$	$p(x_1, y_B, z_A) = 0\ (\frac{1}{9})$
		$p(y_A/x_4) = 0\ (\frac{1}{3})$	$p(x_4, y_A) = 0\ (\frac{1}{9})$	$p(x_4/y_A) = 0\ (\frac{2}{9})$	$p(y_B/z_B) = \frac{2}{3}\ (\frac{5}{9})$	$p[x_1/(y_B, z_B)] = 0\ (0)$	$p(x_1, y_B, z_B) = 0\ (0)$
$p(x_2) = \frac{1}{6}$	$p(y_B) = \frac{1}{2}$	$p(y_B/x_1) = 0\ (\frac{1}{3})$	$p(x_1, y_B) = 0\ (\frac{1}{9})$	$p(x_1/y_B) = 0\ (\frac{2}{9})$		$p[x_2/(y_A, z_A)] = 0\ (0)$	$p(x_2, y_A, z_A) = 0\ (0)$
		$p(y_B/x_2) = 0\ (\frac{1}{3})$	$p(x_2, y_B) = 0\ (\frac{1}{18})$	$p(x_2/y_B) = 0\ (\frac{1}{9})$		$p[x_2/(y_A, z_B)] = 1\ (\frac{1}{2})$	$p(x_2, y_A, z_B) = \frac{1}{6}\ (\frac{1}{9})$
		$p(y_B/x_3) = 1\ (\frac{2}{3})$	$p(x_3, y_B) = \frac{1}{6}\ (\frac{1}{9})$	$p(x_3/y_B) = \frac{1}{3}\ (\frac{2}{9})$		$p[x_2/(y_B, z_A)] = 0\ (0)$	$p(x_2, y_B, z_A) = 0\ (0)$
		$p(y_B/x_4) = 1\ (\frac{2}{3})$	$p(x_4, y_B) = \frac{1}{3}\ (\frac{2}{9})$	$p(x_4/y_B) = \frac{2}{3}\ (\frac{4}{9})$		$p[x_2/(y_B, z_B)] = 0\ (\frac{1}{5})$	$p(x_2, y_B, z_B) = 0\ (\frac{1}{18})$
$p(x_3) = \frac{1}{6}$	$p(z_A) = \frac{1}{2}$	$p(z_A/x_1) = 1$	$p(x_1, z_A) = \frac{1}{3}$	$p(x_1/z_A) = \frac{2}{3}$		$p[x_3/(y_A, z_A)] = 0\ (\frac{1}{5})$	$p(x_3, y_A, z_A) = 0\ (\frac{1}{18})$
		$p(z_A/x_2) = 0$	$p(x_2, z_A) = 0$	$p(x_2/z_A) = 0$		$p[x_3/(y_A, z_B)] = 0\ (0)$	$p(x_3, y_A, z_B) = 0\ (0)$
		$p(z_A/x_3) = 1$	$p(x_3, z_A) = \frac{1}{6}$	$p(x_3/z_A) = \frac{1}{3}$		$p[x_3/(y_B, z_A)] = 1\ (\frac{1}{2})$	$p(x_3, y_B, z_A) = \frac{1}{6}\ (\frac{1}{9})$
		$p(z_A/x_4) = 0$	$p(x_4, z_A) = 0$	$p(x_4/z_A) = 0$		$p[x_3/(y_B, z_B)] = 0\ (0)$	$p(x_3, y_B, z_B) = 0\ (0)$
$p(x_4) = \frac{1}{3}$	$p(z_B) = \frac{1}{2}$	$p(z_B/x_1) = 0$	$p(x_1, z_B) = 0$	$p(x_1/z_B) = 0$		$p[x_4/(y_A, z_A)] = 0\ (0)$	$p(x_4, y_A, z_A) = 0\ (0)$
		$p(z_B/x_2) = 1$	$p(x_2, z_B) = \frac{1}{6}$	$p(x_2/z_B) = \frac{1}{3}$		$p[x_4/(y_A, z_B)] = 0\ (\frac{1}{2})$	$p(x_4, y_A, z_B) = 0\ (\frac{1}{9})$
		$p(z_B/x_3) = 0$	$p(x_3, z_B) = 0$	$p(x_3/z_B) = 0$		$p[x_4/(y_B, z_A)] = 0\ (0)$	$p(x_4, y_B, z_A) = 0\ (0)$
		$p(z_B/x_4) = 1$	$p(x_4, z_B) = \frac{1}{3}$	$p(x_4/z_B) = \frac{2}{3}$		$p[x_4/(y_B, z_B)] = 1\ (\frac{4}{5})$	$p(x_4, y_B, z_B) = \frac{1}{3}\ (\frac{2}{9})$

[a] The numbers in parentheses refer to Example 8.2.

of Table 8.4 illustrate this *noiseless case*, since the conditional probabilities of column 7 are either zero or one and the nonzero joint probabilities of column 8 are equal to the corresponding a-priori probabilities $p(x_i)$. However, if only y_j or z_k is considered separately, then it is only possible to conclude that, of the four possible input symbols, some one out of a set of two was sent. The assignment of the symbols y_j and z_k can be looked upon as a mapping from an X space into a Y space, a Z space, and a $Y \times Z$ product space as shown in Figure 8.7. Obviously only the mapping from X to $Y \times Z$ is one to one.

Now the various information gains can be calculated from Table 8.4. For example,

$$I(x_1; y_A) = \log_2 \frac{p(x_1/y_A)}{p(x_1)}$$

$$= \log_2 \frac{2/3}{1/3} = 1.0 \text{ bits},$$

$$I(x_1; z_A/y_A) = \log_2 \frac{p[x_1/(y_A, z_A)]}{p(x_1/y_A)}$$

$$= \log_2 \frac{1}{2/3} = 0.585 \text{ bits},$$

$$I(x_1; y_A, z_A) = I(x_1; y_A) + I(x_1; z_A/y_A)$$

$$= 1.000 + 0.585 = 1.585 \text{ bits}.$$

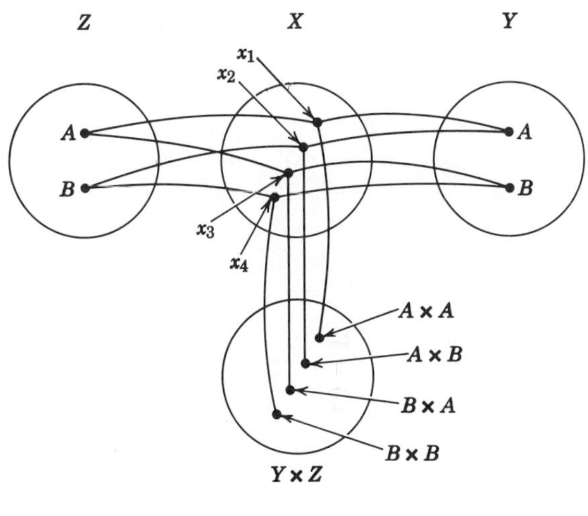

Figure 8.7 *Example* 8.1 *considered as a mapping*

As previously mentioned, once a y_j and a z_k are received, it is clear which symbol has been sent. Therefore the total information gain must be equal to the self-information in the given symbol x_i. For example,

$$\begin{aligned} I(x_1; y_A, z_A) &= I(x_1) = -\log_2 p(x_1) \\ &= \log_2 3 = 1.585 \text{ bits.} \end{aligned}$$

The various information measures are listed in Table 8.5.

Table 8.5. Information Measures for Example 8.1

Source Symbol x_i	Source Entropy (bits) $-\log_2 p(x_i)$	Destination Symbol y_j	z_k	Information Gain (bits) $I(x_i; y_j)$	$I(x_i; z_k/y_j)$	$I(x_i; y_j, z_k)$
x_1	1.585	A	A	1	0.585	1.585
x_2	2.585	A	B	1	1.585	2.585
x_3	2.585	B	A	1	1.585	2.585
x_4	1.585	B	B	1	0.585	1.585

• •

In the example just considered, the communication channel was *noiseless* in the sense that, for each source symbol x_i, a given (unique) pair of destination symbols y_j and z_k was produced. In general, noise and distortion in the channel will cause errors in reception so that the a-posteriori conditional probabilities $p[x_i/(y_j, z_k)]$ have nonzero values over a set of possible x_i. In other words, on the basis of the destination symbols, an observer is not able to determine precisely which source symbol was sent. We now consider such an example.

Example 8.2 (*Noisy Reception*) We will consider the same situation as given in Example 8.1 except that we will introduce an error in the reception of the destination symbol y_j. Suppose this probability of error is $\frac{1}{3}$; that is, one-third of the time B is received when A should be and one-third of the time A is received when B should be. The new situation is shown in Table 8.6

Table 8.6. A Noisy Communication Problem

Source symbol x_i		1	1	1	1	1	1	2	2	2	3	3	3	4	4	4	4	4	4
Destination symbol	y_j	A	A	A	A	B	B	A	A	B	B	B	A	B	B	B	B	A	A
	z_k	A	A	A	A	A	A	B	B	B	A	A	A	B	B	B	B	B	B

Table 8.7. Information Measure for Example 8.2

	Source	Correct Reception					Incorrect Reception				
Source Symbol	Entropy (bits)	Destination Symbol		Information Gain (bits)			Destination Symbol		Information Gain (bits)		
$p(x_i)$	$-\log_2 p(x_i)$	y_j	z_k	$I(x_i; y_j)$	$I(x_i; z_k/y_j)$	$I(x_i; y_j, z_k)$	y_j	z_k	$I(x_i; y_j)$	$I(x_i; z_k/y_j)$	$I(x_i; y_j, z_k)$
x_1	1.585	A	A	0.415	0.848	1.263	B	A	−0.585	1.170	0.585
x_2	2.585	A	B	0.415	1.170	1.585	B	B	−0.585	0.848	0.263
x_3	2.585	B	A	0.415	1.170	1.585	A	A	−0.585	0.848	0.263
x_4	1.585	B	B	0.415	0.848	1.263	A	B	−0.585	1.170	0.585

where multiple listings have been made so that all listings are equiprobable. It can be seen that the probabilities $p(y_j)$ are unchanged since as many A's are changed to B's as B's to A's. This is an example of a symmetric channel in which the errors do not distinguish among the symbols. However, the conditional probabilities $p(y_j/x_i)$ are no longer either zero or unity. For example, $p(y_A/x_1)$ is now only $\frac{2}{3}$ instead of unity. The other $p(y_j/x_i)$ are shown in Table 8.4 in parentheses to make it easier to compare this example to Example 8.1. Similarly, the remaining probabilities change as listed.

The mutual information between x_i and y_j is now changed also. For example,

$$I(x_1; y_A) = \log_2 \frac{p(x_1/y_A)}{p(x_1)} = \log_2 \frac{4}{3} = 0.415 \text{ bits.}$$

In addition we now have to consider

$$I(x_1; y_B) = \log_2 \frac{p(x_1/y_B)}{p(x_1)} = \log_2 \frac{2}{3} = -0.405 \text{ bits.}$$

This last information gain is seen to be negative; in this case the observer has been misled. The new set of information gains is shown in Table 8.7 where we now take into account both correct and incorrect reception.

• •

It should be kept in mind that, when errors are present, an observer will not be able to calculate the information gain even after the reception of all the symbols relating to a given source symbol, since the same series of received symbols may represent several different source symbols. Thus the observer is unable to say which source symbol has been sent and the best that he can do is to compute the information gain with respect to each possible source symbol. In many cases it would be more desirable to have a quantity which is independent of the particular symbols. We shall obtain a number of quantities of this nature in the remainder of this section and in subsequent sections.

The mutual information $I(x_i; y_j)$, as given by Equation 8.5-1, may be expanded to give

$$I(x_i; y_j) = -\log p(x_i) + \log p(x_i/y_j) = I(x_i) - I(x_i/y_j) \quad (8.5\text{-}16)$$

where $I(x_i)$ is the self-information of x_i and $I(x_i/y_j)$ is the self-information required to specify x_i after the ocurrence of y_j. Suppose that a given y_j always occurs in conjunction with a given x_i so that $p(x_i/y_j)$ is unity. Then we have

$$I(x_i, y_j) = I(x_i), \qquad p(x_i/y_j) = 1. \quad (8.5\text{-}17)$$

In this case a given y_j can be identified uniquely with some one x_i; thus the self-information $I(x_i)$ is the mutual information provided by a symbol x_i

about itself. Suppose that x_i and y_j are statistically independent so that $p(x_i/y_j) = p(x_i)$; then

$$I(x_i; y_j) = 0, \qquad p(x_i/y_j) = p(x_i). \tag{8.5-18}$$

For the case where $p(x_i/y_j) < p(x_i)$ it is clear from Equation 8.5-1 or 8.5-16 that the mutual information is negative; that is,

$$I(x_i; y_j) < 0, \qquad p(x_i/y_j) < p(x_i). \tag{8.5-19}$$

We might say in this case that the output y_j is misleading us about the input x_i.

The mutual information $I(x_i; y_j)$ is a random variable just as was the self-information $I(x_i)$. However, two probability spaces X and Y are involved now and several ensemble averages are possible. We define $I(X; y_j)$, the information provided by y_j with respect to the ensemble X, by

$$I(X; y_j) = E_{X/y_j}\{I(x_i; y_j)\} = \sum_{i=1}^{M} p(x_i/y_j)I(x_i; y_j). \tag{8.5-20}$$

In the same way we define $I(x_i; Y)$ as the information provided by x_i with respect to the ensemble Y:

$$I(x_i; Y) = E_{Y/x_i}\{I(x_i; y_j)\} = \sum_{j=1}^{N} p(y_j/x_i)I(x_i; y_j). \tag{8.5-21}$$

It is easily shown that both of these new information measures are non-negative:

$$I(X; y_j) \geq 0 \tag{8.5-22}$$

and

$$I(x_i; Y) \geq 0. \tag{8.5-23}$$

For example, we use the inequality given by Equation 8.4-11 and write

$$I(x_i; y_j) = \frac{1}{\ln b} \ln \frac{p(x_i/y_j)}{p(x_i)} \geq \frac{1}{\ln b}\left[1 - \frac{p(x_i)}{p(x_i/y_j)}\right]. \tag{8.5-24}$$

Now we use Equation 8.5-20 together with this last inequality to obtain

$$I(X; y_j) \geq \frac{1}{\ln b} \sum_i p(x_i/y_j)\left[1 - \frac{p(x_i)}{p(x_i/y_j)}\right] \tag{8.5-25}$$

or

$$I(X; y_j) \geq \frac{1}{\ln b}\left[\sum_i p(x_i/y_j) - \sum_i p(x_i)\right] = 0 \tag{8.5-26}$$

as was to be proved. The proof that $I(x_i; Y) \geq 0$ is identical.

We now proceed to define the *average mutual information* $I(X; Y)$ as a statistical average of $I(x_i; y_j)$ with respect to the joint probability $p(x_i, y_j)$; that is,

$$\boxed{\begin{aligned} I(X; Y) &= E_{XY}\{I(x_i; y_j)\} = \sum_i \sum_j p(x_i, y_j) I(x_i; y_j) \\ &= \sum_i \sum_j p(x_i, y_j) \log \frac{p(x_i/y_j)}{p(x_i)}. \end{aligned}} \tag{8.5-27}$$

This new function $I(X; Y)$ is the first information measure that we have defined which does not depend on the individual symbols x_i or y_j. Thus it is a property of the whole communication system and will turn out to be only the first in a series of similar quantities used as a basis for the characterization of communication systems.

We note in passing that $I(X; Y)$ has a number of expected properties. For example it is nonnegative. This fact follows immediately from Equations 8.5-20 and 8.5-22 since $I(X; y_j) \geq 0$ and

$$\sum_{j=1}^{N} p(y_j) I(X; y_j) = \sum_{i=1}^{M} \sum_{j=1}^{N} p(y_j) p(x_i/y_j) I(x_i; y_j) \geq 0 \tag{8.5-28}$$

or

$$\boxed{I(X; Y) \geq 0.} \tag{8.5-29}$$

If the ensembles *X and Y are statistically independent*, then it follows from Equation 8.5-18 that

$$\boxed{I(X; Y) = 0.} \tag{8.5-30}$$

Also we see from Equations 8.5-4 and 8.5-27 that $I(X; Y)$ is symmetric in X and Y so that

$$\boxed{I(X; Y) = I(Y; X).} \tag{8.5-31}$$

8.6 Joint and Conditional Entropy

We are dealing now with both a source alphabet $\{x_i\}$ of M symbols and a destination alphabet $\{y_j\}$ of N symbols with corresponding probabilities $p(x_i)$ and $p(y_j)$, respectively. We have defined the source entropy $H(X)$ by

Equation 8.2-4. It is obvious that a similar quantity, the destination entropy $H(Y)$, can be defined analogously by

$$H(Y) = -\sum_{j=1}^{N} p(y_j) \log p(y_j). \qquad (8.6\text{-}1)$$

This quantity will have, of course, all of the properties developed in Section 8.4.

In the previous section we have introduced the joint probability $p(x_i, y_j)$ and have used it to define average mutual information $I(X; Y)$. In the same way we define the *joint or system entropy* $H(X, Y)$ by

$$\boxed{H(X, Y) = -\sum_{i=1}^{M} \sum_{j=1}^{N} p(x_i, y_j) \log p(x_i, y_j).} \qquad (8.6\text{-}2)$$

Clearly, if X and Y are *statistically independent* so that

$$p(x_i, y_j) = p(x_i)p(y_j) \qquad (8.6\text{-}3)$$

for all i and j, then Equation 8.6-2 can be written as

$$\boxed{H(X, Y) = H(X) + H(Y).} \qquad (8.6\text{-}4)$$

On the other hand, if X and Y are not independent, then Equation 8.6-2 can be written as

$$\begin{aligned} H(X, Y) &= -\sum_{i=1}^{M} \sum_{j=1}^{N} p(x_i, y_j) \log [p(x_i)p(y_j/x_i)] \\ &= H(X) - \sum_{i=1}^{M} \sum_{j=1}^{N} p(x_i, y_j) \log p(y_j/x_i). \qquad (8.6\text{-}5) \end{aligned}$$

The last term in this equation will be called a *conditional entropy* $H(Y/X)$ defined by

$$\boxed{H(Y/X) = -\sum_{i=1}^{M} \sum_{j=1}^{N} p(x_i, y_j) \log p(y_j/x_i)} \qquad (8.6\text{-}6)$$

so that Equation 8.6-5 becomes

$$\boxed{H(X, Y) = H(X) + H(Y/X).} \qquad (8.6\text{-}7)$$

In the same way

$$\boxed{H(X, Y) = H(Y) + H(X/Y)} \tag{8.6-8}$$

where the conditional entropy $H(X/Y)$ is defined by

$$\boxed{H(X/Y) = -\sum_{i=1}^{M}\sum_{j=1}^{N} p(x_i, y_j) \log p(x_i/y_j).} \tag{8.6-9}$$

The conditional entropies just defined each satisfy an important inequality

$$0 \le H(Y/X) \le H(Y) \tag{8.6-10}$$

and

$$0 \le H(X/Y) \le H(X). \tag{8.6-11}$$

The left sides of these two inequalities are proved exactly as in Equations 8.5-22 and 8.5-23. The right sides may be proved as follows. Let us expand Equation 8.5-27 for the average mutual information as

$$\begin{aligned} I(X; Y) &= \sum_{i=1}^{M}\sum_{j=1}^{N} p(x_i, y_j) \log \frac{p(x_i, y_j)}{p(x_i)p(y_j)} \\ &= \sum_{i=1}^{M}\sum_{j=1}^{N} p(x_i, y_j) \log p(x_i, y_j) \\ &\quad - \sum_{i=1}^{M} p(x_i) \log p(x_i) \\ &\quad - \sum_{j=1}^{N} p(y_j) \log p(y_j) \end{aligned} \tag{8.6-12}$$

or

$$\boxed{I(X; Y) = -H(X, Y) + H(X) + H(Y) \ge 0.} \tag{8.6-13}$$

This last expression can be rewritten in two equivalent forms by using either Equation 8.6-7 or 8.6-8:

$$\boxed{I(X; Y) = H(Y) - H(Y/X) \ge 0} \tag{8.6-14}$$

or

$$\boxed{I(X; Y) = H(X) - H(X/Y) \ge 0.} \tag{8.6-15}$$

From these last two expressions it is apparent that the right sides of the inequalities of Equations 8.6-10 and 8.6-11 are satisfied. It is also clear, say from Equation 8.6-13, that $H(X, Y)$ satisfies the inequality

$$\boxed{H(X, Y) \leq H(X) + H(Y).} \tag{8.6-16}$$

In other words the joint entropy of two ensembles X and Y is a maximum when the ensembles are independent.

We have now defined the five entropies $H(X)$, $H(Y)$, $H(Y/X)$, $H(X/Y)$, and $H(X, Y)$. At this point it may be appropriate to comment briefly on the meaning of the two conditional entropies $H(Y/X)$ and $H(X/Y)$. Let us refer first to Equation 8.6-15. This equation expresses the fact that the average information gained about a message, when a communication is completed, is equal to the average source information less the average uncertainty that still remains about the message. From another point of view, the quantity $H(X/Y)$ is the average additional information needed at the destination after reception to completely specify the message sent. Thus $H(X/Y)$ represents the information lost in the channel. It is frequently called the *equivocation.* Let us now consider Equation 8.6-14. This equation indicates that the information transmitted consists of the difference between the destination entropy and that part of the destination entropy that is not information about the source; thus the term $H(Y/X)$ can be considered a *noise entropy* added in the channel.

8.7 Sources of Information—The Zero-Memory Source

Up to this point we have considered that the information source of Figure 8.1 has associated with it an alphabet $\{X\}$ of size M with elements (or symbols) $x_1, x_2, \ldots, x_M$. To each element x_i we have assigned a probability $p(x_i)$; we have defined an information measure on this element; and we have developed some average properties [such as source entropy $H(X)$] of this measure. In all of our treatment we have been dealing with a single element (or symbol) assumed to be emitted from the source.

In practice we are interested in sending a message set $\{U\}$ as mentioned in Section 8.1. Each member message u_i must, in general, be represented by a sequence of symbols, since it will usually be true that the number of possible messages exceeds the size M of the source alphabet. Thus we are concerned with sequences of symbols and, for our purposes, these sequences replace the original messages u_i which need not be considered further. When the source alphabet $\{X\}$ is of finite size M, we will say that we have a *finite discrete source.* We will be concerned now with the interrelationships existing between symbols in the generated sequences and with the classification of sources according to these interrelationships.

In Section 3.1, a stochastic process x_t, $t \in T$, was defined as an indexed set of random variables where T is the *parameter set* of the process. If the set T is a sequence, then x_t is a stochastic process with discrete parameter (also called a *random sequence* or *series*). One way to look at the output of our *finite discrete source* is that it is a discrete-parameter stochastic process with each possible given sequence one of the ensemble members or realizations of the process. Thus the study of information sources can be reduced to a study of random processes.

The simplest case to consider is that in which the successive symbols (random variables) are statistically independent. Such a source will be called a *zero-memory source*. If the successive symbols obey the same fixed probability law so that the one distribution $p(x_i)$ determines the appearance of each indexed symbol, then the source will be called *stationary*. However, it will usually be apparent from the context whether the source is stationary or nonstationary and we will frequently omit this adjective and refer simply to a zero-memory source.

Let us consider sequences of length n, each member of the sequence being a realization of the random variable x_i with fixed probability distribution $p(x_i)$. Since there are M possible realizations of the random variable and n terms in the sequence, there must be M^n distinct sequences possible of length n. Let us denote the random variable x_i in the jth position by x_{i_j} so that the sequence set (the message set) can be represented by

$$\{U\} = X^n = \{x_{i_1}, x_{i_2}, \ldots, x_{i_n}\}, \qquad i = 1, 2, \ldots, M. \tag{8.7-1}$$

The symbol X^n is sometimes used to represent this sequence set and has been called the *nth extension of the zero-memory source X*. It is clear that the probability of occurrence of a given message u_i is just the product of the probabilities of occurrence of the individual terms in the sequence

$$p\{u_i\} = p(x_{i_1})p(x_{i_2})\cdots p\{x_{_n}\}. \tag{8.7-2}$$

Now the entropy for the extended source X^n is just

$$H(X^n) = -\sum_{X^n} p\{u_i\} \log p\{u_i\} \tag{8.7-3}$$

or, from Equation 8.7-2,

$$H(X^n) = -\sum_{X^n} p\{u_i\} \log p\{x_{i_1}\} - \sum_{X^n} p\{u_i\} \log p\{x_{i_2}\} - \cdots$$
$$- \sum_{X^n} p\{u_i\} \log p\{x_{i_n}\}. \tag{8.7-4}$$

But each term in this expression may be written as

$$-\sum_{X^n} p\{u_i\} \log p\{x_{i_j}\} = -\sum_{X} p(x_i) \log p(x_i) = H(X) \tag{8.7-5}$$

so that

$$\boxed{H(X^n) = nH(X) \text{ information units/sequence}} \tag{8.7-6}$$

as expected. Note that, if base 2 logarithms are used, then $H(X)$ has units of bits/symbol, n is symbols/sequence, and $H(X^n)$ is in units of bits/sequence. For a zero-memory source, all sequence averages of information measure are obtained by multiplying the corresponding symbol measure by the number of symbols in the sequence.

Example 8.3 Suppose that we toss a fair coin 100 times. We assume that the ensemble $R = X^{75}$ describes the results of the first 75 tosses and $S = X^{60}$ the results of the last 60. We have

$$p(x_i) = \tfrac{1}{2}, \qquad i = 1, 2,$$

$$H(X) = \sum_{i=1}^{2} \tfrac{1}{2} \log_2 2 = 1 \text{ bit/symbol}$$

and hence, from Equation 8.7-6,

$$H(R) = 75\, H(X) = 75 \text{ bits/sequence.}$$

In the same way

$$H(S) = 60\, H(X) = 60 \text{ bits/sequence}$$

and

$$\begin{aligned} H(R, S) &= 100 \text{ bits,} \\ H(R/S) &= 40 \text{ bits,} \\ H(S/R) &= 25 \text{ bits.} \end{aligned}$$

It follows from Equation 8.6-13 that the average mutual information of the two sequences is

$$I(R; S) = 35 \text{ bits.}$$

• •

8.8 The Markov Information Source

In the zero-memory source of the previous section, each symbol in the sequence of n symbols occurred independently of every other symbol in the sequence. Such a model is not general enough for most information sources. A constructive way to generalize this model is to assume that the occurrence of a given symbol depends on some number m of immediately preceding symbols. Thus the information source can be considered to produce the mth-order Markov chain of Section 3.12 and is called an *mth-order Markov source.*

For an mth-order Markov source, the m symbols preceding a given symbol position will be called the *state* s_j of the source at that symbol position. It is clear that, if there are M possible symbols x_i, then the mth-order Markov source will have $M^m = q$ possible states s_j making up the *state set*

$$S = \{s_1, s_2, \ldots, s_q), \qquad q = M^m. \tag{8.8-1}$$

At a given time corresponding to one symbol position the source will be in a given state s_j. There will exist a probability $p(s_k/s_j) = p_{jk}$ that the source will move into another state s_k with the emission of the next symbol. The set of all such conditional probabilities is expressed by the *transition matrix* $\boldsymbol{T}$ where

$$\boldsymbol{T} = [p_{jk}] = \begin{bmatrix} p_{11} & p_{12} & \cdots & p_{1q} \\ p_{21} & p_{22} & \cdots & p_{2q} \\ \vdots & & & \\ p_{q1} & p_{q2} & \cdots & p_{qq} \end{bmatrix} \tag{8.8-2}$$

A *Markov matrix* or *stochastic matrix* is any square matrix with non-negative elements such that the row sums are unity. It is clear that $\boldsymbol{T}$ is such a matrix since

$$\sum_{j=1}^{q} p_{ij} = \sum_{j=1}^{q} p(s_j/s_i) = 1, \qquad i = 1, 2, \ldots, q. \tag{8.8-3}$$

Conversely, any stochastic matrix is a possible transition matrix for a Markov source of order m where $q = M^m$ is equal to the number of rows or columns of the matrix. We have already pointed out in Section 3.12 that a Markov chain is completely specified by its transition matrix $\boldsymbol{T}$ and by an *initial distribution vector* $\boldsymbol{\pi}$ giving the probability distribution for the first state occurring. For the zero-memory source, the transition matrix reduces to a stochastic matrix where all the rows are identical and are each equal to the initial distribution vector $\boldsymbol{\pi}$, which is in turn equal to the vector giving the source alphabet a-priori probabilities. Thus, in this case, we have

$$p_{jk} = p(s_k/s_j) = p(s_k) = p(x_k), \qquad k = 1, 2, \ldots, M. \tag{8.8-4}$$

From a slightly more formal point of view we can now define an mth-order Markov source as an initial probability vector $\boldsymbol{\pi}$, a transition matrix $\boldsymbol{T}$, and a *symbol set* or *alphabet* $x = \{x_1, x_2, \ldots, x_M\}$. It will sometimes be convenient to use a graphical representation (called a *state diagram*) of a Markov source. In this graph, each of the $M^m = q$ possible source states s_i are represented by a point, and the possible transitions between states are represented by directed lines between the points.

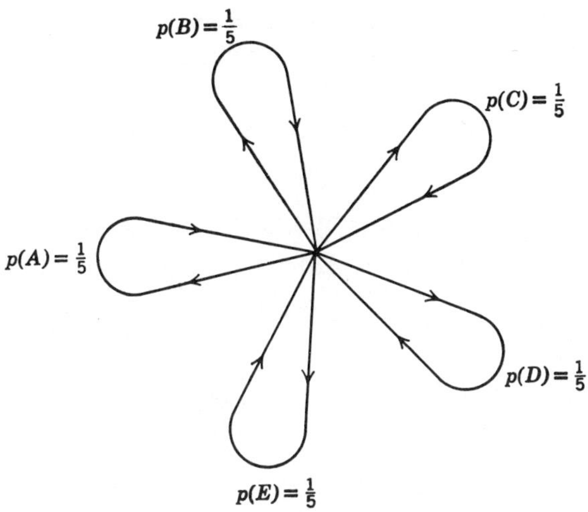

Figure 8.8 *The state diagram for a zero-memory source:*

$$X = \{A, B, C, D, E\}$$
$$p(x_i) = \tfrac{1}{5}$$
$$x_i = A, B, C, D, E$$

Example 8.4 Let the symbols be chosen from the alphabet of five letters A, B, C, D, and E, each with probability $\frac{1}{5}$, successive choices being independent. This is an example of a zero-memory source; a typical sequence produced by this source might be

$$BDCBCECBCADCBDDAAECEEAABB\ldots.$$

Note that there is only one possible state for the source in this example since the probability of a given symbol being produced at any symbol position is always the same. The state diagram is, therefore, degenerate and is shown in Figure 8.8.

• •

For our purposes we shall be interested only in Markov sources which are *ergodic*. A Markov source is ergodic if it is possible for the source to move from any one state s_i to any other state s_j in a finite number of steps. It can be shown [9] that this definition of ergodicity is consistent with that given in Section 3.4 in the sense that an ergodic Markov source will emit a Markov chain which is an ergodic random process. Such sources will emit sequences of symbols which, for very long sequences, are "typical" in that they have the statistical regularity discussed in Section 3.4.

A careful treatment of the ergodicity of a source is beyond the scope of this discussion. However, a few remarks can be made which are intuitively valid. Suppose we have the simple transition matrix

$$\boldsymbol{T} = \begin{bmatrix} p_{11} & p_{12} \\ p_{21} & p_{22} \end{bmatrix} \tag{8.8-5}$$

describing a two-state source with states s_1 and s_2. We will write $\boldsymbol{T}^2$ to indicate the matrix product of $\boldsymbol{T}$ and $\boldsymbol{T}$, so that

$$\boldsymbol{T}^2 = \begin{bmatrix} p_{11} & p_{12} \\ p_{21} & p_{22} \end{bmatrix} \begin{bmatrix} p_{11} & p_{12} \\ p_{21} & p_{22} \end{bmatrix} = \begin{bmatrix} p_{11}^2 + p_{12}p_{21} & p_{11}p_{12} + p_{12}p_{22} \\ p_{21}p_{11} + p_{22}p_{21} & p_{21}p_{12} + p_{22}^2 \end{bmatrix}. \tag{8.8-6}$$

Let us consider now the element in the first row and column of the new matrix $\boldsymbol{T}^2$. This element will be denoted by $p_{11}^{(2)}$ and is

$$p_{11}^{(2)} = p_{11}^2 + p_{12}p_{21}. \tag{8.8-7}$$

But this is just the probability, given that the source is in state s_1, that it will be in state s_1 again after two transitions. The two terms of Equation 8.8-7 correspond to the two ways in which this condition can occur; that is, the state can remain s_1 for two transitions or the state can change from s_1 to s_2 on the first transition and back from s_2 to s_1 on the second. It is clear that, for the general transition matrix $\boldsymbol{T}$, the product $\boldsymbol{T}^2$ has elements $p_{ij}^{(2)}$ which give the conditional probability of transition in two steps to state s_j given state s_i. These elements are

$$p_{ij}^{(2)} = \sum_{k=1}^{q} p_{ik}p_{kj}. \tag{8.8-8}$$

More generally, the matrix $\boldsymbol{T}^r$ gives the transition probabilities between states after r steps and has elements

$$p_{ij}^{(r)} = \sum_{k_1=1}^{q} \cdots \sum_{k_{r-1}=1}^{q} p_{ik_1}p_{k_1k_2}\cdots p_{k_{r-1}j}. \tag{8.8-9}$$

A Markov source with transition matrix $\boldsymbol{T}$ is called *regular* if, for some positive integer r, the matrix $\boldsymbol{T}^r$ has no zero elements. A sufficient condition for a Markov source to be ergodic is that it be regular. It can be seen that, if the transition matrix is regular, then it is always possible to go from any one state to any other state in a finite number of steps, in fact in r steps.

Regular Markov sources have a number of interesting properties. One of the most remarkable and useful is that the regular transition matrix $\boldsymbol{T}$ uniquely defines a vector $\mathbf{P}$ which gives the *stationary* probability distribution of the states s_i *independent* of the initial distribution vector $\boldsymbol{\pi}$. In other words, we can determine the relative frequency of occurrence of the states from the transition matrix T. More specifically, as discussed in Section 3.12, it can be shown [10] that

$$\lim_{n \to \infty} T^n = A \tag{8.8-10}$$

where A is a stochastic matrix with identical rows, each of which is the probability vector $\mathbf{P}$

$$\mathbf{P} = [p_1 \quad p_2 \quad \cdots \quad p_q] \tag{8.8-11}$$

where $p_i = p(s_i)$ is the *stationary* probability distribution of state s_i. We can obtain $\mathbf{P}$ from the relationship [10]

$$\mathbf{P}T = \mathbf{P}. \tag{8.8-12}$$

Also, for *any* initial distribution vector $\boldsymbol{\pi}$, we have

$$\lim_{n \to \infty} \boldsymbol{\pi} T^n = \mathbf{P}. \tag{8.8-13}$$

It is obvious that any zero-memory source such as that of Example 8.4 is an ergodic source since it has only one state.

Example 8.5 Let us consider a first-order Markov source with alphabet

$$X = \{A, B, C\},$$

initial probability vector $\boldsymbol{\pi}$ given by

$$\boldsymbol{\pi} = [\tfrac{9}{27} \quad \tfrac{16}{27} \quad \tfrac{2}{27}],$$

and transition matrix

$$T = \begin{bmatrix} 0 & \frac{4}{5} & \frac{1}{5} \\ \frac{1}{2} & \frac{1}{2} & 0 \\ \frac{1}{2} & \frac{2}{5} & \frac{1}{10} \end{bmatrix}.$$

Thus $P(B/A)$ is $\frac{4}{5}$ and $P(C/A)$ is $\frac{1}{5}$. In this case $M = 3$ and $m = 1$ so that there are $q = M^m = 3$ possible states. The state diagram is given by Figure 8.9. The new symbol produced is the one to which the arrow is pointed. A typical sequence produced by this source might be the following:

ABBABABABABABABBBABBBBBABABABABAB
BBACACABBABBBBABBABACBBBABA.

Of course, the initial probability vector $\boldsymbol{\pi}$ does not enter into the state diagram.

Let us now investigate the ergodicity of the source. We find that

$$T^2 = \begin{bmatrix} \frac{1}{2} & \frac{12}{25} & \frac{1}{50} \\ \frac{1}{4} & \frac{13}{20} & \frac{1}{10} \\ \frac{1}{4} & \frac{16}{25} & \frac{11}{100} \end{bmatrix}$$

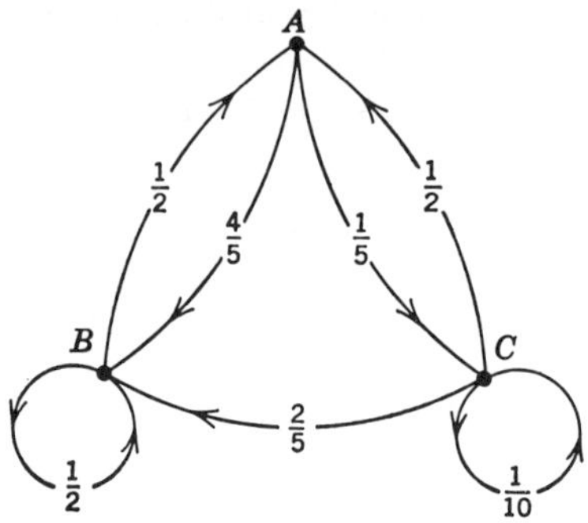

Figure 8.9 *The state diagram for a first-order Markov source*

which has no zero elements. Thus the source is ergodic and it is possible to go from any state to any other state in two transitions, but not in one. This result was apparent from a cursory examination of the state diagram of Figure 8.9.

The stationary-state distribution vector **P** is obtained from Equation 8.8-12. We have

$$[p_1 \quad p_2 \quad p_3]\begin{bmatrix} 0 & \frac{4}{5} & \frac{1}{5} \\ \frac{1}{2} & \frac{1}{2} & 0 \\ \frac{1}{2} & \frac{2}{5} & \frac{1}{10} \end{bmatrix} = [p_1 \quad p_2 \quad p_3]$$

or

$$\tfrac{1}{2}(p_2 + p_3) = p_1,$$
$$\tfrac{4}{5}p_1 + \tfrac{1}{2}p_2 + \tfrac{2}{5}p_3 = p_2,$$
$$\tfrac{1}{5}p_1 + \tfrac{1}{10}p_3 = p_3.$$

These three equations are easily solved simultaneously to find that

$$p_1 = \tfrac{9}{2}p_3, \qquad p_2 = 8p_3.$$

Since **P** is a probability vector

$$\sum_{i=1}^{3} p_i = 1 = p_1 + p_2 + p_3$$

or

$$p_1 = \tfrac{9}{27}, \qquad p_2 = \tfrac{16}{27}, \qquad p_3 = \tfrac{2}{27},$$

or

$$\mathbf{P} = [\tfrac{9}{27} \quad \tfrac{16}{27} \quad \tfrac{2}{27}].$$

Thus the initial distribution vector $\boldsymbol{\pi}$ has been (inadvertently?) chosen to be equal to the stationary-state distribution vector **P**.

• •

Example 8.6 As a final example, consider a second-order Markov source with the same alphabet and initial probability vector as in the previous example. The number of possible states is 9 since

$$q = M^m = 3^2 = 9.$$

The symbol transition probabilities are assumed to be given by

	A	B	C
AA	$\frac{1}{6}$	$\frac{1}{2}$	$\frac{1}{3}$
AB	$\frac{1}{4}$	$\frac{1}{4}$	$\frac{1}{2}$
AC	$\frac{1}{5}$	$\frac{2}{5}$	$\frac{2}{5}$
BA	$\frac{1}{4}$	$\frac{1}{8}$	$\frac{5}{8}$
BB	$\frac{2}{7}$	$\frac{2}{7}$	$\frac{3}{7}$
BC	$\frac{1}{2}$	$\frac{1}{4}$	$\frac{1}{4}$
CA	$\frac{1}{2}$	$\frac{1}{3}$	$\frac{1}{6}$
CB	$\frac{1}{5}$	$\frac{1}{5}$	$\frac{3}{5}$
CC	$\frac{1}{3}$	$\frac{1}{3}$	$\frac{1}{3}$

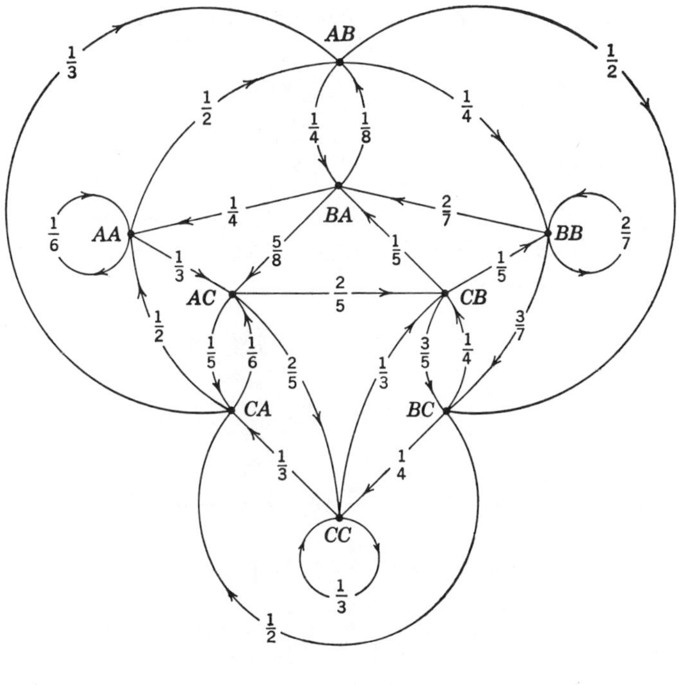

Figure 8.10 *The state diagram for a second-order Markov source*

Thus $p(A/AA) = \frac{1}{6}$. The corresponding transition matrix $\boldsymbol{T}$ is easily obtained from these probabilities and is

$\boldsymbol{T} =$

	AA	AB	AC	BA	BB	BC	CA	CB	CC
AA	$\frac{1}{6}$	$\frac{1}{2}$	$\frac{1}{3}$	0	0	0	0	0	0
AB	0	0	0	$\frac{1}{4}$	$\frac{1}{4}$	$\frac{1}{2}$	0	0	0
AC	0	0	0	0	0	0	$\frac{1}{5}$	$\frac{2}{5}$	$\frac{2}{5}$
BA	$\frac{1}{4}$	$\frac{1}{8}$	$\frac{5}{8}$	0	0	0	0	0	0
BB	0	0	0	$\frac{2}{7}$	$\frac{2}{7}$	$\frac{3}{7}$	0	0	0
BC	0	0	0	0	0	0	$\frac{1}{2}$	$\frac{1}{4}$	$\frac{1}{4}$
CA	$\frac{1}{2}$	$\frac{1}{3}$	$\frac{1}{6}$	0	0	0	0	0	0
CB	0	0	0	$\frac{1}{5}$	$\frac{1}{5}$	$\frac{3}{5}$	0	0	0
CC	0	0	0	0	0	0	$\frac{1}{3}$	$\frac{1}{3}$	$\frac{1}{3}$

It is clear that not all state transitions are possible in one step since the first symbol in each two-symbol state must be the same as the last symbol in the previous state. For higher-order Markov sources, the transition matrices will be even more restricted. In fact, for an Nth-order Markov source, transition in one step is possible only between two states in which the first $N - 1$ symbols of the second sequence are identical to and in the same order as the last $N - 1$ symbols of the first sequence.

The state diagram for this example is shown in Figure 8.10. It will be left to the reader to show that the source is ergodic.

• •

Example 8.7 Let us now look at an example of a nonergodic source. Consider a second-order Markov source with an alphabet of two symbols A and B, a transition matrix given by

$\boldsymbol{T} =$

	AA	AB	BA	BB
AA	1	0	0	0
AB	0	0	$\frac{1}{2}$	$\frac{1}{2}$
BA	$\frac{1}{4}$	$\frac{3}{4}$	0	0
BB	0	0	0	1

and a state diagram shown in Figure 8.11. It is clear from this state diagram that states AA and BB have an unusual property, namely, once either one of these states is attained, it continues for the rest of the sequence. Thus, after enough time, every sequence will become a sequence of A's only or a sequence of B's only. Such sequences will not be "typical."

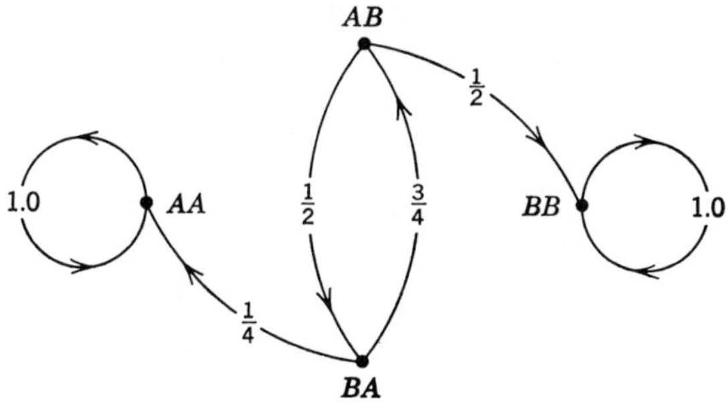

Figure 8.11 The state diagram of a nonergodic Markov source

Let us calculate several of the matrices $\boldsymbol{T}^r$. For $r = 2$, we have

$$\boldsymbol{T}^2 = \begin{bmatrix} 1 & 0 & 0 & 0 \\ \frac{1}{8} & \frac{3}{8} & 0 & \frac{1}{2} \\ \frac{1}{4} & 0 & \frac{3}{8} & \frac{3}{8} \\ 0 & 0 & 0 & 1 \end{bmatrix}.$$

If we continue until we reach $r = 5$, we find

$$\boldsymbol{T}^5 = \begin{bmatrix} 1 & 0 & 0 & 0 \\ \frac{22}{128} & 0 & \frac{9}{128} & \frac{97}{128} \\ \frac{97}{256} & \frac{27}{256} & 0 & \frac{132}{256} \\ 0 & 0 & 0 & 1 \end{bmatrix}$$

and, for $r = 6$,

$$\boldsymbol{T}^6 = \begin{bmatrix} 1 & 0 & 0 & 0 \\ \frac{97}{512} & \frac{27}{512} & 0 & \frac{388}{512} \\ \frac{194}{512} & 0 & \frac{27}{512} & \frac{291}{512} \\ 0 & 0 & 0 & 1 \end{bmatrix}.$$

Several facts are clear; in the first place, the top row of $\boldsymbol{T}^r$ will always contain the elements 1 0 0 0 and the bottom row the elements 0 0 0 1 for any integer value of r. Thus the transition matrix $\boldsymbol{T}$ is *not* regular. In the second place, it can be seen from the state diagram that, if r is large enough, the probability of ending in either state AB or BA must approach zero, no matter what the initial state. Thus the second and third columns of $\boldsymbol{T}^r$ will have only zero elements in the limit as r becomes large. We now have

$$\lim_{r\to\infty} \boldsymbol{T}^r = \begin{bmatrix} 1 & 0 & 0 & 0 \\ p_{21} & 0 & 0 & p_{24} \\ p_{31} & 0 & 0 & p_{34} \\ 0 & 0 & 0 & 1 \end{bmatrix}.$$

• •

8.9 The Entropy of a Markov Source

We have pointed out in the previous section that a Markov source is described by its state s_i from the set of possible states $S = \{s_1, s_2, \ldots, s_q\}$ where $q = M^m$. For each state s_i there exists the set of probabilities $p(s_j/s_i) = p_{ij}$ that, if the source is in state s_i, it will move to s_j on the production of the next symbol x_k. For each state s_i we can define an entropy $H(s_i)$ by

$$\begin{aligned} H(s_i) &= -\sum_{j=1}^{q} p(s_j/s_i) \log p(s_j/s_i) \\ &= -\sum_{k=1}^{M} p(x_k/s_i) \log p(x_k/s_i). \end{aligned} \tag{8.9-1}$$

The source entropy $H(S)$ in information units per symbol is the expected value of $H(s_i)$; that is,

$$\begin{aligned} H(S) &= -\sum_{i=1}^{q} p(s_i)H(s_i) \\ &= -\sum_{i=1}^{q}\sum_{j=1}^{q} p(s_i)p(s_j/s_i) \log p(s_j/s_i) \\ &= -\sum_{i=1}^{q}\sum_{k=1}^{M} p(s_i)p(x_k/s_i) \log p(x_k/s_i) \end{aligned} \tag{8.9-2}$$

where $p(s_i) = p_i$ is the stationary-state probability and is the ith element of the vector $\mathbf{P}$ defined by Equation 8.8-12.

For the zero-memory source, successive symbols are independent and there is only one possible state, so that $p(s_i) = 1$ and $q = 1$. Also each conditional probability is given by

$$p(x_k/s_i) = p(x_k) \tag{8.9-3}$$

so that Equation 8.9-2 becomes

$$H(S) = -\sum_{k=1}^{M} p(x_k) \log p(x_k) \tag{8.9-4}$$

in agreement with the original definition of Equation 8.2-4.

For the first-order Markov source, each state is uniquely determined by one symbol of the source alphabet and we have

$$p(x_k/s_i) = p(x_k/x_i). \tag{8.9-5}$$

Now the source entropy becomes

$$H(S) = -\sum_{i=1}^{M} \sum_{k=1}^{M} p(x_i, x_k) \log p(x_k/x_i). \tag{8.9-6}$$

Let us consider the second equality in Equation 8.9-1. This equation is of the form

$$H(s_i) = -\sum_{k=1}^{M} f(x_k) \log f(x_k) \tag{8.9-7}$$

where $f(x_k)$ is a probability distribution on the symbols x_k. In this case $f(x_k)$ is a conditional probability but it is still true that

$$f(x_k) \geq 0, \qquad k = 1, 2, \ldots, M, \tag{8.9-8}$$

$$\sum_{k=1}^{M} f(x_k) = 1. \tag{8.9-9}$$

Thus $H(s_i)$ has all of the properties developed in Section 8.4 and in particular by Equation 8.4-7,

$$H(s_i) \leq \log M \tag{8.9-10}$$

where M is the number of symbols x_i in the source alphabet. The maximum value $\log M$ occurs when the x_k are equally likely so that

$$f(x_k) = p(x_k/s_i) = \frac{1}{M}, \qquad \text{all } k, i. \tag{8.9-11}$$

Thus Equation 8.9-2 can be written as

$$H(S) \leq \log M \sum_{i=1}^{q} p(s_i) = \log M$$

or

$$\boxed{H(S) \leq \log M.} \tag{8.9-12}$$

The maximum occurs when each conditional probability is given by Equation 8.9-11 and, hence, when successive symbols are independent and each has probability of occurrence of $1/M$.

For a given source, the ratio of the actual entropy $H(S)$ to the maximum value it can have with the same alphabet is called the *relative entropy* of the source. The *redundancy* η of the source is defined as the positive difference between unity and this relative entropy:

$$\eta = 1 - \frac{H(S)}{\log M}. \qquad (8.9\text{-}13)$$

The quantity $\log M$ is sometimes called the *capacity* of the alphabet.

Example 8.8 Let us refer to the first-order Markov source of Example 8.5. The source entropy is given by Equation 8.9-6 and can be written as

$$\begin{aligned} H(S) = &-p(A)[p(A/A)\log_2 p(A/A) + p(B/A)\log_2 p(B/A) \\ &\qquad + p(C/A)\log_2 p(C/A)] \\ &- p(B)[p(A/B)\log_2 p(A/B) + p(B/B)\log_2 p(B/B) \\ &\qquad + p(C/B)\log_2 p(C/B)] \\ &- p(C)[p(A/C)\log_2 p(A/C) + p(B/C)\log_2 p(B/C) \\ &\qquad + p(C/C)\log_2 p(C/C)] \\ = &\tfrac{9}{27}(\tfrac{4}{5}\log_2 \tfrac{5}{4} + \tfrac{1}{5}\log_2 5) + \tfrac{16}{27}(\tfrac{1}{2}\log_2 2 + \tfrac{1}{2}\log_2 2) \\ &+ \tfrac{2}{27}(\tfrac{1}{2}\log_2 2 + \tfrac{2}{5}\log_2 \tfrac{5}{2} + \tfrac{1}{10}\log_2 10) \end{aligned}$$

or

$$H(S) = 0.934 \text{ bits/symbol.}$$

The capacity of the alphabet is $\log_2 3$ or 1.585 bits per symbol. Hence the source redundancy is

$$\eta = 1 - \frac{0.934}{1.585} = 0.411 \text{ bits/symbol.}$$

• •

A more general inequality than that of Equation 8.9-12 can be obtained. Suppose we are given an arbitrary mth-order Markov source with state set S. It will be convenient to define the *adjoint source* S^a as the zero-memory source with the same alphabet X and the same set of symbol probabilities $p(x_i)$. The entropy $H(S^a)$ of this adjoint source is given by Equation 8.9-4 and obeys the inequality of Equation 8.4-7:

$$H(S^a) \leq \log M. \qquad (8.9\text{-}14)$$

We will now show that $H(S)$ is bounded from above by $H(S^a)$. We can see from Equation 8.4-13 that

$$\sum_{i=1}^{q} \sum_{k=1}^{M} p(s_i, x_k) \log \frac{p(s_i)p(x_k)}{p(s_i, x_k)} \leq 0 \tag{8.9-15}$$

with equality if, and only if,

$$p(s_i, x_k) = p(s_i)p(x_k), \qquad \text{all } i, k. \tag{8.9-16}$$

On rearranging the inequality we have

$$\sum_{i=1}^{q} \sum_{k=1}^{M} p(s_i)p(x_k/s_i) \log \frac{1}{p(x_k/s_i)} \leq \sum_{i=1}^{q} \sum_{k=1}^{M} p(s_i, x_k) \log \frac{1}{p(x_k)}. \tag{8.9-17}$$

But the right side of this equation can be written as

$$\sum_{i=1}^{q} \sum_{k=1}^{M} p(s_i, x_k) \log \frac{1}{p(x_k)} = \sum_{k=1}^{M} p(x_k) \log \frac{1}{p(x_k)} = H(S^a) \tag{8.9-18}$$

and, from Equation 8.9-2, the left side is $H(S)$ so that

$$\boxed{H(S) \leq H(S^a) \leq \log M.} \tag{8.9-19}$$

We conclude that the addition of constraints on the symbols in the form of conditional probabilities in the message sequences serves only to decrease the source entropy or information content per symbol.

Suppose we deal with very long sequences C_N of length N symbols from an ergodic mth-order Markov source. Let us consider first the zero-memory case where successive symbols x_i are independent. For a long sequence of length N it is highly probable that each symbol x_i will occur approximately $Np(x_i)$ times in the sequence. The probability $p(C_N)$ of occurrence of a highly probable (typical) sequence is approximately the joint probability that the symbol x_1 occurs $Np(x_1)$ times, the symbol x_2 occurs $Np(x_2)$ times, and so on, so that

$$p(C_N) \approx \prod_{i=1}^{M} [p(x_i)]^{Np(x_i)}. \tag{8.9-20}$$

On taking the logarithm of both sides, we have

$$\log p(C_N) \approx N \sum_{i=1}^{M} p(x_i) \log p(x_i) = -NH(S) \tag{8.9-21}$$

or

$$H(S) \approx -\frac{\log p(C_N)}{N}. \tag{8.9-22}$$

Thus, for a zero-memory source, entropy should be given approximately by this last expression where $p(C_N)$ is the probability of occurrence of a "typical" sequence.

This intuitively reasonable result for a zero-memory source can be extended to ergodic Markov sources and can be stated precisely as the following theorem [11]:

Theorem: *For a given ergodic Markov source with entropy $H(S)$ and for any $\epsilon > 0$, $\delta > 0$, a number N_0 exists such that all sequences C_N of length $N \geq N_0$ fall into one of the following two classes:*

1. *A class of sequences of combined probability less than ϵ.*
2. *The remaining sequences C_N each with probability $p(C_N)$ satisfying the inequality*

$$\left| \frac{-\log p(C_N)}{N} - H(S) \right| < \delta.$$

Although the proof of this theorem is straightforward, it is too long for inclusion here and the reader is referred to other sources [11, 12].

8.10 Noiseless Coding and the Noiseless Discrete Channel

Most of the preceding discussion has emphasized the information source and its properties although both source and destination were included in the concept of mutual information defined in Section 8.5. In this section we begin to consider the properties of the communication channel serving as the link between the source and the destination. It has already been shown that a fundamental part of the description of the information source is a source alphabet $X = \{x_1, x_2, \ldots, x_M\}$ of size M. The source emits sequences of x_i's chosen from this source alphabet. In general an arbitrary channel will not accept and transmit this sequence. Instead the channel will accept a sequence of some other elements a_i chosen from a *code alphabet* A of *size* D where

$$A = \{a_1, a_2, \ldots, a_D\} \tag{8.10-1}$$

with D generally smaller than M. For example, the *binary channel* accepts only two symbols, say 0 and 1, while the information source might emit English text with at least 26 distinct symbols. The elements a_i of the code alphabet are frequently called *code elements* or *code characters* while a given sequence of a_i's may be called a *code word.*

The situation is now describable in terms of Figure 8.12 where an encoder E has been added between the source and channel. The process of *coding* or *encoding* the source consists of associating with each source symbol x_i a given code word, which is just a given sequence of a_i's. Thus the source emits a sequence of x_i's chosen from the source alphabet X and the encoder emits a

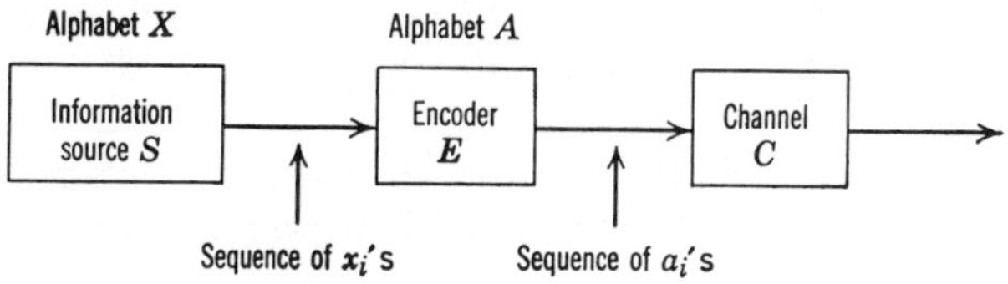

Figure 8.12 *Encoding a source*

sequence of a_i's chosen from the code alphabet A. It will be assumed in all subsequent discussions that the code words are distinct, that is, that each code word corresponds to only one source symbol.

Even though each code word is required to be distinct, sequences of code words may not have this property. An example is code A of Table 8.8 where

Table 8.8. Four Binary Coding Schemes[a]

Source Symbols	Code A	Code B	Code C	Code D
x_1	0	0	0	0
x_2	1	10	01	10
x_3	00	110	011	110
x_4	11	111	0111	1110

[a] Code A is not uniquely decipherable.
Codes B, C, and D are uniquely decipherable.
Codes B and D are instantaneous codes.
Codes C and D are comma codes.

a source of size four has been encoded in binary code with characters 0 and 1. In code A the code words are distinct but sequences of code words are not. For example, the sequence 0011 may correspond to any one of the source sequences $x_1x_1x_2x_2$, $x_1x_1x_4$, $x_3x_2x_2$, or x_3x_4. It is clear that such a code is not *uniquely* decipherable. On the other hand, a given sequence of code words taken from code B will correspond to a distinct sequence of source symbols. An examination of code B will show that in no case is a code word formed by adding characters to another word. We say that no code word is a *prefix* of another. It is clear that this is a *sufficient* (but not necessary) condition for a code to be uniquely decipherable. That it is not necessary can be seen from an examination of code C of Table 8.8. This code is uniquely decipherable even though many of the code words are prefixes of other words. For both C and D any sequence of code words can be decoded by subdividing the sequence of 0's and 1's to the left of every 0 for code C and to the right of every zero for code D. The character 0 is the first (or last) character of every code word and acts as a "comma"; therefore this type of code is called a *comma code.*

In general the channel will require a finite amount of time to transmit each code character. It is apparent then that the code words should be as short as possible in order to maximize information transfer per unit time. The average length L of a code is given by

$$L = \sum_{i=1}^{M} n_i p(x_i) \tag{8.10-2}$$

where n_i is the length (number of code characters) of the code word for the source symbol x_i, and $p(x_i)$ is the probability of occurrence of x_i. Although the average code length cannot be computed unless the set $\{p(x_i)\}$ is given, it is obvious that codes C and D of Table 8.8 will have a greater average length than code B unless $p(x_4) = 0$. From this example we might assume that comma codes are not optimal with respect to minimum average length.

Let us encode the sequence $x_3x_1x_3x_2$ into codes B, C, and D of Table 8.8 as follows:

Code B—110011010,
Code C—011001101,
Code D—110011010.

Suppose that we now start to decode into the original sequence of x_i's. Notice that codes B and D are fundamentally different from code C in that codes B and D may be decoded word by word *without examining subsequent code characters* while code C cannot be so treated. For example, when the first character 0 of code C is received, the observer is unable to tell, until the next character is observed, whether or not this is the *word* for source symbol x_1. Similarly, when the first two characters 01 are received, they may not be interpreted as the code word for x_2 until subsequent code characters are examined. Codes B and D are called *instantaneous codes* while code C is noninstantaneous. The instantaneous codes have the property previously mentioned that no code word is a prefix of another code word.

It may now be considered that the aim of noiseless coding is to produce codes with the following two properties:

1. *Unique decipherability.*
2. *Minimum average length* L for a given source S with alphabet X and probability set $\{p(x_i)\}$.

Codes that have both of these properties will be called *optimal.* It can be shown [13, 14] that if, for a given source S, a code is optimal among instantaneous codes, then it is optimal among all uniquely decipherable codes.

Thus it is sufficient to consider instantaneous codes. It is interesting to point out that a *necessary* property of optimal codes is that source symbols with higher probabilities must have shorter code words, that is,

$$p(x_i) > p(x_j) \Rightarrow n_i \le n_j. \tag{8.10-3}$$

The proof is simple. Suppose the converse is true so that $p(x_i) > p(x_j)$ but $n_i > n_j$. Let L_1 be the resulting average code length. Now produce a new code by interchanging code words i and j. The average code length L_2 of the new code is shorter than L_1 since

$$\begin{aligned} L_2 - L_1 &= n_j p(x_i) + n_i p(x_j) - [n_i p(x_i) + n_j p(x_j)] \\ &= [p(x_i) - p(x_j)][n_j - n_i] < 0. \end{aligned}$$

Let us restrict our attention for the moment to instantaneous codes and attempt to find necessary and sufficient conditions for the existence of such codes for a given source alphabet X of size M. The encoding procedure consists of the assignment of a code word to each of the M source symbols. The code word for the source symbol x_i will be of length n_i or, in other words, will consist of n_i code elements chosen from the code alphabet of size D. For convenience, let us arrange the code words in order of increasing size so that we have the ordered set of M positive integers $n_1, n_2, \ldots, n_M$ where $n_i \le n_{i+1}$.

Observe first that there are at most D instantaneous sequences of length one. If the symbol W_k is used to denote the *actual* number of words of length k, then

$$W_1 \le D. \tag{8.10-4}$$

The number of actual instantaneous two-element words (instantaneously distinguishable from each other and from the W_1 one-element words) must obey the inequality

$$W_2 \le (D - W_1)D \tag{8.10-5}$$

since the first element can be chosen in $(D - W_1)$ ways and the second element in D ways. It is clear that

$$W_3 \le [(D - W_1)D - W_2]D = D^3 - W_1 D^2 - W_2 D \tag{8.10-6}$$

since the first two elements can be chosen in $[(D - W_1)D - W_2]$ ways and the third element in D ways. In the general case, W_k satisfies the inequality

$$W_k \le D^k - W_1 D^{k-1} - \cdots - W_{k-1} D. \tag{8.10-7}$$

This last expression may be divided by D^k and rearranged to yield

$$\sum_{j=1}^{k} W_j D^{-j} \le 1. \tag{8.10-8}$$

This sum is also given by

$$\sum_{j=1}^{k} W_j D^{-j} = \sum_{n_i \le k} D^{-n_i} \tag{8.10-9}$$

since the right side may be written as

$$\sum_{n_i \le k} D^{-n_i} = \underbrace{D^{-1} + \cdots + D^{-1}}_{w_1 \text{ times}} + \underbrace{D^{-2} + \cdots + D^{-2}}_{w_2 \text{ times}} + \cdots + \underbrace{D^{-k} + \cdots + D^{-k}}_{w_k \text{ times}}. \tag{8.10-10}$$

Let k be set equal to n_M, the number of code elements in the longest code word; then Equation 8.10-9 becomes

$$\sum_{j=1}^{n_M} W_j D^{-j} = \sum_{i=1}^{M} D^{-n_i}. \tag{8.10-11}$$

This expression may be combined with the inequality of Equation 8.10-8 to yield a *necessary* condition for the construction of an instantaneous code, namely

$$\boxed{\sum_{i=1}^{M} D^{-n_i} \le 1.} \tag{8.10-12}$$

The condition is also *sufficient* since Equations 8.10-7 and 8.10-12 are equivalent if $k = n_M$ and since Equation 8.10-7 was derived by a procedure that could be used to construct an instantaneous code. Equation 8.10-12 is sometimes called the *Kraft Inequality* [15].

Let us refer again to Table 8.8 and check whether the four codes in that table satisfy Equation 8.10-12. For code A, the inequality is not satisfied since

$$\sum_{i=1}^{4} 2^{-n_i} = 2^{-1} + 2^{-1} + 2^{-2} + 2^{-2} = 1.5. \tag{8.10-13}$$

On the other hand, code A is neither instantaneous nor uniquely decipherable. For code B, the equality is satisfied since

$$\sum_{i=1}^{4} 2^{-n_i} = 2^{-1} + 2^{-2} + 2^{-3} + 2^{-3} = 1. \tag{8.10-14}$$

This is the shortest possible instantaneous code and is therefore optimal.

For codes C and D, we have

$$\sum_{i=1}^{4} 2^{-n_i} = 2^{-1} + 2^{-2} + 2^{-3} + 2^{-4} = \tfrac{15}{16} < 1. \qquad (8.10\text{-}15)$$

This result was to be expected for code D since it is instantaneous and obviously longer than code B.

The Kraft Inequality of Equation 8.10-12 has been shown to be a necessary and sufficient condition for the existence of an instantaneous code. Since instantaneous codes are uniquely decipherable, it is clear that the inequality is also a *sufficient* condition for the existence of a *uniquely decipherable* code. Surprisingly enough, as implied earlier, the inequality is also *necessary*. The proof of this important result is due to McMillan [13] although a simpler formulation has been given later by Karush [14]. The proof is reasonably straightforward but will not be given here.

8.11 The Noiseless Coding Theorem

As previously discussed, it is desirable to produce codes which are not only uniquely decipherable but which also have as small an average length L as possible. A necessary and sufficient inequality for the construction of a uniquely decipherable code has been given by the Kraft Inequality of Equation 8.10-12. Refer now to the inequality developed in Equation 8.4-13 which may be written as

$$\sum_{i=1}^{M} p_i \log p_i \geq \sum_{i=1}^{M} p_i \log q_i \qquad (8.11\text{-}1)$$

where the p_i and q_i are any sets of probabilities satisfying Equation 8.4-12. Note that equality occurs if, and only if, $p_i = q_i$ for all i. Let us identify

$$p_i = p(x_i), \qquad (8.11\text{-}2)$$

$$q_i = \frac{D^{-n_i}}{\sum_{i=1}^{M} D^{-n_i}}. \qquad (8.11\text{-}3)$$

With these substitutions, the left side of Equation 8.11-1 is just the negative of the source entropy so that

$$-H(X) \geq -\sum_{i=1}^{M} n_i p(x_i) \log D - \sum_{i=1}^{M} p(x_i) \log \left(\sum_{i=1}^{M} D^{-n_i} \right). \qquad (8.11\text{-}4)$$

Since Equation 8.10-12 must be satisfied if a uniquely decipherable code is to be constructed, the term $\log\left(\sum_{i=1}^{M} D^{-n_i}\right)$ must be nonpositive. Consequently, with the use of Equation 8.10-2, this last equation becomes

$$H(X) \leq L \log D + (\text{nonpositive number}). \tag{8.11-5}$$

Thus the average code length is bounded from below by

$$\boxed{L \geq \frac{H(X)}{\log D}.} \tag{8.11-6}$$

Note that equality (and minimum code length) occurs if, and only if,

$$p(x_i) = D^{-n_i}, \qquad i = 1, 2, \ldots, M. \tag{8.11-7}$$

Recall that $p(x_i)$ is the a-priori probability of occurrence of x_i; D is the size of the code alphabet; and n_i is the number of elements in the code word for x_i. Except in special cases, Equation 8.11-7 will not be satisfied, and the inequality of Equation 8.11-6 will be strict. As a counterexample, however, refer to code B of Table 8.8 and assume that the source symbols have the a-priori probabilities $p(x_1) = \frac{1}{2}, p(x_2) = \frac{1}{4}, p(x_3) = p(x_4) = \frac{1}{8}$. Then Equation 8.11-7 is satisfied and the average code length L is

$$L = \sum_{i=1}^{4} n_i p(x_i) = \tfrac{1}{2} + \tfrac{2}{4} + \tfrac{3}{8} + \tfrac{3}{8} = 1\tfrac{3}{4} \text{ elements/symbol.}$$

For this case the source entropy $H(X)$ is

$$\begin{aligned} H(X) &= -\sum_{i=1}^{4} p(x_i) \log p(x_i) \\ &= \tfrac{1}{2}\log 2 + \tfrac{1}{4}\log 4 + \tfrac{1}{8}\log 8 + \tfrac{1}{8}\log 8 \\ &= \tfrac{1}{2} + \tfrac{2}{4} + \tfrac{3}{8} + \tfrac{3}{8} = 1\tfrac{3}{4} \text{ bits/symbol} \end{aligned}$$

and, as expected,

$$L = \frac{H(X)}{\log 2} = H(X).$$

Such a code where the equality of Equation 8.11-6 applies will be called *absolutely optimal.*

The relationship given by Equation 8.11-6 places a lower bound on the average length of a uniquely decipherable code, and this lower bound is attained only in the special case where $p(x_i) = D^{-n_i}$ for all i. Let us examine the reasons why the lower bound is not attained in the general case. As pointed out in Equation 8.2-3, the self-information $I(x_i)$ associated with the symbol x_i is

$$I(x_i) = -\log p(x_i). \tag{8.11-8}$$

The maximum information associated with a code element a_j is log D where D is the size of the code alphabet. Thus the minimum number of a's required for the code word for x_i is

$$n_i' = \frac{I(x_i)}{\log D}. \tag{8.11-9}$$

However, the number n_i' is not usually an integer; in fact it is the integer n_i if, and only if, Equation 8.11-7 is satisfied. For the general case, then, the integer n_i is bounded by

$$n_i' \leq n_i < n_i' + 1 \tag{8.11-10}$$

or, from Equation 8.11-9,

$$\frac{I(x_i)}{\log D} \leq n_i < \frac{I(x_i)}{\log D} + 1. \tag{8.11-11}$$

A uniquely decipherable code can be constructed with code words whose lengths $n_1, n_2, \ldots, n_M$ are chosen according to this last inequality. This is the case since the n_i determined by Equation 8.11-11 satisfy the Kraft Inequality of Equation 8.10-12. Only the left side of this last inequality need be checked since the right side will make the sum $\sum_{i=1}^{M} D^{-n_i}$ even smaller. We have

$$\sum_{i=1}^{M} D^{-n_i} \leq \sum_{i=1}^{M} D^{-I(x_i)/\log D} = \sum_{i=1}^{M} D^{\log_D p(x_i)} = \sum_{i=1}^{M} p(x_i) = 1. \tag{8.11-12}$$

Now Equation 8.11-11 may be multiplied by $p(x_i)$ and summed to yield

$$\boxed{\frac{H(X)}{\log D} \leq L \leq \frac{H(X)}{\log D} + 1.} \tag{8.11-13}$$

This relationship* is frequently called the *noiseless coding theorem*. It gives the upper and lower bounds on the average length of a uniquely decipherable code that can be constructed for a source with entropy $H(X)$. It has already been noted that any source with symbol probabilities satisfying Equation 8.11-7 can be encoded into a code of absolutely minimum average length $H(X)/\log D$. Equation 8.11-13 guarantees that, for an arbitrary source, the average code-word length can be constrained to be no more than one code element longer than this absolute minimum.

* Some authors refer to Equation 8.11-6 as the noiseless coding theorem.

Actually the average code length L can be made arbitrarily close to $H(X)/\log D$ by encoding sequences of source symbols rather than considering each symbol separately.

Example 8.9 Consider a source with four symbols x_1, x_2, x_3, and x_4 each with a-priori probabilities as indicated in Table 8.9. The code designated as code 1 is an instantaneous (and hence uniquely decipherable) code which is the same as code B of Table 8.8. In this case the a-priori probabilities $p(x_i)$ are not equal to D^{-n_i} and the average code length L_1 is greater than $H(X)/\log D$ as shown. Suppose now that pairs of symbols $x_i x_j$ are encoded as shown in the table. Code 2 is an instantaneous code to accomplish this purpose. Again the joint probabilities $p(x_i, x_j)$ are not equal to D^{-n_i} but the average code length L_2 has been reduced. It is apparent that the process of encoding "blocks" of source symbols instead of single symbols has reduced the average code length in this particular case. If blocks of length three—that is, sequences of the form $x_i x_j x_k$—are now encoded, it will be found that the

Table 8.9. An Example of Block Coding[a]

Source Symbols x_i	A-priori Probabilities $p(x_i)$	Code 1	D^{-n_i}	Source Symbols $x_i x_j$	A-priori Probabilities $p(x_i, x_j)$	Code 2	D^{-n_i}
x_1	$\frac{1}{3}$	0	$\frac{1}{2}$	x_1x_1	$\frac{1}{9}$	000	$\frac{1}{8}$
				x_1x_2	$\frac{1}{9}$	001	$\frac{1}{8}$
				x_2x_2	$\frac{1}{9}$	010	$\frac{1}{8}$
x_2	$\frac{1}{3}$	10	$\frac{1}{4}$	x_2x_1	$\frac{1}{9}$	0110	$\frac{1}{16}$
				x_1x_3	$\frac{1}{18}$	0111	$\frac{1}{16}$
				x_1x_4	$\frac{1}{18}$	1000	$\frac{1}{16}$
x_3	$\frac{1}{6}$	110	$\frac{1}{8}$	x_2x_3	$\frac{1}{18}$	1001	$\frac{1}{16}$
				x_2x_4	$\frac{1}{18}$	1010	$\frac{1}{16}$
				x_3x_1	$\frac{1}{18}$	1011	$\frac{1}{16}$
x_4	$\frac{1}{6}$	111	$\frac{1}{8}$	x_3x_2	$\frac{1}{18}$	1100	$\frac{1}{16}$
				x_4x_1	$\frac{1}{18}$	1101	$\frac{1}{16}$
				x_4x_2	$\frac{1}{18}$	11100	$\frac{1}{32}$
				x_3x_3	$\frac{1}{36}$	11101	$\frac{1}{32}$
				x_3x_4	$\frac{1}{36}$	11110	$\frac{1}{32}$
				x_4x_3	$\frac{1}{36}$	111110	$\frac{1}{64}$
				x_4x_4	$\frac{1}{36}$	111111	$\frac{1}{64}$

[a] Code 1: $L_1 = 2$ code elements/symbol, $H(X) = 1.918$ bits/symbol,
$1.918 \leq L_1 < 2.918$.
Code 2: $L_2 = 1.944^+$ code elements/symbol, $H(X) = 1.918$ bits/symbol,
$1.918 \leq L_2 \leq 2.918$.

resulting average code length L_3 is less than L_2. By using longer and longer blocks, the average code length can be made to approach $H(X)/\log D$ as closely as desired. This concept will now be developed in somewhat more generality.

• •

It will be assumed in this discussion that the sequences of x_i's emitted by the source are independent, that is, the source is zero memory. Let X^n be the set of all possible x sequences of length n as given by Equation 8.7-1. As pointed out in Section 8.7, there are M^n such sequences. The entropy of this nth extension of the zero-memory source X is given by Equation 8.7-6 and is just

$$H(X^n) = nH(X). \tag{8.11-14}$$

Let the average length of the code words associated with these x sequences be denoted by L_n. Then, in exactly the same way as Equation 8.11-13 was derived, L_n must satisfy

$$\frac{H(X^n)}{\log D} \leq L_n < \frac{H(X^n)}{\log D} + 1. \tag{8.11-15}$$

On substituting Equation 8.11-14 and dividing by n, there results

$$\frac{H(X)}{\log D} \leq L \leq \frac{H(X)}{\log D} + \frac{1}{n} \tag{8.11-16}$$

since $nL = L_n$. It is clear from this last expression that L can be constrained to be as close as desired to $H(X)/\log D$ by increasing n, the "block" length. The disadvantage of such block coding lies in the increased complexity of the encoding procedure and in the necessity of providing storage facilities to allow the x sequence to accumulate before encoding. This discussion has assumed a zero-memory source. Similar results can be proved for sources with memory but the procedures are considerably more complicated.

8.12 The Construction of Codes

The two preceding sections have been concerned with the general properties of codes and with bounds on average code length. In this section the actual construction of codes is considered with particular attention to optimal codes; that is, uniquely decipherable codes of minimum length.

Let us begin with a case in which an absolutely optimal code exists; that is, the source symbol probabilities satisfy Equation 8.11-7. An example that has been discussed previously is shown in Table 8.10. For the binary case ($D = 2$), it is clear that

$$p(x_i) = D^{-n_i}, \qquad i = 1, 2, 3, 4.$$

Note that the code is code *B* of Table 8.8 and, as previously mentioned, is absolutely optimal for the source probabilities of Table 8.10. A constructive procedure for realizing this code is illustrated in the same table. First, the source symbols are arranged in order of decreasing probability. Then, the set of source symbols is divided into two equiprobable groups; the first group consists of x_1 and the second group of x_2, x_3, and x_4. The code element 0 is assigned as the first element in the first group and the code element 1 as the first element in the second group. Next, each of the two groups is subdivided into two equiprobable groups; the code element 0 is assigned as the second element in the first of each subgroup and the code element 1 as the second element in the second of each subgroup. This process is repeated until each subgroup consists of a single source symbol. The resulting code is instantaneous and absolutely optimal if $p(x_i) = D^{-n_i}$ for all i.

Table 8.10. The Construction of an Optimal Code: $D = 2$

Source Symbol x_i	A-priori Probabilities $p(x_i)$	Step 1	Step 2	Step 3	Final Code
x_1	$\frac{1}{2}$	0			0
x_2	$\frac{1}{4}$	1	0		10
x_3	$\frac{1}{8}$	1	1	0	110
x_4	$\frac{1}{8}$	1	1	1	111

Average code length $L = 1\frac{3}{4}$ code elements/symbol.
Source entropy $H(X) = 1\frac{3}{4}$ bits/symbol.

$$L = \frac{H(X)}{\log_2 2} = H(X).$$

For arbitrary code alphabet size ($D \geq 2$), the procedure may be generalized in an obvious fashion as follows:

1. Arrange the M source symbols in order of decreasing probability.
2. Arrange the D code elements in an arbitrary but fixed order, that is, $a_1, a_2, \ldots, a_D$.
3. Divide the set of symbols x_i into D groups with equal probabilities of $1/D$ each. This division is always possible if Equation 8.11-7 is satisfied.
4. Assign the element a_1 as the first digit for symbols in the first group, a_2 for the second, and a_i for the ith group.

5. After the first division each of the resulting groups contains a number of symbols equal to D raised to some integral power, if Equation 8.11-7 is satisfied. Thus a typical group, say group i, contains D^{k_i} symbols where k_i is an integer (which may be zero). This group of symbols can be further subdivided k_i times into D parts of equal probabilities. Each division decides one additional code digit in the sequence. A typical symbol x_i is isolated after q divisions. If it belongs to the i_1 group after the first division, the i_2 group after the second division, and so on, then the code word for x_i will be $a_{i_1}a_{i_2}\cdots a_{i_q}$.

An illustration of the construction of an absolutely optimal code for the case where $D = 3$ is given in Table 8.11. This procedure ensures that source symbols with high probabilities will have short code words and vice versa,

Table 8.11. The Construction of an Optimal Code: D = 3

Source Symbols x_i	A-priori Probabilities $p(x_i)$	Step 1	Step 2	Step 3	Final Code
x_1	$\frac{1}{3}$	1			1
x_2	$\frac{1}{9}$	0	1		0 1
x_3	$\frac{1}{9}$	0	0		0 0
x_4	$\frac{1}{9}$	0	−1		0 −1
x_5	$\frac{1}{27}$	−1	1	1	−1 1 1
x_6	$\frac{1}{27}$	−1	1	0	−1 1 0
x_7	$\frac{1}{27}$	−1	1	−1	−1 1 −1
x_8	$\frac{1}{27}$	−1	0	1	−1 0 1
x_9	$\frac{1}{27}$	−1	0	0	−1 0 0
x_{10}	$\frac{1}{27}$	−1	0	−1	−1 0 −1
x_{11}	$\frac{1}{27}$	−1	−1	1	−1 −1 1
x_{12}	$\frac{1}{27}$	−1	−1	0	−1 −1 0
x_{13}	$\frac{1}{27}$	−1	−1	−1	−1 −1 −1

Average code length L = 2 code elements/symbol.
Source entropy $H(X) = 2 \log_2 3$ bits/symbol.

$$L = \frac{H(X)}{\log_2 3}.$$

since a symbol with probability D^{-n_i} will be isolated after n_i divisions and thus will have n_i elements in its code word as required by Equation 8.11-7.

The code resulting from the process just discussed is sometimes called the *Shannon-Fano* code. It is apparent that the same encoding procedure can be followed whether or not the source probabilities satisfy Equation 8.11-7. The set of symbols x_i are simply divided into D groups with probabilities as nearly equal as possible. The procedure is sometimes ambiguous, however, as an examination of Table 8.12 will show. In that table two possible Shannon-Fano codes are given for a source whose symbol probabilities do not satisfy Equation 8.11-7. The ambiguity arises, of course, in the choice of approximately equiprobable subgroups. As shown, code 1 is slightly shorter than code 2.

Table 8.12. Two Shannon-Fano Codes[a]

Source Symbols x_i	A-priori Probabilities $p(x_i)$	Code 1 Step 1	Code 1 Step 2	Code 1 Step 3	Code 1 Step 4	Code 1 Final Code	Code 2 Step 1	Code 2 Step 2	Code 2 Step 3	Code 2 Step 4	Code 2 Final Code
x_1	0.40	0				0	0	0			00
x_2	0.20	1	0	0		100	0	1			01
x_3	0.12	1	0	1		101	1	0	0		100
x_4	0.08	1	1	0	0	1100	1	0	1		101
x_5	0.08	1	1	0	1	1101	1	1	0		110
x_6	0.08	1	1	1	0	1110	1	1	1	0	1110
x_7	0.04	1	1	1	1	1111	1	1	1	1	1111

[a] Average length of code 1:

$$L_1 = 1(0.40) + 3(0.20) + 3(0.12) + 4(0.08) + 4(0.08) + 4(0.08) + 4(0.04)$$
$$= 2.48 \text{ code elements/symbol.}$$

Average length of code 2:

$$L_2 = 2(0.40) + 2(0.20) + 3(0.12) + 3(0.08) + 3(0.08) + 4(0.08) + 4(0.04)$$
$$= 2.52 \text{ code elements/symbol.}$$

For the general case where Equation 8.11-7 is not satisfied, a procedure due to Huffman [16] guarantees an optimal code, that is, one with minimum average length. This procedure for code alphabet of arbitrary size D may be described as follows:

1. As before, arrange the M source symbols in order of decreasing probability.

2. As before, arrange the code elements in an arbitrary but fixed order, that is, $a_1, a_2, \ldots, a_D$.
3. Combine (sum) the probabilities of the D least likely symbols and reorder the resulting $M - (D - 1)$ probabilities; this step will be called reduction 1. Repeat as often as necessary until there are D ordered probabilities remaining. *Note*: For the binary case ($D = 2$), it will always be possible to accomplish this reduction in $M - 2$ steps. When the size of the code alphabet is arbitrary, the last reduction will result in exactly D ordered probabilities if, and only if,

$$M = D + n(D - 1)$$

where n is an integer. In case this relationship is not satisfied, the proper procedure to follow is to add *dummy* source symbols with zero probability. The entire encoding procedure is followed as before and, at the end, the dummy symbols are thrown away.
4. Start the encoding with the last reduction which consists of exactly D ordered probabilities; assign the element a_1 as the first digit in the code words for all the source symbols associated with the first probability; assign a_2 to the second probability; and a_i to the ith probability.
5. Proceed to the next-to-the-last reduction; this reduction consists of $D + (D - 1)$ ordered probabilities for a net gain of $D - 1$ probabilities. For the D new probabilities, the first code digit has already been assigned and is the same for all of these D probabilities; assign a_1 as the second digit for all source symbols associated with the first of these D new probabilities; assign a_2 as the second digit for the second of these D new probabilities, and so on.
6. The encoding procedure terminates after $1 + n(D - 1)$ steps which is one more than the number of reductions.

As illustrations of the Huffman coding procedure, a binary code is constructed in Table 8.13 and a ternary code in Table 8.14. Note that the binary code is essentially the same (except for a relabeling of code elements) as code 1 of Table 8.12. The ternary code is the same as that shown in Table 8.11. For a proof of the optimality of the Huffman code, the reader is referred to the literature [16, 17].

This treatment of encoding procedures has assumed that the source symbols are emitted independently, that is, that the source is zero memory. As discussed in Sections 8.8 and 8.9, this assumption is not always realistic and a better model may be the Markov source of arbitrary order m. In such a case the actual code construction may become much more complicated than for the zero-memory case [18]. The noiseless coding theorem of Equation 8.11-13 still holds, however, except that $H(X)$ is replaced by the source entropy $H(S)$ of Equation 8.9-2.

Table 8.13. Construction of a Huffman Code ($D = 2$)

Source Symbols x_i	A-priori Probabilities $p(x_i)$	Final Code	Reduction 1	Step 5	Reduction 2	Step 4	Reduction 3	Step 3	Reduction 4	Step 2	Reduction 5	Step 1
x_1	0.40	0	0.40	0	0.40	0	0.40	0	0.40	0	0.60	1
x_2	0.20	111	0.20	111	0.20	111	0.24	10	0.36	11	0.40	0
x_3	0.12	101	0.12	101	0.16	110	0.20	111	0.24	10		
x_4	0.08	1101	0.12	100	0.12	101	0.16	110				
x_5	0.08	1100	0.08	1101	0.12	100						
x_6	0.08	1001	0.08	1100								
x_7	0.04	1000										

Average code length $L = 1(0.40) + 3(0.20) + 3(0.12) + 4(0.08) + 4(0.08) + 4(0.08) + 4(0.04)$

$= 2.48$ code elements/symbol.

Table 8.14. Construction of a Huffman Code ($D = 3$)

Source Symbols x_i	A-priori Probabilities $p(x_i)$	Final Code	Reduction 1	← Step 5	Reduction 2	← Step 4	Reduction 3	← Step 3	Reduction 4	← Step 2	Reduction 5	Step 1
x_1	$\frac{1}{3}$	1	$\frac{1}{3}$	1	$\frac{1}{3}$	1	$\frac{1}{3}$	1	$\frac{1}{3}$	1	$\frac{1}{3}$	1
x_2	$\frac{1}{9}$	−1 1	$\frac{1}{9}$	−1 1	$\frac{1}{9}$	−1 1	$\frac{1}{9}$	−1 1	$\frac{1}{3}$	0	$\frac{1}{3}$	0
x_3	$\frac{1}{9}$	−1 0	$\frac{1}{9}$	−1 0	$\frac{1}{9}$	−1 0	$\frac{1}{9}$	−1 0	$\frac{1}{9}$	−1 1	$\frac{1}{3}$	−1
x_4	$\frac{1}{9}$	−1 −1	$\frac{1}{9}$	−1 −1	$\frac{1}{9}$	−1 −1	$\frac{1}{9}$	−1 −1	$\frac{1}{9}$	−1 0		
x_5	$\frac{1}{27}$	0 −1 1	$\frac{1}{9}$	0 1	$\frac{1}{9}$	0 1	$\frac{1}{9}$	0 1	$\frac{1}{9}$	−1 −1		
x_6	$\frac{1}{27}$	0 −1 0	$\frac{1}{27}$	0 −1 1	$\frac{1}{9}$	0 0	$\frac{1}{9}$	0 0				
x_7	$\frac{1}{27}$	0 −1 −1	$\frac{1}{27}$	0 −1 0	$\frac{1}{27}$	0 −1 1	$\frac{1}{9}$	0 −1				
x_8	$\frac{1}{27}$	0 0 1	$\frac{1}{27}$	0 −1 −1	$\frac{1}{27}$	0 −1 0						
x_9	$\frac{1}{27}$	0 0 0	$\frac{1}{27}$	0 0 1	$\frac{1}{27}$	0 −1 −1						
x_{10}	$\frac{1}{27}$	0 0 −1	$\frac{1}{27}$	0 0 0								
x_{11}	$\frac{1}{27}$	0 1 1	$\frac{1}{27}$	0 0 −1								
x_{12}	$\frac{1}{27}$	0 1 0										
x_{13}	$\frac{1}{27}$	0 1 −1										

$$\text{Average code length} = 1(\tfrac{1}{3}) + 2(\tfrac{1}{9}) + 2(\tfrac{1}{9}) + 2(\tfrac{1}{9}) + 3(\tfrac{1}{27}) + 3(\tfrac{1}{27}) + 3(\tfrac{1}{27}) + 3(\tfrac{1}{27}) + 3(\tfrac{1}{27}) + 3(\tfrac{1}{27}) + 3(\tfrac{1}{27}) + 3(\tfrac{1}{27}) + 3(\tfrac{1}{27})$$

$$= 2 \text{ code elements/symbol.}$$

$$M = \mathrm{D} + n(D - 1) \quad \text{or} \quad 13 = 3 + 2n \Rightarrow n = 5 \text{ reductions.}$$

8.13 Channel Capacity

A number of equivalent expressions were developed in Sections 8.5 and 8.6 for average mutual information $I(X; Y)$ where X was a source alphabet of M symbols and Y was a destination alphabet of N symbols as shown in Figure 8.6. The most useful of these for the present discussion will be Equations 8.6-14 and 8.6-15 which may be combined as

$$I(X; Y) = H(Y) - H(Y/X) = H(X) - H(X/Y) \geq 0. \quad (8.13\text{-}1)$$

It is clear from this expression as well as from Examples 8.1 and 8.2 of Section 8.5 that the average mutual information depends not only on the statistical characteristics of the channel but also on the distribution $p(x_i)$ of the input alphabet X. If the input distribution is varied until Equation 8.13-1 is a maximum for a given channel, the resulting value of $I(X; Y)$ is called the *channel capacity* C of that channel, that is,

$$C = \max_{p(x_i)} I(X; Y). \quad (8.13\text{-}2)$$

It should be noted that, in general, $H(X)$, $H(Y)$, $H(X/Y)$, and $H(Y/X)$ all depend on the input distribution $p(x_i)$. Hence, *in the general case*, it is not a simple matter to maximize Equation 8.13-1 with respect to $p(x_i)$.

All of the measures of information that have been considered in this treatment have involved only probability distributions. Thus, for the model of Figure 8.6, the joint distribution $p(x_i, y_j)$ is sufficient. Suppose the source [and hence the input distribution $p(x_i)$] is known; then it follows from the usual conditional probability relationship

$$p(x_i, y_j) = p(x_i)p(y_j/x_i) \quad (8.13\text{-}3)$$

that only the distribution $p(y_j/x_i)$ is needed for $p(x_i, y_j)$ to be determined. This conditional probability $p(y_j/x_i)$ can then be taken as a description of the information channel connecting the source X and the destination Y. Thus a *discrete, memoryless channel* can be defined as the probability distribution

$$p(y_j/x_i), \qquad x_i \in X \quad \text{and} \quad y_j \in Y, \quad (8.13\text{-}4)$$

or, equivalently, by the *channel matrix* $\boldsymbol{D}$ where

$$\boldsymbol{D} = [p(y_j/x_i)] = \begin{bmatrix} p(y_1/x_1) & p(y_2/x_1) & \cdots & p(y_N/x_1) \\ p(y_1/x_2) & p(y_2/x_2) & \cdots & p(y_N/x_2) \\ \vdots & & & \vdots \\ p(y_1/x_M) & & & p(y_N/x_M) \end{bmatrix}. \quad (8.13\text{-}5)$$

A number of special types of channels are readily distinguished. Some of the simplest and/or most interesting can be listed as follows.

Lossless Channel

A channel will be called *lossless* if $H(X/Y) = 0$ for all input distributions $p(x_i)$ so that Equation 8.13-1 can be written as

$$I(X; Y) = H(X). \tag{8.13-6}$$

Thus the information transfer is equal to the source entropy, and no source information is lost in transmission. Channel capacity is obtained when the source entropy is a maximum, or, from Equation 8.4-8,

$$C = \max_{p(x_i)} H(X) = \log M. \tag{8.13-7}$$

This maximum is obtained when all x_i are equally likely or, in other words, when $p(x_i)$ is a uniform distribution. The lossless channel may be illustrated schematically as in Figure 8.13. The Y alphabet, of size N, may be partitioned into M disjoint sets B_i; each set is uniquely associated with one, and only one,

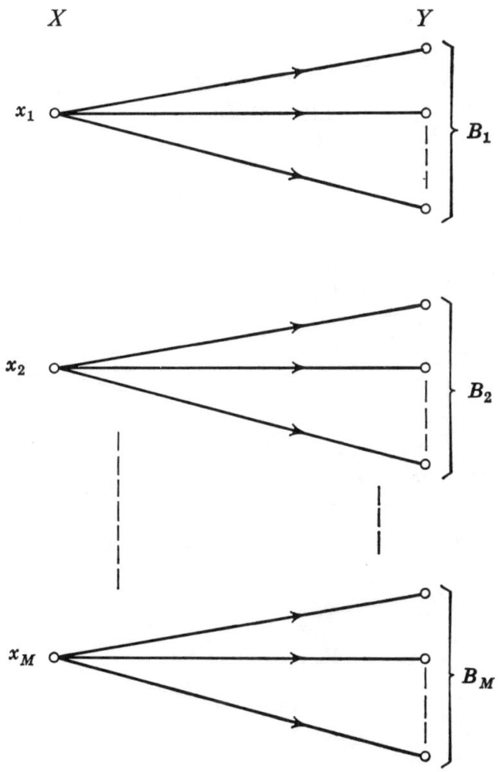

Figure 8.13 *The lossless channel*

member of the input alphabet X. Thus the conditional probabilities $p(x_i/y_j)$ take on only the values zero and unity; specifically

$$p(x_i/y_j) = \begin{cases} 1, & y_j \in B_i, \\ 0, & y_j \notin B_i. \end{cases} \tag{8.13-8}$$

Example 8.1 of Section 8.5 shows a lossless channel although two information destinations rather than one are involved. Refer to column 7 of Table 8.4 where the conditional probabilities $p(x_i/y_j, z_k)$ are all either zero or unity. When a given destination symbol is received in the lossless channel, it is clear which source symbol was sent.

The Deterministic Channel

A channel will be called *deterministic* if $H(Y/X) = 0$ for all input distributions $p(x_i)$ so that Equation 8.13-1 can be written as

$$I(X; Y) = H(Y). \tag{8.13-9}$$

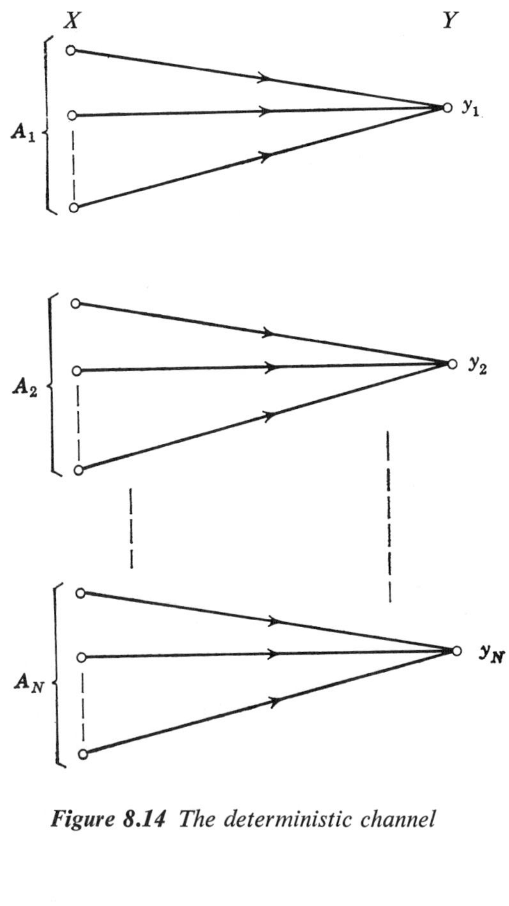

Figure 8.14 *The deterministic channel*

Thus the information transfer is equal to the destination entropy. Channel capacity is obtained when the destination entropy is a maximum, or

$$C = \max_{p(x_i)} H(Y) = \log N. \tag{8.13-10}$$

This maximum is attained when all the y_j are equally likely or, in other words, when $p(y_j)$ is a uniform distribution. The deterministic channel is shown in Figure 8.14. The X alphabet, of size M, may be partitioned into N disjoint sets A_j; each set is uniquely associated with one, and only one, member of the destination alphabet Y. Thus the conditional probabilities $p(y_j/x_i)$ take on only the values zero and unity; specifically

$$p(y_j/x_i) = \begin{cases} 1, & x_i \in A_j, \\ 0, & x_i \notin A_j. \end{cases} \tag{8.13-11}$$

When a given source symbol is sent in the deterministic channel, it is clear which destination symbol will be received.

For the deterministic channel, the channel matrix $\boldsymbol{D}$ of Equation 8.13-5 has one, and only one, nonzero element in each row and this element is unity.

The Noiseless Channel

A channel will be called *noiseless* if it is both lossless and deterministic. In this case, Equation 8.13-1 becomes

$$I(X; Y) = H(X) = H(Y). \tag{8.13-12}$$

Each A set and each B set has only one member; the source and destination alphabets are of the same size, that is, $M = N$; and channel capacity is

$$C = \log M = \log N. \tag{8.13-13}$$

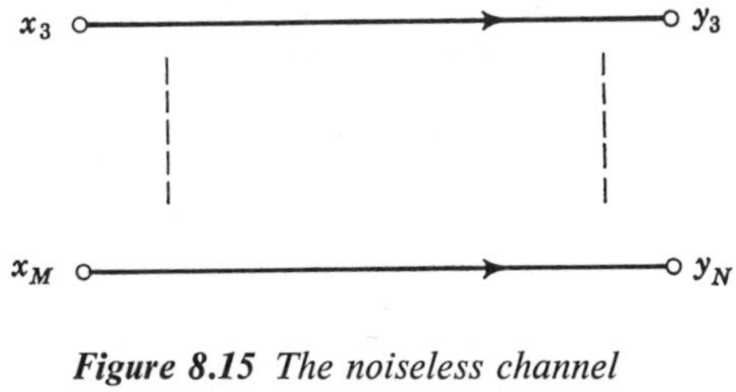

Figure 8.15 The noiseless channel

The noiseless channel is shown in Figure 8.15. The channel matrix $\boldsymbol{D}$ has one, and only one, nonzero element in each row *and* in each column and this element is unity.

The Symmetric Channel

A channel will be called *symmetric* if the rows of the channel matrix $\boldsymbol{D}$ are identical except for permutations *and* if the columns are identical except for permutations. If $\boldsymbol{D}$ is square, then rows and columns are identical except for permutations. The importance of the symmetric channel lies in the fact that the conditional entropy $H(Y/X)$ is independent of the input distribution $p(x_i)$ and depends only on the channel matrix $\boldsymbol{D}$. As a consequence, the determination of channel capacity is greatly simplified.

To show that $H(Y/X)$ is independent of $p(x_i)$, we rearrange Equation 8.6-6 to obtain

$$H(Y/X) = \sum_{i=1}^{M} p(x_i)\left[-\sum_{j=1}^{N} p(y_j/x_i)\log p(y_j/x_i)\right]. \qquad (8.13\text{-}14)$$

The term in brackets has the same value for each x_i since the rows of the channel matrix $\boldsymbol{D}$ are identical except for permutations. Since this term is not a function of the index i, the input distribution $p(x_i)$ sums to unity and $H(Y/X)$ becomes

$$H(Y/X) = -\sum_{j=1}^{N} p(y_j/x_i)\log p(y_j/x_i) \qquad (8.13\text{-}15)$$

as was to be proved. The channel capacity C_s of the symmetric channel now becomes

$$C_s = [\max_{p(x_i)} H(Y)] - H(Y/X) \qquad (8.13\text{-}16)$$

or, since the maximum is attained when the y_j are equally likely,

$$C_s = \log N + \sum_{j=1}^{N} p(y_j/x_i)\log p(y_j/x_i). \qquad (8.13\text{-}17)$$

The case most often treated is the *binary symmetric channel* (BSC) where $M = N = 2$. In this case the channel matrix can be written as

$$\boldsymbol{D} = \begin{bmatrix} p & 1-p \\ 1-p & p \end{bmatrix} \qquad (8.13\text{-}18)$$

and the channel capacity C_{s2} is

$$C_{s2} = \log 2 + p\log p + (1-p)\log(1-p). \qquad (8.13\text{-}19)$$

As a matter of convenience, let us define the function $G(p)$ by

$$G(p) = -[p \log p + (1 - p) \log (1 - p)]. \tag{8.13-20}$$

Note that this expression is mathematically identical to the entropy of a binary source as given in Equation 8.4-1 and plotted in Figure 8.3 using base 2 logarithms. Now Equations 8.13-15 and 8.13-19 can be written as

$$H(Y/X) = G(p) \tag{8.13-21}$$

and

$$C_{s2} = \log 2 - G(p). \tag{8.13-22}$$

If base 2 logarithms are used, this last expression can be plotted as a function of p as shown in Figure 8.16. As expected, the channel capacity is large if p, the probability of correct transmission, is either close to unity or to zero. If $p = \frac{1}{2}$, there is no statistical evidence as to which symbol was sent, and the channel capacity is zero.

Let us consider the BSC in somewhat more detail. Assume that the source symbols are $x_1 = 0$ and $x_2 = 1$ with a-priori probabilities $p(0) = w$ and $p(1) = 1 - w$. The complete system is then shown in Figure 8.17, where the channel matrix corresponds to Equation 8.13-18. The conditional entropy $H(Y/X)$ has already been given by Equation 8.13-21. The destination entropy $H(Y)$ can be found from

$$H(Y) = -\sum_{j=1}^{2} p(y_j) \log p(y_j) \tag{8.13-23}$$

where $p(y_j)$ is given by

$$p(y_j) = \sum_{i=1}^{2} p(x_i) p(y_j/x_i) \tag{8.13-24}$$

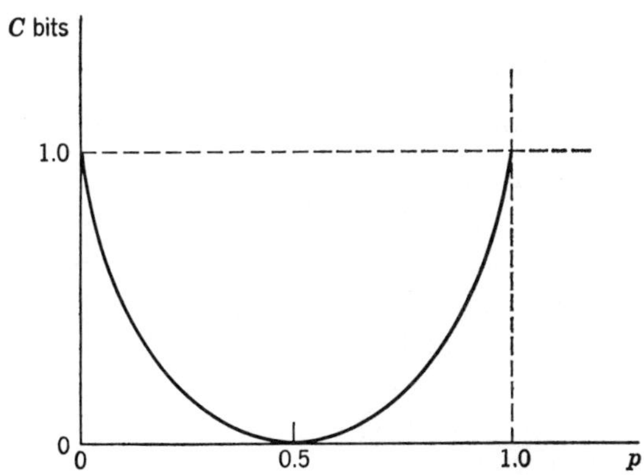

Figure 8.16 *The capacity of the binary symmetric channel*

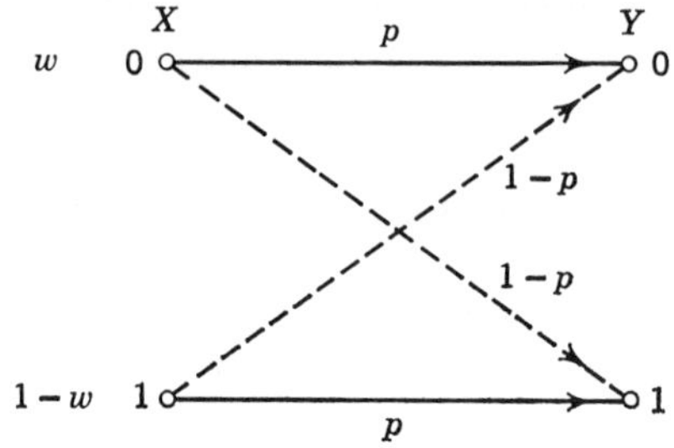

Figure 8.17 *The BSC with source*

or

$$p(y_1) = wp + (1 - w)(1 - p), \tag{8.13-25}$$

$$p(y_2) = w(1 - p) + (1 - w)p. \tag{8.13-26}$$

Now Equation 8.13-1 may be used to write the information transfer as

$$I(X; Y) = G[wp + (1 - w)(1 - p)] - G(p) \tag{8.13-27}$$

where $G(p)$ has already been defined by Equation 8.13-20.

Example 8.10 Let a BSC and source be described by

$$\boldsymbol{D} = \begin{bmatrix} \frac{2}{3} & \frac{1}{3} \\ \frac{1}{3} & \frac{2}{3} \end{bmatrix},$$

$$p(x_1) = w = \tfrac{3}{4}, \qquad p(x_2) = 1 - w = \tfrac{1}{4}.$$

The function $G(p)$ of Equation 8.13-20 is

$$G(\tfrac{2}{3}) = \tfrac{2}{3}\log_2 \tfrac{3}{2} + \tfrac{1}{3}\log_2 3 = 0.918 \text{ bits/symbol}$$

so that the channel capacity is

$$C = \log_2 2 - 0.918 = 0.082 \text{ bits/symbol.}$$

The information transferred is given by Equation 8.13-27:

$$I(X; Y) = G(\tfrac{7}{12}) - G(\tfrac{2}{3}) = 0.979 - 0.918 = 0.061 \text{ bits/symbol.}$$

Thus the channel capacity is somewhat larger than the information transfer in this case. The reason, of course, is that the input distribution $p(x_i)$ was not chosen to maximize $I(X; Y)$. It is fairly obvious from Equation 8.13-27 that the optimal choice is $p(x_1) = p(x_2) = \frac{1}{2}$ not only in this particular example but for the BSC in general.

● ●

If the channel is not symmetric, then the channel capacity is not so easily found. Indeed, all of the calculations become more complicated as the following example indicates.

Example 8.11 Let a binary channel and source be described by

$$\boldsymbol{D} = \begin{bmatrix} \frac{2}{3} & \frac{1}{3} \\ \frac{1}{10} & \frac{9}{10} \end{bmatrix},$$

$$p(x_1) = \tfrac{3}{4}, \qquad p(x_2) = \tfrac{1}{4}.$$

The probabilities $p(y_j)$ may be calculated from Equation 8.13-24 to yield

$$p(y_1) = \tfrac{3}{4}\cdot\tfrac{2}{3} + \tfrac{1}{4}\cdot\tfrac{1}{10} = \tfrac{21}{40},$$

$$p(y_2) = \tfrac{3}{4}\cdot\tfrac{1}{3} + \tfrac{1}{4}\cdot\tfrac{9}{10} = \tfrac{19}{40}.$$

The conditional probabilities $p(x_i/y_j)$ can now be found from

$$p(x_i/y_j) = \frac{p(x_i)p(y_j/x_i)}{p(y_j)},$$

$$p(x_1/y_1) = \tfrac{20}{21}, \qquad p(x_2/y_1) = \tfrac{1}{21},$$

$$p(x_1/y_2) = \tfrac{10}{19}, \qquad p(x_2/y_2) = \tfrac{9}{19}.$$

Now the various entropies can be calculated. Note first that the source entropy $H(X)$ is

$$H(X) = \tfrac{3}{4}\log_2 \tfrac{4}{3} + \tfrac{1}{4}\log_2 4 = 0.811 \text{ bits/symbol}.$$

In the same way, we find

$$H(X/y_1) = -\sum_{i=1}^{2} p(x_i/y_1)\log p(x_i/y_1),$$

$$= \tfrac{20}{21}\log_2 \tfrac{21}{20} + \tfrac{1}{21}\log_2 21 = 0.276 \text{ bits/symbol},$$

$$H(X/y_2) = \tfrac{10}{19}\log_2 \tfrac{19}{10} + \tfrac{9}{19}\log_2 \tfrac{19}{9} = 0.998 \text{ bits/symbol},$$

$$H(X/Y) = \sum_{j=1}^{2} p(y_i)H(X/y_j) = 0.619 \text{ bits/symbol}.$$

Thus the information transfer $I(X;\, Y)$ is given by

$$I(X;\, Y) = H(X) - H(X/Y) = 0.811 - 0.619 = 0.192 \text{ bits/symbol}.$$

However, no simple relationship exists for channel capacity.

• •

The capacity C of a channel has been defined in terms of information transfer. An equally valid way to evaluate a channel is to determine how often it transmits a message incorrectly. For the BSC of Figure 8.17, it is clear that the probability of incorrect transmission is $1 - p$ and of correct transmission is p. For the general case, it may be considerably more difficult to determine these probabilities. Also it seems reasonable that there should be a relationship between the probability of correct transmission and channel capacity; a channel with high probability of correct transmission should tend to have a high capacity. These and related topics will be discussed in the next section.

8.14 Error Probability, Decision Rules, and the Fano Bound

Let us refer to Figure 8.13 which is a schematic representation of a lossless channel. In such a channel it is very clear as to which source symbol x_i has been sent when a given destination symbol y_j is received. The Y set has been partitioned into subsets B_i and each of these subsets is uniquely associated with a given x_i. In the general case of a lossy channel, a similar partitioning of the Y set is necessary in order to decide which x_i was sent when a given y_j is received. More specifically, a *decision scheme* or *decoding scheme* $\mathscr{B}$ is a partitioning of the Y set into M disjoint and exhaustive sets $B_1, B_2, \ldots, B_M$ such that, when a destination symbol y_k falls into set B_i, it is decided that symbol x_i was sent. Implicit in this definition is a *decision rule* $d(y_j)$, which is a function specifying uniquely a source symbol for each destination symbol. Thus, for Example 8.11, a reasonable decision scheme is obviously (?!)

$$B_1 = \{y_1\}, \qquad B_2 = \{y_2\}$$

and the corresponding decision rule is

$$d(y_1) = x_1, \qquad d(y_2) = x_2.$$

At this point, it is instructive to examine the error probability associated with a given decision scheme. Let $p(e/y_j)$ be the probability of error when it is decided that y_j has been received. Then the *total error probability* $p(e)$ is

$$p(e) = \sum_{j=1}^{N} p(y_j)p(e/y_j). \tag{8.14-1}$$

For a given decision scheme $\mathscr{B}$, the conditional error probability $p(e/y_j)$ can be written as

$$p(e/y_j) = 1 - p[d(y_j)/y_j] \tag{8.14-2}$$

where $p[d(y_j)/y_j]$ is the conditional probability $p(x_i/y_j)$ with x_i assigned by the decision rule; that is, for a given decision scheme, $d(y_j) = x_i$. The probability $p(y_j)$ is determined only by the source a-priori probability $p(x_i)$ and by the channel matrix $\boldsymbol{D} = [p(y_j/x_i)]$. Hence, only the term $p(e/y_j)$ in Equation 8.14-1 is a function of the decision scheme. Since Equation 8.14-1 is a sum of nonnegative terms, the error probability is a minimum when each summand is a minimum. Thus the term $p(e/y_j)$ should be a minimum for each y_j. It follows from Equation 8.14-2 that the minimum error scheme is that scheme which assigns a decision rule

$$d(y_j) = x^*, \qquad j = 1, 2, \ldots, N, \tag{8.14-3}$$

where x^* is defined by

$$p(x^*/y_j) \geq p(x_i/y_j), \qquad i = 1, 2, \ldots, M. \tag{8.14-4}$$

In other words, each y_j is decoded as the *a-posteriori most likely* x_i. This scheme, which minimizes the probability of error $p(e)$, is usually called the *ideal observer*.

Example 8.12 Consider the binary channel and source of Example 8.11. The conditional probabilities $p(x_i/y_j)$ have already been calculated and are

$$p(x_1/y_1) = \tfrac{20}{21}, \qquad p(x_2/y_1) = \tfrac{1}{21},$$

$$p(x_1/y_2) = \tfrac{10}{19}, \qquad p(x_2/y_2) = \tfrac{9}{19}.$$

The ideal observer decision scheme is

$$B_1 = \{y_1, y_2\}, \qquad B_2 = \{\varnothing\}$$

where $\varnothing$ is the rule set, with decision rule

$$d(y_1) = x_1, \qquad d(y_2) = x_1!!.$$

The surprising result is that x_1 should be considered sent regardless of whether y_1 or y_2 is received. The error probability is

$$p(e) = \sum_{j=1}^{2} p(y_j)\{1 - p[d(y_j)/y_j]\}$$

or

$$p(e) = \tfrac{21}{40}(1 - \tfrac{20}{21}) + \tfrac{19}{40}(1 - \tfrac{10}{19}) = \tfrac{1}{4}.$$

Suppose that the "reasonable" decision rule

$$d(y_1) = x_1, \qquad d(y_2) = x_2$$

is used instead of the ideal observer. Now the error probability becomes

$$p(e) = \tfrac{21}{40}(1 - \tfrac{20}{21}) + \tfrac{19}{40}(1 - \tfrac{9}{19}) = \tfrac{11}{40}.$$

● ●

This example has indicated that the ideal observer is not always a completely satisfactory decision scheme. It suffers from two major disadvantages:

1. For a given channel $\boldsymbol{D}$, the scheme is defined only for a given input distribution $p(x_i)$. It would be preferable to have a scheme that was insensitive to input distributions.

2. The scheme minimizes average error but does not bound certain errors. For example, as has just been demonstrated, some symbols may always be received incorrectly.

Despite these disadvantages, the ideal observer is a straightforward scheme which does minimize average error. It is also widely used as a standard with which other decision schemes may be compared.

Consider the special case in which the input distribution $p(x_i)$ is uniformly distributed so that all x_i are equally likely:

$$p(x_i) = \frac{1}{M}, \qquad i = 1, 2, \ldots, M. \tag{8.14-5}$$

Now the conditional likelihood $p(x_i/y_j)$ is

$$p(x_i/y_j) = \frac{p(x_i)p(y_j/x_i)}{p(y_j)} = \frac{p(y_j/x_i)}{Mp(y_j)}. \tag{8.14-6}$$

For a given y_j, that input x_i is chosen which makes $p(y_j/x_i)$ a maximum, and the decision rule is

$$d(y_j) = x^{\dagger}, \qquad j = 1, 2, \ldots, N, \tag{8.14-7}$$

where $x^{\dagger}$ is defined by

$$p(y_j/x^{\dagger}) \geq p(y_j/x_i), \qquad i = 1, 2, \ldots, M. \tag{8.14-8}$$

The probability of error becomes

$$p(e) = \sum_{j=1}^{N} p(y_j)\left[1 - \frac{p(y_j/x^{\dagger})}{Mp(y_j)}\right]. \tag{8.14-9}$$

This decoder is sometimes called the *maximum likelihood* decoder or decision scheme.

Example 8.13 Let a channel and source be described by the channel matrix

$$\boldsymbol{D} = \begin{bmatrix} 1 & 0 & 0 \\ 1 & 0 & 0 \\ 0 & \frac{1}{2} & \frac{1}{2} \end{bmatrix} = [p(y_j/x_i)]$$

and the input probabilities

$$p(x_1) = \tfrac{1}{2}, \qquad p(x_2) = \tfrac{1}{4}, \qquad p(x_3) = \tfrac{1}{4}.$$

The various probabilities are easily calculated to be

$$p(y_1) = \tfrac{3}{4}, \qquad p(y_2) = \tfrac{1}{8}, \qquad p(y_3) = \tfrac{1}{8},$$

and

$$p(x_1/y_1) = \tfrac{2}{3}, \qquad p(x_2/y_1) = \tfrac{1}{3}, \qquad p(x_3/y_1) = 0,$$
$$p(x_1/y_2) = 0, \qquad p(x_2/y_2) = 0, \qquad p(x_3/y_2) = 1,$$
$$p(x_1/y_3) = 0, \qquad p(x_2/y_3) = 0, \qquad p(x_3/y_3) = 1.$$

Now the ideal observer decision scheme is

$$B_1 = \{y_1\}, \qquad B_2 = \{\varnothing\}, \qquad B_3 = \{y_2, y_3\},$$

while the associated decision rule is

$$d(y_1) = x_1, \qquad d(y_2) = x_3, \qquad d(y_3) = x_3.$$

The possibility of error of the scheme is

$$\begin{aligned} p(e) &= \sum_{j=1}^{3} p(y_j)\{1 - p[d(y_j)/y_j]\} \\ &= \tfrac{3}{4}(1 - \tfrac{2}{3}) + \tfrac{1}{8}(1 - 1) + \tfrac{1}{8}(1 - 1) \\ &= \tfrac{1}{4}. \end{aligned}$$

Note that this error is just the a-priori probability that x_2 is sent since x_1 and x_3 are always received correctly.

Suppose now that the a-priori probabilities are equal so that

$$p(x_1) = p(x_2) = p(x_3) = \tfrac{1}{3}.$$

In this case

$$p(y_1) = \tfrac{2}{3}, \qquad p(y_2) = \tfrac{1}{6}, \qquad p(y_3) = \tfrac{1}{6}$$

and

$$p(x_1/y_1) = \tfrac{1}{2}, \qquad p(x_2/y_1) = \tfrac{1}{2}, \qquad p(x_3/y_1) = 0,$$
$$p(x_1/y_2) = 0, \qquad p(x_2/y_2) = 0, \qquad p(x_3/y_2) = 1,$$
$$p(x_1/y_3) = 0, \qquad p(x_2/y_3) = 0, \qquad p(x_3/y_3) = 1.$$

Now the ideal observer is also the maximum likelihood decoder and is given by

$$B_1 = \{y_1\}, \qquad B_2 = \{\varnothing\}, \qquad B_3 = \{y_2, y_3\}.$$

When y_1 is received, there is equal probability that x_1 or x_2 was sent; either choice may be made and the error in either case is

$$p(e) = \tfrac{2}{3}(1 - \tfrac{1}{2}) + 0 + 0 = \tfrac{1}{3}$$

which is the a-priori probability of either x_1 or x_2.

• •

The error probability $p(e)$ has been given in terms of the decision rule $d(y_j)$ by Equations 8.14-1 and 8.14-2. As mentioned previously, it would appear that a relationship should exist between this probability and channel capacity. As a matter of convenience in deriving such a relationship, let us reintroduce the notation used previously:

$$d(y_j) = x^*, \tag{8.14-10}$$

$$p[d(y_j)/y_j] = p(x^*/y_j). \tag{8.14-11}$$

Thus x^* is the source element that the decision rule specifies as sent when y_j is received, and $p(x^*/y_j)$ is the a-posteriori probability of that element. The source alphabet X consists of the element x^* and the remaining elements $X - x^*$. With this notation the error probability can be written as

$$p(e) = \sum_{j=1}^{N} p(y_j)[1 - p(x^*/y_j)]. \tag{8.14-12}$$

A convenient place to start the derivation is to consider the probability $p_1(x_i)$ defined by

$$p_1(x_i) = \frac{p(x_i/y_j)}{1 - p(x^*/y_j)}, \qquad x_i \in X - x^*, \tag{8.14-13}$$

with the two properties $p_1(x_i) \geq 0$ and

$$\sum_{X - x^*} p_1(x_i) = 1. \tag{8.14-14}$$

Thus $p_1(x_i)$ is a probability density defined for $M - 1$ terms and its associated entropy satisfies the inequality

$$-\sum_{X - x^*} p_1(x_i) \log p_1(x_i) \leq \log (M - 1) \tag{8.14-15}$$

with equality if, and only if, the $(M - 1)$ terms are all equally likely On multiplying both sides of this equation by $p(y_j)$, summing over Y, and rearranging, we obtain

$$-\sum_{Y} p(y_j) \sum_{X - x^*} p(x_i/y_j) \log p_1(x_i) \leq p(e) \log (M - 1) \tag{8.14-16}$$

where Equation 8.14-12 has been used for $p(e)$. The conditional entropy $H(X/Y)$ may be written as

$$H(X/Y) = -\sum_{Y} p(y_j)\left[\sum_{X - x^*} p(x_i/y_j) \log p(x_i/y_j) + p(x^*/y_j) \log p(x^*/y_j)\right]. \tag{8.14-17}$$

After some rearrangement and the use of Equations 8.14-13 and 8.14-16, this expression becomes

$$H(X/Y) \le p(e)\log(M-1) - \sum_{Y} p(y_j) \times \left\{ p(x^*/y_j)\log p(x^*/y_j) + \sum_{X-x^*} p(x_i/y_j)\log[1 - p(x^*/y_j)] \right\}. \tag{8.14-18}$$

It is now necessary to identify the last two terms on the right side of this last equation. For convenience, define the auxiliary alphabet $Z = (z_1, z_2)$ with joint probabilities $p(z_1, y)$ and $p(z_2, y)$ given by

$$p(z_1, y_j) = p(x^*, y_j), \tag{8.14-19}$$

$$p(z_2, y_j) = p(X - x^*, y_j) \tag{8.14-20}$$

Thus z_1 is formally associated with the element x^* and z_2 with the remaining elements $X - x^*$ of the source alphabet It is clear that

$$p(z_1) = \sum_{Y} p(x^*, y_j) = \sum_{Y} p(y_j)p(x^*/y_j) = 1 - p(e) \tag{8.14-21}$$

and

$$p(z_2) = 1 - p(z_1) = p(e). \tag{8.14-22}$$

Now the entropy of the Z alphabet is

$$\begin{aligned} H(Z) &= -\sum_{k=1}^{2} p(z_k)\log p(z_k) \\ &= -p(e)\log p(e) - [1 - p(e)]\log[1 - p(e)] \\ &= G[p(e)] \end{aligned} \tag{8.14-23}$$

where the function $G(\cdot)$ has been defined previously by Equation 8.13-20. In the same way, the conditional entropy $H(Z/Y)$ is

$$H(Z/Y) = -\sum_{Y} p(y_j) \sum_{k=1}^{2} p(z_k/y_j)\log p(z_k/y_j) \tag{8.14-24}$$

or

$$H(Z/Y) = -\sum_{Y} p(y_j)\left\{ p(x^*/y_j)\log p(x^*/y_j) + \sum_{X-x^*} p(x_i/y_j)\log[1 - p(x^*/y_j)] \right\}. \tag{8.14-25}$$

It has been pointed out previously in Section 8.6, as given by Equation 8.6-10, that any conditional entropy $H(Z/Y)$ satisfies the inequality

$$H(Z/Y) \leq H(Z). \tag{8.14-26}$$

Note now that Equation 8.14-18 can be written as

$$H(X/Y) \leq p(e) \log (M - 1) + H(Z/Y) \tag{8.14-27}$$

or, from Equations 8.14-23 and 8.14-26,

$$\boxed{H(X/Y) \leq G[p(e)] + p(e) \log (M - 1).} \tag{8.14-28}$$

This expression is known as the *Fano Bound*; it relates the error probability $p(e)$ to the channel capacity through Equations 8.13-1 and 8.13-2. The three terms can be interpreted as follows:

1. $H(X/Y)$ is the equivocation. As discussed at the end of Section 8.6, it is the average additional information needed at the destination after reception to completely determine the symbol that was sent.
2. $G[p(e)]$ is the entropy of the binary system with probabilities $p(e)$ and $[1 - p(e)]$. In other words it is the average amount of information needed to determine if the decision rule resulted in an error.
3. Log $(M - 1)$ is the maximum amount of information needed to determine which among the remaining $M - 1$ symbols was sent if the decision rule was incorrect; this information is needed with probability $p(e)$.

As pointed out after Equation 8.14-15, the equality occurs in Equation 8.14-27 if, and only if, the $M - 1$ source symbols that were not selected are equally likely.

8.15 The Noisy Coding Theorem

In Sections 8.13 and 8.14, the concept of channel capacity was defined and some of the properties of discrete, memoryless channels were discussed. It will now be shown that capacity is a fundamental property of an information channel in the sense that it is possible to transmit information through the channel at any rate less than the channel capacity with arbitrarily small probability of error. This result, which will be stated more precisely later, is called the *noisy coding theorem* or *Shannon's fundamental theorem for a noisy channel.*

A proof of this theorem for an arbitrary channel is difficult and quite beyond the level of this treatment. However, a heuristic (i.e., nonrigorous) discussion of the BSC is relatively straightforward and will afford considerable insight to the general problem. Let us begin by stating carefully the assumptions that will be made in order that minimum confusion will exist as to what has been proved.

The Information Source

Let the source consist of a set $U = \{u_1, u_2, \ldots, u_m\}$ of m messages u_k. It will be assumed that these are *independent* and *equally likely*. This last assumption is a strong one, and it may be necessary to form this message set by encoding an original source alphabet $X = \{x_1, x_2, \ldots, x_M\}$ into long sequences; that is, each sent message u_k may consist of a sequence of x_i's so that

$$u_k = x_{i_1} x_{i_2} \cdots x_{i_p}$$

where the length p of the message u_k may be large. Now the source entropy $H(U)$ is given by

$$H(U) = \log m. \tag{8.15-1}$$

Encoding the Message Set U

Each message u_k will be encoded into a code word consisting of a sequence of n binary digits for transmission over the binary channel. The number of possible code words of length n is 2^n; the number of messages u_k is m; thus the inequality

$$2^n \geq m$$

or

$$n \geq \log_2 m \tag{8.15-2}$$

must be satisfied. The code words will be assigned at *random*, that is, the probability that a given n-place code word (of the 2^n possible code words) will be chosen to represent one of the set of m messages is $m/2^n$.

The Channel

It will be assumed that the channel is the BSC of Figure 8.17 with probability p of correct transmission of a binary digit and probability $q = 1 - p$ of incorrect transmission.

The Decision Scheme or Decoder

At the destination, the received message (there are 2^n possible) will be compared with all m of the messages that could have been sent. The message u_k will be considered sent which differs from the received message in the least number of binary digits. As pointed out in problem 16, this decoder is the ideal observer (actually the maximum likelihood decoder since, in this particular case, the source probabilities are equally likely).

Consider a given code word of length n. The probability that exactly r digits will be altered in transmission (while $n - r$ digits are not altered) is just the probability of $n - r$ successes and r failures in n Bernoulli trials. This is the binomial distribution of Section 2.16 and is given by

$$P(r \text{ errors}) = \binom{n}{n-r} p^{n-r} q^r, \qquad r = 0, 1, 2, \ldots, n. \tag{8.15-3}$$

The mean number of errors is

$$E(r) = nq = n(1 - p). \tag{8.15-4}$$

Thus the average number of binary digits altered by noise in the BSC will be nq.

There are $\binom{n}{r}$ code words (out of the total of 2^n) that differ *in exactly r digits* from a given code word. Therefore the number K of code words that differ *in nq digits or less* from a given code word is given by the sum

$$K = \sum_{r=0}^{nq} \binom{n}{r} = \binom{n}{0} + \binom{n}{1} + \cdots + \binom{n}{nq-1} + \binom{n}{nq}. \tag{8.15-5}$$

For $q < \frac{1}{2}$, each term in this series for K is larger than the preceding term so that K is bounded by

$$K \leq (nq + 1)\binom{n}{nq} = (nq + 1)\frac{n!}{(nq)!(n - nq)!}. \tag{8.15-6}$$

For large n, the factorial is given approximately by Stirling's formula

$$n! \approx \sqrt{2\pi}\, n^{n+1/2}\, e^{-n}. \tag{8.15-7}$$

With this substitution, Equation 8.15-6 can be written as

$$K \leq \sqrt{\frac{(nq + 1)^2}{2\pi npq}}\, q^{-nq} p^{-np}. \tag{8.15-8}$$

Thus, on the average, a given sent code word can be received as any one of K code words differing in nq digits or less from the sent word. An error will result if any message u_k, other than the one originally transmitted, is represented by one of this group of K code words.

Recall that the code words were assigned at random from the set of 2^n possible code words. Therefore the expected number of messages that could be changed by errors in the BSC into the original message and hence be confused with this original message is

$$L = K\frac{m}{2^n}. \tag{8.15-9}$$

It follows from Equation 8.15-8 that L is bounded by

$$L \leq m\sqrt{\frac{(nq + 1)^2}{2\pi npq}}\, 2^{-nC_{s2}} \tag{8.15-10}$$

where C_{s2} is the capacity of the BSC and is given by Equation 8.13-19; that is,

$$C_{s2} = 1 + p \log_2 p + q \log_2 q. \tag{8.15-11}$$

For finite m and C_{s2}, the limit of L for large n is zero; that is,

$$\lim_{n\to\infty} L = 0. \tag{8.15-12}$$

Thus, for sufficiently long code words, the probability that a message will be decoded incorrectly approaches zero. Now let the number of messages m be fixed by the relation

$$m = \frac{2^{nC_{s2}}}{n}. \tag{8.15-13}$$

Then Equation 8.15-10 becomes

$$L \leq \sqrt{\frac{(nq+1)^2}{2\pi n^3 pq}} \tag{8.15-14}$$

and, as before,

$$\lim_{n\to\infty} L = 0. \tag{8.15-15}$$

The channel input entropy per message is given by Equation 8.15-1. This entropy per code digit is

$$H'(U) = \frac{\log_2 m}{n} \text{ bits/digit} \tag{8.15-16}$$

or, from Equation 8.15-13,

$$H'(U) = C_{s2} - \frac{\log n}{n} \tag{8.15-17}$$

and

$$\lim_{n\to\infty} H'(U) = C_{s2} \text{ bits/digit}. \tag{8.15-18}$$

Thus, for sufficiently long message sequences, information can be conveyed in the BSC at a rate approaching channel capacity and with arbitrarily small probability of error.

The noisy coding theorem can be stated more precisely as follows: "Consider a discrete memoryless channel with nonzero capacity C; fix two numbers H and ϵ such that

$$0 < H < C \tag{8.15-19}$$

and

$$\epsilon > 0. \tag{8.15-20}$$

Let us transmit m messages $u_1, u_2, \ldots, u_m$ by code words each of length n binary digits. The positive integer n can be chosen so that

$$m \geq 2^{nH}. \tag{8.15-21}$$

In addition, at the destination the m sent messages can be associated with a set $V = \{v_1, v_2, \ldots, v_m\}$ of received messages and with a decision rule $d(v_j) = u_j$ such that

$$p[d(v_j)/v_j] \geq 1 - \epsilon, \qquad (8.15\text{-}22)$$

that is, decoding can be accomplished with a probability of error that does not exceed ϵ." The foregoing discussion of this theorem for the BSC is not a formal proof, of course, but is merely a plausibility argument. However, a rigorous proof is quite involved and somewhat difficult. The interested reader should refer to Ash [17], Feinstein [19], Wolfowitz [20], or Gallagher [21]. There is a converse to the noisy coding theorem [17,20] which states that it is not possible to produce an encoding procedure that allows transmission at a rate greater than channel capacity with arbitrarily small error.

8.16 Error-Correcting Codes

The codes considered in Section 8.12 were designed for minimum length in the case of noiseless transmission. For noisy channels, the coding theorem just discussed guarantees the existence of a code that will allow transmission at any rate less than channel capacity and with arbitrarily small probability of error. However, the theorem does not provide a constructive procedure to devise such codes. Indeed, it implies that very long sequences of source symbols may have to be considered if reliable transmission at rates near channel capacity are to be obtained. In this section, we consider some of the elementary properties of simple *error-correcting codes*; that is, codes that can be used to increase reliability in the transmission of information through noisy channels by correcting at least some of the errors that occur so that overall probability of error is reduced.

The discussion will be restricted to the BSC and the notation of the previous section will be used. Thus, a source alphabet $X = \{x_1, x_2, \ldots, x_M\}$ of M symbols will be used to form a message set U of m messages u_k where $U = \{u_1, u_2, \ldots, u_m\}$. Each u_k will consist of a sequence of the x_i's. Each message u_k will be encoded into a sequence of n binary digits for transmission over the BSC of Figure 8.17. At the destination, there exists a set $V = \{v_1, v_2, \ldots, v_{2^n}\}$ of all possible binary sequences of length n. The inequality $m \leq 2^n$ of Equation 8.15-2 must hold. The problem is to associate with each sent message u_k a received message v_j so that $p(e)$, the overall probability of error, is reduced.

In the discussion of the noisy coding theorem, a decoding scheme was used that examined the received message v_j and identified it with the sent message u_k which differed from it in the least number of binary digits. In all of the discussions of this section, it will be assumed that this decoder is used. Let us define the *distance*, or *Hamming distance*, $d(v_j, v_k)$ between two binary sequences v_j and v_k of length n as that integer which is the number of digits in which v_j and v_k disagree. Thus, if the distance between two sequences is zero, the two sequences are identical; if, on the other hand,

$$v_j = 011010, \qquad v_k = 100011,$$

then the distance $d(v_j, v_k)$ is four. It is easily seen that this distance measure has the following elementary properties:

$$(1)\quad d(v_j, v_k) \geq 0 \text{ with equality if, and only if, } v_j = v_k. \tag{8.16-1}$$

$$(2)\quad d(v_j, v_k) = d(v_k, v_j). \tag{8.16-2}$$

$$(3)\quad d(v_j, v_l) \leq d(v_j, v_k) + d(v_k, v_l). \tag{8.16-3}$$

$$(4)\quad d(v_j, v_k) \leq n. \tag{8.16-4}$$

The decoder which we use is a *minimum-distance* decoder. As mentioned in the previous section and shown in problem 16, the ideal observer decoding scheme is a minimum distance scheme for the BSC.

It is intuitively apparent that the sent messages should be represented by code words that all have the greatest possible distances between them. Let us investigate this matter in more detail by considering all binary sequences of length $n = 3$; there are $2^n = 2^3 = 8$ such sequences, namely

000	001	011	111
	010	110	
	100	101	

It is convenient to represent these as the eight corners of a unit cube as shown in Figure 8.18(a) where the x axis corresponds to the first digit, the y axis to

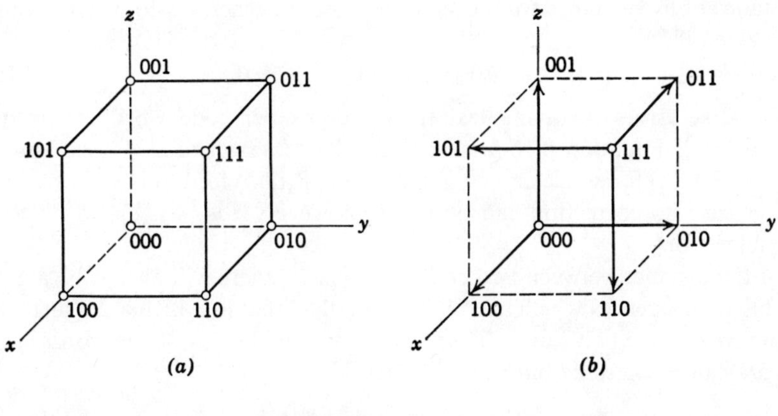

Figure 8.18 *The representation of binary sequences as the corners of an n cube, n = 3: (a) the eight binary sequences of length 3; (b) the shift in sequences 000 and 111 from a single error*

the second, and the z axis to the third. Although direct pictorial representation is not possible, it is clear that binary sequences of length n greater than three can be considered as the corners of the corresponding n cube.

Suppose that all eight binary sequences are used as code words to encode a source. If any binary digit is changed in transmission, an error will result at the destination since the sent message will be interpreted incorrectly as one of the three possible messages that differ in one code digit from the sent message. This situation is illustrated in Figure 8.18(b) for the code words 000 and 111. A change of one digit in each of these code words produces one of three possible other code words as shown.

Figure 8.18(b) suggests that only two code words, say 000 and 111, should be used. The distance between these two words, or any other two words on opposite corners of the cube, is three. If only one digit is changed in the transmission of each of these two code words, then they can be correctly distinguished at the destination by a minimum-distance decoder. If two digits are changed in each word in transmission, then it will not be possible to make this distinction.

It is clear that this reasoning can be extended to sequences containing more than three binary digits. For any $n \geq 3$, single errors in each code word can be corrected. If double errors are to be corrected without fail, it is necessary that there exist at least two code words with a distance between them of at least five; thus, for this case, binary code words of length five or greater must be used.

In the light of the previous discussion, it can be seen that the error-correcting properties of a code depend on the distance $d(v_j, v_k)$ between code words. Specifically, single errors can be corrected if all code words employed are at least a distance of three apart, double errors if the words are at a distance of five or more from each other, and, in general, q-fold errors can be corrected if

$$d(v_j, v_k) \geq 2q + 1, \qquad j \neq k. \tag{8.16-5}$$

Of course errors involving less than q digits per code word can also be corrected if Equation 8.16-5 is satisfied. The case where $n = 3$ is shown in Figure 8.19(a) for single errors and in Figure 8.19(b) for double errors. In the latter case no correction can be made; however, it is possible to detect all double errors.

If the distance between two code words is $2q$, there will always be a group of binary sequences which are "in the middle," that is, which are a distance q from *each* of the two words. Thus, by the proper choice of code words, q-fold errors can be *detected* but not corrected if

$$d(v_j, v_k) = 2q, \qquad j \neq k. \tag{8.16-6}$$

This situation is illustrated in Figure 8.20 for binary sequences of length four. Single errors can be corrected but double errors can be detected only.

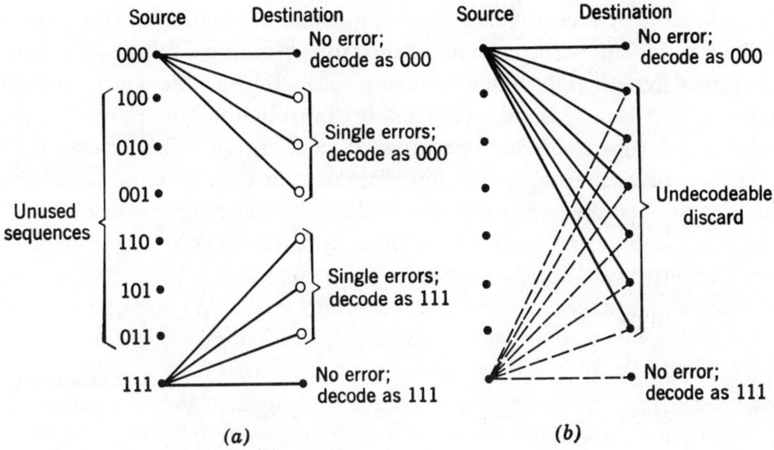

Figure 8.19 Error detection and correction for binary sequences of length 3: (a) error corrections for single errors, $d = 3 \geq 2q + 1 = 3$; (b) failure to correct when double errors can occur, $d = 3 < 2q + 1 = 5$

Let us now consider the maximum number of code words r that can be selected from the set of 2^n possible binary sequences of length n to form a code that will correct all single, double, . . . , q-fold errors. In both of the examples of Figures 8.19 and 8.20, the number of code words selected was two. In fact (see problem 17) there is no single-error correcting code for $n = 3, 4$ containing more than two words. Suppose we consider a given code consisting of the words $\ldots, u_k, u_j, \ldots$. All binary sequences of distance q or less from u_k must "belong" to u_k, and to u_k only, if q-fold errors are to be corrected. Thus, associated with u_k, are all binary sequences of distance $0, 1, 2, \ldots, q$ from u_k. The number of such sequences has been given previously by Equation 8.15-5 and is

$$\binom{n}{0} + \binom{n}{1} + \binom{n}{2} + \cdots + \binom{n}{q} = \sum_{i=0}^{q} \binom{n}{i}. \tag{8.16-7}$$

Since there are r of the code words, the total number of sequences associated with all the code words is $r \sum_{i=0}^{q} \binom{n}{i}$. This number can be no larger than 2^n, the total number of distinct binary sequences of length n. Therefore the following inequality must hold:

$$r \sum_{i=0}^{q} \binom{n}{i} \leq 2^n$$

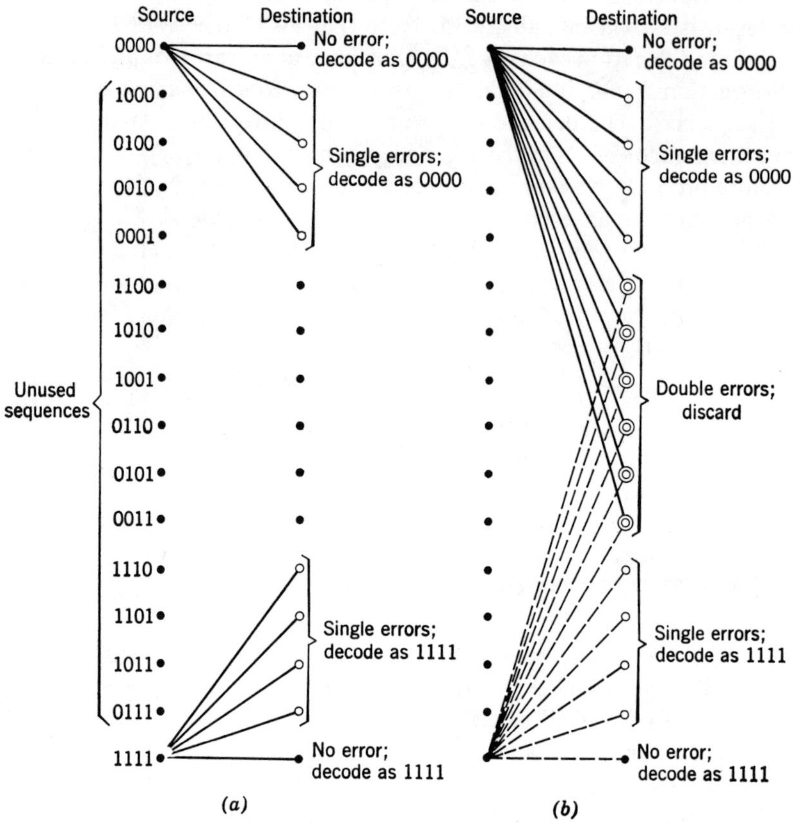

Figure 8.20 *Error detection and correction for binary sequences of length 4: (a) error correction for single errors, $d = 4 \geq 2q + 1 = 3$; (b) error correction for single errors and error detection for double errors, $d = 4 \geq 2q = 4$*

or

$$r \leq \frac{2^n}{\sum_{i=0}^{q} \binom{n}{i}}. \tag{8.16-8}$$

This is a *necessary* upper bound on the number of code words that can be used to correct all errors up to and including q-fold errors. As implied by problem 17, it is not sufficient.

Let us refer to Figure 8.19 which lists the eight possible binary sequences of length three. Suppose we add one binary digit to each sequence in such a way that the total number of 1's in the sequence is *even* (or odd, if you wish). The result is shown in Table 8.15. Note that all of the word sequences of length four differ from each other by a distance of at least two. In accordance with Equation 8.16-6, it should be possible now to *detect single errors* in all eight sequences. The detection method is straightforward. At the receiver, count the number of 1's in the sequence; if the number is odd, a single error (or, more precisely, an odd number of errors) has occurred; if the number is even, no error (or an even number of errors) has occurred. This particular scheme is a good one if single errors only are likely to occur and if detection only (rather than correction) is desired. The extra digit that was added is called a *parity check digit* and the scheme is a very simple example of a *parity check code.*

Table 8.15. A Parity Check Code for Single-Error Detection

Message Digits	Check Digits	Word
000	0	0000
100	1	1001
010	1	0101
001	1	0011
110	0	1100
101	0	1010
011	0	0110
111	1	1111

More generally, in parity check codes, the encoded sequence consists of n binary digits of which only $k < n$ are *information digits* while the remaining $l = n - k$ digits are used for error detection and correction and are called *check digits* or *parity checks.* The example of Table 8.15 is a single-error detecting code but, in general, q-fold errors may be detected and/or corrected. As the number of errors to be detected and/or corrected increases, the number l of check digits must increase. Thus, for fixed word length n, the number of information digits $k = n - l$ will decrease as more and more errors are to be detected and/or corrected. Also the total number of words in the code cannot exceed the right side of Equation 8.16-8 or the number 2^k.

Parity check codes are relatively easy to implement. The simple example given of a single-error detecting code requires only that the number of 1's in each code word be counted. In this light, it is of considerable importance to note that these codes satisfy the noisy coding theorem of Section 8.15. In other words, it is possible to encode a source by parity check coding for transmission over a BSC at a rate approaching channel capacity and with arbitrarily small probability of error [22]. Then, from Equation 8.15-21, we have

$$2^{nH} = 2^k \tag{8.16-9}$$

or H, the rate of transmission, is given by

$$H = \frac{k}{n} \tag{8.16-10}$$

as $n \to \infty$, the probability of error $p(e)$ approaches zero. Thus, in a certain sense, it is sufficient to limit a study of error-correcting codes to the parity check codes. These codes are discussed extensively in several books [18, 22, 23] and a large periodical literature exists. Further discussion is beyond the scope of this treatment.

8.17 Continuous Channels

All of the preceding discussion of this chapter has concerned discrete message distributions and channels. Further, it has been assumed, either implicitly or explicitly, that the time parameter is discrete, that is, that a certain number of messages, symbols, code digits, and so on, are transmitted per unit time. Thus we have been concerned with *continuous-amplitude, discrete-time* channels and with messages that can be modeled as *discrete random processes* with *discrete parameter*. As discussed in Section 3.1, there are three other possibilities depending on whether the process amplitude and the time parameter have discrete or continuous distributions.

In this section, we consider the *continuous-amplitude, discrete-time* channel in which the input messages can be modeled as *continuous random processes* with *discrete parameter*. It will be shown later that continuous-time cases of engineering interest can be treated by techniques that amount to the replacement of the continuous parameter by a discrete parameter. The most straightforward method involves the application of the sampling theorem of Section 7.5 to bandlimited processes. In this case the process is sampled at equispaced intervals of length $1/2W$ seconds where W is the highest frequency of the process. Thus the continuous parameter t is replaced by the discrete parameter $t_k = k/2W$, $k = \ldots, -1, 0, 1, \ldots$.

Let us restrict our attention for the moment, however, to continuous-amplitude, discrete-time situations. The discrete density $p(x_i)$, $i = 1, 2, \ldots, M$, of the source message set is replaced by the continuous density $W_x(x)$ where, in general, $-\infty < x < \infty$, although the range of x may be restricted in particular cases. In the same way, other discrete densities are replaced by continuous densities. For example, the destination distribution $p(y_j)$, $j = 1, 2, \ldots, N$, becomes $W_y(y)$ and the joint distribution $p(x_i, y_j)$ will be $W_2(x, y)$.

A reasonable place to begin the study of continuous distribution is with the definition of source entropy as given by Equation 8.2-4 for the discrete case:

$$H(X) = -\sum_i p(x_i) \log p(x_i). \qquad (8.17\text{-}1)$$

Let us approximate the continuous density $W_x(x)$ by the relationship

$$p(x_i) \approx W_x(x_i)\,\Delta x_i \qquad (8.17\text{-}2)$$

where $p(x_i)$ is approximately the probability that the continuous random variable x with density $W_x(x)$ lies in an interval Δx_i which includes x_i. In the limiting case as $\Delta x_i \to 0$, this relationship will become exact and Equation 8.17-1 will become

$$H(X) = \lim_{\Delta x_i \to 0} \left\{ -\sum_i W_x(x_i)\,\Delta x_i \log [W_x(x_i)\,\Delta x_i] \right\}. \qquad (8.17\text{-}3)$$

The logarithm may be expanded to yield

$$\begin{aligned} H(X) = &- \lim_{\Delta x_i \to 0} \sum_i W_x(x_i) \log [W_x(x_i)]\,\Delta x_i \\ &- \lim_{\Delta x_i \to 0} \sum_i W_x(x_i)\,\Delta x_i \log \Delta x_i. \end{aligned} \qquad (8.17\text{-}4)$$

The first term of this expression may be considered as a limiting form of the integral $\int_{-\infty}^{\infty} W_x(x) \log W_x(x)\,dx$ while the second term tends to infinity; for example, if the intervals Δx_i are equal

$$\begin{aligned} -\lim_{\Delta x_i \to 0} \sum_i W_x(x_i)\,\Delta x \log \Delta x &= -\lim_{\Delta x \to 0} \log \Delta x \int_{-\infty}^{\infty} W_x(x)\,dx \\ &= -\lim_{\Delta x \to 0} \log \Delta x = \infty. \end{aligned} \qquad (8.17\text{-}5)$$

This rather nonrigorous approach suggests that the entropy of a continuous distribution $W_x(x)$ might be defined in analogy with the discrete case by the first term of Equation 8.17-4

$$\boxed{H(X) = -\int_{-\infty}^{\infty} W_x(x) \log W_x(x)\,dx.} \qquad (8.17\text{-}6)$$

This definition is not completely satisfactory for a number of reasons having to do with the properties of this new $H(X)$.

$H(X)$ May be Negative, Positive, or Zero: In the discrete case, it was shown in Equation 8.4-5 that $H(X)$ was nonnegative. This is no longer necessarily true. For example, let $W_x(x)$ be uniformly distributed in the interval $(0, 1/a)$. Then, by Equation 8.17-6, we have

$$H(X) = -\int_0^{1/a} a \log a \, dx = -\log a = \begin{cases} > 0, & a < 1, \\ = 0, & a = 1, \\ < 0, & a > 1. \end{cases} \tag{8.17-7}$$

H(X) Depends on the Coordinate System: In the interest of generality, let us consider the set of random variables $x_1, x_2, \ldots, x_n$ with joint distribution

$$W_x(x_1, x_2, \ldots, x_n) = W_x(\mathbf{x}). \tag{8.17-8}$$

Consider the entropy

$$H(\mathbf{X}) = -\int_{-\infty}^{\infty} \cdots \int_{-\infty}^{\infty} W_x(\mathbf{x}) \log W_x(\mathbf{x}) \, d\mathbf{x} \tag{8.17-9}$$

where $d\mathbf{x} = dx_1 dx_2 \cdots dx_n$. Let us see what happens when we transform to a new coordinate system (set of random variables) $y_1, y_2, \ldots, y_n$ with joint distribution

$$W_y(y_1, y_2, \ldots, y_n) = W_y(\mathbf{y}) \tag{8.17-10}$$

and entropy $H(\mathbf{Y})$ where

$$H(\mathbf{Y}) = -\int_{-\infty}^{\infty} \cdots \int_{-\infty}^{\infty} W_y(\mathbf{y}) \log W_y(\mathbf{y}) \, d\mathbf{y}. \tag{8.17-11}$$

It follows from Section 2.20 and Appendix C that $W_x(\mathbf{x})$ and $W_y(\mathbf{y})$ are related by

$$W_y(\mathbf{y}) = W_x(\mathbf{x}) |J(\mathbf{x}/\mathbf{y})| \tag{8.17-12}$$

where $J(\mathbf{x}/\mathbf{y})$ is the Jacobian of the transformation from $\mathbf{x}$ to $\mathbf{y}$ and is given by Equation C.2-9. This last expression together with

$$d\mathbf{y} = J(\mathbf{y}/\mathbf{x}) \, d\mathbf{x} \tag{8.17-13}$$

may be substituted into Equation 8.17-11 to yield

$$H(\mathbf{Y}) = -\int_{-\infty}^{\infty} \cdots \int_{-\infty}^{\infty} W_x(\mathbf{x}) |J(\mathbf{x}/\mathbf{y})| \log [W_x(\mathbf{x}) |J(\mathbf{x}/\mathbf{y})|] J(\mathbf{y}/\mathbf{x}) \, d\mathbf{x}. \tag{8.17-14}$$

On using the relationship

$$|J(\mathbf{x}/\mathbf{y})| J(\mathbf{y}/\mathbf{x}) = \pm 1 \tag{8.17-15}$$

and expanding the logarithm, we obtain

$$H(\mathbf{Y}) = -\int_{-\infty}^{\infty} \cdots \int_{-\infty}^{\infty} W_x(\mathbf{x}) \log W_x(\mathbf{x}) \, d\mathbf{x} - \int_{-\infty}^{\infty} \cdots \int_{-\infty}^{\infty} W_x(\mathbf{x}) \log |J(\mathbf{x}/\mathbf{y})| \, d\mathbf{x} \tag{8.17-16}$$

or

$$\boxed{H(\mathbf{Y}) = H(\mathbf{X}) - E_{\mathbf{x}}\{\log |J(\mathbf{x}/\mathbf{y})|\}} \tag{8.17-17}$$

where $E_{\mathbf{x}}\{\cdot\}$ indicates the expectation operation with respect to the $\mathbf{x}$ distribution. Thus the entropy of a continuous distribution changes with the coordinate system. As an example, consider the linear transformation

$$y_i = \sum_{j=1}^{n} a_{ij}x_j, \qquad i = 1, 2, \ldots, n. \tag{8.17-18}$$

In this case the Jacobian is

$$|J(\mathbf{y}/\mathbf{x})| = \left\| \begin{matrix} a_{11} & a_{12} & \cdots & a_{1n} \\ a_{21} & a_{22} & \cdots & a_{2n} \\ \vdots & & & \\ a_{n1} & a_{n2} & \cdots & a_{nn} \end{matrix} \right\| = \|a_{ij}\| \tag{8.17-19}$$

where $\|a_{ij}\|$ is the magnitude of the determinant $|a_{ij}|$. Now Equation 8.17-17 becomes

$$H(\mathbf{Y}) = H(\mathbf{X}) + \log \|a_{ij}\|. \tag{8.17-20}$$

In one dimension, where $y = ax$, this becomes

$$H(Y) = H(X) + \log |a|. \tag{8.17-21}$$

H(X) Is Invariant to Translation: Let us consider only the simple translation

$$y = x + b. \tag{8.17-22}$$

The entropy $H(Y)$ is

$$\begin{aligned} H(Y) &= -\int_{-\infty}^{\infty} W_y(y) \log W_y(y)\, dy \\ &= -\int_{-\infty}^{\infty} W_x(y - b) \log W_x(y - b)\, dy. \end{aligned} \tag{8.17-23}$$

After the change of variable $x = y - b$, we have

$$H(Y) = -\int_{-\infty}^{\infty} W_x(x) \log W_x(x)\, dx = H(X) \tag{8.17-24}$$

as expected. Actually this result follows immediately from Equation 8.17-17 since the Jacobian of a translation is unity.

It is clear that joint and conditional entropies can be defined in exact analogy to the discrete case discussed in Section 8.5. If the joint density $W_2(x, y)$ exists and if

$$W_x(x) = \int_{-\infty}^{\infty} W_2(x, y)\, dy \tag{8.17-25}$$

and

$$W_y(y) = \int_{-\infty}^{\infty} W_2(x, y)\, dx, \tag{8.17-26}$$

then the joint entropy $H(X, Y)$ is given by

$$H(X, Y) = -\int_{-\infty}^{\infty}\int_{-\infty}^{\infty} W_2(x, y) \log W_2(x, y)\, dx\, dy \tag{8.17-27}$$

and the conditional entropies $H(X/Y)$ and $H(Y/X)$ are

$$H(X/Y) = -\int_{-\infty}^{\infty}\int_{-\infty}^{\infty} W_2(x, y) \log \frac{W_2(x, y)}{W_y(y)}\, dx\, dy \tag{8.17-28}$$

and

$$H(Y/X) = -\int_{-\infty}^{\infty}\int_{-\infty}^{\infty} W_2(x, y) \log \frac{W_2(x, y)}{W_x(x)}\, dx\, dy. \tag{8.17-29}$$

The average mutual information follows from Equation 8.5-27 or from Equation 8.6-13 and is

$$I(X; Y) = -\int_{-\infty}^{\infty}\int_{-\infty}^{\infty} W_2(x, y) \log \frac{W_x(x)W_y(y)}{W_2(x, y)}\, dx\, dy. \tag{8.17-30}$$

Although the entropy of a continuous distribution can be negative, positive, or zero, the average mutual information $I(X; Y)$ is nonnegative as in the discrete case. Consider two continuous densities $p(x) \geq 0$ and $q(x) \geq 0$ where

$$\int_{-\infty}^{\infty} p(x)\, dx = \int_{-\infty}^{\infty} q(x)\, dx = 1. \tag{8.17-31}$$

It follows from Equation 8.4-11 that

$$\int_{-\infty}^{\infty} p(x) \log_a \frac{p(x)}{q(x)}\, dx \geq \frac{1}{\ln a}\int_{-\infty}^{\infty} p(x)\left[1 - \frac{q(x)}{p(x)}\right] dx = 0 \tag{8.17-32}$$

with equality if, and only if, $p(x) = q(x)$. It can be seen immediately from Equation 8.17-30 that

$$I(X; Y) \geq 0 \tag{8.17-33}$$

with equality when x and y are statistically independent, that is, when

$$W_2(x, y) = W_x(x)W_y(y). \tag{8.17-34}$$

8.18 The Maximization of the Entropy of Continuous Distributions

It was shown in Section 8.4 that the entropy of a discrete distribution is a maximum when the distribution is uniform, that is, when all outcomes are equally likely. In the continuous case, the entropy depends on the coordinate system, and it is possible to maximize this entropy subject to various constraints on the associated density function. The maximization itself is the so-called isoperimetric problem of the calculus of variations and is discussed in some detail in Section F.5. For our purposes, the procedure is as follows: It is desired to find $y = y(x)$ so that the integral

$$I = \int_a^b F(x, y)\, dx \tag{8.18-1}$$

is an extremum subject to the constraints that

$$\begin{aligned} \int_a^b F_1(x, y)\, dx &= c_1 \\ \vdots \qquad & \quad \vdots \\ \int_a^b F_n(x, y)\, dx &= c_n \end{aligned} \tag{8.18-2}$$

where the c_i are preassigned constants and the F_i are functions determined by the problem. The function y is found by solving the equation

$$\frac{\partial F}{\partial y} + \lambda_1 \frac{\partial F_1}{\partial y} + \cdots + \lambda_n \frac{\partial F_n}{\partial y} = 0. \tag{8.18-3}$$

The λ_i are constants (called undetermined multipliers) whose values are found by substituting the y found from Equation 8.18-3 into the set of equations given by Equation 8.18-2. We consider now several of the commonest and most useful entropy maximizations.

The Maximization of $H(X)$ for a Fixed Variance of x: We wish to maximize

$$H(X) = -\int_{-\infty}^{\infty} W_x(x) \log W_x(x)\, dx \tag{8.18-4}$$

subject to the constraints that

$$\int_{-\infty}^{\infty} W_x(x)\, dx = 1 \tag{8.18-5}$$

and

$$\int_{-\infty}^{\infty} x^2 W_x(x)\, dx = \sigma^2. \tag{8.18-6}$$

This problem has been solved as Example F.7 in Section F.5 and the result is

$$W_x(x) = \frac{1}{\sqrt{2\pi}\sigma} e^{-x^2/2\sigma^2}, \qquad -\infty < x < \infty. \tag{8.18-7}$$

Thus, for fixed variance, the normal distribution has the largest entropy. It is clear that

$$\ln W_x(x) = -\frac{1}{2} \ln 2\pi\sigma^2 - \frac{x^2}{2\sigma^2} \tag{8.18-8}$$

and that the entropy in this case is

$$H(X) = \frac{1}{2} \ln 2\pi\sigma^2 \int_{-\infty}^{\infty} W_x(x)\, dx + \frac{1}{2\sigma^2} \int_{-\infty}^{\infty} x^2 W_x(x)\, dx$$

or

$$\boxed{H(X) = \tfrac{1}{2} \ln 2\pi\sigma^2 + \tfrac{1}{2} \ln e = \tfrac{1}{2} \ln 2\pi e\sigma^2.} \tag{8.18-9}$$

This last result will be of considerable use later. Note that the normal distribution need not have a zero mean since entropy is invariant to a translation. For convenience, the natural logarithm has been used and the units of H are nats as discussed in Section 8.3.

The Maximization of $H(X)$ for a Limited Peak Value of x: In this case, the single constraint is

$$\int_{-M}^{M} W_x(x)\, dx = 1 \tag{8.18-10}$$

and $W_x(x)$ is found from the equation

$$\frac{\partial}{\partial W_x}(-W_x \ln W_x + \lambda_1 W_x) = 0 \tag{8.18-11}$$

or

$$-1 - \ln W_x + \lambda_1 = 0$$

or

$$W_x(x) = e^{\lambda_1 - 1} = \text{a constant}. \tag{8.18-12}$$

This result may be used in Equation 8.18-10 to obtain the uniform distribution

$$W_x(x) = \begin{cases} \dfrac{1}{2M}, & |x| \le M, \\ 0, & |x| > M. \end{cases} \tag{8.18-13}$$

The associated entropy is

$$H(X) = -\int_{-M}^{M} \frac{1}{2M} \log \frac{1}{2M}\, dx = \log 2M. \tag{8.18-14}$$

The Maximization of H(X) for x Limited to Nonnegative Values and a Given Average Value: The constraints are

$$\int_0^\infty W_x(x)\, dx = 1 \tag{8.18-15}$$

and

$$\int_0^\infty x W_x(x)\, dx = u \tag{8.18-16}$$

and $W_x(x)$ is found from

$$\frac{\partial}{\partial W_x}[-W_x \ln W_x + \lambda_1 W_x + \lambda_2 x W_x] = 0$$

or

$$-1 - \ln W_x + \lambda_1 + \lambda_2 x = 0$$
$$W_x(x) = e^{\lambda_1 - 1}\, e^{\lambda_2 x}.$$

As before λ_1 and λ_2 may be eliminated through Equations 8.18-15 and 8.18-16 to yield

$$W_x(x) = \begin{cases} 0, & x < 0, \\ \dfrac{1}{u} e^{-x/u}, & x \geq 0. \end{cases} \tag{8.18-17}$$

The entropy associated with this distribution is

$$H(X) = \ln u \int_0^\infty W_x(x)\, dx + \frac{1}{u}\int_0^\infty x W_x(x)\, dx$$

or

$$H(X) = \ln (u) + 1 = \ln (ue). \tag{8.18-18}$$

8.19 Gaussian Signals and Channels

Let us assume that the source symbol x and the destination symbol y are jointly gaussian so that the joint density $W_2(x, y)$ is

$$W_2(x, y) = \frac{1}{2\pi\sigma_x\sigma_y\sqrt{1 - \rho^2}} \exp\left\{-\frac{1}{2(1 - \rho^2)}\left[\left(\frac{x}{\sigma_x}\right)^2 - 2\rho\frac{xy}{\sigma_x\sigma_y} + \left(\frac{y}{\sigma_y}\right)^2\right]\right\} \tag{8.19-1}$$

where σ_x^2 and σ_y^2 are the variances of x and y, respectively, and ρ is the correlation coefficient given by

$$\rho = \frac{E\{xy\}}{\sigma_x\sigma_y}. \tag{8.19-2}$$

The univariate densities of x and y are given, of course, by

$$W_x(x) = \frac{1}{\sqrt{2\pi}\sigma_x} e^{-(1/2)(x/\sigma_x)^2}, \qquad -\infty < x < \infty, \tag{8.19-3}$$

and

$$W_y(y) = \frac{1}{\sqrt{2\pi}\sigma_y} e^{-(1/2)(y/\sigma_y)^2}, \qquad -\infty < y < \infty. \tag{8.19-4}$$

Let us make use of Equation 8.17-30 to calculate the average mutual information $I(X; Y)$. We have

$$\frac{W_2(x, y)}{W_x(x)W_y(y)} = \frac{1}{\sqrt{1-\rho^2}} \exp\left\{-\frac{\rho^2}{2(1-\rho^2)}\left[\left(\frac{x}{\sigma_x}\right)^2 - \frac{2xy}{\rho\sigma_x\sigma_y} + \left(\frac{y}{\sigma_y}\right)^2\right]\right\} \tag{8.19-5}$$

and

$$I(X; Y) = -\frac{1}{2}\ln(1-\rho^2)\int_{-\infty}^{\infty}\int_{-\infty}^{\infty} W_2(x, y)\, dx\, dy$$

$$- \frac{\rho^2}{2(1-\rho^2)}\int_{-\infty}^{\infty}\int_{-\infty}^{\infty} W_2(x, y)\left[\left(\frac{x}{\sigma_x}\right)^2 - \frac{2xy}{\rho\sigma_x\sigma_y} + \left(\frac{y}{\sigma_y}\right)^2\right] dx\, dy. \tag{8.19-6}$$

This expression may be rewritten as

$$I(X; Y) = -\frac{1}{2}\ln(1-\rho^2) - \frac{\rho^2}{2(1-\rho^2)}(1 - 2 + 1)$$

or

$$\boxed{I(X; Y) = -\tfrac{1}{2}\ln(1-\rho^2).} \tag{8.19-7}$$

Thus the average mutual information in two jointly gaussian random variables is a function only of the correlation coefficient ρ and varies from zero to infinity as shown in Figure 8.21.

The noise entropy $H(Y/X)$ can be written as

$$H(Y/X) = H(Y) - I(X; Y) = \tfrac{1}{2}\ln 2\pi e\sigma_y^2 + \tfrac{1}{2}\ln(1-\rho^2)$$

or

$$H(Y/X) = \tfrac{1}{2}\ln 2\pi e\sigma_y^2(1-\rho^2). \tag{8.19-8}$$

Suppose that x and y are jointly gaussian as a result of independent zero-mean gaussian noise n being added in the channel to the gaussian input x so that

$$y = x + n. \tag{8.19-9}$$

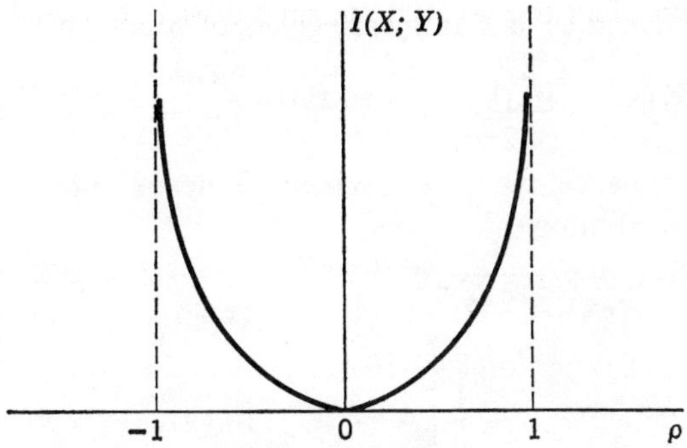

Figure 8.21 The average mutual information in two jointly gaussian random variables

In this case the correlation coefficient ρ becomes

$$\rho = \frac{E\{x^2 + nx\}}{\sigma_x \sigma_y} = \frac{\sigma_x^2}{\sigma_x \sigma_y} = \frac{\sigma_x}{\sigma_y} \tag{8.19-10}$$

and the noise entropy is

$$\boxed{H(Y/X) = \tfrac{1}{2} \ln 2\pi e \sigma_n^2} \tag{8.19-11}$$

where σ_n^2 is the noise variance given by

$$\sigma_n^2 = E\{n^2\} = \sigma_y^2 - \sigma_x^2. \tag{8.19-12}$$

In this situation, Equation 8.19-7 can be rewritten as

$$I(X; Y) = -\frac{1}{2} \ln \left(1 - \frac{\sigma_x^2}{\sigma_y^2}\right) \tag{8.19-13}$$

or

$$I(X; Y) = \frac{1}{2} \ln \left(1 + \frac{\sigma_x^2}{\sigma_n^2}\right). \tag{8.19-14}$$

It is conventional to define the signal power as $S = \sigma_x^2$ and the noise power as $N = \sigma_n^2$ and to rewrite this last expression as

$$\boxed{I(X; Y) = \frac{1}{2} \ln \left(1 + \frac{S}{N}\right)} \tag{8.19-15}$$

where S/N is the signal-to-noise power ratio.

Let us now consider the n-variate normal distribution of Equation 2.22-13

$$W_n(\mathbf{x}) = W_n(x_1, x_2, \ldots, x_n) = \frac{|\sigma^{ij}|^{1/2}}{(2\pi)^{n/2}} e^{-(1/2)\mathbf{x}'\Lambda^{-1}\mathbf{x}} \tag{8.19-16}$$

where, for convenience, we have assumed all means to be zero. The entropy $H(\mathbf{X})$ of this distribution is

$$H(\mathbf{X}) = -\int_{-\infty}^{\infty} \cdots \int_{-\infty}^{\infty} W_n(\mathbf{x}) \ln \frac{|\sigma^{ij}|^{1/2}}{(2\pi)^{n/2}} \, d\mathbf{x} + \frac{1}{2}\int_{-\infty}^{\infty} \cdots \int_{-\infty}^{\infty} W_n(\mathbf{x})\mathbf{x}'\Lambda^{-1}\mathbf{x} \, d\mathbf{x}. \tag{8.19-17}$$

The last integral in this expression can be written as

$$E_{\mathbf{x}}\{\mathbf{x}'\Lambda^{-1}\mathbf{x}\} = E_{\mathbf{x}}\left\{\sum_{i=1}^{n}\sum_{j=1}^{n} x_i\sigma^{ij}x_j\right\} = \sum_{i=1}^{n}\sum_{j=1}^{n} \sigma^{ij}E_{\mathbf{x}}\{x_ix_j\} = \sum_{i=1}^{n}\sum_{j=1}^{n} \sigma^{ij}\sigma_{ij}. \tag{8.19-18}$$

But the elements of the covariance matrix and its inverse satisfy the relationship

$$\sum_{j=1}^{n} \sigma^{ij}\sigma_{ij} = 1 \tag{8.19-19}$$

so that Equation 8.19-18 becomes

$$E_{\mathbf{x}}\{\mathbf{x}'\Lambda^{-1}\mathbf{x}\} = n. \tag{8.19-20}$$

Therefore the entropy $H(\mathbf{X})$ becomes

$$H(\mathbf{X}) = -\ln \frac{|\sigma^{ij}|^{1/2}}{(2\pi)^{n/2}} + \ln (e^{n/2})$$

or

$$\boxed{H(\mathbf{X}) = \ln [(2\pi e)^{n/2}|\sigma^{ij}|^{-1/2}].} \tag{8.19-21}$$

Note that for $n = 1$, the determinant $|\sigma^{ij}|$ is just $1/\sigma^2$ and Equation 8.18-9 results. For $n = 2$, $|\sigma^{ij}|$ is given by Equation 2.22-11 as

$$|\sigma^{ij}| = \frac{1}{\sigma_1^2\sigma_2^2(1 - \rho^2)} \tag{8.19-22}$$

and $H(\mathbf{X})$ becomes

$$H(\mathbf{X}) = \ln [2\pi e\sigma_1\sigma_2(1 - \rho^2)^{1/2}]. \tag{8.19-23}$$

The results of Equation 8.19-21 can be used to find the average mutual information in two gaussian vectors. The vector $\mathbf{x}$ has been defined previously by Equation 2.22-1. In the same way, define $\mathbf{y}$ by

$$\mathbf{y} = \begin{bmatrix} y_1 \\ y_2 \\ \vdots \\ y_m \end{bmatrix} \tag{8.19-24}$$

where m and n are not necessarily equal. Denote the inverses of the covariance matrices of $\mathbf{x}$ and $\mathbf{y}$ by $[\sigma_x^{ij}]$ and $[\sigma_y^{ij}]$, respectively, and that of the vector

$$\mathbf{z} = \begin{bmatrix} \mathbf{x} \\ \mathbf{y} \end{bmatrix} \tag{8.19-25}$$

by $[\sigma^{ij}]$. Then it follows from Equations 8.6-13 and 8.19-21 that the average mutual information in $\mathbf{x}$ and $\mathbf{y}$ is given by

$$\begin{aligned} I(\mathbf{X}; \mathbf{Y}) = &\ln [(2\pi e)^{n/2}|\sigma_x^{ij}|^{-1/2}] + \ln [(2\pi e)^{m/2}|\sigma_y^{ij}|^{-1/2}] \\ &- \ln [(2\pi e)^{(n+m)/2}|\sigma^{ij}|^{-1/2}] \end{aligned}$$

or

$$\boxed{I(\mathbf{X}; \mathbf{Y}) = \frac{1}{2} \ln \frac{|\sigma^{ij}|}{|\sigma_x^{ij}|\,|\sigma_y^{ij}|}.} \tag{8.19-26}$$

In analogy with Section 8.13, channel capacity C for the continuous-amplitude, discrete-time channel is

$$C = \max_{W_x(x)} I(X; Y) = \max_{W_x(x)} [H(Y) - H(Y/X)]. \tag{8.19-27}$$

Suppose the channel consists of an additive noise which is a sequence of independent gaussian random variables n each with zero mean and variance σ_n^2. In this case the conditional probability $W(y/x)$ at each time instant is normal with variance σ_n^2 and mean equal to the particular realization of x. The noise entropy $H(Y/X)$ is given by Equation 8.19-11 and Equation 8.19-27 becomes

$$C = \max_{W_x(x)} [H(Y)] - \tfrac{1}{2} \ln 2\pi e \sigma_n^2. \tag{8.19-28}$$

If the input power is fixed at σ_x^2, then the output power is fixed at $\sigma_y^2 = \sigma_x^2 + \sigma_n^2$ and $H(Y)$ is a maximum if $y = x + n$ is a sequence of independent gaussian random variables. The value of $H(Y)$ is

$$H(Y) = \tfrac{1}{2} \ln 2\pi e(\sigma_x^2 + \sigma_n^2) \tag{8.19-29}$$

and the channel capacity becomes

$$\boxed{C = \frac{1}{2}\ln\left(1 + \frac{\sigma_x^2}{\sigma_n^2}\right) = \frac{1}{2}\ln\left(1 + \frac{S}{N}\right)} \tag{8.19-30}$$

where S/N is the signal-to-noise power ratio. Note that the input x is a sequence of independent gaussian random variables and this last equation is identical to Equation 8.19-15. Thus, for additive independent gaussian noise and an input power limitation, the discrete-time, continuous-amplitude channel has a capacity given by Equation 8.19-30. This capacity is realized when the input is an independent sequence of independent, identically distributed gaussian random variables.

8.20 Bandlimited Channels

In this section we consider messages that can be modeled as continuous random processes with continuous parameters. The channels that transmit these messages will be called *amplitude-continuous, time-continuous* channels. Specifically, we shall restrict attention to signals $x(t)$ which are at least wide-sense stationary with autocorrelation function

$$R_x(\tau) = E\{x(t)x(t + \tau)\} \tag{8.20-1}$$

and power spectral density

$$\varphi_x(\omega) = \int_{-\infty}^{\infty} R_x(\tau)\, e^{-j\omega\tau}\, d\tau. \tag{8.20-2}$$

The process will be assumed to have zero mean and to be *bandlimited* $(-2\pi W, 2\pi W)$ in the sense that the power spectral density $\varphi_x(\omega)$ vanishes outside this interval, that is,

$$\varphi_x(\omega) = 0, \qquad |\omega| > 2\pi W. \tag{8.20-3}$$

It is shown in Section 7.5 that a deterministic signal $f(t)$ bandlimited $(-2\pi W, 2\pi W)$ admits a *sampling representation*, that is, can be written as Equation 7.5-14. It can also be shown, as was done in Section 4.6 for the Fourier series, that the random process $x(t)$ admits the same expansion

$$x(t) = \sum_{k=-\infty}^{\infty} x\left(\frac{k}{2W}\right) \frac{\sin(2\pi Wt - k\pi)}{(2\pi Wt - k\pi)}. \tag{8.20-4}$$

The right side of this expression converges mean-square to the process $x(t)$. The proof is straightforward. Since $\varphi_x(\omega)$ vanishes outside $(-2\pi W, 2\pi W)$, its Fourier transform $R_x(\tau)$ has a sampling representation

$$R_x(\tau) = \sum_{k=-\infty}^{\infty} R_x\left(\frac{k}{2W}\right) \frac{\sin(2\pi Wt - k\pi)}{(2\pi Wt - k\pi)}. \tag{8.20-5}$$

On defining the partial sum as in Equation 4.6-11, we find, after some manipulation,

$$\lim_{N\to\infty} E\{|x(t) - x_N(t)|^2\} = R_x(0) - 2R_x(0) + R_x(0) = 0 \qquad (8.20\text{-}6)$$

and $x_N(t)$ converges mean-square to $x(t)$. Thus the process $x(t)$ with continuous time parameter t may be represented by the process $x(k/2W)$, $k = \ldots, -2, -1, 0, 1, 2, \ldots,$ with discrete-time parameter $t_k = k/2W$. For bandlimited signals or channels it is sufficient, therefore, to consider the discrete-time case and to relate the results to continuous time through Equation 8.20-4.

Suppose the continuous-time process $x(t)$ has a spectrum $\varphi_x(\omega)$ which is *flat and bandlimited* as shown in Figure 3.9(*b*) where $\omega_b = 2\pi W$. The corresponding autocorrelation function $R_x(\tau)$ is shown in the same figure and can be written as

$$R_x(\tau) = 2N_0W\frac{\sin 2\pi W\tau}{2\pi W\tau}. \qquad (8.20\text{-}7)$$

This function passes through zero at intervals of $1/2W$ so that

$$R_x\left(\frac{k}{2W}\right) = 0, \qquad k = \ldots, -2, -1, 1, 2, \ldots. \qquad (8.20\text{-}8)$$

Thus, samples spaced $k/2W$ apart are *uncorrelated if the power spectral density is flat and bandlimited* $(2\pi W, 2\pi W)$. *If the process is gaussian, the samples are independent.* This implies that continuous-time, bandlimited $(-2\pi W, 2\pi W)$ gaussian channels, where the noise has a flat spectrum, have a capacity C given by Equation 8.19-30

$$C = \frac{1}{2}\ln\left(1 + \frac{S}{N}\right) \text{ nats/sample.} \qquad (8.20\text{-}9)$$

Here N is the variance of the additive, flat, bandlimited gaussian noise and S is $R_x(0)$, the fixed variance of the input signal. The units of Equation 8.20-9 are on a per sample basis. Since there are $2W$ samples per unit time, the capacity per unit time can be written as

$$\boxed{C' = W\ln\left(1 + \frac{S}{N}\right) \text{ nats/second.}} \qquad (8.20\text{-}10)$$

8.21 Concluding Remarks

The fundamental ideas developed in this chapter were the notion of *channel capacity* and the *noisy coding theorem.* Although specific discussions were frequently limited to the BSC, the concept of capacity can be applied, in a general way, to all communication channels. Similarly, the idea expressed by the noisy coding theorem—namely, that channel noise places fundamental limits on the *rate* but not the *accuracy* of information transfer—has an equally broad applicability.

This discussion has been limited to zero-memory channels. Although these provide satisfactory models for many physical communication systems, channels with memory occur in a wide variety of physical situations. Results for such channels are considerably more complicated although those for the zero-memory case often may be extrapolated to yield satisfactory, if not optimal, solutions.

The noisy coding theorem sets an upper limit (the channel capacity) on the rate of information transfer while guaranteeing the *existence* of codes that allow transmission at this rate with arbitrarily small probability of error. On the other hand, the theorem does not provide a deterministic procedure for the construction of such codes and, to date, their production proceeds more or less independently. Although a very extensive literature exists on error-correcting codes, little direct connection exists between this area and information theory itself.

PROBLEMS

1. Prove whether or not the following functions are concave or convex:
   (a) $\log x$.
   (b) $x \log x$.
   (c) $\dfrac{\log x}{x}$.
   (d) $x^2 \log x$.
2. Show that $\log_a b = \log_a c \log_c b$.
3. You are watching a horse race with eight participants. The results of the race are flashed on the board in the order: *win*, *place*, and *show*.
   (a) What is the total information transmitted if all entries are equally likely to win?
   (b) What is the information provided by each of the signs: *win*, *place*, and *show*?
   (c) If the order of transmittal is *show*, *place*, *win*, what information is provided by each?
4. Show that the Markov source of Example 8.6 is regular and hence, ergodic. Find the stationary-state distribution vector $\mathbf{P}$.
5. Prove the left sides of the inequalities of Equations 8.6-10 and 8.6-11.

6. A first-order Markov source with alphabet $x = \{A, B\}$ has a transition matrix $\boldsymbol{T}$ given by

$$\boldsymbol{T} = \begin{array}{c|cc} & A & B \\ \hline A & 1-p & p \\ B & q & 1-q \end{array}$$

   (a) Prove that the stationary source probabilities are $p(A) = q/(p + q)$ and $p(B) = p/(p + q)$.
   (b) Draw the state diagram and find the source entropy $H(S)$.
   (c) Plot $H(S)$ versus p for $p = q$; for $q = 1$, $p \neq q$.
7. Find the entropy and redundancy for the sources of Examples 8.4 and 8.6.
8. In Example 8.7 find the values of p_{21}, p_{24}, p_{31}, and p_{34} in the limit as $r \to \infty$.
9. Consider a source that emits independent symbols at the rate of two per second. The possible symbols are chosen from the alphabet A, B, C, D, and E.
   (a) Calculate the maximum source entropy per symbol and per second.
   (b) Calculate the source entropy per symbol and per second for the following symbol probabilities:

$$P(A) = 0.5,$$
$$P(B) = 0.3,$$
$$P(C) = 0.1,$$
$$P(D) = 0.05,$$
$$P(E) = 0.05.$$

10. A source emits the two symbols A and B such that $P(A) = 0.95$, $P(B) = 0.05$.
   (a) Encode by the Fano method message groups of one, two, three, and four symbols and calculate the code efficiency in each case. Calculate the average information per code digit.
   (b) Encode the message group of three symbols by the Huffman method and calculate the code efficiency.
11. For the following source alphabet X find the Shannon, Fano, and Huffman binary codes.

Symbol	*Probability*	*Symbol*	*Probability*
x_1	$p(x_1) = 0.22$	x_5	$p(x_5) = 0.12$
x_2	$p(x_2) = 0.16$	x_6	$p(x_6) = 0.09$
x_3	$p(x_3) = 0.15$	x_7	$p(x_7) = 0.08$
x_4	$p(x_4) = 0.13$	x_8	$p(x_8) = 0.05$

12. Find the capacity of each of the channels described by the following matrices:

(a) $\boldsymbol{D} = [p(y_j/x_i)] = \begin{bmatrix} \frac{3}{8} & \frac{3}{8} & \frac{1}{8} & \frac{1}{8} \\ \frac{1}{8} & \frac{1}{8} & \frac{3}{8} & \frac{3}{8} \end{bmatrix}$.

(b) $\boldsymbol{D} = \begin{bmatrix} \frac{1}{2} & \frac{1}{2} \\ \frac{1}{2} & \frac{1}{2} \end{bmatrix}$.

(c) $\boldsymbol{D} = \begin{bmatrix} \frac{1}{2} & \frac{1}{2} \\ \frac{1}{3} & \frac{2}{3} \end{bmatrix}$.

13. Find the capacity of the channel described by the following matrix:

$$\boldsymbol{D} = [p(y_j/x_i)] = \begin{bmatrix} 1 & 0 \\ 1 & 0 \\ 0 & 1 \end{bmatrix}.$$

For what input distribution is this capacity achieved?

14. The *binary erasure channel* (BEC) has two inputs $\{0, 1\}$ and three outputs $\{0, e, 1\}$ where e indicates an erasure, that is, it cannot be determined at the destination whether 0 or 1 was sent. The BEC can be represented by the channel matrix

$$\boldsymbol{D} = [p(y_j/x_i)] = \begin{bmatrix} p & 1-p & 0 \\ 0 & 1-p & p \end{bmatrix}.$$

Find the channel capacity.

15. A channel is described by the matrix

$$\boldsymbol{D} = [p(y_j/x_i)] = \begin{bmatrix} \frac{1}{2} & \frac{3}{8} & \frac{1}{8} \\ \frac{1}{8} & \frac{1}{2} & \frac{3}{8} \\ \frac{3}{8} & \frac{1}{8} & \frac{1}{2} \end{bmatrix}.$$

Let the a-priori probabilities of the x_i be

$$p(x_1) = \tfrac{2}{3}, \qquad p(x_2) = p(x_3) = \tfrac{1}{6}.$$

Find the ideal observer decision scheme and calculate the probability of error for the scheme.

16. For the BSC, prove that the ideal observer chooses as correct the member of the source set that differs in the least number of binary digits from the received message.

17. Show that, for $n = 3$ and 4, it is not possible to construct a single-error correcting code with more than two words. Show that, as a consequence, Equation 8.16-8 is not a sufficient condition.

18. For a code word length of $n = 5$, devise the single-error correcting code that has the largest number of code words. Show that this code satisfies Equation 8.16-8. Repeat for double-error correction.
19. Find a distribution, other than the uniform, for which the associated entropy is $+\infty$.

REFERENCES

1. J. R. Pierce, *Symbols, Signals, and Noise*, Harper, New York, 1961, Chap. II.
2. C. Cherry, *On Human Communication, A Review, A Survey, and A Summary*, M.I.T. Press, Cambridge, Mass., 1957.
3. R. V. L. Hartley, "Transmission of Information," *Bell Syst. Tech. J.*, **7**, 1928, 535–563.
4. N. Wiener, *Cybernetics*, M.I.T. Press, Cambridge, Mass., 1948.
5. N. Wiener, *The Extrapolation, Interpolation, and Smoothing of Stationary Time Series*, M.I.T. Press, Cambridge, Mass., 1949.
6. C. E. Shannon, "A Mathematical Theory of Communication," *Bell Syst. Tech. J.*, **27**, 1948, 379–423, 623–656.
7. C. E. Shannon, "Communication in the Presence of Noise," *Proc. IRE*, **37**, 1949, 10–21.
8. P. M. Woodward and I. L. Davies, "Information Theory and Inverse Probability in Telecommunication," *Proc. IEE* (London), **99**, Pt. III, 1952, 37.
9. W. Feller, *An Introduction to Probability Theory and Its Applications*, Vol. I, 3rd ed., Wiley, New York, 1968.
10. S. Karlin, *A First Course in Stochastic Processes*, Academic, New York, 1966.
11. A. I. Khinchin, *Mathematical Foundations of Information Theory*, Dover, New York, 1957, pp. 17–22.
12. B. McMillan, "The Basic Theorems of Information Theory," *Ann. Math. Stat.*, **24**, 1953, 196–219.
13. B. McMillan, "Two Inequalities Implied by Unique Decipherability," *IRE Trans. on Information Theory*, IT-2, December 1956, 115–116.
14. J. Karush, "A Simple Proof of an Inequality of McMillan," *IRE Trans. on Information Theory*, IT-7, April 1961, 118.
15. L. G. Kraft, "A Device for Quantizing, Grouping, and Coding Amplitude Modulated Pulses," M. S. Thesis, Department of Electrical Engineering, Massachusetts Institute of Technology, 1949.
16. D. A. Huffman, "A Method for the Construction of Minimum Redundancy Codes," *Proc. IRE*, **40**, 1952, 1098–1101.
17. R. Ash, *Information Theory*, Wiley (Interscience), New York, 1965, Chap. 2.
18. R. M. Fano, *Transmission of Information*, M.I.T. Press, Cambridge, Mass., 1961, Chap. 4.
19. A. Feinstein, *Foundations of Information Theory*, McGraw-Hill, New York, 1958.
20. J. Wolfowitz, *Coding Theorems of Information Theory*, Prentice-Hall, Englewood Cliffs, N. J., 1961.
21. R. G. Gallagher, "A Simple Derivation of the Coding Theorem and Some Applications," *IEEE Trans. on Information Theory*, IT-11, January 1965, 3–18.
22. R. Ash *op. cit.*, Chaps. 4 and 5.
23. W. W. Peterson, *Error-Correcting Codes*, M.I.T. Press, Cambridge, Mass., 1961.

ADDITIONAL READING

1. General Books

(1949) Shannon, C. E., and W. Weaver. *The Mathematical Theory of Communication*, Univ. of Illinois Press, Urbana, Ill.

(1952) Bell, D. A. *Information Theory and Its Engineering Applications*, Pitman, London.

(1953) Goldman, S. *Information Theory*, Prentice-Hall, Englewood Cliffs, N. J.

(1956) Brillouin, L. *Science and Information Theory*, Academic, New York.

(1957) Khinchin, A. I. *Mathematical Foundations of Information Theory*, Dover, New York.

(1958) Feinstein, A. *Foundations of Information Theory*, McGraw-Hill, New York.

(1959) Kullback, S. *Information Theory and Statistics*, Wiley, New York.

Meyer-Eppler, W. *Grundlagen and Anwendungen der Informations Theorie*, Springer-Verlag, Berlin.

(1961) Fano, R. M. *Transmission of Information*, M.I.T. Press, Cambridge, Mass.

Reza, F. M. *An Introduction to Information Theory*, McGraw-Hill, New York.

Wolfowitz, J. *Coding Theorems of Information Theory*, Prentice-Hall, Englewood Cliffs, N. J.

Vitushkin, A. G. *Theory of the Transmission and Processing of Information* (trans. by R. Feinstein), Pergamon, London.

(1963) Abramson, N. *Information Theory and Coding*, McGraw-Hill, New York.

Fey, P. *Informationstheorie*, Akademie-Verlag, Berlin.

Schwartz, L. S. *Principles of Coding, Filtering, and Information Theory*, Spartan, Baltimore, Md.

(1964) Pinsker, M. S. *Information and Information Stability of Random Variables and Processes* (trans. and ed. by R. Feinstein), Holden-Day, San Francisco.

Raisbeck, G. *Information Theory*, M.I.T. Press, Cambridge, Mass.

(1965) Ash, R. *Information Theory*, Wiley (Interscience), New York.

Billingsley, P. B. *Ergodic Theory and Information*, Wiley, New York.

(1968) Jelinek, F. *Probabilistic Information Theory*, McGraw-Hill, New York.

Gallager, R. *Information Theory and Reliable Communication*, Wiley, New York.

2. Published Bibliographies

The following very extensive bibliographies are available. Note that the titles are misleading in that "information theory" is used as a synonym for what we have called "statistical communication theory." An appreciable part of each bibliography, however, is devoted to those topics of information measure and coding treated here in Chapter 8.

(1953) Stumpers, F. L. "A Bibliography of Information Theory, Communication Theory —Cybernetics," *IRE Trans. on Information Theory*, PGIT-2, November.

(1955) Stumpers, F. L. "A Bibliography of Information Theory, First Supplement," *IRE Trans. on Information Theory*, IT-1, September, 31–47.

(1957) Stumpers, F. L. "A Bibliography of Information Theory, Second Supplement," *IRE Trans. on Information Theory*, IT-3, June, 150–166.

(1960) Stumpers, F. L. "A Bibliography on Information Theory, Third Supplement," *IRE Trans. on Information Theory*, IT-6, March, 25–51.

(1961) Elias, P., A. Gill, R. Price, N. Abramson, P. Swerling, and L. A. Zadeh. "Progress in Information Theory in the U.S.A., 1957–1960," *IRE Trans. on Information Theory*, IT-7, July, 128–144.

(1963) Zadeh, L. A., N. Abramson, A. V. Balakrishnan, D. Braverman, M. Eden, E. A. Feigenbaum, T. Kailath, R. M. Lerner, J. Massey, G. E. Meuller, W. W. Peterson, R. Price, G. Sebestyen, D. Slepian, A. J. Thomasian, and G. L. Turin. "Report on Progress in Information Theory in the U.S.A., 1960–1963," *IEEE Trans. on Information Theory*, IT-9, October, 221–264.

In addition to these bibliographies, the book by Cherry [Reference 2] contains nearly 400 selected references.

Appendix A

The Riemann-Stieltjes Integral

A.1 The Riemann Integral

A fundamental problem in the elementary calculus [1, 2] is the determination of the area under a positive curve $f(x)$ in some interval $[a, b]$* such as is shown in Figure A.1. The usual way to approach this problem is to subdivide the interval into m segments $[x_i - x_{i-1}]$ and to form the sums

$$\sum_{i=1}^{m} f(\xi_i)[x_i - x_{i-1}] \tag{A.1-1}$$

where ξ_i is a point in the ith interval. In other words, any arbitrary set of ξ_i and x_i produce a partition of $[a, b]$ given by

$$a = x_0 < \xi_1 < x_1 < \xi_2 < x_2 < \cdots < \xi_m < x_m = b. \tag{A.1-2}$$

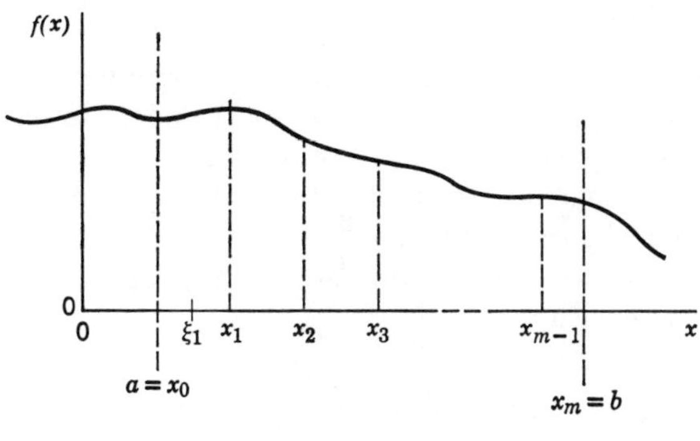

Figure A.1 Determination of the area under a curve

* The notation $[a, b]$ will be used to denote a *closed interval*; that is, one that includes the endpoints a and b. The *open interval* will be denoted by (a, b).

The sum represented by Equation A.1-1 has a different value for each partition of $[a, b]$, of course, but each such value is an approximation to the area between $f(x)$ and the x axis in the interval $[a, b]$. It is intuitively apparent that, if $f(x)$ is reasonably well behaved, then the sum becomes a better and better approximation to the area as the number m increases and as the length $[x_i - x_{i-1}]$ of the longest interval decreases. For sufficiently well-behaved $f(x)$ a unique limit exists which is the *Riemann definite integral* of elementary calculus. More precisely, the Riemann integral of $f(x)$ with respect to x over the interval $[a, b]$ is defined as

$$\int_a^b f(x)\,dx = \lim_{\max|x_i - x_{i-1}| \to 0} \sum_{i=1}^{m} f(\xi_i)[x_i - x_{i-1}] \qquad \text{(A.1-3)}$$

for any arbitrary sequence of partitions of the form of Equation A.1-2.

At this point a word about notation is in order. It is clear that, in ordinary summations,

$$\sum_{i=1}^{n} a_i = \sum_{j=1}^{n} a_j = a_1 + a_2 + \cdots + a_n. \qquad \text{(A.1-4)}$$

The symbols i and j are each called an "index of summation" and the sum does not depend on the notation used for this index. In the same way

$$\int_a^b f(x)\,dx = \int_a^b f(y)\,dy \qquad \text{(A.1-5)}$$

and the definite integral depends on a, b, and f but not on the "variables of integration" x and y.

A.2 Functions of Bounded Variation

A function $g(x)$ is called *increasing* (or, more correctly, *nondecreasing*) in $[a, b]$ if $g(x_2) \geq g(x_1)$ for any pair of values x_1 and x_2 in $[a, b]$ for which $x_2 > x_1$. Similarly, $g(x)$ is called *decreasing* (or *nonincreasing*) in $[a, b]$ if $g(x_2) \leq g(x_1)$. A function which is either an increasing function or a decreasing function is called *monotonic.*

A function $g(x)$ is of *bounded variation* [3] in $[a, b]$ if, and only if, there exists a number M such that

$$\sum_{i=1}^{m} |g(x_i) - g(x_{i-1})| < M \qquad \text{(A.2-1)}$$

for all partitions

$$a = x_0 < x_1 < x_2 < \cdots < x_m = b \qquad \text{(A.2-2)}$$

of the interval. Alternately, $g(x)$ is of bounded variation if, and only if, it can be written in the form

$$g(x) = g_1(x) - g_2(x) \tag{A.2-3}$$

where the functions $g_1(x)$ and $g_2(x)$ are bounded and nondecreasing in $[a, b]$. If the function $g(x)$ is bounded and has a finite number of relative maxima and minima, and discontinuities in the finite open interval (a, b), then it is of bounded variation in (a, b). This last statement is known as the *Dirichlet Conditions* [3, 4] and will turn out to be important in the treatment of Fourier series in Appendix D.

Where realizable signals are concerned, if $g(t)$ is of bounded variation in some interval, then it is bounded and its high-frequency components must make a limited total contribution to its frequency spectrum.

A.3 The Riemann-Stieltjes Integral

The Riemann-Stieltjes integral [3, 5, 6] of the function $f(x)$ with respect to the function $g(x)$ over the interval $[a, b]$ is defined as

$$\int_a^b f(x)\, dg(x) = \lim_{\max|x_i - x_{i-1}| \to 0} \sum_{i=1}^{m} f(\xi_i)[g(x_i) - g(x_{i-1})] \tag{A.3-1}$$

for any arbitrary sequence of partitions of the form of Equation A.1-2. A sufficient condition for the existence of the limit given by Equation A.3-1 is that $g(x)$ be of *bounded variation* and $f(x)$ be *continuous* on $[a, b]$ (or that $f(x)$ be of bounded variation and $g(x)$ be continuous). It is apparent that, for the case where

$$g(x) = \alpha x, \tag{A.3-2}$$

the Riemann-Stieltjes integral reduces to the ordinary Riemann integral.

Some idea of the geometric meaning of the Riemann-Stieltjes integral may be obtained by referring to Figure A.2. It is apparent that the Riemann-Stieltjes sum

$$\sum_{i=1}^{m} f(\xi_i)[g(x_i) - g(x_{i-1})] \tag{A.3-3}$$

is not an approximation to the area under $f(x)$ except when $g(x)$ is a straight line through the origin with unit slope. Rather, the sum is weighted by the values that $g(x)$ assumes in the interval $[a, b]$. As mentioned in the preceding paragraph, one of the functions f or g need not be particularly well behaved as far as continuity is concerned, but only of bounded variation. For example, suppose that $u_-(x)$ is the unit step function given by

$$u_-(x) = \begin{cases} 0, & x < 0, \\ 1, & x \geq 0. \end{cases} \tag{A.3-4}$$

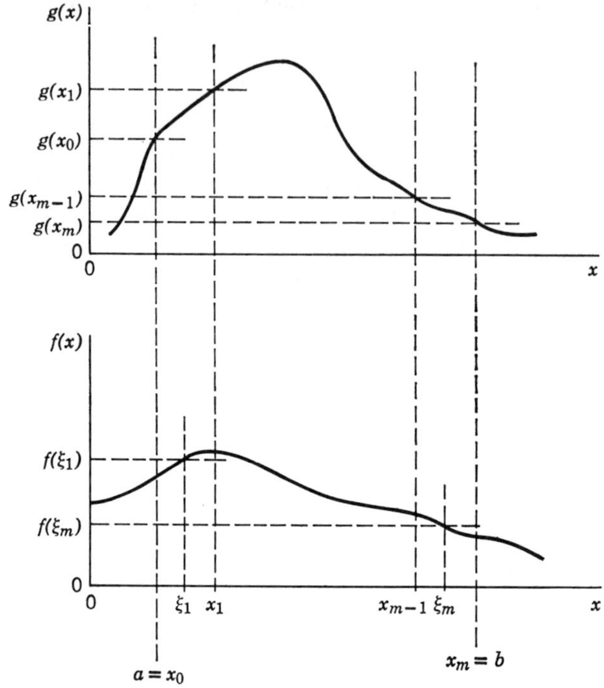

Figure A.2 *The Riemann-Stieltjes integral: the first term in the Riemann-Stieltjes sum is* $f(\xi_1)[g(x_1) - g(x_0)]$*; the mth term in the Riemann-Stieltjes sum is* $f(\xi_m)[g(x_m) - g(x_{m-1})]$

Then the Riemann-Stieltjes integral may be used to write the sum

$$\sum_k \alpha_k f(k) = \int_{-\infty}^{\infty} f(x)\, d \sum_k \alpha_k u_-(x - k). \tag{A.3-5}$$

In particular, let $f(x)$ be a density function which may be either discrete, continuous, or mixed. Its cumulative distribution function $F(a)$ is

$$F(a) = P(x \leq a) = \int_{-\infty}^{a} dF(x) \tag{A.3-6}$$

and the expression

$$\int_{-\infty}^{\infty} dF(x) = 1 \tag{A.3-7}$$

replaces both Equations 2.10-6 and 2.10-8.

The Riemann-Stieltjes integral, if it exists, has a number of properties analogous to those of the Riemann integral as well as some for which there is no obvious analogy:

1. $\int_a^b (\alpha_1 f_1 + \alpha_2 f_2)\, dg = \alpha_1 \int_a^b f_1\, dg + \alpha_2 \int_a^b f_2\, dg.$ (A.3-8)

2. $\int_a^b f\, d(\alpha_1 g_1 + \alpha_2 g_2) = \alpha_1 \int_a^b f\, dg_1 + \alpha_2 \int_a^b f\, dg_2.$ (A.3-9)

3. $\int_a^b f\, dg = -\int_b^a f\, dg.$ (A.3-10)

4. If g is constant in $[a, b]$, $\int_a^b f\, dg = 0.$ (A.3-11)

5. If f is constant in $[a, b]$, $\int_a^b f\, dg = f[g(b) - g(a)].$ (A.3-12)

6. $\int_a^b f\, dg = f(a)[k - g(a)] + f(b)[g(b) - k],$ (A.3-13)

   if g is the constant k in (a, b) and f is continuous at a and b.

7. $\int_a^b f(x)\, dg(x) = f(b)g(b) - f(a)g(a) - \int_a^b g(x)\, df(x)$ (A.3-14)

   from integration by parts.

8. $\int_a^b f(x)\, dg(x) = \int_a^b f(x)g'(x)\, dx,$ (A.3-15)

   if $g(x)$ is continuously differentiable on $[a, b]$.

REFERENCES

1. R. Courant, *Differential and Integral Calculus*, Vol. I, Interscience, New York, 1937.
2. S. Lang, *A First Course in Calculus*, Addison-Wesley, Reading, Mass., 1964.
3. T. M. Apostal, *Mathematical Analysis*, Addison-Wesley, Reading, Mass., 1957.
4. R. V. Churchill, *Fourier Series and Boundary Value Problems*, McGraw-Hill, New York, 1941.
5. H. Kostelman, *Modern Theories of Integration*, Dover, New York, 1960.
6. S. Hartman and J. Mikusinski, *The Theory of Lebesgue Measure and Integration*, Pergamon, New York, 1961.

Appendix B

The Dirac Delta Function

B.1 Step Functions

As mentioned in Appendix A, it is often convenient to introduce the concept of a *step function* [1], a function that changes its value only on a discrete set of discontinuities. The commonest step function considered is the *unit-step function* which may be defined in at least three ways:

1. *Symmetrical unit-step function* $u(x)$*:*

$$u(x) = \begin{cases} 0, & x < 0, \\ \frac{1}{2}, & x = 0, \\ 1, & x > 0. \end{cases} \tag{B.1-1}$$

2. *Asymmetrical unit-step function* $u_-(x)$*:*

$$u_-(x) = \begin{cases} 0, & x < 0, \\ 1, & x \geq 0. \end{cases} \tag{B.1-2}$$

3. *Asymmetrical unit-step function* $u_+(x)$*:*

$$u_+(x) = \begin{cases} 0, & x \leq 0, \\ 1, & x > 0. \end{cases} \tag{B.1-3}$$

Each of these is illustrated in Figure B.1. It is apparent that $u_+(x)$ is continuous on the left and that $u_-(x)$ is continuous on the right at $x = 0$ (see Section 2.10).

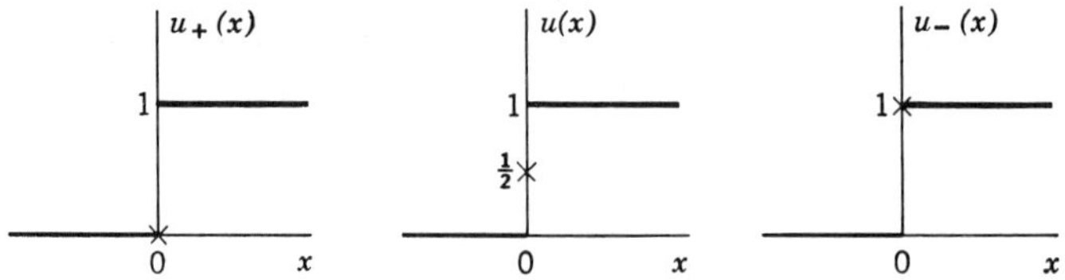

Figure B.1 *Unit-step functions*

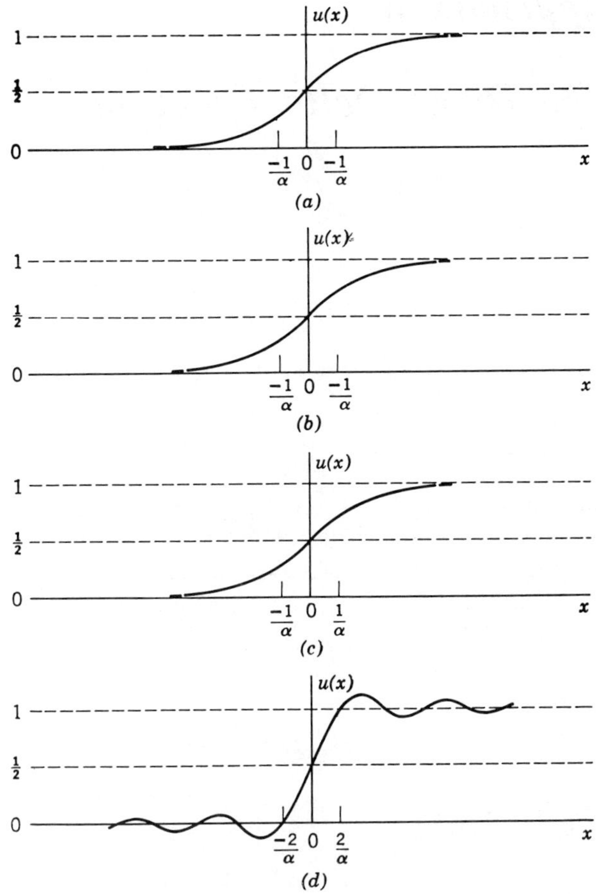

Figure B.2 *Approximations to the unit-step function:*

$$(a)\quad u(x) = \lim_{\alpha \to \infty} \left[\frac{1}{2} + \frac{1}{\pi} \text{arc tan } (\alpha x)\right]$$

$$(b)\quad u(x) = \lim_{\alpha \to \infty} \frac{1}{2} [\text{erf } (\alpha x) + 1]$$

$$(c)\quad u(x) = \lim_{\alpha \to \infty} (2^{-e^{-\alpha x}})$$

$$(d)\quad u(x) = \lim_{\alpha \to \infty} \frac{1}{\pi} \int_{-\infty}^{\alpha x} \frac{\sin y}{y} \, dy$$

The unit-step function $u(x)$ may be represented as a limit of many continuous functions, for example,

1. $u(x) = \lim_{\alpha \to \infty} \left[\frac{1}{2} + \frac{1}{\pi} \text{arc tan } (\alpha x)\right]$. (B.1-4)

2. $u(x) = \lim_{\alpha \to \infty} \frac{1}{2}[\text{erf}(\alpha x) + 1].$ (B.1-5)

where the error function erf (αx) has already been defined by Equation 2.18-12 as

$$\text{erf}(\alpha x) = \frac{2\alpha}{\sqrt{\pi}} \int_0^{\alpha x} e^{-\alpha^2 y^2}\, dy.$$

3. $u(x) = \lim_{\alpha \to \infty} [2^{-e^{-\alpha x}}].$ (B.1-6)

4. $u(x) = \lim_{\alpha \to \infty} \frac{1}{\pi} \int_{-\infty}^{\alpha x} \frac{\sin y}{y}\, dy.$ (B.1-7)

These functions are illustrated in Figure B.2. As a matter of interest we point out also that

$$u(t) = \frac{1}{2} + \frac{1}{2\pi} \int_{-\infty}^{\infty} \frac{\sin \omega t}{\omega}\, d\omega. \tag{B.1-8}$$

Another step function of some interest is the *signum function*

$$\text{sgn}\, x = \begin{cases} -1, & x < 0, \\ 0, & x = 0, \\ 1, & x > 0. \end{cases} \tag{B.1-9}$$

It is clear that

$$\text{sgn}\, x = 2u(x) - 1. \tag{B.1-10}$$

B.2 The Dirac Delta Function

The Dirac delta function $\delta(x)$ [2] is of considerable utility in engineering problems [3, 4] even though it is mathematically pathological. It is defined in terms of its "sampling" property with regard to an arbitrary function $f(x)$ for $a < b$:

$$\int_a^b f(x)\delta(x - k)\, dx = \begin{cases} 0, & \text{for } k < a \text{ or } k > \text{b}, \\ \frac{1}{2}f(k + 0), & \text{for } k = a, \\ \frac{1}{2}f(k - 0), & \text{for } k = b, \\ \frac{1}{2}[f(k - 0) + f(k + 0)], & \text{for } a < k < b. \end{cases} \tag{B.2-1}$$

In the usual case where (a, b) is $(-\infty, \infty)$ and where $f(x)$ is continuous, Equation B.2-1 becomes simply

$$\int_{-\infty}^{\infty} f(x)\delta(x - k)\, dx = f(k). \tag{B.2-2}$$

Note that $\delta(x)$ is not a function by the usual definitions since Equation B.2-2 implies that

$$\delta(x) = 0, \qquad x \neq 0, \tag{B.2-3}$$

$$\int_{-\infty}^{\infty} \delta(x)\, dx = \int_{-\epsilon}^{\epsilon} \delta(x)\, dx = 1 \tag{B.2-4}$$

which are not consistent relationships. A more general approach to the definition of "function" can be made through the *theory of distributions* [5, 6] in order to justify definitions such as Equation B.2-1. For our purposes we will regard the delta function as a formal representation of the operation defined by Equation B.2-1 and will not attempt to further justify its use.

A comparison of Equations A.3-5 and B.2-2 indicates that the use of the delta function could be avoided by employing the Riemann-Stieltjes integral:

$$\int_{-\infty}^{\infty} f(x)\, du(x-k) = \int_{-\infty}^{\infty} f(x)\delta(x-k)\, dx = f(k). \tag{B.2-5}$$

This relation also suggests that, in a formal sense, the delta function can be regarded as the "derivative" of the unit-step function:

$$\frac{d}{dx} u(x) = \delta(x). \tag{B.2-6}$$

In the same way, the rth "derivative" $\delta^{(r)}(x)$ of the delta function is defined for $a < b$ by

$$\int_a^b f(x)\delta^{(r)}(x-k)\, dx$$
$$= \begin{cases} 0, & \text{for } k < a \text{ and } k > b, \\ \frac{1}{2}(-1)^r f^{(r)}(k+0), & \text{for } k = a, \\ \frac{1}{2}(-1)^r f^{(r)}(k-0), & \text{for } k = b, \\ \frac{1}{2}(-1)^r [f^{(r)}(k-0) + f^{(r)}(k+0)], & \text{for } a < k < b, \end{cases} \tag{B.2-7}$$

where $f^{(r)}(x)$ is the rth derivative of $f(x)$ with respect to x. Again, for the infinite interval and continuous $f(x)$, this becomes

$$\int_{-\infty}^{\infty} f(x)\delta^{(r)}(x-k)\, dx = (-1)^r f^{(r)}(k). \tag{B.2-8}$$

A number of formal relationships involving the delta function may be obtained directly from the defining relationship of Equation B.2-1. For example:

1. $\delta(x) = \delta(-x).$ (B.2-9)

2. $\delta(\alpha x) = \dfrac{1}{\alpha}\delta(x), \qquad \alpha > 0.$ (B.2-10)

3. $f(x)\delta(x - k) = \frac{1}{2}[f(k - 0) + f(k + 0)]\delta(x - k).$ (B.2-11)

4. $x\,\delta(x) = 0.$ (B.2-12)

5. $\delta(x^2 - \alpha^2) = \dfrac{1}{2\alpha}[\delta(x - \alpha) + \delta(x + \alpha)], \qquad \alpha > 0.$ (B.2-13)

6. $\displaystyle\int_{-\infty}^{\infty} \delta(\alpha - x)\delta(x - \beta)\,dx = \delta(\alpha - \beta).$ (B.2-14)

Approximation of Delta Functions

As in the case of the unit-step functions, the Dirac delta function may be approximated by a number of continuous functions, for example,

1. $\delta(x) = \lim\limits_{\alpha\to\infty} \delta_1(x, \alpha) = \lim\limits_{\alpha\to\infty} \dfrac{\alpha}{\pi(1 + \alpha^2x^2)},$ (B.2-15)

2. $\delta(x) = \lim\limits_{\alpha\to\infty} \delta_2(x, \alpha) = \lim\limits_{\alpha\to\infty} \dfrac{\alpha}{\sqrt{\pi}} e^{-\alpha^2x^2},$ (B.2-16)

3. $\delta(x) = \lim\limits_{\alpha\to\infty} \delta_3(x, \alpha) = \lim\limits_{\alpha\to\infty} \dfrac{\alpha}{\pi}\dfrac{\sin \alpha x}{\alpha x},$ (B.2-17)

which are shown in Figure B.3 with α the parameter. These are approximations in the sense that, when $\delta_k(x, \alpha)$ is used for $\delta(x)$ in Equations B.2-1 or B.2-2, then the limit of the integral is unchanged; that is,

$$\lim_{\alpha\to\infty} \int_{-\infty}^{\infty} f(x)\delta_j(x - k, \alpha)\,dx = f(k), \qquad j = 1, 2, 3, \qquad \text{(B.2-18)}$$

whenever $f(x)$ is continuous. Note that $\delta_1(x, \alpha)$, $\delta_2(x, \alpha)$, and $\delta_3(x, \alpha)$ are the derivatives with respect to x of Equations B.1-4, B.1-5, and B.1-7 respectively. The delta function may be approximated also by discontinuous functions. Perhaps the commonest such approximation is by the pulse of unit area shown in Figure B.3(d) and given by

$$\delta(x) = \lim_{\alpha\to 0} \delta_4(x, \alpha) = \lim_{\alpha\to 0} \frac{u(x + \alpha) - u(x - \alpha)}{2\alpha}. \qquad \text{(B.2-19)}$$

A useful relationship, called *Dirichlet's Integral Formula*, may be obtained from an application of the delta function approximation of Equation B.2-17 to Equation B.2-1:

$$\lim_{\alpha\to\infty} \frac{1}{\pi}\int_{-\infty}^{\infty} f(x)\frac{\sin \alpha(k - x)}{k - x}\,dx = \tfrac{1}{2}[f(k - 0) + f(k + 0)], \qquad \text{(B.2-20)}$$

$$-\infty < k < \infty.$$

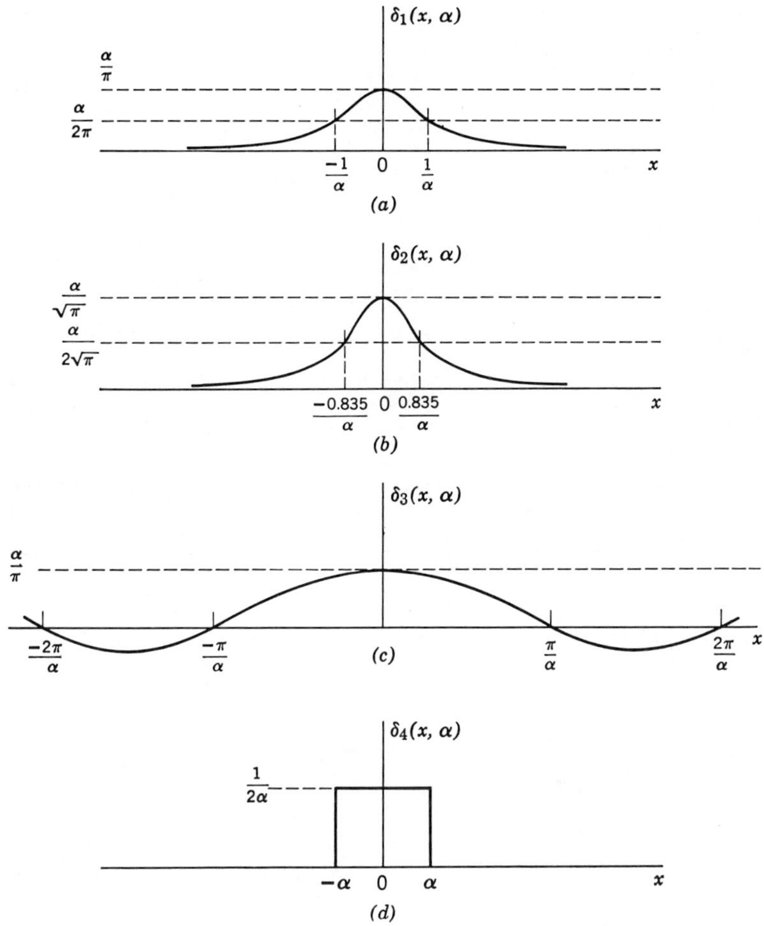

Figure B.3 *Approximations to the Dirac delta function:*

$$(a) \quad \delta(x) = \lim_{\alpha \to \infty} \frac{\alpha}{\pi(1 + \alpha^2 x^2)}$$

$$(b) \quad \delta(x) = \lim_{\alpha \to \infty} \frac{\alpha}{\sqrt{\pi}} e^{-\alpha^2 x^2}$$

$$(c) \quad \delta(x) = \lim_{\alpha \to \infty} \frac{\alpha}{\pi} \frac{\sin \alpha x}{\alpha x}$$

$$(d) \quad \delta(x) = \lim_{\alpha \to \infty} \frac{u(x + \alpha) - u(x - \alpha)}{2\alpha}$$

This expression is of considerable value in the treatment of Fourier integrals and in the derivation of the concept of power spectral density.

The "derivatives" of the delta function may be approximated by successive differentiation of the various approximating forms $\delta_k(x, \alpha)$, $k = 1, 2, 3, 4$, or by other means [3].

REFERENCES

1. T. M. Apostal, *Mathematical Analysis*, Addison-Wesley, Reading, Mass., 1957.
2. P. A. M. Dirac, *Principles of Quantum Mechanics*, Oxford Univ. Press, Oxford, England, 1935.
3. H. S. Carslaw and J. C. Jaeger, *Operational Methods in Applied Mathematics*, Dover, New York, 1963.
4. W. Kaplan, *Operational Methods for Linear Systems*, Addison-Wesley, Reading, Mass., 1962.
5. L. Schwartz, *Theorie des Distributions*, Vols. I and II, Hermann, Paris, 1950–1951.
6. M. J. Lighthill, *An Introduction to Fourier Analysis and Generalized Functions*, Cambridge Univ. Press, Cambridge, England, 1958.

Appendix C

The Transformation of Coordinates

C.1 Introduction

A situation which arises frequently in engineering problems is that involving a change in variables in an integral. In the one-dimensional case, it is desired to change from the variable x to the variable y in the integral

$$I_1 = \int f(x)\,dx \tag{C.1-1}$$

where

$$y = h(x) \tag{C.1-2}$$

is a one-to-one continuously differentiable transformation and

$$x = h^{-1}(y) \tag{C.1-3}$$

is the inverse transformation. The change is straightforward since

$$I_1 = \int f[h^{-1}(y)]\,dx = \int g(y)\frac{dx}{dy}\,dy \tag{C.1-4}$$

where, for convenience, $g(y)$ is written for $f[h^{-1}(y)]$.

Example C.1 You are given the definite integral

$$I_1 = \int_0^a dx = a.$$

If y is related to x through

$$y = x^3,$$

evaluate the integral by transforming to the y-coordinate system. We have

$$\frac{dy}{dx} = 3x^2 = 3y^{2/3}$$

and

$$I_1 = \int_0^{a^3} \tfrac{1}{3} y^{-2/3}\, dy = y^{1/3}\Big|_0^{a^3} = a$$

as expected.

• •

C.2 Transformation in Two Dimensions

In two dimensions the problem can be considered as a transformation of a differential area from one coordinate system x_1, x_2 to another system y_1, y_2. As before, suppose y_1 and y_2 are related to x_1 and x_2 through the one-to-one continuously differentiable transformations

$$y_1 = h_1(x_1, x_2) \tag{C.2-1}$$

$$y_2 = h_2(x_1, x_2). \tag{C.2-2}$$

Then a function $f(x_1, x_2)$ can be written as

$$f(x_1, x_2) = g(y_1, y_2) \tag{C.2-3}$$

where

$$g(y_1, y_2) = f[h_1^{-1}(y_1, y_2), h_2^{-1}(y_1, y_2)] \tag{C.2-4}$$

and the inverse functions h_1^{-1} and h_2^{-1} are found by solving Equations C.2-1 and C.2-2 simultaneously. It remains to find the function J in the equality

$$\iint_R f(x_1, x_2)\, dx_1\, dx_2 = \iint_{R'} g(y_1, y_2) J\, dy_1\, dy_2 \tag{C.2-5}$$

where R is some region of integration in the x_1x_2-plane and R' is the corresponding region in the y_1y_2-plane. A differential area $dx_1\, dx_2$ is shown in Figure C.1. First, we examine dx_1, holding x_2 constant so that

$$dx_1 = \frac{\partial x_1}{\partial y_1} dy_1 + \frac{\partial x_1}{\partial y_2} dy_2, \tag{C.2-6}$$

$$0 = \frac{\partial x_2}{\partial y_1} dy_1 + \frac{\partial x_2}{\partial y_2} dy_2. \tag{C.2-7}$$

These two equations may be solved simultaneously for dx_1 to yield

$$dx_1 = \frac{[(\partial x_1/\partial y_1)(\partial x_2/\partial y_2)] - [(\partial x_1/\partial y_2)(\partial x_2/\partial y_1)]}{\partial x_2/\partial y_2} dy_1. \tag{C.2-8}$$

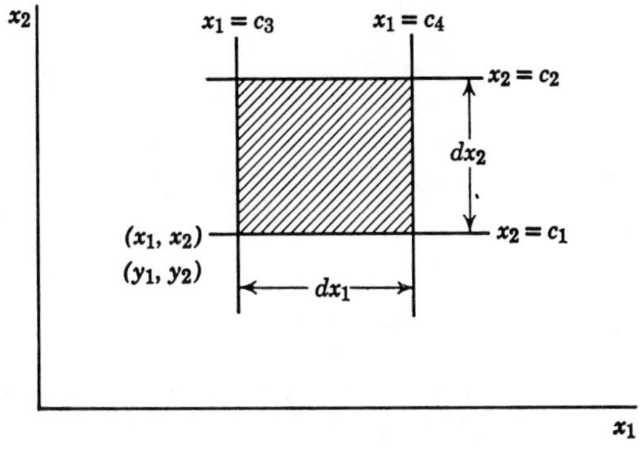

Figure C.1 Transformation of coordinates

We introduce the notation

$$J\left(\frac{x_a, x_b, x_c, \ldots, x_n}{y_a, y_b, y_c, \ldots, y_n}\right) = \begin{vmatrix} \frac{\partial x_a}{\partial y_a} & \frac{\partial x_a}{\partial y_b} & \cdots & \frac{\partial x_a}{\partial y_n} \\ \frac{\partial x_b}{\partial y_a} & \frac{\partial x_b}{\partial y_b} & \cdots & \frac{\partial x_b}{\partial y_n} \\ \vdots & & & \\ \frac{\partial x_n}{\partial y_a} & \frac{\partial x_n}{\partial y_b} & \cdots & \frac{\partial x_n}{\partial y_n} \end{vmatrix} \tag{C.2-9}$$

where the symbol | | indicates the determinant. Now Equation C.2-8 can be written as

$$dx_1 = \frac{J(x_1, x_2/y_1, y_2)}{J(x_2/y_2)}\, dy_1 \tag{C.2-10}$$

and the left side of Equation C.2-5 may be written as

$$\iint f(x_1, x_2)\frac{J(x_1, x_2/y_1, y_2)}{J(x_2/y_2)}\, dy_1\, dx_2.$$

Holding y_1 constant we obtain

$$dx_2 = \frac{\partial x_2}{\partial y_2}\, dy_2 = J\left(\frac{x_2}{y_2}\right) dy_2 \tag{C.2-11}$$

or

$$\iint_R f(x_1, x_2)\, dx_1\, dx_2 = \iint_{R'} g(y_1, y_2) J\left(\frac{x_1, x_2}{y_1, y_2}\right) dy_1\, dy_2. \tag{C.2-12}$$

The function $J(x_1, x_2/y_1, y_2)$ is called the *Jacobian of the transformation* of Equations C.2-1 and C.2-2. It is apparent from the development that

$$\iint_{R'} g(y_1, y_2)\, dy_1\, dy_2 = \iint_{R} f(x_1, x_2) J\left(\frac{y_1, y_2}{x_1, x_2}\right) dx_1\, dx_2 \qquad \text{(C.2-13)}$$

and that

$$J\left(\frac{x_1, x_2}{y_1, y_2}\right) J\left(\frac{y_1, y_2}{x_1, x_2}\right) = 1. \qquad \text{(C.2-14)}$$

Example C.2 Let us transform from rectangular coordinates x, y to polar coordinates r, θ where

$$x = r\cos\theta,$$

$$y = r\sin\theta.$$

Then the Jacobian of the transformation is

$$J\left(\frac{x, y}{r, \theta}\right) = \begin{vmatrix} \dfrac{\partial x}{\partial r} & \dfrac{\partial x}{\partial \theta} \\[2ex] \dfrac{\partial y}{\partial r} & \dfrac{\partial y}{\partial \theta} \end{vmatrix} = \begin{vmatrix} \cos\theta & -r\sin\theta \\ \sin\theta & r\cos\theta \end{vmatrix},$$

$$J\left(\frac{x, y}{r, \theta}\right) = r,$$

and

$$\iint_{R} f(x, y)\, dx\, dy = \iint_{R'} f(r\cos\theta, r\sin\theta) r\, dr\, d\theta.$$

• •

C.3 Extension to the *n*-Dimensional Case

The procedures of the previous section may be extended to n dimensions in a straightforward fashion. The problem is to obtain the volume transformation

$$dx_1\, dx_2 \cdots dx_n = J\, dy_1\, dy_2 \cdots dy_n. \qquad \text{(C.3-1)}$$

In general,

$$dx_i = \sum_{j=1}^{n} \frac{\partial x_i}{\partial y_j} dy_j, \qquad i = 1, 2, \ldots, n. \qquad \text{(C.3-2)}$$

First examine dx_1, holding $x_2, x_3, \ldots, x_n$ constant:

$$dx_1 = \sum_{j=1}^{n} \frac{\partial x_1}{\partial y_j} dy_j, \tag{C.3-3}$$

$$0 = \sum_{j=1}^{n} \frac{\partial x_2}{\partial y_j} dy_j, \tag{C.3-4}$$

$$\vdots$$

$$0 = \sum_{j=1}^{n} \frac{\partial x_n}{\partial y_j} dy_j. \tag{C.3-5}$$

This set of equations may be solved for dx_1 to yield

$$dx_1 = \frac{J(x_1, x_2, \ldots, x_n/y_1, y_2, \ldots, y_n)}{J(x_2, x_3, \ldots, x_n/y_2, y_3, \ldots, y_n)} dy_1. \tag{C 3-6}$$

Now transform dx_2 holding $y_1, x_3, \ldots, x_n$ constant:

$$dx_2 = \frac{J(x_2, x_3, \ldots, x_n/y_2, y_3, \ldots, y_n)}{J(x_3, x_4, \ldots, x_n/y_3, y_4, \ldots, y_n)} dy_2. \tag{C 3-7}$$

Finally

$$dx_n = J\left(\frac{x_n}{y_n}\right) dy_n. \tag{C 3-8}$$

The transformation of integrals is now

$$\iint \cdots \int_R f(x_1, x_2, \ldots, x_n)\, dx_1\, dx_2 \cdots dx_n$$

$$= \iint \cdots \int_{R'} g(y_1, y_2, \ldots, y_n) J\left(\frac{x_1, x_2, \ldots, x_n}{y_1, y_2, \ldots, y_n}\right) dy_1\, dy_2 \cdots dy_n \tag{C 3-9}$$

as expected.

REFERENCES

1. T. M. Apostal, *Calculus*, Vol. II, Blaisdell, New York, 1962.
2. R. Courant, *Differential and Integral Calculus*, Vol. II, Interscience, New York, 1936.
3. H. Jeffreys and B. S. Jeffreys, *Methods of Mathematical Physics*, Cambridge Univ. Press, Cambridge, England, 1950.

Appendix D

Fourier Series and the Fourier and Laplace Transforms

D.1 Introduction

In many problems involving signal analysis and processing, a fundamental step in the solution will be the representation of the signal in some appropriate form. There are many possible representations. For example, the signal $f(t)$ can be written as the sum of an *even* function $f_1(t) = f_1(-t)$ and an *odd* function $f_2(t) = -f_2(-t)$:

$$f(t) = f_1(t) + f_2(t) \tag{D.1-1}$$

where the even function is determined by

$$f_1(t) = \tfrac{1}{2}[f(t) + f(-t)] \tag{D.1-2}$$

and the odd function by

$$f_2(t) = \tfrac{1}{2}[f(t) - f(-t)]. \tag{D.1-3}$$

The two representations that seem to be encountered most commonly are the *power series*

$$f(x) = f(0) + \frac{f^{(1)}(0)}{1!}x + \frac{f^{(2)}(0)}{2!}x^2 + \cdots + \frac{f^{(n)}(0)}{n!}x^n + \cdots \tag{D.1-4}$$

and the *trigonometric series*

$$\begin{aligned} f(x) = a_0 &+ a_1 \cos x + a_2 \cos 2x + \cdots \\ &+ b_1 \sin x + b_2 \sin 2x + \cdots . \end{aligned} \tag{D.1-5}$$

However, many others arise, particularly in the solution of boundary value problems, and a large literature exists on the subject [1, 2, 3].

D.2 Periodic Functions

A function $f(x)$ is called *periodic* with period T if there exists a constant $T > 0$ for which

$$f(t + T) = f(t) \tag{D.2-1}$$

for all t and $t + T$ in the domain of definition of $f(t)$. A function that is periodic with period T repeats itself every T seconds.

Example D.1 The functions sin t, cos t, tan t, cot t, and so on, are all periodic with period 2π in $(-\infty, \infty)$. The functions sin kt, cos kt, tan kt, cot kt, and so on, are all periodic with period $2\pi/k$ in $(-\infty, \infty)$.

● ●

The sum, difference, product, or quotient of two periodic functions of period T is a function of period T. Also it is apparent that a periodic function of period T is also periodic with period nT, $n = 2, 3, 4, \ldots$, that is,

$$\begin{aligned} f(t) &= f(t + T) = f(t + 2T) = f(t + 3T) = \cdots \\ &= f(t - T) = f(t - 2T) = f(t - 3T) = \cdots. \end{aligned} \tag{D.2-2}$$

If a periodic function with period T is integrable over any interval of length T, then it is integrable over any other interval of length T; in fact,

$$\int_a^{a+T} f(t)\, dt = \int_b^{b+T} f(t)\, dt, \qquad \text{for all } a, b. \tag{D.2-3}$$

Periodic functions were studied extensively by the French mathematician J. B. Fourier (1768–1830) who applied trigonometric series to the solution of many problems of mathematical physics.

D.3 Fourier Series

A series of the form

$$\sum_{k=0}^{n} (a_k \cos kx + b_k \sin kx)$$

will be called a trigonometric polynomial of order n. As mentioned in Section D.1, such series are of fundamental importance in the solution of the boundary value problems of mathematical physics. One of the lasting achievements of nineteenth-century mathematics was to show that an arbitrary periodic function could be represented by such series.

We begin with the representation of a function $f(t)$ by the trigonometric polynomial

$$f(t) = \frac{a_0}{2} + \sum_{k=1}^{n} (a_k \cos k\omega_1 t + b_k \sin k\omega_1 t) \tag{D.3-1}$$

where $\omega_1 = 2\pi/T$. This representation is periodic with period T. If it is to be useful, it is necessary to find a general means for calculating the coefficients a_k and b_k. Let us multiply both sides of Equation D.3-1 by $\cos m\omega_1 t$ and integrate from $-T/2$ to $T/2$ in t. The result is

$$\begin{aligned}\int_{-T/2}^{T/2} f(t) \cos m\omega_1 t\, dt &= \frac{a_0}{2} \int_{-T/2}^{T/2} \cos m\omega_1 t\, dt \\ &\quad + \sum_{k=1}^{n} a_k \int_{-T/2}^{T/2} \cos k\omega_1 t \cos m\omega_1 t\, dt \\ &\quad + \sum_{k=1}^{n} b_k \int_{-T/2}^{T/2} \sin k\omega_1 t \cos m\omega_1 t\, dt. \qquad \text{(D.3-2)}\end{aligned}$$

This expression may be simplified by noting that

$$\begin{aligned}\int_{-T/2}^{T/2} \cos k\omega_1 t \cos m\omega_1 t\, dt &= \tfrac{1}{2} \int_{-T/2}^{T/2} [\cos (k + m)\omega_1 t + \cos (k - m)\omega_1 t]\, dt \\ &= \begin{cases} \delta_{km} T/2 & k \text{ and } m \neq 0, \\ T, & k = m = 0, \end{cases} \qquad \text{(D.3-3)}\end{aligned}$$

where δ_{km} is the Kronecker delta defined by

$$\delta_{km} = \begin{cases} 0, & k \neq m, \\ 1, & k = m. \end{cases} \qquad \text{(D.3.4)}$$

In the same way, we have

$$\begin{aligned}\int_{-T/2}^{T/2} \sin k\omega_1 t \ \cos m\omega_1 t\, dt &= \tfrac{1}{2} \int_{-T/2}^{T/2} [\sin (k + m)\omega_1 t + \sin (k - m)\omega_1 t]\, dt \\ &= 0, \qquad \text{for all } k \text{ and } m. \qquad \text{(D.3-5)}\end{aligned}$$

It is clear from Equations D.3-3 and D.3-5 that only one term will be nonzero in Equation D.3-2 and that

$$a_k = \frac{2}{T} \int_{-T/2}^{T/2} f(t) \cos k\omega_1 t\, dt, \qquad k = 0, 1, \ldots. \qquad \text{(D.3-6)}$$

Now, if we multiply both sides of Equation D.3-1 by $\sin k\omega_1 t$ and follow the same procedure as before, we obtain an expression for the coefficient b_k:

$$b_k = \frac{2}{T} \int_{-T/2}^{T/2} f(t) \sin k\omega_1 t\, dt, \qquad k = 0, 1, \ldots. \qquad \text{(D.3-7)}$$

These expressions for a_k and b_k are called the *Fourier coefficients* of order k of the function $f(t)$, and the resulting representation of Equation D.3-1 is called the nth-order *Fourier polynomial* of the function $f(t)$. As $n \to \infty$, the series becomes the *Fourier series* representation of $f(t)$.

Example D.2 If a sine wave $A \sin \omega_1 t$ is passed through an ideal half-wave rectifier, the resulting waveform is shown in Figure D.1. Let us find the Fourier series representation. From Equation D.3-6 ,we have

$$\begin{aligned} a_k &= \frac{2}{T}\int_0^{T/2} A \sin \omega_1 t \cos k\omega_1 t \, dt \\ &= \frac{A}{T}\int_0^{T/2} [\sin (1+k)\omega_1 t + \sin (1-k)\omega_1 t] \, dt \\ &= \frac{A}{T}\left[-\frac{\cos (1+k)\omega_1 t}{(1+k)\omega_1} - \frac{\cos (1-k)\omega_1 t}{(1-k)\omega_1}\right]_0^{T/2} \\ &= \begin{cases} 0, & k \text{ odd}, \\ \dfrac{2A}{\pi}\left(\dfrac{1}{1-k^2}\right), & k \text{ even}. \end{cases} \end{aligned}$$

From Equation D.3-7,

$$\begin{aligned} b_k &= \frac{2}{T}\int_0^{T/2} A \sin \omega_1 t \sin k\omega_1 t \, dt \\ &= \frac{A}{T}\int_0^{T/2} [\cos (1-k)\omega_1 t - \cos (1+k)\omega_1 t] \, dt \\ &= \frac{A}{T}\left[\frac{\sin (1-k)\omega_1 t}{(1-k)\omega_1} - \frac{\sin (1+k)\omega_1 t}{(1+k)\omega_1}\right]_0^{T/2} \\ &= \begin{cases} \dfrac{A}{2}, & k = 1, \\ 0, & k \neq 1. \end{cases} \end{aligned}$$

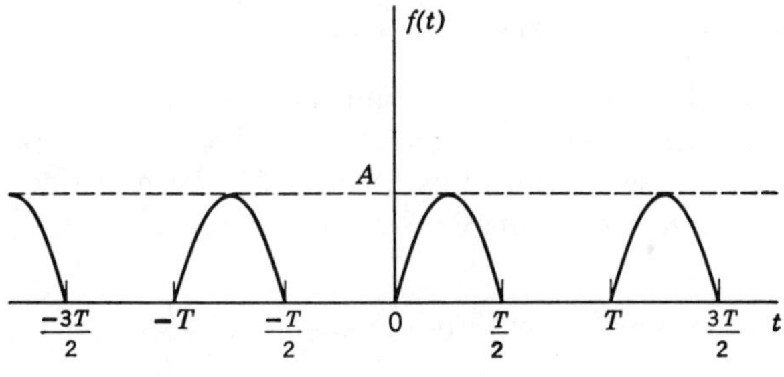

Figure D.1 *A half-wave rectified sine wave:*

$$f(t) = \begin{cases} A \sin \omega_1 t, & nT \le t \le (2n+1)\dfrac{T}{2} \\ 0, & (2n+1)\dfrac{T}{2} \le t \le (n+1)T \end{cases}$$

$$\omega_1 = \frac{2\pi}{T}, \qquad n = \ldots, -3, -2, -1, 0, 1, 2, 3, \ldots$$

Thus the Fourier series is

$$f(t) = \frac{A}{\pi} + \frac{A}{2}\sin\omega_1 t + \frac{2A}{\pi}\sum_{\substack{n\ \text{even}\\ n\geq 2}} \frac{\cos k\omega_1 t}{1-k^2}.$$

• •

An examination of Equations D.3-6 and D.3-7 shows that if $f(t)$ is an even function [that is, if $f(t) = f(-t)$], then $b_k = 0$, and if $f(t)$ is an odd function [that is, if $f(t) = -f(-t)$], then $a_k = 0$. Thus an even function of period T can be represented by a *Fourier cosine series*

$$f(t) = \frac{a_0}{2} + \sum_{k=1}^{\infty} a_k \cos k\omega_1 t \tag{D.3-8}$$

where

$$a_k = \frac{4}{T}\int_0^{T/2} f(t)\cos k\omega_1 t\, dt, \qquad k = 0, 1, \ldots. \tag{D.3-9}$$

In the same way, an odd function of period T can be represented by a *Fourier sine series*

$$f(t) = \sum_{k=1}^{\infty} b_k \sin k\omega_1 t \tag{D.3-10}$$

where

$$b_k = \frac{4}{T}\int_0^{T/2} f(t)\sin k\omega_1 t\, dt, \qquad k = 1, 2, \ldots. \tag{D.3-11}$$

D.4 Some Properties of the Fourier Series [4, 5, 6, 7]

In the previous discussion we have said that the Fourier series is used to "represent" a function $f(t)$, but we have not defined precisely what we mean by this term "represent." Until this situation has been clarified it would be more appropriate to write Equation D.3-1 as

$$f(t) \sim \frac{a_0}{2} + \sum_{k=1}^{n} (a_k \cos k\omega_1 t + b_k \sin k\omega_1 t) \tag{D.4-1}$$

since the meaning of the equality has not been established.

Let $f_n(t)$ be any trigonometric polynomial given by

$$f_n(t) = \frac{\alpha_0}{2} + \sum_{k=1}^{n} (\alpha_k \cos k\omega_1 t + \beta_k \sin k\omega_1 t) \tag{D.4-2}$$

where the α_k and β_k are as yet undetermined. We attempt to approximate a function $f(t)$ by this polynomial and investigate the mean-squared difference ϵ^2 given by

$$\epsilon^2 = \frac{1}{T}\int_{-T/2}^{T/2} [f(t) - f_n(t)]^2 \, dt \tag{D.4-3}$$

or by

$$\epsilon^2 = \frac{1}{T}\int_{-T/2}^{T/2} [f^2(t) - 2f(t)f_n(t) + f_n^2(t)] \, dt. \tag{D.4-4}$$

On substituting Equation D.4-2 in this last expression and rearranging, we obtain

$$\epsilon^2 = \left\{\frac{1}{T}\int_{-T/2}^{T/2} f^2(t)\, dt - \left[\frac{a_0^2}{4} + \sum_{k=1}^{n} (a_k^2 + b_k^2)\right]\right\}$$
$$+ \frac{1}{2}(\alpha_0 - a_0)^2 + \sum_{k=1}^{n} [(\alpha_k - a_k)^2 + (\beta_k - b_k)^2] \tag{D.4-5}$$

where the a_k and b_k are given by Equations D.3-6 and D.3-7. It is clear that the terms in curly brackets are independent of the choice of the coefficients α_k and β_k and that the remaining terms are nonnegative; therefore the mean-squared difference ϵ^2 is least when these remaining terms are zero, or when

$$\alpha_k = a_k, \qquad k = 0, 1, 2, \ldots, n, \tag{D.4-6}$$

$$\beta_k = b_k, \qquad k = 1, 2, 3, \ldots, n. \tag{D.4-7}$$

Thus, of all trigonometric polynomials of order n, the Fourier polynomial gives the least mean-square approximation of a function $f(t)$.

Suppose that a Fourier polynomial of order n has in fact been used to approximate $f(t)$. The mean-squared difference is given by Equation D.4-5 as

$$\epsilon^2 = \frac{1}{T}\int_{-T/2}^{T/2} f^2(t)\, dt - \left[\frac{a_0^2}{4} + \sum_{k=1}^{b} (a_k^2 + b_k^2)\right] \geq 0 \tag{D.4-8}$$

and is nonnegative from the original definition. Thus we obtain the *Bessel Inequality*

$$\frac{a_0^2}{2} + \sum_{k=1}^{n} (a_k^2 + b_k^2) \leq \frac{1}{T}\int_{-T/2}^{T/2} f^2(t)\, dt. \tag{D.4-9}$$

The right side of this equation is independent of n and is, in fact, just the normalized signal energy. Since the series on the left is a sum of nonnegative terms and is bounded, it must converge in the limit as $n \to \infty$, or, in other words, the general term $(a_k^2 + b_k^2)$ must approach zero in the limit. Since both a_k^2 and b_k^2 are nonnegative,

$$\lim_{k\to\infty} a_k = \lim_{k\to\infty} b_k = 0, \tag{D.4-10}$$

provided only that $f(t)$ is both integrable and square integrable in $[-T/2, T/2]$, that is, that

$$\int_{-T/2}^{T/2} f(t)\, dt < \infty \tag{D.4-11}$$

and

$$\int_{-T/2}^{T/2} f^2(t)\, dt < \infty. \tag{D.4-12}$$

These last conditions are necessary for Equations D.4-3 and D.4-4 to be meaningful. It can be shown [5, 6] that in the limit as $n \to \infty$, the equality sign holds in Equation D.4-9 for any $f(t)$ that is bounded and integrable in $[-T/2, T/2]$. The resulting relation is often called *Parseval's Theorem*:

$$\frac{a_0^2}{4} + \sum_{k=1}^{\infty} (a_k^2 + b_k^2) = \frac{1}{T}\int_{-T/2}^{T/2} f^2(t)\, dt. \tag{D.4-13}$$

Let us now investigate briefly the point-by-point convergence of the Fourier polynomial $f_n(t)$ to the function $f(t)$ in the interval $[-T/2, T/2]$. It will be assumed that $f(t)$ satisfies the *Dirichlet Conditions* of Appendix A; that is, that it is bounded and has a finite number of relative maxima and minima and discontinuities in the interval $[-T/2, T/2]$.

We will need to make use of two definite integrals [5, 6, 7] known as *Dirichlet's Integrals*, which are

$$\lim_{\alpha \to \infty} \int_0^a f(x) \frac{\sin \alpha x}{\sin x}\, dx = \frac{\pi}{2} f(+0), \tag{D.4-14}$$

$$\lim_{\alpha \to \infty} \int_a^b f(x) \frac{\sin \alpha x}{\sin x}\, dx = 0 \tag{D.4-15}$$

where $0 < a < b < \pi$ and $f(x)$ satisfies the Dirichlet Conditions in (a, b). We shall not prove these relationships, but they can be justified in a heuristic fashion as follows. The left side of Equation D.4-14 can be written as

$$\lim_{\alpha \to \infty} \int_0^a \pi \left[f(x) \frac{x}{\sin x}\right]\left(\frac{\alpha}{\pi} \frac{\sin \alpha x}{\alpha x}\right) dx.$$

As α becomes larger and larger the second expression in the integrand approaches a delta function $\delta(x)$ (from Equation B.2-17) and Equation D.4-14 becomes

$$\pi \int_0^a \left[f(x) \frac{x}{\sin x}\right] \delta(x)\, dx = \frac{\pi}{2} f(+0) \tag{D.4-16}$$

from Equation B.2-1. In the same way, Equation D.4-14 follows since the interval (a, b) does not include the origin.

The kth term in the Fourier polynomial $f_n(t)$ can be written as

$$a_k \cos k\omega_1 t + b_k \sin k\omega_1 t = \left[\frac{2}{T}\int_{-T/2}^{T/2} f(x) \cos k\omega_1 x \, dx\right] \cos k\omega_1 t + \left[\frac{2}{T}\int_{-T/2}^{T/2} f(x) \sin k\omega_1 x \, dx\right] \sin k\omega_1 t = \frac{2}{T}\int_{-T/2}^{T/2} f(x) \cos k\omega_1 (x - t) \, dx \tag{D.4-17}$$

where x is a variable of integration. Since [7]

$$\frac{1}{2} + \sum_{k=1}^{n} \cos kz = \frac{\sin (n + \frac{1}{2})z}{2 \sin \frac{1}{2}z}, \tag{D.4-18}$$

then $f_n(t)$ can be written as

$$f_n(t) = \frac{2}{T}\int_{-T/2}^{T/2} f(x) \frac{\sin \omega_1 (n + \frac{1}{2})(x - t)}{2 \sin (\omega_1/2)(x - t)} \, dx \tag{D.4-19}$$

or, with a change of variable $y = (\omega_1/2)(x - t)$, as

$$f_n(t) = \frac{1}{\pi}\int_{-(1/2)(\pi + \omega_1 t)}^{(1/2)(\pi - \omega_1 t)} f\left(t + \frac{2y}{\omega_1}\right) \frac{\sin (2n + 1)y}{\sin y} \, dy. \tag{D.4-20}$$

This integral is a function of t through the limits. We consider the range of t to be $[-T/2, T/2]$, which corresponds to a range of $[-\pi, \pi]$ for the variable of integration y. There are three points in this latter interval in which $\sin y$ is zero, namely $-\pi, 0, \pi = y$. If we consider values of t for which $-T/2 < t < T/2$, then, of these three points, only $y = 0$ is included. From Equations D.4-14 and D.4-15, we see that Equation D.4-20 becomes

$$\lim_{n \to \infty} f_n(t) = \tfrac{1}{2}[f(t - 0) + f(t + 0)] \tag{D.4-21}$$

for $-T/2 < t < T/2$. If $f(t)$ is continuous at the point t, then the limit of $f_n(t)$ converges to $f(t)$ at that point.

We now consider the point $t = T/2$. Now Equation D.4-20 becomes

$$f_n(t) = \frac{1}{\pi}\int_{-\pi}^{0} f\left(\frac{T}{2} + \frac{2y}{\omega_1}\right) \frac{\sin (2n + 1)y}{\sin y} \, dy. \tag{D.4-22}$$

Now $\sin y$ is zero at $y = 0, -\pi$. For convenience Equation D.4-22 can be written as

$$f_n(t) = \frac{1}{\pi}\int_{-\pi}^{-\pi+\epsilon} f\left(\frac{T}{2} + \frac{2y}{\omega_1}\right) \frac{\sin (2n + 1)y}{\sin y} \, dy + \frac{1}{\pi}\int_{-\epsilon}^{0} f\left(\frac{T}{2} + \frac{2y}{\omega_1}\right) \frac{\sin (2n + 1)y}{\sin y} \, dy \tag{D.4-23}$$

since the expression will be evaluated only for large n. In the limit, this expression can be found from Equations D.4-14 and D.4-15 to be

$$\lim_{n\to\infty} f_n(t) = \frac{1}{2}\left[f\left(\frac{T}{2}+0\right) + f\left(\frac{T}{2}-0\right)\right], \qquad t = \frac{T}{2}. \tag{D.4-24}$$

The same result is found for the limit when $t = -T/2$.

Thus it has been shown that the Fourier series converges to the value of the function at points of continuity in $-T/2 \leq t \leq T/2$ and to the mean of the discontinuity at points of discontinuity in that interval.

It is clear that, for any continuous function $f(t)$, the Bessel Inequality of Equation D.4-9 becomes, in the limit, Parseval's Theorem given by Equation D.4-13. This is true since, for continuous $f(t)$, the integrand of Equation D.4-3 can be made arbitrarily small for all $-T/2 < t < T/2$ by making n large enough. The resulting error ϵ^2, which is the difference between the two sides of the Bessel Inequality, then becomes arbitrarily small. Furthermore, it is intuitively apparent that Parseval's Theorem will hold for any function that is bounded and integrable in $[-T/2, T/2]$. The references [5, 6, 7] should be consulted for further details.

D.5 The Complex Form for the Fourier Series

The Fourier series expansion for $f(t)$ in $[-T/2, T/2]$ may be rearranged to give

$$\begin{aligned} f(t) &= \frac{a_0}{2} + \sum_{k=1}^{\infty} (a_k \cos k\omega_1 t + b_k \sin k\omega_1 t) \\ &= \frac{1}{2}\sum_{k=-\infty}^{k=-1} (a_k + jb_k)\, e^{jk\omega_1 t} + \frac{a_0}{2} + \frac{1}{2}\sum_{k=1}^{\infty} (a_k - jb_k)\, e^{jk\omega_1 t} \end{aligned} \tag{D.5-1}$$

since

$$a_{-k} = a_k \tag{D.5-2}$$

and

$$b_{-k} = -b_k. \tag{D.5-3}$$

If we define a new set of Fourier coefficients c_k by

$$c_k = \tfrac{1}{2}(a_k - jb_k), \qquad k > 0, \tag{D.5-4}$$

then we note that

$$c_{-k} = \tfrac{1}{2}(a_k + jb_k) \tag{D.5-5}$$

and that Equation D.5-1 can be written as

$$f(t) = \sum_{k=-\infty}^{\infty} c_k\, e^{jk\omega_1 t}. \tag{D.5-6}$$

This is called the *complex form of the Fourier series.* Furthermore the coefficients c_k can be obtained from Equations D.3-6 and D.3-7 as

$$c_k = \frac{1}{T}\int_{-T/2}^{T/2} f(t)\, e^{-jk\omega_1 t}\, dt. \tag{D.5-7}$$

This expression could be found alternately from Equation D.5-6 by noting the orthogonality relationship

$$\frac{1}{T}\int_{-T/2}^{T/2} e^{jk\omega_1 t}\, e^{-jm\omega_1 t}\, dt = \delta_{km} \tag{D.5-8}$$

and following the procedure of Section D.3. It is sometimes more convenient to use this complex form rather than the usual trigonometric series. In the complex notation, Parseval's Theorem (Equation D.4-13) becomes

$$\frac{1}{T}\int_{-T/2}^{T/2} f^2(t)\, dt = \sum_{k=-\infty}^{\infty} |c_k|^2. \tag{D.5-9}$$

D.6 The Gibbs Phenomenon

The trigonometric polynomial $f_n(t)$ does not approach the function $f(t)$ uniformly over any interval containing a point of discontinuity of $f(t)$. The form of the deviation of $f_n(t)$ from $f(t)$ is known as the *Gibbs Phenomenon* [7, 8]. A detailed discussion is beyond the scope of this text [see Reference 8] but a brief treatment will be given in terms of a specific example.

Example D.3 Find the Fourier series for the "square wave" given by

$$f(t) = \begin{cases} -1, & -\dfrac{T}{2} < t < 0, \\ 1, & 0 < t < \dfrac{T}{2}. \end{cases}$$

This signal is shown in Figure D.2(a). It is an odd function and can be represented by the Fourier sine series of Equation D.3-10. The coefficients b_k are given by

$$b_k = \frac{4}{T}\int_0^{T/2} \sin k\omega_1 t\, dt = \frac{2}{k\pi}(1 - \cos k\pi)$$

$$= \begin{cases} 0, & \text{k even}, \\ \dfrac{4}{\pi k}, & k \text{ odd}. \end{cases}$$

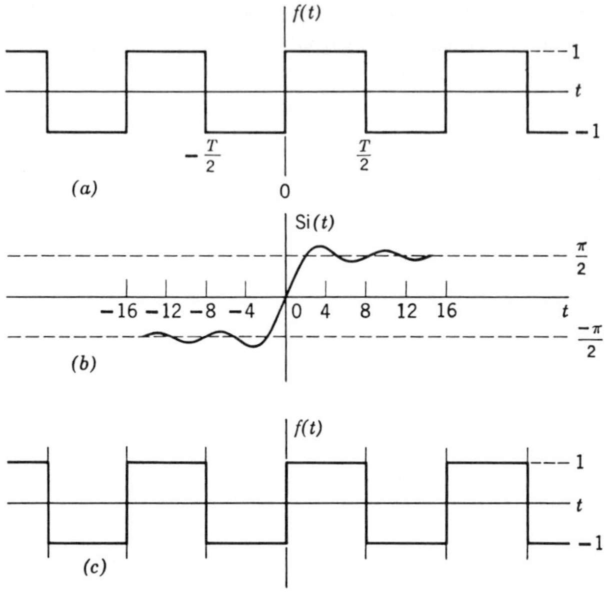

Figure D.2 *A square wave and the Gibbs Phenomenon: (a) a square wave; (b) the function Si(t); (c) the Gibbs Phenomenon*

Thus the Fourier series is

$$f(t) = \frac{4}{\pi} \sin \omega_1 t + \frac{4}{3\pi} \sin 3\omega_1 t + \frac{4}{5\pi} \sin 5\omega_1 t + \cdots.$$

• •

The Fourier polynomial $f_n(t)$ (which is the nth partial sum of the Fourier series) is given by Equation D.4-19. Let us restrict ourselves to the square wave we have just considered so that Equation D.4-19 becomes

$$f_n(t) = \frac{2}{T} \int_0^{T/2} \frac{\sin \omega_1 (n + \frac{1}{2})(x - t)}{2 \sin (\omega_1/2)(x - t)} \, dx - \frac{2}{T} \int_{-T/2}^{0} \frac{\sin \omega_1 (n + \frac{1}{2})(x - t)}{2 \sin (\omega_1/2)(x - t)} \, dx. \tag{D.6-1}$$

On substituting $(\omega_1/2)(x - t) = y$ in the first integral and $(\omega_1/2)(x - t) = -y$ in the second and rearranging, we obtain

$$f_n(t) = \frac{1}{\pi} \int_{-\omega_1 t/2}^{\omega_1 t/2} \frac{\sin (2n + 1)y}{\sin y} \, dy + \frac{1}{\pi} \int_{(\pi + \omega t_1)/2}^{(\pi - \omega_1 t)/2} \frac{\sin (2n + 1)y}{\sin y} \, dy. \tag{D.6-2}$$

We desire to investigate these two integrals for large n in the vicinity of $t = 0$ where the discontinuity occurs. The second integral, for small t, is evaluated over an interval in the vicinity of $y = \pi/2$ where its integrand has a magnitude near unity. The first integral, on the other hand, is evaluated in the vicinity of $y = 0$ and its integrand has a large value there which can be found by taking the ratio of the derivatives of the numerator and denominator to obtain

$$\left.\frac{(2n + 1)\cos(2n + 1)y}{\cos y}\right|_{y=0} = 2n + 1. \tag{D.6-3}$$

Thus, for large n, the second integral in Equation D.6-2 can be neglected and the partial sum in the vicinity of the discontinuity at the origin is given approximately, for large n, by

$$f_n(t) = \frac{1}{\pi}\int_{-\omega_1 t/2}^{\omega_1 t/2} \frac{\sin(2n + 1)y}{\sin y}\, dy. \tag{D.6-4}$$

In the vicinity of $y = 0$, $\sin y$ can be replaced by y and Equation D.6-4 becomes approximately

$$f_n(t) \approx \frac{2}{\pi}\int_{0}^{\omega_1 t/2} \frac{\sin(2n + 1)y}{y}\, dy \tag{D.6-5}$$

or, with the change in variable $(2n + 1)y = x$,

$$f_n(t) \approx \frac{2}{\pi}\int_{0}^{[(n+(1/2)]\omega_1 t} \frac{\sin x}{x}\, dx = \frac{2}{\pi}\,\mathrm{Si}\,[(n + \tfrac{1}{2})\omega_1 t] \tag{D.6-6}$$

where the *sine integral* Si (t) is defined by

$$\mathrm{Si}\,(t) = \int_0^t \frac{\sin x}{x}\, dx.$$

It is tabulated in many references [9, 10]. This function is zero for $t = 0$ and approaches $\pi/2$ as $t \to \infty$ and $-\pi/2$ as $t \to -\infty$ as shown in Figure D.2(b).

Equation D.6-6 explains the behavior of $f_n(t)$ in the vicinity of the origin for large n and for the "square-wave" example. As the number of terms n in the Fourier approximation increases, the form of the curve of Figure D.2(b) does not change but the scale is compressed and the oscillations occur more rapidly. Thus the successive maximum and minimum values do not change but they are crowded in toward the origin and, for large n, the approximation $f_n(t)$ would tend to appear as shown in Figure D.2(c). In the general case there will be similar overshoots at all discontinuities. The subject is discussed in detail in [8] and a number of examples are given there.

D.7 The Fourier Integral Theorem

The frequency spectrum of a periodic signal $f(t)$ with period T can be obtained by considering the complex Fourier series representation of Equation D.5-6

$$f(t) = \sum_{k=-\infty}^{\infty} c_k \, e^{jk\omega_1 t} \tag{D.7-1}$$

where, as before, the Fourier coefficient is obtained from Equation D.5-7

$$c_k = \frac{1}{T}\int_{-T/2}^{T/2} f(x)\, e^{-jk\omega_1 x}\, dx \tag{D.7-2}$$

and

$$\omega_1 = \frac{2\pi}{T}. \tag{D.7-3}$$

In the frequency domain, the coefficient c_k is a measure of the signal magnitude and relative phase angle at the frequency $k\omega_1$. It should be emphasized that c_k generally is complex and thus both its magnitude and angle must be considered. For the square wave of Figures D.2 and D.3 the coefficients of the complex Fourier series are given by

$$c_k = \tfrac{1}{2}(a_k - jb_k) = \begin{cases} 0, & k \text{ even}, \\ -\dfrac{j2}{\pi k}, & k \text{ odd}. \end{cases} \tag{D.7-4}$$

The magnitude and angle of this *line spectrum* is plotted in Figure D.3.

One approach to the problem of representing a nonperiodic signal might be to use a Fourier series expansion in the interval $[-T/2, T/2]$ and allow $T \to \infty$. It is apparent that, when this is done, the spectral line spacing $\omega_1 = 2\pi/T$ will become smaller and smaller. In the limit the function $f(t)$ is defined for all t and the line spectrum becomes a continuous function of frequency. The result is the *Fourier integral representation*, which will now be derived. Many detailed treatments are available in the literature [5, 6, 8, 11].

With the substitution of Equation D.7-2 into Equation D.7-1, we have

$$f(t) = \sum_{k=-\infty}^{\infty} \left[\frac{1}{T}\int_{-T/2}^{T/2} f(x)\, e^{-jk\omega_1 x}\, dx\right] e^{jk\omega_1 t}. \tag{D.7-5}$$

We now consider the limiting case where $T \to \infty$ (or $\omega_1 \to 0$) in such a way that $k\omega_1 \to \infty$ as $k \to \infty$. After the substitutions

$$k\omega_1 = \omega = k\,\Delta\omega, \tag{D.7-6}$$

Equation D.7-5 may be written as

$$f(t) = \frac{1}{2\pi}\sum_{k=-\infty}^{\infty} \left[\int_{-T/2}^{T/2} f(x)\, e^{-j\omega x}\, dx\right] e^{j\omega t}\,\Delta\omega. \tag{D.7-7}$$

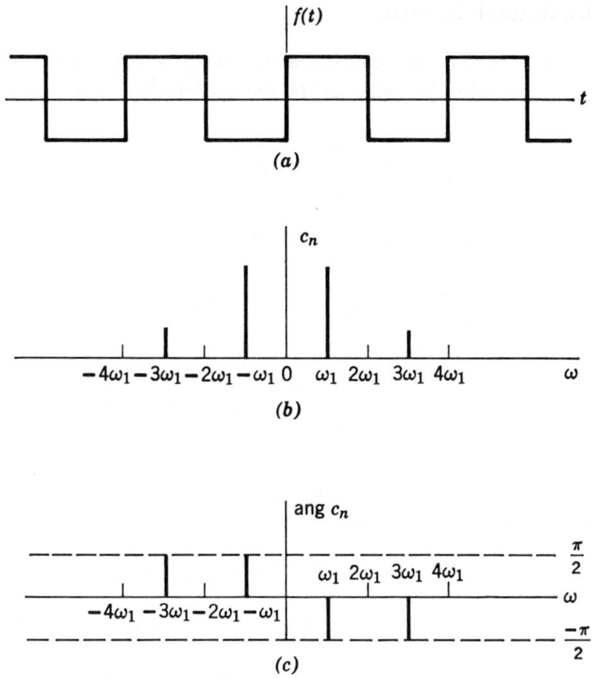

***Figure D.3** Line spectrum of a square wave: (a) square wave; (b) line spectrum magnitude; (c) line spectrum angle*

In the limit as $T \to \infty$, it is reasonable to expect that this equation will approach

$$f(t) = \frac{1}{2\pi} \int_{-\infty}^{\infty} F(\omega)\, e^{j\omega t}\, d\omega \tag{D.7-8}$$

where

$$F(\omega) = \int_{-\infty}^{\infty} f(x)\, e^{-j\omega x}\, dx. \tag{D.7-9}$$

However, it should be emphasized that this procedure is not rigorous since the limit given is not the definition of the integral and since $F(\omega)$ is itself a function of the limiting operation. This point will be discussed in the next paragraph. Equations D.7-8 and D.7-9 are called a *Fourier-transform pair*, relating two functions $f(t)$ and its *Fourier transform* $F(\omega)$. If Equation D.7-9 is substituted into Equation D.7-8, the result is one form of the *Fourier Integral Theorem*:

$$f(t) = \frac{1}{2\pi} \int_{-\infty}^{\infty} \left[\int_{-\infty}^{\infty} f(x)\, e^{-j\omega x}\, dx \right] e^{j\omega t}\, d\omega. \tag{D.7-10}$$

We proceed now to verify this theorem somewhat more carefully.

Let us consider the function

$$f_A(t) = \frac{1}{2\pi}\int_{-A}^{A}\left[\int_{-\infty}^{\infty} f(x)\, e^{-j\omega x}\, dx\right] e^{j\omega t}\, d\omega. \tag{D.7-11}$$

Interchanging the order of integration, we have

$$f_A(t) = \int_{-\infty}^{\infty} f(x)\left[\frac{1}{2\pi}\int_{-A}^{A} e^{j\omega(t-x)} d\omega\right] dx$$

or

$$f_A(t) = \int_{-\infty}^{\infty} f(x)\,\frac{\sin A(t-x)}{\pi(t-x)}\, dx. \tag{D.7-12}$$

The function $[\sin A(t-x)]/[\pi(t-x)]$ is one of the approximating forms for the delta function $\delta(t-x)$ as given by Equation B.2-17 and shown in Figure B.3. Thus Equation D.7-12 becomes in the limit

$$\lim_{A\to\infty} f_A(t) = \int_{-\infty}^{\infty} f(x)\,\delta(t-x)\, dx$$

$$= \tfrac{1}{2}[f(t-0) + f(t+0)] \tag{D.7-13}$$

and the Fourier Integral Theorem of Equation D.7-10 should be written as

$$\tfrac{1}{2}[f(t-0) + f(t+0)] = \frac{1}{2\pi}\int_{-\infty}^{\infty}\int_{-\infty}^{\infty} f(x)\, e^{-j\omega x}\, dx\, e^{j\omega t}\, d\omega \tag{D.7-14}$$

which reduces to Equation D.7-10 if $f(t)$ is continuous at t.

The use of the delta function in the verification of the Fourier Integral Theorem has tended to obscure the restrictions which must be placed on $f(t)$ for the theorem to hold. A careful examination of Equation D.7-12 as given in [8] will establish these restrictions. Sufficient conditions are that $f(t)$ satisfy the Dirichlet Conditions of Appendix A and that $f(t)$ be absolutely convergent, that is,

$$\int_{-\infty}^{\infty} |f(t)|\, dt < \infty. \tag{D.7-15}$$

The Fourier-transform pair of Equations D.7-8 and D.7-9 is identical to that given in Chapter 3 as Equations 3.6-7 and 3.6-9 and has been shown to be valid for functions that are continuous and absolutely integrable. If the functions are not continuous, then the transform is equal to the mean value of the discontinuity as given by Equation D.7-14. As mentioned at the beginning of Section 3.6, Fourier integral theory can be developed using the concepts of "square integrability" and "convergence in the mean." The procedures are similar to those already used in this appendix and the results are Equations 3.6-3 and 3.6-4. A detailed discussion is given in Chapter 3 of [11].

D.8 Elementary Properties of Fourier Transforms

The Fourier-transform pair $f(t)$ and $F(\omega)$ are related through Equations D.7-8 and D.7-9, or, equivalently, through the Fourier Integral Theorem of Equation D.7-10. Some of the more elementary properties of the transform $F(\omega)$ follow directly from Equation D.7-9. It will be convenient sometimes to denote the Fourier transform of a time function by $\mathscr{F}(\ )$ so that Equation D.7-9 can be written as

$$F(\omega) = \mathscr{F}[f(t)]. \tag{D.8-1}$$

Since integration is a linear operation, it follows that the Fourier transformation is also *linear*. Thus if the time functions $a(t)$ and $b(t)$ have Fourier transforms $A(\omega)$ and $B(\omega)$, respectively, it follows that

$$\mathscr{F}[ca(t) + db(t)] = cA(\omega) + dB(\omega) \tag{D.8-2}$$

where c and d are arbitrary constants. The Fourier transform is also *symmetric*; that is, if $f(t)$ is real

$$\mathscr{F}[f(t)] = F(\omega) \Rightarrow \mathscr{F}[f(-t)] = F(-\omega) = F^*(\omega) \tag{D.8-3}$$

where $F^*(\omega)$ is the complex conjugate of $F(\omega)$. If $f(t)$ is an *even* function, then Equation D.7-9 can be written as

$$F(\omega) = \int_{-\infty}^{\infty} f(x) \cos \omega x \, dx = 2\int_{0}^{\infty} f(x) \cos \omega x \, dx \tag{D.8-4}$$

while, if $f(t)$ is an *odd* function,

$$F(\omega) = -j\int_{-\infty}^{\infty} f(x) \sin \omega x \, dx = -2j\int_{0}^{\infty} f(x) \sin \omega x \, dx. \tag{D.8-5}$$

Suppose that the time function $f(t)$ with Fourier transform $F(\omega)$ is translated by an amount t_0 to form the new function $f(t - t_0)$. The Fourier transform of $f(t - t_0)$ is

$$\mathscr{F}[f(t - t_0)] = \int_{-\infty}^{\infty} f(x - t_0)\, e^{-j\omega x}\, dx. \tag{D.8-6}$$

Let us make the linear change of variable $y = x - t_0$ so that

$$\begin{aligned}\mathscr{F}[f(t - t_0)] &= \int_{-\infty}^{\infty} f(y)\, e^{-j\omega y}\, e^{-j\omega t_0}\, dy \\ &= e^{-j\omega t_0}\int_{-\infty}^{\infty} f(y)\, e^{-j\omega y}\, dy \\ &= e^{-j\omega t_0} F(\omega).\end{aligned} \tag{D.8-7}$$

Thus translation in the time domain by an amount t_0 corresponds to multiplication in the frequency domain by the factor $e^{j\omega t_0}$. In the same way, it is easy to show that

$$\mathscr{F}[f(t)\, e^{j\omega_0 t}] = F(\omega - \omega_0). \tag{D.8-8}$$

The derivative $[df(t)]/dt$ may be transformed as

$$\mathscr{F}\left[\frac{df(t)}{dt}\right] = \int_{-\infty}^{\infty} \frac{df(t)}{dt}\, e^{-j\omega t}\, dt.$$

Integration by parts yields

$$\mathscr{F}\left[\frac{df(t)}{dt}\right] = f(t)\, e^{-j\omega t}\Big|_{-\infty}^{\infty} + j\omega \int_{-\infty}^{\infty} f(t)\, e^{-j\omega t}\, dt$$

or

$$\mathscr{F}\left[\frac{df(t)}{dt}\right] = j\omega F(\omega) \tag{D.8-9}$$

if $f(\infty) = f(-\infty) = 0$. This last condition must hold in any case if $f(t)$ is to be absolutely integrable and, hence, transformable. Successive applications of the same procedure yields

$$\frac{d^n}{dt^n}[f(t)] = (j\omega)^n F(\omega). \tag{D.8-10}$$

The relationships that have been developed in this section as well as some others that will be used in the text are summarized in Table 2, Appendix I. Some of the more commonly encountered Fourier-transform pairs are also given in that table as well as in Table 3.1 of Chapter 3.

D.9 The Laplace Transform

In a wide variety of problems, particularly those involving the analysis of linear systems with nonrandom inputs, it is necessary or desirable to assume that the driving forces are applied at some fixed time, say $t = 0$. In such cases, a given driving force or signal is of interest only for positive time since its behavior for negative time is irrelevant to the problem. The Laplace transform $F_l(s)$ of a time function $f(t)$ is defined as

$$F_l(s) = \int_0^{\infty} f(t)\, e^{-st}\, dt \tag{D.9-1}$$

where s is a complex variable with real part σ and imaginary part $j\omega$ so that

$$s = \sigma + j\omega. \tag{D.9-2}$$

Note that integration in the time domain takes place only over the interval $(0, \infty)$. Thus any two different time functions $f_1(t)$ and $f_2(t)$ that are identical for positive time will have the same Laplace transform.

It has already been pointed out in Section D.7 that a sufficient condition for the existence of the Fourier transform $F(\omega)$ of $f(t)$ is that $f(t)$ be absolutely integrable as given by Equation D.7-15. In the same way the Laplace transform $F_l(s)$ exists if

$$\int_0^\infty |f(t)|\, e^{-\sigma t}\, dt < \infty \tag{D.9-3}$$

for some real number σ. Most functions encountered in engineering satisfy this condition. Indeed many functions which do not have Fourier transforms will have Laplace transforms due to the rapid decay of $e^{-\sigma t}$. For example, for all n,

$$\lim_{x\to\infty} x^n\, e^{-x} = 0$$

and

$$\int_0^\infty x^n\, e^{-x}\, dx = \Gamma(n+1)$$

where $\Gamma(n)$ is the gamma function. On the other hand, the Fourier transform of x^n does not exist for $n > 0$, and for $n = 0$ can only be expressed as a delta function.

The inversion of Equation D.9-1 to find $f(t)$ from $F_l(s)$ is not as simple as for the Fourier transform. It can be shown [5, 12] that $f(t)$ is given by

$$f(t) = \frac{1}{2\pi j}\int_{\sigma_1 - j\infty}^{\sigma_1 + j\infty} F(s)\, e^{st}\, ds \tag{D.9-4}$$

where σ_1 is a number greater than the σ of Equation D.9-3. This expression for $f(t)$ is a contour integral in the complex s plane and may be difficult to evaluate. The usual procedure is to avoid the inverse transformation and build up a table of Laplace-transform pairs $f(t)$ and $F_l(s)$ from the direct transform of Equation D.9-1. The function $f(t)$ given by Equation D.9-4 corresponds to the $f(t)$ of Equation D.9-1 for positive time and is zero for negative time.

It can be shown in a straightforward manner that the Laplace transform has many of the same properties as the Fourier transform. A number of these as well as some common transform pairs are listed in Table 3, Appendix I. An examination of this table shows that there are certain differences between the two types of transforms. For example, the lower limit in the Laplace transform expression is $t = 0$. Thus, the Laplace transform $\mathscr{L}(\ )$ of the derivative of a function becomes

$$\mathscr{L}\left[\frac{df(t)}{dt}\right] = \int_0^\infty \frac{df(t)}{dt}\, e^{-st}\, dt = e^{-st} f(t)\Big|_0^\infty + s\int_0^\infty f(t)\, e^{-st}\, dt$$
$$= sF_l(s) - f(0).$$

The initial condition $f(0)$ enters into the transform of the derivative. In the solution of ordinary differential equations by the use of the Laplace transform, the initial conditions associated with the equations are automatically included.

It is clear that the Laplace and Fourier transforms are distinct. Any function $f(t)$ that is nonzero for negative time cannot be represented uniquely as an inverse Laplace transform. On the other hand, many functions that have Laplace transforms will not have Fourier transforms. However, if

$$f(t) = 0, \qquad t < 0,$$

and

$$\int_0^{\infty} |f(t)|\, dt < \infty,$$

the number σ in Equation D.9-1 can be set equal to zero and

$$\mathcal{F}[f(t)] = \mathcal{L}[f(t)]_{s=j\omega}.$$

D.10 Signal Energy and the Power Spectrum

Consider the two time functions $f_1(t)$ and $f_2(t)$ with Fourier transforms $F_1(\omega)$ and $F_2(\omega)$, respectively. Equation D.7-8 may be used to write

$$\int_{-\infty}^{\infty} f_1(t) f_2(t)\, dt = \frac{1}{(2\pi)^2} \int_{-\infty}^{\infty} \int_{-\infty}^{\infty} F_1(x)\, e^{jxt}\, dx \int_{-\infty}^{\infty} F_2(y)\, e^{jyt}\, dy\, dt. \tag{D.10-1}$$

This expression may be rearranged to give

$$\int_{-\infty}^{\infty} f_1(t) f_2(t)\, dt = \frac{1}{2\pi} \int_{-\infty}^{\infty} \int_{-\infty}^{\infty} F_1(x) F_2(y) \left[\frac{1}{2\pi} \int_{-\infty}^{\infty} e^{jt(x+y)}\, dt\right] dx\, dy. \tag{D.10-2}$$

The integral in brackets may be evaluated from pair number 7 of Table 3.1 and is $\delta(x + y)$. Now Equation D.10-2 becomes

$$\int_{-\infty}^{\infty} f_1(t) f_2(t)\, dt = \frac{1}{2\pi} \int_{-\infty}^{\infty} F_1(x) F_2(-x)\, dx = \frac{1}{2\pi} \int_{-\infty}^{\infty} F_1(-x) F_2(x)\, dx. \tag{D.10-3}$$

For the special case where $f_1(t) = f_2(t) = f(t)$, this reduces to

$$\int_{-\infty}^{\infty} f^2(t)\, dt = \frac{1}{2\pi} \int_{-\infty}^{\infty} |F(\omega)|^2\, d\omega. \tag{D.10-4}$$

This last equation gives the signal energy expressed either in the time or frequency domain. It is one of the many forms of *Parseval's Theorem* which has already been given in terms of the coefficients of a Fourier series by Equations D.4-13 and D.5-9. The quantity $|F(\omega)|^2$ is sometimes called the *power spectrum* of $f(t)$ but it is more conventional to define power spectrum somewhat differently as in Equation 7.8-9.

REFERENCES

1. P. M. Morse and H. Feshbach, *Methods of Theoretical Physics*, Vols. I and II, McGraw-Hill, New York, 1953.
2. H. Margenau and G. M. Murphy, *The Mathematics of Physics and Chemistry*, Vol. I, Van Nostrand, Princeton, N. J., 1956.
3. W. Magnus and F. Oberhettinger, *Formulas and Theorems for the Special Functions of Mathematical Physics*, Chelsea, New York, 1949.
4. R. V. Churchill, *Fourier Series and Boundary Value Problems*, McGraw-Hill, New York, 1941.
5. W. Kaplan, *Operational Methods for Linear Systems*, Addison-Wesley, Reading, Mass., 1962.
6. T. M. Apostol, *Mathematical Analysis*, Addison-Wesley, Reading, Mass., 1957.
7. Z. Zygmund, *Trigonometric Series*, Dover, New York, 1955.
8. H. S. Carslaw, *Introduction to the Theory of Fourier's Series and Integrals*, Dover, New York, 1930.
9. E. Jahnke and F. Emde, *Tables of Functions*, Dover, New York, 1945.
10. *Handbook of Mathematical Tables*, Chemical Rubber, Cleveland, Ohio, 1962.
11. E. C. Titchmarsh, *Introduction to the Theory of Fourier Integrals*, Oxford Univ. Press, Oxford, England, 1948.
12. H. S. Carslaw and J. C. Jaeger, *Operational Methods in Applied Mathematics*, Dover, New York, 1963.

Appendix E

Some Inequalities Including Schwarz's Inequality

E.1 Introduction

There are several inequalities of considerable utility in many areas of analysis. We discuss and develop these briefly in the following sections. We begin with an elementary inequality that will be useful in later proofs.

Let a, b be two real nonnegative numbers

$$a > 0, \qquad b > 0, \tag{E.1-1}$$

and let x be a real variable. Form the function

$$y = F(x; a, b) = xa + (1 - x)b - a^x b^{1-x}. \tag{E.1-2}$$

Then it is apparent that

$$F(0; a, b) = F(1; a, b) = 0, \tag{E.1-3}$$

that is, the curve of y versus x passes through (0, 0) and (1, 0). Furthermore, on differentiating twice with respect to x, we have

$$\frac{d^2F(x; a, b)}{dx^2} = -\left(\log \frac{b}{a}\right)^2 a^x b^{1-x} \begin{cases} < 0, & a \neq b, \\ = 0, & a = b. \end{cases} \tag{E.1-4}$$

Note also that, for $a = b$,

$$F(x; a, a) = 0. \tag{E.1-5}$$

Thus, for $a = b$, the curve of y versus x is the x axis and, for $a \neq b$, it is concave downward and passes through the points (0, 0) and (1, 0). In other words

$$F(x; a, b) > 0, \qquad 0 < x < 1, \qquad a \neq b. \tag{E.1-6}$$

It follows from Equations E.1-5 and E.1-6 and from the definition of $F(x; a, b)$ as given by Equation E.1-1 that

$$\boxed{\begin{gathered} a^x b^{1-x} \leq xa + (1-x)b \\ \text{for} \\ a > 0, \quad b > 0, \quad 0 < x < 1 \end{gathered}} \tag{E.1-7}$$

which is the result desired.

E.2 Hölder's Inequality

Let $f(t)$ and $g(t)$ be functions of the real variable t in the interval (a, b). Assume that

$$\int_a^b |f(t)|^p \, dt < \infty \tag{E.2-1}$$

and

$$\int_a^b |g(t)|^q \, dt < \infty \tag{E.2-2}$$

where p and q are real numbers related by

$$\frac{1}{p} + \frac{1}{q} = 1 \tag{E.2-3}$$

and $p > 1$. We note first that, for any t, either

$$|g(t)| \leq |f(t)|^{p-1} \tag{E.2-4}$$

and hence

$$|f(t)g(t)| \leq |f(t)|^p \tag{E.2-5}$$

or else

$$|g(t)| > |f(t)|^{p-1}, \tag{E.2-6}$$

$$|f(t)| \leq |g(t)|^{1/(p-1)} = |g(t)|^{q-1} \tag{E.2-7}$$

and hence

$$|f(t)g(t)| \leq |g(t)|^q. \tag{E.2-8}$$

It follows from Equations E.2-1 and E.2-2 that

$$\boxed{\int_a^b |f(t)g(t)| \, dt < \infty} \tag{E.2-9}$$

which is the first result desired.

Let us assume that neither $f(t)$ nor $g(t)$ vanishes so that

$$f(t)g(t) \neq 0. \tag{E.2-10}$$

In Equation E.1-6 we let

$$x = \frac{1}{p}, \qquad 1 - x = \frac{1}{q} \tag{E.2-11}$$

and

$$a = \frac{|f(t)|^p}{\int_a^b |f(t)|^p \, dt}, \tag{E.2-12}$$

$$b = \frac{|g(t)|^q}{\int_a^b |g(t)|^q \, dt}. \tag{E.2-13}$$

Now Equation E.1-7 becomes, for each t,

$$\frac{|f(t)g(t)|}{\left[\int_a^b |f(t)|^p \, dt\right]^{1/p}\left[\int_a^b |g(t)|^q \, dt\right]^{1/q}} \leq \frac{|f(t)|^p}{p\int_a^b |f(t)|^p \, dt} + \frac{|g(t)|^q}{q\int_a^b |g(t)|^q \, dt}. \tag{E.2-14}$$

We now integrate both sides with respect to t in the interval (a, b) to obtain

$$\frac{\int_a^b |f(t)g(t)| \, dt}{\left[\int_a^b |f(t)|^p \, dt\right]^{1/p}\left[\int_a^b |g(t)|^q \, dt\right]^{1/q}} \leq \frac{1}{p} + \frac{1}{q} = 1. \tag{E.2-15}$$

On rearranging, we have *Hölder's Inequality*:

$$\boxed{\begin{gathered}\int_a^b |f(t)g(t)| \, dt \leq \left[\int_a^b |f(t)|^p \, dt\right]^{1/p}\left[\int_a^b |g(t)|^q \, dt\right]^{1/q} < \infty \\ \frac{1}{p} + \frac{1}{q} = 1.\end{gathered}} \tag{E.2-16}$$

A weaker inequality may be written which will be more convenient for our purposes. Since

$$\left|\int_a^b f(t)g(t) \, dt\right| \leq \int_a^b |f(t)g(t)| \, dt, \tag{E.2-17}$$

$$\boxed{\begin{gathered}\left|\int_a^b f(t)g(t) \, dt\right| \leq \left[\int_a^b |f(t)|^p \, dt\right]^{1/p}\left[\int_a^b |g(t)|^q \, dt\right]^{1/q}, \\ \frac{1}{p} + \frac{1}{q} = 1.\end{gathered}} \tag{E.2-18}$$

Hölder's Inequality for sums is proved in a similar way. Let f_n and g_n be sequences of numbers such that

$$\sum_n |f_n|^p < \infty \tag{E.2-19}$$

and

$$\sum_n |g_n|^q < \infty \tag{E.2-20}$$

where p and q satisfy Equation E.2-3. By an argument identical to that used in arriving at Equation E.2-9, we conclude that

$$\sum_n |f_n g_n| < \infty. \tag{E.2-21}$$

As before we assume that neither f_n nor g_n vanishes so that

$$f_n g_n \neq 0. \tag{E.2-22}$$

In Equation E.1-7, we let

$$x = \frac{1}{p}, \qquad 1 - x = \frac{1}{q} \tag{E.2-23}$$

and

$$a = \frac{|f_n|^p}{\sum_n |f_n|^p}, \tag{E.2-24}$$

$$b = \frac{|g_n|^q}{\sum_n |g_n|^q}. \tag{E.2-25}$$

After summing both sides of the resulting expression and using Equation E.2-3, we obtain Hölder's Inequality for sums:

$$\boxed{\begin{gathered} \sum_n |f_n g_n| \leq \left(\sum_n |f_n|^p\right)^{1/p} \left(\sum_n |g_n|^q\right)^{1/q}, \\ \frac{1}{p} + \frac{1}{q} = 1 \end{gathered}} \tag{E.2-26}$$

or, since

$$\left|\sum_n f_n g_n\right| \leq \sum_n |f_n g_n|, \tag{E.2-27}$$

a weaker but sometimes more useful form is

$$\boxed{\begin{gathered} \left|\sum_n f_n g_n\right| \leq \left(\sum_n |f_n|^p\right)^{1/p} \left(\sum_n |g_n|^q\right)^{1/q}, \\ \frac{1}{p} + \frac{1}{q} = 1. \end{gathered}} \tag{E.2-28}$$

E.3 Minkowski's Inequality

Suppose that two functions $f_1(t)$ and $f_2(t)$ satisfy Equation E.2-1 where $f(t)$ is $f_1(t)$ or $f_2(t)$ with $p > 1$. It is clear that, for each t,

$$|f_1(t) \pm f_2(t)|^p \leq \begin{cases} |2f_1(t)|^p, & \text{for } f_1(t) \geq f_2(t), \\ |2f_2(t)|^p, & \text{for } f_1(t) < f_2(t). \end{cases} \tag{E.3-1}$$

Hence

$$\int_a^b |f_1(t) \pm f_2(t)|^p \, dt \leq 2^p \int_a^b |f_1(t)|^p \, dt + 2^p \int_a^b |f_2(t)|^p \, dt < \infty \tag{E.3-2}$$

for finite p. We now note that

$$\int_a^b |f_1(t) \pm f_2(t)|^p \, dt \leq \int_a^b |f_1(t) \pm f_2(t)|^{p-1} |f_1(t)| \, dt + \int_a^b |f_1(t) \pm f_2(t)|^{p-1} |f_2(t)| \, dt. \tag{E.3-3}$$

Let us apply Hölder's Inequality to each of the integrals on the right with

$$g(t) = |f_1(t) + f_2(t)|^{p-1}, \tag{E.3-4}$$

$$f(t) = |f_k(t)|, \qquad k = 1, 2. \tag{E.3-5}$$

The result is

$$\int_a^b |f_1(t) \pm f_2(t)|^p \, dt \leq \left[\int_a^b |f_1(t)|^p \, dt\right]^{1/p} \left[\int_a^b |f_1(t) \pm f_2(t)|^{(p-1)q} \, dt\right]^{1/q} + \left[\int_a^b |f_2(t)|^p \, dt\right]^{1/p} \left[\int_a^b |f_1(t) \pm f_2(t)|^{(p-1)q} \, dt\right]^{1/q}. \tag{E.3-6}$$

Note that $(p - 1)q = p$ and that $1 - 1/q = 1/p$. Thus, on dividing both sides of this last inequality by the common term on the right side, we obtain Minkowski's Inequality:

$$\left[\int_a^b |f_1(t) \pm f_2(t)|^p \, dt\right]^{1/p} \leq \left[\int_a^b |f_1(t)|^p \, dt\right]^{1/p} + \left[\int_a^b |f_2(t)|^p \, dt\right]^{1/p}, \quad p \geq 1. \tag{E.3-7}$$

Actually we have proved the inequality only for $p > 1$, but the case $p = 1$ is a trivial result of the triangle inequality

$$|x \pm y| \leq |x| + |y|. \tag{E.3-8}$$

Minkowski's Inequality for sums is proved in a similar way and is

$$\left(\sum_n |f_{1n} \pm f_{2n}|^p\right)^{1/p} \le \left(\sum_n |f_{1n}|^p\right)^{1/p} + \left(\sum_n |f_{2n}|^p\right)^{1/p}, \quad p \ge 1. \tag{E.3-9}$$

E.4 Schwarz's Inequality

This is a special case of Hölder's Inequality obtained by placing $p = q = 2$ in Equation E.2-18 or E.2-28 to obtain

$$\left|\int_a^b f(t)g(t)\, dt\right| \le \left[\int_a^b |f(t)|^2\, dt\right]^{1/2} \left[\int_a^b |g(t)|^2\, dt\right]^{1/2} \tag{E.4-1}$$

or

$$\left|\sum_n f_n g_n\right| \le \left(\sum_n |f_n|^2\right)^{1/2} \left(\sum_n |g_n|^2\right)^{1/2}. \tag{E.4-2}$$

Since this inequality is relatively easy to prove directly, it will be instructive to do so.

We assume that $f(t)$ and $g(t)$, functions of the real variable t, satisfy Equations E.2-1 and E.2-2 with $p = q = 2$. Hence they satisfy Equation E.2-9. Now form the integral

$$\int_a^b [\lambda f^*(t) + g^*(t)][\lambda f(t) + g(t)]\, dt = \lambda^2 A + \lambda(B + B^*) + C = k(\lambda) \ge 0 \tag{E.4-3}$$

where λ is a real variable, the symbol * means complex conjugate, and

$$A = \int_a^b |f(t)|^2\, dt \ge 0, \tag{E.4-4}$$

$$B = \int_a^b f^*(t)g(t)\, dt, \tag{E.4-5}$$

$$C = \int_a^b |g(t)|^2\, dt \ge 0. \tag{E.4-6}$$

The integral of Equation E.4-3 exists, is real, and is a nonnegative function of λ, say $k(\lambda)$. Since $k(\lambda)$ is nonnegative, it must have no real roots except possibly a double root. From the quadratic formula, then, we must have

$$\left(\frac{B + B^*}{2}\right)^2 = (\text{Re } B)^2 \le AC \tag{E.4-7}$$

or

$$(B + B^*)^2 \leq 4AC. \tag{E.4-8}$$

On substituting for A, B, B^*, and C, we obtain

$$\left\{\int_a^b [f^*(t)g(t) + f(t)g^*(t)]\,dt\right\}^2 \leq 4\int_a^b |f(t)|^2\,dt \int_a^b |g(t)|^2\,dt. \tag{E.4-9}$$

This is the form of Schwarz's Inequality that is appropriate for complex functions $f(t)$ and $g(t)$, where, for example,

$$f(t) = f_1(t) + jf_2(t), \tag{E.4-10}$$

$$g(t) = g_1(t) + jg_2(t), \tag{E.4-11}$$

and $f_1(t)$, $f_2(t)$, $g_1(t)$, and $g_2(t)$ are real. For the case where $f(t)$ and $g(t)$ are real, then

$$f^*(t)g(t) + f(t)g^*(t) = 2f(t)g(t) \tag{E.4-12}$$

and Equation E.4-1 follows immediately.

Note that equality is obtained [aside from the trivial case where $f(t)$ or $g(t)$ is zero] when the double root exists in Equation E.4-3; that is, when

$$\lambda f(t) + g(t) = \lambda f^*(t) + g^*(t) = 0. \tag{E.4-13}$$

Since λ is real, then $f(t)$ and $g(t)$ are linearly related. Looking at the problem from a slightly different point of view, we see that there is a real value of λ for which Equation E.4-3 is zero and for which its first derivative with respect to λ vanishes; that is,

$$2\lambda A + (B + B^*) = 0 \tag{E.4-14}$$

or

$$\lambda = -\frac{B + B^*}{2A} = -\frac{\int_a^b [f^*(t)g(t) + f(t)g^*(t)]\,dt}{2\int_a^b |f(t)|^2\,dt}. \tag{E.4-15}$$

Equation E.4-15 holds if and only if

$$g(t) = -\lambda f(t). \tag{E.4-16}$$

This last relationship is equivalent to Equation E.4-13.

REFERENCES

1. E. C. Titchmarsh, *The Theory of Functions*, Oxford Univ. Press, Oxford, England, 1939.
2. M. E. Munroe, *Introduction to Measure and Integration*, Addison-Wesley, Reading, Mass., 1953.
3. N. Dunford and J. T. Schwartz, *Linear Operators, Part I: General Theory*, Interscience, New York, 1958.
4. A. C. Zaanen, *Linear Analysis*, North-Holland, Amsterdam, Holland, 1960.

Appendix F

The Calculus of Variations

F.1 Maxima and Minima of a Function

We begin by reviewing some of the elementary notions of maxima and minima [1, 2]. Suppose we are given the function $y = f(x)$, assumed to be single valued and representable by Taylor's expansion:

$$f(x + h) = f(x) + \frac{h}{1!} f^{(1)}(x) + \frac{h^2}{2!} f^{(2)}(x) + \cdots + \frac{h^n}{n!} f^{(n)}(x) + \cdots \quad \text{(F.1-1)}$$

where

$$f^{(i)}(x) = \frac{d^i f(x)}{dx^i}. \quad \text{(F.1-2)}$$

Consider the function at the point $x = a$ and form the difference

$$\Delta(h) = f(a + h) - f(a) = hf^{(1)}(a) + \frac{h^2}{2!} f^{(2)}(a) + \cdots. \quad \text{(F.1-3)}$$

Assume that $f(a)$ is either a maximum or a minimum as shown in Figure F.1. Then, for sufficiently small h, the sign of $\Delta(h)$ must equal the sign of $\Delta(-h)$. However, for small h, the difference $\Delta(h)$ of Equation F.1-3 is given approximately by

$$\Delta(h) \approx hf^{(1)}(a). \quad \text{(F.1-4)}$$

Thus, for the sign of $\Delta(h)$ to be unchanged when h changes sign, it follows that

$$\boxed{f^{(1)}(a) = 0} \quad \text{(F.1-5)}$$

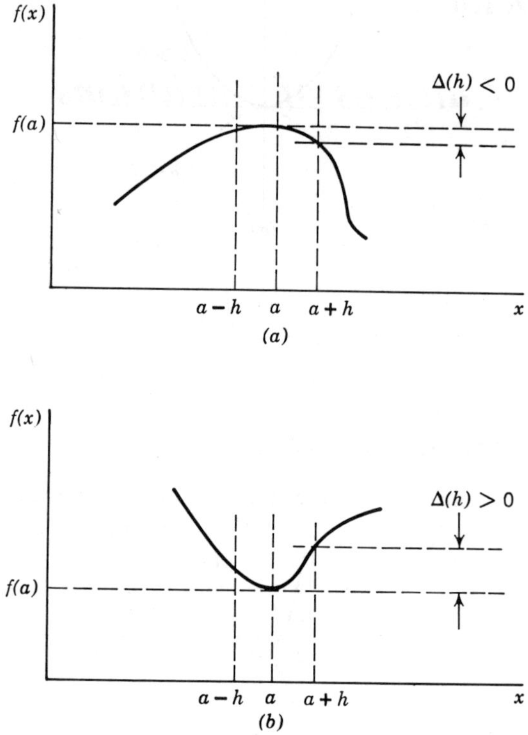

Figure F.1 *Maxima and minima of f(x): (a) a maximum; (b) a minimum*

is a *necessary* condition for $f(x)$ to have a maximum or minimum at $x = a$. We say that $f(x)$ is *stationary* at the point $x = a$ if Equation F.1-5 is satisfied. Not all stationary points are maxima or minima, however.

Suppose that Equation F.1-5 is satisfied. Then, for small h,

$$\Delta(h) \approx \frac{h^2}{2!} f^{(2)}(a). \tag{F.1-6}$$

It is apparent from Figure F.1 that $\Delta(h) < 0$ for a maximum and $\Delta(h) > 0$ for a minimum. Therefore, if the second derivative is nonzero,

$$f^{(2)}(a) = \begin{cases} < 0, & \text{at a maximum,} \\ > 0, & \text{at a minimum.} \end{cases} \tag{F.1-7}$$

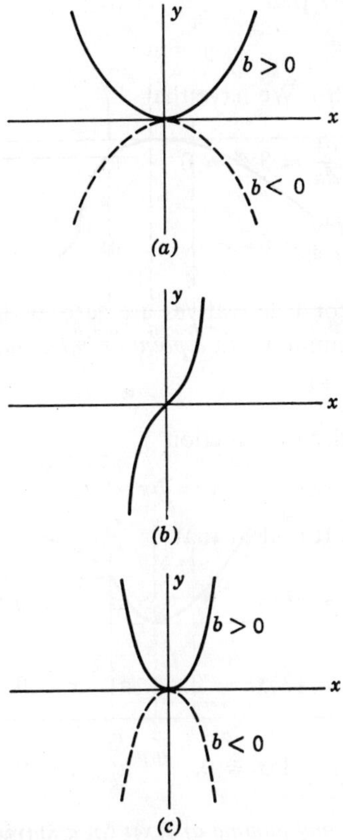

Figure F.2 *Some functions with stationary values:* (*a*) *the function* $y = bx^2$; (*b*) *the function* $y = x^3$; (*c*) *the function* $y = bx^4$

Example F.1 Let us consider the function

$$y = bx^2$$

where b is a real number. This curve is plotted in Figure F.2(*a*). It is clear that

$$\frac{dy}{dx} = 2bx = 0, \quad \text{at} \quad x = 0,$$

and that

$$\frac{d^2y}{dx^2} = 2b = \begin{cases} < 0, & b < 0 \text{ (max)}, \\ > 0, & b > 0 \text{ (min)}. \end{cases}$$

• •

Example F.2 Suppose that

$$y = x^3$$

as shown in Figure F.2(*b*). We have that

$$\frac{dy}{dx} = 3x^2 = 0, \qquad \text{at} \quad x = 0,$$

$$\frac{d^2y}{dx^2} = 6x = 0, \qquad \text{at} \quad x = 0.$$

Here both first and second derivatives are zero at the origin. The function has no maximum or minimum, but a *point of inflection* at the origin.

• •

Example F.3 Consider the function

$$y = bx^4$$

shown in Figure F.2(*c*). It is clear that

$$\frac{dy}{dx} = 4bx^3 = 0, \qquad \text{at} \quad x = 0,$$

$$\frac{d^2y}{dx^2} = 12bx^2 = 0, \qquad \text{at} \quad x = 0,$$

$$\frac{d^3y}{dx^3} = 24bx = 0, \qquad \text{at} \quad x = 0,$$

$$\frac{d^4y}{dx^4} = 24b = \begin{cases} < 0, & b < 0 \text{ (max)}, \\ > 0, & b > 0 \text{ (min)}. \end{cases}$$

In this example, the first three derivatives are zero at the origin; nevertheless the function has either a maximum or a minimum there. Note that the appropriate approximation to the difference $\Delta(h)$ is

$$\Delta(h) \approx \frac{h^4}{4!} f^{(4)}(0).$$

• •

It is apparent that Taylor's Theorem provides the basis for a general statement of *sufficient* conditions for a maximum or a minimum when the *necessary* condition has been fulfilled; that is, when the first derivative vanishes. We consider the difference $\Delta(h)$ as approximated by the first nonvanishing term of the Taylor series. If this term contains an odd power of h, we have a point of inflection. If this term contains an even power of h, we have a maximum if the derivative of corresponding order is negative and a minimum if the derivative is positive.

F.2 Maxima and Minima of Functions of Two or More Variables

In this section we will consider the function $f(x, y)$ of the two variables x and y. Extension to the case of more than two variables will be obvious. As before, we begin with Taylor's expansion of f:

$$\begin{aligned} f(x+h, y+k) &= f(x, y) + \left(h\frac{\partial f}{\partial x} + k\frac{\partial f}{\partial y}\right) \\ &+ \frac{1}{2!}\left(h^2\frac{\partial^2 f}{\partial x^2} + 2hk\frac{\partial^2 f}{\partial x\,\partial y} + k^2\frac{\partial^2 f}{\partial y^2}\right) + \cdots \\ &+ \frac{1}{n!}\left[\sum_{i=0}^{n}\binom{n}{i}h^i k^{n-i}\frac{\partial^n f}{\partial x^i\,\partial y^{n-i}}\right] + \cdots \qquad \text{(F.2-1)} \end{aligned}$$

where $\binom{n}{i}$ is the binomial coefficient given by

$$\binom{n}{i} = \frac{n!}{i!(n-i)!} \qquad \text{(F.2-2)}$$

Let us form the difference $\Delta(h, k)$ given by

$$\Delta(h, k) = f(a+h, b+k) - f(a, b) \qquad \text{(F.2-3)}$$

and note that, for small h and k, it is given approximately by

$$\Delta(h, k) \approx \left(h\frac{\partial f}{\partial x} + k\frac{\partial f}{\partial y}\right)_{x=a,\ y=b}. \qquad \text{(F.2-4)}$$

It is clear that, for $\Delta(h, k)$ to have the same sign independently of the signs of h and k, it is necessary that

$$\boxed{\frac{\partial f}{\partial x} = 0, \qquad \frac{\partial f}{\partial y} = 0, \qquad \text{at}\quad x = a,\ y = b.} \qquad \text{(F.2-5)}$$

These are necessary conditions for $f(x, y)$ to have a maximum or a minimum at (a, b). If these conditions are satisfied then, for small h and k,

$$\Delta(h, k) \approx \frac{1}{2!}\left(h^2\frac{\partial^2 f}{\partial x^2} + 2hk\frac{\partial^2 f}{\partial x\,\partial y} + k^2\frac{\partial^2 f}{\partial y^2}\right)_{x=a,\ y=b}. \qquad \text{(F.2-6)}$$

This expression may be written as

$$\begin{aligned} \Delta(h, k) &\approx \frac{1}{2!}(h^2A + 2hkB + k^2C) \\ &\approx \frac{1}{2!}\frac{(Ah+Bk)^2 + (AC-B^2)k^2}{A}. \qquad \text{(F.2-7)} \end{aligned}$$

Since $(Ah + Bk)^2$ is always nonnegative, it is clear that the sign of $\Delta(h, k)$ is independent of the signs of h and k if

$$D = AC - B^2 \geq 0 \tag{F.2-8}$$

or

$$\boxed{\frac{\partial^2 f}{\partial x^2}\frac{\partial^2 f}{\partial y^2} \geq \left(\frac{\partial^2 f}{\partial x\,\partial y}\right)^2 \geq 0, \quad \text{at} \quad x = a,\ y = b.} \tag{F.2-9}$$

Thus, when this last equation is satisfied,

$$f(a, b) = \begin{cases} \text{max} & \text{if } \dfrac{\partial^2 f}{\partial x^2} < 0, \quad \dfrac{\partial^2 f}{\partial y^2} < 0 \quad \text{at } x = a, y = b, \\ \text{min} & \text{if } \dfrac{\partial^2 f}{\partial x^2} > 0, \quad \dfrac{\partial^2 f}{\partial y^2} > 0 \quad \text{at } x = a, y = b, \end{cases} \tag{F.2-10}$$

if $f(a, b)$ has a minimum or maximum at $x = a$, $y = b$. In other words, the function $f(x, y)$ is

$$\left.\begin{matrix}\text{max} \\ \text{min}\end{matrix}\right\} \text{ at } x = a, y = b \qquad \text{if } h^2A + 2hkB + k^2C \begin{cases} < 0, \\ > 0. \end{cases} \tag{F.2-11}$$

Example F.4 Find the maxima and minima of the surface

$$f(x, y) = \frac{1}{2c}\left(\frac{x^2}{a^2} - \frac{y^2}{b^2}\right).$$

Equation F.2-5 becomes

$$0 = \frac{\partial f}{\partial x} = \frac{x}{ca^2}, \qquad 0 = \frac{\partial f}{\partial y} = -\frac{y}{cb^2}.$$

These partial derivatives vanish when $x = y = 0$. The second derivatives are

$$\frac{\partial^2 f}{\partial x^2} = \frac{1}{ca^2}, \qquad \frac{\partial^2 f}{\partial x\,\partial y} = 0, \qquad \frac{\partial^2 f}{\partial y^2} = -\frac{1}{cb^2}.$$

Note that the condition of Equation F.2-8 cannot be satisfied since $\partial^2 f/\partial x^2$ and $\partial^2 f/\partial y^2$ have opposite signs. In other words

$$D = AC - B^2 = -\frac{1}{a^2b^2c^2} < 0.$$

The function $f(x, y)$ is a hyperbolic paraboloid. It has a *saddle point* or *minimax* point at the origin. These are points where the first partial derivatives vanish but where $D < 0$.

• •

F.3 Lagrange's Method of Undetermined Multipliers

It will not always be true that the several variables involved are independent. A common problem is to find the maxima or minima of a function $u(x, y, z)$ where the variables x, y, and z are related through

$$\phi(x, y, z) = K, \quad \text{a constant.} \tag{F.3-1}$$

If u is to have a stationary value, then the total differential du must be zero:

$$du = 0 = \frac{\partial u}{\partial x}\,dx + \frac{\partial u}{\partial y}\,dy + \frac{\partial u}{\partial z}\,dz. \tag{F.3-2}$$

It follows from Equation F.3-1 that the total differential $d\phi$ must also be zero:

$$d\phi = 0 = \frac{\partial \phi}{\partial x}\,dx + \frac{\partial \phi}{\partial y}\,dy + \frac{\partial \phi}{\partial z}\,dz. \tag{F.3-3}$$

Now suppose we multiply $d\phi$ by some parameter λ and add the product to du. We have

$$du + \lambda\,d\phi = 0. \tag{F.3-4}$$

Since Equation F.3-4 involves arbitrary differentials dx, dy, and dz, we conclude that

$$\begin{aligned} \frac{\partial u}{\partial x} + \lambda\frac{\partial \phi}{\partial x} &= 0, \\ \frac{\partial u}{\partial y} + \lambda\frac{\partial \phi}{\partial y} &= 0, \\ \frac{\partial u}{\partial z} + \lambda\frac{\partial \phi}{\partial z} &= 0. \end{aligned} \tag{F.3-5}$$

These last equations together with Equation F.3-1 give four relations in the four unknowns; namely, λ and the values of x, y, and z determining the stationary point location (x, y, z).

More generally, to determine the stationary values of the function $u(x_1, x_2, \ldots, x_n)$ where the n variables x_i are subject to the m constraints

$$\phi_i(x_1, x_2, \ldots, x_n) = K_i, \qquad i = 1, 2, \ldots, m, \tag{F.3-6}$$

we form the function

$$F = u + \sum_{i=1}^{m} \phi_i \lambda_i. \tag{F.3-7}$$

We find the set of λ_i and the values of $x_1, x_2, \ldots, x_n$ which determine the stationary point $(x_1, x_2, \ldots, x_n)$ from the m equations given by Equation F.3-6 and from the n equations

$$\frac{\partial F}{\partial x_j} = 0, \qquad j = 1, 2, \ldots, n. \tag{F.3-8}$$

Example F.5 We desire to find the volume of the largest parallelepiped that can be inscribed in the ellipsoid

$$\frac{x^2}{a^2} + \frac{y^2}{b^2} + \frac{z^2}{c^2} = 1 = \phi(x, y, z).$$

Since the ellipsoid is symmetrically located with respect to the origin, the same will be true of the inscribed parallelepiped with volume

$$u(x, y, z) = 8xyz. \tag{A}$$

Equation F.3-5 becomes

$$8yz + \frac{2\lambda}{a^2} x = 0, \tag{B}$$

$$8xz + \frac{2\lambda}{b^2} y = 0, \tag{C}$$

$$8xy + \frac{2\lambda}{c^2} z = 0. \tag{D}$$

Each of these last three equations may be multiplied by x, y, and z respectively and the result summed to give

$$3u + 2\lambda\phi = 3u + 2\lambda = 0,$$

$$2\lambda = -3u.$$

This last expression for λ may be substituted in Equation B which, after multiplication by x, becomes

$$u\left(1 - \frac{3}{a^2} x^2\right) = 0,$$

$$x = \frac{a}{\sqrt{3}}.$$

In the same way

$$y = \frac{b}{\sqrt{3}},$$

$$z = \frac{c}{\sqrt{3}}$$

and the volume u is

$$u = \frac{8abc}{3\sqrt{3}}.$$

• •

F.4 The Fundamental Problem of the Calculus of Variations

The fundamental problem of the calculus of variations [3, 4, 5] may be stated as follows. Let us define the quantity I by the definite integral

$$I = \int_{x=a}^{x=b} F(x, y, y')\, dx \tag{F.4-1}$$

where $F(\mu, \nu, \sigma)$ is a known function, y' is dy/dx, and $y = y(x)$ is an unknown function of x which is to be chosen to make I a minimum. The problem is to find $y(x)$.

Consider functions $\mathbf{y}$ which are "close" to y in the following sense. Let $\mathbf{y}$ be given by

$$\mathbf{y} = y + \gamma \delta y \tag{F.4-2}$$

where γ is a real variable and δy is an *arbitrary* function of x such that its first two derivatives are continuous in (a, b) and

$$\delta y(a) = \delta y(b) = 0. \tag{F.4-3}$$

The function δy is called the *variation* of y. For small γ and δy, $\mathbf{y}$ is "close" to y in the interval (a, b) and, in any case, coincides with it at the endpoints. The situation is illustrated in Figure F.3. It follows from Equation F.4-2 that

$$\mathbf{y}' = y' + \gamma \delta y'. \tag{F.4-4}$$

The expressions of Equation F.4-2 and F.4-4 for $\mathbf{y}$ and $\mathbf{y}'$ may be substituted for y and y' in Equation F.4-1 to yield a new function $I(\gamma)$ where $I(0) = I$:

$$I(\gamma) = I(0) + \delta I = \int_a^b F(x, y + \gamma\delta y, y' + \gamma \delta y')\, dx. \tag{F.4-5}$$

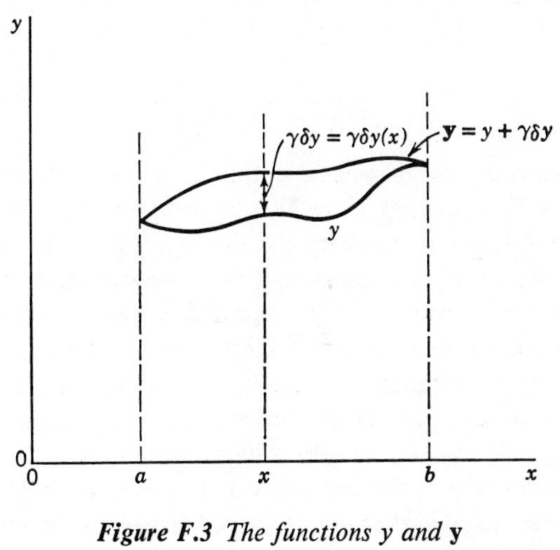

Figure F.3 The functions y and **y**

We want to pick $y(x)$ so that $I(0)$ is a maximum or minimum (or, more generally, a stationary point). A *necessary* condition has been established in Section F.1 and is that

$$\boxed{\left.\frac{dI(\gamma)}{d\gamma}\right|_{\gamma=0} = 0.} \tag{F.4-6}$$

Suppose we expand $F(x, \mathbf{y}, \mathbf{y}')$ in a Taylor series. We have

$$I(\gamma) = \int_a^b \left[F(x, y, y') + \gamma\delta y \frac{\partial F}{\partial y} + \gamma\delta y' \frac{\partial F}{\partial y'} + \text{terms in } \gamma^2, \gamma^3, \ldots\right] dx \tag{F.4-7}$$

or

$$\frac{dI(\gamma)}{d\gamma} = \int_a^b \left[\delta y \frac{\partial F}{\partial y} + \delta y' \frac{\partial F}{\partial y'} + \text{terms in } \gamma, \gamma^2, \ldots\right] dx. \tag{F.4-8}$$

We now let $\gamma = 0$ and obtain Equation F.4-6:

$$\left.\frac{dI(\gamma)}{d\gamma}\right|_{\gamma=0} = 0 = \int_a^b \left[\delta y \frac{\partial F}{\partial y} + \delta y' \frac{\partial F}{\partial y'}\right] dx. \tag{F.4-9}$$

The second term in the integrand may be integrated by parts to yield

$$\begin{aligned}\int_a^b \delta y' \frac{\partial F}{\partial y'}\, dx &= \delta y \left.\frac{\partial F}{\partial y'}\right|_a^b - \int_a^b \delta y \frac{d}{dx}\left(\frac{\partial F}{\partial y'}\right) dx \\ &= -\int_a^b \delta y \frac{d}{dx}\left(\frac{\partial F}{\partial y'}\right) dx \end{aligned} \tag{F.4-10}$$

since $\delta y(b) = \delta y(a) = 0$ from Equation F.4-3. Combining Equations F.4-9 and F.4-10, we have

$$0 = \int_a^b \delta y \left[\frac{\partial F}{\partial y} - \frac{d}{dx}\left(\frac{\partial F}{\partial y'}\right)\right] dx. \tag{F.4-11}$$

The foregoing is a typical integral obtained in problems in the calculus of variations. We will argue that, since δy is an *arbitrary* variation in y, the term in brackets must be zero in order for the integral to be zero. The proof [4, Chap. 3] depends on showing that, if the bracketed term is not zero, then there is a particular continuously twice-differentiable function δy_1 which vanishes at the endpoints such that the integral is not zero. Such a function δy_1 can always be constructed, for example, by defining it to be zero everywhere except in some region where the bracketed term is nonzero. If δy_1 has the same sign in that region, then the integral is obviously nonzero, a contradiction. The reader who is further interested is referred to [4, Chap. 3].

Thus we have concluded that the bracketed term must be zero. Therefore, a *necessary* condition for the definite integral I of Equation F.4-1 to have an extreme value is that

$$\boxed{\frac{\partial F}{\partial y} - \frac{d}{dx}\left(\frac{\partial F}{\partial y'}\right) = 0.} \tag{F.4-12}$$

This expression is called the *Euler-Lagrange (differential) Equation.* The problem is easily extended to the case where F is a function of more variables, some of which may be independent [5]. It should be emphasized that Equation F.4-12 is necessary but not sufficient for a minimum. In many cases it will be apparent from geometrical or physical considerations whether the result gives a maximum, a minimum, or neither.

Note that the total derivative dG/dx of a function $G(x, y, y')$ may be written as

$$\frac{dG}{dx} = \frac{\partial G}{\partial x} + \frac{\partial G}{\partial y}\frac{dy}{dx} + \frac{\partial G}{\partial y'}\frac{dy'}{dx} = \frac{\partial G}{\partial x} + \frac{\partial G}{\partial y}y' + \frac{\partial G}{\partial y'}y''. \tag{F.4-13}$$

Thus Equation F.4-12 may be written explicitly as

$$\boxed{\frac{\partial F}{\partial y} - \frac{\partial^2 F}{\partial x\,\partial y'} - \frac{\partial^2 F}{\partial y\,\partial y'}y' - \frac{\partial^2 F}{\partial y'^2}y'' = 0.} \tag{F.4-14}$$

Example F.6 Show that the shortest curve $y(x)$ joining two fixed endpoints $x = a$ and $x = b$ is a straight line.

The length of the curve is

$$I = \int_a^b ds = \int_a^b (dx^2 + dy^2)^{1/2} = \int_a^b (1 + y'^2)^{1/2}\,dx$$

and is to be a minimum. Therefore the function $F = (1 + y'^2)^{1/2}$ must satisfy

$$\frac{\partial F}{\partial y} - \frac{d}{dx}\left(\frac{\partial F}{\partial y'}\right) = 0$$

or

$$0 - \frac{y''}{(1 + y'^2)^{3/2}} = 0.$$

From this last expression, we conclude that

$$y'' = 0,$$
$$y' = c_1,$$
$$y = c_1x + c_2,$$

where the constants c_1 and c_2 are determined by the fixed endpoints.

• •

F.5 Isoperimetric Problems

Variational situations frequently arise where additional constraints are imposed. One of the commonest of these is to find that $y = y(x)$ which yields a stationary value of

$$I = \int_a^b F(x, y, y')\, dx \tag{F.5-1}$$

subject to the condition that

$$I_1 = \int_a^b F_1(x, y, y')\, dx = c \tag{F.5-2}$$

where c is a constant. Such problems are called *isoperimetric* since one of the best-known problems of this type is that of finding the closed curve of fixed perimeter with the largest area.

Lagrange's method of undetermined multipliers, as discussed in Section F.3, may be applied directly to this class of problems. We consider the integral

$$I_0 = \int_a^b (F + \lambda F_1)\, dx = I + \lambda I_1 \tag{F.5-3}$$

where the parameter λ is a Lagrange multiplier. We find the $y(x)$ that makes I_0 stationary since I_0 will be stationary if I is stationary. But a necessary condition is just the *Euler-Lagrange Condition* with the function F replaced by $F_0 = F + \lambda F_1$:

$$\boxed{\frac{\partial F_0}{\partial y} - \frac{d}{dx}\left(\frac{\partial F_0}{\partial y'}\right) = 0.} \tag{F.5-4}$$

We have two equations, F.5-1 and F.5-4, in the two unknowns $y(x)$ and λ. A formal proof proceeds exactly as in Section F.4.

Example F.7 Suppose we want to maximize the expression

$$I = \int_{-\infty}^{\infty} y(x) \log y(x)\, dx \tag{A}$$

subject to the conditions that

$$I_1 = 1 = \int_{-\infty}^{\infty} y(x)\, dx, \tag{B}$$

$$I_2 = \sigma^2 = \int_{-\infty}^{\infty} x^2 y(x)\, dx \tag{C}$$

where σ^2 is a given constant. If the function $y(x)$ is a probability density function, then I is called the *entropy* of the density and we are asking for that density with fixed variance σ^2 which yields a maximum entropy. Such problems arise in information theory and are treated in Chapter 8.

We form the expression

$$F_0 = F + \lambda_1 F_1 + \lambda_2 F_2 = y \log y + \lambda_1 y + \lambda_2 x^2 y$$

where λ_1 and λ_2 are Lagrangian multipliers. We now apply Equation F.5-4 which, for our problem, reduces to

$$\frac{\partial F_0}{\partial y} = 0 = 1 + \log y + \lambda_1 + \lambda_2 x^2$$

or

$$y = e^{-(\lambda_1+1)} e^{-\lambda_2 x^2}. \tag{D}$$

We find λ_1 and λ_2 by substituting Equation D into Equations B and C. From Equation C we have

$$\sigma^2 = e^{-(\lambda_1+1)} \int_{-\infty}^{\infty} x^2 e^{-\lambda_2 x^2} dx.$$

This expression may be integrated by parts to yield

$$\sigma^2 = \frac{e^{-(\lambda_1+1)}}{2\lambda_2} \int_{-\infty}^{\infty} e^{-\lambda_2 x^2} dx.$$

But the use of Equation B reduces this to

$$\sigma^2 = \frac{1}{2\lambda_2}, \qquad \lambda_2 = \frac{1}{2\sigma^2}.$$

We substitute in Equation B and have

$$e^{\lambda_1+1} = \int_{-\infty}^{\infty} e^{-x^2/2\sigma^2} dx.$$

But the integral is just $\sqrt{2\pi}\sigma$ from the known properties of the normal distribution (see Section 2.18). Therefore the function $y(x)$ is

$$y(x) = \frac{1}{\sqrt{2\pi}\sigma} e^{-x^2/2\sigma^2}, \qquad -\infty < x < \infty,$$

the normal distribution with zero mean and variance σ^2.

• •

F.6 Isoperimetric Problems Involving More General Functionals

The problems considered in the previous sections involved functionals* formed by integrating a certain differential expression in the argument function. More general classes of functionals often will be encountered in variational problems. In such cases the general variational techniques that were used to find the *Euler-Lagrange Equation* must be followed.

A specific problem of this more general type is the following which we will encounter frequently in the optimization of linear systems. It is desired to find the function $\phi(t)$ that yields a stationary value of the expression

$$I = \int_a^b \int_a^b K(s, t)\phi(s)\phi(t)\, ds\, dt \tag{F.6-1}$$

subject to one or both of the constraints

$$I_1 = \int_a^b [\phi(t)]^2\, dt \tag{F.6-2}$$

and

$$I_2 = \int_a^b \phi(t)f(t)\, dt. \tag{F.6-3}$$

Here $K(s, t)$ is a given continuous symmetric function and $f(t)$ is a given continuous function. We form

$$I_0 = I + \lambda_1 I_1 + \lambda_2 I_2$$

where λ_1 and λ_2 are Lagrangian multipliers. We replace ϕ by $\phi + \gamma\delta\phi$ and form $I(\gamma)$. We then perform the operations

$$\left.\frac{\partial I(\gamma)}{\partial \gamma}\right|_{\gamma=0} = 0 \tag{F.6-4}$$

and obtain

$$0 = 2\int_a^b \delta\phi(t)\left[\int_a^b K(s, t)\phi(s)\, ds + \lambda_1\phi(t) + \frac{\lambda_2}{2} f(t)\right] dt. \tag{F.6-5}$$

We conclude, therefore, since $\delta\phi(t)$ is arbitrary, that

$$\boxed{0 = \int_a^b K(s, t)\phi(s)\, ds + \lambda_1\phi(t) + \frac{\lambda_2}{2} f(t).} \tag{F.6-6}$$

This is an integral equation in the unknown $\phi(t)$. It may be solved and then λ_1 and λ_2 may be eliminated through Equations F.6-2 and F.6-3.

* A functional is a quantity that depends on one or more functions rather than on a number of distinct variables. The domain of a functional is a set of admissible functions rather than a region of coordinate space. The length I of the curve of Example F.6 is a functional of the argument function $y(x)$. The integral $\int_a^b y(s, t)y(t, r)\, dt$ is a functional of the argument function $y(s, t)$.

The reader should recognize that we have already encountered special forms of Equation F.6-6. With condition I_2 missing we obtain

$$0 = \int_a^b K(s, t)\phi(s)\, ds + \lambda_1\phi(t) \tag{F.6-7}$$

which is the homogeneous integral equation encountered in Chapter 4 (Equation 4.6-43) determining the characteristic orthogonal functions in the Karhunen-Loeve expansion of a random process. With condition I_1 missing we obtain the integral equation encountered in Chapter 5 (Equation 5.4-14 or 5.10-18) for the matched filter and for the least mean-squared-error linear filter.

A wide class of extremum problems are amenable to this general approach [3, 4, 5].

REFERENCES

1. R. Courant, *Differential and Integral Calculus*, Vols. I and II, Interscience, New York, 1937.
2. S. Lang, *A First Course in Calculus*, Addison-Wesley, Reading, Mass., 1964.
3. R. Courant and D. Hilbert, *Methods of Mathematical Physics*, Vol. I, Interscience, New York, 1953, Chap. V.
4. R. Weinstock, *Calculus of Variations*, McGraw-Hill, New York, 1952.
5. I. M. Gelfand and S. V. Fomin, *Calculus of Variations* (trans. from Russian by R. A. Silverman), Prentice-Hall, Englewood Cliffs, N. J., 1963.

Appendix G

A Tabulation of Integrals Useful in the Calculation of Mean-Squared Error

We consider integrals of the form [1]

$$I_n = \frac{1}{2\pi}\int_{-\infty}^{\infty} \frac{F(j\omega)}{H(j\omega)H(-j\omega)}\, d\omega, \qquad n = 1, 2, \ldots, \tag{G-1}$$

where $F(j\omega)$ is even in ω and is given by

$$F(j\omega) = b_0(j\omega)^{2n-2} + b_1(j\omega)^{2n-4} + \cdots + b_{n-1} \tag{G-2}$$

and where $H(j\omega)$ is given by

$$H(j\omega) = a_0(j\omega)^n + a_1(j\omega)^{n-1} + \cdots + a_n. \tag{G-3}$$

Furthermore $H(\lambda)$, $\lambda = \omega + j\sigma$, has no lower half-plane roots. The values of I_n for $n = 1, 2, 3, \ldots, 7$ are

$$I_1 = \frac{b_0}{2a_0a_1}. \tag{G-4}$$

$$I_2 = \frac{-b_0 + (a_0b_1/a_2)}{2a_0a_1}. \tag{G-5}$$

$$I_3 = \frac{-a_2b_0 + a_0b_1 - (a_0a_1b_2/a_3)}{2a_0(a_0a_3 - a_1a_2)}. \tag{G-6}$$

$$I_4 = \frac{b_0(-a_1a_4 + a_2a_3) - a_0a_3b_1 + a_0a_1b_2 + (a_0b_3/a_4)(a_0a_3 - a_1a_2)}{2a_0(a_0a_3^2 + a_1^2a_4 - a_1a_2a_3)}. \tag{G-7}$$

$$I_5 = \frac{M_5}{2a_0\,\Delta_5}. \tag{G-8}$$

$$\begin{aligned} M_5 = {} & b_0(-a_0a_4a_5 + a_1a_4^2 + a_2^2a_5 - a_2a_3a_4) + a_0b_1(-a_2a_5 + a_3a_4) \\ & + a_0b_2(a_0a_5 - a_1a_4) + a_0b_3(-a_0a_3 + a_1a_2) \\ & + (a_0b_4/a_5)(-a_0a_1a_5 + a_0a_3^2 + a_1^2a_4 - a_1a_2a_3). \end{aligned}$$

$$\begin{aligned} \Delta_5 = {} & a_0^2a_5^2 - 2a_0a_1a_4a_5 - a_0a_2a_3a_5 + a_0a_3^2a_4 \\ & + a_1^2a_4^2 + a_1a_2^2a_5 - a_1a_2a_3a_4. \end{aligned}$$

$$I_6 = \frac{M_6}{2a_0\,\Delta_6}. \tag{G-9}$$

$$\begin{aligned} M_6 = {} & b_0(-a_0a_3a_5a_6 + a_0a_4a_5^2 - a_1^2a_6^2 + 2a_1a_2a_5a_6 + a_1a_3a_4a_6 - a_1a_4^2a_5 \\ & \qquad - a_2^2a_5^2 - a_2a_3^2a_6 + a_2a_3a_4a_5) \\ & + a_0b_1(-a_1a_5a_6 + a_2a_5^2 + a_3^2a_6 - a_3a_4a_5) \\ & + a_0b_2(-a_0a_5^2 - a_1a_3a_6 + a_1a_4a_5) \\ & + a_0b_3(a_0a_3a_5 + a_1^2a_6 - a_1a_2a_5) \\ & + a_0b_4(a_0a_1a_5 - a_0a_3^2 - a_1^2a_4 + a_1a_2a_3) \\ & + (a_0b_5/a_6)(a_0^2a_5^2 + a_0a_1a_3a_6 - 2a_0a_1a_4a_5 - a_0a_2a_3a_5 + a_0a_3^2a_4 \\ & \qquad - a_1^2a_2a_6 + a_1^2a_4^2 + a_1a_2^2a_5 - a_1a_2a_3a_4). \end{aligned}$$

$$\begin{aligned} \Delta_6 = {} & a_0^2a_5^3 + 3a_0a_1a_3a_5a_6 - 2a_0a_1a_4a_5^2 - a_0a_2a_3a_5^2 - a_0a_3^3a_6 \\ & + a_0a_3^2a_4a_5 + a_1^3a_6^2 - 2a_1^2a_2a_5a_6 - a_1^2a_3a_4a_6 + a_1^2a_4^2a_5 \\ & + a_1a_2^2a_5^2 + a_1a_2a_3^2a_6 - a_1a_2a_3a_4a_5. \end{aligned}$$

$$I_7 = \frac{M_7}{2a_0\,\Delta_7}. \tag{G-10}$$

$$M_7 = b_0m_0 + a_0b_1m_1 + a_0b_2m_2 + \cdots + a_0b_6m_6.$$

$$\begin{aligned} m_0 = {} & a_0^2a_6a_7^2 - 2a_0a_1a_6^2a_7 - 2a_0a_2a_4a_7^2 + a_0a_2a_5a_6a_7 + a_0a_3a_5a_6^2 \\ & + a_0a_4^2a_5a_7 - a_0a_4a_5^2a_6 + a_1^2a_6^3 + 3a_1a_2a_4a_6a_7 - 2a_1a_2a_5a_6^2 \\ & - a_1a_3a_4a_6^2 - a_1a_4^3a_7 + a_1a_4^2a_5a_6 + a_2^3a_7^2 - 2a_2^2a_3a_6a_7 \\ & - a_2^2a_4a_5a_7 + a_2^2a_5^2a_6 + a_2a_3a_4^2a_7 - a_2a_3a_4a_5a_6 - a_2a_3^2a_6^2. \end{aligned}$$

$$\begin{aligned} m_1 = {} & a_0a_4a_7^2 - a_0a_5a_6a_7 - a_1a_4a_6a_7 + a_1a_5a_6^2 - a_2^2a_7^2 + 2a_2a_3a_6a_7 \\ & + a_2a_4a_5a_7 - a_2a_5^2a_6 - a_3^2a_6^2 - a_3a_4^2a_7 + a_3a_4a_5a_6. \end{aligned}$$

$$\begin{aligned} m_2 = {} & a_0a_2a_7^2 - a_0a_3a_6a_7 - a_0a_4a_5a_7 + a_0a_5^2a_6 - a_1a_2a_6a_7 + a_1a_3a_6^2 \\ & + a_1a_4^2a_7 - a_1a_4a_5a_6. \end{aligned}$$

$$\begin{aligned} m_3 = {} & -a_0^2a_7^2 + 2a_0a_1a_6a_7 + a_0a_3a_4a_7 - a_0a_3a_5a_6 - a_1^2a_6^2 - a_1a_2a_4a_7 \\ & + a_1a_2a_5a_6. \end{aligned}$$

$$\begin{aligned} m_4 = {} & a_0^2a_5a_7 - a_0a_1a_4a_7 - a_0a_1a_5a_6 - a_0a_2a_3a_7 + a_0a_3^2a_6 + a_1^2a_4a_6 \\ & + a_1a_2^2a_7 - a_1a_2a_3a_6. \end{aligned}$$

$$
\begin{aligned}
m_5 = {} & a_0^2a_3a_7 - a_0^2a_5^2 - a_0a_1a_2a_7 - a_0a_1a_3a_6 + 2a_0a_1a_4a_5 + a_0a_2a_3a_5 \\
& - a_0a_3^2a_4 + a_1^2a_2a_6 - a_1^2a_4^2 - a_1a_2^2a_5 + a_1a_2a_3a_4.
\end{aligned}
$$

$$
\begin{aligned}
m_6 = {} & (1/a_7)(a_0^2a_1a_7^2 - 2a_0^2a_3a_5a_7 + a_0^2a_5^3 - 2a_0a_1^2a_6a_7 + a_0a_1a_2a_5a_7 \\
& + 3a_0a_1a_3a_5a_6 - 2a_0a_1a_4a_5^2 + a_0a_2a_3^2a_7 - a_0a_2a_3a_5^2 - a_0a_3^3a_6 \\
& + a_0a_3^2a_4a_5 + a_1^3a_6^2 + a_1^2a_2a_4a_7 - 2a_1^2a_2a_5a_6 - a_1^2a_3a_4a_6 \\
& + a_1^2a_4^2a_5 - a_1a_2^2a_3a_7 + a_1a_2^2a_5^2 + a_1a_2a_3^2a_6 - a_1a_2a_3a_4a_5).
\end{aligned}
$$

$$
\begin{aligned}
\Delta_7 = {} & -a_0^3a_7^3 + 3a_0^2a_1a_6a_7^2 + a_0^2a_2a_5a_7^2 + 2a_0^2a_3a_4a_7^2 - 3a_0^2a_3a_5a_6a_7 \\
& - a_0^2a_4a_5^2a_7 + a_0^2a_5^3a_6 - 3a_0a_1^2a_6^2a_7 - 3a_0a_1a_2a_4a_7^2 \\
& + a_0a_1a_2a_5a_6a_7 + 3a_0a_1a_3a_5a_6^2 - a_0a_1a_3a_4a_6a_7 + 2a_0a_1a_4^2a_5a_7 \\
& - 2a_0a_1a_4a_5^2a_6 - a_0a_2^2a_3a_7^2 + 2a_0a_2a_3^2a_6a_7 + a_0a_2a_3a_4a_5a_7 \\
& - a_0a_2a_3a_5^2a_6 - a_0a_3^3a_6^2 - a_0a_3^2a_4^2a_7 + a_0a_3^2a_4a_5a_6 + a_1^3a_6^3 \\
& + 3a_1^2a_2a_4a_6a_7 - 2a_1^2a_2a_5a_6^2 - a_1^2a_3a_4a_6^2 - a_1^2a_4^3a_7 \\
& + a_1^2a_4^2a_5a_6 + a_1a_2^3a_7^2 - 2a_1a_2^2a_3a_6a_7 - a_1a_2^2a_4a_5a_7 \\
& + a_1a_2^2a_5^2a_6 + a_1a_2a_3^2a_6^2 + a_1a_2a_3a_4^2a_7 - a_1a_2a_3a_4a_5a_6.
\end{aligned}
$$

REFERENCES

H. M. James, N. B. Nichols, and R. S. Phillips, *Theory of Servomechanisms*, MIT Rad. Lab. Series, Vol. 25, McGraw-Hill, New York, 1947.

Appendix H

Hilbert Transforms, Analytic Signals, and Correlation Functions

H.1 The Hilbert Transform

We consider a real function $f(t)$ and its Fourier transform $F(\omega)$ related by

$$F(\omega) = \int_{-\infty}^{\infty} f(t)\, e^{-j\omega t}\, dt \tag{H.1-1}$$

and

$$f(t) = \frac{1}{2\pi}\int_{-\infty}^{\infty} F(\omega)\, e^{j\omega t}\, d\omega. \tag{H.1-2}$$

In general the frequency function $F(\omega)$ is complex and may be represented from Equation H.1-1 as

$$\begin{aligned} F(\omega) &= \int_{-\infty}^{\infty} f(t)[\cos \omega t - j \sin \omega t]\, dt \\ &= a(\omega) - jb(\omega) \end{aligned} \tag{H.1-3}$$

where the functions $a(\omega)$ and $b(\omega)$ are given by

$$a(\omega) = \int_{-\infty}^{\infty} f(t) \cos \omega t\, dt, \tag{H.1-4}$$

$$b(\omega) = \int_{-\infty}^{\infty} f(t) \sin \omega t\, dt \tag{H.1-5}$$

and are real if $f(t)$ is real. Also

$$F(-\omega) = F^*(\omega) = a(\omega) + jb(\omega) \tag{H.1-6}$$

where the asterisk represents the complex conjugate. Thus, on inverting Equations H.1-3 and H.1-6, we have

$$a(\omega) = \frac{1}{2}[F(\omega) + F(-\omega)], \qquad \text{(H.1-7)}$$

$$b(\omega) = \frac{j}{2}[F(\omega) - F(-\omega)]. \qquad \text{(H.1-8)}$$

It follows from Equation H.1-2 that $f(t)$ may be written as

$$\begin{aligned} f(t) &= \frac{1}{2\pi}\int_{-\infty}^{0} F(\omega)\, e^{j\omega t}\, d\omega + \frac{1}{2\pi}\int_{0}^{\infty} F(\omega)\, e^{j\omega t}\, d\omega \\ &= \frac{1}{2\pi}\int_{0}^{\infty} F(-\omega)\, e^{-j\omega t}\, d\omega + \frac{1}{2\pi}\int_{0}^{\infty} F(\omega)\, e^{j\omega t}\, d\omega \end{aligned}$$

or, from Equations H.1-7 and H.1-8,

$$f(t) = \frac{1}{\pi}\int_{0}^{\infty} a(\omega)\cos \omega t\, d\omega + \frac{1}{\pi}\int_{0}^{\infty} b(\omega)\sin \omega t\, d\omega. \qquad \text{(H.1-9)}$$

This last equation with $a(\omega)$ and $b(\omega)$ defined by Equations H.1-4 and H.1-5 is a *positive frequency* form of Fourier's integral formula. This last integral can be written formally as

$$\lim_{s\to 0} f(t, s) = f(t) = \lim_{s\to 0}\frac{1}{\pi}\int_{0}^{\infty} [a(\omega)\cos \omega t + b(\omega)\sin \omega t]\, e^{-s\omega}\, d\omega. \qquad \text{(H.1-10)}$$

Now let us define the function $g(r)$ by

$$g(r) = \frac{1}{\pi}\int_{0}^{\infty} [a(\omega) - jb(\omega)]\, e^{jr\omega}\, d\omega \qquad \text{(H.1-11)}$$

where r is a complex variable given by

$$r = t + js. \qquad \text{(H.1-12)}$$

Notice that the function $f(t, s)$ defined by Equation H.1-10 is just the real part of $g(r)$. If we denote the imaginary part by $h(r)$, we have

$$g(r) = f(r) + jh(r) = f(t, s) + jh(t, s) \qquad \text{(H.1-13)}$$

where $h(t, s)$ is given by

$$h(t, s) = -\frac{1}{\pi}\int_{0}^{\infty} [b(\omega)\cos \omega t - a(\omega)\sin \omega t]\, e^{-s\omega}\, d\omega. \qquad \text{(H.1-14)}$$

Let us define a function $f_1(t)$ by

$$f_1(t) = -h(t, 0) = \frac{1}{\pi}\int_{0}^{\infty} [b(\omega)\cos \omega t - a(\omega)\sin \omega t]\, d\omega. \qquad \text{(H.1-15)}$$

This integral is called the *allied integral* of Fourier's integral and the function $f_1(t)$ is called the *quadrature function* or *Hilbert transform* of $f(t)$.

Let us now find the relationship between the frequency functions $F_1(\omega)$ and $F(\omega)$, the Fourier transforms of $f_1(t)$ and $f(t)$, respectively. Equation H.1-15 may be rewritten, with the aid of Equations H.1-7 and H.1-8 as

$$f_1(t) = \frac{j}{2\pi}\int_0^\infty [F(\omega) - F(-\omega)] \cos \omega t \, d\omega$$

$$- \frac{1}{2\pi}\int_0^\infty [F(\omega) + F(-\omega)] \sin \omega t \, d\omega$$

$$= \frac{j}{2\pi}\int_0^\infty F(\omega)\, e^{j\omega t} \, d\omega - \frac{j}{2\pi}\int_0^\infty F(-\omega)\, e^{-j\omega t} \, d\omega$$

or, after a linear change in variable in the last integral,

$$f_1(t) = \frac{1}{2\pi}\int_0^\infty jF(\omega)\, e^{j\omega t} \, d\omega - \frac{1}{2\pi}\int_{-\infty}^0 jF(\omega)\, e^{j\omega t} \, d\omega. \qquad \text{(H.1-16)}$$

Also

$$f_1(t) = \frac{1}{2\pi}\int_{-\infty}^\infty F_1(\omega)\, e^{j\omega t} \, d\omega$$

since $f_1(t)$ and $F_1(\omega)$ are a Fourier-transform pair. It is apparent, therefore, from the last two equations that $F_1(\omega)$ and $F(\omega)$ are related by

$$\boxed{F_1(\omega) = jF(\omega) \operatorname{sgn} \omega} \qquad \text{(H.1-17)}$$

where sgn x is the *signum* function given by

$$\operatorname{sgn} x = \begin{cases} 1, & x > 0, \\ 0, & x = 0, \\ -1, & x < 0. \end{cases} \qquad \text{(H.1-18)}$$

The transform pair $f(t)$ and $f_1(t)$ given by Equations H.1-9 and H.1-15 usually are written in one of several alternative forms which show more explicitly the relationship between $f(t)$ and $f_1(t)$. Let us consider Equation H.1-15 and eliminate $a(\omega)$ and $b(\omega)$ by the use of Equations H.1-4 and H.1-5:

$$f_1(t) = \frac{1}{\pi}\int_0^\infty \int_{-\infty}^\infty [f(x) \sin \omega x \cos \omega t - f(x) \cos \omega x \sin \omega t] \, dx \, d\omega$$

or

$$f_1(t) = \frac{1}{\pi}\int_0^\infty \int_{-\infty}^\infty f(x) \sin \omega(x - t) \, dx \, d\omega \qquad \text{(H.1-19)}$$

since $\sin \omega(x - t) = \sin \omega x \cos \omega t - \cos \omega x \sin \omega t$. The last equation may be written formally as

$$f_1(t) = \lim_{\lambda \to \infty} \frac{1}{\pi}\int_0^\lambda \int_{-\infty}^\infty f(x) \sin \omega(x - t) \, dx \, d\omega. \qquad \text{(H.1-20)}$$

On interchanging the order of integration and integrating with respect to λ, we obtain

$$f_1(t) = \lim_{\lambda \to \infty} \frac{1}{\pi} ⨍_{-\infty}^{\infty} \frac{1 - \cos \lambda(x - t)}{x - t} f(x)\, dx \qquad \text{(H.1-21)}$$

where the notation $⨍$ indicates the *Cauchy principal value* of the integral.*

There is a theorem in analysis called the *Reimann-Lebesgue Lemma* [1] which states that, for a function $k(x)$ belonging to $L_1(-\infty, \infty)$; that is, where

$$\int_{-\infty}^{\infty} |k(x)|\, dx < \infty, \qquad \text{(H.1-22)}$$

then it is true that

$$\lim_{\lambda \to \infty} \int_{-\infty}^{\infty} k(x) \cos \lambda x\, dx = 0, \qquad \text{(H.1-23)}$$

$$\lim_{\lambda \to \infty} \int_{-\infty}^{\infty} k(x) \sin \lambda x\, dx = 0. \qquad \text{(H.1-24)}$$

It is apparent then, that if $f(t)$ behaves well enough, the cosine part of Equation H.1-21 is zero, that is,

$$\lim_{\lambda \to \infty} ⨍_{-\infty}^{\infty} \frac{\cos \lambda(x - t)}{x - t} f(x)\, dx = 0 \qquad \text{(H.1-25)}$$

and Equation H.1-21 reduces to

$$\boxed{f_1(t) = -\frac{1}{\pi} ⨍_{-\infty}^{\infty} \frac{f(x)}{t - x}\, dx} \qquad \text{(H.1-26)}$$

* The *Cauchy principal value* of an integral is defined as follows: Given a function $g(x)$ such that

$$\lim_{x \to a} g(x) = \infty, \qquad b < a < c,$$

then the Cauchy principal value of the integral

$$\int_b^c q(x)\, dx$$

is written as

$$⨍_b^c q(x)\, dx$$

and is given by

$$⨍_b^c q(x)\, dx = \lim_{\delta \to 0} \left[\int_b^{a-\delta} q(x)\, dx + \int_{a+\delta}^{c} q(x)\, dx \right].$$

where we emphasize that the Cauchy principal value of the integral must be taken. This last expression is the usual expression for $f_1(t)$, the Hilbert transform of $f(t)$.

The inverse transform relationship is obtained in the same way, making use of the relationship between $F(\omega)$ and $F_1(\omega)$ as given by Equation H.1-17. Thus we write

$$F_1(\omega) = \alpha(\omega) - j\beta(\omega) \tag{H.1-27}$$

and note that, since $a(\omega)$ and $\alpha(\omega)$ are even and $b(\omega)$ and $\beta(\omega)$ are odd, it follows from Equation H.1-17 that

$$\alpha(\omega) = b(\omega), \tag{H.1-28}$$

$$\beta(\omega) = -a(\omega). \tag{H.1-29}$$

We have

$$\begin{aligned} F_1(\omega) &= \int_{-\infty}^{\infty} f_1(t) \cos \omega t \, dt - j \int_{-\infty}^{\infty} f_1(t) \sin \omega t \, dt \\ &= \alpha(\omega) - j\beta(\omega) = b(\omega) + ja(\omega) \end{aligned}$$

so that we can relate $f_1(t)$ to $a(\omega)$ and $b(\omega)$ by

$$a(\omega) = -\int_{-\infty}^{\infty} f_1(t) \sin \omega t \, dt, \tag{H.1-30}$$

$$b(\omega) = \int_{-\infty}^{\infty} f_1(t) \cos \omega t \, dt. \tag{H.1-31}$$

Now, as before, we may substitute these expressions in Equation H.1-9 to obtain

$$f(t) = -\frac{1}{\pi} \int_0^{\infty} \int_{-\infty}^{\infty} f_1(x) \sin \omega(x - t) \, dx \, d\omega. \tag{H.1-32}$$

A comparison of this equation with Equation H.1-19 and the subsequent steps shows that

$$\boxed{f(t) = \frac{1}{\pi} \fint_{-\infty}^{\infty} \frac{f_1(x)}{t - x} \, dx} \tag{H.1-33}$$

is the inverse transform corresponding to Equation H.1-26.

To show that the Cauchy principal value may exist in Equation H.1-21 or H.1-26, assume that $f(x)$ may be written in the form

$$f(x) = f(t) + (x - t)h(t, x) \tag{H.1-34}$$

where $h(t, x)$ is well behaved in the vicinity of $t = x$. Then Equation H.1-26 becomes

$$f_1(t) = -\frac{1}{\pi} f(t) \fint_{-\infty}^{\infty} \frac{dx}{t - x} + \frac{1}{\pi} \int_{-\infty}^{\infty} h(t, x)\, dx$$

$$= -\frac{1}{\pi} f(t) \lim_{\delta \to 0} \left[-\log (t - x) \Big|_{-\infty}^{t-\delta} - \log (t - x) \Big|_{t+\delta}^{\infty} \right] + \frac{1}{\pi} \int_{-\infty}^{\infty} h(t, x)\, dx$$

$$= -\frac{1}{\pi} f(t) \lim_{\substack{\delta \to 0 \\ x \to \infty}} \left[\log \left(\frac{-\delta}{\delta} \right) - \log \frac{t - x}{t + x} \right] + \frac{1}{\pi} \int_{-\infty}^{\infty} h(t, x)\, dx$$

or, finally,

$$f_1(t) = \frac{1}{\pi} \int_{-\infty}^{\infty} h(t, x)\, dx \tag{H.1-35}$$

which may be finite. We now perform a typical calculation of a transform.

Example H.1 Let $f(t)$ be the symmetrical unit pulse given by

$$f(t) = \begin{cases} 1, & |t| \le a, \\ 0, & |t| > a. \end{cases}$$

Then the Hilbert transform $f_1(t)$ must be considered for three ranges of t.

1. $-a \le t \le a$:

$$f_1(t) = -\frac{1}{\pi} \fint_{-a}^{a} \frac{dx}{t - x}$$

$$= -\frac{1}{\pi} \lim_{\delta \to 0} \left(\int_{-a}^{t-\delta} \frac{dx}{t - x} - \int_{t+\delta}^{a} \frac{dx}{x - t} \right)$$

$$= \frac{1}{\pi} \lim_{\delta \to 0} [\log \delta - \log (t + a) + \log (a - t) - \log (\delta)]$$

$$= \frac{1}{\pi} \lim_{\delta \to 0} \left[\log \frac{\delta(a - t)}{\delta(t + a)} \right] = \frac{1}{\pi} \log \frac{a - t}{a + t}.$$

2. $t < -a$:

$$f_1(t) = \frac{1}{\pi} \int_{-a}^{a} \frac{dx}{x - t} = \frac{1}{\pi} \log (x - t) \Big|_{-a}^{a} = \frac{1}{\pi} \log \frac{a - t}{-a - t}.$$

3. $t > a$:

$$f_1(t) = -\frac{1}{\pi} \int_{-a}^{a} \frac{dx}{t - x} = \frac{1}{\pi} \log (t - x) \Big|_{-a}^{a} = \frac{1}{\pi} \log \frac{t - a}{t + a}.$$

A general expression may be written which includes the three ranges of t:

$$f_1(t) = \frac{1}{\pi} \log \left| \frac{t - a}{t + a} \right|.$$

• •

A list of some of the common Hilbert-transform pairs is given in Table H.1.

Table H.1. Common Hilbert-Transform Pairs

	$f(t) = \frac{1}{\pi} \int_{-\infty}^{\infty} \frac{f_1(x)}{t - x}\, dx$	$f_1(t) = -\frac{1}{\pi} \int_{-\infty}^{\infty} \frac{f(x)}{t - x}\, dx$
1	$\begin{cases} 1, & \lvert t\rvert \le a \\ 0, & \lvert t\rvert > a \end{cases}$	$\frac{1}{\pi} \log \left\lvert \frac{t-a}{t+a} \right\rvert$
2	$\frac{1}{1 + t^2}$	$-\frac{t}{1 + t^2}$
3	$\cos t$	$-\sin t$
4	$\sin t$	$\cos t$
5	$\frac{\sin t}{t}$	$-\frac{2}{t} \sin^2 \left(\frac{t}{2}\right) = \frac{\cos t - 1}{t}$
6	$\delta(t)$	$-\frac{1}{\pi t}$

H.2 Causal Signals and Physically Realizable Systems

We say that a signal $f(t)$ is *causal* if it vanishes for negative time and that a linear system is *physically realizable* or *causal* if its impulse response $f(t)$ vanishes for negative time. In this section we will develop the fact that the real and imaginary parts of the Fourier transforms of causal functions are Hilbert-transform pairs.

It was shown in Equation H.1-17 that the Fourier transforms of Hilbert pairs $f(t)$ and $f_1(t)$ were related by

$$F_1(\omega) = (j \operatorname{sgn} \omega) F(\omega). \tag{H.2-1}$$

From a system point of view, the time function $f_1(t)$ with Fourier transform $F_1(\omega)$ could be considered the result of passing the time function $f(t)$ with Fourier transform $F(\omega)$ through the linear filter with complex transfer function $H(\omega)$ given by

$$H(\omega) = j \operatorname{sgn} \omega. \tag{H.2-2}$$

The impulse response $h(t)$ of this filter is

$$h(t) = \frac{1}{2\pi} \int_{-\infty}^{\infty} j \operatorname{sgn} \omega\, e^{j\omega t}\, d\omega$$

or

$$h(t) = \lim_{\alpha \to 0} \frac{1}{2\pi} \int_{-\infty}^{\infty} j \operatorname{sgn} \omega\, e^{-\alpha|\omega|}\, e^{j\omega t}\, d\omega = -\frac{1}{\pi t}. \tag{H.2-3}$$

Thus, $f_1(t)$ is given by the convolution

$$f_1(t) = -f(t)*\frac{1}{\pi t} = -\frac{1}{\pi}\fint_{-\infty}^{\infty} \frac{f(x)}{t-x}\,dx \tag{H.2-4}$$

in agreement with Equation H.1-26, as expected. The unrealizable filter $h(t)$ could be considered to function as a *Hilbert "transformer"* as shown in Figure H.1.

The Hilbert transformation is seen to be a linear operation and corresponds to a filtering in which the amplitudes of the spectral components are unchanged but the phases are advanced by $\pi/2$ for positive frequencies and retarded by $\pi/2$ for negative frequencies. Since two successive Hilbert trans-

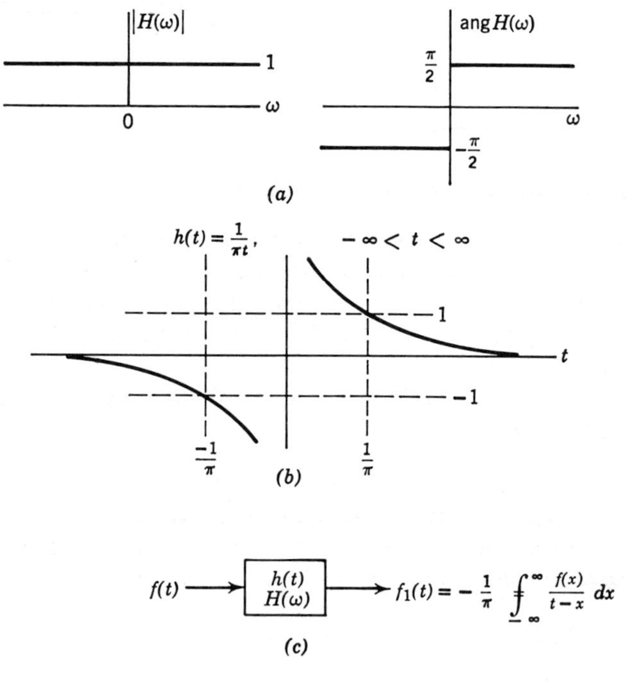

Figure H.1 *The Hilbert "transformer": (a) the complex transfer function $H(\omega)$; (b) the impulse response $h(t)$; (c) the "transformer."*

$$H(\omega) = j \operatorname{sgn} \omega$$

$$|H(\omega)| = |\operatorname{sgn} \omega| = 1$$

$$\operatorname{ang} H(\omega) = \frac{\pi}{2} \operatorname{sgn} \omega$$

formations would reverse the phases of all components, we would expect that the result would be the negative of the original function; that is,

$$f(t) = -\left(-\frac{1}{\pi t}\right)*f_1(t)$$

or

$$f(t) = \frac{1}{\pi}\fint_{-\infty}^{\infty} \frac{f_1(x)}{t - x}\, dx$$

in agreement with Equation H.1-33.

Now consider a function $f(t)$ which vanishes for negative time; that is,

$$f(t) = 0, \qquad t < 0. \tag{H.2-5}$$

Let us further assume that $f(t)$ contains no singularities (delta functions) at the origin. Let us write $f(t)$ as the sum of an even part $f_e(t)$ and an odd part $f_o(t)$

$$f(t) = f_e(t) + f_o(t) \tag{H.2-6}$$

where

$$f_e(t) = \tfrac{1}{2}[f(t) + f(-t)], \tag{H.2-7}$$

$$f_o(t) = \tfrac{1}{2}[f(t) - f(-t)]. \tag{H.2-8}$$

It follows from Equation H.2-5 that, for $t < 0$, $f_e(t) = -f_o(t)$ and, for $t > 0$, $f_e(t) = f_o(t)$. In other words,

$$f_e(t) = f_o(t) \operatorname{sgn} t, \tag{H.2-9}$$

$$f_o(t) = f_e(t) \operatorname{sgn} t. \tag{H.2-10}$$

After transforming, these relationships become, from analogy with Equations H.2-2 and H.2-4,

$$F_e(\omega) = \frac{1}{2\pi} F_o(\omega)*\frac{2}{j\omega} = \frac{-j}{\pi}\fint_{-\infty}^{\infty} \frac{F_o(\nu)}{\omega - \nu}\, d\nu \tag{H.2-11}$$

and

$$F_o(\omega) = \frac{1}{2\pi} F_e(\omega)*\frac{2}{j\omega} = \frac{-j}{\pi}\fint_{-\infty}^{\infty} \frac{F_e(\nu)}{\omega - \nu}\, d\nu. \tag{H.2-12}$$

But $F(\omega)$ and $f(t)$ are related by

$$F(\omega) = \operatorname{Re} F(\omega) + j \operatorname{Im} F(\omega) = \int_0^{\infty} [f(t) \cos \omega t - jf(t) \sin \omega t]\, dt$$

and, from Equation H.2-6, we have

$$F(\omega) = F_e(\omega) + F_o(\omega). \tag{H.2-13}$$

Since the transform of an even function is real and the transform of an odd function is purely imaginary, it follows that

$$F_e(\omega) = \operatorname{Re} F(\omega), \tag{H.2-14}$$

$$F_o(\omega) = j \operatorname{Im} F(\omega). \tag{H.2-15}$$

Hence Equations H.2-11 and H.2-12 can be written as

$$\operatorname{Re} F(\omega) = \frac{1}{\pi} \int_{-\infty}^{\infty} \frac{\operatorname{Im} F(\nu)}{\omega - \nu} \, d\nu \tag{H.2-16}$$

and

$$\operatorname{Im} F(\omega) = -\frac{1}{\pi} \int_{-\infty}^{\infty} \frac{\operatorname{Re} F(\nu)}{\omega - \nu} \, d\nu. \tag{H.2-17}$$

As previously stated in Section 7.3, the real and imaginary parts of the Fourier transform of a causal function are a Hilbert-transform pair.

Example H.2 Consider the low-pass RC filter with impulse response

$$f(t) = u(t)\, e^{-\alpha t}$$

where $u(t)$ is the unit-step function. This is a physically realizable network with complex transfer function

$$F(\omega) = \int_{0}^{\infty} e^{-\alpha t}\, e^{-j\omega t}\, dt$$

$$= \frac{1}{\alpha + j\omega} \cdot \frac{\alpha - j\omega}{\alpha - j\omega} = \frac{\alpha}{\alpha^2 + \omega^2} + j \frac{-\omega}{\alpha^2 + \omega^2}.$$

Thus the real and imaginary parts of $F(\omega)$ are given by

$$\operatorname{Re} F(\omega) = \frac{\alpha}{\alpha^2 + \omega^2},$$

$$\operatorname{Im} F(\omega) = \frac{-\omega}{\alpha^2 + \omega^2}.$$

These do indeed constitute a Hilbert-transform pair as shown by comparison with pair number 2 of Table H.1 with α replaced by unity and ω by t.

• •

H.3 Single-Sideband Systems

Let $a(t)$ be a low-pass function with Fourier transform $A(\omega)$ and Hilbert transform $a_1(t)$. We form the new function $b(t)$ given by

$$b(t) = a(t) \cos \omega_0 t + a_1(t) \sin \omega_0 t. \tag{H.3-1}$$

The spectrum (Fourier transform) of $b(t)$ is

$$B(\omega) = \int_{-\infty}^{\infty} b(t)\, e^{-j\omega t}\, dt. \tag{H.3-2}$$

On substituting Equation H.3-1 we obtain

$$B(\omega) = \int_{-\infty}^{\infty} a(t)\left(\frac{e^{j\omega_0 t} + e^{-j\omega_0 t}}{2}\right) e^{-j\omega t}\, dt$$

$$+ \int_{-\infty}^{\infty} a_1(t)\left(\frac{e^{j\omega_0 t} - e^{-j\omega_0 t}}{2j}\right) e^{-j\omega t}\, dt$$

$$= \frac{1}{2}\int_{-\infty}^{\infty} a(t)\, e^{-jt(\omega - \omega_0)}\, dt + \frac{1}{2}\int_{-\infty}^{\infty} a(t)\, e^{-jt(\omega + \omega_0)}\, dt$$

$$+ \frac{1}{2j}\int_{-\infty}^{\infty} a_1(t)\, e^{-jt(\omega - \omega_0)}\, dt - \frac{1}{2j}\int_{-\infty}^{\infty} a_1(t)\, e^{-jt(\omega + \omega_0)}\, dt$$

or it follows from Equation H.1-17 that

$$B(\omega) = \tfrac{1}{2}A(\omega - \omega_0) + \tfrac{1}{2}A(\omega + \omega_0)$$
$$+ \tfrac{1}{2}[A(\omega - \omega_0)\operatorname{sgn}(\omega - \omega_0) - A(\omega + \omega_0)\operatorname{sgn}(\omega + \omega_0)]. \tag{H.3-3}$$

This last expression may be evaluated for three ranges of ω:

1. $\omega > \omega_0$; $\operatorname{sgn}(\omega - \omega_0) = 1$, $\operatorname{sgn}(\omega + \omega_0) = 1$:

$$B(\omega) = \tfrac{1}{2}A(\omega - \omega_0) + \tfrac{1}{2}A(\omega + \omega_0) + \tfrac{1}{2}A(\omega - \omega_0) - \tfrac{1}{2}A(\omega + \omega_0)$$
$$= A(\omega - \omega_0).$$

2. $-\omega_0 < \omega < \omega_0$; $\operatorname{sgn}(\omega - \omega_0) = -1$, $\operatorname{sgn}(\omega + \omega_0) = 1$:

$$B(\omega) = \tfrac{1}{2}A(\omega - \omega_0) + \tfrac{1}{2}A(\omega + \omega_0) - \tfrac{1}{2}A(\omega - \omega_0) - \tfrac{1}{2}A(\omega + \omega_0)$$
$$= 0.$$

3. $\omega < -\omega_0$; $\operatorname{sgn}(\omega - \omega_0) = -1$, $\operatorname{sgn}(\omega + \omega_0) = -1$:

$$B(\omega) = \tfrac{1}{2}A(\omega - \omega_0) + \tfrac{1}{2}A(\omega + \omega_0) - \tfrac{1}{2}A(\omega - \omega_0) + \tfrac{1}{2}A(\omega + \omega_0)$$
$$= A(\omega + \omega_0).$$

We note that the transform of $a(t)\cos\omega_0 t$ is given by the first two terms of Equation H.3-3; that is,

$$\mathscr{F}[a(t)\cos\omega_0 t] = \tfrac{1}{2}A(\omega - \omega_0) + \tfrac{1}{2}A(\omega + \omega_0). \tag{H.3-4}$$

This is the so-called *double-sideband* spectrum of the amplitude-modulated signal $a(t)$ with carrier frequency ω_0 radians per second. The signal $b(t)$ is the *single-sideband* equivalent of $a(t)$. The spectra $A(\omega)$, $\mathscr{F}[a(t)\cos\omega_0 t]$, and $B(\omega)$ are illustrated in Figure H.2 for an arbitrary low-pass function $A(\omega)$.

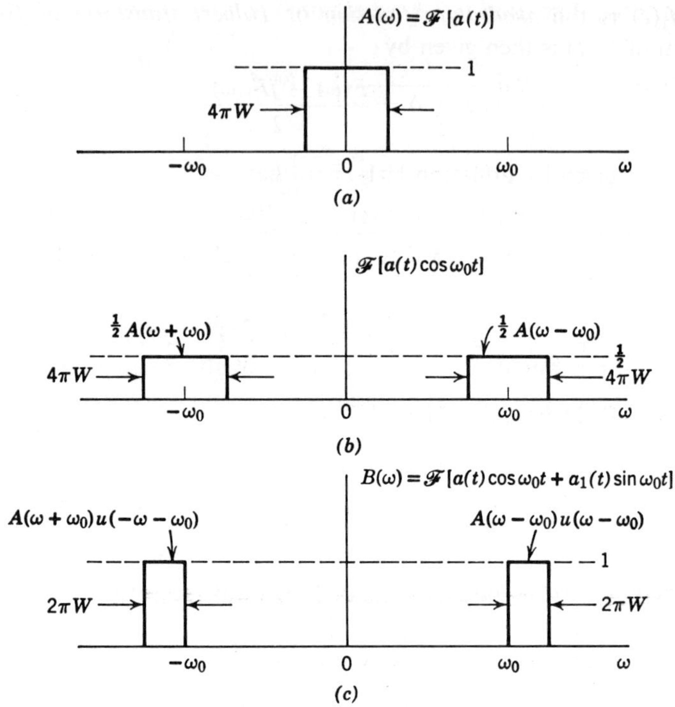

Figure H.2 *Double- and single-sideband amplitude modulation:* (*a*) *a low-pass signal;* (*b*) *the associated double-sideband AM signal;* (*c*) *the associated single-sideband AM signal*

Note that the two lower-frequency sidebands of $a(t)\cos\omega_0 t$ are missing in $b(t)$. The spectrum $B(\omega)$ may be written as

$$B(\omega) = A(\omega - \omega_0)u(\omega - \omega_0) + A(\omega + \omega_0)u(-\omega - \omega_0) \qquad \text{(H.3-5)}$$

where $u(\cdot)$ is the unit-step function.

H.4 Analytic Signals

In Section 7.4, we discussed briefly the *analytic signal* $f_+(t)$ defined in such a way as to have only positive frequency components. In the present discussion, let us approach the problem slightly differently and define the analytic signal $f_+(t)$ associated with $f(t)$ by the relation

$$f_+(t) = \frac{f(t) - jf_1(t)}{\sqrt{2}} \qquad \text{(H.4-1)}$$

where $f_1(t)$ is the *quadrature function* or *Hilbert transform* of $f(t)$. The spectrum of $f_+(t)$ is then given by

$$F_+(\omega) = \frac{F(\omega) - jF_1(\omega)}{\sqrt{2}}. \tag{H.4-2}$$

But $F_1(\omega)$ is given by Equation H.1-17 so that

$$F_+(\omega) = \frac{F(\omega) + F(\omega)\,\text{sgn}\,\omega}{\sqrt{2}} = \begin{cases} 0, & \omega < 0, \\ \sqrt{2}F(\omega), & \omega > 0, \end{cases}$$

or

$$F_+(\omega) = \sqrt{2}\,F(\omega)u(\omega) \tag{H.4-3}$$

where $u(\omega)$ is the unit-step function. Thus the analytic signal $f_+(t)$ has only a positive frequency spectrum as previously noted and is given by the transform relationship

$$f_+(t) = \frac{1}{\sqrt{2}\pi}\int_0^\infty F(\omega)\,e^{j\omega t}\,d\omega \tag{H.4-4}$$

which was the definition given in Section 7.4. Since $\sqrt{2}\,F(\omega)u(\omega)$ is not an even function, $f_+(t)$ is complex for real $f(t)$. The spectrum $F_+(\omega)$ is called the *positive frequency spectrum* of $f(t)$. Also, from Parseval's relation, we have

$$\int_{-\infty}^\infty f^2(t)\,dt = \frac{1}{2\pi}\int_{-\infty}^\infty |F(\omega)|^2\,d\omega = \mathscr{E} \tag{H.4-5}$$

the total energy in $f(t)$. But, from Equation H.4-3,

$$\int_{-\infty}^\infty |f_+(t)|^2\,dt = \frac{1}{2\pi}\int_{-\infty}^\infty |F_+(\omega)|^2\,d\omega = \frac{1}{\pi}\int_0^\infty |F(\omega)|^2\,d\omega = \mathscr{E}. \tag{H.4-6}$$

From a system point of view, the analytic signal $f_+(t)$ can be considered as the output of a linear filter with input $f(t)$ and impulse response $h(t)$ given by

$$h(t) = \frac{\delta(t)}{\sqrt{2}} + \frac{j}{\sqrt{2}\pi t} \tag{H.4-7}$$

where $\delta(t)$ is the Dirac delta function. Then the output $f_+(t)$ is the convolution $f(t)*h(t)$ or

$$\begin{aligned} f_+(t) &= \int_{-\infty}^\infty f(x)h(t-x)\,dx \\ &= \frac{1}{\sqrt{2}}\int_{-\infty}^\infty f(x)\delta(t-x)\,dx + \frac{j}{\sqrt{2}}\frac{1}{\pi}\fint_{-\infty}^\infty \frac{f(x)}{t-x}\,dx \\ &= \frac{f(t) - jf_1(t)}{\sqrt{2}} \end{aligned}$$

as originally defined.

Example H.3 Let $f(t)$ be given by $\cos \omega t$. Then the Hilbert transform is obtained from Table H.1 and is $-\sin \omega t$. The analytic signal is

$$f_+(t) = \frac{\cos \omega t + j \sin \omega t}{\sqrt{2}} = \frac{1}{\sqrt{2}} e^{j\omega t}.$$

Thus the analytic signal corresponding to the sinusoid $\cos \omega t$ is the phasor $(1/\sqrt{2})\, e^{j\omega t}$, a complex function.

• •

H.5 Instantaneous Envelope, Phase, and Frequency

It will often be convenient to represent the arbitrary (real) signal $f(t)$ as an amplitude- and angle-modulated sinusoid of the form

$$f(t) = A(t) \cos \theta(t) \tag{H.5-1}$$

where $A(t)$ is called the *envelope function* and $\theta(t)$ the *phase function* of $f(t)$. The problem is to agree on a unique definition for $A(t)$ and $\theta(t)$. It is apparent that the problem is solved if $\theta(t)$ is defined since $A(t)$ is then given by

$$A(t) = \frac{f(t)}{\cos \theta(t)}. \tag{H.5-2}$$

Let us begin by considering the function

$$g(t) = B \cos \omega t \tag{H.5-3a}$$

where B and ω are constants. The Hilbert transform $g_1(t)$ of this function is given in Table H.1 as

$$g_1(t) = -B \sin \omega t \tag{H.5-3b}$$

and, from Equations H.5-3a and H.5-3b,

$$\tan \omega t = -\frac{g_1(t)}{g(t)}$$

or

$$\omega t = \arctan \left[-\frac{g_1(t)}{g(t)} \right]. \tag{H.5-4}$$

It follows from Equation H.5-3a that B is given by

$$B = \frac{g(t)}{\cos \omega t} = \frac{g(t)}{\cos (\arctan \{-[g_1(t)/g(t)]\})}. \tag{H.5-5}$$

For this trivial case we have uniquely defined ωt and B by Equations H.5-4 and H.5-5 respectively. Furthermore, the frequency ω is given by the time derivative of Equation H.5-4 or

$$\omega = \frac{d}{dt} \left\{ \arctan \left[-\frac{g_1(t)}{g(t)} \right] \right\}. \tag{H.5-6}$$

Let us now return to Equation H.5-1 and, by analogy with the last paragraph, define $A(t)$ and $\theta(t)$. We consider $f(t)$ and its Hilbert transform $f_1(t)$. The *phase function* $\theta(t)$ is defined as in Equation H.5-4 by

$$\boxed{\theta(t) = \arc\tan\left[-\frac{f_1(t)}{f(t)}\right].} \tag{H.5-7}$$

The *envelope function* $A(t)$ is defined as in Equation H.5-5 by

$$\boxed{A(t) = \frac{f(t)}{\cos\left(\arc\tan\{-[f_1(t)/f(t)]\}\right)}.} \tag{H.5-8}$$

We now define the *instantaneous frequency* ω_i, by analogy with Equation H.5-6, to be the time derivative of $\theta(t)$, or

$$\boxed{\omega_i = \frac{d}{dt}\,\theta(t) = \frac{d}{dt}\left\{\arc\tan\left[-\frac{f_1(t)}{f(t)}\right]\right\}.} \tag{H.5-9}$$

We have the relationship

$$\frac{d}{dx}(\arc\tan u) = \frac{1}{1+u^2}\frac{du}{dx};$$

therefore Equation H.5-9 can be written in the alternate form

$$\omega_i = \frac{f_1(t)[df(t)/dt] - f(t)[df_1(t)/dt]}{[f(t)]^2 + [f_1(t)]^2}. \tag{H.5-10}$$

We illustrate the procedure with several examples.

Example H.4 Let $f(t)$ be given by

$$f(t) = \sin\omega t$$

with Hilbert transform $f_1(t)$ obtained from Table H.1 as

$$f_1(t) = \cos\omega t.$$

The phase function $\theta(t)$ is obtained from Equation H.5-7 as

$$\theta(t) = \arc\tan\left(-\frac{\cos\omega t}{\sin\omega t}\right) = \arc\tan(-\cot\omega t)$$

$$= \arc\tan\left[\tan\left(\omega t - \frac{\pi}{2}\right)\right] = \omega t - \frac{\pi}{2}$$

and the envelope function $A(t)$ is given by Equation H.5-8 as

$$A(t) = \frac{\sin \omega t}{\cos(\omega t - \pi/2)} = 1$$

so that we have

$$\sin \omega t = A(t) \cos \theta(t) = \cos\left(\omega t - \frac{\pi}{2}\right)$$

as expected. The instantaneous frequency ω_i is

$$\omega_i = \frac{d}{dt}\theta(t) = \omega.$$

• •

Example H.5 We now consider the somewhat less trivial case where

$$f(t) = \frac{1}{1 + t^2}$$

with Hilbert transform

$$f_1(t) = -\frac{t}{1 + t^2}.$$

The phase function $\theta(t)$ becomes

$$\theta(t) = \arctan\left(\frac{t}{1 + t^2}\frac{1 + t^2}{1}\right) = \arctan t.$$

The envelope function is

$$A(t) = \frac{1/(1 + t^2)}{\cos(\arctan t)}.$$

The instantaneous frequency ω_i is

$$\omega_i = \frac{d}{dt}\theta(t) = \frac{d}{dt}(\arctan t) = \frac{1}{1 + t^2}.$$

• •

H.6 Correlation Functions and Hilbert Transforms

The correlation functions associated with random processes were first defined in Section 3.3. If $a(t)$ is a random process, then the *autocorrelation function* $R_{aa}(t, t + \tau)$ was defined as

$$R_{aa}(t, t + \tau) = E\{a(t)a(t + \tau)\} \tag{H.6-1}$$

where $E\{\ \}$ is the expectation operator. Similarly, for two random processes $a(t)$ and $\hat{a}(t)$, the *cross-correlation* function $R_{a\hat{a}}(t, t + \tau)$ was defined as

$$R_{a\hat{a}}(t, t + \tau) = E\{a(t)\hat{a}(t + \tau)\}. \tag{H.6-2}$$

If the process $a(t)$ is at least *wide-sense stationary*, then Equation H.6-1 becomes

$$R_{aa}(t, t + \tau) = R_{aa}(\tau) \tag{H.6-3}$$

a function of the time difference τ only. Also if $a(t)$ and $\hat{a}(t)$ are jointly at least wide-sense stationary, Equation H.6-2 becomes

$$R_{a\hat{a}}(t, t + \tau) = R_{a\hat{a}}(\tau). \tag{H.6-4}$$

We showed in Section H.3 that a single-sideband AM signal $b(t)$ may be written as

$$b(t) = a(t) \cos \omega_0 t + \hat{a}(t) \sin \omega_0 t \tag{H.6-5}$$

if $\hat{a}(t)$ is the Hilbert transform of $a(t)$. It will often happen in the study of modulation systems that equations of the form of H.6-5 will occur where $a(t)$ is a random process. Then the correlation function of the new process $b(t)$ will involve the correlation function of the Hilbert transform $\hat{a}(t)$ and the cross-correlation function of $a(t)$ and $\hat{a}(t)$. Note that $\hat{a}(t)$ is simply a new random process defined by the linear operation of Equation H.1-26. We now proceed to develop some of these relationships.

Let us assume that $a(t)$ is a random process which is at least wide-sense stationary with mean $u = 0$ and autocorrelation function $R_{aa}(\tau)$ given by

$$R_{aa}(\tau) = E\{a(t)a(t + \tau)\} \tag{H.6-6}$$

where, as for all autocorrelation functions,

$$R_{aa}(\tau) = R_{aa}(-\tau). \tag{H.6-7}$$

If $\hat{a}(t)$ is the Hilbert transform of $a(t)$, what is the autocorrelation function $R_{\hat{a}\hat{a}}(t, t + \tau)$? We begin with

$$\begin{aligned} R_{\hat{a}\hat{a}}(t, t + \tau) &= E\{\hat{a}(t)\hat{a}(t + \tau)\} \\ &= E\left\{\left[-\frac{1}{\pi}\int_{-\infty}^{\infty} \frac{a(x)}{t - x}\, dx\right]\left[-\frac{1}{\pi}\int_{-\infty}^{\infty} \frac{a(y)}{t + \tau - y}\, dy\right]\right\} \end{aligned}$$

from Equation H.1-26. Let us interchange the order of integration and expectation and write

$$R_{\hat{a}\hat{a}}(t, t + \tau) = \frac{1}{\pi^2}\int_{-\infty}^{\infty} \frac{1}{t - x}\int_{-\infty}^{\infty} \frac{R_{aa}(y - x)}{t + \tau - y}\, dy\, dx. \tag{H.6-8}$$

After the linear change in variable $y - x = u$ in the inner integral we have

$$R_{\hat{a}\hat{a}}(t, t + \tau) = -\frac{1}{\pi}\int_{-\infty}^{\infty} \frac{\hat{R}_{aa}(t + \tau - x)}{t - x}\, dx \tag{H.6-9}$$

where $\hat{R}_{aa}(\eta)$ is the Hilbert transform of $R_{aa}(\eta)$ and is given in the usual way by

$$\hat{R}_{aa}(\eta) = -\frac{1}{\pi}\int_{-\infty}^{\infty} \frac{R_{aa}(u)}{\eta - u}\, du. \tag{H.6-10}$$

After another change in variable $t + \tau - x = y$ in Equation H.6-9 we obtain

$$\boldsymbol{R}_{\hat{a}\hat{a}}(t, t + \tau) = \frac{1}{\pi} \fint_{-\infty}^{\infty} \frac{\hat{R}_{aa}(y)}{\tau - y}\, dy = R_{\hat{a}\hat{a}}(\tau) \qquad \text{(H.6-11)}$$

so that the process $\hat{a}(t)$ is at least wide-sense stationary. The fact could have been assumed immediately since the Hilbert transformation is a linear time-invariant operation. From the evenness (Equation H.6-7) of the autocorrelation function, Equation H.6-11 becomes

$$R_{\hat{a}\hat{a}}(\tau) = R_{\hat{a}\hat{a}}(-\tau) = \frac{1}{\pi} \fint_{-\infty}^{\infty} \frac{\hat{R}_{aa}(y)}{\tau - y}\, dy. \qquad \text{(H.6-12)}$$

But this last equation is just the inverse transform corresponding to Equation H.6-10 or

$$R_{aa}(\tau) = \frac{1}{\pi} \fint_{-\infty}^{\infty} \frac{\hat{R}_{aa}(y)}{\tau - y}\, dy = R_{\hat{a}\hat{a}}(\tau) \qquad \text{(H.6-13)}$$

so that, finally,

$$\boxed{R_{\hat{a}\hat{a}}(\tau) = R_{aa}(\tau).} \qquad \text{(H.6-14)}$$

The autocorrelation function of the Hilbert transform of a wide-sense stationary process is equal to the autocorrelation function of the process.

In the same way, the various cross-correlation properties can be developed. We write

$$\begin{aligned} R_{a\hat{a}}(\tau) &= E\{a(t)\hat{a}(t + \tau)\} \\ &= -\frac{1}{\pi} \fint_{-\infty}^{\infty} \frac{E\{a(t)a(y)\}}{t + \tau - y}\, dy \\ &= -\frac{1}{\pi} \fint_{-\infty}^{\infty} \frac{R_{aa}(x)}{\tau - x}\, dx = \hat{R}_{aa}(\tau), \end{aligned} \qquad \text{(H.6-15)}$$

$$\boxed{R_{a\hat{a}}(\tau) = \hat{R}_{aa}(\tau).} \qquad \text{(H.6-16)}$$

Also

$$\begin{aligned} R_{\hat{a}a}(\tau) &= -\frac{1}{\pi} \fint_{-\infty}^{\infty} \frac{E\{a(y)a(t + \tau)\}}{t - y}\, dy \\ &= \frac{1}{\pi} \fint_{-\infty}^{\infty} \frac{R_{aa}(x)}{\tau - x}\, dx = -\hat{R}_{aa}(\tau) \end{aligned} \qquad \text{(H.6-17)}$$

or

$$\boxed{R_{\hat{a}a}(\tau) = -\hat{R}_{aa}(\tau).} \qquad \text{(H.6-18)}$$

Finally, we have

$$\hat{R}_{aa}(-\tau) = -\frac{1}{\pi} \fint_{-\infty}^{\infty} \frac{R_{aa}(-x)}{-\tau - x} dx$$

$$= \frac{1}{\pi} \fint_{-\infty}^{\infty} \frac{R_{aa}(x)}{\tau - x} dx = -\hat{R}_{aa}(\tau), \tag{H.6-19}$$

$$\boxed{\hat{R}_{aa}(-\tau) = -\hat{R}_{aa}(\tau).} \tag{H.6-20}$$

REFERENCE

1. E. C. Titchmarsh, *Introduction to the Theory of Fourier Integrals,* Oxford Univ. Press, Oxford, England, 1948, p. 11.

Appendix I

Useful Tables

Table I.1. The Normal Distribution

x	$f(x)$	$F(x)$	x	$f(x)$	$F(x)$
0.00	0.3989	0.5000	1.40	0.1497	0.9192
0.05	0.3984	0.5199	1.50	0.1295	0.9332
0.10	0.3970	0.5398	1.60	0.1109	0.9452
0.15	0.3945	0.5596	1.70	0.0940	0.9554
0.20	0.3910	0.5793	1.80	0.0790	0.9641
0.25	0.3867	0.5987	1.90	0.0656	0.9713
0.30	0.3814	0.6179	2.00	0.0540	0.9772
0.35	0.3752	0.6368	2.10	0.0440	0.9821
0.40	0.3683	0.6554	2.20	0.0355	0.9861
0.45	0.3605	0.6736	2.30	0.0283	0.9893
0.50	0.3521	0.6915	2.40	0.0224	0.9918
0.55	0.3429	0.7088	2.50	0.1075	0.9938
0.60	0.3332	0.7257	2.60	0.0136	0.9953
0.65	0.3230	0.7422	2.70	0.0104	0.9965
0.70	0.3123	0.7580	2.80	0.00791	0.99744
0.75	0.3011	0.7734	2.90	0.00595	0.99813
0.80	0.2897	0.7881	3.00	0.00443	0.99865
0.85	0.2780	0.8023	3.20	0.00238	0.99931
0.90	0.2661	0.8159	3.40	0.00123	0.99966
0.95	0.2541	0.8289	3.60	0.00061	0.99984
1.00	0.2420	0.8413	3.80	0.00029	0.99993
1.10	0.2179	0.8643	4.00	0.00013	0.99997
1.20	0.1942	0.8849	4.50	0.000016	0.999996
1.30	0.1714	0.9032	5.00	0.0000015	0.9999997

$$f(x) = \frac{1}{\sqrt{2\pi}} e^{-x^2/2}, \qquad F(x) = \frac{1}{\sqrt{2\pi}} \int_{-\infty}^{x} e^{-y^2/2}\, dy$$

Table I.2. Fourier-Transform Properties and Pairs

A. Elementary Properties of the Fourier Transform

Property	$f(t)$	$F(\omega)$
Linearity	$ca(t) + db(t)$	$cA(\omega) + dB(\omega)$
Time translation	$f(t - t_0)$	$e^{-j\omega t_0}F(\omega)$
Frequency translation	$f(t)\, e^{j\omega_0 t}$	$F(\omega - \omega_0)$
Scaling	$f(at)$	$\frac{1}{\lvert a\rvert} F\left(\frac{\omega}{a}\right)$
Differentiation	$\frac{d^n}{dt^n}[f(t)]$	$[j\omega]^n F(\omega)$
Modulation	$f(t) \cos \omega_0 t$	$\frac{1}{2}F(\omega - \omega_0) + \frac{1}{2}F(\omega + \omega_0)$
Convolution	$a(t) * b(t)$	$A(\omega)B(\omega)$

B. Fourier-Transform Pairs

Number	$f(t)$	$F(\omega)$
1	$u(t - a) - u(t - b), \quad a < b$	$\frac{e^{-j\omega a} - e^{-j\omega b}}{j\omega}$
2	$u(t + a) - u(t - a), \quad a > 0$	$2a \frac{\sin a\omega}{a\omega}$
3	$e^{-bt}u(t), \quad b > 0$	$\frac{1}{b + j\omega}$
4	$e^{-b\lvert t\rvert}, \quad b > 0$	$\frac{2b}{b^2 + \omega^2}$
5	$\frac{1}{b^2 + t^2}$	$\frac{\pi}{b} e^{-b\lvert\omega\rvert}, \quad b > 0$
6	1	$2\pi\delta(\omega)$
7	$\delta(t)$	1
8	$\frac{1}{\sqrt{2\pi}} e^{-t^2/2}$	$e^{-\omega^2/2}$

Table I.3. Laplace-Transform Properties and Pairs

A. Elementary Properties of the Laplace Transform

Property	$f(t)$	$F_l(s)$
Linearity	$ca(t) + db(t)$	$cA_l(s) + dB_l(s)$
Time translation	$f(t - t_0)u(t - t_0), \quad t_0 \geq 0$	$e^{-t_0 s}F_l(s)$
Frequency translation	$e^{-at}f(t)$	$F_l(s + a), \quad a \geq 0$
Scaling	$f(at), \quad a \geq 0$	$\frac{1}{a} F_l\left(\frac{s}{a}\right)$
Convolution	$\int_0^t a(\tau)b(t - \tau)\, d\tau$	$A_l(s)B_l(s)$

B. Laplace-Transform Pairs

Number	$f(t)$	$F_l(s)$
1	$u(t - a) - u(t - b), \quad 0 \leq a < b$	$\frac{e^{-sa} - e^{-sb}}{s}$
2	$u(t)$	$\frac{1}{s}$
3	$e^{-bt}, \quad b > 0$	$\frac{1}{b + s}$
4	t	$\frac{1}{s^2}$
5	t^n	$\frac{n!}{s^{n+1}}$
6	$t\, e^{-bt}, \quad b > 0$	$\frac{1}{(b + s)^2}$
7	$\cos bt$	$\frac{s}{b^2 + s^2}$
8	$\sin bt$	$\frac{b}{s^2 + b^2}$
9	$e^{-at} \cos bt, \quad a > 0$	$\frac{a + s}{b^2 + (s + a)^2}$
10	$e^{-at} \sin bt, \quad a > 0$	$\frac{b}{b^2 + (s + a)^2}$
11	$\frac{df}{dt}$	$sF_l(s) - f(0)$
12	$\frac{d^2f}{dt^2}$	$s^2F_l(s) - sf(0) - f^{(1)}(0)$
13	$\int_0^t f(x)\, dx$	$\frac{F_l(s)}{s}$

Table I.4. Logarithms to the Bases 2, e, and 10

x	$\log_2 x$	$\log_e x$	$\log_{10} x$	x	$\log_2 x$	$\log_e x$	$\log_{10} x$
1	0	0	0	26	4.7004	3.2581	1.4150
2	1.0000	0.6932	0.3010	27	4.7549	3.2958	1.4314
3	1.5850	1.0986	0.4771	28	4.8074	3.3320	1.4472
4	2.0000	1.3863	0.6021	29	4.8580	3.3673	1.4624
5	2.3219	1.6094	0.6990	30	4.9069	3.4012	1.4771
6	2.5850	1.7918	0.7782	31	4.9542	3.4340	1.4914
7	2.8074	1.9459	0.8451	32	5.0000	3.4657	1.5051
8	3.0000	2.0794	0.9031	33	5.0444	3.4965	1.5185
9	3.1699	2.1972	0.9542	34	5.0875	3.5264	1.5314
10	3.3219	2.3026	1.0000	35	5.1293	3.5554	1.5441
11	3.4594	2.3979	1.0414	36	5.1699	3.5835	1.5563
12	3.5850	2.4849	1.0792	37	5.2095	3.6109	1.5862
13	3.7004	2.5650	1.1139	38	5.2479	3.6376	1.5798
14	3.8074	2.6391	1.1461	39	5.2854	3.6636	1.5911
15	3.9069	2.7081	1.1761	40	5.3219	3.6889	1.6021
16	4.0000	2.7726	1.2041	50	5.6439	3.9120	1.6990
17	4.0875	2.8332	1.2304	60	5.9069	4.0943	1.7782
18	4.1699	2.8904	1.2553	70	6.1293	4.2485	1.8451
19	4.2479	2.9444	1.2788	80	6.3219	4.3820	1.9031
20	4.3219	2.9957	1.3010	90	6.4919	4.4998	1.9542
21	4.3923	3.0445	1.3222	100	6.6439	4.6051	2.0000
22	4.4594	3.0910	1.3424	120	6.9069	4.7874	2.0792
23	4.5236	3.1355	1.3617	140	7.1273	4.9416	2.1461
24	4.5850	3.1781	1.3802	160	7.3219	5.0752	2.2041
25	4.6439	3.2189	1.3979	180	7.4919	5.1930	2.2553
				200	7.6439	5.2983	2.3010

Table I.5. Entropy of a Binary Source

p	$-p \log_2 p$	$-(1 - p) \log_2 (1 - p)$	H
0	0	0	0
0.01	0.0664	0.0144	0.0808
0.02	0.1129	0.0285	0.1414
0.04	0.1858	0.0565	0.2423
0.06	0.2435	0.0839	0.3274
0.08	0.2915	0.1107	0.4022
0.10	0.3322	0.1368	0.4690
0.12	0 3671	0.1623	0.5294
0.14	0.3971	0.1871	0.5842
0.16	0.4230	0.2113	0.6343
0.18	0.4453	0.2348	0.6801
0.20	0.4644	0.2575	0.7219
0.22	0.4806	0.2796	0.7602
0.24	0.4941	0.3009	0.7950
0.26	0.5053	0.3214	0.8267
0.28	0.5142	0.3412	0.8554
0.30	0.5211	0.3602	0.8813
0.32	0.5260	0.3784	0.9044
0.34	0.5292	0.3956	0.9248
0.36	0.5306	0.4121	0.9427
0.38	0.5304	0.4276	0.9580
0.40	0.5288	0.4422	0.9710
0.42	0.5257	0.4558	0.9815
0.44	0.5212	0.4684	0.9896
0.46	0.5154	0.4800	0.9954
0.48	0.5083	0.4905	0.9988
0.50	0.5000	0.5000	1.0000

$H = H(p) = -p \log_2 p - (1 - p) \log_2 (1 - p)$ bits

Note: $H(p) = H(1 - p)$; $0 \leq p \leq \frac{1}{2}$.

Index